Building an Effective Security Program for Distributed Energy Resources and Systems

Building an Effective Security Program for Distributed Energy Resources and Systems

Understanding Security for Smart Grid and Distributed Energy Resources and Systems

Volume 1

Mariana Hentea

This edition first published 2021

Registered Office
John Wiley & Sons, Inc., 111 River Street, Hoboken, NJ 07030, USA

Editorial Office
111 River Street, Hoboken, NJ 07030, USA

For details of our global editorial offices, customer services, and more information about Wiley products visit us at www.wiley.com.

Wiley also publishes its books in a variety of electronic formats and by print-on-demand. Some content that appears in standard print versions of this book may not be available in other formats.

Library of Congress Cataloging-in-Publication Data

Names: Hentea, Mariana, author.
Title: Building an effective security program for distributed energy resources and systems: understanding security for smart grid and distributed energy resources and systems / Mariana Hentea.
Description: Hoboken, NJ : Wiley, 2021. | Includes bibliographical references and index.
Identifiers: LCCN 2020045336 (print) | LCCN 2020045337 (ebook) | ISBN 9781118949047 (cloth) | ISBN 9781119070429 (adobe pdf) | ISBN 9781119070436 (epub)
Subjects: LCSH: Smart power grids–Security measures.
Classification: LCC TK3105 .H45 2021 (print) | LCC TK3105 (ebook) | DDC 621.31068/4–dc23
LC record available at https://lccn.loc.gov/2020045336
LC ebook record available at https://lccn.loc.gov/2020045337

Cover Design: Wiley
Cover Image: © Henrik5000/Getty Images

Set in 9/11pt STIXTwoText by SPi Global, Pondicherry, India

SKY10025935_033121

To my husband, Toma, and our children, Irina and Marius, for their love and patience.
"There is no doubt that it is around the family and the home that
all the greatest virtues. . . are created, strengthened and maintained." (Winston Churchill)

Contents

Foreword

"Just because something doesn't do what you planned it to do doesn't mean it's useless." (Thomas A. Edison, US Inventor)

Environmental policies, energy rising costs, and technology innovations are challenging many assumptions that were used to build current electric utility infrastructure, which has been evolving for more than a century. The power grid is the most complex man-made system that allows access to electricity, a fundamental enabler for the economy. While access to electricity is the greatest engineering achievement of the twentieth century, the grid of today does not have the attributes necessary to meet the demands of the twenty-first century and beyond.

The Smart Grid paradigm promises to improve the power grid reliability and enable sustainability and customer choice. To meet the power grid concerns, utilities around the world are investing in distributed energy resources (DERs). However, different utilities have different reasons and business drivers for investing in DERs management. Besides grid reliability, the increasingly rapid adoption of DERs is driven by other factors such as to meet the world's energy efficiency and greenhouse gas emission goals. With these drivers for investing in DERs and Smart Grids, cybersecurity solutions are imperative for reliable energy delivery. In highly connected world via Internet and with an increasing sophistication of threats, it is unrealistic to assume energy delivery systems are isolated or immune from compromise.

To achieve the interoperability of Smart Grid devices and systems, it is required that standards and protocols align policy, business, and technology in a manner that would enable all electric resources, including demand-side resources, to contribute to an efficient, reliable electricity network. There is a need to understand that ensuring cybersecurity and privacy of the information is more than conformance to standards.

Security and privacy needs for Smart Grid and DERs, strategies, security requirements, risk management, security and privacy design, and countermeasures as well as standards and best industry practice recommendations are discussed in this book of three volumes:

Understanding Security for Smart Grid and Distributed Energy Resources and Systems (Vol 1)
Building Security Program for Smart Grid and Distributed Energy Resources and Systems (Vol 2)
Effective Security Program for Smart Grid and Distributed Energy Resources and Systems: An Engineering Approach (Vol 3)

The aim of this three-volume book is building security and privacy programs to support the development of Smart Grid Systems and DER systems that are reliable, secure, resilient, and flexible. The cybersecurity problem becomes a very complex problem for the Smart Grid system, defined also a system of systems. The basic concepts, approaches, and frameworks are described in this three-book set. Smart Grid and DERs security and privacy issues are gradually introduced and discussed from many perspectives.

The sequence is starting with introductory topics for the security and privacy programs and Smart Grid and DER needs (Volume 1), followed by more advanced and detailed functions of the programs as well as discussion of Smart Grid and DER characteristics, vulnerabilities, threats, potential risks (Volume 2), to efficient and effective security and programs that include monitoring, reporting, and control based on security measurements and security metrics as well as intelligent decision-making (Volume 3). Each volume and some topics can be used independently for limited purposes.

These books include information about strategies, security requirements, risk management, security design, and countermeasures as well as regulations, standards, and best practice recommendations. The focus is on describing the most specific issues of Smart Grid and DERs including building security and privacy program blocks to handle several aspects of the security and privacy risks for the Smart Grid and DER systems. These books demonstrate how to blend Engineering techniques with standards and best security practices. Finally, a perspective on the future DER systems cannot be discussed without taking a look at the vision on the future Smart Grids and research needs.

The information provided in this three-volume book could be used to educate current workforce, future graduates, academic/research, and regulators to understand the complex cybersecurity domain in the context of the various paradigms (e.g. Smart Grid, convergence of security by design and privacy by design) and emerging technologies (e.g. Internet of Things, wireless technologies, big data analytics, machine learning, intelligent control, and decision-making).

Preface Volume 1

Understanding Security for Smart Grid and Distributed Energy Resources and Systems

"If you want to find the secrets of the universe, think in terms of energy, frequency and vibration." (Nikola Tesla, US Inventor)

The emergence of Smart Grid paradigm and distributed energy resources (DERs) applications requires innovation and deployment of new technologies, processes, and policies. DERs are typically smaller electricity generation or storage units located in a community, business, or home. They can serve consumers' energy needs locally and can provide support for the grid. All points of the power grid infrastructure will come under challenge, so it is critical that we fix the process and trust issues in DERs and future Smart Grid technologies.

The more sophisticated technologies and devices become, the greater the danger of them being stolen or adapted for misuse. The growing popularity of wireless technology used in several computing systems may have finally attracted enough hackers to make the potential for serious security threats a reality. In fact, the number and types of mobile threats – including viruses, spyware, malicious downloadable applications, phishing, and spam – have spiked in recent months. One can argue that device makers and wireless service providers have long focused on communications and other services, with security remaining an afterthought.

There is a growing concern about the security and safety of the control systems in terms of vulnerabilities, lack of protection, and awareness. In the past, control systems were isolated from other Information Technology (IT) systems. Historically, IT teams and industrial control systems or operational technology (OT) teams have been organized vertically based on the technology stack they managed. Connection to the Internet is new (early 1990s) and debatable among specialists. However, even without any connection to the Internet, these systems are still vulnerable to external or internal attackers that can exploit vulnerabilities in private communication networks and protocols, software such as operating systems, custom and vendor software, data storage software, databases, and applications.

Therefore, the increasing cyber attacks to energy sector and critical infrastructure are National concerns that require better security and privacy protection, an educated work force of Engineers in the area of security and privacy issues, and Security Professionals in the area of industrial control systems, particularly developing and implementing security protection for emerging Smart Grid applications and DER systems.

The security frameworks and initiatives surrounding the Smart Grid technology hence need to be provided and applied in a time-critical fashion before larger implementations of Smart Grid roll out without good designs. Additionally, the electrical power community needs to critically consider applications of such frameworks to legacy power grid implementations to avoid security add-ons that could be costly and inefficient.

While no single solution can be applied today to protect the power grid, this book (Volume 1), **Understanding Security for Smart Grid and Distributed Energy Resources and Systems**, provides an introduction of the fundamental concepts of cybersecurity, Smart Grid, DERs, power systems, and energy sector as a critical infrastructure. It discusses strategies, approaches, methods, frameworks, and standards that could help current work force in the electrical sector and power product manufacturers to:

- Understand the security problem as it applies to the power grid, energy sector, and electricity subsector.
- Understand the cybersecurity terms and evolution of terms.
- Understand the Smart Grid concepts, DERs, and system needs for protection against intentional or unintentional threats.
- Construct new engineering approaches to cybersecurity such as integrated organizational cooperation, strategic and tactical methods to be implemented, and increasing standards compliance requirements as well as fostering public trust that security is a high priority to those who provide these critical energy resources.
- Define trust in a dynamic, collaborative environment and understand what it means to provide trust throughout an interaction.
- Use a common framework for security policies and support of interoperability, ensuring security, and continuity.
- Recognize the importance of standards in the development of Smart Grid technologies and DER systems to develop a framework that includes protocols and model standards for information security management.
- Describe relevant cybersecurity standards or best practices that can be used for the specific applications.
- Understand the scope and limitations of the security controls.
- Identify the capability of the components or system to be updated to meet future cybersecurity requirements or technologies.

The key topics discussed in the book include:

- Smart Grid paradigm, DERs and systems, scope of security and privacy, computing and information systems for business and industrial applications, critical Smart Grid systems, overview of Smart Grid cybersecurity standards, and key players in Smart Grid standards development.
- Cybersecurity concepts and cybersecurity evolution, cybersecurity for electrical sector as a National Priority, emerging technologies, the needs for Smart Grid cybersecurity, solutions, security, and privacy programs.
- Principles of cybersecurity, characteristics of information, critical security characteristics of information and systems, information security models.
- Applying security principles to Smart Grid and DERs, Smart Grid infrastructure and technologies by considering IT systems infrastructure versus industrial control systems infrastructure with their differences and similarities including the IT and Operational convergence trends.
- Smart Grid vulnerabilities, threats, recent cyber attacks, security controls, and cybersecurity challenges.
- Critical infrastructure, critical infrastructure interdependencies, energy sector as a component of critical infrastructure, information security frameworks (NIST Cybersecurity Framework and NIST Privacy framework – generic frameworks), terrorism challenges addressing security of control systems, emerging technologies, and impacts to cybersecurity.
- Characteristics of Smart Grid and DER systems, power system services and operations, energy management system, electrical utilities evolution, Smart Grid conceptual models (NIST conceptual model, IEEE model, European Union conceptual model), power and smart devices, and Smart Grid key technologies.
- Analysis of power system characteristics (e.g. stability, partial stability), analysis of DER impacts, addressing issues (e.g. cybersecurity, reliability, resiliency, cyber-physical systems), Smart Grid interoperability dimensions, interoperability framework, and addressing cross-cutting issues.
- Distributed energy systems, DER technologies and security challenges, establishing information security governance, and examples of Smart Grid applications and cybersecurity expectations.

- Security management as a broad field of management, security management components and tasks, security program definition and functions, security management process, asset management, physical security and safety, security versus safety, information security management infrastructure, models and frameworks for information security management, privacy program functions, and approaches for building a security program and privacy program.
- Security management for Smart Grid systems – strategic, tactical, and operational views, unified view of security management based on risk management for both IT systems and control systems, systemic security management – comparison and discussion of models, efficient and effective management solutions, security models for electrical sector – electricity subsector cybersecurity capability maturity model (ES-CM2), NIST framework, etc., implementation challenges on achieving security governance, and ensuring information assurance, certification, and accreditation.

The topics discussed in this book help to educate the Security Professionals, Power Control Engineers, management, regulators, service providers, and inform the public at large about the Smart Grid paradigm, DERs, and needs for Security and Privacy protection. Also, the book may be used to educate future graduates (e.g. engineers, computer science, IT graduates, business, and law) to gain skills and more knowledge on understanding and managing the security and privacy risks of Smart Grid and DERs as well as approaches for defining and maintaining a security and privacy program. For example, Law students can use the material from the book to understand the cybersecurity issues for critical infrastructure problems. Also, they can learn about the current regulations, the power and consumers' needs for new regulations in the future.

Research and academia communities could use the book to have a broader view of the cybersecurity problems for Smart Grid, critical infrastructure and energy sector.

Acknowledgments

Although I am the sole author of this three-volume book, the content is the product of my work experience and learning from discussions with colleagues and friends about various topics and projects at work, interactions with researchers at conferences and workshops, meetings and presentations provided by professional societies, my published research works, presentations and talks at conferences, teaching courses in the university, leading research projects with students, meetings with IEEE members, etc.

Besides these, I have been inspired by Dr. Martha Evens' strength and dedication to seek new work and educate others. Dr. Martha Evens encouraged me to pursue a doctoral degree in Artificial Intelligence, after I accomplished an MS in computer science at Illinois Institute of Technology, Chicago, IL, USA. Still after several decades, Dr. Evens (now emeritus professor) provided advice on how to manage the writing of this book. She always encouraged me to pursue my own research interests.

The chosen topic – cybersecurity for the Smart Grid and distributed energy resources – is the result of my own decision, after I learned about threats to power grid and the need for providing more information on security matters to engineers.

I thank Dr. Simone Taylor for reading my book proposal and offering the opportunity to publish this book. My thanks also go to reviewers, Antony Sami, Brett Kurzman, Kari Capone, Sarah Lemore, and the team of editors and managers from Wiley. Their support and advice in completing the writing task are very much appreciated.

28 November 2019 *Mariana Hentea*

Part I

Understanding Security and Privacy Problem

1

Security

1.1 Introduction

Over a short period of time, people and businesses have come to depend greatly upon computer technology and automation in many different aspects of their lives. Computers are involved in managing and operating public utilities, banking, e-commerce and other financial institutions, medical equipment and healthcare services, government offices, military defense systems, and almost every possible business and day-to-day activities of the people. This level of dependence and the extent of Internet technology integration made security necessary discipline as stated by the Organisation for Economic Cooperation and Development (OECD) in [OECD 2006]:

> Security must become an integral part of the daily routine of individuals, businesses and governments in their use of Internet Communication Technologies (ICTs) and conduct of online activities.

Security is the condition of being protected against danger and loss. In general usage, security is similar to safety. Security means that something is not only secure but also it has been secured.

There are various definitions of security provided by different dictionaries (e.g. security is freedom from danger; safety) (see more definitions in Appendix A), but all of them basically agree on some components, and they miss this point: they do not translate readily into information technology (IT) terms. In the IT sector, there is an acceptance that there is no pure risk-free state, whatever it is done (or not done), but it carries a risk.

Therefore, the definitions should not be considered as absolute descriptions of the word security in the real world because they individually describe a practically impossible goal. In order to describe security in a more realistic way, by combining the definitions provided by two dictionaries, new definitions are suggested (e.g. [Fragkos 2005]).

Thus, the definition of security is understood as the capability of a system to protect its resources and to perform to its design goals. However, definitions may differ among users, standards organizations, and industries. Also, several concepts and definitions for security and many related terms have evolved in time to reflect emerging trends. Some other terms are used such as information security and cybersecurity. In a computing context, the term security implies cybersecurity [TechTarget]. Information security was first brought to the public's attention by the release of the first guidelines to protect the security of information systems in 1992 [OECD 1992].

Ten years later, the OECD reviewed the guidelines to take into account the generalized adoption of Internet technologies, which enabled the openness and interconnection of formerly closed and isolated information systems. The need to develop a culture of security and greater awareness was initiated in 2002 by OECD [OECD 2002] for OECD members and nonmembers alike; it was adopted by United Nations in 2002 [UN 2002]. The OECD document [OECD 2002] emphasizes the need to take into

Building an Effective Security Program for Distributed Energy Resources and Systems: Understanding Security for Smart Grid and Distributed Energy Resources and Systems, Volume 1, First Edition. Mariana Hentea.

account the emergence of the open Internet and the generalization of interconnectivity. These guidelines apply to all participants in the new information society.

Security is, therefore, currently a widespread and growing concern that covers all areas of society: business, domestic, financial, government, and so on. Often security has different meanings to different people. There are several definitions and terms that sometimes make the security an ambiguous field. For example, in the energy sector, energy security refers to the uninterrupted availability of energy sources at an affordable price [IEA 2016]. To a power engineer, security means that power flows between utilities are open. Another view of security is a three-legged stool consisting of physical security, information technology (IT) security, and industrial control systems (ICS) security [Weiss 2010].

Security has a wide base and addresses specific issues regarding computers, networks, communication devices, data, information, people, organizations, and governments. Users must have confidence that information systems operate as intended without unanticipated failures or problems. Also, users must have confidence that information is handled timely, accurately, confidentially, and reliably.

Following this document [OECD 2002], OECD published more technical guidelines and recommendations for the implementation and management of security [OECD 2003], [USCIB 2004], [OECD 2005], [OECD 2008] including privacy [OECD 2016]. Revisions of the guidelines are reported in [OECD 2012a], [OECD 2012c].

On 17 September 2015 the OECD Council adopted the Recommendation on Digital Security Risk Management [OECD 2015], which replaces the 2002 guidelines. The [OECD 2015] document provides guidance for a new generation of national strategies on the management of digital security risk aimed to optimize the economic and social benefits expected from digital openness. The recommendation calls on governments, public, and private organizations to adopt an approach to digital security risk management that builds trust and takes advantage of the open digital environment for economic and social prosperity. As described in this document, digital security implies that security is approached from at least four different perspectives, each stemming from a different culture and background, recognized practices, and objectives:

- Technology that is focusing on the functioning of the digital environment (often called information security, computer security, or network security by experts).
- Law enforcement and, more generally, legal aspects (e.g. cybercrime).
- National and international security, including aspects such as the role of information and communication technologies (ICTs) with respect to intelligence, conflict prevention, warfare, etc.
- Economic and social prosperity, encompassing wealth creation, innovation, growth, competitiveness, and employment across all economic sectors, as well as aspects such as individual liberties, health, education, culture, democratic participation, science, and other dimensions of well-being in which the digital environment is driving progress.

The continuous growth of cybersecurity threats and attacks including the increasing sophistication of the malware is impacting the security of energy sector and other critical infrastructures. The energy industry includes electricity sector that provides the production and delivery of power to consumers through a grid connection.

Currently, cybersecurity is a widespread and growing concern for the energy sector. In addition, the energy market shows the presence of emerging Smart Grid phenomena, which introduce new security concerns. In the context of this book, security has a wide base and addresses specific issues regarding power grid and Smart Grid with its related technologies such as Internet of things, cyber–physical systems, industrial control systems, communication networks, computers, information, organization, and people, and others.

1.2 Smart Grid

The Smart Grid is evolving from the traditional electrical grid. An electrical grid (also referred to as an electricity grid or electric grid) is an interconnected network for delivering electricity from suppliers to consumers. It consists of generating stations that produce electrical power, high-voltage transmission lines that carry power from distant sources to demand centers, and distribution lines that connect individual customers. The US electric power system has provided highly reliable electricity for more than a century.

1.2.1 Traditional Power Grid Architecture

The traditional architecture (see Figure 1.1) is based on large-scale generation remotely located from consumers, hierarchical control structures with minimal feedback, limited energy storage, one-way control, and passive loads.

As illustrated in Figure 1.1, the electricity sector is composed of four distinct functions: generation, transmission, distribution, and system operations. Once electricity is generated, it is generally sent through high-voltage, high-capacity transmission lines to local electricity distributors. Once there, electricity is transformed into a lower voltage and sent through local distribution lines for consumption by industrial plants, businesses, and residential consumers.

Because electric energy is generated and consumed almost instantaneously, the operation of an electric power system requires that a system operator constantly balance the generation and consumption of power. Figure 1.2 shows additional functional systems (transmission system, system operations, distribution system) and substation connected to different customers (offices, residential customers, and industrial customers). Information including basic definitions of terms and concepts related to the electrical power grid can be also found in the references and glossaries included in Appendix B.

1.2.1.1 Key Players

In the US electric sector, the key players include utilities and system operators [GAO 2011]:

- Utilities own and operate electricity assets, which may include generation plants, transmission lines, distribution lines, and substations including structures often seen in residential and commercial areas that contain technical equipment such as switches and transformers to ensure smooth, safe flow of current and voltage. Utilities may be owned by investors, municipalities, and individuals (as in cooperative utilities).
- System operators are sometimes affiliated with a particular utility or sometimes independent and responsible for managing the electricity flows in multiple utility areas. The system operators manage and control the generation, transmission, and distribution of electric power using control systems, IT information systems, and network-based systems that monitor and control sensitive processes and physical functions, including opening and closing circuit breakers (see definitions in Appendix B). Therefore, the effective functioning of the electricity industry is highly dependent on these control systems.

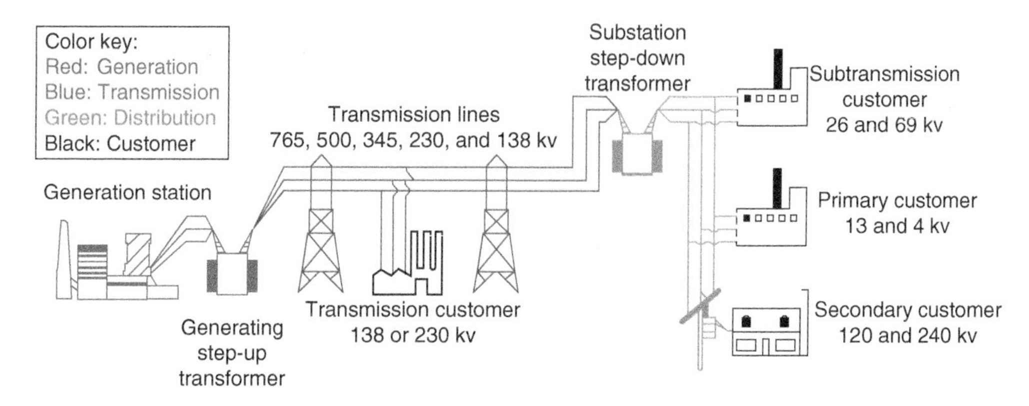

Figure 1.1 Traditional electricity delivery system. *Source:* [DOE 2015a]. Public Domain.

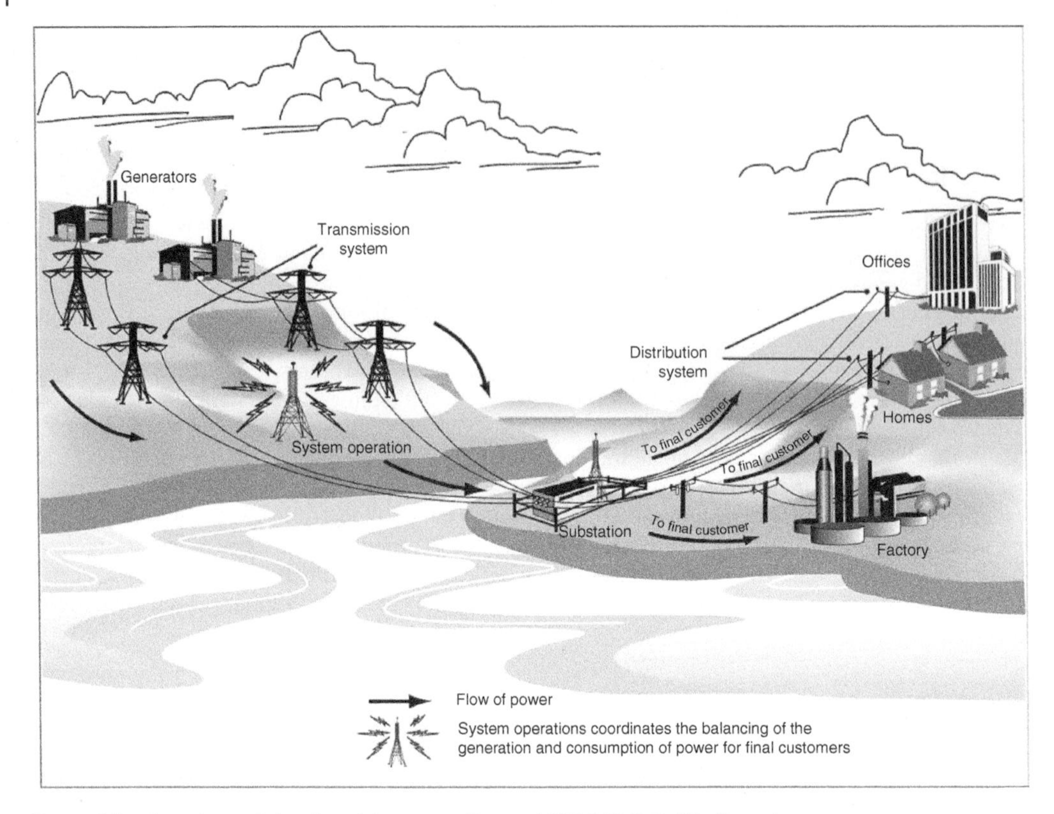

Figure 1.2 Functions of the electricity sector. *Source:* [GAO 2011]. Public Domain.

However, for many years, the US electricity network lacked opportunities such as [GAO 2011]:

- Adequate technologies (e.g. sensors) to allow system operators to monitor how much electricity was flowing on distribution lines.
- Communication networks to further integrate parts of the electricity grid with control centers.
- Computerized control devices to automate system management and recovery.

1.2.1.2 Electric Grid Design of the Future

As the electric grid transitions from the traditional design to the design of the future, new features and technologies must be incorporated. Increasing communications and computing capabilities are transforming power grid from the traditional centralized model to an integrated hybrid centralized/decentralized system. Therefore, society and the power industry in particular are challenged by the transformation of the power grid, as introduced by Nikola Tesla about 120 years ago, into a Smart Grid.

Figure 1.3 depicts an electric power grid that is evolving to include more distributed control, two-way flows of electricity and information, more energy storage, and new market participants including consumers as energy producers.

1.2.2 Smart Grid Definitions

The definition of a Smart Grid is broad and encompasses many aspects of electric grid operation and management. A Smart Grid is an improved electrical power grid, a network of transmission lines, substations, transformers, and more that deliver electricity from suppliers to consumers by using two-way digital technology to communicate with end loads and appliances at industrial, commercial, and

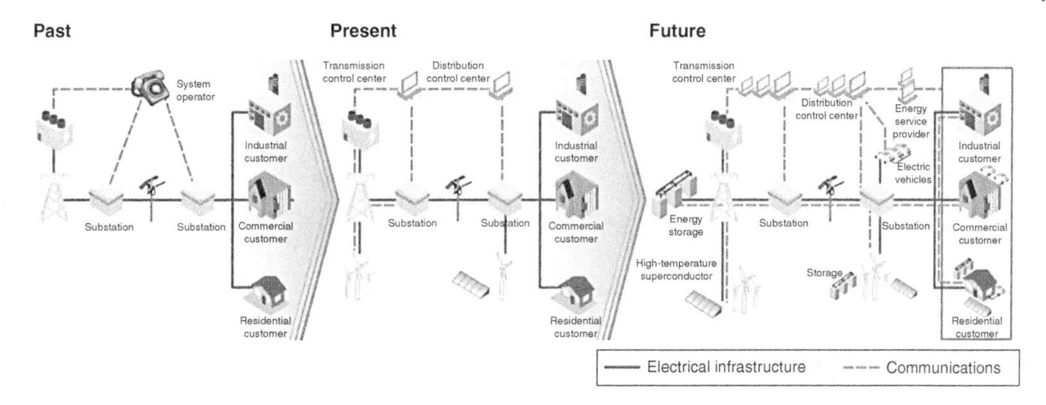

Figure 1.3 Evolution of the electric power grid. *Source:* [DOE 2015a]. Public Domain.

residential premises to save energy and reduce capital and operational cost by improving reliability, security, and efficiency of current power grid. The Smart Grid enables greater use of electricity generated from renewable resources.

Smart Grids are typically described as electricity systems complemented by communication networks, monitoring and control systems, smart devices, and end-user interfaces [OECD 2010], [OECD 2009].

Another Smart Grid definition blends both functions and components [OECD 2012b] and refers to an electricity network that uses digital and other advanced technologies to monitor and manage the transport of electricity from all generation sources to meet the varying electricity demands of end users. Smart Grids coordinate the needs and capabilities of all generators, grid operators, end users, and electricity market stakeholders to operate all parts of the system as efficiently as possible, minimizing costs and environmental impacts while maximizing system reliability, resilience, and stability [IEA 2011].

The Smart Grid is a vision of the future electricity delivery infrastructure that improves network efficiency and resilience while empowering consumers and addressing energy sustainability concerns [Gartner IT].

The SmartGrids Platform was started by the Directorate-General for Research of the European Commission in 2005 [SmartGrids 2006]. This initiative aims at boosting the competitive situation of the European Union in the field of electricity networks, especially smart power grids. The establishment of a European Technology Platform (ETP) in this field was for the first time suggested by the industrial stakeholders and the research community at the first International Conference on the Integration of Renewable and Distributed Energy Resources [Conference 2004].

Although there is no formal definition of a Smart Grid based on its features proposed in the literature, the Smart Grid may be considered as a power grid in which modern sensors, communication links, and computational power are used to improve the efficiency, stability, and flexibility of the system [Rihan 2011].

The 2006 report of European Commission [SmartGrids 2006] describes the vision of the "Future: operation of system will be shared between central and distributed generators. Control of distributed generators could be aggregated to form microgrids or 'virtual' power plants to facilitate their integration both in the physical system and in the market." Figure 1.4 shows how the concept of SmartGrids works.

SmartGrids was a new concept for electricity networks across Europe. The initiative aims to respond to the rising challenges and opportunities, bringing benefits to all users, stakeholders, and companies. Also, the Advisory Council of the technology platform SmartGrids proposed new ways for Europe to move forward on improving the efficiency of the generation, transmission, and distribution of electricity. By using cleaner energy resources (e.g. solar, wind), the SmartGrids aims to benefit the European economy and help improve consumers' needs.

Figure 1.5 illustrates the vision for the future electricity networks. As presented in the [SmartGrids 2006] report, a proportion of the electricity generated by large conventional plants will be displaced by distributed generation, renewable energy sources, demand response and demand-side management, and energy storage.

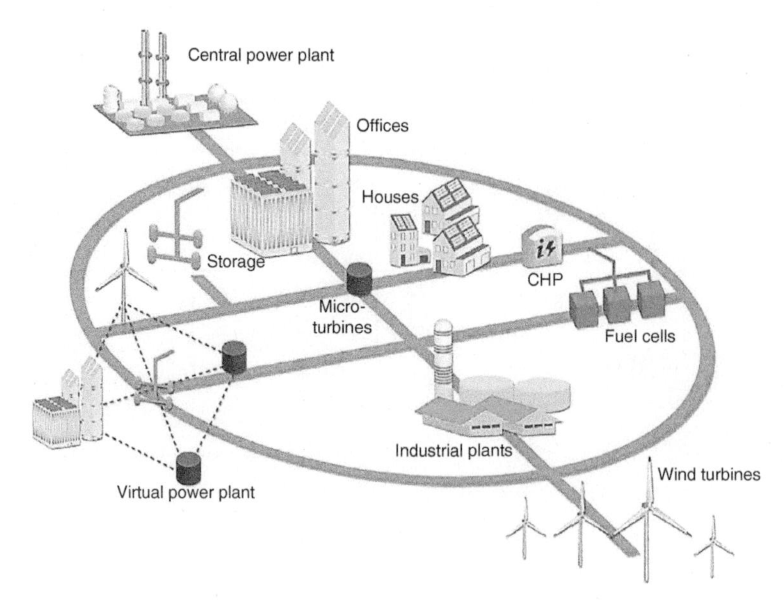

Figure 1.4 SmartGrids concept. *Source:* [SmartGrids 2006]. © European Communities, 2006.

Figure 1.5 Future network vision. *Source:* [SmartGrids 2006]. © European Communities, 2006.

Figure 1.5 (Continued)

Figure 1.6 depicts the layout of the modernized electrical grid with voltages and depictions of electrical lines that are typical of Germany and other European systems. A vision of future developments in Europe is depicted at VISION 2050 site (see [VISION 2050]). More European reference publications are available at [ETIP SNET].

The development of Smart Grid in the United States is a result of Title XIII of the Energy Independence and Security Act of 2007 [EISA 2007], which provided legislative support for Department of Energy's (DOE) Smart Grid activities and reinforced its role in leading and coordinating national grid modernization efforts [Mandates 2007].

The smart power grid delivers electricity from suppliers (e.g. central power plant, distributed generation resources such as wind turbines, microturbines, etc.) to consumers using two-way digital technology to communicate with end loads and appliances at industrial, commercial, and residential premises to save energy, reduce capital and operational cost by improving efficiency, and increase reliability and transparency. Also, the Smart Grid includes control systems, intelligent devices, and communication networks that keep track of electricity flowing in the grid.

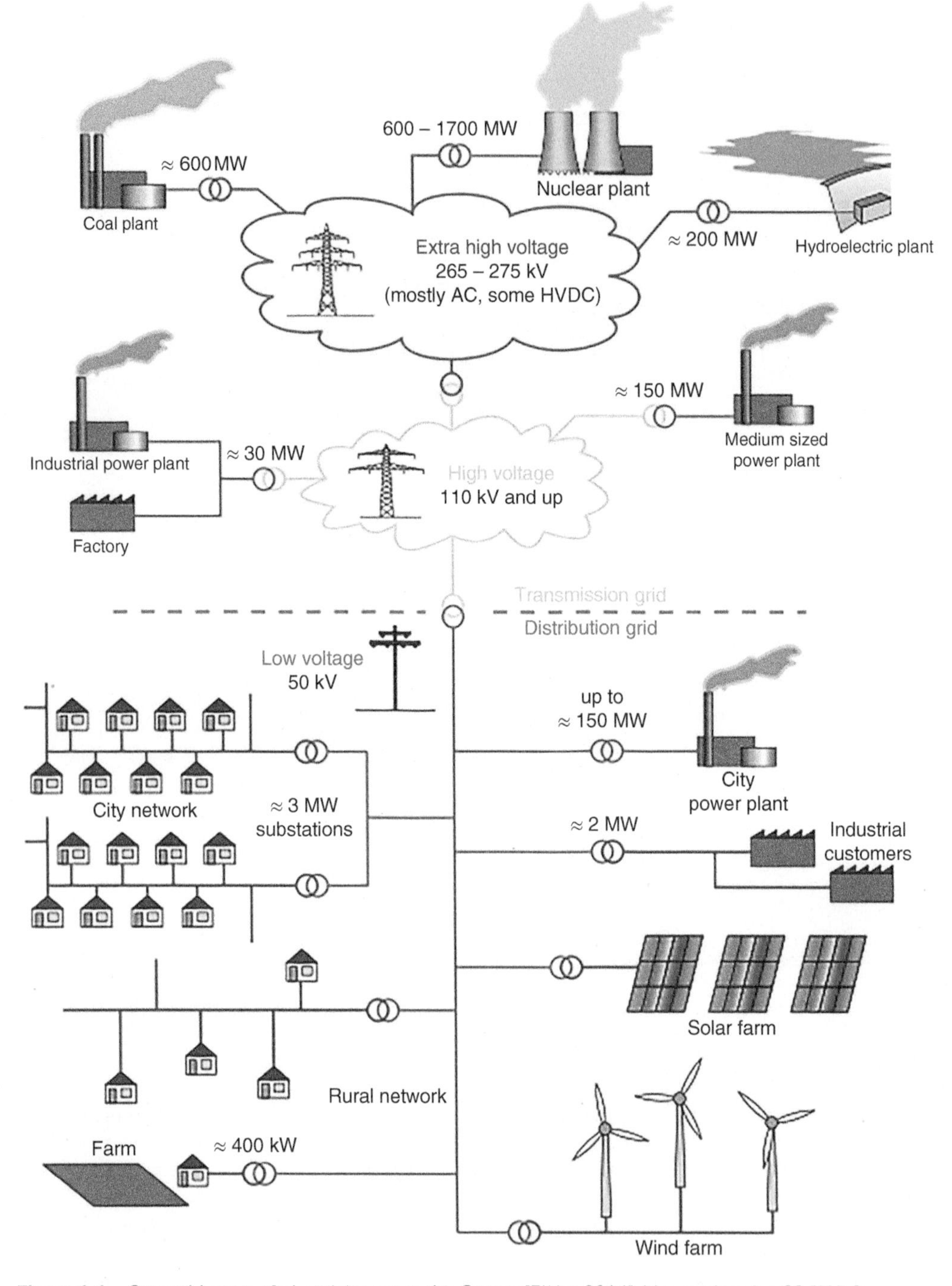

Figure 1.6 General layout of electricity networks. *Source:* [ElNet 2014]. Licensed under CC BY 3.0.

Until this point in time, research on the Smart Grid has revolutionized the way the energy trade will be performed in the near future. The Smart Grid concept challenged the majority of the energy trade stakeholders regarding several aspects of the current setting: infrastructure should be reengineered, and new legislation should be developed, and new business models should be implemented, and so forth. Motivations for Smart Grid developments in the United States are also described in [DOE 2009].

As we have seen in the above definitions, Smart Grid and similar names such as intelligent grid, modern grid, future grid, modernized grid, and so on are all being used to describe a digitized and intelligent version of the current power grid.

1.2.3 Drivers for Change

Examples of drivers for change in the electric power system in the United States include:

- Integration of Smart Grid technologies (see definition in Appendix B) for managing complex power systems, driven by the availability of advanced technologies that can better manage progressively challenging loads.
- Growing expectations for a resilient and responsive power grid in the face of more frequent and intense weather events, cyber and physical attacks, and interdependencies with natural gas and water systems.

Smart Grid technologies and applications encompass a diverse array of modern communications, sensing, control, information, and energy technologies that are already being developed, tested, and deployed throughout the grid. These technologies are divided into three basic categories [NSF 2011]:

- Advanced ICTs (including sensors and automation capabilities) that improve the operation of transmission and distribution systems.
- Advanced metering solutions, which improve on or replace legacy metering infrastructure.
- Technologies, devices, and services that access and leverage energy usage information, such as smart appliances that can use energy data to start operating when energy is cheaper or renewable energy is available.

The Smart Grid vision increases the use of IT systems, networks, and two-way communication to automate actions that system operators formerly had to perform manually.

Thus, the grid modernization is an ongoing process, and initiatives have commonly involved installing advanced metering infrastructure (AMI) (smart meters) in homes and commercial buildings that enable two-way communication between the utility and customer. Other initiatives include adding smart components to provide the system operator with more detailed data on the conditions of the transmission and distribution systems and better tools to observe the overall condition of the grid (referred to as wide area situational awareness). These components include advanced smart switches on the distribution system that communicate with each other to reroute electricity around a troubled line and high-resolution, time-synchronized monitors, called phasor measurement units, on the transmission system. Concepts such as smart loads, smart generation distribution, smart electric vehicles (EVs), smart buildings, and smart switch are enabling a smarter grid [DOE 2015a]. Figure 1.7 illustrates a Smart Grid configuration, with common smart components (smart meter, smart appliances, phasor measurement unit, wind turbines, electric vehicle, smart switch, two-way communication lines), although utilities making Smart Grid investments may opt for alternative configurations depending on cost, customer needs, and local conditions.

To deliver electricity more cost effectively in response to consumer needs and at the same time with less damage to the climate, the Smart Grid uses distributed energy resources (DERs), advanced communication, and control technologies.

1.2.4 Smart Grid Communication Infrastructure

Communication infrastructure is the backbone of the communication system upon which various broadcasting and telecommunication services are operated. The infrastructure is the core component that connects upstream production, such as voice, data, and audiovisual services, with downstream consumers. In basic terms, communication infrastructure involves technology, products, and network connections that allow for the transmission of communications over large distances. According to [P2030 2011], the facilitation of Smart Grid consists of these following aspects: power engineering, communication

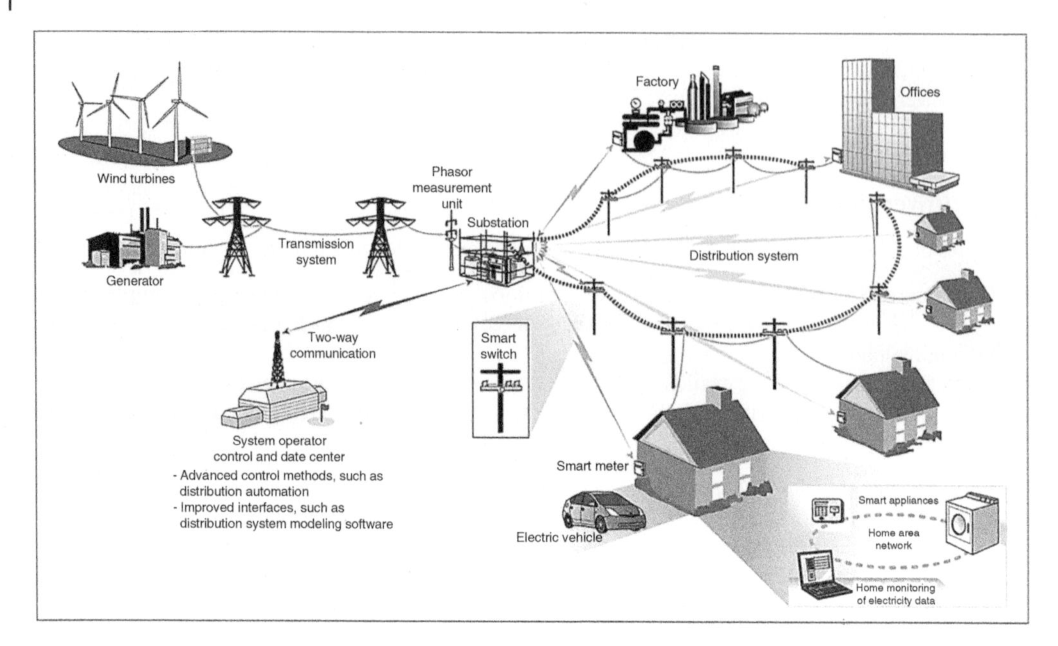

Figure 1.7 Common Smart Grid components. *Source:* [GAO 2011]. Public Domain.

technology, and information technology. A Smart Grid is characterized by the bidirectional connection of electricity and information flows to create an automated, widely distributed delivery network.

A communication infrastructure is an essential part to the success of the emerging Smart Grid [Yan 2013]. Through a communication infrastructure, a Smart Grid can improve power reliability and quality to eliminate electricity blackout.

As described in [Chen 2010], Smart Grid supports two-way power flow and information flow to reach optimal electric power operation. Smart Grid shall consequently collect all kinds of information of electricity generation (centralized or distributed), consumption (instantaneous or predictive), storage (or conversion to energy in other forms), and distribution through the communication infrastructure. Then, the optimization of electricity utilization can be realized through appropriate information technology such as grid or cloud computing to allow appropriate actions in the entire Smart Grid through communication infrastructure again.

Communication infrastructure is a complex ecosystem of separate yet interconnected systems. It consists of a variety of networks, including the broader Internet, cellular networks, optical backhaul networks, and local area networks. A scalable and pervasive communication infrastructure is crucial in both construction and operation of a Smart Grid [Yan 2013].

For the purpose of planning and organization of the diverse, expanding collection of interconnected networks that will compose the Smart Grid, NIST adopted the approach of dividing the Smart Grid into seven logical domains, known as Smart Grid Conceptual Reference Model. The model includes the following domains [NIST SP1108r1]:

- Generation – Includes traditional generation sources and DERs; may also store energy for later distribution; generation includes coal, nuclear, and large-scale hydrogeneration usually attached to transmission; DERs are associated with customer and distribution domains providing generation and storage and with service provider aggregated energy resources.
- Transmission – Carriers of bulk electricity over long distances; may also store and generate electricity.
- Distribution – Distributors of electricity to and from customers; may also store and generate electricity.
- Customers – End users of electricity (residential, commercial, and industrial); may also generate, store, and manage the use of energy.
- Operations – Managers of the movement of electricity.

- Markets – Operators and participants in electricity markets.
- Service providers – Organizations providing services to electrical customers and to utilities.

Figure 1.8 illustrates the interaction of roles in different Smart Grid domains through secure communication. It shows communications (blue lines) and energy/electricity flows (yellow lines) connecting each domain and how they are interrelated. Each individual conceptual domain is itself composed of important Smart Grid elements that are connected to each other through two- way communications and energy/electricity paths. These connections are the basis of the future, intelligent and dynamic power electricity grid.

Each domain and its subdomains encompass Smart Grid actors and applications (see description in Table 1.1). Actors include devices, systems, or programs that make decisions and exchange information necessary for performing applications; smart meters, solar generators, and control systems represent examples of devices and systems. Applications, on the other hand, are tasks performed by one or more actors within a domain. For example, corresponding applications may be home automation, solar energy generation and energy storage, and energy management (see more information in [NIST SP1108r3]). Appendix D includes more information about each domain including graphical representation of interactions with other domains.

The Smart Grid Roadmap [NIST SP1108r1] describes the conceptual reference model as a tool for discussing the characteristics, uses, behavior, and other elements of Smart Grid domains and the relationships among these elements. The model is a tool for identifying the standards and protocols needed to ensure interoperability and cybersecurity and defining and developing architectures for systems and subsystems within the Smart Grid.

Interoperability of a Smart Grid is the ability of diverse systems to work together, use the compatible parts, exchange information or equipment from each other, and work cooperatively to perform tasks. It enables integration, effective cooperation, and two-way communications among the many interconnected elements of the Smart Grid.

The viewpoint depicted in the diagram provides a high-level, overarching logical architecture representation of a few major relationships that existing applications have to Smart Grid domains.

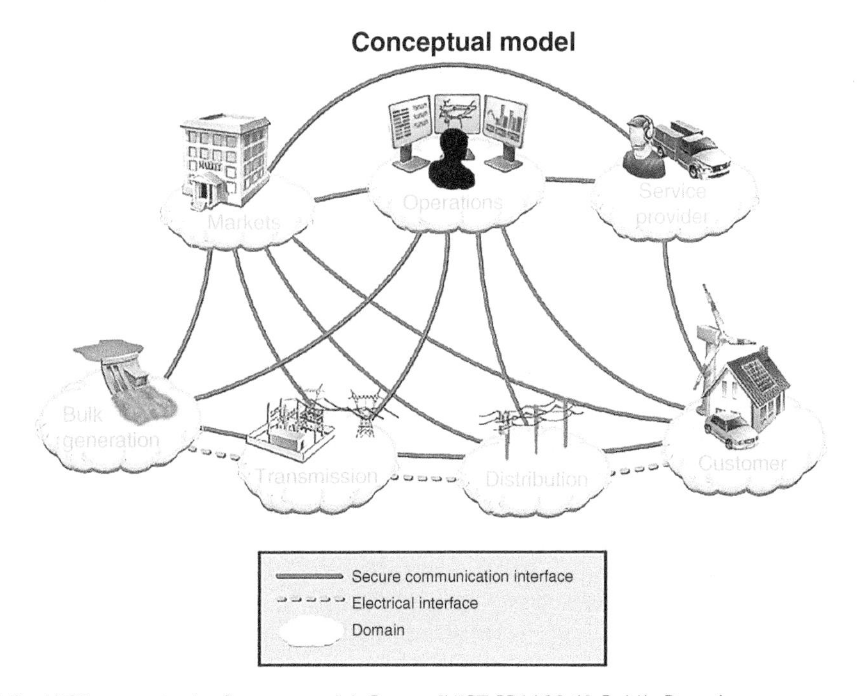

Figure 1.8 NIST conceptual reference model. *Source:* [NIST SP1108r1]. Public Domain.

Table 1.1 Domains and actors in the Smart Grid conceptual model.

	Domain	Roles/services in the domain
1	Customer	The end users of electricity. May also generate, store, and manage the use of energy. Traditionally, three customer types are discussed, each with its own domain: residential, commercial, and industrial
2	Markets	The operators and participants in electricity markets
3	Service providers	The organizations providing services to electrical customers and to utilities
4	Operations	The managers of the movement of electricity
5	Generation	The generators of electricity. May also store energy for later distribution. This domain includes traditional generation sources (traditionally referred to as generation) and distributed energy resources (DER). At a logical level, generation includes coal, nuclear, and large-scale hydrogeneration usually attached to transmission. DER (at a logical level) is associated with customer and distribution domain provided generation and storage and with service provider aggregated energy resources
6	Transmission	The carriers of bulk electricity over long distances. May also store and generate electricity
7	Distribution	The distributors of electricity to and from customers. May also store and generate electricity

Source: [NIST SP1108r1]. Public Domain.

This diagram suggests what their possible communication paths could be in a Smart Grid. It is also a useful way to identify potential intra- and inter-domain interactions between existing and new applications, along with capabilities enabled by these interactions.

To enable the Smart Grid functionality, the actors in a particular domain often interact with actors in other domains. Actors are devices, systems, or programs that make decisions and exchange information necessary for executing applications within the Smart Grid. Actors have the capability to make decisions and to exchange information with other actors. Organizations may have actors in more than one domain. Figure 1.9 illustrates a composite view of actors within domains of the conceptual reference model.

Information about the actors associated with the NIST conceptual reference model is provided in [NISTIR 7628r1], [NIST SP1108r3]. Table 1.2 includes a list of selected actors and their description.

Figure 1.10 shows a high-level view of information network and existing applications mapped to Smart Grid domains. An information network is a collection, or aggregation, of interconnected computers, communication devices, and other information and communication technologies. Technologies in a network exchange information and share resources.

Smart Grid consists of many different types of networks, not all of which are shown in the diagram. The Smart Grid is a network of many systems and subsystems, as well as a network of networks. That is, many systems with various ownership and management boundaries are interconnected to provide end-to-end services between and among stakeholders as well as between and among intelligent devices.

The communication infrastructure includes different types of networks, technologies, and services that are organized based on criteria such as geography, topology, business purpose, technology, ownership, etc. Understanding these networks is crucial to perform analysis of requirements for security and quality of service considerations.

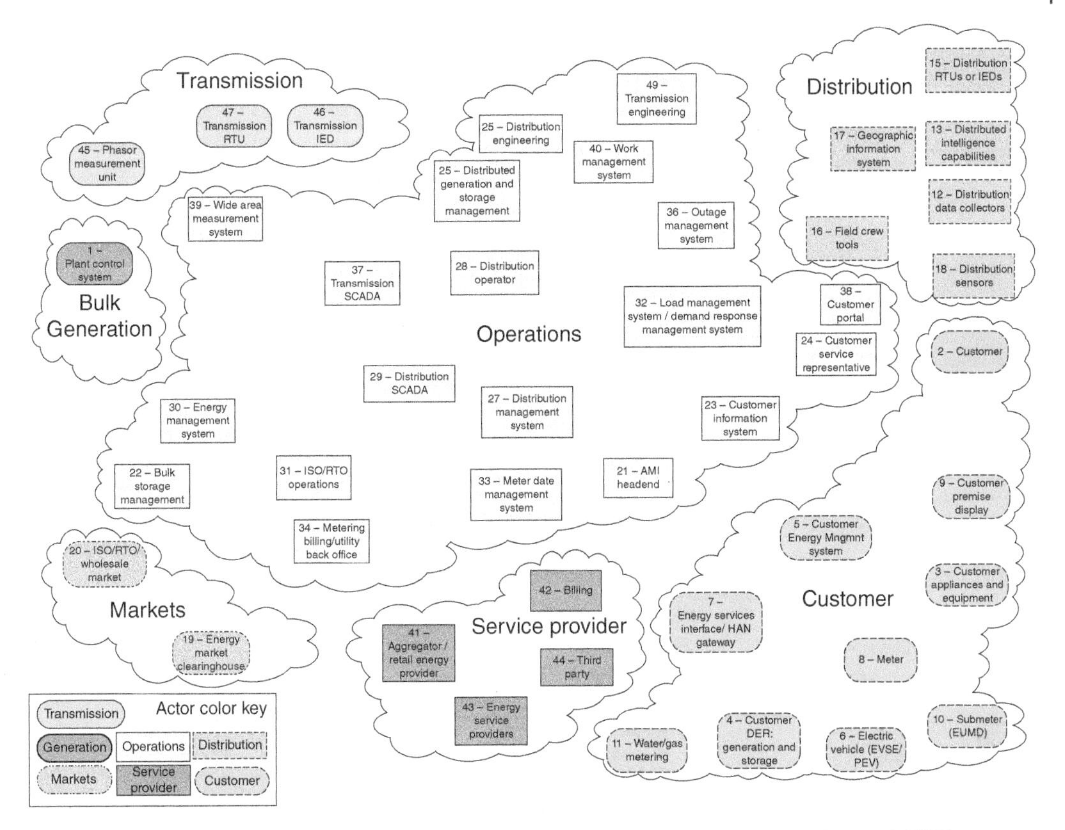

Figure 1.9 View of the actors within domains of NIST conceptual reference model. *Source:* [NISTIR 7628r1]. Public Domain.

The networks include the enterprise bus that connects control center applications to markets and generators and with each other; wide area networks that connect geographically distant sites; field area networks that connect devices, such as intelligent electronic devices (IEDs) that control circuit breakers and transformers; and premises networks that include customer networks as well as utility networks within the customer domain. These networks may be implemented using public (e.g. the Internet) and nonpublic networks in combination. Both public and nonpublic networks require implementation and maintenance of appropriate security and access control to support the Smart Grid.

Given that the Smart Grid will not only be a system of systems but also a network of information networks, a thorough analysis of network and communications requirements for each network is needed. This analysis should differentiate among the requirements pertinent to different Smart Grid applications, actors, and domains.

Figure 1.11 illustrates a general architecture for Smart Grid communication infrastructure, which includes home area networks (HAN), business area networks (BAN), neighborhood area networks (NAN), data centers, and substation automation integration systems.

As illustrated in this pictorial view, the authors of this work [Baimel 2016] summarize the communication infrastructure of the Smart Grid as based on three types of networks to include HAN, NAN, and WAN described as follows:

1.2.4.1 HAN

HAN is deployed and operated within a small area (tens of meters), usually a house or a small office; HAN has relatively low transmission data rate compared with other two networks, hundreds of bits per

Table 1.2 Actor descriptions for the logical reference model.

Actor no	Acronym	Actor	Actor description
1	DCS	Plant control system – distributed control system	A local control system at a bulk generation plant. This is sometimes called a distributed control system (DCS)
2		Customer	An entity that pays for electrical goods or services. A customer of a utility, including customers who provide more power than they consume
3		Customer appliances and equipment	A device or instrument designed to perform a specific function, especially an electrical device, such as a toaster, for household use. An electric appliance or machinery that may have the ability to be monitored, controlled, and/or displayed
4	DER	Distributed energy resources (customer generation and storage)	Energy generation resources, such as solar or wind, used to generate and store energy (located on a customer site) to interface to the controller (home area network/business area network [HAN/BAN]) to perform an energy-related activity
5	EMS	Energy management system	An application service or device that communicates with devices in the home. The application service or device may have interfaces to the meter to read usage data or to the operations domain to get pricing or other information to make automated or manual decisions to control energy consumption more efficiently. The EMS may be a utility subscription service, a third-party offered service, a consumer-specified policy, a consumer-owned device, or a manual control by the utility or consumer
6	PEV/ EVSE	Plug-in electric vehicle/electric vehicle service element	PEV is a vehicle propelled by an electric motor and powered by a rechargeable battery. It can be recharged using an external power source. When the external power source is the power grid, the EV is connected through the EVSE that provides power and communication
7	HAN Gateway	Home Area Network Gateway	An interface between the distribution, operations, service provider, and customer domains and the devices within the customer domain
8		Meter	Point-of-sale device used for the transfer of product and measuring usage from one domain/ system to another
9		Customer Premise Display	A device that displays usage and cost data to the customer on location
11		Water/gas metering	A point-of-sale device used for the transfer of product (water and gas) and measuring usage from one domain/system to another
12		Distribution data collector	A data concentrator collecting data from multiple sources and modifying/transforming it
13		Distributed intelligence capabilities	Advanced automated/intelligence application that operates in a normally autonomous mode from the centralized control system to increase reliability and responsiveness

15	RTU/ IED	Distribution remote terminal unit/intelligent electronic device	Receives data from sensors and power equipment and can issue control commands, such as tripping circuit breakers, if voltage, current, or frequency anomalies are identified, RTUs and/or IEDs can raise/lower voltage levels to maintain the desired voltage range
17	GIS	Geographic information system	A spatial asset management system that provides utilities with asset information and network connectivity for advanced applications
18		Distribution sensor	A device that measures a physical process and converts it into a signal that can be read by an observer or by an instrument
19		Energy Market Clearinghouse	Wide area energy market operation system providing high-level market signals for distribution companies (ISO/RTO and utility operations)
21	AMI	Advanced metering infrastructure headend	This system manages the information exchanges between third-party systems or systems not considered headend, such as the meter data management system (MDMS) and the AMI network
22		Bulk storage management	Provides management for energy storage connected to the bulk power system
23	CIS	Customer Information System	Enterprise-wide software applications that allow companies to manage aspects of their relationship with a customer
25		Distributed generation and storage management	Distributed generation is the process of generating electricity from many small local energy sources. Storage management enables the efficient integration of distributed generation sources into the grid
27	DMS	Distribution management systems	A suite of application software that supports electric system operations. Example applications include topology processor, online three-phase unbalanced distribution power flow, contingency analysis, study mode analysis, switch order management, short-circuit analysis, Volt/VAR management, and loss analysis. These applications provide operations staff and engineering personnel additional information and tools to help accomplish their objectives
29	SCADA	Supervisory control and data acquisition	A supervisory computerized system that that gathers and processes data and applies operational controls for distribution side systems used to control dispersed assets
30	EMS	Energy management system	A system used by electric grid operators to monitor, control, and optimize the performance of the generation and/or transmission system
32	LMS/ DRMS	Load management systems/ demand response management system	An LMS issues load management commands to appliances or equipment at customer locations in order to decrease load during peak or emergency situations. The DRMS issues pricing or other signals to appliances and equipment at customer locations in order to request customers (or their preprogrammed systems) to decrease or increase their loads in response to the signals

(*Continued*)

Table 1.2 (Continued)

Actor no	Acronym	Actor	Actor description
33	MDMS	Meter data management system	System that stores meter data (e.g. energy usage, energy generation, meter logs, meter test results) and makes data available to authorized systems. This system is a component of the customer communication system. This may also be referred to as a billing meter
34		Metering/billing/utility back office	Back office utility systems for metering and billing
36		Outage management system	An OMS is a computer system used by operators of electric distribution systems to assist in outage identification and restoration of power. Major functions usually found in an OMS include: Listing all customers who have outages Prediction of location of fuse or breaker that opened upon failure Prioritizing restoration efforts and managing resources based upon criteria such as location of emergency facilities, size of outages, and duration of outages Providing information on extent of outages and number of customers impacted to management, media, and regulators Estimation of restoration time Management of crews assisting in restoration Calculation of crews required for restoration
39	WAMS	Wide area measurement system	Communication system that monitors all phase measurements and substation equipment over a large geographical base that can use visual modeling and other techniques to provide system information to power system operators
41		Aggregator/retail energy provider	Any marketer, broker, public agency, city, county, or special district that combines the loads of multiple end-use customers in facilitating the sale and purchase of electric energy, transmission, and other services on behalf of these customers
42		Billing	An entity that performs the function of generating an invoice to obtain payment from the customer
43	ESP	Energy service provider	Provides retail electricity, natural gas, and clean energy options, along with energy efficiency products and services
44		Third party	A third party providing a business function outside of the utility
45	PMU	Phasor measurement unit	A device that measures the electrical parameters of an electricity grid with respect to universal time (UTC) such as phase angle, amplitude, and frequency to determine the state of the system
48		Security/network/system management	An entity that monitors and configure the security, network, and system devices

Source: [NISTIR 7628r1]. Public Domain.

second (bps). In a typical implementation, a HAN consists of a broadband Internet connection that is shared between multiple users through a wired or wireless modem. It enables the communication and sharing of resources between computers, mobile, and other devices over a network connection. In Smart Grid implementation, all smart home devices that consume energy and smart meters can be connected to HAN. The device-based data is acquired and transmitted through HAN to the smart meters. HAN allows more efficient home energy management. HAN can be implemented by ZigBee or Ethernet technologies.

1.2.4.2 NAN

NAN is deployed and operated within area of hundred meters, which is actually a few urban buildings. Several HANs can be connected to one NAN, and they transmit data of energy consumed by each house to the NAN. The NAN delivers this data to local data centers for storage. This data storage is important for charging the consumers and data analysis for energy generation demand pattern recognition. The NAN has up to 2 Kbps transmission data rate. The NAN can be implemented by PLC, Wi-Fi, and cellular technologies.

1.2.4.3 WAN

WAN is deployed and operated within vast area of tens of kilometers and consists of several NANs and local distribution companies (LDCs). LDC is a distribution company that maintains the portion of the utility supply grid that is closest to the residential and small commercial consumer. Moreover, the communication of all Smart Grid's components, including operator control center, main and renewable energy generation, transmission and distribution, is based on WAN. The WAN has very high transmission data rate up to few Gbps. The WAN can be implemented by Ethernet networks, WiMAX, 3G/LTE, and microwave transmission.

Several published research works are focused on Smart Grid communication infrastructure. This work [Chen 2010] presents a communication infrastructure for Smart Grid based on hierarchical electricity distribution. The *power* system is typically *hierarchical* and is divided by functional areas depending on the voltage levels. These levels are the transmission network, the high/middle voltage (HV/MV) substations, the MV *distribution* network, the MV/LV transformer substations, and the low voltage (LV) network.

This work [Yan 2013] approaches Smart Grid communication infrastructure as a system of systems that is extremely complex. As a consequence, modeling, analysis, and design of a suitable communication infrastructure should meet many new challenges. The models to be used must be capable of accounting for uncertainty as a way to simulate emerging behavior. The control system and particularly communication infrastructure must be designed to manage uncertainty and inconsistencies to be resilient or gracefully degrade when necessary. Finally, the performance metric must be adjusted to the new nature of the power system.

The authors of this work [Yan 2013] explore the challenges for Smart Grid system. Since a Smart Grid system may have over millions of consumers and devices, the demand for reliability and security is extremely critical. Through a communication infrastructure, a Smart Grid can improve power reliability and quality to eliminate electricity blackout. Security is a challenging issue since the ongoing Smart Grid systems are facing increasing vulnerabilities as more and more automation, remote monitoring, controlling, and supervision entities are interconnected. The authors argue that a Smart Grid built on the technologies of sensing, communications, and control technologies offer a very promising future for utilities and users. Summarizing the basic requirements of communication infrastructure in Smart Grid paradigm highlights that efficiency, reliability, and security of interconnected devices and systems are critical to enabling Smart Grid communications.

However, sensors can be used in many Smart Grid applications, but their limitations and challenges need to be accounted. For example, sensors for substation condition motoring can be simple or complex, depending on the parameters to be monitored and monitoring techniques for transformer condition monitoring. Sensors are also used on other power equipment such as surge arresters, circuit

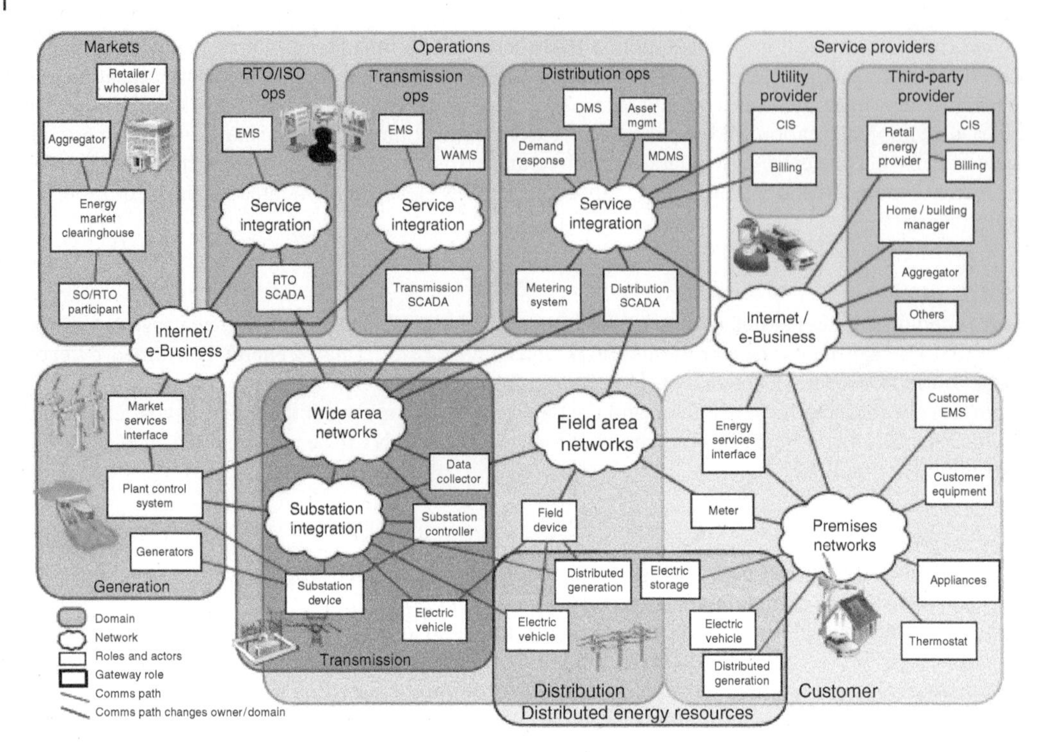

Figure 1.10 Legacy application types within NIST conceptual domains. *Source:* [NISTIR 7628]. Public Domain.

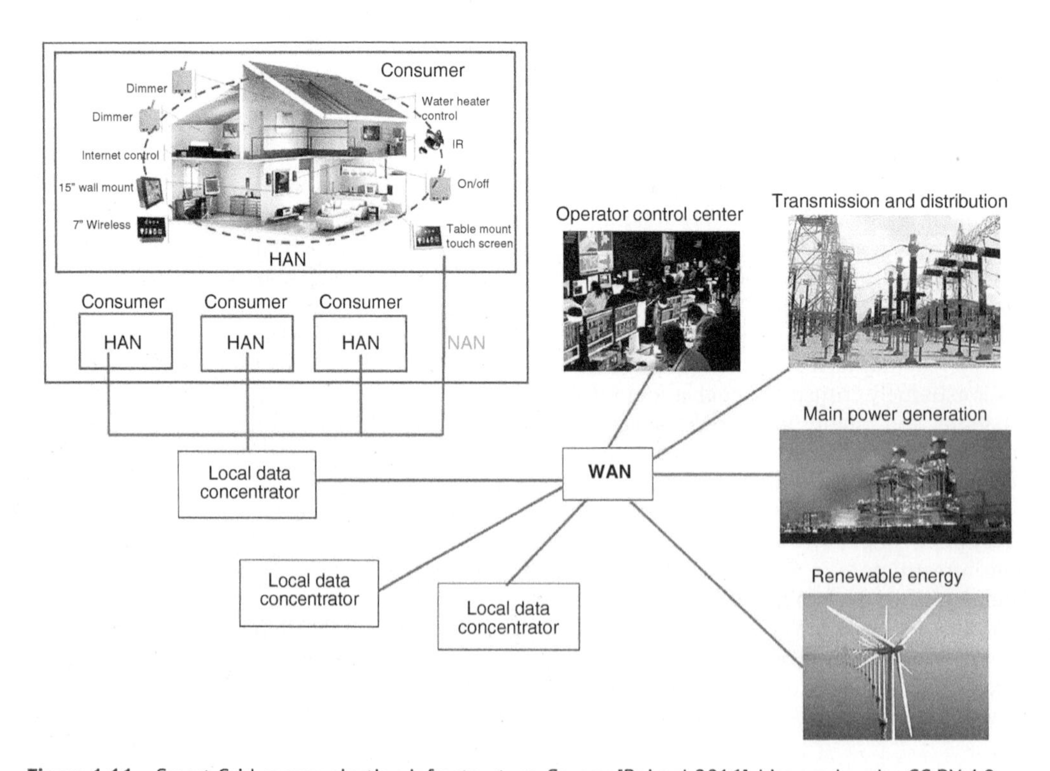

Figure 1.11 Smart Grid communication infrastructure. *Source:* [Baimel 2016]. Licensed under CC BY 4.0.

breakers, instrument transformers, capacitor banks, station batteries, etc. While there are benefits of using the sensor technology, there should be a strategy how to use in a secure way.

There should be an understanding of the benefits and challenges of several technologies that are currently available to be used for Smart Grid implementation. This work [Baimel 2016] provides a brief introduction on the advantages and disadvantages of some common technologies used for Smart Grid communication infrastructure:

- Although ZigBee technology offers advantages (e.g. low cost, relatively small bandwidth use), its operation in unlicensed frequency of 868 MHz and 2.4 GHz could have interference with other Wi-Fi, Bluetooth, and microwave signals.
- Advantages of the cellular networks include already existing infrastructure with wide area of deployment, high rates of data transfer, and available security algorithms that are already implemented in the cellular communications, but major disadvantage is sharing cellular networks with other users, so they are not fully dedicated to the Smart Grid communications; their use be a serious problem in case of emergency state of the grid.
- Advantages of the PLC are already established as widespread infrastructure that reduces installation costs; disadvantages are presence of higher harmonics in the power lines that interfere with communication signals and limited frequency of communication.

Therefore, with each technology used for Smart Grid implementation, there should be more vigilance in early identification of the security challenges of that technology in the environment to be used such that a decision for risk mitigation in terms of acceptance, transfer, or avoidance must be made before and during the design phase. As highlighted in [Baimel 2016], the success of the Smart Grid depends directly on reliable, robust, and secure communication system with high data rate capability.

1.2.5 Secure Energy Infrastructure

Figure 1.12 depicts the relationship between secure energy infrastructure and Smart Grid characteristics to include power system reliability, power quality, integrated communication, and compatible devices and appliances.

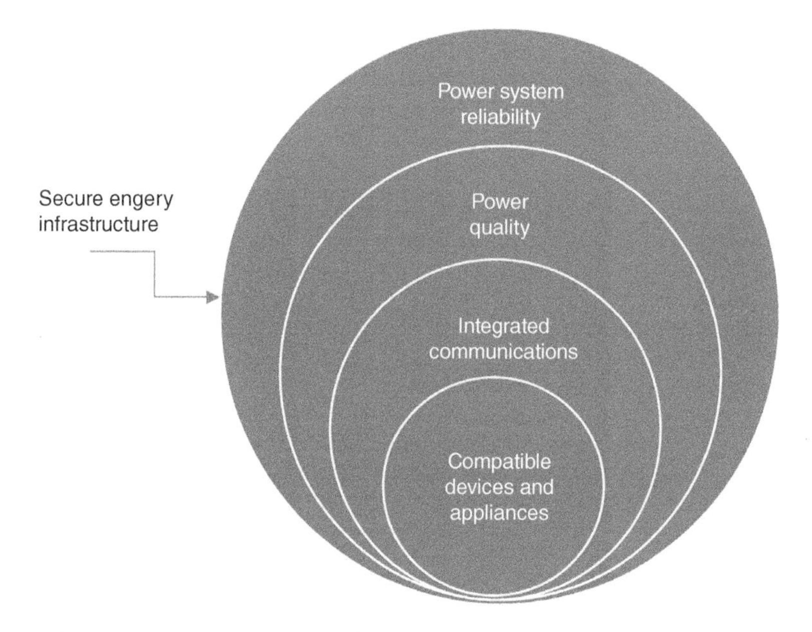

Figure 1.12 Relationships on secure energy infrastructure.

Besides employing security controls to protect systems or networks, this work [Liu 2018] highlights the need to ensure that networks are deployed in configurations that allow them to have resilience in spite of cyber attacks being present. Network resilience is the ability of the network to withstand harm or to return to an acceptable operational condition after it has been harmed by external perturbations. From a network perspective, adversarial resilience happens through two basic mechanisms:

- Properly setting up the connections between entities in a network to withstand threats and allow network protocols to have an avenue for redirecting communications when connections are broken.
- Through feedback mechanisms that call forth additional redundancy in the network's design to repurpose components to meet new challenges.

The resilience requires a network topological structure that can be adapted to provide enhanced resilience to attacks against a network. The authors argue that the performance and resilience of communication networks are intimately tied to their underlying topological structure. Networks whose topological graphs do not exhibit sufficient interconnection are brittle and can be easily broken by network attacks that break links or remove nodes from the network. Representing a network as a graph, the authors explore the notion of network connectivity as quantified by the Fiedler value. The magnitude of this value reflects how well connected the overall graph is. It has been used in analyzing the robustness and synchronizability of networks. The Fiedler value can be used by an adversary to determine locations where to attack a communication network and, conversely, how it can be used by network defense protocols to strengthen the network, thereby providing resilience to network attacks.

Due to the distributed nature of these DER systems, it becomes necessary to properly interconnect them with the rest of the Smart Grid network [ENISA 2016]. These communication networks must cover all the devices from the DER systems in order to enable remote control from the utility and operation centers and ensure their efficient interaction within the whole grid. Furthermore, as they are distributed, these communications will need to be secured to protect them from attacks and accesses from unauthorized users.

1.3 Distributed Energy Resources

Distributed energy resources or DERs are dispatchable energy generation and storage technologies, typically up to 10 MWs in size, that are interconnected to the distribution grid to provide electric capacity and/or energy to a customer or a group of customers and potentially export the excess to the grid for economical purposes. Distributed generation (DG), distributed resources (DR), DER, or dispersed power (DP) is the use of small-scale power generation technologies located close to the load being served. DERs may be subdivided into:

- DG, which can be classified as:
  - Renewable generation, e.g. small wind power systems, solar photovoltaic (PV), fuel cells, biofuel generators and digesters, nuclear, hydro, and other generation resources using renewable fuel
  - Nonrenewable generation, e.g. microturbines, small combustion turbines, diesel, natural gas and dual-fueled engines, etc.
- Distributed storage, which can be classified as electric storage, mechanical, electrochemical, thermal, etc.
  - Electric storage, e.g. battery systems and uninterruptible power supplies (UPS), flywheel, superconducting magnetic energy storage (SMES), etc.
  - Thermal storage – these convert electric power to thermal (cooling/heating) energy for later use, e.g. ice-based air cooling system and high-capacity brick-based air heating system.
- Plug-in electric vehicles (PEVs) and hybrid vehicles (PEV/PHEV) may also be considered as DER; the EVs may be used to supply stored electric energy back to the grid.

DERs are small, modular, energy generation and storage technologies that provide electric capacity or energy where it is needed. A unified and generally accepted definition of DER does not exist, so many definitions have evolved from different perspectives over time.

The current trend to a utility infrastructure containing DERs (also called DG), energy storage systems, and central control may offer future relief from grid-transmitted and facility-caused disturbances. The financial impacts of less than adequate power quality and reliability and the consequent demand for solutions, coupled with the unique characteristics of DERs, offer an opportunity for more rapid deployments of DERs as primary and/or backup power supplies. DERs have the potential to overcome utility and site-specific supply problems and to help compensate for other developing limitations of the existing distribution grid.

A DER is a device that produces electricity and is connected to the electrical system, either behind the meter in the customer's premises or on the utility's primary distribution system. A DER can utilize a variety of energy inputs including, but not limited to, liquid petroleum fuels, biofuels, natural gas, sun, wind, and geothermal energy. Electricity storage devices can also be classified as DERs.

DERs are smaller power sources that can be aggregated to provide power necessary to meet regular demand. As the electricity grid continues to modernize, DERs based on storage and advanced renewable technologies can help facilitate the transition to a smarter grid [EPRI 2015].

1.3.1 DER Characteristics

The following includes a list of key characteristics for DERs:

- Generate electric power.
- Provide energy according to the consumer demand.
- Run on fuel or use previous stored energy can be dispatched for a given period of time.
- Small-scale power generation sources located close to where electricity is used.
- Provide an alternative to or an enhancement of the traditional electric power grid.
- Systems can be quickly developed.
- Costly efficient alternative to the construction of large central power plants and high-voltage transmission lines [Capehart 2014].
- Offer consumers the potential for lower cost, higher service reliability, high power quality, increased energy efficiency, and energy independence.
- Reduce emissions.
- Expand options for energy services.
- Operate in the grid-connected mode or islanding mode [Feng 2012].
- Dispatched remotely by utility or locally by individual owners.
- Controlled locally or under the control of a distribution management system (DMS) or load aggregator [Taylor 2012].
- Various stakeholders such as energy companies, equipment suppliers, regulators, energy users, and financial and supporting companies.

However, DERs are used in several ways and in different systems.

1.3.2 DER Uses

Behind-the-meter DER may be bundled with regular load and managed alongside the demand response (DR) resources – such as a residential rooftop PV solar panel. But often, DER is treated separately in part due to its control capabilities. In addition to a regular retail tariff, behind-the-meter DER may be subject to net metering or feed-in tariff, where excess generation can be exported to the grid at an established or a dynamic price.

Similar to DR resources, DER assets can be registered and enrolled into a DR program. DR is defined as changes in electric use by demand-side resources from their normal consumption patterns in response to changes in the price of electricity or to incentive payments designed to induce lower electricity use at times of high wholesale market prices or when system reliability is jeopardized [FERC 2012].

Furthermore, DER assets are typically required to meet additional technical requirements and certification for grid interconnection.

Depending on the size of a DER and its export capabilities, submetering and telemetry capabilities may be required to monitor the impact of the DER operation on the distribution grid reliability and power quality. Also, renewable resources may receive renewable energy credits (RECs) and may also qualify as a must-run resource, e.g. wind power in some regions. These capabilities need to be incorporated into pricing and control signals associated with DER operation.

Also, DER systems can be interconnected with many other systems in the Smart Grid. The major systems interconnected with DER systems are microgrids, distribution system, and synchrophasor system. These systems such as microgrid and synchrophasor system along with DER help in providing high-quality energy with increased efficiency and reliability where consumer can produce and manage their energy usage. Distribution system helps in connecting these independent generation units with the main power grid.

As the penetration rate of renewable energy increases, besides issues on how to connect renewable to power grid and operate the renewable or how to build storage plants, other issues have to be addressed. These challenges include:

- Implementation of priority applications as identified by Federal Energy Regulatory Commission (FERC) [FERC 2009]:
  - Demand and response.
  - Wide area situational awareness, which means to know continually what is going on in space and time in a dynamic environment, with awareness of potential threats, opportunities, and the range and implications of potential actions and options.
  - Energy storage.
  - Electric transportation.
- A consistent approach for integrating the communication backbone for providing business and control information between different systems, entities, consumers, and service providers [EPRI 2007].
- Interoperability and standard interfaces between systems as well as interfaces with DER networks and systems [EPRI 2007].

However, these additional requirements exceed the capabilities of the existing grid. They cannot be achieved by simply modifying the current supervisory control and data acquisition (SCADA) network. Therefore, the Smart Grid communication network must incorporate new design features that also accommodate other two major requirements:

- Integration of time-dependent renewable resources.
- Controlling the load that is affected by the dynamic consumer demand and response.

Other examples of the new technologies and applications include:

- Utility-scale renewable sources that feed energy into the transmission system.
- Distributed and renewable energy resources that feed into the distribution system.
- PEVs, which will potentially create large load increases in some sections of the grid.
- New demand-side management techniques that will give consumers interactive ways to participate in Smart Grid markets.
- Storage technologies that allow introducing some latency between the generation and consumption cycles to help compensate for the time-varying nature of renewable resources such as wind or solar energy.

Also, FERC [FERC 2009] identifies two crosscutting priorities, namely, cybersecurity and communication and coordination across intersystem interfaces.

1.3.3 DER Systems

DER systems include a combination of technologies and energy options and can be used in several ways [Renewable Energy], [DER Systems]. The effective use of grid-connected DER may require DER with more reliable capabilities such as power electronic interfaces, communications, and control devices for efficient dispatch and operation [Taylor 2012].

DER systems include generation and storage systems, both renewable and nonrenewable. Renewable systems may include PV systems, wind turbines, biofuel systems, fuel cells, battery storage systems,

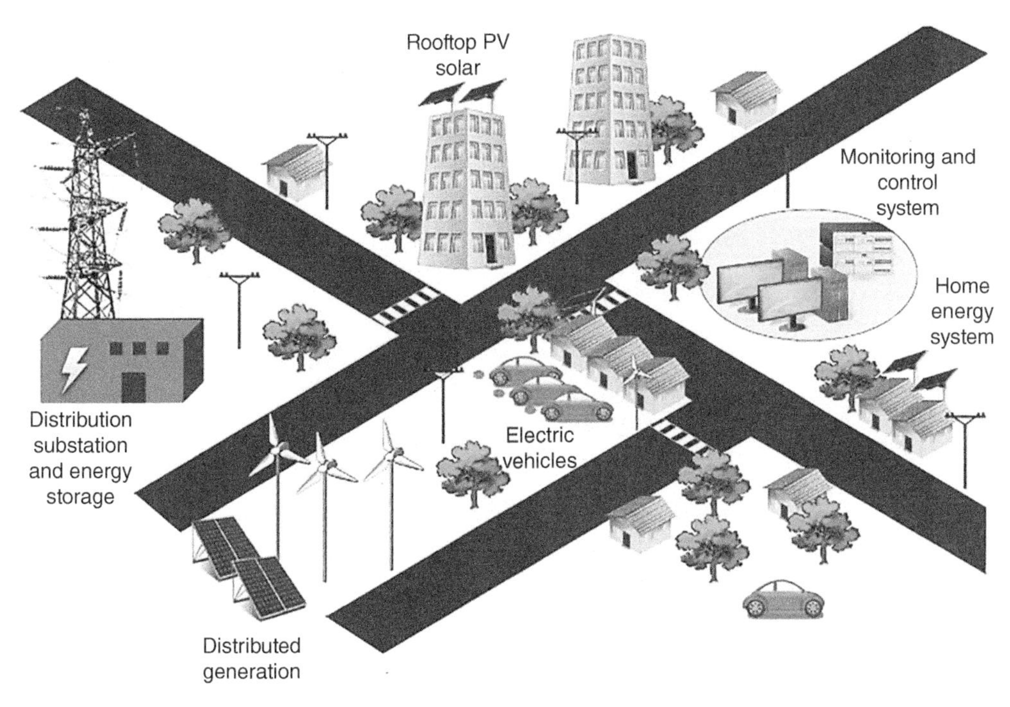

Figure 1.13 DER locations scenario. *Source:* [ENISA 2015b]. Public Domain.

electric and thermal storage systems, cogeneration systems, and small hydro plants [DER Systems]. Nonrenewable systems may include diesel generators, gas turbine generators, and others [NREL 2014]. Figure 1.13 illustrates the different DER systems that can provide energy to the Smart Grid, utility-scale providers, or homes.

Each type of DER system has its own unique characteristics, but in general, each DER system can be treated as a small- to medium-sized source of electric power. EVs can sometimes act as DER systems. Since EVs also have different purposes, they are identified as separate from the other types of DER systems in [EPRI 2013].

Security concerns related to DERs expand too including many areas, from the smallest entity (device) to the highest entity (Smart Grid). Another issue is the dispersed responsibility for security control. Utilities do not typically have direct organizational control over these DER systems and often need to operate through DER owners, commercial retail energy providers, aggregators, virtual power plant (VPP) managers, and other third parties [ENISA 2015b]. Thus, key drivers of DERs developments include microgrids and VPP [DOE 2015b].

1.3.4 Microgrid

A group of interconnected loads and DERs within clearly defined electrical boundaries that acts as a single controllable entity with respect to the grid is called a microgrid. The microgrid can connect and disconnect from the grid to enable it to operate in both grid-connected and island modes. The concept of the microgrid is rapidly evolving beyond backup power to include islanding capabilities for critical infrastructure and the management of DERs (e.g. batteries, renewables, and EVs) in conjunction with building or industrial loads.

Most microgrids operating today are single-customer microgrids and focus on integrating traditional generation resources (e.g. CHP and diesel generators) with new technologies such as renewable generation and electric energy storage systems. Customized communication and control technologies were developed to enable these resources to act as a single entity with respect to the grid.

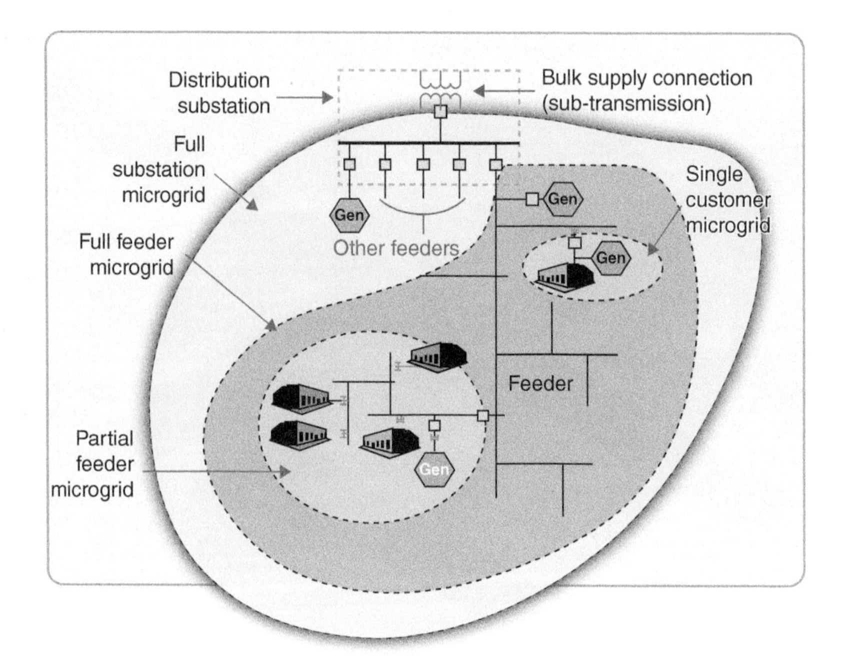

Figure 1.14 Alternative microgrid configurations. *Source:* [DOE 2015b]. Public Domain.

Figure 1.14 shows the concept of a nested microgrid with a naming convention according to their configuration that is used to differentiate them. Microgrids can exist in multiple configurations: independently, networked along a feeder, or nested within another. Microgrids help with:

- Reliability by serving as a grid resource in grid-connected mode and switching to island mode on detecting a contingency, thus improving reliability metrics
- Energy security by ensuring critical loads can be served for sustained periods of time during catastrophic events such as hurricanes or attacks
- Environment by operating in a manner to maximize electricity produced from renewable resources to reduce overall emissions

1.3.5 Virtual Power Plant

A VPP is an operating concept where a group of DERs that are geographically disperse (e.g. associated with different utility meters, residing on different feeders, or not having clearly defined electrical boundaries) is aggregated and coordinated to act as a single entity (see Figure 1.15). This technology concept can include any combination of individual grid-enabled customer resources (e.g. DG, EVs, and energy storage) or integrated resources (e.g. smart buildings and microgrids). Additionally, the control of the

Figure 1.15 VPP schematic view. *Source:* [OECD 2012b]. © 2012, OECD.

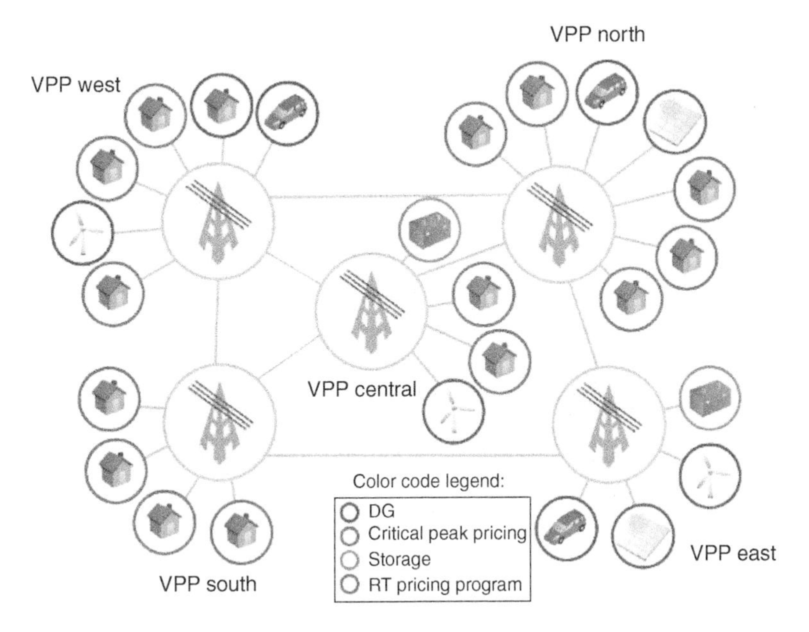

Figure 1.16 Illustration of connected virtual power plants. *Source:* [DOE 2015b]. Public Domain.

aggregated resources is accomplished through a mix of strategies and signals that can involve markets [Zurborg 2010].

The main components of a VPP are interconnected control and management devices as well as an integrated software management system. A software platform, such as a distributed energy resource management system (DERMS), can be used to implement the VPP concept [Saadeh 2015].

The concept of connected VPP is shown in Figure 1.16. An integrated and interoperable DRMS with various utility DMSs makes it possible to use VPP for purposes such as:

- Optimize distribution system conditions
- Aggregate various DR to provide services to the bulk transmission system

Advances in communication, modeling, and controls are needed to use this technology more broadly. Examples of information systems for Smart Grid and DER systems are further discussed in the book.

1.4 Scope of Security and Privacy

Defining the scope of security in the context of IT is not a trivial task because security is complex. Security is defined as the policies, practices, and technology that must be in place for an organization to use information for supporting business activities with a reasonable assurance of safety [Volonino 2004]. There are numerous corporate assets to protect. The protection involves managerial policies, technologies, and legal and ethical issues.

Therefore, in the context of this book, security has a wide base and addresses specific issues regarding Smart Grids, DERs, power grids, network communications, computers, technologies, information, organizations, critical infrastructures, and people.

Security relates to several areas such as physical security, information security, operation security, communication security, national security, network security, disaster recovery, laws and ethics, management, etc. These areas are interrelated and they affect each other. All technology, hardware, people, and procedures are woven together into a security fabric [Harris 2005]. In fact, the so-called information society is increasingly dependent on a wide range of software systems whose mission is critical, such as telecommunications, control systems, financial systems, and power systems. The potential losses that are faced by businesses and organizations that rely on all these systems, both hardware and software, therefore signify that it is crucial for information systems to be properly secured from the outset.

Security is a condition of the system being protected from unintended or unauthorized access, change, or destruction [IIC 2015]. Therefore, a product, system, or service is considered to be secure to the extent that its users can rely on its functions (or will function) in the intended way [ENISA Glossary].

1.4.1 Security for the Smart Grid

Often security has different meanings to different people (e.g. security professional, control engineer, management). There are several definitions and terms that sometimes make the security an ambiguous field. With the Smart Grid one can associate also the word security, which is one of few Smart Grid terms that is overworked and overloaded (e.g. assigned multiple definitions). Such definitions range all the way from ensuring reliability, keeping the lights on, to protecting the confidentiality of customer information [Nordell 2012]. There are many facets of security that one needs to understand such as security as reliability, security as communication reliability, and security as information protection.

1.4.1.1 Security as Reliability

Security as reliability is one aspect that is used by power engineers to describe power system reliability as the ability of the bulk power system to withstand unexpected disturbances such as short circuits or unanticipated loss of system elements due to natural causes [Nordell 2012]. Traditional power engineering practice views reliability as security.

Electric system reliability is the degree to which the performance of the elements of the electrical system results in power being delivered to consumers within accepted standards and in the amount desired. Reliability in this context encompasses two concepts, adequacy and security:

- Adequacy implies that there are sufficient generation and transmission resources installed and available to meet projected electrical demand plus reserves for contingencies.
- Security implies that the system will remain intact operationally (e.g. will have sufficient available operating capacity) even after outages or other equipment failure. The degree of reliability may be measured by the frequency, duration, and magnitude of adverse effects on consumer service.

Among reasons for developing an interconnected electric utility system is also the improvement in the reliability of services to customers when individual generating plant reliability was (and still is) much less than 100%. Currently, the security focus of the industry has expanded to include withstanding disturbances caused by man-made physical or cyber attacks.

1.4.1.2 Security as Communication Reliability

Security as communication reliability is used to describe the reliability for power system communication, which has several facets, including the probability that a given message will be lost entirely, the use of redundant communication paths and automatic failover to protect against message loss, the expected time delay (latency) in delivering a message, and the expected variability of that time delay (jitter) [Nordell 2012]. It also involves how competing messages may (or may not) be given priority when communication channels are saturated. This latter parameter is known as quality of service (QoS) and has long been practiced in the world of telephony, but it is a relatively new concept for power system engineers.

1.4.1.3 Security as Information Protection

Security as information protection involves measures taken to ensure the anonymity of electronic information, both in transit and when stored on digital systems; of primary importance is information related to protecting personal information related to utility customers and information about the electric power system that may be of interest to parties who wish to harm the utility or its customers [Nordell 2012].

The four interrelated dimensions to energy security are described as physical, cyber, supply, and conflict-related as defined in [DOE 2015a]:

- Physical security risks are related to damage to energy supply, storage, and delivery infrastructures, such as the electric grid, pipeline networks, and rail and marine systems.
- Cybersecurity risks are related to the compromise of ICT-based controls that operate and coordinate energy supply, delivery, and end-use systems.
- Supply security risks are related to price shocks and international supply disruptions of energy commodities, critical materials, and/or equipment.
- Conflict-related security risks are associated with unrest in foreign countries linked to, or impacting, energy.

Therefore, multiple definitions of security need to be explored to find some common thread that can help ensure the success of the pursuit of a smarter electrical grid while maintaining security – in all of its various meanings [Nordell 2012].

Grid security and the privacy of people including consumers are of vital importance in the energy sector. If there is any compromise of the personal data or security of the power service, it can undermine everything. An incident would not only create a breach of privacy or security, but it might also compromise the potential future markets the technology might have been able to create if the service had been secure.

1.4.2 Privacy

Similar to security, privacy has many definitions for use on different contexts, cultures, and jurisdictions. One definition is provided as [Dictionary 1994]:

> The condition of being secluded from others; secrecy.

Generally, privacy means a state in which an individual is not observed or disturbed by others.

Privacy refers to protection of personal data. Personal data means any information relating to an identified or identifiable individual (data subject) [Shei 2013].

In the Internet and Web context, where users exchange private data via Web or email with organizations or other users, sometimes unknown users, users experience many concerns:

- What personal information can be shared with whom.
- Whether and how one can share information anonymously.

Thus, users are concerned with privacy as it relates to personally identifiable information (PII). This is associated with collection, ownership, access control, integrity control, distribution, modifications, repurposing, reconstruction, and disposition of relating to an individual.

In some situations, an individual might choose to withhold their identity to be publicly unknown or anonymous. In protecting the PII, one option is anonymity. Anonymity is a result of not having identifying characteristics (such as a name or description of physical appearance) disclosed. More concepts and principles related to privacy are available at [OECD 2016]. Therefore, privacy rights are defined in constitutional and common law. Privacy laws deal with the regulation of personal information about individuals that can be collected, stored, and used by governments and other public as well as private organizations.

There is not one universal, internationally accepted definition of privacy; it can mean many things to different individuals. At its most basic, privacy can be seen as the right to be left alone. Privacy terms are defined differently among various industries, groups, countries, and even individuals. Furthermore, privacy should not be confused, as it often is, with being the same as confidentiality, and personal information is not the same as confidential information. Confidential information is information for which access should be limited to only those with a business need to know and that could result in compromise to a system, data, application, or other business function if inappropriately shared.

Additionally, privacy can often be confused with security. Although there may be significant overlap between the two, they are also distinct concepts. There can be security without having privacy, but there cannot be privacy without security; it is one of the elements of privacy.

1.4.2.1 Privacy in the Smart Grid

It is important to understand that privacy considerations with respect to a Smart Grid include examining the rights, values, and interests of individuals; it involves the related characteristics, descriptive information, and activities [NISTIR 7628r1]. Thus, data privacy is impacted by the practices of customers who supply personal data and all entities that gather or handle that data.

Also, new energy usage data collected outside of smart meters, such as from home energy management systems (EMS), is also created through applications of Smart Grid technologies. As those data items become more specific and are made available to additional individuals, the complexity of the associated privacy issues increases as well.

Another perspective on privacy is described as consisting of four dimensions [NISTIR 7628r1]:

- Privacy of personal information involves the right to control when, where, how, to whom, and to what extent an individual shares his/her own personal information, as well as the right to access personal information given to others, to correct it, and to ensure it is safeguarded and disposed of appropriately.
- Privacy of the person is the right to control the integrity of one's own identity and body (physical requirements, health problems, and required medical devices).
- Privacy of personal behavior is the right to keep any knowledge of their activities, and their choices, from being shared with others.
- Privacy of personal communications is the right to communicate without undue surveillance, monitoring, or censorship.

Privacy as a strategy for Smart Grid applications should include all four dimensions [NISTIR 7628r1]. Most Smart Grid entities directly address the personal information dimension, but the other dimensions are not included. There is a gap in the laws and regulations. Therefore, the other dimensions should also be considered in the Smart Grid context because new types of energy use data may be created and communicated. Unique electric signatures for consumer electronics and appliances could be compared against some common appliance usage profiles to develop detailed, time-stamped activity reports within personal dwellings. Charging station information might reveal the detailed whereabouts of an EV/PEV/PHEV. This data did not exist before the application of Smart Grid technologies. Smart Grid applications may reveal details (energy usage patterns or other type of activities), either explicitly or implicitly, about an individual's household dwelling or other type of premises.

Although many of the types of data items accessible through the Smart Grid are not new, there is now the possibility that other parties, entities, or individuals will have access to those data items, and there are now many new uses for and ways to analyze the collected data, which may raise substantial privacy concerns. The reputation of an energy service provider might also be impacted by gaps in customer data privacy protection.

1.4.3 The Need for Security and Privacy

Security has a wide base and addresses specific issues regarding computers, information, and organizations. The continuous growth of cybersecurity threats and attacks including the increasing sophistication of the malware is impacting the security of critical infrastructure, ICS, power grids, EMS, and SCADA control systems [Hentea 2007].

There is a growing concern about the security and safety of the control systems in terms of vulnerabilities, lack of protection, and awareness.

Besides security concerns, computer systems including control systems raise the issue of safety causing harm and catastrophic damage when they fail to support applications as intended. Therefore, information security management principles, processes, and security architecture need to be applied to smart power grid systems without exception [ENISA 2015a].

Smart Grid technologies and applications create new security and privacy risks and concerns in unexpected ways. Concerns of privacy of consumers and people are of vital importance in the energy sector. If there is any compromise of the personal data or security of the power service, it can undermine many services and applications. An incident would not only create a breach of privacy or confidentiality, integrity, or availability of the information, but it might also compromise the potential future markets the technology might have been able to create if it the service had been secure. Therefore, the vulnerability of the power system is not mainly a matter of electric system or physical system, but it is also a matter of cybersecurity. Attacks (such as attacks upon the power system, attacks by the power system, and attacks through the power system) to the Smart Grid applications could bring huge damage to the economy and public safety.

In complex interactive systems like Smart Grid whose elements are tightly coupled, the likelihood of targeted attack as well as failures from erroneous operations and natural disasters and accidents is quite high. Vulnerabilities and attacks can be at different levels – software controlling or controlled device, application, storage, data access, LAN, enterprise, private communication links, and public PSTN and Internet-based communications.

The destruction of power grid systems and assets would have a debilitating impact on energy security, economic security, public health, or safety. With a system that handles power generation, transmission, and distribution, security responsibility extends beyond the traditional walls of the data center. An intruder can, intentionally or unintentionally, cause a power line to be energized that would endanger lives. Similarly, a power line may be de-energized in such a way as to cause damage to transmission and control systems and possibly endanger the safety of employees and the public. Therefore, each organization should develop its own policy to protect assets, employees, and general public who are at risk when human (intentional or unintentional) threats or natural disasters occur.

Security controls (called also safeguards, measures, or countermeasures) are needed to ensure protection of an organization assets (tangible and intangible) and people as well as safety of people. Tangible assets are physical assets that include power equipment, computers, devices, facilities, and supplies. Intangible assets include data, information, reputation, intellectual property, copyrights, trade secrets, business strategies, and any other information valuable to any organization.

It is recognized that as new capabilities are included in the Smart Grid, potential privacy issues may occur [NISTIR 7628]. A privacy policy framework for the Smart Grid and for smart homes is suggested in [GridWise 2011]. This framework is limited and addresses only consumer privacy issues that arise from the collection, use, and retention of such data no matter from what source it is collected.

1.5 Computing and Information Systems for Business and Industrial Applications

Business firms and other organizations rely on information systems to carry out and manage their operations, interact with their customers and suppliers, and compete in the marketplace. An information system is composed of many components that include hardware, software, data, people, procedures, and networks necessary to use information as a resource in an organization [Whitman 2011]. An information system is an organized assembly of computing and communication resources and procedures – e.g. equipment and services, together with their supporting infrastructure, facilities, and personnel – that create, collect, record, process, store, transport, retrieve, display, disseminate, control, or dispose information to accomplish a specified set of functions [RFC 4949].

An information system aims to support operations, management, and decision making. In a broad sense, the term is used to refer not only to the ICT that an organization uses but also to the way in which people interact with this technology in support of business processes. At the center of any computing or information system is data and information.

1.5.1 Information System Classification

Figure 1.17 shows different types of information systems. Distinct types of information systems include the following [O'Brien 2011]:

- Management information systems are used to analyze and facilitate strategic and operational activities. They utilize systems to generate information to improve efficiency and effectiveness of decision making, providing critical information tailored to the information needs of executives (e.g. decision support).
- Operations support information systems refer to the interconnected databases and applications implemented for the ongoing central maintenance of electronic operational data. Operations information system process data generated by business operations, maintain records about the exchanges, handle routine (yet critical) tasks, monitor, and control industrial processes.

As depicted in this diagram, both categories Operations support and Management Support Systems categories include specialized processing systems such expert systems (expert advice to decision makers), knowledge management systems (manage organizational knowledge), strategic information systems (support competitive advantage), and functional business systems (support business functions).

Also, information systems include specialized systems such as industrial/process control systems, telephone switching and private branch exchange (PBX) systems, and environmental control systems [CNSSI 4009].

Process control systems, SCADA systems, distributed control systems (DCS), and other smaller control system configurations including skid-mounted programmable logic controllers (PLC) are often found in the industrial sectors and critical infrastructures. These are also known under a general term, industrial control system (ICS) [NIST SP800-82r2]. SCADA systems and DCS monitor the flow of electricity from generators through transmission and distribution lines. These electronic systems enable efficient operation and management of electric systems through the use of automated data collection and equipment control. A related term is industrial control network, which is a system of interconnected equipment used to monitor and control physical equipment in industrial environments [Galloway 2012].

Industrial control system is a generalized term referring to a system of electronic components that control the physical operations of machines. Automated or operator-entered commands can be issued to machines, either locally in-plant or remote station control devices, often referred to as field devices. The machines may transmit sensor data back to the controller for monitoring and automated operational functions. ICSs are typically used to operate the infrastructure in industries such as electrical, water, oil and gas, discrete manufacturing, and chemical including experimental and

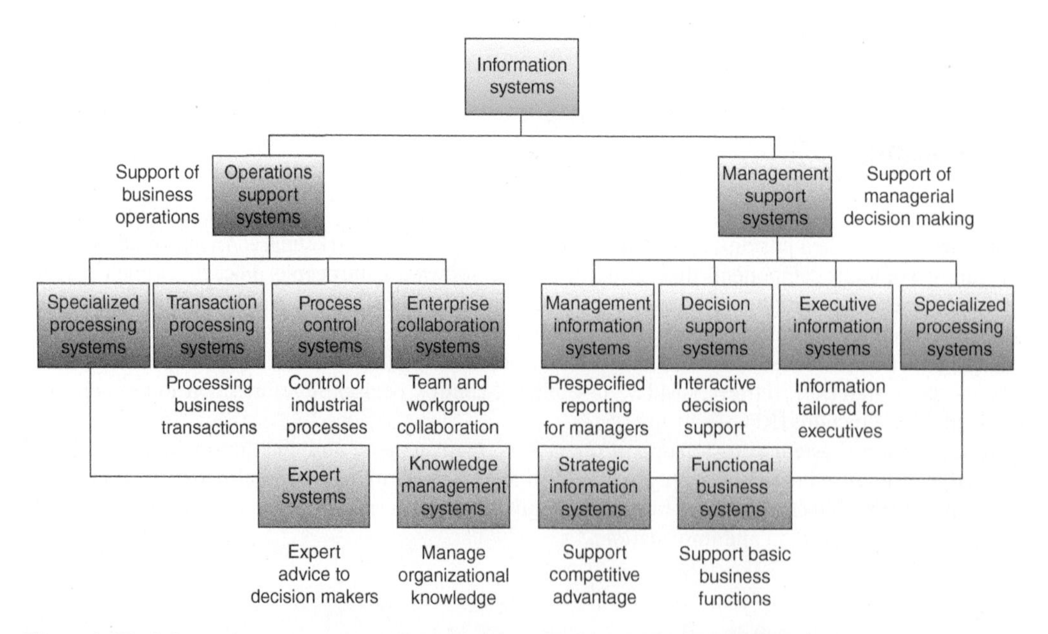

Figure 1.17 Information systems classification. *Source:* [O'Brien 1999]. © 1999, McGraw-Hill.

research facilities such as nuclear fusion laboratories. The reliable operation of modern infrastructures including Smart Grid depends on computerized systems and SCADA systems.

1.5.2 Information Systems in Power Grids

The power grid transmits and distributes electrical power generated from primary fossil or renewable energy resources, such as coal or wind. Computers and networks manage, monitor, protect, and control the continuous real-time delivery of electrical power. Figure 1.18 depicts a view of key characteristics of Smart Grid compared with traditional power grid.

A Smart Grid system may include IT, which is a discrete system of electronic information resources organized for the collection, processing, maintenance, use, sharing, dissemination, or disposition of information. A Smart Grid system may also consist of operational technologies (OT) or ICS, which comprise several types of operational and control systems, including SCADA systems, DCS, and other control system configurations such as skid-mounted PLC that are often found in the industrial sectors and critical infrastructures [NISTIR 7628r1].

The Smart Grid has great potential for driving innovation in the ways electricity is produced, managed, and consumed. As described in [OECD 2012b], applications of ICTs and especially the opportunities provided by the Internet can help sustain electricity supply while limiting environmental

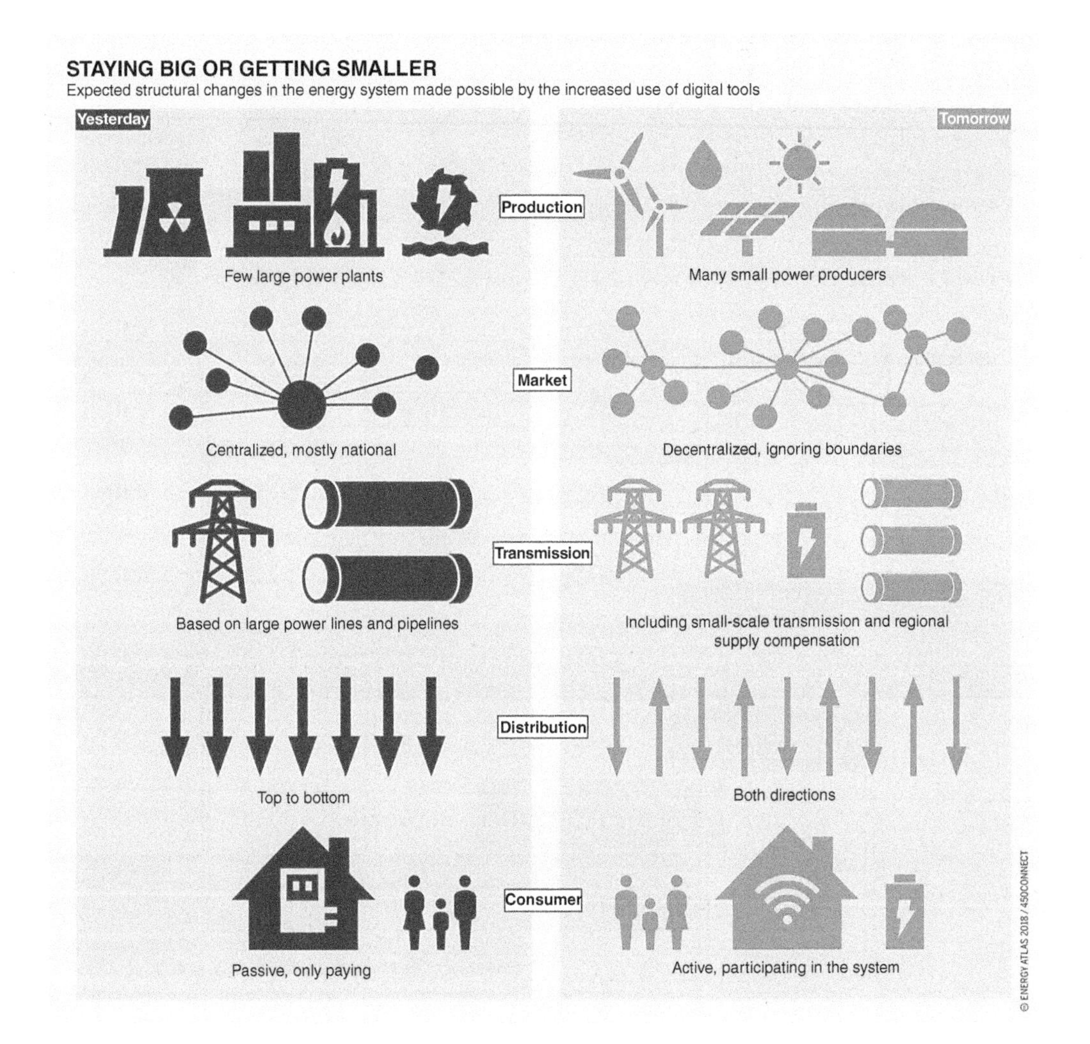

Figure 1.18 Smart Grid characteristics vs. traditional power system. *Source:* [Bartz/Stockmar]. Licensed under CC BY-SA 4.0.

impacts. In a Smart Grid, ICTs estimate the operational state with the vision of surviving a cyber incident while sustaining critical energy delivery functions and optimized power flow for economic and efficient generation dispatch. ICTs are seen as promoting a wider integration of renewable energy sources, promoting low-carbon transport options including EVs, and inducting structural shifts in electricity consumption [OECD 2012d].

The Smart Grid is a particular application area expected to help tackle a number of structural challenges that global energy supply and demand are facing. Technologies and the use of data enable improved and more accurate information about the availability, price, and environmental impacts of energy, thereby empowering producers and consumers to make more informed energy conservation choices.

The Internet especially gives rise to a new generation of businesses providing services around electricity, adding further value and innovation to the energy sector value chain. The transition to a modern grid requires the adoption of advanced technologies, such as smart meters, automated feeder switches, fiber-optic and wireless networks, storage, and other new hardware. These devices require a new communication and control layer to manage a changing mix of supply resources and provide new services.

Figure 1.19 shows a simplified view of the energy sector value chain. However, information gaps may affect the Smart Grid outcomes.

Figure 1.20 depicts an overview of the ICTs for different domains (generation, transmission, distribution, customer) and customer areas (industrial, building, residential).

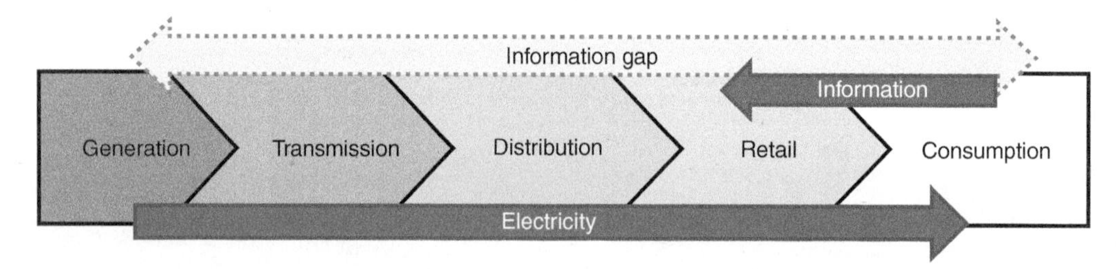

Figure 1.19 Stylized electricity sector value chain. *Source:* [OECD 2012b]. © 2012, OECD.

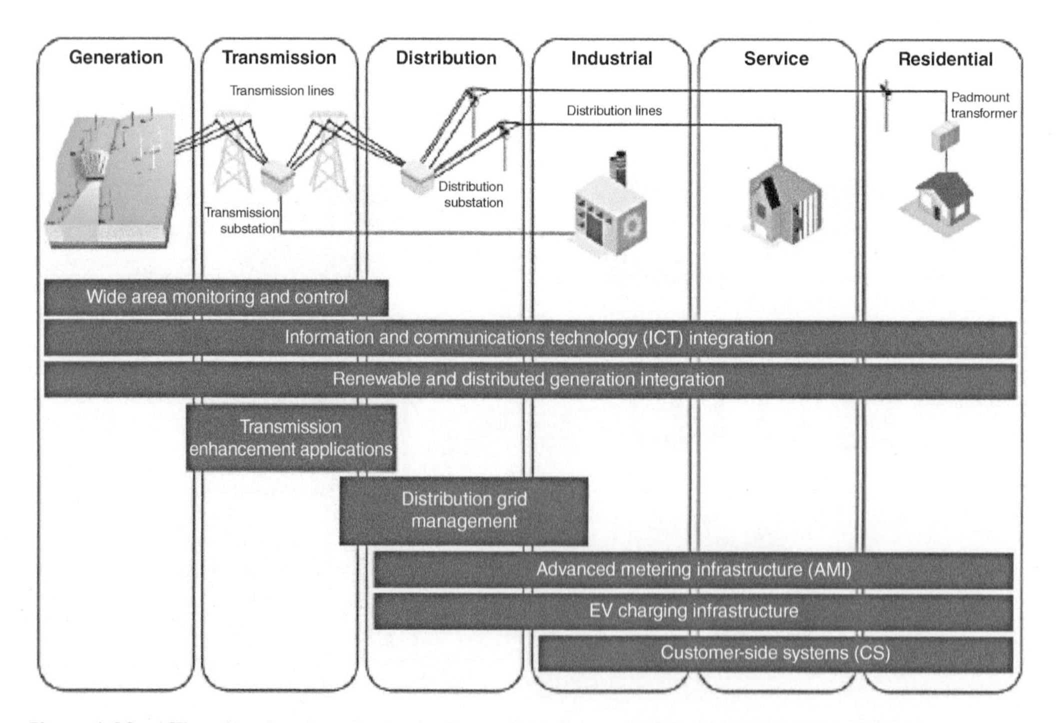

Figure 1.20 ICT application domains in the Smart Grid. *Source:* [OECD 2012b]. © 2012, OECD.

ICTs and Internet applications are essential to modernization of the electric power grid in every domain. A summary of electricity challenges and ICTs addressing these challenges is shown in Table 1.3.

Table 1.3 Electricity sector challenges and potential ICT applications.

Electricity sector challenges	ICT applications
Generation	
Renewable energy generation	Smart meters Vehicle-to-grid (V2G) and grid-to-vehicle (G2V)
Distributed, small-scale electricity generation	Virtual power plants Vehicle-to-grid (V2G) and grid-to-vehicle (G2V) Smart meters
Transport (transmission and distribution)	
Transmission and distribution grid management	Sensor-based networks Embedded systems and software Integrated software systems and application programming interfaces (APIs) Smart meters Communication protocols, including machine-to-machine communications (M2M)
Storage	
Storage capacities (physical and logical)	V2G, G2V and vehicle-to-home (V2H) Smart meters End-user interfaces
Retail	
Dynamic and real-time pricing for electricity consumption and distributed generation	Smart meters End-user interfaces
Consumption	
Electricity conservation and energy efficiency	End-user interfaces Smart meters Electricity data intelligence
Demand management (automated)	End-user interfaces Smart meters Communication protocols including M2M Smart building technologies Smart electronic devices Data centers and cloud computing
Integration of electric vehicles and renewable sources	End-user interfaces Smart meters V2G, G2V Communication protocols including M2M Integrated software systems and APIs
Facilitate access to electricity in developing countries (electrification)	

Source: [OECD 2012b]. © 2012, OECD.

1.5.3 DER Information Systems

As DERs are used in the Smart Grid, a brief overview of DER systems helps security engineers and control engineers to understand the scope of DER tasks before planning for security protection. Examples of DER information systems (called also DER systems) that are used for DER operations and other tasks include the following:

- Wide area situational awareness encompasses monitoring and display of power system components and performance across interconnections and over large geographic areas in near real time. The goals of situational awareness are to understand and ultimately optimize the management of power network components (including DERs), behavior, and performance, as well as to anticipate, prevent, or respond to problems before disruptions can arise.
- Cybersecurity management encompasses measures to ensure the protection of information, control systems, IT systems, and telecommunication infrastructures that support DERs.
- Network management includes monitoring, reporting, and control of communications; for example, DER systems use a variety of public and private communication networks, both wired and wireless; the identification of performance metrics and core operational requirements of different applications is also critical to the Smart Grid.
- DR management and automation is focused on maximizing performance of DERs and PEVs; automation of distribution systems becomes increasingly more important to the efficient and reliable operation of the overall power system; the benefits of distribution management include increased reliability, reductions in peak loads, and improved capabilities for managing distributed sources of renewable energy.
- Energy storage management systems include means of storing energy, directly or indirectly, data acquisition, controls of communications, and security protections.
- Electric transportation refers, primarily, to enabling large-scale integration of PEVs to increase use of renewable sources of energy.
- Database management systems and the communication-associated networks that create a two-way network between advanced meters and utility business systems, enabling collection and distribution of information to customers and other parties, such as the competitive retail supplier or the utility itself; customers may be consumers and producers of energy (owners of renewable and storage systems).

The following are examples of DER information systems that can be included in the category of management information systems:

- DR and consumer energy efficiency encompasses mechanisms and incentives for utilities, business, industrial, and residential customers to cut energy use during times of peak demand or when power reliability is at risk; DR is necessary for optimizing the balance of power supply and demand.
- Information security management system supports the approach to developing, implementing, and improving the effectiveness of an organization's information security with regard to the management of risk.

Other categories of information systems may include:

- End-user computing systems to support the direct hands on use of computers by end users for operational and managerial applications.
- Business information systems to support the operational and managerial applications of the basic business functions of the organization, for example, integrated systems in Smart Grid.
- Strategic information systems to support the operational and managerial applications of the basic business functions of the organization.

1.6 Integrated Systems in a Smart Grid

OT represents a broad category of applications that utilities use for operations that include safe and reliable generation and delivery of energy. OT includes power equipment (e.g. operating gear, oil circuit breakers, solid-state relays, and many devices in between) and control room applications such as SCADA systems that monitor the network, reaching out to devices as complex as substation gateways or to devices as simple as sensors. OT is often applied within a mission critical framework.

If OT is more focused in operations, IT systems are just the opposite. IT systems allow machines to exchange information directly with humans, usually within a second or longer. Examples include improved enterprise resource planning (ERP), geographic information systems (GIS), and customer relationship management (CRM) systems, along with office-based productivity tools and mobile computing devices, which have now permeated the utility workplace. Until recently, the growth in IT stood independent of the hidden OT equipment, serving and protecting the grid. The interconnection of a wide array of traditional and emerging utility business and operation information systems such as EMS, DMS, SCADA, market management system (MMS), AMI, Customer Information System (CIS), and outage management system (OMS) allows the utilities, customers, and other service providers not only to monitor energy usage and grid status with higher precision and accuracy but also to analyze and activate distributed energy resources and storage options and construct pricing responses to appropriately reorient consumption patterns to balance available power generation capacity and demand continuously – all in real time. These activities require secure automated information exchange, analysis, and intelligent decision making distributed throughout the grid.

Therefore, the picture is starting to change. The Smart Grid is transforming utility operations and pushing IT across its traditional boundary into OT so rapidly that is blurring the distinction between the two categories (see Figure 1.21). The dynamics of IT and OT integration and how utilities can leverage this convergence for smarter, more cost-effective, and more reliable operation are discussed in [Meyers 2013].

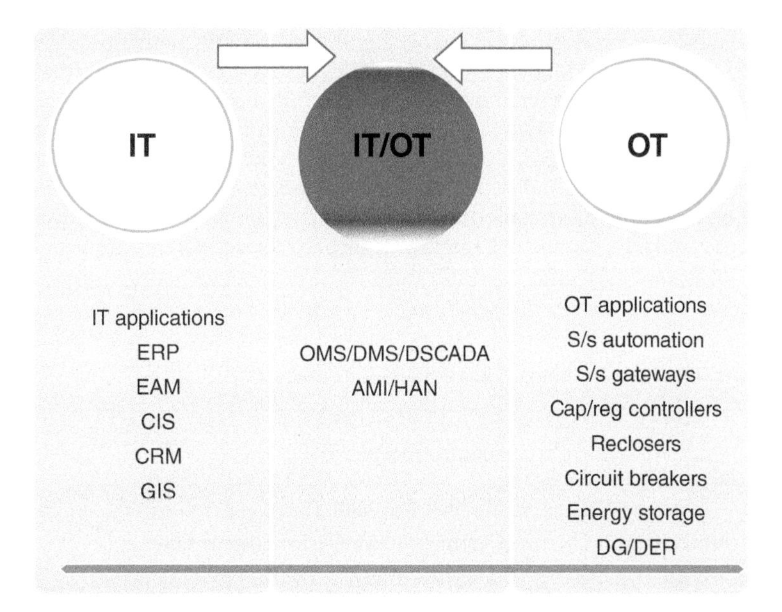

Figure 1.21 OT and IT integration. ERP, Enterprise Resources Planning; EAM, Enterprise Asset Management; GIS, Geographic Information Systems; CRM, Customer Relationship Management; CIS, Customer Information Systems; OMS, Outage Management System; DMS, Distribution Management Systems; DG, Distributed Generation; DER, Distributed Energy Resource; HAN, Home Area Network; AMI, Advanced Metering Infrastructure; DSCADA, Distribution Supervisory Control And Data Acquisition (DSCADA)

Smart meters and home area network technologies are also helping to blur the lines between the energy supply and energy distribution domains. For example, the OMS at many sites has migrated away from the enterprise and toward the operation domain.

1.6.1 Trends

The Smart Grid is redefining technology norms, and the modernization is characterized by the following trends:

- The continuous growth in OT deployment.
- The continuous implementation of IT by the utility to model, monitor, and manage its distribution system.
- An urgent requirement for utilities to integrate their IT and OT networks.

However, the extensive digital technologies that enable significant improvements to new energy systems also increase the attack surface for cyber intrusion. For power system and supply security reasons, DER systems have to include ancillary services that are commonly seen on traditional power systems or bulk generation systems, so as to ensure compatibility with older and legacy devices and systems.

1.6.2 Characteristics

The characteristics of the future grid are distinctly different from those of the current power system as shown in Table 1.4. As shown in this table, there are more challenges including security that have to be addressed by Smart Grid systems.

The extensive digital technologies that enable significant improvements to new energy systems also increase the attack surface for cyber intrusion. The increasingly significant role of information systems and growing dependence on managing the power flow (generation, transmission, and distribution), markets, customers, financial, and trade needs call for special efforts to foster confidence in computing systems and information systems for business and industrial applications. Since the emergence of Internet and World Wide Web technologies, the control systems were integrated with the business and IT systems and became more exposed to cyber threats. Although specific threats target control systems and intelligent devices, these systems are also exposed to the

Table 1.4 Comparison of key attributes of current and future systems.

Current system	Future paradigm
Monolitic	Modular and agile
Centralized generation	Centralized and distributed generation
Decisions driven by cost	Decisions driven by cost and sustainability
Vulnerable to catastrophic events	Contained events
Limited energy choices	Personalized energy options
Vulnerable to new threats	Inherently secure against threats

Source: [DOE 2015b]. Public Domain.

same cyberspace threats as any business system because they share the common vulnerabilities with the traditional IT systems.

1.7 Critical Smart Grid Systems

The availability and reliability of computing and information systems for business and power grid applications are dependent on the secure operations of ICSs and other infrastructures. The following sections include a brief introduction of key systems and security concerns. Any attack on any of these systems can propagate on other systems too.

1.7.1 Industrial Control Systems

A control system is a device or set of devices to manage, command, direct, or regulate the behavior of other devices or systems. ICSs are typically used to operate the infrastructure in industries such as electrical, water, oil and gas, and chemical including experimental and research facilities such as nuclear fusion laboratories. SCADA systems, DCS, and other smaller control system configurations including skid-mounted PLC are often found in the industrial sectors and critical infrastructures. These are also known under a general term, ICS. The reliable operation of modern infrastructures depends on computerized systems and SCADA systems.

In the past, control systems were isolated from other IT systems. Connection to the Internet is new (early 1990s) and debatable among specialists. Many experts agree that exposing control systems to the public PSTN and Internet carries unacceptable risk. However, even without any connection to the Internet, these systems are still vulnerable to external or internal attackers that can exploit vulnerabilities in private communication network and protocol, software such as operating systems, custom and vendor software, data storage software, databases, and applications.

Control systems are exposed to the same cyberspace threats like any business system because they share the common vulnerabilities with the traditional IT systems. In complex interactive systems like Smart Grid whose elements are tightly coupled, likelihood of targeted attack as well as failures from erroneous operations and natural disasters and accidents is quite high. Vulnerabilities and attacks could be at different levels – software controlling or controlled device, application, storage, data access, LAN, enterprise, private communication links, and public PSTN and Internet-based communications.

1.7.2 SCADA Systems

SCADA system is a common process automation system that is used to gather data from sensors and instruments located at remote sites and to transmit data at a central site for either control or monitoring purposes. The collected data is usually viewed on one or more SCADA host computers located at the central or master site. Based on information received from remote stations, automated or operator-driven supervisory commands can be pushed to remote station control devices, which are often referred to as field devices. Generally, a SCADA system includes the following components:

- Instruments that sense process variables.
- Operating equipment connected to instruments.
- Local processors that collect data and communicate with the site's instruments and operating equipment called PLC, remote terminal unit (RTU), intelligent electronic device (IED), or programmable automation controller (PAC).
- Short-range communications between local processors, instruments, and operating equipment.

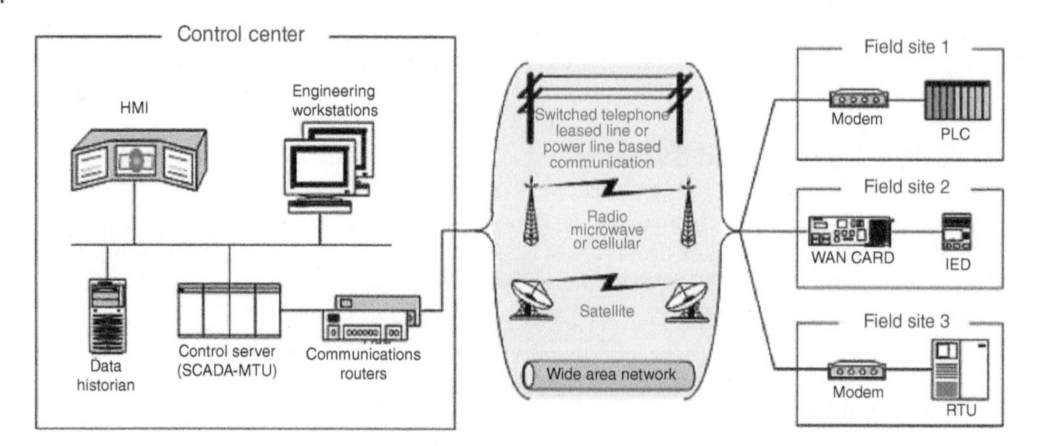

Figure 1.22 SCADA general diagram. *Source:* [NIST SP800-82r2]. Public Domain.

- Host computers as central point of human monitoring and control of the processes, storing databases, and display of statistical control charts and reports. Host computers are also known as master terminal unit (MTU), the SCADA server, or a PC with human–machine interface (HMI).
- Long-range communications between local processors and host computers using wired and/or wireless network connections.

Figure 1.22 shows a typical SCADA system with HMI installed in a control center and connected via communication network to monitored field sites.

SCADA system is a category of control systems used to monitor or control processes such as chemical, transport, water supply, power generation and distribution, and gas and oil supply. A control system is a device or set of devices to manage, command, direct, or regulate the behavior of other devices or systems. However, SCADA is not a full control system, but rather focuses on the supervisory level. Usually, SCADA systems involve a human-in-the-loop control and decision-making processes.

The architecture of a SCADA system consists of one of more MTUs that are used by engineers in a control station to monitor and control a large number of RTUs located in field or industrial plants. An MTU is a general-purpose computer or server running SCADA utility programs and RTUs are generally small dedicated devices designed for rough field or industrial environment. One or more SCADA MTUs retrieve real-time analog and status data from RTUs, store to data historian, and analyze the data. MTUs automatically send control commands to the RTUs or enable the engineers to do so manually.

Vulnerability discovery techniques and appropriate engineering activities are required to ensure security, reliability, and safety of plants that use SCADA control systems.

These systems evolved from static to dynamic systems. The increased connectivity to Internet and mobile device technology has also a major impact on control system architectures. Modern products are often based on component architectures using commercial off-the-shelf (COTS) product elements as units. Security and safety of the SCADA control systems in terms of vulnerabilities, lack of protection, and awareness are discussed in [Hentea 2007]. Information security management principles and processes need to be applied to SCADA systems without exception.

1.7.3 Energy Management Systems

Although not unique definition is used for EMS, one definition is more common:

> Computer systems used by operators of electric to monitor, control, and optimize the performance of the power grid (generation, transmission, distribution).

The functionality of any EMS is supported by a combination of two systems:

- SCADA system for monitoring and control.
- Advanced applications for optimization.

Energy management may be supported by distributed energy management technologies that include energy storage devices and various methods for reducing overall electrical load. The technology references as SCADA/EMS or EMS/SCADA are common. In [Weiss 2010], EMS is described as a SCADA system with additional applications.

EMS may exclude the monitoring and control functions, more specifically referring to a suite of power applications. In other circumstances, EMS refers to a system of hardware and software components for an automated control and monitoring of the heating, ventilation, and lighting needs of a building or group of buildings such as university campuses, office buildings, or factories [Panke 2002]. Most of these EMS also provide facilities for the reading of electricity, gas, and water meters. The data obtained from these EMS systems can then be used to produce trend analysis and annual consumption forecasts.

The increasing number of SCADA/EMS systems operating in the electric industry, the growing number of market participants, and the development of complex market models relying more on IT technologies contribute to raising the interdependency between the operation of the power grid and the operation of the wholesale electric market.

1.7.4 Advanced Meter Systems

The AMI enable measurement of detailed, time-based information and frequent collection and transmittal of such information to various parties. Advanced metering systems are composed of hardware, software, communications, consumer energy displays and controllers, and applications (such as customer-associated systems, meter data management, and supplier business systems).

AMI typically refers to the full measurement and collection system that includes meters at the customer site; communication networks between the customer and a service provider, such as an electric, gas, or water utility; and data reception and management systems that make the information available to the service provider [EPRI 2007], [UCAIUG 2008].

Smart Grid is defined as system of systems, implying also a network of information networks. Although some characteristics are similar to IT systems, AMI systems have characteristics that differ from traditional information systems. Many of these differences stem from the fact that AMI systems are integrated into the physical power grid. In some cases, adversely impacting an AMI system can pose significant risk to the health and safety of human lives and serious damage to the environment, as well as serious financial issues such as production losses.

Unlike automatic meter reading, AMI enables two-way communications with the meter. Smart meters enable two-way communication between the meter and the central system. A smart meter is an electronic device that records consumption of electric energy, natural gas, or water in intervals of an hour or less and communicates that information at least daily back to the utility for monitoring and billing purposes. Each meter must be able to reliably and securely communicate the information collected to some central location.

In addition to communication with the headend network, smart meters may need to be part of a home area network, which can include a customer display and a hub to interface one or more meters with the headend. Technologies for this network vary from country to country. Many security concerns center on the inherent hacking weakness of wireless technology, combined with the remotely controllable software incorporated into smart meters.

Security of the devices and information and other issues associated with meter data transmission from the customer meters to the AMI host system need to be addressed to ensure that only authorized devices provide and receive meter data.

1.8 Standards, Guidelines, and Recommendations

In the development of Smart Grid, requirements may be developed based on standards, guidelines, and recommendations defined as follows:

- Standards are documents, established by consensus, that provide rules, guidelines, or characteristics for products, activities, or their results; once adopted by an industry, they can be enforced, and an organization may comply.
- Guidelines are a set of recommended practices produced by a recognized authoritative source representing subject matter experts and community consensus or internally by an organization; the guidelines may be adopted, but there are no compliance requirements.
- Recommendations are a set of advices and good practices.

1.8.1 Overview of Various Standards

Standards are being talked about all over the world, and it is a confusing world full of new acronyms, players, consultants, and companies. While overwhelming at times and competitive at others, standards organizations are all working toward the same goal. The goal is for vendor interoperability and collaboration without the installation of custom systems. Standards help in maximizing safety, compatibility, and quality of processes and products.

Technical standards are the result of a standardization process, and they can be classified as:

- De facto standards that are followed by informal convention or dominant usage.
- De jure standards that are part of legally binding contracts, laws, or regulations.
- Voluntary standards that are published and available for people to consider for use.

A technical standard is a formal document of specifications that define the capabilities of a product for a particular use including the function and performance of a device or system. It is usually a document that establishes uniform engineering or technical criteria, methods, processes, and practices. A standard or a group of standards can be used to support a function.

Standardization is greatly increased when companies release new products to market. Compatibility is important for products to be successful; this allows consumers to use their new items along with what they already own. Also, standards may include safety issues following these guidelines [ISO/IEC 51].

The existence of a published standard does not necessarily imply that it is useful or correct. Just because an item is stamped with a standard number does not, by itself, indicate that the item is fit for any particular use. The people who use the item or service (engineers, trade unions, etc.) or specify it (building codes, government, industry, etc.) have the responsibility to consider the available standards, specify the correct one, enforce compliance, and use the item correctly. Therefore, key attributes need to be considered.

1.8.2 Key Standard Attributes and Conformance

Key standard attributes should include:

- Enable interoperability – Standards are critical to enabling interoperable systems and components; mature, robust standards are the foundation of mass markets for the millions of components that will have a role in the future Smart Grid.
- Enable innovation – Standards enable innovation where thousands of companies may construct individual components.
- Enable consistency – Standards also enable consistency in system management and maintenance over the life cycles of components.

The electricity industry is facing multiple competing standards, guidelines, and recommendations. The Smart Grid is no exception.

1.8.3 Smart Grid Standards

Collaboration on standards for the Smart Grid is a must, and the power industry needs to embrace the work the standards organizations are doing. Participation is encouraged; but more important are the conversations that occur at appropriate levels of abstraction. It is also important for all standards organizations to work at the level that suits the application. For example, when specifying a protocol, the bits and bytes are important; however, when specifying a business integration framework, the bits and bytes are less important than workflow, which takes precedence. This is a delicate balance but one that the industry will need to develop and embrace [Vos 2009].

In the Smart Grid, the evidence of the essential role of standards is growing. A Congressional Research Service report, for example, cited the ongoing deployment of smart meters as an area in need of widely accepted standards. The US investment in smart meters predicted to be at least $40 billion for a million new smart meters over a period of five years [Kaplan 2009], [ON World 2009], [WH 2016].

The principle of using open standards is highly encouraged to support interoperability and therefore help popularize new technologies [Microsoft 2002]. An open standard is a standard that is publicly available and has various rights to use associated with it. It may also include an explanation of how it was designed and why (e.g. open process). There is no single definition, and interpretations vary with usage [Open Standard]. Among known standards organizations, only the IETF and ITU-T explicitly refer to their standards as open standards that allow reasonable and nondiscriminatory patent licensing fee requirements. However, those in the open-source software community think that an open standard is only open if it can be freely adopted, implemented, and extended.

1.8.3.1 Key Players in Smart Grid Standards Development

The major key players involved in the development of standards supporting the Smart Grid include:

- NIST – Private agency (former federal technology agency).
- ISA – International organization of engineers in control of industrial processes.
- SGIP (Smart Grid Interoperability Panel) – Different areas.
- UCA International User Group – Suppliers of electric/gas/water utility systems.
- GridWise Alliance – Transform the US electric grid to achieve a sustainable energy future.
- EPRI/EEI – Research and innovation in many areas.
- ZigBee Alliance – Low-power radio inside buildings.
- Wi-Fi – Interoperability of wireless products.
- IEEE P2030 – Standards and guidelines for Smart Grid applications.
- IRENA – International renewable integration.
- ISO – International standards; technology standards.
- IEC – Active in many areas of the Smart Grid.
- IETF – Internet standards in many areas of the Smart Grid.
- ITU-T – Active in many areas of the telecommunications and Smart Grid.
- ANSI – Working on meters.
- NAESB – North America industry interoperability standards for gas and electricity.
- NEMA – Electrical equipment manufacturers.
- NRECA – Electric cooperative utilities.
- State legislatures.
- Federal/state regulators.
- FERC – US Federal Energy Regulatory Commission.
- NERC – Reliability of US interconnected systems including portions of Canada and Mexico.
- ITU-T – Communications and many Smart Grid applications.
- OASIS – Cross-domain standards for services to enable machine-based scheduling of human-centric activities.
- ASHRAE – HVAC and refrigeration standards.
- IETF – Internet standards for the Smart Grid.
- SAE – Communication between PEV and the electric power grid.
- OpenADR Alliance – Standards for DR implementations.
- Bacnet – Standards for commercial buildings and integration with the Smart Grid.
- OPC – Standards for open connectivity of ICSs and process control.

One of the most important side benefits of the Smart Grid is the work being performed by government and industry groups in collaboration. Developing interoperability standards plays a key role in supporting grid modernization.

The work of the National Institute of Standards and Technology (NIST) and industry associations such as the International Electrotechnical Commission (IEC), the Electric Power Research Institute (EPRI), and the Smart Grid Interoperability Panel (SGIP) and trade groups like the GridWise Alliance (GWA) and GridWise Architecture Council (GWAC) all contribute to establishing the definitions and specifications for connecting grid devices. These groups have enabled rapid progress forward in the development of the Smart Grid. Processes are already in place to close the gaps in current standards. Most grid-focused interoperability projects that adhere to the current standards can now move forward with a high degree of confidence. The most active of these groups include the GridWise Architecture Council and NIST's SGIP.

1.8.3.1.1 GridWise Architecture Council The GridWise Architecture Council includes members from different domains of Smart Grid technology that is sponsored by DOE. Although NIST has been assigned the primary responsibility to coordinate development of a standards framework for information management to achieve interoperability of Smart Grid devices and systems, the Energy Independence and Security Act of 2007 (EISA) requires that NIST consult with GWAC to define the standards and set up investment grants.

The GridWise Architecture Council has enormous influence in the development of the Smart Grid framework and the GWAC stack, adapted from the OSI layered stack, which helped to stimulate innovation in the computer industry.

1.8.3.1.2 NIST Smart Grid Interoperability Panel The NIST initiated the SGIP in 2009 to support NIST in fulfilling its responsibility, under the EISA, to coordinate standard development for the Smart Grid. Since January 2013, SGIP (http://sgip.org) entered a new phase – self-sustaining entity with the majority of funding to come from industry stakeholders. The NIST SGIP is the way NIST interacts with the electricity industry and other stakeholders. They are working on Smart Grid standards, developing priority action plans, and designing the testing and certification standards. SGIP developed the Smart Grid conceptual model and cybersecurity requirements [NISTIR 7628r1] including recommendations for security solutions. Specific NIST activities include:

- Identifying existing applicable standards.
- Addressing and solving gaps where a standard extension or new standard is needed.
- Identifying overlaps where multiple standards address some common information.

NIST maintains an active role and continues to support SGIP's mission to provide a framework for coordinating all Smart Grid stakeholders in an effort to accelerate standard harmonization and advance the interoperability of Smart Grid devices and systems. The catalog of standards (http://sgip.org/Catalog-of-Standards) is a compendium of standards and practices considered to be relevant for the development and deployment of a robust and interoperable Smart Grid. The catalog is expected to be a larger compilation that can support the FERC, but it is independent of FERC decision making.

The SGIP has several priority-specific committees and working groups. NIST maintains an active presence in these groups. Among these groups, we mention the cybersecurity (SGCC) group and domain expert working groups (DEWGs). The SGCC working group identifies and analyzes security requirements and develops a risk mitigation strategy to ensure the security and integrity of the Smart Grid. DEWGs perform analyses and provide expertise in specific application domains including distributed renewables, generation, and storage.

Once there is, in the judgment of the FERC, sufficient consensus concerning the standards developed under NIST's oversight, FERC is directed to adopt such standards and protocols as may be necessary to ensure Smart Grid functionality and interoperability in interstate transmission of electric power and regional and wholesale electricity markets [EISA 2007]. The law delegates to the FERC the responsibility of defining what sufficient consensus and adopts means in the context of the standards.

Recognizing the needs of the energy sector, FERC identified four functional priorities for the development of key interoperability standards for the following areas:

- Demand and response.
- Wide area situational awareness.
- Energy storage.
- Electric transportation.

Also, FERC identifies two crosscutting priorities, system security (cybersecurity and physical security) and intersystem communication, a common semantic framework (e.g. agreement as to meaning and software models) for enabling effective communication and coordination across inter-system interfaces.

On 22 November 2013, FERC approved Version 5 of the critical infrastructure protection standards (CIP Version 5), which represents significant progress in mitigating cyber risks to the bulk power system. In 2014, NERC initiated a program to help industry transition directly from the currently enforceable CIP Version 3 standards to CIP Version 5. The goal of the transition program is to improve industry's understanding of the technical security requirements for CIP Version 5, as well as the expectations for compliance and enforcement.

While NERC-CIP Version 5 of standards was released on 22 November 2013, organizations must transition all high- and medium-impact BES to NERC-CIP v5 on 1 April 2016. Low-impact BES systems can wait until 1 April 2017. However, there is no clear cybersecurity strategy as many CIP standards were made inactive and many standards are pending enforcement. It is recommended to visit [NERC CIP] portal for the most current standards and recent activities.

1.8.3.2 How to Use Standards

One of the predominant topics of the emerging Smart Grid is standardization [Uslar 2013]. Education on how to use standards is rarely the focus of curricula in colleges and universities. Guidelines and books may be useful in getting help for using the standards. A comprehensive introduction to Smart Grid standards and their applications for developers, consumers, and service providers is provided in [Sato 2015]. The authors consider the need for standards interoperability and integration in the Smart Grid. The authors claim a methodology for understanding and identification of the fundamental standards needed by developers for DER, electric storage, and E-mobility/plug-in vehicles. However, many standards may not be applicable forever, but they could become obsolete in a short period of time or could change continuously, or new standards could emerge. Therefore, the methodology to select a new standard is needed.

An introductory textbook for people trying to get firsthand and condensed knowledge on Smart Grid standardization with a focus on ICT as well as to have a reference textbook dealing with the various standards to be applied in Smart Grids is a motivation for the authors of this book [Uslar 2013].

Other criteria may be useful too. For example, it is better to use a mature standard. A mature standard is a standard that has been in use for sufficient time that most of its initial faults and inherent problems have been identified and removed or reduced by further development [NIST SP1108r3].

1.8.4 Cybersecurity Standards

Cybersecurity standards enable organizations to practice safe security techniques and to reduce the number of successful cybersecurity attacks. In general, the standards provide outlines as well as specific techniques for implementing cybersecurity functions. Appendix J includes a list of most common acronyms used in the book.

Cybersecurity guidance is provided by national and international organizations. Standards are continuously developed and revised by different organizations, forums, and associations that are:

- International – e.g. IEC, ISA, ISO, ITU, IETF, IEEE.
- Consortium – e.g. SAE, OGC, ZigBee Alliance, HomePlug Alliance, Wi-Fi Alliance, HomeGrid Forum, OASIS, ISF.
- Regional and National – e.g. NIST, ANSI, NEMA, ASHRAE, NAISB.

DOE is working with NIST to enable manufacturers of products to use current cybersecurity guidance. In 2012, the DOE published a guideline for risk management process [DOE 2012]. In the United States, NIST published standards that are mandatory for federal agencies as well as special publications that provide guidance for information system security for private industries. Examples of alliances include:

- ZigBee.
- Wi-Fi.
- HomePlug.
- Powerline.
- Z-Wave.

Current activities in ICS security are supported by many standards, programs, organizations, forum, and associations such as:

- American Gas Association (AGA) Standard 12, Cryptographic Protection of SCADA Communications.
- American Petroleum Institute (API) Standard 1164, Pipeline SCADA Security.
- Center for Control System Security at Sandia National Laboratories (SNL).
- Chemical Sector Cyber Security Program.
- Chemical Industry Data Exchange (CIDX).
- DHS Control Systems Security Program (CSSP).
- DHS CSSP Recommended Practices.
- DHS Process Control Systems Forum (PCSF).
- Electric Power Research Institute (EPRI).
- Institute of Electrical and Electronics Engineers (IEEE).
- Institute for Information Infrastructure Protection (I3P).
- International Electrotechnical Commission (IEC) Technical Committees 65 and 57.
- ISA99 Industrial Automation and Control Systems Security Standards.
- ISA100 Wireless Systems for Automation.
- International Council on Large Electric Systems (CIGRE).
- LOGI2C – Linking the Oil and Gas Industry to Improve Cyber Security.
- National SCADA Test Bed (NSTB).
- NIST 800 Series Security Guidelines.
- NIST Industrial Control System Security Project.
- NIST Industrial Control Security Testbed.
- North American Electric Reliability Council (NERC).
- SCADA and Control Systems Procurement Project.
- US-CERT Control Systems Security Center (CSSC).

2

Advancing Security

2.1 Emerging Technologies

While the term security (or cybersecurity) is broadly defined and understood, there is a trend about the multidisciplinary aspects of the concept and more specifically about the need to advance technical security. While the technical view about is unilateral, we consider that advancing security for Smart Grid is also needed because of the emerging technologies. Although the world of emerging technologies in Smart Grid is almost incomprehensible, we provide an overview and introduction to these related technologies: Internet of Things (IoT), Internet of Everything (IoE), and cyber–physical systems (CPS). We discuss how these technologies impact security of Smart Grid systems and how security controls should be increased at higher levels.

2.1.1 Internet of Things

In simple terms, the IoT refers to the networked interconnection of everyday objects. The IoT is a general evolution of the Internet from a network of interconnected computers to a network of interconnected objects [IntSoc 2015a].

Since the term IoT was first coined by the Auto-ID center in 1999 [AUTO-ID 1999], the development of the underlying concepts has ever increased its pace [Santucci 2010]. Nowadays, the IoT presents a strong focus of research with various initiatives working on the (re)design, application, and usage of standard Internet technologies in the IoT technology.

Despite being a buzzword, IoT technology denotes a trend where a large number of embedded devices employ communication services offered by communication protocols. The embedded electronics, software, sensors, and network connectivity enable the objects to collect and exchange data. Many of these devices, often called smart objects, are not directly operated by humans but exist as components spread out in the environment [RFC 7452]. Such devices have been used in the industry for decades, usually in the form of non- Internet Protocol (IP)/proprietary protocols that are connected to IP-based networks by way of protocol translation gateways.

The IoT is defined as a global infrastructure for the information society, enabling advanced services by interconnecting (physical and virtual) things based on existing and evolving interoperable information and communication technologies [ITU-T 2012]. The IoT refers broadly to the extension of network connectivity and computing capability to objects, devices, sensors, and items not ordinarily considered to be computers [IntSoc 2015a]. The IoT technology assumes the interconnection of highly heterogeneous networked entities and networks following a number of communication patterns such as human to human (H2H), human to thing (H2T), thing to thing (T2T), or thing to things (T2Ts).

With many definitions, but similar concepts, there is no single, universally accepted definition for the term. Different definitions are used by various groups to describe or promote a particular view of what IoT means and its most important attributes. Some definitions specify the concept of the Internet, while

Building an Effective Security Program for Distributed Energy Resources and Systems: Understanding Security for Smart Grid and Distributed Energy Resources and Systems, Volume 1, First Edition. Mariana Hentea.

others do not [IntSoc 2015a], and others define the connection of things on a multipoint basis [ABI Research 2015a]. The various definitions of IoT emphasize different aspects of the IoT phenomenon from different focal points and use cases, but there are concerns such as the following:

- The disparate definitions could be a source of confusion in dialogue on IoT issues, particularly in discussions between stakeholder groups or industry segments.
- Different perspectives that could be factored into discussions create a vulnerable technology that may not be able to deal with several threats (e.g. economic, cyber, natural, etc.).

Some fuzziness still exists in these definitions, but one argues that every physical object has a virtual component that can produce and consume services and collaborate toward a common goal [Roman 2011]. Things have identities and virtual personalities operating in smart spaces using intelligent interfaces to connect and communicate within social, environmental, and user contexts [EC-EPoSS 2008]. These characteristics enable IoT to extend anywhere, anyhow, anytime computing to anything, anyone, any service [EC-EPoSS 2008], [Roman 2011]. In the IoT paradigm, everything real becomes virtual, which means that each person and thing has a locatable, addressable, and readable counterpart on the Internet.

2.1.1.1 Characteristics of Objects

The interconnected objects have main characteristics [Roman 2011]:

- Existence – Things exist in the physical world with the aid of specific technologies, such as an embedded communication device becoming a virtual persona.
- Sense of self – Things have, either implicitly or explicitly, an identity that describes them and can process information, make decisions, and behave autonomously.
- Connectivity – Things can initiate communication with other entities and an element in their surroundings; a remote entity can locate and access them.
- Interactivity – Things can interoperate and collaborate with a wide range of heterogeneous entities, whether human or machine or real or virtual such that they produce and consume a wide variety of services.
- Dynamicity – Things can interact with other things at any time, any place, and in any way; they can enter and leave the network at will, no need to be limited to a single physical location, and can use a range of interface types.

An optional sixth characteristic is environmental awareness. Sensors can enable a thing to perceive physical and virtual data about its environment, such as water radiation or network overhead (this characteristic is optional because not all things will exhibit it, such as an object enhanced with a lower-end radio-frequency identification (RFID) tag).

2.1.1.2 Technologies

The vision of the IoT has evolved due to a convergence of multiple technologies, ranging from wireless communication to the Internet and from embedded systems to microelectromechanical systems (MEMS). This means that the traditional fields of embedded systems, wireless sensor networks, control systems, automation (including home and building automation), and others all have made contributions to the development of IoT technology (see technology roadmap; Figure 2.1).

Energy-efficient microcontrollers act as brains due to objects with embedded intelligence. A sensor is a special device that perceives certain characteristics of the real world and transfers them into a digital representation.

Sensor technology provides objects with sensory receptors, and RFID provides a way for them to distinguish one another, much as people recognize a face. Finally, low-energy wireless technology, such as specified in the IEEE 802.15.4 standard, supplies the virtual counterparts of voice and hearing. Multiple applications already use these and other technologies, such as machine-to-machine (M2M) communication, virtual worlds, and robotics.

The IoT is a vision that encompasses and surmounts several technologies at the confluence of nanotechnology, biotechnology, information technology (IT), and cognitive sciences [Santucci 2010]. However, to be a virtual being, an IoT object needs only enough technology to realize its role and complete its mission.

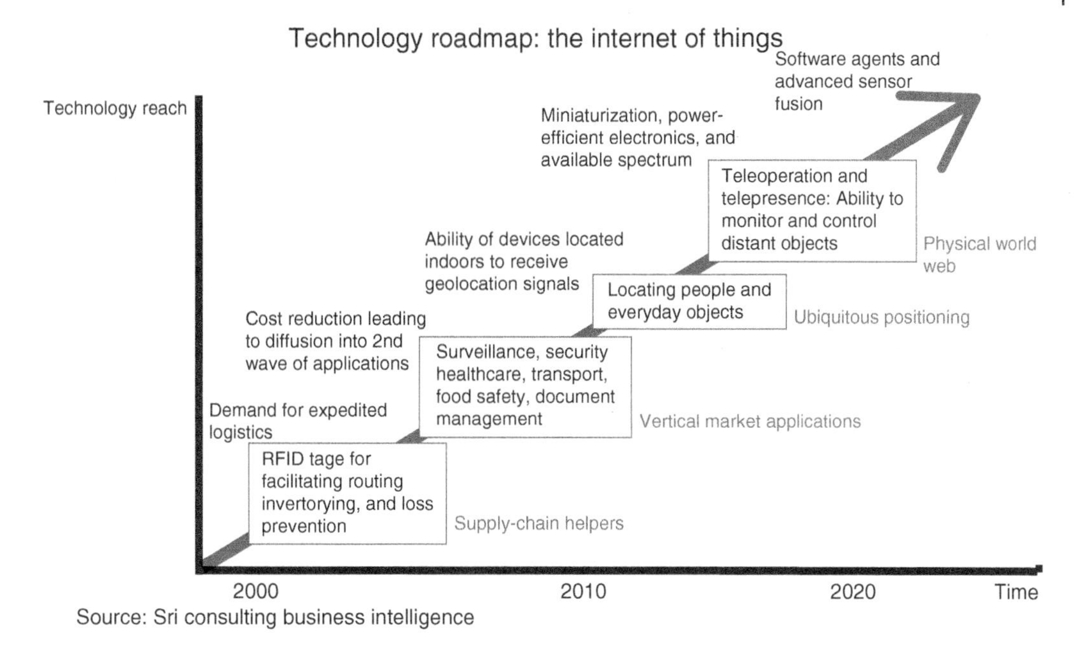

Figure 2.1 Technology road map. *Source:* [WikiTech]. Public Domain.

The combination of various technologies has enabled objects to exhibit these characteristics, allowing them to become virtual beings. IoT technology mixes with social aspects, big data, and cloud computing and as an enabler for CPS. As a result of these technologies and the cross integration of all the technologies, a new level of control and feedback mechanisms is being developed in order to enhance the processes.

2.1.1.3 IoT Applications

IoT technology is expected to offer advanced connectivity of devices, systems, and services that goes beyond M2M communications and covers a variety of protocols, domains, and applications.

By 2025, Internet nodes may reside in everyday things – food packages, furniture, paper documents, and more [SRI BI 2008]. The emergence of various applications includes building and industrial automation and cars that can interconnect millions of objects for sensing things like power quality, tire pressure, and temperature and that can actuate engines and lights. This trend quickly made it of the utmost importance to extend the IP protocol suite for these networks [Borman 2010].

It is predicted that there will be between 20 billion of IoT-connected devices [Gartner 2015]. The IoT is expected to grow to 50 billion connected devices in 2020 [Cisco 2011] and to provide valuable information to consumers, manufacturers, and utility providers. IoT is one of the platforms of today's Smart Grid, smart city, and smart energy management systems.

The most common and relevant IoT applications to Smart Grid and DER systems include energy management, distribution automation, building and home automation, and infrastructure management.

2.1.1.3.1 Energy Management Integration of sensing and actuation systems, connected to the Internet, is likely to optimize energy consumption as a whole. It is expected that IoT devices will be integrated into all forms of energy-consuming devices (switches, power outlets, bulbs, televisions, etc.) and be able to communicate with the utility supply company in order to effectively balance power generation and supply.

IoT technologies offer consumers, manufacturers, and utility providers new ways to manage devices and ultimately conserve resources and save money by using smart meters, home gateways, smart plugs, and connected appliances.

The introduction of electric smart plugs, in-home displays, and smart thermostats has given consumers a choice of which household devices they want to monitor. Additionally, the IoT delivers more data for manufacturers and utility providers to reduce costs through diagnostics and neighborhood-wide meter reading capabilities [Monnier 2013].

Besides home-based energy management, the IoT is especially relevant to the Smart Grid since it provides systems to gather and act on energy and power-related information in an automated fashion with the goal to improve the efficiency, reliability, economics, and sustainability of the production and distribution of electricity.

Another application is a city energy consumption service to monitor the energy consumption of the whole city [Zanella 2014]. This service enables authorities and citizens to get a clear and detailed view of the amount of energy required by different services (public lighting, transportation, traffic lights, control cameras, heating/cooling of public buildings, etc.). This service helps identify the main energy consumption sources and to set priorities in order to optimize their consumption behavior.

2.1.1.3.2 Distribution Automation The Smart Grid provides systems to gather and act on energy and power-related information in an automated fashion with the goal of improving the efficiency, reliability, economics, and sustainability of the production and distribution of electricity. Using advanced metering infrastructure (AMI) devices connected to the Internet backbone, electric utilities can not only collect data from end-user connections but also manage other distribution automation devices like transformers and reclosers. Figure 2.2 shows an application involving a smart meter with capabilities as an intranet of things and a smart meter as part of the IoT.

In the intranet-of-things scenario (left), the meter interacts only with SCADA system. In the IoT scenario (right), the meter interacts with the SCADA system, household members, other houses, and emergency personnel.

2.1.1.3.3 Building and Home Automation IoT devices are being used to monitor and control the mechanical, electrical, and electronic systems used in various types of buildings (e.g. public and private, industrial, institutional, or residential). Home automation systems, like other building automation systems, are typically used to control lighting, heating, ventilation, air conditioning, appliances, communication systems, entertainment, and home security devices to improve convenience, comfort, energy efficiency, and security.

2.1.1.3.4 Infrastructure Management Monitoring and controlling operations of offshore wind farms is a key application of the IoT [IoT 2015]. IoT applications have greater scope and flexibility,

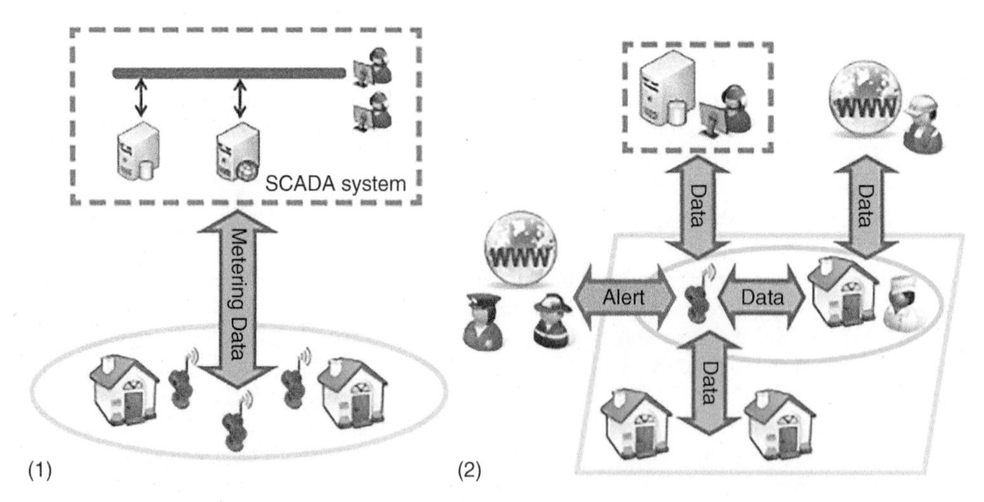

Figure 2.2 A smart meter application in two scenarios. *Source:* [Roman 2011]. © 2011, IEEE.

when they are able to interact not only with objects in other scenarios and domains but also with real and virtual entities.

2.1.1.4 IoT Security and Privacy

Several significant obstacles remain to fulfilling the IoT vision, chief among them security [Roman 2011]. One aspect is that IoT technology is being developed rapidly without appropriate consideration of the profound security challenges involved including the regulatory changes that might be necessary. In particular, as the IoT spreads widely, cyber attacks are likely to become an increasingly physical (rather than simply virtual) threat.

IoT developments point to future threat opportunities and risks that will arise when people can remotely control, locate, and monitor even the most mundane devices and articles to the extent that everyday objects become information security risks. Appliances such as refrigerator may be hacked and used to send spam messages [Starr 2014]. Although email application has not disturbed the power grid, it is wise to think about the danger of using a smart appliance as a platform for launching cyber attacks on Smart Grid systems.

The IoT technology is not only a human tool; instead it should be considered as an active agent because it already influences moral decision making, which in turn affects human agency, privacy, and autonomy. There are concerns regarding the impact of IoT on consumer privacy, because the Big Data focus on collecting everything and keep it around forever will have security impacts.

The IoT devices could distribute those risks far more widely than the Internet has to date. Massively parallel sensor fusion may undermine social cohesion if it proves to be fundamentally incompatible with Fourth Amendment guarantees against unreasonable search [SRI BI 2008].

The IoT and Smart Grid technologies will together be aggressively integrated into the developed world's socio-economic fabric with little, if any public or governmental oversight [Tracy 2015]. Therefore, the technology needs to gain consumers' trust due to privacy concerns [Bachman 2015]. However, more regulations for the manufacturers may benefit the protection of the consumers.

Perceived as creepy new wave of the Internet [Halpern 2014] or as a disruptive technology [SRI BI 2008], in order to have a widespread adoption of any object identification system, there is a need to have a technically sound solution to guarantee privacy and the security of the customers among some other issues [Ishaq 2013]. The challenge is to prevent the growth of malicious models or at least to mitigate and limit their impact [Roman 2011].

IoT applications are impacted by the security features that should make attacks significantly more difficult or even impossible [EC-EPoSS 2008]. The selection of security features and mechanisms will continue to be determined by the impact on business processes, and trade-offs will be made between chip size, cost, functionality, interoperability, security, and privacy.

Some argue that is imperative that companies and governments need to capture these trends to ensure that the IoT is recognized as useful [EC-EPoSS 2008]. In addition, education and information are critical for the success of the IoT technology because privacy concerns about the misuse of information are high and final users do not clearly see the advantages and disadvantages of the widespread adoption of this technology. IoT security is not only a spectrum of device vulnerability but also unique security and privacy concerns of systems using these devices.

Another recommendation is that effective and appropriate security solutions can be achieved only if the participants involved with these devices apply a collaborative security approach as described in [IntSoc 2015b].

2.1.1.5 Challenges

A number of potential challenges may stand in the way of the IoT vision – particularly in the areas of security; privacy; interoperability and standards; legal, regulatory, and rights issues; and the inclusion of emerging economies [IntSoc 2015a]. As concluded in this report, there is a need to address IoT challenges and maximize its benefits while reducing its risks.

From a broader perspective, the IoT can be perceived as a vision with technological and societal implications [ITU-T 2012]. While such extreme interconnection will bring unprecedented convenience and economy, there are many challenges in terms of security and privacy risks that require novel approaches to ensure its safe and ethical use [Roman 2011]. Due to the complex nature of connected devices, their integration with other services, and the general insensitivity of hardware engineers to security issues, security is a technical and a cultural problem that regulators have little power to directly enforce. To make matters worse, even though the US Federal Trade Commission (FTC) recognizes the problem, it can do little to protect consumers as the IoT grows [Clearfield 2013].

While IoT technology can benefit several stakeholders, if the proper technology standards and policies are not in place, the backlash could easily stifle innovation [Palermo 2014].

Over the next 10–15 years, the IoT is likely to develop fast and shape a newer information society and knowledge economy, but the direction and pace with which developments will occur are difficult to forecast [Santucci 2010]. In order to reap the full benefits of such a technological disruption, resolutions are recommended to address these challenges in Europe [Santucci 2011]:

- Mobilize a critical mass of research and innovation effort for the creation of new products, processes, and services.
- Develop a new definition of privacy for a changed world.
- Protect the different building blocks of the IoT, considering how these blocks will work together and what kind of interoperable security mechanisms must be created, and to assure a certain level of security during the cooperation among IoT multiple actors, especially human beings, machines, and objects.
- Develop ethics for the IoT by promoting an important dialogue between computer scientists and the broader public and by bridging the digital divide between those with access to technology and those without.

Another challenge is the big data trend. The IoT connects everything with everyone in an integrated global network. People, machines, natural resources, production lines, logistics networks, consumption habits, recycling flows, and virtually every other aspect of economic and social life will be linked via sensors and software to the IoT platform, continually feeding big data to every node – businesses, homes, vehicles – moment to moment, in real time. Big data, in turn, will be processed with advanced analytics, transformed into predictive algorithms, and programmed into automated systems to improve thermodynamic efficiencies, dramatically increase productivity, and reduce the marginal cost of producing and delivering a full range of goods and services to near zero across the entire economy.

Another concern regarding IoT technologies pertains to the environmental impacts of the manufacture, use, and eventual disposal of all these semiconductor-rich devices. Modern electronics are replete with a wide variety of heavy metals and rare earth metals, as well as highly toxic synthetic chemicals. This makes them extremely difficult to recycle properly. Electronic components are often simply incinerated or dumped in regular landfills, thereby polluting soil, groundwater, surface water, and air.

However, the next evolutionary stage of the IoT is touted by the IoE technology.

2.1.2 Internet of Everything (IoE)

The Internet of Everything is a technology concept that sees previously unconnected objects and processes converging with the ones that are digital first by their nature. This all-encompassing convergence of physical and digital domains is set to disrupt individual organizations and entire industries like nothing before [ABI Research 2015b]. ABI Research defines the IoE market as a combination of the IoT, Internet of Digital, and Internet of Humans. Value at stake drives in the connection of everything.

Figure 2.3 represents an infographic of IoE components and subcomponents as perceived by the markets a few years ago. According to [ABI Research 2015b], the market is expected to grow more than 40 billion devices on the IoE by 2020.

According to ABI Research market tracker of 2019 for IoE, the forecast for 2023 is that IoT will represent 54% of IoE, a higher percentage than Internet of Digital evaluated at 43%.

Figure 2.4 shows a chart representing the ABI Research market forecast of IoE for 2023 year, where:

- Internet of Digital includes PCs and digital home and mobile devices.
- Internet of Humans includes wearable computing.

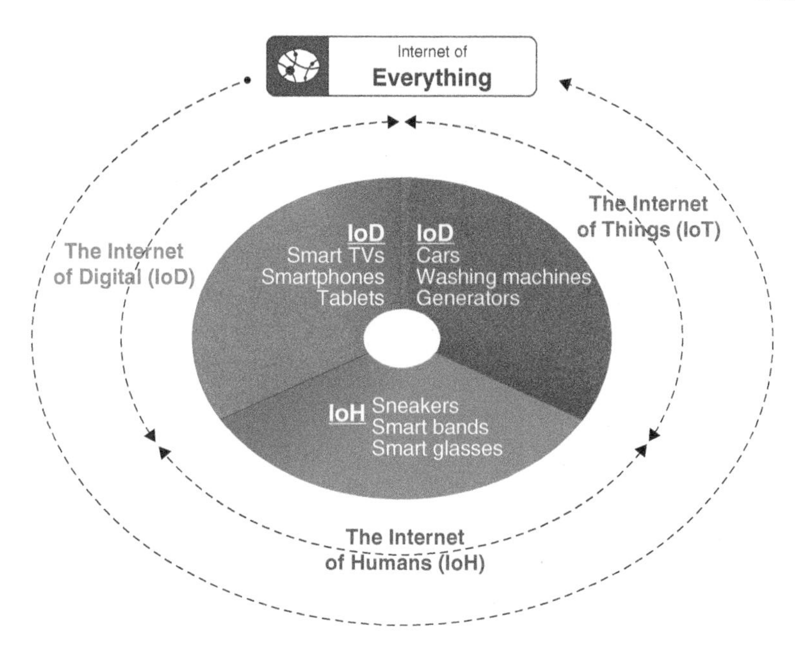

Figure 2.3 Infographic of Internet of Everything. *Source*: [ABI Research 2015a]. Printed based on courtesy of ABI Research Internet of Everything Market.

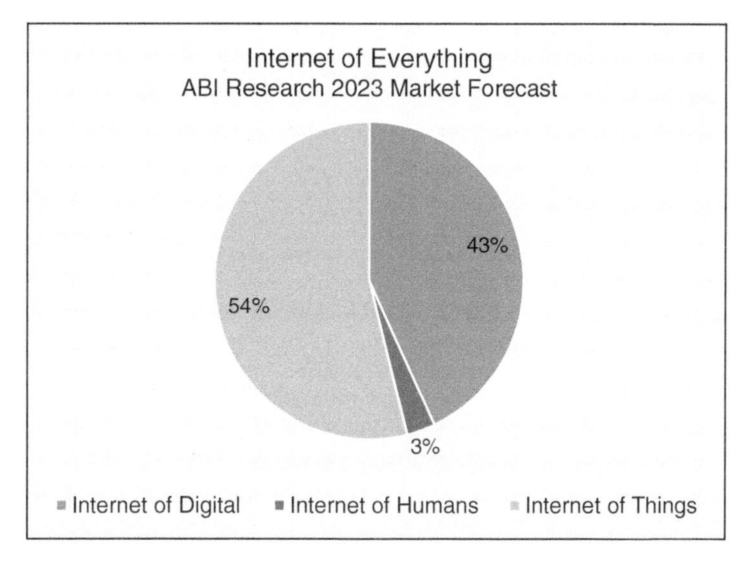

Figure 2.4 ABI Research Market Forecast for Internet of Everything. *Source:* Based on courtesy of ABI Research Internet of Everything Market Tracker, 2019.

- IoT includes utilities and industrial IoT, smart cities and buildings, retail advertising and supply chain, connected car, and smart home.

Figure 2.5 shows the overall concept of IoE, a networked connection of people, processes, data, and things. Examples of things include devices, cars, generators, and washing machines. In the context of IoE, cybersecurity is moving away from the traditional centralized view to a decentralized approach whereby security happens as close as possible to the endpoint. Specifically, in IoE, identity must extend

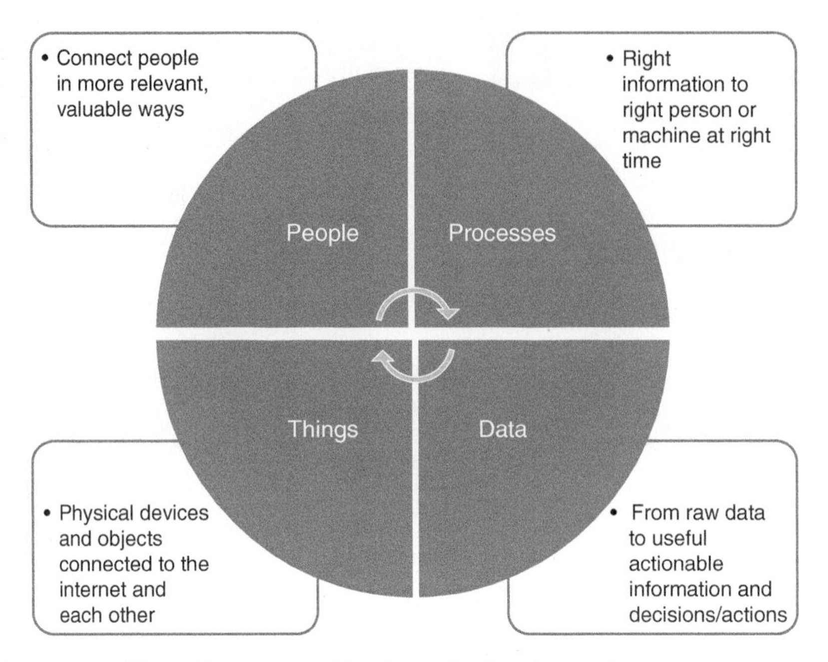

Figure 2.5 Internet of Everything connected in a large distributed network.

beyond conventional identity. However, creating a unified identity that addresses users in both physical world and virtual world is still a challenge.

On one hand, the IoE is being enabled by advancements with standardized, ultra-low-power wireless technologies (see Figure 2.6). The likes of Bluetooth and ZigBee have proven instrumental in driving sensor and node implementations, while Wi-Fi technology and cellular connectivity serve as a backbone for transferring the collected data to the cloud.

Consumers in particular expect to integrate all connected devices on a single home network – specifically on Wi-Fi networks. They may embrace a range of connected applications such as home security, smart energy, and in-vehicle infotainment. Wi-Fi enables human interaction with the IoE [Wi-Fi 2014].

An enormous range of device manufacturers, service providers, and software makers stand to build entirely new businesses addressing the IoE opportunity. However, this opportunity brings more security and privacy concerns for the consumers and businesses that support various applications.

Networked embedded systems have emerged under various names such as Internet/Web of Things/Objects, Internet of Everything, smart objects, Cooperating Objects, Industry 4.0, the Industrial Internet, cyber–physical systems, M2M, the Internet of Everything, the Smarter Planet, TSensors (Trillion Sensors), or the Fog (like the cloud, but closer to the ground). The vision is of a technology that deeply connects our physical world with our information world, although one may argue on the differences and their focus. In this research paper [Karnouskos 2011], the author argues that does not really differentiate when it refers to them as an amalgamation of computational and physical properties.

However, the trend is toward integration and interaction of the physical world with the information world. Recently, the community has come to understand that the principal challenges in embedded systems stem from their interaction with physical processes, and not from their limited resources. Since CPS emerge as a distinct category, the following section is an overview of this concept.

2.1.3 Cyber–Physical Systems

The term cyber–physical systems was coined by Helen Gill at the National Science Foundation in the United States in 2006 [CPS 2006]. CPS comprise interacting digital, analog, physical, and human components engineered for function through integrated physics and logic [NIST CPS].

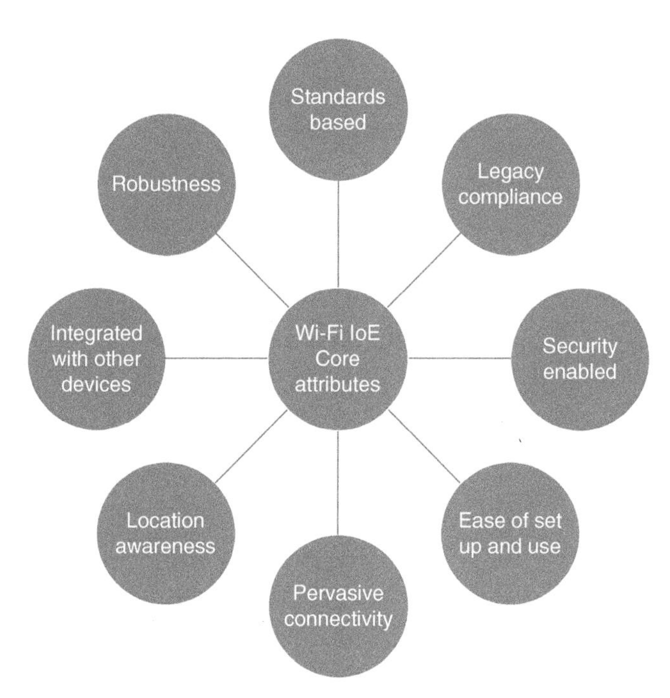

Figure 2.6 A perspective of the Wi-Fi Internet of Everything – core attributes.

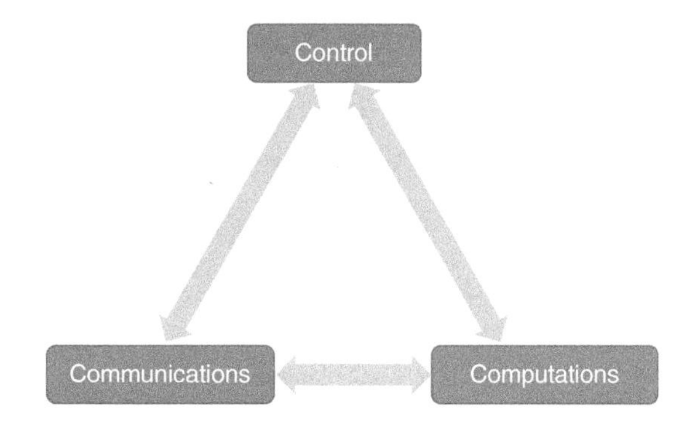

Figure 2.7 Simple cyber-physical representation.

A CPS is an integration of computation with physical processes whose behavior is defined by both cyber and physical parts of the system [Lee 2015a]. The authors argue that it is not sufficient only to separately understand the physical components and the computational components, but we must also understand their interaction. CPS is about the intersection, not the union, of the physical and the cyber [Lee 2010], [Lee 2015a], [Lee 2015b]. The embedded computers and networks monitor and control the physical processes, usually with feedback loops where physical processes affect computations and vice versa. This work [Lee 2015a] provides methodology and techniques for designing CPS. Figure 2.7 is a simple representation of a CPS with the components, computation, communication, and control that interact continuously.

CPS are heterogeneous blends by nature. They combine computation, communication, and physical dynamics [Lee 2015a]. The authors envision that several CPS applications may be based on a structure to include three main parts:

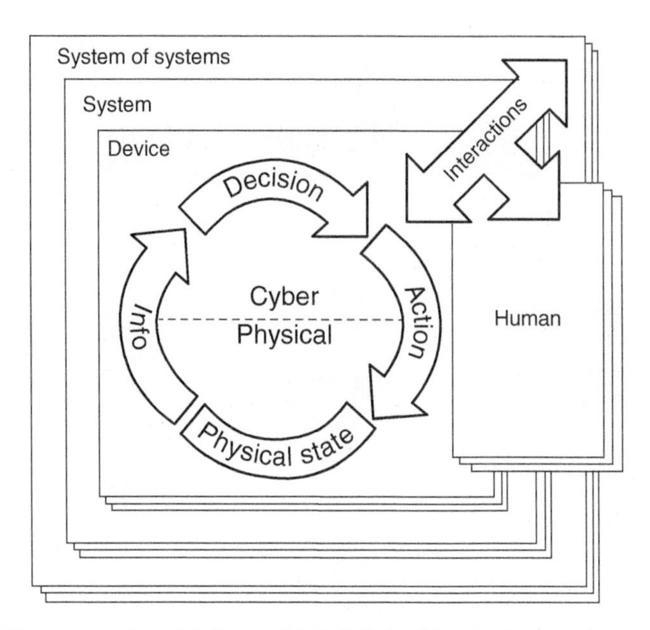

Figure 2.8 NIST CPS conceptual model. *Source:* [NIST SP1500-201]. Public Domain.

- First, the physical plant is the physical part of a CPS, not realized with computers or digital networks; it can include mechanical parts, biological or chemical processes and human operators.
- Second, there are one or more computational platforms, which consist of sensors, actuators, one or more computers, and (possibly) one or more operating systems (OS).
- Third, there is a network fabric, which provides the mechanisms for the computers to communicate; the platforms and the network fabric form the cyber part of the CPS.

A more detailed and current definition of CPS is provided by NIST [NIST SP1500-201]:

> Cyber–physical systems integrate computation, communication, sensing, and actuation with physical systems to fulfill time-sensitive functions with varying degrees of interaction with the environment, including human interaction.

A CPS conceptual model is shown in Figure 2.8. This CPS representation highlights the potential interactions of devices and systems in a system of systems (SoS) (e.g. a CPS infrastructure).

As shown in Figure 2.8, CPS may be as simple as an individual device (a device that has an element of computation and interacts with the physical world through sensing and actuation), or a CPS can consist of one or more cyber–physical devices that form a system or can be an SoS, consisting of multiple systems that consist of multiple devices. This pattern is recursive and depends on one's perspective (e.g. a device from one perspective may be a system from another perspective). Ultimately, a CPS must contain the decision flow together with at least one of the flows for information or action. The information flow represents digitally the measurement of the physical state of the physical world, while the action flow impacts the physical state of the physical world. This allows for collaborations from small and medium scale up to city/nation/world scale.

The scope of CPS is very broad by nature; there are large number and variety of domains, services, applications, and devices. Also, CPS controls have a variety of levels of complexity ranging from automatic to autonomic. CPS go beyond conventional product, system, and application design traditionally conducted in the absence of significant or pervasive interconnectedness. There are many differences that characterize CPS from traditional systems. Examples of characteristics are listed in Table 2.1.

The CPS will provide the foundation of our critical infrastructure, form the basis of emerging and future smart services, and improve our quality of life in many areas [NIST CPS].

Table 2.1 CPS characteristics.

Characteristic	Description	Remarks
Cyber and physical	Combination of cyber and physical components	
Connectedness	Generally involves sensing, computation and actuation	Involves combination of IT and OT with associated timing constraints
System of systems (SoS)	May bridge multiple purposes and time and data domains	Different time domains may reference different time scales or have different granularities or accuracies Time scale: a system of unambiguous ordering of events
Emergent behaviors	Open nature of CPS composition	Understanding a behavior that cannot be reduced to a single CPS subsystem, but comes about through the interaction of possibly many CPS subsystems
Methodology	A methodology needed to ensuring interoperability, managing evolution, and dealing with emergent effects	Example: NIST 1500-201 framework
Repurposed	Other purpose use beyond applications that were their basis of design	
Application enabler	Enabling cross-domain applications	
Trustworthiness concern	Potential impact on the physical world	Urgent need for emphasis on security, privacy, safety, reliability, resilience, and assurance for pervasive interconnected devices and infrastructures
Broad range of platform and algorithm complexity	Accommodate a variety of computational models	
Variety of modes of communication	From stand-alone systems to highly networked systems	May use legacy protocols or anything up to more object exchange protocols
Heterogeneity	Wide range of heterogeneous devices (sensors, controllers, control schemes, input sources, platforms, etc.)	Complexity associated with the sensing and control loop(s) with feedback that are central to CPS must be well addressed in any design
Co-design	Design of the hardware and the software jointly to inform tradeoffs between the cyber and physical components of the system	
Typically a time-sensitive component	Timing is a central architectural concern	A bound may be required on a time interval, e.g. the latency between when a sensor measurement event occurred and the time at which the data was made available to the CPS

(*Continued*)

Table 2.1 (Continued)

Characteristic	Description	Remarks
Interaction with the operating environment	CPS measure and sense and then calculate and act upon their environment, typically changing one or more of the observed properties (thus providing closed-loop control)	
Typically a human environment	CPS environment typically includes humans and humans function	Architecture must support a variety of modes of human interaction: human as CPS controller or partner in control; human as CPS user; human as the consumer of CPS output; and human as the direct object of CPS to be measured and acted upon

Source: Adapted from [NIST SP1500-201].

2.1.4 Cyber–Physical Systems Applications

The vision is that CPS could improve many existing systems, such as robotic manufacturing systems; electric power generation and distribution; process control in chemical factories; distributed computer games; transportation of manufactured goods; heating, cooling, and lighting in buildings; people movers such as elevators; and bridges that monitor their own state of health. The impact of such improvements on safety, energy consumption, and the economy is potentially enormous. So modern businesses rely on CPS to accurately sync the real-world status on backend systems and processes.

CPS can be found extensively in multiple domains including the electricity sector [Parolini 2012]. CPS is seen as an integral part of the Smart Grid as discussed by Karnouskos in [Karnouskos 2011], [Karnouskos 2012]. The perspective of this researcher is that the Smart Grid will have to heavily depend on CPS that are able to monitor, share, and manage information and actions on the business as well as the physical power grid. Many traditional parts of the Smart Grid are increasingly CPS dominated. In generation, CPS control the connection to the network as well as the operational aspects in the electricity generation side such as solar and wind parks, hydro facilities, etc.

CPS involve traditional IT as in the passage of data from sensors to the processing of those data in computation. CPS also involve traditional operational technology (OT) for control aspects and actuation. The combination of these IT and OT worlds along with associated timing constraints is a particularly new feature of CPS.

Figure 2.9 depicts the use of CPS for smart transportation [Ling 2015]. The components of the CPS are a collection of computing devices communicating with one another and interacting with the physical world via sensors and actuators in a feedback loop as described in [Lee 2015a].

Figure 2.10 is a simple representation of a CPS with the components, computation, communication, and control that interact with the cyber and physical world. Generally, the structure for a CPS includes physical plant, computational platforms, and the network fabric. An application may use two networked platforms with their own sensors and/or actuators. The embedded computers interact with a physical plant through sensors and actuators and with each other through a network fabric. The action taken by the actuators affects the data provided by the sensors through the physical plant.

As described in [Lee 2015b], the design of CPS, therefore, requires understanding the joint dynamics of computers, software, networks, and physical processes. The author argues that it is this study of joint dynamics that sets this CPS discipline apart. CPS is a discipline that combines engineering models and methods from mechanical, environmental, civil, electrical, biomedical, chemical, aeronautical, and industrial engineering with the models and methods of computer science. Therefore, there are theoretical and practical challenges in the design of CPS applications; among them security is an alarming concern that requires imperative investment.

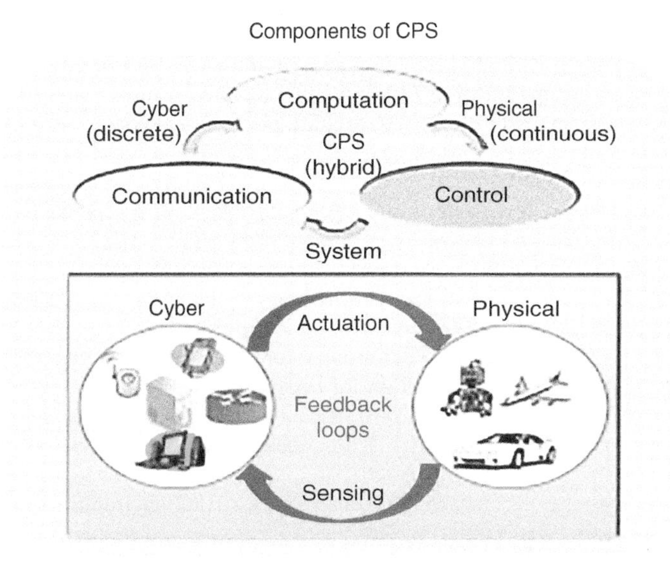

Figure 2.9 Components of CPS for smart transportation. *Source:* [Ling 2015]. © 2016, IEICE.

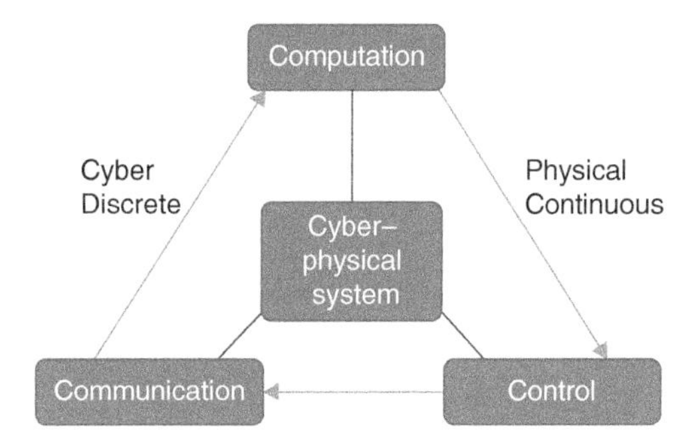

Figure 2.10 Cyber–physical system – simple structure.

2.2 Cybersecurity

The term cybersecurity is associated with the security of the cyberspace, which was coined in a science fiction novel [Gibson 1984], as a futuristic computer network that people use by plugging their minds into it or the electronic medium of computer networks, in which online communication takes place. However, there are many definitions for cybersecurity and cyberspace that evolved over time. Many cyber terms are coming into vogue, and a few organizations have tried to include significant definitions that allow us to make useful distinctions when compared with existing terms. Thus, when searching for definitions of certain security concepts and terms, we find identical definitions (one glossary references another glossary), similar definitions, or definitions that are too short or too long, or missing.

2.2.1 Cybersecurity Definitions

The following is a sequence of definitions for cybersecurity and cyberspace as provided in known glossaries.

Cybersecurity is the ability to protect or defend the use of cyberspace from cyber attacks [CNSSI 4009].

Cybersecurity is the ability to protect or defend the use of cyberspace from cyber attacks [NISTIR 7298r2].

Cybersecurity is the activity or process, ability or capability, or state whereby information and communication systems and the information contained therein are protected from and/or defended against damage, unauthorized use or modification, or exploitation [NICCS 2016].

Cyberspace is a global domain within the information environment consisting of interdependent IT infrastructures, telecommunication networks and computer processing systems, and embedded processors and controllers [CNSSI 4009].

Cyberspace is a global domain within the information environment consisting of interdependent IT infrastructures, telecommunication networks and computer processing systems, and embedded processors and controllers [NISTIR 7298r2].

Cyberspace is the interdependent network of IT infrastructure that includes the Internet, telecommunication networks, computer systems, and embedded processors and controllers [NICCS 2016].

As shown above, we found identical definitions for both terms in these glossaries [CNSSI 4009], [NISTIR 7298r2] and a similar definition for cyberspace in the glossary [NICCS 2016]. While these terms are not defined in [ISO/IEC 27000], [RFC 4949]], the International Telecommunication Union approved the overview of cybersecurity as described in [ITU-T 2008], which is not really a concise definition.

In common usage, the term cyberspace refers also to the virtual environment of information and interactions between people [WH 2009]. However, a new term, cyber ecosystem, is encompassing more entities. It is defined as the interconnected information infrastructure of interactions among persons, processes, data, and information and communication technologies, along with the environment and conditions that influence those interactions [NICCS 2016].

2.2.2 Understanding Cybersecurity Terms

Cybersecurity is the ability to protect or defend the use of cyberspace from cyber attacks [CNSSI 4009]. Further, a cybersecurity attack is defined as an attack via cyberspace for the purpose of disrupting, disabling, or destroying a computing environment/infrastructure [CNSSI 4009]. However, this definition excludes the possibility of physical attacks, unintentional human errors, and natural disasters that can also disrupt a computing environment/infrastructure. Physical attacks may be realized without using the cyberspace, but still causing harm to cyberspace. Often two definitions are combined into one definition. For example, the cybersecurity definition [CNSSI 4009] is concatenated with another definition (measures taken to protect a computer or computerized system [IT and OT] against unauthorized access or attack) to make the cybersecurity definition provided by the US Department of Energy (DOE) [DOE 2014a].

However, no unique definition for cybersecurity is available across the Internet [Franscella 2013]. As pointed out in [Vacca 2012], no formal accepted definition of cybersecurity currently exists. On the use of cybersecurity versus cyber security, the communities agreed on using the word cybersecurity [Franscella 2013].

Often the cybersecurity is covering all security dimensions from technology to economic and social, legal, law enforcement, human rights, national security, warfare, international stability, intelligence, and other aspects. The widespread use of this term often masks the broad and complex nature of the subject matter [OECD 2015].

When comparing cybersecurity with information security, some people regard these concepts as overlapping, being the same thing [ENISA 2015a]. Others may view information security as focused on protecting specific individual systems and the information within organizations, while cybersecurity is seen as being focused on protecting the infrastructure and networks of critical information infrastructures.

Information security is defined as measures adopted to prevent the unauthorized use, misuse, modification, or denial of use of knowledge, facts, data, or capabilities [Maiwald 2004]. This term is defined in [NISTIR 7298r2] as the protection of information and information systems from unauthorized access, use, disclosure, disruption, modification, or destruction in order to provide confidentiality, integrity, and availability. Although this definition may seem more focused (implying security goals such as confidentiality, integrity, and availability), it is still not accurate because protection measures should also provide for non-repudiation and other attributes of the information.

Although there is no universally accepted nor straightforward definition of cybersecurity or information security [ENISA 2015a], we need to understand the differences among these various definitions and views.

The recommendations of the [OECD 2015] document introduce the concept of digital security risk (see definition in Appendix A) that requires a response fundamentally different in nature from other categories of risk needs to be countered. To that effect, the term cybersecurity and more generally the prefix cyber that helped convey this misleading sense of specificity do not appear in the recommendation. Digital security risk is dynamic in nature. It includes aspects related to the digital and physical environments, the people involved in the activity, and the organizational processes supporting it.

The abundance of definitions for security terms is the result of various aspects and attributes that an interested party may want to emphasize in the definition of a concept. Also, many security- and privacy-related concepts and terms evolved as the security paradigms changed in time, particularly in the way IT security was addressed. Appendix A includes a table showing different definitions for common security terms as provided by known standards and glossaries.

This is an indication of the development of a field where a foundation for defining the basic concepts is still evolving. However, it is necessary to have more consistent definitions among related and dependent terms. An appropriate balance between comprehensive and extended definitions is needed also for promoting terms that are useful to users and general public, not only to security experts and researchers. These terms are needed in communicating, writing, and understanding news and documents dealing with security policies, directives, instructions, and guidance.

Often, the lack of knowledge of the definitions or lack of unique definitions prompts for defining these terms in each industry. For example, DOE published a glossary of concepts including a set of cybersecurity terms in [DOE 2014a]. Several terms are taken from other documents, or they are adapted for the energy sector use. There is a problem when these dictionaries are not continuously updated; when new terms may appear, some terms could become obsolete or be changed in the referenced glossary. Therefore, one solution is to check the definitions and their maintenance status of these terms. The security team needs to agree on the basic terms to avoid language confusion and avoid rolling out ambiguous activities.

Since some security terms do not have common definitions or new updates emerge, we recommend previewing the definition of the most current dictionaries of security terms and concepts as defined by known standard organizations such as the International Organization for Standardization (ISO)/IEC, the Internet Engineering Task Force (IETF), and International Society of Automation (ISA). Often the glossary adopted by an organization may need to be revised. Definitions of related security terms (cybersecurity, threat, vulnerability, asset, countermeasure, exposure, security service, etc.) are also available in published guides maintained by security professionals such as [Harris 2013], [Krutz 2004]. Figure 2.11 shows a visual representation of the relationships among different security concepts (terms). Definitions of the terms are provided in [CC 2.3] (see also Appendix A).

In addition, security and privacy concepts have to be understood by users, security designers, and managers; otherwise misunderstanding creates confusion or ambiguity in communication that undermines the successful implementation of security and privacy programs.

The assets may have vulnerabilities that may be exploited by a threat agent leading to risk that can damage the asset. The owner of the assets wants to minimize the risk and uses countermeasures (controls or safeguards). Applying the right countermeasure can eliminate the vulnerability and exposure and thus reduce the risk. One issue is that eliminating the threat agent may not be possible, but it is possible to protect the asset and prevent the threat agent from exploiting vulnerabilities within the asset's environment.

These terms and definitions of security terms continue to change and evolve with technology developments, emerging new technologies, and research trends. This work [Von Solms 2013] discusses the similarities and differences between these terms: cybersecurity, information security, and communications security. The authors argue that cybersecurity goes beyond the boundaries of traditional information security to include not only the protection of information resources but also that of other assets, including the reference to the human factor. Figure 2.12 illustrates graphically the relationships among these concepts.

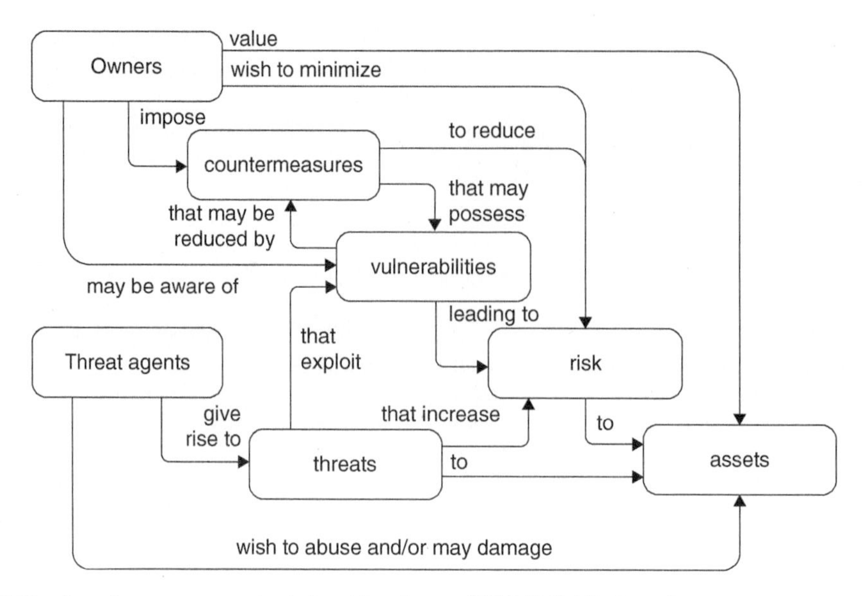

Figure 2.11 Security concepts and relationships. *Source:* [CC 2.3]. Public Domain.

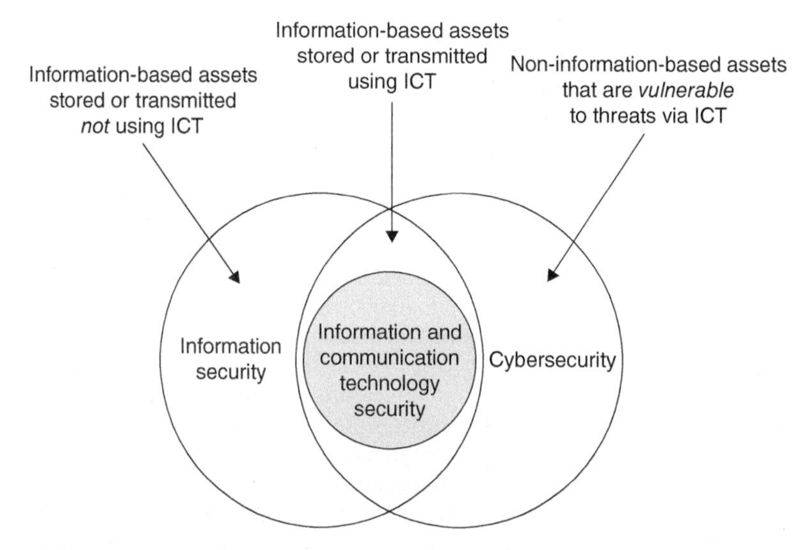

Figure 2.12 Information security and cybersecurity relationship. *Source:* [Von Solms 2013]. © 2013, Elsevier.

This work [Craigen 2014] is another attempt to provide a new definition for the term cybersecurity from a multidisciplinary perspective as follows:

> Cybersecurity is the organization and collection of resources, processes, and structures used to protect cyberspace and cyberspace-enabled systems from occurrences that misalign de jure from de facto property rights.

However, the definition is missing the point that cybersecurity is a field of research, an industry, and a societal issue. There are many different theoretical and interpretational aspects that could or even should be considered when discussing cybersecurity as a concept and a term.

Appendix A includes several definitions promoted by organizations and glossaries including DOE. Although there is no universally accepted nor straightforward definition of cybersecurity and other related terms, we need to understand these definitions and views.

2.2.3 Cybersecurity Evolution

In the past, before Internet technologies became the mainstream technology, there were few risks and limited definitions for security and security expertise. Security evolved from protecting a file, an application, or a computer to protecting a larger area that comprises many computers, networks, organizations, and people.

The security field evolved from an obscure term known initially only to military and governments to include organizations of all kinds, the public, and the globe. For some time, few professionals were involved in security matters. Today organizations and governments are continuously searching for better security professionals to protect their information and their other resources. A review of the terms and definitions for cybersecurity is well documented in [Bay 2016].

Security definitions evolved from simple terms like computer security, IT security, and information security to more recent terms identified as cyber security or cybersecurity, the last term winning, although the cyber security term is still used in some publications [Franscella 2013].

As we observed earlier, security terms are differently defined in many books and guidances; therefore we use the terms security, cybersecurity, and information security in this book based on the well-known standards. We acknowledge that there are subtleties in these definitions. NIST guidelines for the Smart Grid use the term cybersecurity (e.g. [NISTIR 7628], [NISTIR 7628r1]). However, we discuss the information security based on definitions included in standards. The ISO/IEC definition is as preservation of information attributes such as confidentiality, integrity, availability, authenticity, accountability, non-repudiation, and reliability [ISO/IEC 27000]. Another defines security as a property of a system by which confidentiality, integrity, availability, accountability, authenticity, and reliability are achieved [ISO 15443].

We also discuss security in the context of an environment determining the setting and circumstances of all interactions and influences with the system of interest [ISO/IEC 42010]. Other issues that need to be understood and managed include the interdependence of cybersecurity and reliability of the power grid.

A cyber attack on devices that protect and control the power grid could result in power disruption or damaged equipment. Similarly, physical attacks on power equipment or cyber infrastructure may impact the information, energy system, and energy services. Security is a system condition that results from the establishment and maintenance of measures to protect the system [RFC 4949]. Therefore, the installation of security controls should avoid interfering with critical energy delivery functions.

Although safety is defined as freedom from risk that is not tolerable [ISO/IEC 51] and safety issues are being the objective of dedicated departments within an organization, we need to discuss it in the context of cybersecurity. Safety is the condition of the system operating without causing unacceptable risk of physical injury or damage to the health of people, either directly or indirectly as a result of damage to property or to the environment [IIC 2015]. For example, inappropriate security controls (e.g. electronic locks to computer facilities without capabilities to open doors or windows) may harm people (working in these facilities) that need to escape when there is a natural disaster, a power down, or a fire.

Smart Grid cybersecurity must address not only deliberate attacks, such as from disgruntled employees, industrial espionage, and terrorists, but also inadvertent compromises of the information infrastructure due to user errors, equipment failures, and natural disasters.

2.3 Advancing Cybersecurity

Security problems for Smart Grid and DER systems require solutions and developments that go beyond all practices (e.g. focused on vulnerability management, reactive strategies, obfuscation, depth of defense, perimeter security, etc.) that proved unsuccessful in the face of new threats. Therefore, it is imperative to focus on more advanced methods and more research to find solutions to unsolved problems [Wulf 2001].

The availability and reliability of computing and information systems for business and power grid applications are dependent on the secure operations of industrial control systems (ICSs) and other infrastructures.

2.3.1 Contributing Factors to Cybersecurity Success

Recently, industry experts and researchers have participated more often in international forums and standards organizations that promote security techniques, security technologies, and standards. One need is a new security model for IT [Wang 2011]. Therefore, securing the Smart Grid and its applications requires the development of more comprehensive definitions and cybersecurity models.

As more technologies penetrate the power grid and help integrate more variable renewable sources of electricity and facilitate the greater use of electric vehicles and energy storage, there are challenges to implementing cybersecurity to ensure the safety and reliability of the Smart Grid.

Although standards and guidelines have been identified to support the implementation of minimum security measures that set a baseline for cybersecurity across the energy sector, many security challenges require solutions such as:

- Advancing cybersecurity and privacy design.
- Understanding interdependencies.
- Open systems view.

2.3.2 Advancing Cybersecurity and Privacy Design

Progress in cybersecurity for DER applications depends on achieved more quantitative and more visually understanding of the performance of the DER communications and control components, data and information characteristics, cyber infrastructure, business objectives, application requirements, security architecture design principles, data traffic patterns, vulnerabilities, and threats. Therefore, it is necessary to build cybersecurity and privacy by design into all DER systems and processes from the beginning. This saves on system life cycle development costs and protects organizations from expensive system modifications to meet the evolving threats.

Security by design and privacy by design are not new buzzwords; they are old principles to be applied by developers during development cycle. Security cannot be added to any system or application associated with power grid as an afterthought. There is a need to start from scratch, at the very beginning of any system development or technology integration, and consider privacy and security requirements in all design, test, and implementation criteria. Strategic consideration of these issues can make a huge difference in the confidence and protection that the overall system provides. This is necessary whether the design effort is focusing on DER applications, silicon chips, DER components, network components, end-user devices, architecture, or the system as whole.

2.3.2.1 Understanding Interdependencies

Information technologies contribute to raising the interdependency between the operation of the power grid (including generation, transmission, and distribution) and the operation of the wholesale electricity market. The electric market and the power system become more closely tight every day. The operation of one depends on the continuous and reliable operation of the other. In addition, the vulnerability of the power system is not mainly a matter of electric system or physical system, but is more a matter of cybersecurity. Attacks (such as attacks upon the power system, attacks by the power system, and attacks through power system) to the Smart Grid infrastructures could bring huge damages on the economy and public safety.

Control systems such as SCADA are highly interconnected with IT systems within electric industry and with external infrastructures and economic sectors. Historically, control system security meant locating and identifying problems in a closed-loop system; now unauthorized intrusion or attacks are evolving issues that have to be addressed.

The interdependencies are manifested at different levels. Security dependencies can occur and have all sorts of side effects. Risk assessment and management in large-scale systems such as smart power grid requires an understanding of how and to what degree the systems are interdependent. Instances of interdependencies with other infrastructures are reported in [Amin 2003].

The smart power grid infrastructure is characterized by interdependencies (physical, cyber, geographical, and logical) and complexity (collections of interacting components). Cyber interdependencies are a result of the pervasive computerization and automation of infrastructures. There is a need for developing tools and techniques that allow a critical infrastructure such as the power grid to self-heal in response to threats, failures, natural disasters, or other perturbations. Also, other scenarios have to be considered. For example, there is a cascading effect due to interdependencies of electric infrastructure with other infrastructures such as gas, telecommunications, transportation, financial, etc.

2.3.2.2 Open Systems

The SCADA obscurity approach used in SCADA systems is debatable; it has proven that it does not work anymore. Even to this day, many SCADA systems are perceived as either invulnerable to cyber attacks or uninteresting to potential hackers (security by obscurity principle). The obscurity principle implies use of concealment for a design, implementation, etc. to provide security. A system relying on the security through obscurity principle may have theoretical or actual security vulnerabilities, but its owners or designers believe that the flaws are not known and that attackers are unlikely to find them. If the strength of the program's security depends on the ignorance of the user, a knowledgeable user can defeat that security mechanism.

The principle of open design states that the security of a mechanism should not depend on the secrecy of its design or implementation [Bishop 2005]. Designers and implementers of security must not depend on secrecy of the details of their design and implementation to ensure security. A methodology based on economic analysis of the obscurity principle and open systems paradigm for determining when obscurity does not help security (there is no security through obscurity) and when the open paradigm affects security (loose lips sink ships) is described in [Swire 2004]. The proposed model provides a systematic way to identify the costs and benefits of disclosure for security.

Another example is the home area network (HAN), which is enabled by open and interoperable standards. The use of open and interoperable standards is key to accessibility, availability, innovation, and widespread adoption. Standards provide:

- Cybersecurity that protects systems and data.
- Interoperable components that protect investments in technology and enable growth in the HAN ecosystem.
- Competition among consumer products companies, which drives down costs while increasing choices for consumers.
- Reduced maintenance and support costs caused by proprietary solutions.
- A common understanding of information exchange.

Interoperability of DERs with HAN is key to advancing Smart Grid applications. For purposes of the HAN specifications, a DER is a HAN device with functionality that measures and communicates its full energy production. DERs generate electricity, which may provide for all or a portion of the premises' electrical needs. A DER may be interconnected to the utility electric distribution system, and any net energy flowing on to the electric grid may be recorded in a separate channel on the AMI meter. Additional information about AMI components and open Smart Grid can be found in [UCAIUG], [NETL 2008].

The DER production may also be managed by an EMS that optimizes the premises energy consumption. Also, open standards are key enablers for the success of the IoT, as it is for any kind of M2M communication.

One crucial aspect is how to protect privacy in open systems. Personal privacy can also be compromised when information is disclosed in open systems. Solutions require analysis of several factors. Compelling goals such as accountability, economic growth, free speech, and privacy should be included in any overall decision about whether to disclose information [Swire 2004].

In addition, a designer has to consider regulations on privacy policy, corporate responsibility, and user trust because compliance is required by several committees (e.g. UN Human Rights Watch Group). A committee established in the United States in August 2013 has the task to review policies and regulations on Intelligence and Communications Technologies to support commitment to privacy and civil liberties and maintain the public trust in the United States.

2.4 Smart Grid Cybersecurity: A Perspective on Comprehensive Characterization

Two major views of Smart Grid security include cybersecurity and physical security [P2030 2011], [DOE 2015a]. Although there are differences on these views, there are common aspects and many interdependencies between these views that require a unified view of security. Another perspective is that power grid information security and protection has aspects of both ICS and IT systems. Although both ICS and IT systems require information security services to combat malicious attacks, the specifics of how these services are used for the power grid depend upon appropriate risk assessment and risk control. Distinct types of attacks targeting ICS and IT systems as well as different performance requirements of these systems determine a specific priority order of the security services implemented for each system. In addition, the Smart Grid trends toward the integration of the operational and business systems require a unified view of security based on risk management instead of applying the old approach of separate techniques for IT and ICS systems. A unified approach based on risk management techniques is described in [Ray 2010].

Therefore, solutions for designing cybersecurity into the Smart Grid with the vision of surviving a cyber incident while sustaining critical energy delivery functions must be carefully engineered and require novel solutions [Hawk 2014], [Ray 2010].

2.4.1 Forces Shaping Cybersecurity

Cybersecurity progress requires understanding of all its aspects as a vector quantity including the forces shaping its evolution as identified in [Agresti 2010]:

- Rebranding exercise means cybersecurity is replacing information security and information assurance as the term of choice; this move promises an engagement with the public.
- Organizational imperative means that offering information assurance in cyberspace, although infeasible, might be required because of assets ownership changes in contracts and service agreements with outside parties.
- Cyberspace domain redefinition means becoming part of a virtual world that:
  - Refers to the virtual environment of information and interactions between people [FERC 2009]
  - Includes the physical world such as CPS and IoT [NSF 2014], [Lee 2010], [IoT], [EC-EPoSS 2008]

  In this context, the Smart Grid and process control systems are types of CPS [CPS 2014], and smart meters and home automation are examples of the IoT applications.
- National defense priority requires securing cyberspace, which is a matter of survival for any nation; one strategy is providing new interaction models such as engaging the public and private sectors to collaborate on the protection of critical infrastructures.

Figure 2.13 is a view of these forces categorized as style (rebranding) and substance (organizational imperative, cyberspace domain, and national defense priority). These forces require us to consider new tasks to include engaging the public, externalizing (globalization) the organization, understanding the virtual domain, and establishing protection strategies for critical infrastructures. Due to cyberspace prominence and collaboration, securing cyberspace must be a personal, organizational, national, and global priority.

Energy sector users and organizations need to succeed in understanding these forces including disturbances created by emerging technologies and trends.

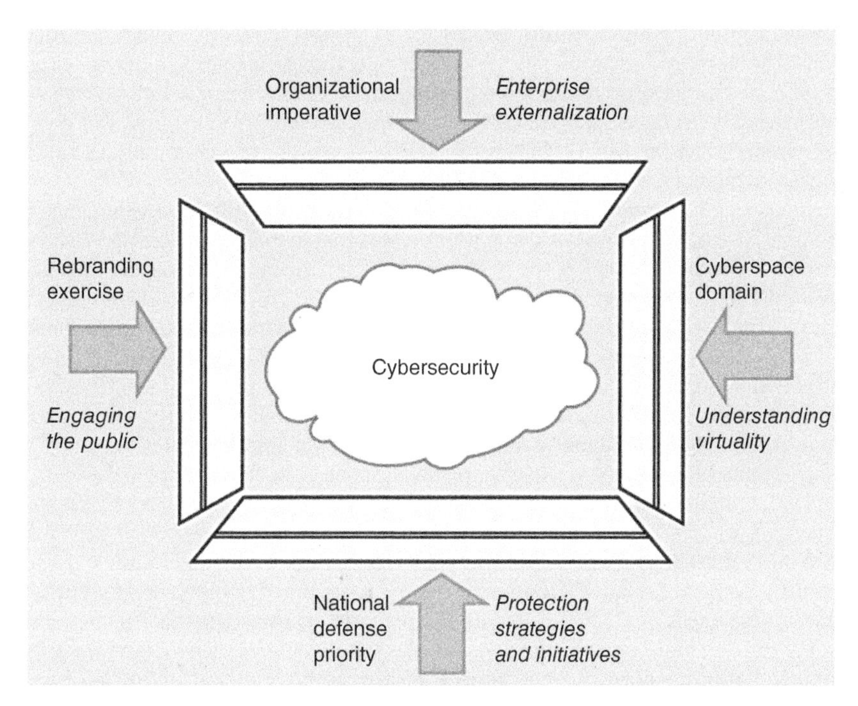

Figure 2.13 Forces shaping cybersecurity. *Source:* [Agresti 2010]. © 2010, IEEE.

2.4.2 Smart Grid Trends

Examples of trends that impact Smart Grid cybersecurity include the following:

- Mobile computing refers to workforce dependence on being mobile and pervasive computing that has generated several devices that assist with this mobility. Forrester Research referred to mobile computing as the empowered movement, since companies are empowering their employees with modern consumer-oriented technologies to better serve their customers [Forrester 2010]. Due to the decreasing cost of computing and the ubiquity of smartphone usage, applications have been developed to be used in home automation to control or remotely monitor a thermostat for air conditioning or a switch for lights.
- Future Internet and its services are driven by users' needs, and new technologies enable these services. Examples of future Internet services for Smart Grid include:
  - Energy consumer demand and response
  - Distributed energy storage with guarantees
  - Personalized energy consumer profile
  - Control of consumer's appliances
- Web as ubiquitous computer refers to the convergence of mobile smart devices, cloud computing, and software as a service, which enables Web with anytime and anywhere computing capabilities [Pendyala 2009]; the applications are moving from the local PC machine to the ubiquitous computer. One issue is that users cannot keep up with frequent software updates and configuring Wi-Fi security settings.
- Embedded systems surround us in many forms from cars to cell phones, video equipment to MP3 players, and dishwashers to home thermostats. However, security for these systems is an open and more difficult problem than security for desktop and enterprise computing. Even a washing machine can be used as a platform to launch distributed denial-of-service (DoS) attacks against the public, an organization, or the government.

- Data-intensive computing and real-time processing of massive data streams are required by more applications. The North American electric power grid operations generate 15 terabytes of raw data per year, and estimates for analytic results from control, market, maintenance, and business operations exceed 45 Tbytes/day. As developers add new high-resolution sensors to the grid, this data volume is increasing rapidly. Data-intensive problems challenge conventional computing architectures with demanding CPU, memory, and I/O requirements.
- Network changes and adoption of new services are determined by the increasing amounts of data collected from sensors, home devices, and power devices that demands reliable and faster communication networks. Profound changes are beginning to occur in public networks, data centers, and enterprise networks such as upgrades of major carrier backbones to higher data rates and the replacement of infrastructure based on old technologies for core networking/transport. Adoption of new services is facilitated by the migration to new protocols (e.g. IPv6, SIP, MIP (MIPv4, MIPv6)) and the emergence of Web services.
- Home networking in a step toward the next-generation unified home networking technology enables operation over all types of in-home wiring (phone line, power line, coaxial cable, and Cat-5 cable) using a single transceiver with few programmable parameters to connect home devices. Thus, new threats and vulnerabilities may occur to smart meters and DER devices installed for customer energy management.
- Virtualization is being adopted as a standard for businesses, but the tools and technologies for addressing the security issues are relatively immature or not consolidated to offer sound security solutions. Although the advantages of virtualization are not disputed (examples include reductions in energy costs that are causing more organizations to consider virtual environments), the protection of a virtual computer hardware platform, OS, storage device, or computer network resources requires attention.
- Virtual organization entity is formed whenever a developer creates an application or a workflow that features autonomous services owned by multiple organizations, each of which shares some proprietary services and part of its own knowledge. However, virtual organization introduces security concerns. For example, traditional access control methods based on the identity of each user in a virtual organization do not scale as the number of users and services increase, especially when the population of users and services is highly dynamic as in the Smart Grid environment.
- Deperimeterization is a process that causes the boundaries between systems, which means the disappearing of boundaries between systems and organizations, to disappear; in this process, they become connected and fragmented at the same time. The most obvious problem is how to reorganize the security. Deperimeterization implies not only that the border of the organization's IT infrastructure becomes blurred but also that the border of the organization' accountability fades.
- Global challenges influence the nature of the organization, and scope of information processing has evolved; managing information security is not just enforcing restrictions to maintain information security services such as confidentiality, integrity, availability, and non-repudiation. In the new millennium, there are demands for more responsibility, integrity of people, trustworthiness, and ethicality [Dhillon 2001]. The most relevant global challenges include poor software quality, weaknesses of protocols, and services.
- Internet has become a critical infrastructure because societies and economies are converging on the Internet, and the distinction between physical and virtual worlds is blurring.

2.5 Security as a Personal, Organizational, National, and Global Priority

The increased use of information systems is generating many benefits, but it has created an ever larger gap between the need to protect systems and the degree of protection. Society, including business, public services, and individuals, has become very dependent on technologies that are not yet sufficiently dependable. Often, information systems are vulnerable to attacks upon or failures of information systems. Certain information systems, both public and private, such as those used in power grid, military or defense installations, nuclear power plants, hospitals, transport systems, and securities exchanges, offer fertile ground for antisocial behavior or terrorism. However, users need to have confidence that information systems will operate as intended without unanticipated failures or problems to personal security and privacy.

2.5.1 Security as Personal Priority

The right to security of the person is guaranteed by Article 3 of the Universal Declaration of Human Rights [UN 1948] that reads:

> Everyone has the right to life, liberty and security of person.

The Article 12 of this document reads:

> No one shall be subjected to arbitrary interference with his privacy, family, home or correspondence, nor to attacks upon his honour and reputation. Everyone has the right to the protection of the law against such interference or attacks.

With the proliferation of computers, the right to privacy of the individual is threatened by the use of information due to the new technology and the Internet. As a result, privacy policies (such as those described in [OECD 1980]) have been adopted in many countries around the world. Other publications assess the impact of technology on the private lives of people (e.g. [Britz 1996]).

2.5.2 Protection of Private Information

In 2010, the OECD celebrated the 30th anniversary of the guidelines for protection of privacy and flows of personal data [OECD 1980] through a series of events and papers such as [OECD 2011]. This report presents an overview of other documents that were published after the guidelines (e.g. [OECD 2002], [OECD 1999]). Also, it provides an analysis of the importance of the privacy guidelines document of 1980. The report shows that there were still privacy risks for organizations, individuals, and society even 30 years after the publication of the guidelines in 1980. Examples of risks and issues include the following [OECD 2011]:

- Certain risks associated with privacy have increased as a result of the shift in scale and volume of personal data flows and the ability to store data indefinitely.
- Definition of personal data in the guidelines is broad (any information related to an identified or identifiable individual). Given the current power of analytics and the apparent limitations of anonymization techniques, this means vast amounts of data potentially now fall under the scope of privacy regimes.
- An increasing economic value of personal data gives rise to concerns related to the security of personal data, unanticipated uses, monitoring, and trust.
- Organizations often retain large amounts of personal data for various purposes.
- High-profile data breaches have shone a light on the challenges of safeguarding personal data; concepts of data controller and data processor raise new concerns.
- An increasing concern that the long-standing territorial/regional approaches to data protection may no longer be sufficient as the world increasingly moves online and data is available everywhere, at any time.
- Uncertainty over questions of applicable law, jurisdiction, and oversight on the global nature of data flows; some organizations may not always be able or willing to tailor their services to meet the specific needs of each jurisdiction.
- Differences that remain among various national and regional approaches to data protection, which are more noticeable when applied to global data flows.

- Increasing difficulty for individuals to understand and make choices related to the uses of their personal data; the uses of personal data are becoming increasingly complex and nontransparent to individuals.
- Advances in technology and changes in organizational practices, which have transformed occasional transborder transfers of personal data into a continuous multipoint global flow.

As a result of this environment, the security of personal data has become an issue of concern to governments, businesses, and citizens [OECD 2011]. The report shows that the volume of personal data being transferred over public networks and retained by organizations has changed the risk profile, potentially exposing larger quantities of data in a single data breach. A data breach is a loss, unauthorized access to, or disclosure of personal data as a result of a failure of the organization to effectively safeguard the data. Data breaches can be attributed to both internal and external factors as discussed in [OECD 2011]:

- Internal factors such as errors or deliberate malicious activity on the part of employees as well as errors or malicious activity on the part of third parties that are involved in processing personal data on behalf of organizations; the risk of potential harm is from identity theft to individuals and from the misuse of their personal data; organizations are impacted too – a substantial financial cost in recovering from the breach and fixing problems within the organization to prevent a recurrence; may be subject to legal actions, including private actions or fines levied by various authorities, where allowed; costs to the organization's reputation; loss of trust or confidence, which can have serious financial consequences.
- External factors include intrusion from outside threat agents (e.g. malware); both organizations and individuals' home computers and other devices are also at risk.

Other developments of recent years include:

- A focus on finding common approaches to privacy protection at a global level, such as the development of international standards, as a response to the borderless nature of data flows, concerns around impediments to those flows, and the different cultural and legal traditions that have shaped the implementation of the privacy guidelines over the past 30 years.
- Finding global solutions and a better understanding of different cultures' views of privacy and the social and economic value of transborder data flows may help to achieve this goal.
- Seeking consensus on developing privacy protections in increasing numbers of countries besides OECD members.
- Increased support from the global privacy community and commitment within international organizations, governments, and privacy enforcement authorities to addressing current challenges.

Many activities and policies for cybersecurity and privacy are supported by the Department of Homeland Security (DHS) in the United States [DHS 2016a]. One example is the policy of 2008 that declares the Fair Information Practice Principles (FIPPs) as the foundation and guiding principles of the DHS's privacy program. FIPPs are time-tested and universally recognized principles that form the basis of the Privacy Act of 1974 and dozens of other federal privacy and information protection statutes. Also, a recent Executive Order [WH 2013] directs DHS to issue an annual report using the FIPPs to assess the Department's cyber operations under the Executive Order.

2.5.3 Protecting Cyberspace as a National Asset

In the light of the risk and potential consequences of cyber events, strengthening the security and resilience of cyberspace has become an important homeland security mission in the United States [DHS 2015]. However, emerging cyber threats require engagement from the entire American community to create a safer cyber environment – from government and law enforcement to the private sector and, most importantly, members of the public. Cybersecurity is a shared responsibility as pointed by DHS [DHS 2016b].

A framework for protecting the US infrastructure is described in [CERT 2003]. As pointed out in this document, securing cyberspace is an extraordinarily difficult strategic challenge that requires a coordinated and focused effort from our entire society – the federal government, state and local governments, the private sector, and the American people. The cornerstone of America's cyberspace security strategy is and will remain a public–private partnership. The strategic objectives are to:

- Prevent cyber attacks against America's critical infrastructures.
- Reduce national vulnerability to cyber attacks.
- Minimize damage and recovery time from cyber attacks that do occur.

Also, strategies that the United States can use for cyberspace protection are described to include the following objectives: establish a comprehensive strategy, maintain strong deterrents, strengthen public–private partnerships, avoid bureaucratic overreach, and forge an international consensus. These strategies can help policy makers make better-informed decisions about how to properly defend the country from threats [Peritz 2010].

In the same time, it is recognized that the perimeter of information systems and networks is increasingly blurred and that, as a consequence, the management of risks and the protection measures should extend to the more global ecosystem level.

The analysis report [OECD 2012a] reveals the success of the guiding principles of [OECD 2002] to create a framework for security in an open digital world where participants reduce risk before accepting it, instead of avoiding risk by limiting interconnectivity. These guidelines have been adopted by OECD members and non-OECD members. Responding to cybersecurity challenges has become a national policy priority in many countries. Gaps in the 2002 guidelines and new cybersecurity challenges are further analyzed in this report [OECD 2012c]. This report highlights many issues such as the following:

- New national strategies to strengthen cybersecurity are pursuing a double objective: driving further economic and social prosperity by using the full potential of the Internet as a new source of growth and platform for innovation and protecting cyberspace-reliant societies against cyber threats.
- Governments are developing comprehensive approaches integrating all facets of cybersecurity into holistic frameworks covering economic, social, educational, legal, law enforcement, technical, diplomatic, military, and intelligence-related aspects. The result is the elevation of this overall subject matter as a government policy priority and a higher degree of governmental coordination to develop strategies.
- The scope of most strategies generally covers all information systems and networks, including critical information infrastructures that are not connected to the Internet.
- Strategies generally lay out a narrative that varies across countries and leads to the introduction of various key objectives and concepts.
- Most strategies recognize that cyberspace is largely owned and operated by the private sector and that policies should be based on public–private partnerships, which may include business, civil society, and academia. However, they place variable emphasis on this aspect.
- While cybersecurity strategies share common concepts, there are still differences such as the concepts of cybersecurity and cyberspace that are not used by all countries.
- Although strategies share fundamental values, some concepts are specific to some countries, such as the economic aspects of cybersecurity, the need for dynamic policies, and the emergence of sovereignty considerations.
- Most strategies also stress the importance of the international dimension of cybersecurity and the need for better alliances and partnerships with like-minded countries or allies, including capacity building of less developed countries; all countries support the establishment of stronger international mechanisms at the policy and the operational levels. In this respect, policy makers need to:
  - Overcome complex coordination and cooperation challenges, internally across governmental bodies and with nongovernmental stakeholders, both at the domestic and international levels
  - Develop and implement action plans according to their strategies in a variety of areas such as critical information infrastructure protection, research and development, skills and jobs, economic incentives, cybersecurity exercises, etc.
- Although the protection of critical information infrastructures is generally included in the scope of cybersecurity strategies, the issue of cross-border interdependencies is rarely addressed at a strategic level.

A national cybersecurity plan is not only a strategic framework for nation's approach to cybersecurity; it is also a tool to improve the security and resilience of national infrastructures and services. Although there are many – and considerably different – definitions, a cybersecurity strategy has proven to be an instrument that helps governments manage the efforts of all involved parties in order to tackle risks related to cyber issues at a national level [ENISA 2015a]. Therefore, international and regional cooperation is needed at strategic levels to include the development of contingency and response plans in advance as well as the importance of regional and international exercises.

2.6 Cybersecurity for Electrical Sector as a National Priority

The electric grid, as government and private experts describe it, is the glass jaw of American industry. Energy-related risks to national security can broadly be categorized into physical, cyber, economic, and conflict related, though significant overlaps among these categories exist.

Cybersecurity and industry experts have expressed concern that, if not implemented securely, Smart Grid systems will be vulnerable to attacks that could result in widespread loss of electrical services essential to maintaining the national economy and security in the United States [GAO 2012]. Besides describing cyber threats to critical infrastructures, which include the electric grid, the report discusses key challenges to securing Smart Grid systems and networks.

Often, reports on cyber threats to electrical sector are published by media. A reporter describes a threat and the impacts of threat realization as follows:

> If an adversary lands a knockout blow, it could black out vast areas of the continent for weeks; interrupt supplies of water, gasoline, diesel fuel and fresh food; shut down communications; and create disruptions of a scale that was only hinted at by Hurricane Sandy and the attacks of Sept. 11. [Wald 2013]

Although this scenario sounds like a piece of science fiction, the reporter is warning the public about the fragility of the electric system that is tightly integrated that a collapse in one spot, whether by error or intent, can set off a cascade of power failure.

Another warns about the electric utility industry lacking adequate protection, and a major cyber threat to critical infrastructures is from the electric utilities [Weiss 2013]. While facts and impacts are not yet encountered by the electrical sector as described in these publications, we need to understand the dangers of cyber threats including other issues. Energy technologies must be robust and resistant to these vulnerabilities.

At least we have to consider that in today's highly connected world, with an increasingly sophisticated cyber threat, it is unrealistic to assume energy delivery systems are isolated or immune from compromise [Hawk 2014]. The grid is essential for almost everything, but it is mostly controlled by investor-owned companies or municipal or regional agencies. That expertise involves running 5 800 major power plants and 450 000 miles of high-voltage transmission lines, monitored and controlled by a staggering mix of devices installed over decades.

Some utilities use their own antique computer protocols and are probably safe from hacking – what the industry calls security through obscurity [Hawk 2014]. Also, cybersecurity in the IT/OT systems for the Smart Grid continues to be a significant topic and has been made even more critical by the convergence of IT/OT [Meyers 2013]. This convergence has enabled an enabled a new range of consumer-based OT, most of which is beyond the reach or control in the traditional utility. Therefore, an IT/OT-converged approach allows utility personnel to deploy each grid modernization application project as a part of a connected whole.

Cybersecurity is a serious and ongoing security, safety, and economic challenge for the electricity sector. The critical role of cybersecurity in ensuring the effective operation of the Smart Grid is documented in legislation and in the DOE energy sector plans (e.g. [DOE 2011]).

Securing the grid is one pillar of the framework set forth in the policy of Energy Independence and Security Act of 2007, and the Recovery Act of the Federal Government [EISA 2007] states:

> It is the policy of the United States to support the modernization of the Nation's electricity transmission and distribution system to maintain a reliable and secure electricity infrastructure that

can meet future demand growth and to achieve each of the following, which together characterize a Smart Grid:

1) Increased use of digital information and controls technology to improve reliability, security, and efficiency of the electric grid
2) Dynamic optimization of grid operations and resources, with full cybersecurity.

Security of grid implies safety and protection of assets, organization, consumers, and public from threats (intentional and unintentional) including natural disasters. Cybersecurity for the Smart Grid needs to support both the reliability of the grid and the security (and privacy) of the information that is generated, processed, transmitted, stored, or disposed. Defined in broad terms, cybersecurity for the power industry covers all issues involving automation and communications that affect the operation of electric power systems, the functioning of the utilities that manage them, and the business processes that support the customer base [NISTIR 7628r1].

2.6.1 Need for Cybersecurity Solutions

Cybersecurity solutions for energy infrastructure are imperative for reliable energy delivery. While reliability remains a fundamental principle of grid modernization efforts, reliability requires cybersecurity [Hawk 2014], [P2030 2011]. As the need for cybersecurity increases, this work [Hawk 2014] discusses energy sector partnerships that are designing cybersecurity for the Smart Grid with the vision of surviving a cyber incident while sustaining critical energy delivery functions.

A recently released document [DOE 2014b] provides guidance and requirements for cybersecurity features for the supply chain vendors and manufacturers of equipment, devices, and software used in power systems. Also, NIST's three-volume document [NISTIR 7628r1] provides guidance to organizations for cybersecurity and privacy strategies, architecture, requirements, supportive analyses, and references.

Ensuring a resilient electric grid is particularly important since it is arguably the most complex and critical infrastructure that other sectors depend upon to deliver essential services. Figure 2.14 is a schematic representation of electricity sector interdependencies with other sectors of the economy. Each

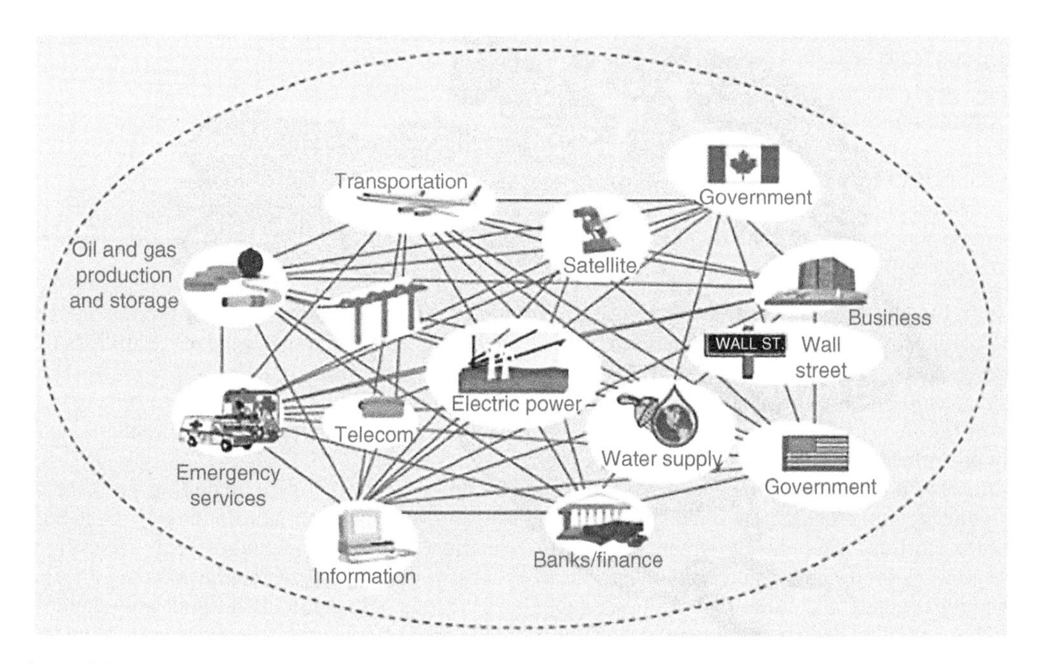

Figure 2.14 Interdependencies across the economy. *Source:* [DHS 2010]. Public Domain.

infrastructure depends on other infrastructures to function successfully. The potential impact of the increasing threats in the electricity sector is amplified by the connectivity between information systems, the Internet, and other infrastructures, creating opportunities for attackers to disrupt the electricity sector and other critical services such as banks, government, transportation, etc. [GAO 2012].

Over the past two decades, the roles of the electricity sector stakeholders have shifted: generation, transmission, and delivery functions have been separated into distinct markets; customers have become generators using distributed generation technologies; and vendors have assumed new responsibilities to provide advanced technologies and improve security. These changes have created new responsibilities for all stakeholders in ensuring the continued security and resilience of the electric power grid.

2.6.2 The US Plans

In the United States, the Federal Energy Regulatory Commission (FERC) defines polices for Smart Grid cybersecurity. Cybersecurity is briefly understood as encompassing measures to ensure the confidentiality, integrity, and availability of the electronic information communication systems and the control systems necessary for the management, operation, and protection of the Smart Grid's energy, IT, and telecommunication infrastructures [FERC 2009]. DOE supports the administration's strategic comprehensive approach to cybersecurity for the power grid. Also, DOE works closely with the DHS, industry, and other government agencies on an ongoing basis to reduce the risk of energy disruptions due to cyber attacks.

The DOE envisions a robust, resilient energy infrastructure in which continuity of business and services is maintained through secure and reliable information sharing, effective risk management programs, coordinated response capabilities, and trusted relationships between public and private security partners at all levels of industry and government [DOE 2010]. While the DOE cybersecurity roadmap provides a foundation for the development and adoption of interoperability and cybersecurity standards, the updated roadmap of 2011 [DOE 2011] goes on to recognize the advances in cybersecurity and other technology including the evolving needs of the energy sector such as the following:

- Providing a broader focus on energy delivery systems, including control systems, Smart Grid technologies, and the interface of cyber and physical security – where physical access to system components can impact cybersecurity.
- Building on successes and addressing gaps require new priorities to be identified such as enhancing vulnerability disclosure between government, researchers, and industry; addressing gaps to further advance technologies.
- Advancing threat capabilities by implementing enhanced security capabilities to protect energy delivery systems against threats that are becoming increasingly innovative, complex, and sophisticated.
- Emphasizing a culture of security that includes training people for developing and implementing the best available security policies, procedures, and technologies tailored to the energy delivery systems operational environment.

In its broadest sense, cybersecurity for the power industry covers all issues involving automation and communications that affect the operation of electric power systems, the functioning of the utilities that manage them, and the business processes that support the customer base.

Actions to develop the Smart Grid architecture include the coordinated advancement of standards across the electric power system, including device characteristics, communication requirements, security, and other system aspects [DOE 2015a].

Implementation of cybersecurity can occur through a variety of mechanisms, including use of standards and recommendations, enforcement of regulations, and voluntary compliance in response to business incentives. The energy sector, specifically electrical sector organizations, can use several mechanisms for designing and implementation of security and protection of energy systems. In addition, utilities, vendors, consultants, national laboratories, higher education institutions, governmental entities, and other organizations continuously contribute and participate in the standards and guidance of the electricity sector.

Also, energy systems and networks cross the national borders, making international collaboration a necessary component of the sector's efforts to develop standards to secure the energy infrastructure.

2.7 The Need for Security and Privacy Programs

A global survey was conducted on security governance, specifically on how boards of directors and senior management are governing the security of their organizations' information, applications, and networks. The survey respondents included 75% participants from critical infrastructure companies and represented [Westby 2012]:

- Energy and utilities companies.
- Financial sector.
- Healthcare.
- Industrials.
- IT and telecommunication companies.

The survey reveals issues related to security pasture of compared industries as follows:

- Boards still are not undertaking key oversight activities related to cyber risks, such as reviewing budgets, security program assessments, and top-level policies; assigning roles and responsibilities for privacy and security; and receiving regular reports on breaches and IT risks.
- Utilities are one of the least prepared organizations when it comes to risk management [Westby 2012].
- Utilities/energy sector and the industrial sector came in last in numerous areas – surprising is that these companies are part of critical infrastructure.
- All industry sectors surveyed are not properly assigning privacy responsibilities.
- Energy/utilities and IT/telecom respondents indicated that their organizations never (0%) rely upon insurance brokers to provide outside risk expertise, while the industrials sector relies upon them 100%.

Another report [GAO 2011] reveals that several security issues are missing including:

- An effective mechanism for sharing information on cybersecurity and other issues.
- Cybersecurity awareness.
- Security features built into Smart Grid systems.
- Metrics to measure cybersecurity.

In addition, the vulnerability of the power system is not mainly a matter of electric system or physical system, but is also a matter of cybersecurity. Attacks (such as attacks upon the power system, attacks by the power system, and attacks through power system) to the Smart Grid infrastructures could bring huge damages on the economy and public safety.

Smart Grid technologies and applications like smart meters, smart appliances, or customer energy management systems create new privacy risks and concerns in unexpected ways. Concerns of privacy of consumers and people are of vital importance in the energy sector. If there is any compromise of the personal data or security of the power service, it can undermine many services and applications. An incident would not only create a breach of privacy or confidentiality, integrity, or availability of the information, but it might also compromise the potential future markets the technology might have been able to create if it the service had been secure. Therefore, information security management principles, processes, and security architecture need to be applied to smart power grid systems without exception. All these objectives need to be included in the security program.

2.7.1 Security Program

Cybersecurity implies the implementation of security measures (safeguards) to ensure protection of an organization assets (tangible and intangible), people, and safety. Tangible assets are physical assets that include power equipment, computers, devices, facilities, and supplies. Intangible assets include data, information, reputation, intellectual property, copyrights, trade secrets, business strategies, and any other information valuable to an organization.

The destruction of power grid systems and assets would have a debilitating impact on energy security, economic security, public health, or safety. With a system that handles power generation, transmission, and distribution, security responsibility extends beyond the traditional walls of the data center. An intruder can, intentionally or unintentionally, cause a power line to be energized that would endanger lives. Similarly, a power line may be de-energized in such a way as to cause damage to transmission and control systems and possibly endanger the safety of employees and the public. Therefore, each organization should develop its own policy to protect assets, employees, and general public who are at risk when human (intentional or unintentional) threats or natural disasters occur. Each organization should develop its own cybersecurity strategy for the implementation of a security program. Cybersecurity must address not only deliberate attacks launched by disgruntled employees, agents of industrial espionage, and terrorists but also inadvertent compromises of the information infrastructure due to user errors, equipment failures, and natural disasters [NISTIR 7628].

Security program is a plan or outline that must cover security governance, planning, prevention, operations, incident response, and business continuity. Variants of Smart Grid implementations have already been rolled out in various jurisdictions across the United States as well as the rest of the world for several years. The window of opportunity to integrate security into the Smart Grid from the beginning is shrinking fast. However, it is also necessary to understand the interdependency and mutual vulnerability of the wholesale electric grid and the wholesale electric market in maintaining the security and stability of the smart power grid. Market participants require to ensure protection of their critical cyber assets and to support an appropriate security program.

A security program needs to be built using the security engineering approach. This requires focus on building systems to remain dependable in the face of malice, error, or mischance [Anderson 2008]. Also, the successful implementation of a security program requires certain basic functions that should be included in any budget allocation [Whitman 2014].

2.7.2 Privacy Program

As new capabilities are included in the Smart Grid, potential new privacy concerns will emerge for which no legal mitigation currently exists. A significant number of privacy breaches occur not because of an attack but through noncompliance with privacy policy or having no policy. For example, a laptop that has a copy of PII data becomes a privacy breach if the laptop is improperly disposed of, lost, or stolen. Hence, measures for protection of privacy have to be designed and implemented too. Thus, a privacy program should be planned, designed, implemented, and maintained. Factors that should be considered in design of a security program include the following:

- Privacy rights continue to evolve by legislation, litigation, and regulation, and the data gathered will be subject to the relevant jurisdiction(s).
- Anonymization
  If private information is not properly anonymized, even data like electrical appliance usage or electric vehicle charging schedules may constitute a privacy violation. In electrical sector, the ownership and rights associated with PII varies by jurisdiction. In some jurisdictions, the person owns their data, while in other jurisdictions, ownership is less clear. For example, a utility that gathers contact and other information for billing purposes may be restricted in use of the PII for any other purposes without consent of the customer – possession of the data is not the same as ownership.
- Technologies and capabilities
  The advancing of technologies such as data mining and pattern recognition can be used on identifying the identity of persons when customer data and energy data is analyzed. Recognizing electric signatures of smart appliances and developing detailed, time-stamped activity reports, utilities, or third-party service providers can determine lifestyle details that could be legitimately characterized as PII in most jurisdictions.

- Dedicated privacy group with its own management

  Although in many organizations, security group is supporting the privacy requirements, the future commands for more responsibility and accountability for the implementation of data privacy specifically in smaller-size enterprises, and need for establishment of a dedicated privacy group with its own management [Shei 2013]. The organizations have to understand that security is only one aspect of privacy and privacy protection implies organization and business decisions.

Ensuring privacy requires a bundle of technologies, policies, culture, regulations, and harmony between many business units from security to legal to human resources to employees [Shei 2013]. Examples of guidelines and recommendations for the protection of privacy data and harmonization of disparities in national privacy regulations are documented in [OECD 2013].

Currently, many countries, organizations, and associations support efforts to empower and educate people to protect their privacy, control their digital footprint, and make the protection of privacy and data a great priority in their lives. In the United States, National Cyber Security Alliance mandates that [NCSA 2014]:

> Everyone – from home computer users to multinational corporations – needs to be aware of the personal data others have entrusted to them and remain vigilant and proactive about protecting it.

This document [NISTIR 7628r1] provides definitions, requirements, safeguards, and use case impacts of privacy breaches. Privacy considerations with respect to the Smart Grid include four aspects: privacy of personal information, privacy of the person, privacy of personal behavior, and privacy of personal communications.

A privacy policy framework for the Smart Grid and for smart homes is suggested in [GridWise 2011]. This framework is limited and addresses only consumer privacy issues that arise from the collection, use, and retention of such data no matter from what source it is collected.

In this book, we do not focus on engineering a privacy program, although some approaches used in engineering the security program could be used for building a privacy program.

2.8 Standards, Guidelines, and Recommendations

A revised NIST document [NISTIR 7628r1] promotes a new cybersecurity framework to protect the Smart Grid. A current list of standards is available. Many accelerated standards and guidelines are focused on topics such as:

- Metering
- Data usage information
- Electric vehicles
- Pricing
- Demand response
- Substation communication
- Energy storage
- Renewables.

2.8.1 Electricity Sector Guidance

In the United States, the DOE envisions a robust, resilient energy infrastructure in which continuity of business and services is maintained through secure and reliable information sharing, effective risk management programs, coordinated response capabilities, and trusted relationships between public and private security partners at all levels of industry and government [DOE 2015c].

Within the electricity subsector, the FERC is focused on the development of key standards to achieve interoperability and functionality of Smart Grid systems and devices [FERC 2009]. FERC certified the North American Electric Reliability Corporation (NERC) as the Electric Reliability Organization that is responsible for developing reliability standards, subject to FERC oversight, review, and approval.

NERC developed the critical infrastructure protection (CIP) standards [NERC CIP], which FERC approved in 2008. The NERC CIP standards suite is composed of a whole family of standards that are continuously revised and changed. These standards were originally devised and implemented to prevent big blackouts – so they are considered both rigorous and heavily enforced only for bulk power systems (generation and transmission).

However, NERC cybersecurity standards and supplementary documents are often similar to guidance applicable to federal agencies [GAO 2011] and do not apply to all power grid functions. In addition, the standards adoption by the electric power industry is lacking coordination and a consistent approach in monitoring industry compliance with voluntary standards. FERC is responsible for regulating aspects of the electric power industry, which includes adopting cybersecurity and other standards it deems necessary to ensure Smart Grid functionality and interoperability.

2.8.2 International Collaboration

An essential element of Smart Grid developments around the globe is coordination for the development of international standards. As the United States and other nations construct their Smart Grids, use of international standards ensures the broadest possible market for Smart Grid suppliers.

NIST is devoting considerable resources and multilateral engagement with other countries to cooperate in the development of international standards for the Smart Grid. In addition, NIST and the International Trade Administration (ITA) have partnered with the DOE to establish the International Smart Grid Action Network (ISGAN), a multinational collaboration of 23 countries and the European Union.

ISGAN complements the Global Smart Grid Federation, a global stakeholder organization, which serves as an association of associations to bring together leaders from Smart Grid stakeholder organizations around the world. This organization supports Smart Grid solutions emerging to address the economic, policy, and regulatory challenges of variable renewables. Similarly, the Clean Energy Solutions Foundation (https://cleanenergysolutions.org) helps governments design and adopt policies and programs that support the deployment of clean energy technologies. Regulatory policies around the globe promote renewable electricity standards (recommendations, good practices, design considerations) to accelerate renewable energy deployment.

However, cybersecurity for the IT/OT systems for the Smart Grid continues to be a significant topic and has been made even more critical by the convergence of IT/OT [Meyers 2013]. This convergence has enabled a new range of consumer-based OT, most of which is beyond the reach or control in the traditional utility. Therefore, an IT/OT-converged approach allows utility personnel to deploy each grid modernization application as a part of a connected whole.

In addition, diverse IoT-based applications call for different deployment scenarios and requirements, which have usually been handled in a proprietary implementation. However, since the IoT is connected to the Internet, most of the devices comprising IoT services need to operate utilizing standardized technologies. The Internet Protocol for Smart Objects (IPSO) Alliance (www.ipso-alliance.org) promotes the IoT. Prominent standardization bodies, such as IETF (www.ietf.org), the Institute of Electrical and Electronics Engineers (IEEE) (www.ieee.org), and European Telecommunications Standards Institute (ETSI) (www.etsi.org), are working on developing protocols, systems, architectures, and frameworks to enable the IoT devices to interoperate.

In respect to privacy, international standards bodies are currently working on establishing standards to assist organizations in better protecting personal data. Examples include:

- The ISO is working on technical standards for a Privacy Framework and Privacy Reference Architecture.
- Regional standards organizations, such as the American National Standards Institute (ANSI) and the European Committee for Standardization (CEN), are involved in data protection standards; CEN\

ISSS reported to the European Commission in 2003 on the utility of standards in enforcing the directive. Their work continues in setting standards for networks, biometrics, identity and authentication, cryptographic protocols, security management, deidentification of health information, data storage, and other standards that have a bearing on privacy architectures.
- The ETSI produces standards for information and communication technologies.
- United Nations Internet Governance Forum (IGF) and the regional IGFs are increasing the privacy discussions. In 2009 and 2010, the IGF program included a main session on security, openness, and privacy as well as numerous workshops devoted to privacy issues.

The need for open standards is a trend that can increase standards adoption toward improving security and privacy of Smart Grid. However, cyber threats can evolve faster than standards. To safeguard vital interests like electrical grids, it is also needed collaboration and information sharing among federal, state, local governments, and industry [NISTIR 7628r1].

References Part 1

[ABI Research 2015a] ABIResearch. (2015a). *Internet of Everything*. https://www.abiresearch.com/pages/what-is-internet-everything/

[ABI Research 2015b] ABIResearch. (2015b). *Recent Research*. https://www.abiresearch.com/market-research/service/m2m-iot-ioe/

[Agresti 2010] Agresti, W. (2010). The Four Forces Shaping Cybersecurity. *Computer*, *43*(2), 101–104. https://doi.org/10.1109/MC.2010.53

[Amin 2003] Amin, M. (2003). *Complex interactive networks/systems initiative: final summary report*. https://massoud-amin.umn.edu/sites/massoud-amin.umn.edu/files/2020-03/cinsi-final_overview_report_ver6.pdf

[Anderson 2008] Anderson, R.J. (2008). *Security Engineering: A Guide to Building Dependable Distributed Systems* (2nd Edition). Wiley.

[AUTO-ID 1999] AUTO-ID. (1999). *The Leading Academic Research Network on the Internet of Things*. https://www.autoidlabs.org/

[Bachman 2015] Bachman, K. (2015, January 28). *FTC sees privacy threats in the 'Internet of Things'* [News]. POLITICO Magazine. http://www.politico.com/story/2015/01/internet-of-things-ftc-114666.html#ixzz3jwp4D7fl

[Baimel 2016] Baimel, D., Tapuchi, S., Baimel, N. (2016). Smart grid communication technologies. *Journal of Power and Energy Engineering*, *4*(8), Paper ID 69361. CC BY 4.0. http://dx.doi.org/10.4236/jpee.2016.48001

[Bartz/Stockmar] Bartz/Stockmar. (2018, May 24). *Characteristics of a Smart Grid (Right) Versus the Traditional System (Left)*. CC BY-SA 4.0. https://en.wikipedia.org/wiki/Smart_grid#/media/File:Staying_big_or_getting_smaller.jpg

[Bay 2016] Bay, M. (2016). What is cybersecurity? *French Journal for Media Research*, *6*. https://frenchjournalformediaresearch.com:443/lodel-1.0/main/index.php?id=988

[Bishop 2005] Bishop, M. (2005). *Introduction to Computer Security*. Addison-Wesley Professional.

[Borman 2010] Bormann, C., Vasseur, J., Shelby, Z. (2010). The internet of things. *the IETF Journal*, *6*(2), 1, 4–5.

[Britz 1996] Britz, J.J. (1996). Technology as a threat to privacy: ethical challenges to the information profession. *Microcomputers for Information Management*, *13*(3), 75–93.

[Capehart 2014] Capehart, B.L., Brombley, M.R. (2014). *Automated Diagnostics and Analytics for Buildings*. Fairmont Press.

Building an Effective Security Program for Distributed Energy Resources and Systems: Understanding Security for Smart Grid and Distributed Energy Resources and Systems, Volume 1, First Edition. Mariana Hentea.

[CC 2.3] Common Criteria. (2005). *Common Criteria for Information Technology Security Evaluation. Part 1: Introduction and General Model Version 2.3* [CCMB-2005-08-001]. https://www.commoncriteriaportal.org/files/ccfiles/ccpart1v2.3.pdf

[CERT 2003] CERT. (2003). *The National Strategy to Secure Cyberspace*. U.S. Government via Department of Homeland Security. https://us-cert.cisa.gov/sites/default/files/publications/cyberspace_strategy.pdf

[Chen 2010] Chen, K-C., Yeh, P-C., Hsieh, H-Y., Chang, S-C. (2010). Communication infrastructure of smart grid. *2010 4th International Symposium on Communications, Control and Signal Processing (ISCCSP)* (pp. 1-5), Limassol, Cyprus, (March 2010). https://doi.org/10.1109/ISCCSP.2010.5463330

[Cisco 2011] Cisco. (2011, April). *Internet of Things How the Next Evolution of the Internet is Changing Everything* [White Paper]. https://www.cisco.com/c/dam/en_us/about/ac79/docs/innov/IoT_IBSG_0411FINAL.pdf

[Clearfield 2013] Clearfield, C. (2013, September 18). *Why the FTC can't regulate the Internet of Things* [News]. Forbes. http://www.forbes.com/sites/chrisclearfield/2013/09/18/why-the-ftc-cant-regulate-the-internet-of-things/#231c8d853ae4

[CNSSI 4009] CNSSI. (2010). *Committee on National Security Systems (CNSS) Information Assurance (IA) Glossary*. Committee on National Security Systems (CNSS) Instruction No. 4009. *Revised in 2015.

[Conference 2004] Cordis. (2004). *International Conference on the Integration of Renewable Energy Sources and Distributed Energy Resources*, Brussels, Belgium, (December 2004). https://cordis.europa.eu/event/id/22548-international-conference-on-the-integration-of-renewable-energy-sources-and-distributed-energy

[CPS 2006] Cyberphysicalsystems. (2006). *Cyber-Physical Systems*. Cyber Physical Systems Organization. http://cyberphysicalsystems.org/Cyber-Physical_Systems.html

[CPS 2014] Wikipedia. (2014, May 16). *Cyber-Physical System*. http://en.wikipedia.org/wiki/Cyber-physical_system

[Craigen 2014] Craigen, D., Diakun-Thibault, N., Purse, R. (2014). Defining cybersecurity. *Technology Innovation Management Review*, *4*(10), 13–20. https://timreview.ca/article/835

[DER Systems] CPSENERGY. (n.d.). *DER systems typical distributed energy resources systems*. https://cpsenergy.com/en/construction-and-renovation/distributed-generation/distributed-energy-resources.html

[Dhillon 2001] Dhillon, G. (2001). Challenges on managing information security in the new millennium. In G. Dhillon (Ed.), *Information Security Management: Global Challenges in the New Millennium* (pp. 1–8). IGI Global. https://doi.org/10.4018/978-1-878289-78-0.ch001

[DHS 2010] DHS. (2010). *Energy sector-specific plan an annex to the national infrastructure protection plan*. Department of Homeland Security. https://www.dhs.gov/xlibrary/assets/nipp-ssp-energy-2010.pdf

[DHS 2015] DHS. (2015, September 22). *Cybersecurity overview*. Department of Homeland Security. https://www.dhs.gov

[DHS 2016a] DHS. (2016a, January 16). *Cybersecurity and Privacy*. Department of Homeland Security. https://www.dhs.gov/cybersecurity-and-privacy

[DHS 2016b] DHS. (2016b). *Cyber Safety*. Department of Homeland Security. https://www.dhs.gov/cyber-safety

[Dictionary 1994] Dictionary. (1994). *American Heritage Dictionary* (third edition). Bantam Doubleday Dell Publishing Group, Inc.

[DOE 2009] DOE. (2009). *The Smart Grid: an introduction.* U.S. Department of Energy. http://energy.gov/sites/prod/files/oeprod/DocumentsandMedia/DOE_SG_Book_Single_Pages%281%29.pdf

[DOE 2010] DOE. (2010). *DOE energy sector specific plan: an annex to the national infrastructure protection plan.* U.S. Department of Energy. http://energy.gov/oe/downloads/energy-sector-specific-plan-annex-national-infrastructure-protection-plan

[DOE 2011] DOE. (2011, September 2011). *Roadmap to achieve energy delivery systems cybersecurity.* Energy Sector Control Systems Working Group. http://energy.gov/sites/prod/files/Energy%20Delivery%20Systems%20Cybersecurity%20Roadmap_finalweb.pdf

[DOE 2012] DOE. (2012, May). *Electricity subsector cybersecurity risk management process* [DOE/OE-0003]. U.S. Department of Energy. http://energy.gov/sites/prod/files/Cybersecurity%20Risk%20Management%20Process%20Guideline%20-%20Final%20-%20May%202012.pdf

[DOE 2014a] DOE. (2014a). *Cybersecurity Capability Maturity Model (ES-C2M2) Glossary.* U.S. Department of Energy. http://energy.gov/sites/prod/files/2014/02/f7/ES-C2M2-v1-1-Feb2014.pdf

[DOE 2014b] DOE. (2014b, April). *Cybersecurity procurement language for energy delivery systems.* U.S. Department of Energy. http://www.energy.gov/oe/downloads/cybersecurity-procurement-language-energy-delivery-april-2014

[DOE 2015a] DOE. (2015a). *Quadrennial technology review an assessment of energy technologies and research opportunities* (September 2015). U.S. Department of Energy. http://energy.gov/sites/prod/files/2015/09/f26/Quadrennial-Technology-Review-2015_0.pdf

[DOE 2015b] DOE. (2015b). *Quadrennial technology review 2015 chapter 3: Enabling modernization of the electric power system technology assessments.* U.S. Department of Energy. http://www.energy.gov/sites/prod/files/2015/09/f26/QTR2015-3D-Flexible-and-Distributed-Energy_0.pdf

[DOE 2015c] DOE. (2015c). *Energy sector cybersecurity framework implementation guidance* (January 2015). U.S. Department of Energy. http://energy.gov/sites/prod/files/2015/01/f19/Energy%20Sector%20Cybersecurity%20Framework%20Implementation%20Guidance_FINAL_01-05-15.pdf

[EC-EPoSS 2008] EC-EPoSS. (2008). Internet of things in 2020 A Roadmap for the future [Workshop Report]. *The European Technology Platform on Smart Systems Integration (EPoSS)* (September 2008). http://old.sztaki.hu/~pbakonyi/bme/kieg/Internet-of-Things_in_2020_EC-EPoSS_Workshop_Report_2008_v3.pdf

[EISA 2007] EISA. (2007). *U.S. Energy Independence and Security Act of 2007 Public Law No: 110-140 Title XIII, Sec. 1301.* http://www.gpo.gov/fdsys/pkg/BILLS-110hr6enr/pdf/BILLS-110hr6enr.pdf

[ElNet 2014] MBizon. (2014). *Electric Network.* Wikimedia. CC BY 3.0. https://commons.wikimedia.org/wiki/File:Electricity_Grid_Schematic_English.svg

[ENISA 2015a] ENISA. (2015a, February 23). *Methodologies for the identification of critical information infrastructure assets and services.* European Union Agency for Network and Information Security. http://doi.org/10.2824/38100

[ENISA 2015b] ENISA. (2015b). *Communication network interdependencies in smart grids – annexes. Annex C methodology for the identification of critical communication networks links and components.* European Union Agency for Cybersecurity. DOI: 978-92-9204-140-3

[ENISA 2016] ENISA. (2016). *Communication network interdependencies in smart grids – annexes, methodology for the identification of critical communication networks links and components* (January 2016). https://www.enisa.europa.eu/topics/critical-information-infrastructures-and-services/communication-network-interdependencies-in-smart-grids-annexes

[ENISA Glossary] ENISA. (2015). *Glossary*. European Union Agency for Cybersecurity. https://www.enisa.europa.eu/topics/threat-risk-management/risk-management/current-risk/risk-management-inventory/glossary

[EPRI 2007] EPRI. (2007). *Advanced metering infrastructure*. EPRI. www.epri.com

[EPRI 2013] EPRI. (2013). *Cyber security for DER systems 1.0*. EPRI. https://smartgrid.epri.com/doc/der%20rpt%2007-30-13.pdf

[EPRI 2015] EPRI. (2015). *Distribution Modeling Guidelines: Recommendations for System and Asset Modeling for Distributed Energy Resource Assessments* (3002006115). https://www.epri.com/research/products/3002006115

[ETIP-SNET] ETIP-SNET. (n.d.). *Europe reference publications*. https://www.etip-snet.eu/publications/reference-publications

[Feng 2012] Feng, X., Stoupis, J., Mohagheghi, S., Larsson, M. (2012). Introduction to smart grid applications. In L.T. Berger, I. Krzysztof (Eds.), *Smart Grid Applications, Communications, and Security* (pp. 3–48). Wiley.

[FERC 2009] FERC. (2009). *Smart Grid Policy. Final Rule Federal Register, Rules and Regulations, 74*(142). http://www.gpo.gov/fdsys/pkg/FR-2009-07-27/html/E9-17624.htm

[FERC 2012] FERC. (2012). *Assessment of demand response and advanced metering* [Staff Report]. Federal Energy Regulatory Commission. https://www.ferc.gov/sites/default/files/2020-05/12-20-12-demand-response.pdf

[Forrester 2010] Schadler, T. (2010, November 3). Welcome to the Empowered Era [Blog]. *Forrester*. http://blogs.forrester.com/ted_schadler/10-11-03-welcome_to_the_empowered_era

[Fragkos 2005] Fragkos, G., Blyth, A. (2005). Security threat assessment across large network infrastructures. In C. Johnson (Ed.), *First Workshop on Safeguarding National Infrastructures* (pp. 22–29). http://www.dcs.gla.ac.uk/~johnson/infrastructure/Infrastructure_Proceedings.PDF

[Franscella 2013] Franscella, J. (2013, July 17). Cybersecurity vs. Cyber Security: When, Why and How to Use the Term [Blog]. *SecurityWeek*. http://www.infosecisland.com/blogview/23287-Cybersecurity-vs-Cyber-Security-When-Why-and-How-to-Use-the-Term.html

[Galloway 2012] Galloway, B., Hancke, G.P. (2012). Introduction to industrial control networks. *IEEE Communications Surveys & Tutorials, 15*(2), 860–880. https://doi.org/10.1109/SURV.2012.071812.00124

[GAO 2011] GAO. (2011). *ELECTRICITY GRID MODERNIZATION Progress Being Made on Cybersecurity Guidelines, but Key Challenges Remain to be Addressed* (GAO-11-117). United States Government Accountability Office. https://www.gao.gov/new.items/d11117.pdf

[GAO 2012] GAO. (2012). *Cybersecurity challenges in securing the modernized electricity grid*. United States Government Accountability Office (GAO) (GAO-12-507T). http://www.gao.gov/products/GAO-12-507T

[Gartner 2015] Gartner. (2015, November 10). *Gartner says 6.4 billion connected "Things" will be in use in 2016, up 30 percent from 2015* [Press Release]. Gartner. https://www.gartner.com/en/newsroom/press-releases/2015-11-10-gartner-says-6-billion-connected-things-will-be-in-use-in-2016-up-30-percent-from-2015

[Gartner IT] Gartner. (n.d.). *IT Glossary, Smart Grid*. http://www.gartner.com/it-glossary/smart-grid

[Gibson 1984] Gibson, W. (1984). *Neuromancer*. The Penguin Putnam, *Inc*.

[GridWise 2011] GridWise. (2011). *GridWise Alliance Policy position on data access & privacy issues, final draft* (September 2011). State and Legislative Policy Working Group. http://www.gridwise.org/documents/GWA_SLPWG_PrivacyWhitePaper_9_12_11_FinalR.pdf

[Halpern 2014] Halpern, S. (2014, November 20). *The Creepy New Wave of the Internet [News]*. The New York Review. http://www.nybooks.com/articles/archives/2014/nov/20/creepy-new-wave-internet

[Harris 2005] Harris, S. (2005). *All in One CISSP Exam Guide* (Third Edition). McGrawHill.

[Harris 2013] Harris, S. (2013). *All in One CISSP Exam Guide* (Sixth Edition). McGrawHill.

[Hawk 2014] Hawk, C., Kaushiva, A. (2014). Cybersecurity and the smarter grid. *The Electricity Journal, 27*(8), 84–95. https://doi.org/10.1016/j.tej.2014.08.008

[Hentea 2007] Hentea, M. (2007). SCADA evolution and security issues escalation. *The Newsletter for Information Protection Professionals Computer Security Alert* (pp. 6–7). Computer Security Institute.

[IEA 2011] IEA. (2011, April). *Technology roadmap – smart grids.* IEA. https://www.iea.org/reports/technology-roadmap-smart-grids

[IEA 2016] IEA. (2016). *Energy Security News and Events.* IEA. https://www.iea.org/past-events?year=2016

[IIC 2015] IIC. (2015). *Industrial Internet Vocabulary Version 1.0* (tech-vocab.tr.001). Industrial Internet Consortium. https://www.iiconsortium.org/pdf/Industrial-Internet-Vocabulary.pdf

[IntSoc 2015a] IntSoc. (2015a, October). *The Internet of Things: An overview understanding the issues and challenges of a more connected world.* Internet Society. https://www.internetsociety.org/resources/doc/2015/iot-overview

[IntSoc 2015b] IntSoc. (2015b, April 12). *Collaborative Security: An approach to tackling internet security issues.* Internet Society. https://www.internetsociety.org/collaborativesecurity/approach/

[IoT] Wikipedia. (2014). *Internet of Things.* http://en.wikipedia.org/wiki/Internet_of_Things

[IoT 2015] Wordpress. (2015, January). *Iot in Nutshell.* https://iotunleashed.wordpress.com/iot-in-nutshell/about/

[Ishaq 2013] Ishaq, I., Carels, D., Teklemariam, G.K., Hoebeke, J., Van den Abeele, F., De Poorter, E., Moerman, I., Demeester, P. (2013). IETF Standardization in the Field of the Internet of Things (IoT): A Survey. *J. Sens. Actuator Netw., 2(2)*, 235–287. https://doi.org/10.3390/jsan2020235

[ISO 15443] ISO TR 15443-1:2012 *Information Technology – Security Techniques – Security Assurance Framework – Part 1: I Introduction and Concepts.*

[ISO/IEC 51] *ISO/IEC Guide 51:2014(en) Safety aspects — Guidelines for their inclusion in standards.*

[ISO/IEC 42010] *ISO/IEC 42010:2011 Systems and Software Engineering – Architecture Description.*

[ISO/IEC 27000] *ISO/IEC 27000:2018 Information Technology — Security Techniques — Information Security Management Systems — Overview and Vocabulary* (Edition 5).

[ITU-T 2008] *ITU-T X.1205* (04/2008) *Overview of Cybersecurity.* International Telecommunications Union for the Telecommunication Standardization Sector. https://www.itu.int/rec/T-REC-X.1205-200804-I

[ITU-T 2012] *ITU-T Y.4000/Y.2060* (06/2012) *Overview of the Internet of Things.* International Telecommunications Union for the Telecommunication Standardization Sector. http://www.itu.int/ITU-T/recommendations/rec.aspx?rec=y.2060 *renumbered as ITU-T Y.4000 in 2016

[Kaplan 2009] Kaplan, S.M. (2009). Smart Grid Electrical Power Transmission: Background and Policy Issues. *The Capital Net Government Series* (pp. 1-42).

[Karnouskos 2011] Karnouskos, S. (2011). Cyber-Physical Systems in the SmartGrid. *2011 9th IEEE International Conference on Industrial Informatics* (pp. 20-23), Caparica, Lisbon, Portugal, (July 2011). https://doi.org/10.1109/INDIN.2011.6034829

[Karnouskos 2012] Karnouskos, S. (2012). Cyber-physical systems in the smart grid – potential and challenges (Presentation). *SAP Research EU-US Joint Open Workshop On Cyber Security of ICS and Smart Grids*, Amsterdam, The Netherlands, (October 2012).

[Krutz 2004] Krutz, R.L., Vines R.D. (2004). *The CISSP Prep Guide* (Second edition). Wiley.

[Lee 2010] Lee, E.A. (2010). CPS foundations. *Proceedings of the 47th Design Automation Conference (DAC)* (pp. 737–742), Anaheim, California, USA, (June 2010). ACM. https://doi.org/10.1145/1837274.1837462

[Lee 2015a] Lee, E.A., Seshia, S.A. (2015a). *Introduction to Embedded Systems A Cyber-Physical Systems Approach* (Second Edition). http://leeseshia.org/index.html

[Lee 2015b] Lee, E.A. (2015b). The past, present and future of cyber-physical systems: a focus on models. *Sensors, 15*(3), 4837–4869. https://doi.org/ 10.3390/s150304837

[Ling 2015] Ling, S.G., Tan, Y., Lim, Y. (2015). Design and multiobjective optimization of efficient vehicle management framework for cyber-physical intelligent transport systems. *Proceedings of the 2015 IEICE Communications Society Conference, BS-6-22 Communications 2 - Network and Service Design, Control and Management*, Sendai, Japan, (September 2015).

[Liu 2018] Liu, Y., Trappe, W., Garnaev, A. (2018). Applications of Graph Connectivity to Network Security. In P.M. Djurić, C. Richard (Eds.), *Cooperative and Graph Signal Processing Principles and Applications* (pp. 445–467). Academic Press.

[Maiwald 2004] Maiwald, E. (2004). *Fundamentals of Network Security*. McGraw-Hill/Technology Education.

[Mandates 2007] Energy. (2007). *Smart Grid*. U.S. Department of Energy. http://energy.gov/oe/technology-development/smart-grid

[Meyers 2013] Meyers, J. (2013). *How the convergence of IT and OT enables smart grid development* (998-2095-08-13-13R0). Schneider Electric. http://cdn.iotwf.com/resources/10/How-the-Convergence-of-IT-and-OT-Enables-Smart-Grid-Development_2013.pdf

[Microsoft 2002] Microsoft. (2002). *Internet & Networking Dictionary*. Microsoft Press.

[Monnier 2013] Monnier, O. (2013). *A Smarter grid with the internet of things* [White Paper]. https://www.se.com/us/en/download/document/998-2095-08-13-13R0_EN/

[NCSA 2014] NCSA. (2014). *Online Safety Basics Learn how to protect yourself, your family and devices with these tips and resources*. National Cyber Security Alliance. https://staysafeonline.org/stay-safe-online/online-safety-basics/

[NERC CIP] NERC. (n.d.). *CIP Standards*. North American Electric Reliability Corporation. http://www.nerc.com/pa/Stand/Pages/CIPStandards.aspx

[NETL 2008] NETL. (2008, February). *Advanced metering infrastructure*. National Energy Technology Laboratory for the U.S. Department of Energy Office of Electricity Delivery and Energy Reliability. U.S. Department of Energy. https://www.smartgrid.gov/document/netl_modern_grid_strategy_powering_our_21st_century_economy_advanced_metering_infrastructur

[NICCS 2016] NICCS. (2016). *Explore Terms: A Glossary of common cybersecurity terminology*. National Initiative for Cybersecurity Careers and Studies (NICCS). https://niccs.us-cert.gov/about-niccs/cybersecurity-glossary#key

[NIST CPS] NIST. (n.d.). *Cyber-Physical Systems*. https://www.nist.gov/el/cyber-physical-systems

[NIST SP800-82r2] *NIST Special Publication (SP) 800-82r2 Guide to Industrial Control Systems (ICS) Security Supervisory Control and Data Acquisition (SCADA) systems, Distributed Control Systems (DCS), and other control system configurations such as Programmable Logic Controllers (PLC)* (Revision 2, May 2015). http://dx.doi.org/10.6028/NIST.SP.800-82r2

[NIST SP1108r1] *NIST Special Publication (SP) 1108r1 NIST Framework and Roadmap for Smart Grid Interoperability Standards* (Release 1.0, January 2010). https://doi.org/10.6028/NIST.sp.1108

[NIST SP1108r3] *NIST Special Publication (SP) 1108r3 NIST Framework and Roadmap for Smart Grid Interoperability Standards* (Release 3.0, September 2014). http://dx.doi.org/10.6028/NIST.SP.1108r3

[NIST SP1500-201] *NIST Special Publication (SP) 1500-201 Framework for Cyber-Physical Systems* (Volume 1, Overview Version 1.0, June 2017). https://doi.org/10.6028/NIST.SP.1500-201

[NISTIR 7628] *NIST Interagency Report (NISTIR) 7628* Guidelines *for Smart Grid Cyber Security* (August 2010).
[NISTIR 7628r1] *NIST Interagency Report (NISTIR) 7628r1 Guidelines for Smart Grid Cybersecurity* (Revision 1, September 2014). http://dx.doi.org/10.6028/NIST.IR.7628r1
[NISTIR 7298r2] *NIST Interagency Report (NISTIR) 7298r2 Glossary of Key Information Security Terms* (Revision 2, May 2013). *Superseded by [NISTIR 7298r3].
[Nordell 2012] Nordell, D.E. (2012). Terms of protection: The Many Faces of Smart Grid Security. *IEEE Power & Energy Magazine*, *10*(1), 18–23. https://doi.org/10.1109/MPE.2011.943194
[NREL 2014] NREL. (2014). *2014 Renewable Energy Data Book.* National Renewable Energy Laboratory (NREL). https://www.nrel.gov/docs/fy16osti/64720.pdf
[NSF 2011] NSF. (2011, June). *A Policy framework for the 21st century grid: enabling our secure energy future.* U.S. National Science Foundation. https://www.ourenergypolicy.org/wp-content/uploads/2011/12/2011_06_WhiteHouse_PolicyFramework21stCenturyGrid.pdf
[NSF 2014] NSF. (2014). *Cyber-Physical Systems Research Initiative.* U.S. National Science Foundation (NSF). https://www.nsf.gov/pubs/2014/nsf14542/nsf14542.htm
[O'Brien 1999] O'Brien, J.A. (1999). *Management Information Systems: Managing Information Technology in the Internetworked Enterprise* (4 Edition). Irwin/McGraw Hill.
[O'Brien 2011] O'Brien, J.A., Marakas, G.M. (2011). *Management Information Systems* (Tenth Edition). McGraw-Hill/Irwin.
[OECD 1980] OECD. (1980). *Guidelines on the Protection of Privacy and Transborder Flows of Personal Data.* Organisation for Economic Co-operation and Development. https://www.oecd.org/internet/ieconomy/oecdguidelinesontheprotectionofprivacyandtransborderflowsofpersonaldata.htm
[OECD 1992] OECD. (1992). *Recommendation of the Council Concerning Guidelines for the Security of Information Systems.* Organisation for Economic Co-operation and Development. https://www.oecd.org/internet/ieconomy/oecdguidelinesforthesecurityofinformationsystems1992.htm
[OECD 1999] OECD. (1999). *Guidelines for Consumer Protection in the Context of Electronic Commerce.* Organisation for Economic Co-operation and Development. https://www.oecd.org/sti/consumer/oecdguidelinesforconsumerprotectioninthecontextofelectroniccommerce1999.htm
[OECD 2002] OECD. (2002, July 25). *Guidelines for the Security of Information Systems and Networks: Towards a Culture of Security.* http://www.oecd.org/sti/ieconomy/oecdguidelinesforthesecurityofinformationsystemsandnetworkstowardsacultureofsecurity.htm
[OECD 2003] OECD. (2003, July). *Implementation Plan for the OECD Guidelines for the Security of Information Systems and Networks: Towards a Culture of Security.* Organisation for Economic Co-operation and Development. http://www.oecd.org/sti/ieconomy/15582260.pdf
[OECD 2005] OECD. (2005, December). *The Promotion of a Culture of Security for Information Systems and Networks in OECD Countries* (DSTI/ICCP/REG(2003)8/FINAL). Organisation for Economic Co-operation and Development. http://www.oecd.org/internet/ieconomy/35884541.pdf
[OECD 2006] OECD. (2006). *Security of Information Systems and Networks.* Organisation for Economic Co-operation and Development. http://www.oecd.org/sti/ieconomy/37418730.pdf
[OECD 2008] OECD. (2008, June 18). *The Seoul Declaration for the Future of the Internet Economy Recommendation No. 147.* OECD Publishing. https://doi.org/10.1787/230445718605
[OECD 2009] OECD. (2009, December). *Smart Sensor Networks: Technologies and Applications for Green Growth* (DSTI/ICCP/IE(2009)4/FINAL). Organisation for Economic Co-operation and Development. https://www.oecd.org/sti/44379113.pdf
[OECD 2010] OECD. (2010). *Greener and Smarter ICTs, the Environment and Climate Change.* Organisation for Economic Co-operation and Development. www.oecd.org/dataoecd/27/12/45983022.pdf

[OECD 2011] OECD. (2011). *Thirty Years after the OECD Privacy Guidelines*. Organisation for Economic Co-operation and Development. http://www.oecd.org/sti/ieconomy/49710223.pdf

[OECD 2012a] OECD. (2012a, November 16). *The Role of the 2002 Security Guidelines: Towards Cybersecurity for an Open and Interconnected Economy* (DSTI/ICCP/REG (2012)8/FINAL). OECD Publishing. http://dx.doi.org/10.1787/5k8zq930xr5j-en

[OECD 2012b] OECD. (2012b, January 10). *ICT Applications for the Smart Grid: Opportunities and Policy Implications. OECD Digital Economy Papers* (No. 190) (DSTI/ICCP/REG(2011)12/FINAL). OECD Publishing. http://dx.doi.org/10.1787/5k9h2q8v9bln-en

[OECD 2012c] OECD. (2012c). *Cybersecurity Policy Making at a Turning Point. Analyzing a New Generation of National Cybersecurity Strategies for the Internet Economy*. Organisation for Economic Co-operation and Development. http://www.oecd.org/sti/ieconomy/cybersecurity%20policy%20 making.pdf

[OECD 2012d] OECD. (2012d, October 4). *OECD Internet Economy Outlook 2012*. Organisation for Economic Co-operation and Development. http://www.oecd.org/sti/ieconomy/ oecd-internet-economy-outlook-2012-9789264086463-en.htm

[OECD 2013] OECD. (2013). *The OECD Privacy Framework*. Organisation for Economic Co-operation and Development. https://www.oecd.org/sti/ieconomy/oecd_privacy_framework.pdf

[OECD 2015] OECD. (2015). *Digital Security Risk Management for Economic and Social Prosperity: OECD Recommendation and Companion Document*. OECD Publishing. http://dx.doi. org/10.1787/9789264245471-en

[OECD 2016] OECD. (2016). *Information Security and Privacy*. Organisation for Economic Co-operation and Development. http://www.oecd.org/sti/ieconomy/informationsecurityandprivacy.htm

[ON World 2009] Onworld. (2009, June 17). 100 Million New Smart Meters within the Next Five Years [Press release]. http://www.onworld.com/html/newssmartmeter.htm

[Open Standard] Wikipedia. (n.d.). *Open Standard*. Wikipedia. https://en.wikipedia.org/wiki/ Open_standard

[P2030 2011] *IEEE Std 2030-2011- IEEE Guide for Smart Grid Interoperability of Energy Technology and Information Technology Operation with the Electric Power System (EPS), End-Use Applications, and Loads.*

[Panke 2002] Panke, R.A. (2002). *Energy Management Systems and Direct Digital Control*. The Fairmont Press.

[Palermo 2014] Palermo, F. (2014, July 7). *Internet of things done wrong stifles innovation*. Information Week. http://www.informationweek.com/strategic-cio/executive-insights-and-innovation/ internet-of-things-done-wrong-stifles-innovation/a/d-id/1279157?page_number=1

[Parolini 2012] Parolini, L., Sinopoli, B., Krogh, B.H.,Wang, Z. (2012). A cyber–physical systems approach to data center modeling and control for energy efficiency. *Proceedings of the IEEE* (Vol. 100, No 1, pp. 254–268). IEEE. https://doi.org/10.1109/JPROC.2011.2161244

[Pendyala 2009] Pendyala, V.S., Shim, S.S.Y. (2009). The Web as the ubiquitous computer. *Computer,42*(9), 90–92.

[Peritz 2010] Peritz, A.J., Sechrist, M. (2010, September). *Protecting Cyberspace and the US National Interest*. https://projects.csail.mit.edu/ecir/wiki/images/c/c5/Cyber-vital-national-interest.pdf

[Ray 2010] Ray, P.D., Harnoor, R., Hentea, M. (2010). Smart power grid security: a unified risk management approach. In D.A. Pritchard, L.D. Sanson (Eds.), *44th Annual 2010 IEEE International Carnahan Conference on Security Technology Proceedings* (pp. 276–285), San Jose, California, USA, (October 2010). https://doi.org.10.1109/CCST.2010.5678681

[Renewable Energy] EPRI. (2014, October 22). *EPRI Smart Grid Demonstration Initiative Final Update* (3002004652). https://www.epri.com/#/pages/product/3002004652/?lang=en-US

[RFC 4949] *IETF Request for Comments (RFC) 4949 Internet Security Glossary* (Version 2, August 2007).

[RFC 7452] *IETF Request for Comments (RFC) 7452 Architectural Considerations in Smart Object Networking* (March 2015).

[Rihan 2011] Rihan, M., Ahmad, M., Beg, M.S. (2011). Developing smart grid in India: background and progress. *2011 IEEE PES Conference on Innovative Smart Grid Technologies - Middle East,* (pp. 1-6), Jeddah, Saudi Arabia, (December 2011). https://doi.org/10.1109/ISGT-MidEast.2011.6220788

[Roman 2011] Roman, R., Najera, P., Lopez, J. (2011). Securing the internet of things. *Computer, 44*(9), 51–58. https://doi.org/10.1109/MC.2011.291

[Saadeh 2015] Saadeh, O. (2014). Distributed energy resource management systems: technologies, deployments and opportunities. *GTM Research* (7 October 2014). http://www.greentechmedia.com/research/report/distributed-energy-resource-management-systems-2014?utm_source=GTM%20Research&utm_medium=Email&utm_campaign=DERMS2014

[Santucci 2010] Santucci, G. (2010). The Internet of Things: Between the revolution of the internet and the metamorphosis of objects. In H. Sundmaeker, P. Gille-min, P. Friess, P.S. Woelffle (Eds.), *Vision and challenges for realizing the internet of things* (pp. 11–24). CERP-IoT - Cluster of European Research Projects on the Internet of Things. http://cordis.europa.eu/fp7/ict/enet/documents/publications/iot-between-the-internet-revolution.pdf

[Santucci 2011] Santucci, G. (2011, February 4). *The Internet of Things: A Window to our future.* https://www.theinternetofthings.eu/gerald-santucci-internet-things-window-our-future

[Sato 2015] Sato, T., Kammen, D.M., Bin, D., Macuha, M., Zhou, Z., Wu, J., Tariq, M., Asfaw, S.A. (2015). *Smart Grid Standards: Specifications, Requirements, and Technologies.* Wiley.

[Shei 2013] Shei, H. (2013, October 1). *Understand the state of data security and privacy: 2013 to 2014* [Report]. Forrester. http://mobility-sp.com/images/gallery/FORRESTER-Understand-The-State-Of-Data-Security-And-Privacy-2013-To-2014.pdf

[SmartGrids 2006] EuropeanCommission. (2006). *European SmartGrids Technology Platform Vision and Strategy for Europe's Electricity Networks of the Future* (EUR 22040). http://ec.europa.eu/research/energy/pdf/smartgrids_en.pdf

[SRI BI 2008] SRIBI. (2008, April). *Disruptive civil technologies six technologies with potential impacts on us interests out to 2025* (cr 2008-07). http://www.fas.org/irp/nic/disruptive.pdf

[Starr 2014] Starr, M. (2014). *Fridge caught sending spam emails in botnet attack* [News]. CNET. http://www.cnet.com/news/fridge-caught-sending-spam-emails-in-botnet-attack

[Swire 2004] Swire, P.P. (2004). A model for when disclosure helps security: what is different about computer and network security? *Journal on Telecommunications and High Technology Law 163.* http://dx.doi.org/10.2139/ssrn.53178

[Taylor 2012] Taylor, J. (2012). Part I – What we discovered: bulk system reliability assessment and the smart grid. *EE Online.* https://electricenergyonline.com/energy/magazine/681/article/PART-I-WHAT-WE-DISCOVERED-Bulk-System-Reliability-Assessment-and-the-Smart-Grid.htm

[TechTarget] TechTarget. (n.d.). *Cybersecurity.* http://whatis.techtarget.com/definition/cybersecurity

[Tracy 2015] Tracy, J.F. (2015, January 30). Internet of Things and smart grid to fully eviscerate privacy [Blog].*Memoryholeblog.*http://memoryholeblog.com/2015/01/30/internet-of-things-and-smart-grid-to-fully-eviscerate-privacy/

[UCAIUG 2008] UCAIUG. (2008). *AMI system requirements V1.01* [UCAIUG: AMI-SEC-ASAP]. UCA International User Group. http://energy.gov/sites/prod/files/oeprod/DocumentsandMedia/14-AMI_System_Security_Requirements_updated.pdf

[UCAIUG] UCAIUG. (n.d.). *AMI components, open smart grid, shared documents.* Open Smart Grid User Group. http://osgug.ucaiug.org/Shared%20Documents/Forms/AllItems.aspx

[UN 1948] UN. (1948, December 10*). Universal Declaration of Human Rights (UDHR) General Assembly resolution 217 A*. United Nations. http://www.un.org/en/universal-declaration-human-rights/index.html

[UN 2002] UN. (2002, December 20). *Resolution Concerning the Creation of a Global Culture of Cybersecurity A/RES/57/239*. United Nations. https://digitallibrary.un.org/record/482184

[USCIB 2004] USCIB. (2004). *The Year In Review Highlights of 2003-2004 Accomplishments*. United States Council for International Business. https://www.uscib.org/docs/year_in_review_2004.pdf

[Uslar 2013] Uslar, M., Specht, M., Dänekas, C., Trefke, J., Rohjans, S., González, J.M., Rosinger, C., Bleiker, R. (2013). *Standardization in Smart Grids Introduction to IT-Related Methodologies, Architectures and Standards*. Springer-Verlag. https://doi.org/10.1007/978-3-642-34916-4

[Yan 2013] Yan, Y., Qian, Y., Sharif, H., Tipper, D. (2013). A Survey on Smart Grid Communication Infrastructures: Motivations, Requirements and Challenges. *IEEE Communications Surveys & Tutorials*, *15*(1), 5–20. https://doi.org/10.1109/SURV.2012.021312.00034

[Vacca 2012] Vacca, J.R. (2012). *Computer and Information Security Handbook* (2nd Edition). Morgan Kaufmann.

[VISION 2050] ETIP-SNET. (n.d.) *Europe VISION 2050*. https://www.etip-snet.eu/etip-snet-vision-2050/

[Volonino 2004] Volonino, L., Robinson, S.R. (2004). *Principles and Practice of Information Security Protecting Computers from Hackers and Lawyers*. Pearson Prentice Hall.

[Von Solms 2013] Von Solms, R., Van Niekerk, J. (2013). From information security to cyber security. *Computers & Security*, *38*, 97–102. https://doi.org/10.1016/j.cose.2013.04.004

[Vos 2009] Vos, A. (2009). Demand Response Management Systems: The Next Wave of Smart Grid Software. *Electric Energy Online* (September/October 2009). http://www.electricenergyonline.com/show_article.php?article=444

[Wald 2013] Wald, M.L. (2013, August 16). *As worries over the power grid rise, a drill will simulate a knockout blow* [News]. New York Times. http://www.nytimes.com/2013/08/17/us/as-worries-over-the-power-grid-rise-a-drill-will-simulate-a-knockout-blow.html

[Wang 2011] Wang, R. (2011, July 13). *Coming to Terms with the Consumerization of IT*. Harvard Business Review. https://hbr.org/2011/07/coming-to-terms-with-the-consu

[Weiss 2010] Weiss, J. (2010). *Protecting Industrial Control Systems from Electronic Threats*. Momentum Press.

[Weiss 2013] Weiss, J. (2013, June 4). *A major cyber threat to critical infrastructures is from the electric utilities [Blog]. CONTROL. Unfettered Blog*. http://www.controlglobal.com/blogs/unfettered/a-major-cyber-threat-to-critical-infrastructures-is-from--the-electric-utilities

[Westby 2012] Westby, J.R. (2012, May 16). *Governance of enterprise security: CyLab 2012 report how boards & senior executives are managing cyber risks* [Report]. https://www.cmu.edu/news/stories/archives/2012/may/may16_cylabreport.html

[WH 2009] WhiteHouse. (2009, May 29). *Cyberspace policy review: assuring a trusted and resilient information and communications infrastructure*. U.S. White House. https://www.energy.gov/cio/downloads/cyberspace-policy-review-assuring-trusted-and-resilient-information-and-communications

[WH 2013] WhiteHouse. (2013). *Executive order on improving critical infrastructure cybersecurity executive order no. 13,636, 78 Fed. Reg. 11739, 19 February 2013*. US. White House. www.gpo.gov/fdsys/pkg/FR-2013-02-19/pdf/2013-03915.pdf

[WH 2016] WhiteHouse. (2016, February 25). *Fact sheet: The Recovery act made the largest single investment in clean energy in history, driving the deployment of clean energy, promoting energy efficiency, and supporting manufacturing* [Report]. US. White House. https://www.whitehouse.gov/the-press-office/2016/02/25/fact-sheet-recovery-act-made-largest-single-investment-clean-energy

[Whitman 2011] Whitman, M.E., Mattord, H.J. (2011). *Principles of Information Security* (Fourth edition). Cengage Learning.

[Whitman 2014] Whitman, M.E., Mattord, H.J. (2014). *Management of Information Security* (Fourth edition). Cengage Learning.

[Wi-Fi 2014] Wi-Fi. (2014). *Connect your life: Wi-Fi® and the Internet of Everything*. Wi-Fi Alliance. https://www.wi-fi.org/

[WikiTech] Wikimedia. (2008, April 4). *Technology roadmap. The Internet of Things*. SRI Consulting Business Intelligence/National Intelligence Council. https://commons.wikimedia.org/wiki/File:Internet_of_Things.svg

[Wulf 2001] Wulf, W.A. (2001, October 10). *Cyber security: Beyond the Maginot line. Statement before the House Science Committee U.S* [Speech]. U.S. House of Representatives National Academy of Engineering. http://www.nae.edu/News/SpeechesandRemarks/CyberSecurityBeyondtheMaginotLine.aspx

[Zanella 2014] Zanella, A., Bui, N., Castellani, A., Vangelista, L., Zorzi, M. (2014). Internet of things for smart cities. *IEEE Internet of Things Journal*, *1*(1), 22–32. https://doi.org/ 10.1109/JIOT.2014.2306328

[Zurborg 2010] Zurborg, A. (2010). Unlocking customer value: The virtual power plant. W*orldPow*er *2010* (pp.1-5). https://www.energy.gov/sites/prod/files/oeprod/DocumentsandMedia/ABB_Attachment.pdf

Part II

Applying Security Principles to Smart Grid

3

Principles of Cybersecurity

3.1 Introduction

The global information society is borderless, unconstrained by distance or time. Today, the significance of computer and communication technologies, economically, socially, and politically, is widely accepted. They are key technologies not only in their own right but also as conduits for and components of other goods, services, and activities. Economies, politics, and societies are based less on geography and physical infrastructure than previously and increasingly more dependent on information and information systems (IS).

IS benefit organizations, governments, and public on every aspect of life. They are widely used by government administrations, business organizations, and research institutions. They are critical to the provision of healthcare, energy, transport, and communications. Expanding uses and benefits of IS is an indication of the present situation and that new technological developments are arising to augment the capabilities of IS. Expanded use of IS offers possibilities of greater access to resources, experience, learning, and participation in cultural and civic life.

However, users need to have confidence that IS operate as intended without unanticipated failures or problems. Otherwise, the systems and their underlying technologies may not be used to the extent possible, and further growth and innovation may be inhibited. Loss of confidence may stem equally from outright malfunction or from functioning that does not meet expectations. Clear, uniform, predictable rules have to be in place to ease and encourage growth and use of IS with adequate security. One goal, therefore, is the protection of individuals and organizations from harm resulting from failures of security. All individuals and organizations potentially rely on the proper functioning of IS.

In the absence of sufficient security, IS and, more generally, information and communication technologies cannot not be used to their full potentials. Lack of security or lack of confidence in the security of IS can impede IS development including use of new information and communication technologies. Therefore, the interests of those relying on information should be protected from harm resulting from failures. IS have to provide the information when the users need it and make use with confidence.

Security of IS considers the protection of information and of the systems that manage it, against a wide range of threats in order to ensure business continuity, minimize risks, and maximize the return on investment and business opportunities. Defending IS from unauthorized access and disruption can be achieved with the practice of information security, sometimes shortened to InfoSec. So, building protection is the objective of information security.

Building an Effective Security Program for Distributed Energy Resources and Systems: Understanding Security for Smart Grid and Distributed Energy Resources and Systems, Volume 1, First Edition. Mariana Hentea.

3.2 Information Security

Information security is the protection of information and IS. It encompasses all infrastructures that facilitate its use – processes, systems, services, and technology. Basically, information security is about measures adopted to prevent the unauthorized use, misuse, modification, or denial of use of knowledge, facts, data, or capabilities [Maiwald 2004].

3.2.1 Terminology

Information security, like most technical subjects, uses a complex web of terminology that is evolving.

The terms information security, IT security, network security, computer security, and information assurance are common terms used to describe these practices. These terms are interrelated and share the common goals of protecting the information. Often, they are used interchangeably, but there are significant differences between them. These differences lie primarily in the approach to the subject, the methodologies used, and the areas of concentration.

The following are common definitions that highlight these differences:

- Information security is concerned with the protection of IS.
- IT security is a subset of information security and is concerned with the security of electronic systems, including computer, voice, and data networks.
- Computer security can focus on correct operation of a computer system.
- Information assurance focuses on the reasons for assurance that information is protected, and it is thus reasoning about information security.

In order to protect its systems and operations, an organization may need multiple layers of security in place [Whitman 2014]:

- Physical security addresses the issues related to protection of physical items or areas.
- Personal security involves the protection of individuals.
- Operations security focuses on protection of activities or specific operation.
- Communications security encompasses the protection of communication media, technology, and content.
- Network security is the protection of networking components, connections, and contents.
- Information security is the protection of information and its critical elements, including the systems and hardware that use, store, and transmit that information.

3.2.2 Information Security Components

Figure 3.1 shows that information security includes the broad areas of information security management, policy, computer and data security, and network security.

Although many approaches and methods can be used to secure systems, certain intrinsic expectations must be met whether the system is small or large or owned by a government agency or by a private corporation. These intrinsic expectations are described as generally accepted system security principles.

3.2.3 Security Principles

The principles address computer security from a very high-level viewpoint. Based on the OECD guidelines [OECD 1992], NIST developed a set of security principles intended to help when creating new systems, practices, or policies. The eight principles originally appeared as the elements of computer security in [NIST SP800-14] (currently is replaced by other standards). Also, these principles are included in other standards. They refer to the following:

- Security supports the mission of the organization.
- Security is an integral element of sound management.

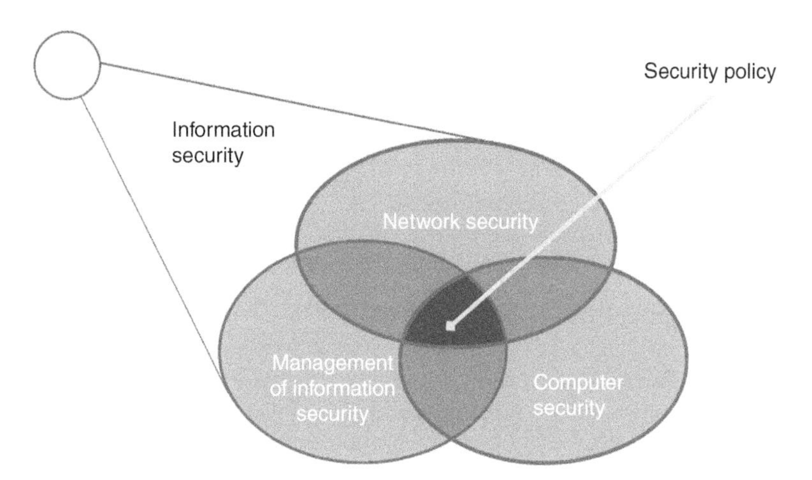

Figure 3.1 Components of information security. *Source:* Adapted from [Whitman 2014]. © 2014, Delmar Learning, a part of Cengage Inc.

- Security should be cost effective.
- System owners have security responsibilities outside their own organizations.
- Security responsibilities and accountability should be made explicit.
- Security requires a comprehensive and integrated approach.
- Security should be periodically reassessed.
- Security is constrained by societal factors.

Management, internal auditors, users, system developers, and security practitioners can use these principles to gain an understanding of the foundation of security. The foundation begins with generally accepted system security principles and continues with common practices that are used in securing systems.

Also, each of the principles applies to various practices. The nature of the relationship between the principles and the practices varies. In some cases, practices are derived from one or more principles; in other cases, practices are constrained by principles. For example, the risk management practice is directly derived from the cost-effectiveness principle. However, the comprehensive and reassessment principles place constraints on the risk management practice. The NIST document uses the terms information technology security and computer security interchangeably. The terms refer to the entire spectrum of information technology including application and support systems. In the following section, we provide a brief introduction of key terms related to information security.

3.3 Security-Related Concepts

Anderson [Anderson 2008] points that many of the terms in security engineering are straightforward, but some are misleading, or even controversial. The concepts threat, vulnerability, risk, and exposure are often interchanged even though they have different meanings [Harris 2013] or misused with different meanings. Also, the terms controls, countermeasures, and safeguards are used interchangeably.

Although we try to use the most current or common definitions identified from the standards and recognized glossaries, we may be using some terms with different meanings (as used by security professionals). Therefore, a control engineer or security engineer should develop sensitivity to the different meaning that common words acquire in different applications [Anderson 2008]. Often, control engineers need to adjust their terminology and security concepts to the evolving security field.

3.3.1 Basic Security Concepts

The following is a brief introduction of the basic security concepts: risk, vulnerability, threat, attack, exposure, consequence, and control. We include definitions as provided by known standards and dictionaries as follows:

> Risk is effect of uncertainty on objectives [ISO/IEC 27000], adopted from [ISO Guide 73]; risk is an expectation of loss expressed as the probability that a particular threat will exploit a particular vulnerability with a particular harmful result [RFC 4949].
>
> Vulnerability is a weakness of an asset or group of assets that can be exploited by one or more threats [ISO/IEC 27000].
>
> Control is measure that is modifying risk [ISO/IEC 27000], adopted from [ISO Guide 73]; countermeasure is an action, device, procedure, or technique that meets or opposes (e.g. counters) a threat, a vulnerability, or an attack by eliminating or preventing it, by minimizing the harm it can cause, or by discovering and reporting it so that corrective action can be taken [RFC 4949].

Security consequence is the outcome of an event such that [ENISA Glossary]:

- There can be more than one consequence from one event; consequences can range from positive to negative.
- Consequences can be expressed qualitatively or quantitatively [ISO Guide 73].

Exposure is the potential loss to an area due to the occurrence of an adverse event [ENISA Glossary]; exposure is a type of threat action whereby sensitive data is directly released to an unauthorized entity (unauthorized disclosure); this type of threat action includes the following subtypes [RFC 4949]:

- Deliberate exposure – intentional release of sensitive data to an unauthorized entity.
- Scavenging – searching through data residue in a system to gain unauthorized knowledge of sensitive data.
- Human error – exposure due to human action or inaction that unintentionally results in an entity gaining unauthorized knowledge of sensitive data.
- Hardware or software error – exposure due to system failure that unintentionally results in an entity gaining unauthorized access.

3.3.2 The Basis for Security

One has to understand that risk is the basis for security. Risk is the probability that something could happen. In information security, it could be the probability of a threat to a system, the probability of a vulnerability being discovered, or the probability of equipment or software malfunction. More precisely, security risk is defined as the probability of a threat agent exploiting a vulnerability to cause harm to a computer, network, system, or company and the resulting business impact. A security risk is any event that could result in the compromise of organizational assets, e.g. the unauthorized use, loss, damage, disclosure, or modification of organizational assets for the profit, personal interest, or political interests of individuals, groups, or other entities constitutes a compromise of the asset and includes the risk of harm to people.

Vulnerabilities can cause an exposure or potential damage due to an attack from a threat. An exposure is an instance when an object is open to possible damage. Risk is the potential for loss that requires protection. If there is no risk, there is no need for security. When risk is examined, the vulnerabilities and threats must be identified. Threats without vulnerabilities pose no risk. Likewise, vulnerabilities without threats pose no risk. Security controls need to be implemented to mitigate the risk.

A new term, called digital security risk, is expression used to describe a category of risk related to the use, development, and management of the digital environment in the course of any activity. This risk can result from the combination of threats and vulnerabilities in the digital environment [OECD 2015]. The risk includes both cyber and physical aspects, the people involved in the activity, and the organizational processes supporting it. Figure 3.2 shows relationships among security terms as an example of threat realization.

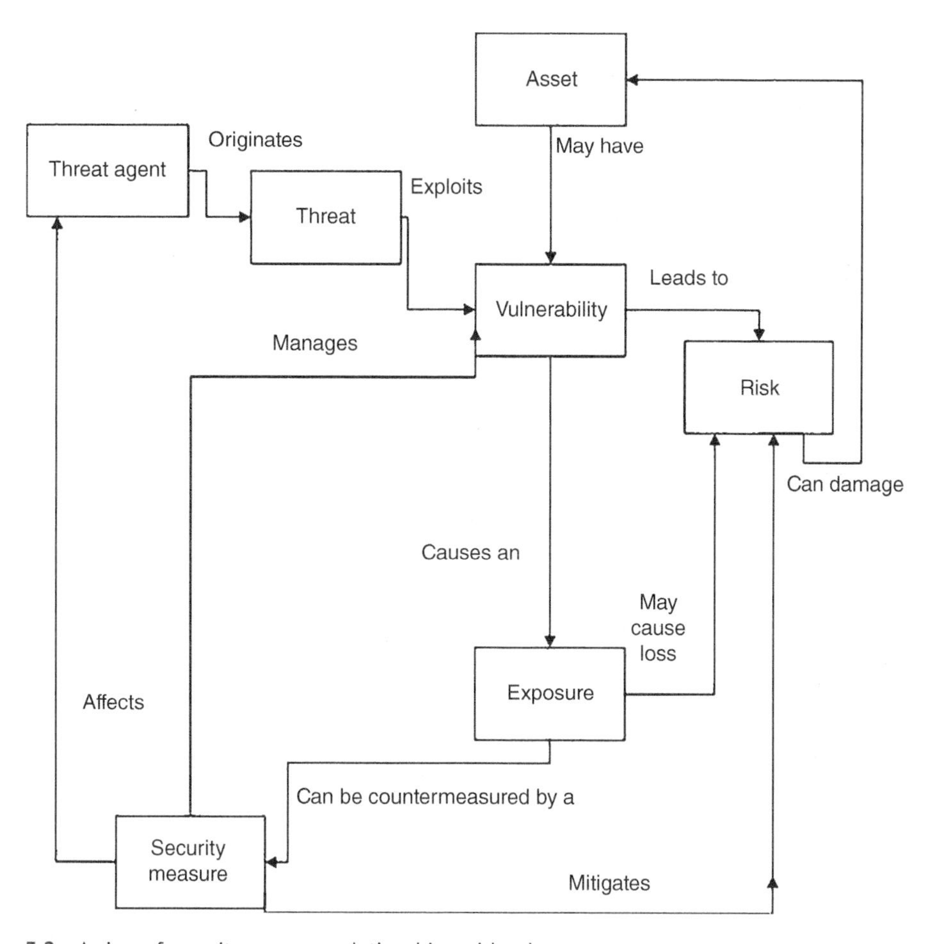

Figure 3.2 A view of security measure relationships with other concepts.

Compromise of organizational assets may adversely affect the enterprise and its business units and their clients. As such, consideration of security risk is a vital component of risk management [Talbot 2009].

3.4 Characteristics of Information

At the center of any IS is data and information. The term data is the collection of raw facts. Data generally means a representation of facts suitable for communication, interpretation, or processing by human beings or by automatic means. Data refers to the lowest abstract or a raw input that when processed or arranged makes meaningful output. It is the group or chunks that represent quantitative and qualitative attributes pertaining to variables.

Information is the meaning assigned to data by means of conventions or transformations applied to that data. Information is usually the processed outcome of data, for example, a set of related data that form a message. Information can be a mental stimulus, perception, representation, or even an instruction. Information is an asset that, like other important business assets, is essential to an enterprise's business. It can exist in many forms. It can be printed or written on paper, stored electronically, transmitted by post or by using electronic means, shown on films, or spoken in conversation.

Good information is that which is used and which creates value. Experience and research show that information can be transformed and has numerous characteristics.

3.4.1 Data Transformation

Observations and recordings are done to obtain data, while analysis is done to obtain information. Understanding the differences between data and information is important when performing a process or task. Data and information are increasingly recognized as enterprise assets.

Data is a collection of values assigned to base measures, derived measures, and/or indicators [ISO/IEC 15939]. Data includes raw facts (e.g. symbols, numbers, or other representation of facts). Examples of data in Smart Grid include sensor data, configuration, command, and control data, financial data, personal data, images, audit logs, encrypted data, etc.

Information is organized or processed data that is timely (e.g. inferences from the data are drawn within the time frame of applicability) and accurate (e.g. with regard to the original data) [Turban 2007]. Information is data in context that within a certain context has a particular meaning; it can be facts, events, things, processes, or ideas including concepts [ISO/IEC 2382].

Further transformation of information results into knowledge, which is further refined in wisdom coined in [Ackoff 1989] as a result of data and information filtration, reduction, and transformation as Figure 3.3 shows these stages.

Knowledge is information that is contextual, relevant, and actionable [Turban 2007]. Wisdom is knowledge in context, knowledge applied in the course of actions [IQ/DQ Glossary]. Wisdom is on top of data–information–knowledge–wisdom (DKIW) hierarchy (pyramid), also called DIKW model (see Figure 3.4).

The model shows wisdom as an ideal state, at the pinnacle of the pyramid. Although this model is directed primarily to the management of organizations, it has been used for IS too. It captures certain insights useful to personnel and organizational management. While it is argued by other scholars that the model has some limitations and wisdom is almost nonexistent [Bernstein 2009], the model is the basis of other models that are used by intelligent decision support systems in agent-mediated environments that are trying to improve decision making in specific applications such as simulation and command and control. The model can be used for similar Smart Grid applications such as situational awareness.

Rapid and converging technological advances in computing, communications, and consumer electronics have caused an explosion in the generation, processing, storage, transmission, and consumption of enormous amounts of power data and information. It is expected that these trends continue to accelerate and to have profound effects on delivery and uses of information, knowledge, and wisdom.

Another view of data transformation is shown in Figure 3.5. This diagram shows in a more realistic perspective the inclusion of noise and metaknowledge layers for the representation of knowledge as

Data ➔Information➔ Knowledge➔Wisdom

Figure 3.3 Data transformation.

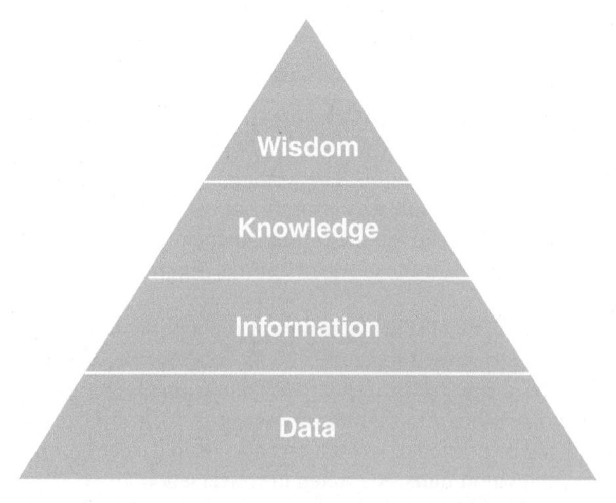

Figure 3.4 DIKW pyramid: data, information, knowledge, and wisdom. *Source:* [Longlivetheux]. Licensed under CC BY-SA 4.0.

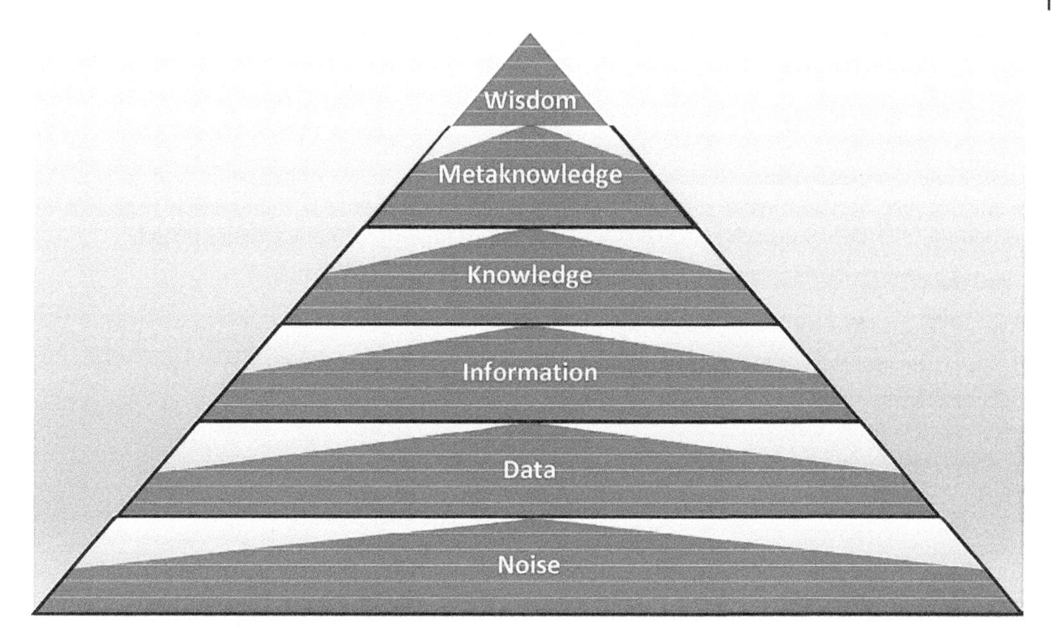

Figure 3.5 The pyramid of knowledge.

used in expert systems [Giarratano 2005]. Knowledge representation is key to the development of expert systems. Expert systems are designed for knowledge representation based on rules of logic called inferences. Elements of knowledge, knowledge representation, methods of representing knowledge, and definitions of knowledge are presented in this work.

One definition of knowledge is [Giarratano 2005]:

1) The fact or condition of knowing something with familiarity gained through experience or association.
2) Acquaintance with or understanding of a science, art, or technique.

Other concepts are defined as follows:

Wisdom is using knowledge in a beneficial way.
Metaknowledge are rules about knowledge.
Knowledge are rules about using information.
Data is potentially useful information.
Noise is no apparent information.

Within power grid applications, data is collected from different sources such:

- Measurements about power flow such as voltage, current, power, etc.
- Events that occur within a device or between interfaces.
- Resources that are affected by those events.
- Agents who participate in the events.

Data (measurements, events, facts) can be transformed to derive useful information and knowledge about the power grid state, security state, etc.

Fundamental to deliver energy services and better quality energy to consumers is transforming data into information that is valued and used (in good and bad ways):

- Data may have important uses for energy conservation, for those customers with the ability to load shift.
- Data has possible value to utility business enterprises.
- Data can also be compiled for various discriminatory, anticompetitive, and/or illegal uses.
- Privacy protections have been circumvented by user error, disgruntled employees, and hackers.

However, the value of data comes from the characteristics it possesses.

3.4.2 Data Characteristics

A characteristic is also called attribute or property (see attribute definition in [ISO/IEC 27000]). Other aspects include frequency of collection, uses, purpose, security, cost, and data quality. The three levels at which information can be used are strategic, tactical, and operational.

The suite of characteristics (known also as properties, attributes, or dimensions) of data is not an exhaustive list because many attributes are continuously defined in dictionaries (e.g. [DAMA Dictionary], [IQ/DQ Glossary], [ASQ Glossary]). The most important characteristics include:

- Accuracy
- Availability
- Authenticity
- Confidentiality
- Completeness
- Correctness
- Consistency
- Dependability
- Integrity
- Possession
- Precision
- Privacy
- Relevance
- Timeliness
- Trustworthiness
- Uniqueness
- Usability
- Utility.

When a characteristic of information changes, the value of that information either increases or, more commonly, decreases. Also, some characteristics of information affect value to users more than others do.

Although security professionals and end users may share some information, there it may not be the same understanding of information characteristics for all, and some conflicts can occur. For example, a process control engineer is more concerned of data availability and timeliness than data confidentiality because information loses all value when it is delivered too late. In other situations, end users may accept a delay of a tenth of a second in the computation of the data as a minor or unnecessary annoyance. Therefore, an organization needs to focus on these characteristics that are part of data management and data governance.

Data management is the development, execution, and supervision of plans, policies, programs, and practices that control, protect, deliver, and enhance the value of data and information assets.

Data governance is a set of processes that ensures that important data assets are formally managed throughout the enterprise. Examples of data management functions include data collection, data development, data security management, data quality management, data warehousing, etc. [DAMA Framework].

Among the many challenges utilities currently face in operating the electric grid, the most prominent data management functions include [Durai 2012]:

- Limited number of digital sensors that can collect raw data from the grid.
- Limited number of communications to existing digital sensors to permit collection of data.
- Limited analytic capabilities for assessing grid condition data so that actionable information can be inferred.
- Limited integration of information from various places in the grid to provide an integrated quality view.

Table 3.1 shows a set of characteristics and definitions for data; the table includes some characteristics that are suitable to other entities (e.g. product or system).

Table 3.1 Examples of data characteristics.

Data Attribute	Definition
Accuracy	Degree of conformity of a measure to a standard or a true value. Level of precision or detail [IQ/DQ Glossary] General term used to describe the closeness of a measurement to the true value [ISO 5725-1]
Accountability	The property of a system or system resource that ensures that the actions of a system entity may be traced uniquely to that entity, which can then be held responsible for its actions [RFC 4949] Property that ensures that the actions of an entity may be traced uniquely to that entity Note 1 to entry: [ISO/IEC 2382]
Accessibility	Data items that are easily obtainable and legal to access with strong protections and controls built into the process
Anonymity	The quality or state of not being named or identified
Anonymous	Pertaining to a data object that has no explicit data type declaration Note 1 to entry: anonymous: term and definition standardized by ISO/IEC [ISO/IEC 2382]
Availability	The property of being accessible and usable upon demand by an authorized entity [RFC 4949] Property of being accessible and usable upon demand by an authorized entity [ISO/IEC 27000]
Auditability	The level to which transactions can be traced and audited through a system [ISACA Glossary]
Authenticity	The property of being genuine and able to be verified and be trusted [RFC 4949] Property that an entity is what it is claims to be [ISO/IEC 27000]
Confidentiality	The property that data is not disclosed to system entities unless they have been authorized to know the data [RFC 4949] Property that information is not made available or disclosed to unauthorized individuals, entities, or processes [ISO/IEC 27000]
Coherence	Adequacy to be reliably combined in different ways and for various uses. Fully coherent data are logically consistent – internally, over time, and across products and programs
Control	Property to exercise authority or influence over
Completeness	A characteristic of information quality measuring the degree to which all required data is known [IQ/DQ Glossary]
Compliance	Adherence to, and the ability to demonstrate adherence to, mandated requirements defined by laws and regulations, as well as voluntary requirements resulting from contractual obligations and internal policies [ISACA Glossary]
Correctness	The property of a system that is guaranteed as the result of formal verification activities [RFC 4949] Conforming to an approved or conventional standard, conforming to or agreeing with fact, logic, or known truth [IQ/DQ Glossary]
Consistency	A measure of information quality expressed as the degree to which a set of data is equivalent in redundant or distributed databases [IQ/DQ Glossary] The condition of adhering together, the ability to be asserted together without contradiction [IQ/DQ Glossary]

(Continued)

Table 3.1 (Continued)

Data Attribute	Definition
Credibility	Refers to the objective and subjective components of the believability of a source or message (https://en.wikipedia.org/wiki/Credibility)
Criticality	An attribute assigned to an asset that reflects its relative importance or necessity in achieving or contributing to the achievement of mission/business goals [NIST SP800-160]
Currency	The quality or state of information of being up-to-date or not outdated [IQ/DQ Glossary]
Integrity	The property that data has not been changed, destroyed, or lost in an unauthorized or accidental manner [RFC 4949] Property of accuracy and completeness [ISO/IEC 27000]
Non-repudiation	Ability to prove the occurrence of a claimed event or action and its originating entities [AHIMA]
Possession	The ownership or control of information, as distinct from confidentiality
Precision	A characteristic of information quality measuring the degree to which data is measure of the ability to distinguish between nearly equal values Note 1 to entry: Example: Four place numerals are less precise than six place numerals; nevertheless, a properly computed four place numeral may be more accurate than an improperly computed six place numeral Note 2 to entry: precision: term and definition standardized by [ISO/IEC 2382] Nown to the right level of granularity [IQ/DQ Glossary]
Privacy	Freedom from intrusion into the private life or affairs of an individual when that intrusion results from undue or illegal gathering and use of data about that individual Note 1 to entry: privacy: term and definition standardized by [ISO/IEC 2382]
Relevance	The extent to which data are useful for the purposes for which they were collected [AHIMA] Concerned with whether the available information sheds light on the issues that are important to users. Assessing relevance is subjective and depends upon the varying needs of users
Reliability	Reliability is a property of something and means consistency; something is reliable if it behaves consistently or produces consistent results
Sensitivity	Measure of importance assigned to information by the information owner to denote its need for protection Note 1 to entry: term and definition standardized by [ISO/IEC 2382]
Timeliness	Concept of data quality that involves whether the data is up-to-date and available within a useful time frame; timeliness is determined by manner and context in which the data are being used
Uniqueness	Limited to a single outcome or result; without alternative possibilities
Utility	The usefulness of information to its intended consumers, including the public. (OMB 515) [IQ/DQ Glossary] Quality or state of having value for some purpose or end

3.4.3 Data Quality

In the preceding section, we introduced general characteristics of data as well as specific characteristics for different uses of information. One specific goal in data collection and processing is data quality that is applied when analyzing data, information, processes, goods, software, hardware, and tasks. The need for data quality is the impact on strategies, planning, and decision making [Roebuck 2011].

Data quality refers to the level of quality of data. Data quality can be defined as the degree to which a set of characteristics of data fulfills requirements [ISO 9000]. Data quality refers to whether data is properly collected, accurate, and fit for its intended purpose.

Terms like characteristics, attributes, and dimensions are commonly used by practitioners and academics to analyze the quality of information. Often, the characteristics term is preferred to dimensions term in analyzing the information quality [English 2009]. Examples of characteristics include completeness, validity, accuracy, consistency, availability, and timeliness. One definition of data quality is:

> The simultaneous presence of consistency, completeness, accuracy, standards compliance and time-stamping [GS1].

Data are of high quality if they are fit for their intended uses in operations, decision making, and planning [Juran 1999]. A Smart Grid application needs to dictate specific requirements for the quality of the data. For example, data collected from sensors is used to monitor lines and substation equipment. Massively deployed sensors collect end-user energy consumption data, weather data, and equipment condition, and operational status is collected to perform real-time rating in the context of actual distribution and transmission line flows. This data has to be accurate, consistent, timely, and available in the real-time constraints of each application (e.g. monitoring) to generate information about the status of the grid or predict what will happen next and develop optimal control strategies. Thus, time is more critical than a regular data reporting. So future improvements to data collection may be needed. Enhancements to communication protocol and algorithms to minimize data collection time are described in [Uludag 2016].

Fact is that often IS practitioners use the term information quality synonymously with data quality. Although characteristics of data quality may be common to information quality, it is recommended to make a distinction between these concepts in certain specifications. Quite often data professionals use more complex models for information quality. The quality of data is typically evaluated on multiple characteristics – those aspects or features of the data used to define, evaluate, and manage its quality. However, categories of information characteristics provide a better understanding of information uses in a context.

3.4.4 Information Quality

Information quality is a measure of the value that the information provides to the user of that information such that meets the requirements of users. Quality is often perceived as subjective, and it can then vary among users and among uses of the information. Quality of information is described by both quantitative and qualitative measures.

Information quality has characteristics that can be grouped in categories. One classification includes categories such as intrinsic (e.g. accuracy), contextual (e.g. completeness), representational (e.g. compatibility), and accessibility (e.g. accessibility) [Wang 1996]. Thus, more important is to mention the user's confidence in meeting specific information requirements. The higher information quality, the higher is user's confidence. To guarantee confidence in that particular information meets some context and specific quality requirements, a process called information quality assurance has to be considered and applied.

3.4.5 System Quality

Software products have to support system quality that is based on a set of structured set of characteristics and subcharacteristics defined in [ISO/IEC 25010]. The set of characteristics include functionality, reliability, usability, efficiency, maintainability, and portability.

Besides the set of characteristics included in this standard, the system quality characteristics for Smart Grid (e.g. control systems) have to include other characteristics such as dependability and security, survivability, trustworthiness, and resilience.

In addition, a new electricity service culture based on *Six Sigma* quality program will better serve the customers. The term *Six Sigma* originated (by Bill Smith at Motorola in the United States) from terminology associated with statistical modeling of manufacturing processes. It is a process that defines, measures, analyzes, improves, and controls existing processes that fall below the Six Sigma specification. The maturity of a manufacturing process can be described by a *Sigma* rating indicating its yield or the percentage of defect-free products it creates. A *Six Sigma* process is one in which 99.99966% of all opportunities to serve a customer are statistically expected to be free of errors (approximately 3.4 defective features per million opportunities).

3.4.6 Data Quality Characteristics Assigned to Systems

Different industry and academic sources have proposed different versions of characteristics and definitions for data quality or information quality. The following are examples of data quality characteristics as they may be used for different systems:

A) Data quality characteristics for a financial IS may include:
- Accuracy
- Completeness
- Consistency
- Uniqueness
- Timeliness.

B) Data quality characteristics for information security assurance may include:
- Consistency
- Completeness
- Accuracy
- Time stamping
- Timeliness
- Accessibility.

C) Data quality characteristics for project management may include [HP PPM]:
- Accuracy
- Completeness
- Consistency
- Timeliness.

Data quality characteristics for Smart Grid may be defined to include characteristics and support the functions of different categories of applications such as the following:

A) Control systems need real-time data and characteristics may include:
- Accessibility
- Accuracy
- Availability
- Integrity
- Time stamping.

B) Business systems data characteristics may include:
- Accuracy
- Completeness
- Consistency
- Compliance
- Uniqueness
- Timeliness.

C) Critical applications data characteristics may include:
- Availability
- Timeliness
- Accessibility
- Compliance.

3.5 Information System Characteristics

Information system comprises the combination of strategic, managerial, and operational activities involved in gathering, processing, storing, distributing, and using information and its related technologies. Information systems are distinct from information technology (IT) in that an information system has an IT component that interacts with the process components.

The International Council on Systems Engineering (INCOSE) (www.incose.org) recognizes the system-level qualities, properties, characteristics, functions, behavior, and performance as described in the most current version of INCOSE Handbook, which aligns more with ISO/IEC standards (e.g. [ISO/IEC 15288], [INCOSE SEH]).

Quality of data, information, information system, and service are common goals in the development of any system to support users and business objectives. Thus, the information system quality is defined as how well the system supports the business process. From a business point of view, it is crucial to understand the business process and its objectives in order to understand what role the IS has in the particular process. The quality attributes are often used to evaluate an IS. Quality attributes are nonfunctional requirements used to evaluate the performance of a system. A list of system quality attributes is provided in Table 3.2. Quality defined as a grade of excellence is a very common term in the manufacturing of a product. It is defined in [ISO 84024] as:

> The totality of features and characteristics of a product or service that bear on its ability to satisfy stated or implied needs.

Thus, quality is viewed in terms of seven dimensions (or characteristics) [OECD Glossary]:

- Relevance
- Credibility
- Accuracy
- Credibility
- Timeliness
- Accessibility
- Interpretability
- Coherence.

Quality is viewed as a multifaceted concept. The quality characteristics of most importance depend on user perspectives, needs, and priorities, which vary across groups of users. An important user is a customer, and quality is fundamentally about looking at products through the eyes of the customer [Buckman]. Juran's quality trilogy is quality planning, quality control, and quality improvement [Juran 1999], [Paton 1999].

In addition, quality of an information system depends on all its components such as hardware, information, software, processes, etc. IT components such as hardware, software, databases, networks, and other related components are used to build IS. Defining what information quality is within an information system and depends greatly on whether dimensions are being identified for the software products as producers of information, processing, storage, communication, and maintenance procedures because one critical component of an IS is software.

Uncertainties may be met and confidence is fostered by building consensus about use of IS. Accepted procedures and rules are needed to provide conditions to increase the reliability of IS. Developers, operators, and users of IS deserve reassurance as to their rights and obligations, including responsibility for system failures.

3.5.1 Software Quality

Software quality is the degree to which software possesses a desired combination of attributes [IEEE 1061]. Recently, software quality attributes in software engineering are described in this standard [ISO/IEC 25010] based on two models:

Table 3.2 System development – quality attributes.

Quality attribute	Description
Availability	Ability of a functional unit to be in a state to perform a required function under given conditions at a given instant of time or over a given time interval assuming that the required external resources are provided Note 2 to entry: The availability defined here is an intrinsic availability where external resources other than maintenance resources do not affect the availability of the functional unit. Operational availability, on the other hand, requires that the external resources be provided [ISO/IEC 2382] Proportion of time that the system is functional and working. It can be measured as a percentage of the total system downtime over a predefined period. Availability will be affected by system errors, infrastructure problems, malicious attacks, and system load Reliability, maintainability and availability ability of a functional unit to be in a state to perform a required function under given conditions at a given instant of time or over a given time interval, assuming the required external resources are provided Note 1 to entry: The term used in [IEV 191-02-5] is availability performance and the definition is the same, with additional notes Note 2 to entry: The availability defined here is an intrinsic availability where external resources other than maintenance resources do not affect the availability of the functional unit. Operational availability, on the other hand, requires that the external resources be provided Property of data or of resources being accessible and usable on demand by an authorized entity Note 1 to entry: term and definition standardized by [ISO/IEC 2382]
Conceptual Integrity	Consistency and coherence of the overall design. This includes the way that components or modules are designed, as well as factors such as coding style and variable naming Data integrity is property of data whose accuracy and consistency are preserved regardless of changes made Note 1 to entry: term and definition standardized by [ISO/IEC 2382]
Critical	Very important device, computer system, process, etc. that if compromised by an incident could have high financial, health, safety, or environmental impact to an organization [IEC 62443-2-1]
Dependability	In Systems Engineering, dependability is a measure of a system's availability, reliability, and its maintainability. In software engineering, dependability is the ability to provide services that can defensibly be trusted within a time period. This may also encompass mechanisms designed to increase and maintain the dependability of a system or software Dependability of an item is the ability to perform as and when required [IEC/TC56] Note 1: Dependability includes availability, reliability, recoverability, maintainability, and maintenance support performance, and, in some cases, other characteristics such as durability, safety and security Note 2: Dependability is used as a collective term for the time related quality characteristics of an item
Flexibility	Ability of a system to adapt to varying environments and situations, and to cope with changes in business policies and rules. A flexible system is one that is easy to reconfigure or adapt in response to different user and system requirements

Interoperability	Ability of diverse components of a system or different systems to operate successfully by exchanging information, often by using services. An interoperable system makes it easier to exchange and reuse information internally as well as externally Capability to communicate, execute programs, or transfer data among various functional units in a manner that requires the user to have little or no knowledge of the unique characteristics of those units Note 1 to entry: term and definition standardized by [ISO/IEC 2382]
Maintainability	Ability of a functional unit, under given conditions of use, to be retained in, or restored to, a state in which it can perform a required function when maintenance is performed under given conditions and using stated procedures and resources [ISO/IEC 2382] Ability of a system to undergo changes to its components, services, features, and interfaces as may be required when adding or changing the functionality, fixing errors, and meeting new business requirements
Manageability	How easy it is to manage the application, usually through sufficient and useful instrumentation exposed for use in monitoring systems and for debugging and performance tuning
Modifiability	Measure of the ease with which changes can be made to a program Note 1 to entry: modifiability: term and definition standardized by [ISO/IEC 2382]
Modularity	Measure of the extent to which a program is composed of modules, such that a change to one module has minimal impact on other modules Note 1 to entry: term and definition standardized by [ISO/IEC 2382]
Performance	An indication of the responsiveness of a system to execute any action within a given time interval. It can be measured in terms of latency or throughput. Latency is the time taken to respond to any event. Throughput is the number of events that take place within a given amount of time
Portability	Capability of a program to be executed on various types of data processing systems without converting the program to a different language and with little or no modification Note 1 to entry: terms and definition standardized by [ISO/IEC 2382]
Reliability	The ability of a system to perform a required function under stated conditions for a specified period of time [RFC 4949], [ISA 62443-1-1] Property of consistent intended behavior and results [ISO/IEC 27000] The ability of an item to perform as required, without failure, for a given time interval, under given conditions [IEC/TC56] Note 1: The time interval duration may be expressed in units appropriate to the item concerned, e.g. calendar time, operating cycles, distance run, etc., and the units should always be clearly stated Note 2: Given conditions include aspects that affect reliability, such as: mode of operation, stress levels, environmental conditions, and maintenance Ability of a system to remain operational over time. Reliability is measured as the probability that a system will not fail to perform its intended functions over a specified time interval

(Continued)

Table 3.2 (Continued)

Quality attribute	Description
Resiliency	The ability to recover from a hardware failure, power outage or other interruption [PC Encyclopedia]
Reusability	Capability for components and subsystems to be suitable for use in other applications and in other scenarios Reusability minimizes the duplication of components and the implementation time
Scalability	Ability of a system to function well when there are changes to the load or demand. Typically, the system will be able to be extended over more powerful or more numerous servers as demand and load increase
Security	Property of a system by which confidentiality, integrity, availability, accountability, authenticity, and reliability are achieved [ISO TR 15443-1], [IIC 2015] Means to protect from disclosure or loss of information, and the possibility of a successful malicious attack. A secure system aims to protect assets and prevent unauthorized modification of information A condition that results from the establishment and maintenance of protective measures that enable an enterprise to perform its mission or critical functions despite risks posed by threats to its use of information systems. Protection measures may involve a combination of deterrence, avoidance, prevention, detection, recovery, and correction that should form part of the enterprise's risk management approach [CNSSI 4009] Note 1: The [CNSSI 4009] definition focuses on security as an organizational enterprise objective Note 2: The engineering perspective views security as a complex quality factor that is composed of multiple quality sub-factors. The most prevalent sub-factors are confidentiality, integrity, and availability. Additionally, the integrity sub-factor can be further divided into hardware, software, data, and communications integrity. Other security relevant quality sub-factors include, but are not limited to, privacy and non-repudiation. There are also quality sub-factors that generally have been considered only by the system safety engineering, such as continuity, resiliency, and fault tolerance, that are now being assessed in terms of susceptibility to malicious intent and the resultant impact on the mission/business; all motivated by mission assurance concerns that span the entire spectrum of incidental and accidental misuse through to attack by an advanced persistent threat The Systems Security Engineering perspective ensures that all security relevant quality sub-factors are satisfied by the engineered system and that the system achieves mission/business security objectives such as that defined in [CNSSI 4009] Information system security is a system characteristic and a set of mechanisms that span the system both logically and physically. The five security goals are integrity, availability, confidentiality, accountability, and assurance [NIST SP800-30r1]
Supportability	How easy it is for operators, developers, and users to understand and use the application, and how easy it is to resolve errors when the system fails to work correctly
Survivability	The ability of a system to remain in operation or existence despite adverse conditions, including natural occurrences, accidental actions, and attacks [RFC 4949] A system is survivable if it complies with its survivability specification [Knight 2003]

Testability	Measure of how easy it is to create test criteria for the system and its components, and to execute these tests in order to determine if the criteria are met. Good testability makes it more likely that faults in a system can be isolated in a timely and effective manner
Trustworthiness	The attribute of an entity that provides confidence to others of the qualifications, capabilities, and reliability of that entity to perform specific tasks and fulfill assigned responsibilities [CNSSI 4009]
	Trustworthiness is an attribute associated with an entity that reflects confidence that the entity will meet its requirements
	Trustworthiness, from the security perspective, reflects confidence that an entity will meet its security requirements while subjected to disruptions, human errors, and purposeful attacks that may occur in the environments of operation
Trustworthy	Trustworthy system is a system that not only is trusted, but also warrants that trust because the system's behavior can be validated in some convincing way, such as through formal analysis or code review [RFC 4949]
Usability	How well the application meets the requirements of the user and consumer by being intuitive, easy to localize and globalize, and able to provide good access for users and a good overall user experience
	A characteristic of an information environment to be user friendly in all its aspects (easy to learn, use, and remember) [IQ/DQ Glossary]

- Quality in use model composed of five characteristics grouped as efficiency, effectiveness, satisfaction, freedom from risk, and context, some of which are further subdivided into subcharacteristics.
- Product quality model (internal and external) composed of eight characteristics subdivided into subcharacteristics. Examples of characteristics include:
  - Functionality suitability (functional completeness, functional correctness, etc.)
  - Performance efficiency (e.g. channel capacity, throughput, response time, latency, etc.)
  - Compatibility (e.g. interoperability, connectivity, hardware/software compatibility, operating system compatibility, etc.)
  - Reliability (e.g. availability, system operational time, system down time, etc.)
  - Security (confidentiality, integrity, non-repudiation, accountability, authenticity)
  - Usability (accessibility, attractiveness, etc.)
  - Maintainability (modularity, modifiability [stability], etc.)
  - Portability (e.g. adaptability, usability, etc.)

The standard [ISO/IEC 25010] includes 8 product quality characteristics for product quality model (in contrast to [ISO/IEC 9126] standard's six) and 31 subcharacteristics and includes 2 new characteristics: security and compatibility with their subcharacteristics. The characteristics defined by both models are relevant to all software products, systems, computer systems, and IS. It is important to understand that some characteristics may be measures, metrics, or indicators. There is a distinction among these terms and cannot be used interchangeably [Bundschuh 2008]. Although the scope of the product quality model is intended to be software and computer systems, many of the characteristics are also relevant to wider systems and services. Complementary to these models is the [ISO/IEC 25012] standard for data quality.

Essential to Smart Grid functionality is interoperability, which is defined as the ability of software and hardware on multiple machines from multiple vendors to communicate meaningfully [Comer 2006]. Interoperability describes the goal of internetworking, namely, to define an abstract, hardware independent networking environment that makes it possible to build distributed communications. Internet applications exhibit a high degree of interoperability.

It is suggested that diversification toward sustainable energy sector products such as DER services and infrastructures can be achieved through market mechanisms, e.g. for transparency and access to information for all value chain participants [OECD 2012].

Another challenge to power grid is the quality of SCADA software. It must be delivered with high assurance of its performance, robustness, reliability, and security [Hale 1998]. This entails using a software engineering process that employs formal techniques to model, test, and analyze software. Tools and methods to achieve the assurance level of SCADA software may include evaluation criteria [CC] and proof-carrying code [PCC]. Thus, survivability profiles can be generated to specify what should happen to a collection of software in the event of a disaster.

3.5.2 System Quality Attributes

Table 3.2 includes a list of quality attributes in terms of system development.

Key issues for availability may include the following:

- A physical tier such as the database server or application server can fail or become unresponsive, causing the entire system to fail.
- Security vulnerabilities can allow denial-of-service (DoS) attacks, which prevent authorized users from accessing the system.
- Inappropriate use of resources can reduce availability. For example, resources acquired too early and held for too long cause resource starvation and an inability to handle additional concurrent user requests.
- Bugs or faults in the application can cause a system-wide failure.
- Frequent updates, such as security patches and user application upgrades, can reduce the availability of the system.
- A network fault can cause the application to be unavailable.

3.6 Critical Information Systems

Critical is defined as a condition of a system resource such that denial of access to, or lack of availability of, that resource would jeopardize a system user's ability to perform a primary function or would result in other serious consequences, such as human injury or loss of life [RFC 4949].

Critical computer systems can be defined as those in which a certain class of failure is to be avoided if it all possible failure can result in significant economic losses, physical damage, or threats to human life. Depending on the class of failure, critical systems may be:

- Safety critical – Can lead to loss of life, serious personal injuries, or damage to the environment.
- Mission critical – Can lead to an inability of the overall system, such as loss of critical infrastructure or data.
- Business critical – Can lead to significant economic costs such as loss of business or reputation.
- Security critical – Can lead to loss of sensitive data.

In order to understand the performance of the information security establishment, a separate and independent framework, the information assurance (IA) framework, must be defined and used.

The following are examples of critical DER systems or Smart Grid systems:

- Safety critical – For example, fault operation of electrochemical batteries may result in spills that cause harm.
- Mission critical – For example, loss of data or unavailability of utility DER management system (DERMS) for distribution (level 4) due to a threat may impact the reliability of power grid.
- Business critical – For example, any threat that causes modification or loss of DR signals during real-time communication can cause disruption to DR applications and impact businesses as well as the reliability of load and demand response functions.
- Security critical – For example, a cyber attack causing the failure of a DER SCADA system (level 3 or level 4) defined as a control point that provides monitoring and control, automatic generation control, real-time power system modeling, and real-time inter-utility data exchange can cause DER SCADA system unavailability, loss of data, commands, service malfunction.

An equally critical facet of information protection is protection of information and commands used to control the power system. Critical power system monitoring and control actions as well as confidential customer information demand a higher level of security protection.

3.6.1 Critical System Characteristics

For critical systems, it is usually the case that the most important system property is the dependability of the system [Sommerville 2004]. The dependability of a system reflects the user's degree of trust in that system. It reflects the extent of the user's confidence that it will operate as users expect and that it will not fail in normal use. The dependability of a system equates to its trustworthiness. A dependable system is a system that is trusted by its users. Usefulness and trustworthiness are not the same thing. A system does not have to be trusted to be useful. There are four principal dimensions or attributes to dependability to include availability, reliability, safety, and security [Sommerville 2004]. As illustrated in the diagram below (Figure 3.6), there are more dimensions.

Other dimensions of dependability include the following:

- Repairability reflects the extent to which the system can be repaired in the event of a failure.
- Maintainability reflects the extent to which the system can be adapted to new requirements.
- Survivability reflects the extent to which the system can deliver services while under hostile attack.
- Error tolerance reflects the extent to which user input errors can be avoided and tolerated.

Systems that are not dependable and are unreliable, unsafe, or insecure may be rejected by their users. The costs of system failure may be very high. Undependable systems may cause information loss with a high consequent recovery cost. Failures may be caused by:

- Hardware failure – Hardware fails because of design and manufacturing errors or because components have reached the end of their natural life.
- Software failure – Software fails due to errors in its specification, design, or implementation.
- Operational failure – Human operators make mistakes, now perhaps the largest single cause of system failures.

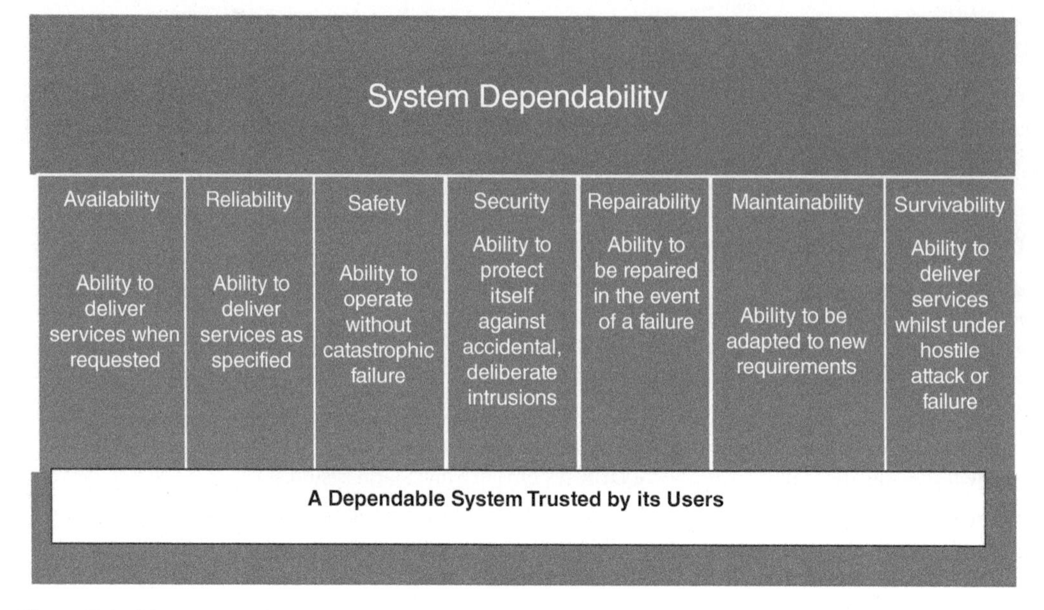

Figure 3.6 Dimensions of dependability.

It is argued that different facets of dependability are suitable for different systems [Knight 2000]. For example, highly reliable operation is usually needed for an embedded control system, highly available operation is usually needed in a SCADA system, and a high level of safety is needed for a power protection system. A system might achieve one facet of dependability but not others. Such a system that fails frequently but only for very brief periods has high availability but low reliability. Many systems are built to operate this way intentionally because it is a cost-effective approach to providing service if reliability (in the formal sense) is not required but availability is [Knight 2000].

However, events that disrupt critical IS are inevitable and must be dealt with in some way. The continued provision of some form of service is more than desirable – in many cases it is essential. A notion of dependability for these systems is defined as survivability because the current facets (reliability, availability, etc.) do not provide the necessary concepts [Knight 2000]. They do not include the notion of degraded service and the associated spectrum of factors that affect the choice of degraded service as an explicit requirement.

The survivability of critical infrastructure systems, such as electric power distribution and telecommunications, has become a major concern of the US government and will garner increasing concern from private industry. These systems should continue to provide acceptable service levels to customers in the face of disturbances – natural, accidental, or malicious [Sullivan 1999], [Ellison 1997].

Many critical systems have requirements for reduced or alternate service under some circumstances. By defining it precisely, system owners can state exactly what the user of a system can expect over time in terms of the provision of service, and system designers have a precise statement of the dependability that the system has to achieve and can design accordingly. A system is survivable if it complies with its survivability specification [Knight 2003]. A survivable system has the ability to continue to provide service (possibly degraded or different) in a given operating environment when various events cause major damage to the system or its operating environment. Service quality, e.g. exactly what a system should do when damaged, seems to be something that should be determined by simple guidelines.

Survivability needs to specify the various and different forms of tolerable service that the system provides given the notion that circumstances might force a change in service to the user [Knight 2003]. A tolerable (but not necessarily preferred) service is one that combines functions that work harmoniously and provide value to the user. The set of tolerable services are the different forms of service that the system must be capable of providing. At any one time, just one member of the set would be operating – the others represent the changed or degraded service definitions.

Creation → Consumption → Disposition

Figure 3.7 Information life cycle.

3.6.2 Information Life Cycle

Information protection must be presented throughout the information life cycle (creation, consumption, and disposition) (Figure 3.7). When information is created, it could very often be at a data format and carry less value to the holder and users of such data. As it is developed, interpreted, and placed in a framework, it becomes consumable information. Information at this stage has the highest value as its consumption by users realizes immediate value.

As the current information is replaced by newer and better quality information, the current information becomes less current and less useful. Such information will eventually be archived or disposed. The worthiness of the information becomes less valuable. However, information could be in other states such as processing, storage, and transmission, and these stages require security protection too.

3.6.3 Information Assurance

A design framework such as IA is usually followed to ensure the protection of the information as the value of information is changing. The value of information produces a different cost of protection. This means a different level of IA specifications needs to be deployed, for example:

- The communication of information from one stage to another should be atomic.
- Any failure of a proper transfer of information should be backtracked to its original state before the transfer to ensure information integrity.

When information is created, IA mostly concerns the completeness and quality of information including the proper operations of the states within the computing systems. Assurance refers to confidence that a claim has been or will be achieved [IEEE 15026-1]. When the scope of claim is security, assurance is the measure of confidence that the security features, practices, procedures, and architecture of an information system accurately mediate and enforce the security policy [CNSSI 4009]. IA focuses on the reasons for assurance that information is protected and is thus reasoning about information security. Data characteristics for information security assurance may include:

- Integrity
- Availability
- Authenticity
- Non-repudiation
- Confidentiality.

Therefore, the following security concerns should be addressed during information life cycle:

- Ownership and confidentiality at the creation stage.
- Integrity, confidentiality, and availability of the information at the consumption stage.
- Proper storage, accessibility, and integrity of the information in the disposition stage.

Techniques can be used to assure that the quality of the information at each stage is within the specification of the IA framework that addresses security concerns. A piece of information should be stopped from passing through to the next stage if the information does not meet the requirements of IA's specification. This could be the result of irregularities of data being exposed or vulnerabilities causing data to be in an unprotected state. With the proper operations of these techniques, this important mechanism of the IA framework would help prevent the occurrence of human error. For example, by reviewing the performance of techniques and how effective they are in stopping the exposure of personal data, the IA framework can be used as a guide for developing a proactive approach to information privacy protection.

3.6.4 Critical Security Characteristics of Information

While cybersecurity as distinct term has various ambiguous definitions, the information security has its own nuances. As it is illustrated below, there are many definitions for information security that could create confusion or conflicting interpretations.

Many information security definitions are identified in different standards or dictionaries. Thus, such definitions may include the specific characteristics (see examples of characteristics in Table 3.1). The following are common definitions:

> Information security (InfoSec) is concerned with the protection of characteristics that include confidentiality, integrity, availability, and non-repudiation and other characteristics of information regardless of the form the information may take as electronic, print, or other forms.

> Information security is preservation of confidentiality, integrity, and availability of information. Note: In addition, other properties, such as authenticity, accountability, non-repudiation, and reliability can also be involved [ISO/IEC 27000].

> Information security (InfoSec) refers to measures that implement and assure security services in information systems, including in computer systems and in communication systems [RFC 4949].

Commonly, information security breach is described as affecting one or more of these common data (information) security attributes (confidentiality, integrity, and availability), known as CIA triad. However, this approach is not sufficient; a broader perspective to include other attributes is needed for analysis to identify the root cause of security intrusions and security compromises.

3.7 Information Security Models

Information security models are methods used to authenticate security policies as they are intended to provide a precise set of rules that a computer can follow to implement the fundamental security concepts, processes, and procedures contained in a security policy. These models can be abstract or intuitive [CYBRARY Models]. Another [Whitman 2014] defines a security model as a collection of security rules that represents the implementation of a security policy. There is an evolving suite of models that have to be understood since old models may still be used in some places.

3.7.1 Evolving Models

The initial CIA triad principles (mainly used by federal agencies in the United States) is only a limited view of security that tends to ignore other important factors such as physical security, organizations, people, and environment. The CIA triad (confidentiality, integrity, and availability) has for several decades been serving as a conceptual model of computer security and, later, InfoSec. Originated in 1975, a wide range of security-related material is based on the CIA triad, despite the facts that the adequacy of the CIA triad has been questioned [Cherdantseva 2013b]. Figure 3.8 shows the security attributes (CIA triad) as they relate to information and information system.

The objective of security of IS is the protection of the interests of those relying on IS from harm resulting from failures of availability, confidentiality, and integrity. This model refers to protection of the basic security characteristics. This model was developed by computer industry, and it has been the standard since the development of mainframe computers.

According to these authors [Cherdantseva 2013b], issues such inadequacy of CIA triad as a complete set of security goals and no coverage of new threats that emerge in the collaborative deperimeterized environment have been reported in the system engineering and InfoSec literature.

Responding to the need for a theoretical foundation for modeling the information, a new model independent of technology and organization is proposed [McCumber 1991]. Since 1991, this model known as McCumber cube (see Figure 3.9) has been also used as the standard for the evaluation of IS security. The model provides a graphical representation of the architectural approach widely used in information security.

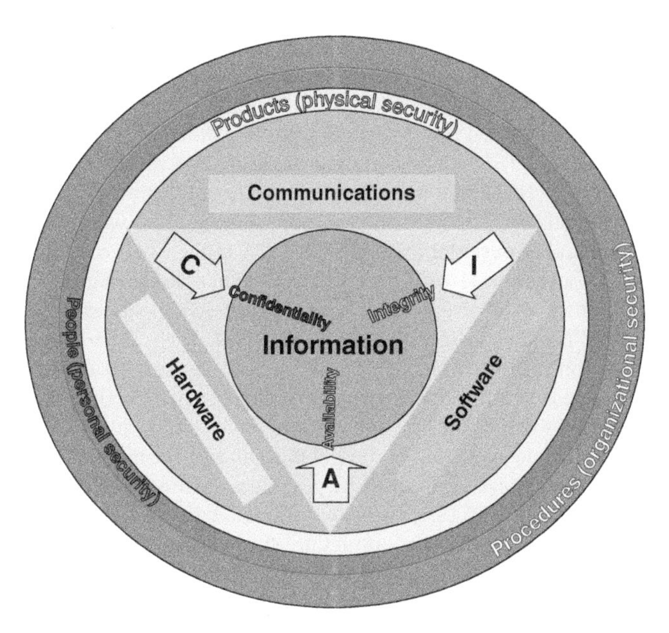

Figure 3.8 CIA information security model. *Source:* [JohnManuel 2009]. Licensed under CC BY SA-3.0.

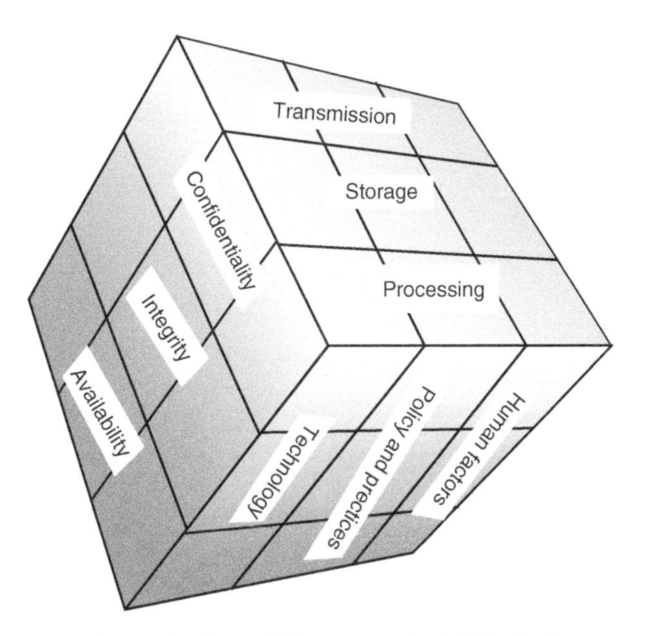

Figure 3.9 McCumber cube. *Source:* [McCumber]. Licensed under CC BY-SA 3.0.

The McCumber cube has three dimensions:

- Critical information characteristics (confidentiality, integrity, and availability).
- Information states (processing, storage, transmission).
- Security measures (technology, policy and practices, human factors [education, training, and awareness]).

The model can be used in the development of IS and as an evaluation tool. Organizations must consider the interconnectedness of all the different factors that impact IA programs. If extrapolated, the 3 dimensions of each axis become a $3 \times 3 \times 3$ cube with 27 cells representing areas that must be addressed to secure IS. To ensure system security, each of the 27 areas must be properly addressed during the security process. For example, the intersection between technology, integrity, and storage requires a control or safeguard that addresses the need to use technology to protect the integrity of information while in storage. What is commonly left out of such a model is the need for guidelines and policies that provide direction for the practices and implementations of technologies [Whitman 2014].

In 1994, the model was adopted by the National Security Telecommunications and Information Systems Security Committee (NSTISSC) to become the National Training Standard for Information Systems Security Professionals (NSTISSI No. 4011). Later, this standard was renamed as CNSS model. Figure 3.10 shows a graphical representation of the NSTISSC model. While the NSTISSC model covers the three dimensions of information security, it omits discussion of detailed guidelines and policies that direct the implementation of controls. Another weakness of using this model with too limited an approach is to view it from a single perspective.

Therefore, the information security evolved from this model because it no longer adequately addresses the constantly changing needs of the users as well as evolving threats. Further analysis of the NSTISSC standard leads to a modified model for IA as described in [Maconachy 2001]. The authors solidify the definitions of InfoSec and IA and show that both INFOSEC and IA are measures to protect systems and the information resident in those systems. These new services ensure the authenticity and non-repudiation characteristics besides confidentiality, integrity, and availability. So, the needs of business included another property of information that is the foundation of non-repudiation security service. The non-repudiation service is a requirement for many applications such as E-commerce, banking, etc.

IA expanded the model to include two new security services, authentication, and non-repudiation. Thus, this definition includes other characteristics besides information characteristics defined by CIA triad model and McCumber model. Figure 3.11 shows the modified view of the model for IA where the

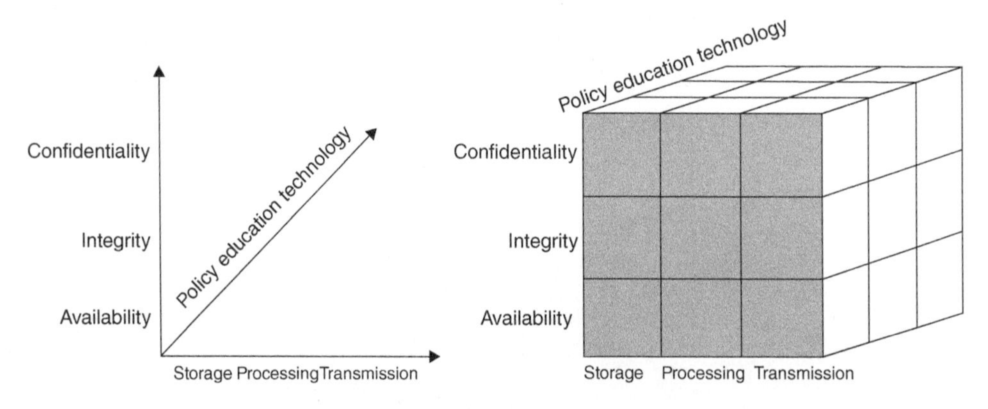

Figure 3.10 NSTISSC model. *Source:* [NSTISSI 4011]. Public Domain.

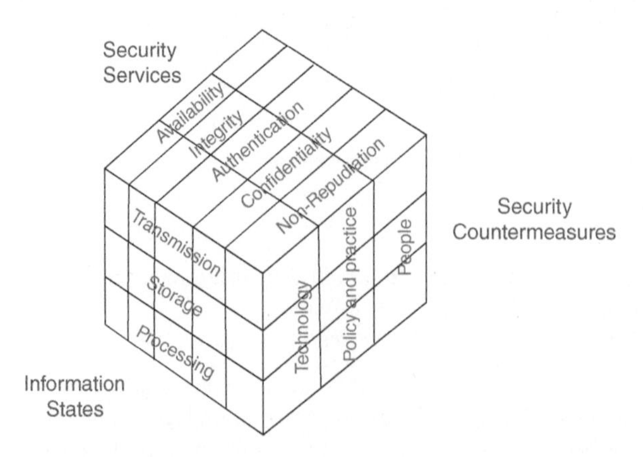

Figure 3.11 The information assurance model. *Source:* [Maconachy 2001]. © 2001, IEEE.

model accounts for three of for dimensions of IA, but security services account for more services than McCumber original model.

Basically, these CIA properties can be ensured by security services such as confidentiality, integrity, and availability services. A security service is a processing or communication service that is provided by a system to give a specific kind of protection to system resources. Security services implement security policies, which are implemented by security mechanisms. A security mechanism is a set of policy rules (or principles) that direct how a system (or an organization) provides security services to protect sensitive and critical system resources (see definitions in Appendix A).

However, other attributes (non-repudiation and authenticity) are being considered in [Maconachy 2001]; so, the authentication and non-repudiation services are included. Currently, more security services may be used to countermeasure the threats and mitigate the risks. Examples include accountability, reliability, privacy, and anonymity. Safety can also be defined to be the control of recognized hazards to achieve an acceptable level of risk.

Besides additional security services (authentication and non-repudiation), another proposed model includes time as a fourth dimension of the McCumber model [Maconachy 2001]. This work argues that time is needed as the fourth dimension: time is not a causal agent of change, but a confounding change agent [Maconachy 2001]. Time may be viewed in three ways:

- At any given time the access to data may be either accessible online or offline. Risk mitigation, as opposed to risk avoidance, takes on a different urgency depending upon connectivity.
- Time relates to IA because at any given time the state of information and IS is in flux. Well-executed systems will include the IA model during all phases of the system development life cycle.
- The human side of the time line leads to career progression; the learning activities over time produce an enhancement to a system security state.

This model, as shown in Figure 3.12, highlights the need to look at information security as evolving in time as new requirements emerge and look broadly at all components instead of focusing only on one component.

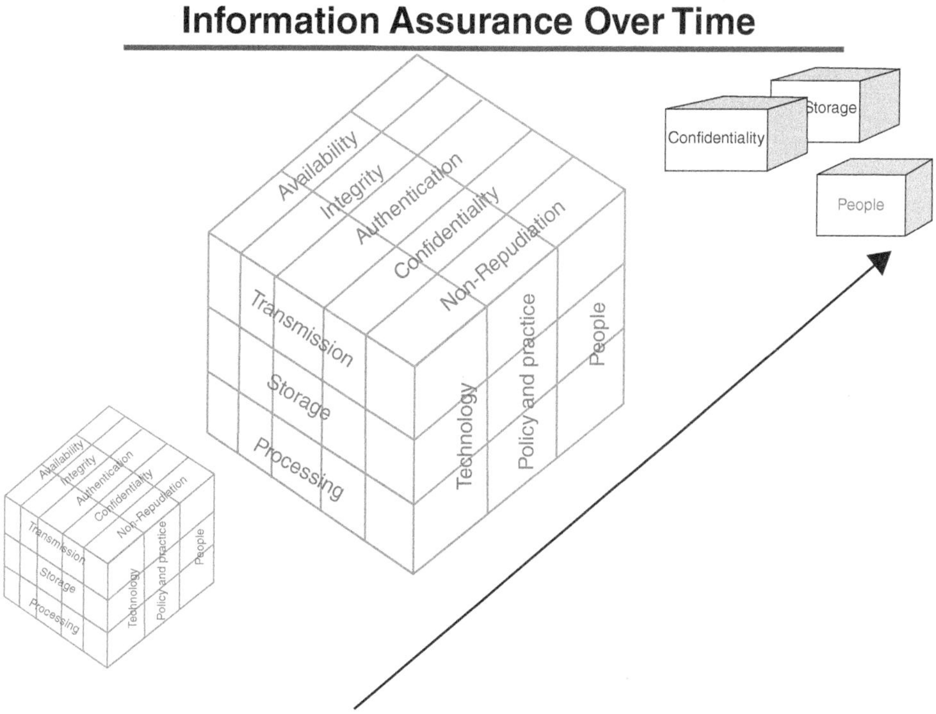

Figure 3.12 IA model with time dimension. *Source:* [Maconachy 2001]. © 2001, IEEE.

Another alternative model to the classic CIA triad is proposed by Parker [Parker 2002]. His model, called the six atomic elements of information, is also known as Parkerian hexad (term coined by M. E. Kabay [Kabay 2000]), and it includes the following characteristics:

- Confidentiality
- Possession or control
- Integrity
- Authenticity
- Availability
- Utility.

These attributes of information are atomic in that they are not broken down into further constituents; they are nonoverlapping in that they refer to unique aspects of information. Any information security breach can be described as affecting one or more of these fundamental attributes of information. The CIA triad (confidentiality, integrity, and availability) is one of the core principles of information security, and it is only a starting point. These attributes support a limited view of security that tends to ignore some additional important factors [Perrin 2008].

The CIA triad principles are used mainly by federal agencies in the United States and need to include another property of information that is the foundation of non-repudiation security service. The non-repudiation service is a requirement for many applications such as E-commerce, banking, etc. However, there is no wide use of Parkerian hexad model among security professionals, although the model is based on a broader perspective of information characteristics. The Parkerian model is expanded in [Whitman 2014] to include one more characteristic, accuracy.

An analysis of the existing information security models finds issues that are discussed in [Cherdantseva 2013b]:

- Lack of an agreed-upon set of security goals (which are interchangeably referred to in the literature as security attributes, properties, fundamental aspects, information criteria, critical information characteristics, and basic building blocks).
- Consideration of time within the models is too generic and does not have pragmatic value.
- Although protection of information outside the organization's perimeter is mentioned in some works, it is not explicitly addressed.
- Many security issues are caused by wrong security decisions being taken on the basis of incomplete knowledge or misunderstanding of the security domain.

Referring to the information assurance and security (IAS) knowledge area [Cherdantseva 2013a], which incorporates the knowledge acquired by both InfoSec and IA, a reference model is described in [Cherdantseva 2013b].

3.7.2 RMIAS Model

The recent reference model of information assurance and security (RMIAS) that was proposed includes a set of relevant security characteristics agreed by security professionals and academics [Cherdantseva 2012], [Cherdantseva 2013b].

A reference model is intended to provide an even higher level of commonality, with definitions that should apply to all IS. A reference model (e.g. RMIAS model) is an abstract framework for understanding significant relationships among the entities of some environment. A reference model consists of a minimal set of unifying concepts, axioms, and relationships within a particular problem domain and is independent of specific standards, technologies, implementations, or other concrete details [OASIS 2006].

The RMIAS model (see Figure 3.13) defines security characteristics as goals and includes more characteristics than any other model. The set of suggested characteristics include:

- Confidentiality
- Integrity
- Availability
- Accountability
- Authenticity and trustworthiness

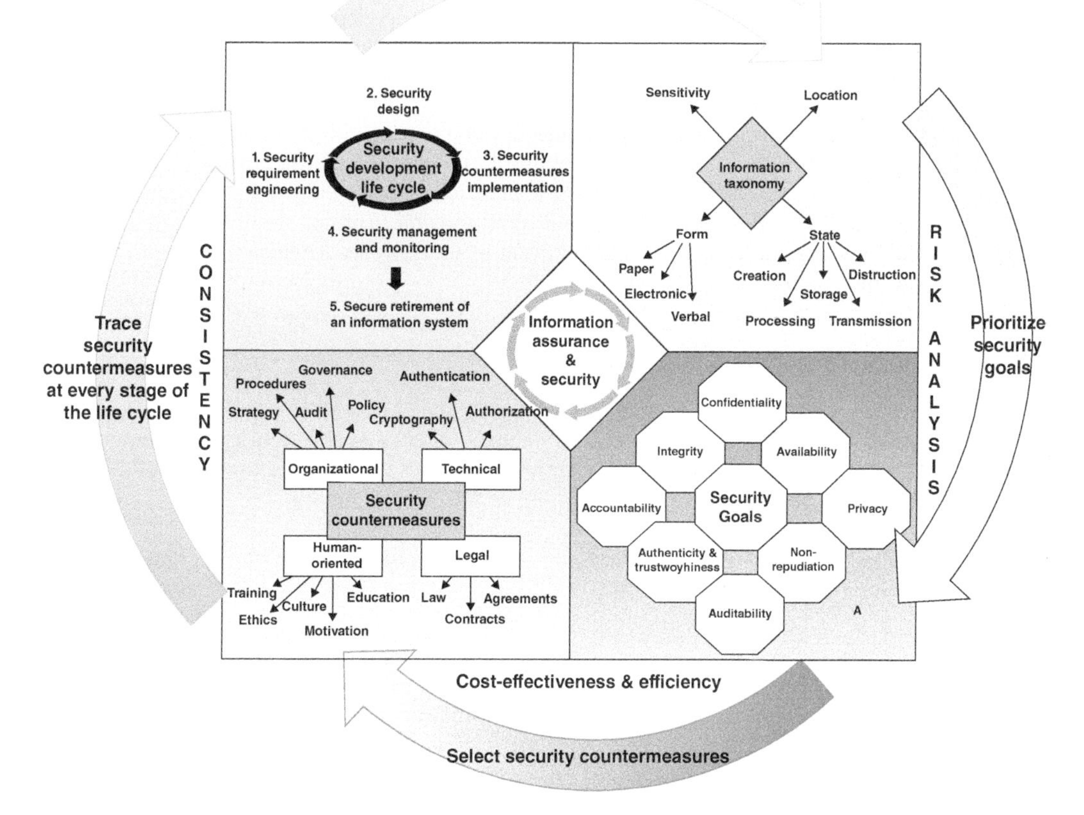

Figure 3.13 RMIAS model. *Source:* [Wikilubina]. Licensed under CC BY-SA 3.0.

- Non-repudiation
- Auditability
- Privacy.

The RMIAS model, depicted in Figure 3.13, has four dimensions:

- Security development life cycle dimension illustrates the progression of IS security during development.
- Information taxonomy dimension describes the nature of information being protected.
- Security goals dimension outlines a broadly applicable list of security goals; a security goal is a desirable ability of an IS to resist a specific category of threats;
  the goal-based approach is recommended, which operates at a higher level of abstraction than a threat-based approach.
- Security countermeasures dimension categorizes countermeasures available for information protection; a security countermeasure can be used to achieve one or more security goals and to mitigate risks to information and vulnerabilities in a system.

These four dimensions are deemed compulsory and sufficient for an understanding of the IAS domain at the chosen high level of abstraction. They do not overlap and do not duplicate each other. The visual representation based on reference model helps a wider audience in understanding the IAS domain knowledge. A formal RMIAS model based on ontology languages is planned. One feature could be the development of machine-readable security policies [Cherdantseva 2013b].

One issue with the RMIAS model is that someone looking at the diagram (on the left, upper corner) has to understand that there is no separation of security development life cycle from the system development life cycle. Indeed, the security development life cycle is correlated with the system development life cycle. There is a correlation of where security tasks (e.g. security requirements engineering) should be included during the activities of a system development (including security in a system development method [Hansche 2004]).

Although security has its own reference model, by the end is the whole system that has to provide end-to-end security. What matters is the integration of security in the IS to meet the requirements for the utility of the information and usability of the IS.

Utility of information is the quality or state of having value for some purpose; information has value when it can serve a purpose. If information is available, but it is not in a format meaningful to the end user, it is not useful. Usability defines how well the application meets the requirements of the user and consumer by being intuitive, easy to localize and globalize, and able to provide good access for users and a good overall user experience. However, the selection of different security countermeasures is also based on the information security goals and risks.

3.7.3 Information Security Goals

According to [OECD 2002], [ISO/IEC 27000], [IUT-T 2008], and [RFC 4949], the primary goal of information security is to guarantee safety of the information stipulated by the security attributes such as confidentiality, integrity, and availability. These are characteristics that describe the utility of the information as follows:

> Confidentiality is the property data that is not disclosed to system entities unless they have been authorized to know the data [RFC 4949].
>
> Integrity is the property that data has not been changed, destroyed, or lost in an unauthorized or accidental manner [RFC 4949].
>
> Availability is the property of being accessible and usable upon demand by an authorized entity [RFC 4949].

The suite of security characteristics is expanded in other information security definitions. For example, [ISO/IEC 27000] model includes non-repudiation, authenticity, accountability, and other aspects such as reliability, privacy, and anonymity. Safety can also be defined to be the control of recognized hazards to achieve an acceptable level of risk [Wiki Safety]. Table 3.3 is a list of characteristics identified as information security goals by different models.

In practice, applications may need to implement protection for a mix of attributes that are referenced by different models. However, certain conflicts between attributes may occur, and resolution is required. In addition, if a product is certified for security based on a model, there appropriate counter will have to be applied using a risk management approach.

A security goal is the desirable security-related property of information or desirable ability of an IS. A study of the evolution of security goals from the 1960s to 2012 highlights issues such as [Cherdantseva 2012], [Cherdantseva 2013b]:

- The set of security goals is neither fixed, nor given.
- Security goals are changing over time in response to the evolution of society, business needs, and ICT.
- There is not an agreed-upon set of goals and authors associate a wide range of goals with InfoSec/IA.
- The same goals are referred to by different names.
- The goals with the same name have conflicting definitions in different sources (various communities input some specific meaning into particular terms).
- Security countermeasures are not distinguished from goals.
- Lack of clarity as to which component of an IS a goal applies to (e.g. integrity may refer to either information or system integrity, or both).
- An InfoSec professional should be alert to the rapid evolution of a valid collection of security goals and should be ready to conduct, regularly, an insightful analysis of security issues posed by emerging technologies.

Table 3.3 Information security goals.

Data attribute	Information security models references
Accountability	ISO/IEC 27000:2014, RMIAS, Whitman, Parkerian Hexad
Accuracy	Whitman
Availability	ISO/IEC 27000:2014, RMIAS, Whitman, Parkerian Hexad, NSTISSC, McCumber, C.I.A. Triad
Authenticity	ISO/IEC 27000:2014, RMIAS, Whitman, Parkerian Hexad, NSTISSC
Auditability	RMIAS
Confidentiality	ISO/IEC 27000:2014, RMIAS, Whitman, Parkerian Hexad, NSTISSC, McCumber, C.I.A. Triad
Control	Parkerian Hexad
Integrity	ISO/IEC 27000:2014, RMIAS, Whitman, Parkerian Hexad, NSTISSC, McCumber, C.I.A. Triad
Non-repudiation	ISO/IEC 27000:2014, RMIAS, NSTISSC
Possession	Whitman, Parkerian Hexad
Privacy	RMIAS
Reliability	ISO/IEC 27000:2014
Trustworthiness	RMIAS
Utility	Whitman, Parkerian Hexad

The set of goals, outlined in the RMIAS, as well as any other set, requires regular revision over time to incorporate goals addressing newly emerging threats. It highlights the fact that the set of goals is not fixed. It may be revised to reflect future changes. Table 3.4 outlines the set of security goals included in RMIS model [Cherdantseva 2013b]. It shows their concise definitions and the applicability of goals to the components of an IS.

An IS defined as a socio-technical system that encompasses six components: information (data), people, business processes (procedures), and ICT (hardware, software, and networks).

3.8 Standards, Guidelines, and Recommendations

Smart Grid technology is not a single silver bullet but a collection of existing and emerging standards-based, interoperable technologies working together.

Guidelines and recommendations for integrating DERs and Smart Grid applications are provided by DOE, EPRI, SANDIA, etc. For example, the DOE Roadmap of 2011 [Roadmap 2011] is an evolving document of [Roadmap 2006] on key issues regarding cybersecurity. It recognizes the cybersecurity and other technology advances and the evolving needs of the energy sector while other groups such as Electronic Privacy Information Center [EPIC] defines a framework that supports the cybersecurity needs of asset owners and operators and strategies for increasing the resilience of energy delivery systems.

The IT, telecommunications, and energy sector-specific plans (SSPs) are published and updated annually. In addition, standards published by NIST, IEEE, ISA, and ISO/IEC are available to organizations, managers, designers, and software manufacturers on complying with requirements for Smart Grid security. It is important to know the continuous development of ISO 27k series of standards [ISO 27k].

NIST is charged with overseeing the identification and selection of hundreds of standards that are required to implement the Smart Grid in the United States. Standards have already been selected for inclusion in NIST's Smart Grid catalog [NIST Wiki].

The Smart Grid Interoperability Panel (SGIP) Cybersecurity Committee (SGCC), which is led and managed by the NIST Information Technology Laboratory (ITL), Computer Security Division, is involved

Table 3.4 RMIS model security goals.

Security Goal	Definition	Information	People	Processes	Hardware	Software	Networks
Accountability[a]	A system ensures to hold users and processes responsible for their actions (e.g. misuse of information)		X	X			
Auditability[b]	A system ensures to conduct persistent, non bypassable monitoring of all actions performed by humans or machines within the system		X	X	X	X	X
Authenticity and Trustworthiness	A system has been determined to provide confidence to other entities to operate within defined levels of risk despite intentional and unintentional threats	X	X	X	X	X	X
Availability	A system ensures that information is accessible and usable upon demand by an authorized entity	X	X	X	X	X	X
Confidentiality	A system ensures that information is not made available or disclosed to unauthorized individuals, entities, or processes	X	X	X	X	X	X
Integrity	A system ensures completeness, accuracy and absence of unauthorized modifications in all its components	X	X	X	X	X	X
Non-repudiation	A system ensures proof (with legal validity) the occurrence of a claimed event or action and its originating entities	X	X	X			
Privacy	A system ensures the privacy of personal data	X	X				
Security attributes		Components of Information Systems					

[a] Accountability is the property that ensures that the actions of an entity can be traced solely to this entity. Accountability guarantees that all operations carried out by individuals, systems or processes can be identified (identification) and that the trace to the author and the operation is kept (traceability).
[b] A computer security audit is a manual or systematic measurable technical assessment of a system or application. Manual assessments include interviewing staff, performing security vulnerability scans, reviewing application and operating system access controls, and analyzing physical access to the systems. Automated assessments, or CAAT's, include system generated audit reports or using software to monitor and report changes to files and settings on a system. Systems can include personal computers, servers, mainframes, network routers, switches [Wiki Audit].
Source: Adapted from [Cherdantseva 2013b]. © 2013, IEEE.

in addressing the critical cybersecurity problems in the areas of advanced metering infrastructure (AMI) security requirements, cloud computing, supply chain, and privacy recommendations related to emerging standards. Their results will provide foundational cybersecurity guidance, cybersecurity reviews of standards and requirements, outreach, and foster collaborations in the crosscutting issue of cybersecurity in the Smart Grid.

The document [DHS 2011b] includes recommendations specifically designed to provide various industry sectors the framework needed to develop sound security standards, guidelines, and best practices. It provides decision makers with a common catalog (framework) from which to select security controls for control systems. This catalog should be viewed as a collection of recommendations to be considered and judiciously employed, as appropriate, when reviewing and developing cybersecurity standards for control systems.

The objective is to advance the development and standardization of cybersecurity, including privacy, policies, measures, procedures, and resiliency in the Smart Grid in 2016. The new technical idea is to adapt existing cybersecurity best practice methodologies and tools and to understand how to apply them in the electric sector while identifying gaps and unique requirements for the grid that require new methodologies and tools.

The SGIP Smart Grid Cybersecurity Committee [SGIP SGCC] aims to address these challenges through collaborations with federal agencies, academia, and industry, through the evaluation of cybersecurity policies and measures in industry standards, and through the development of guidance documents.

3.8.1 SGIP Catalog of Standards

The SGIP Catalog of Standards (CoS) is a compendium of standards and practices considered to be relevant for the development and deployment of a robust and interoperable Smart Grid, and the catalog contains more information than the brief description in the [NIST SP1108r3] document. The relevancy means the standard facilitates interoperability related to the integration of Smart Grid devices or systems. The CoS includes recommended standards for Smart Grid implementations in the United States. Table 3.5 is a list of organizations that support standards relevant to Smart Grid interoperability.

However, there are many other standard organizations involved in standards for power systems and Smart Grid; examples include:

- BSI – British Standards Institution.
- CEN – European Committee for Standardization.
- CENELEC – European Committee for Electrotechnical Standardization.
- DIN – Deutsches Institut für Normung (German Standards Institute).
- ETSI – European Telecommunications Standards Institute.
- AFNOR – Association Française de Normalisation (French Standardization Association).
- UL – Underwriters Laboratories.

However, with the plethora of standards available in the electric industry that can be applied to a Smart Grid domain, there could be gaps. A gap analysis is recommended to identify existing standards against actual Smart Grid requirements and find solutions to fill the gap. A current technical framework of enabling technologies and smart solutions is documented in the book by [Borlase 2017]. The book describes the role of technology developments and coordinated standards in Smart Grid, including various initiatives and organizations helping to drive the Smart Grid effort.

3.8.2 Cybersecurity Standards for Smart Grid

The list of standards needed to secure the Smart Grid is not an exhaustive list. One can find it helpful to check the identified standards in these works [Cleveland 2013], [Cleveland 2016] [ENISA 2012].

The European Network for Information Security Agency (ENISA) maintains the document Smart Grid Security Related Standards Guidelines and Regulatory Documents [ENISA 2012].

Table 3.5 List of organizations developing relevant standards for Smart Grid.

Organization	Name	Recognized – International, Regional, National (USA)	Examples of protocols, standards
ANSI	American National Standards Institute	USA	ANSI C12 Suite
ASHRAE	American Society of Heating, Refrigeration, and Air Conditioning Engineers	USA	BacNet
CEA	Consumer Electronics Association	USA	Lon Protocol Suite
DHS	Department of Homeland Security (DHS)	USA	DHS Cyber Security Procurement Language for Control Systems
GWAC	GridWise Architecture Council	USA	GWAC Interoperability Model
IEC	International Electrotechnical Commission	International	IEC 60870, IEC 61850, IEC 62351
IEEE	Institute of Electrical and Electronics Engineers	International	IEEE 1815 (DNP3), IEEE C37.118.1-2011, IEEE 1547
IETF	Internet Engineering Task Force	International	RFC 6272, IPv4/IPv6
ISO	International Organization for Standardization	International	ISO 9000
ITU	International Telecommunication Union	International	ITU-T G.9972
NAESB	North American Energy Standards Board	USA	REQ-21 Energy Services Provider Interface (ESPI)
NEMA	National Electrical Manufacturers Association	USA	SG-AMI 1-2009
NERC	North American Electric Reliability Corporation	Regional	NERC CIP 002-014
NIST	National Institute of Standards and Technology	USA	NIST Interagency Report NISTIR 7628
NRECA	National Rural Electric Cooperative Association	USA	MultiSpeak
OASIS	Organization for the Advancement of Structured Information Standard	International	EMIX
OGC	Open Geospatial Consortium	International	GML
OPC-UA	Open Platform Communications United Architecture	USA	SOA Architecture
OpenADR	OpenADR Alliance	International	OpenADR 2.0
SAE	Society of Automotive Engineers	International	SAEJ2847/1
UCAIug	Utility Communications Architecture International Users Group	International	OpenHAN
W3C	World Wide Web Consortium	International	HTML, XML
ZigBee	ZigBee Alliance	International	Smart Energy Profile 2.0

ENISA defines the included documents as follows:

- Standards – documents intended for defining new security mechanisms or frameworks focusing on interoperability or certification aspects.
- Guidelines – recommended security good practices, technical reports on specific topics and any worksheet supporting activities such as risk analysis, security requirements definition for Smart Grid components, Smart Grid components assessment from a security perspective, etc.
- Regulatory documents are either security guidelines or standards that are considered mandatory from a legal perspective of because it is de facto standard for an industrial association (e.g. DSO operators).

The two most regarded standards that specifically deal with IT security are the ISO 27001 [ISO/IEC 27001] standard (successor of the ISO 17799) and some of the NIST SP800 standard series. These documents describe other security qualities beyond confidentiality, integrity, and availability; the latter three are generally considered as the foundational security qualities. Additional attributes like authenticity, accountability, and non-repudiation can be considered as subsets of the foundational qualities.

For example, the [ISO/IEC 27001] standard specifies the requirements for establishing, implementing, operating, monitoring, reviewing, maintaining, and improving formalized information security management systems (ISMS) within the context of the organization's overall business risks. This standard can be used by all organizations, regardless of type, size, and nature. It specifies requirements for the implementation of information security controls customized to the needs of individual organizations or parts thereof. Several other standards of ISO 27k series are needed to implement the cybersecurity. The categories of [ISO/IEC 27000] series of standards include requirements, guidelines, sector-specific guidelines, control, and vocabulary.

4

Applying Security Principles to Smart Grid

4.1 Smart Grid Security Goals

The security goals are the basis of a security model, and they are selected based on the business needs for an application, systems, or organization [Dhillon 2007]. The Edison Electric Institute asserts top priorities of power industry as protection of electric grid and ensuring a reliable supply of power. Key to the success of this effort is the ability to provide measures capable of protecting the evolving intelligent network against interruption, exploitation, compromise, or outright attack of cyber assets, whether the attack vector is physical, cyber, or both [Edison].

Cybersecurity in the Smart Grid includes both power and cyber system technologies and processes in information technology (IT) and power system operations and governance. According to NIST Cybersecurity document [NISTIR 7628r1], these technologies and processes need to provide the protection required to ensure confidentiality, integrity, and availability of the Smart Grid cyber infrastructure, including control systems, sensors, and actuators. However, the definition of security goals (attributes) needs to be more inclusive.

The Reference Model of Information Assurance & Security model (RMIAS model) is a comprehensive model that could satisfy the information security requirements for a broad range of Smart Grid systems. Further, control systems have specific needs that require a different set of security attributes (e.g. reliability is needed, while confidentiality or accountability can be excluded).

Cybersecurity needs to be appropriately addressed to the combined power system and IT communication system domains to maintain the reliability of the Smart Grid and privacy of consumer information. Cybersecurity in the Smart Grid must include a balance of both power and cyber system technologies and processes in IT and power system operations and governance. Poorly applied practices from one domain that are applied into another may degrade reliability.

New emerging technologies and increasing risks due to threats and vulnerabilities require Smart Grid systems to adopt a comprehensive security approach (e.g. RMIAS model) and a unified risk management approach to both IT-traditional systems and control systems (as described in [Ray 2010]). The unified view means that IT-specific approaches are blended with control-specific approaches, and not a sum of two approaches. Also, differences on goals need to be addressed accordingly. For example, a traditional IT-focused understanding of cybersecurity is that it is the protection required to ensure confidentiality, integrity, and availability of the electronic information communication system. In general, for IT systems, the priority for the security objectives is confidentiality first, then integrity, and availability. For industrial control systems (ICS), including power systems, the priorities of the security objectives are availability first, integrity second, and then confidentiality. Also, the information security goals need to be expanded to include more attributes based on the needs of the applications.

Information survivability is regarded as critical by electric utility companies to ensuring the continued delivery of electric power to the nation. To realize this goal, one way is the development of

Building an Effective Security Program for Distributed Energy Resources and Systems: Understanding Security for Smart Grid and Distributed Energy Resources and Systems, Volume 1, First Edition. Mariana Hentea.

survivable supervisory control and data acquisition (SCADA) software by using trusted software development methodologies [Hale 1998].

There is a growing concern about the security of the SCADA systems in terms of vulnerabilities, lack of protection, and awareness. Vulnerability discovery techniques and appropriate engineering activities are required to ensure security, reliability, and safety of plants that use SCADA systems. Therefore, information security management principles and processes need to be applied to SCADA systems without exception.

4.2 DER Information Security Characteristics

Information security characteristics for Smart Grid and distributed energy resource (DER) systems must be determined and selected based on the needs of each business and application. It is suggested a set of security characteristics to include:

- Accountability
- Authenticity
- Availability
- Integrity
- Confidentiality
- Non-repudiation
- Reliability
- Trustworthiness
- Privacy
- Anonymity.

Traditionally, in electrical sector, security for information systems (known IT systems) is focused to ensure the confidentiality, integrity, and availability of the information and communication systems. In the context of Smart Grid, information security needs have to be analyzed, and the set of characteristics should be expanded to include other characteristics.

An information system may include both control systems and IT communication systems to maintain the functionality of the Smart Grid. Also, physical, communication, emission, computer, and network security together make up the concept of information security. No one security type can protect a business; therefore, the concept of total information security is critical to the success of the electrical enterprise. Information security in the Smart Grid applications must include a balance of both power and cyber system technologies and processes in IT and power system operations and governance. Poorly applied practices from one domain that are applied into another may degrade the reliability of the power grid.

4.2.1 Information Classification

After identifying all important information (or data) for any application, it should be properly classified. An organization handles lots of data that is created, used, and maintained. Information classification is required before an organization can decide what protection measures are necessary to be implemented. The primary purpose of data classification is to indicate the level of availability, integrity, and confidentiality protection. Information classification ensures data is protected in the most cost-effective manner. Each classification should have separate handling requirements and procedures pertaining to how data is accessed, used, and destroyed. In addition, layers of responsibility should be defined.

4.2.2 Information Classification Levels

The reason to classify information (data) is to organize it according to its sensitivity to loss, disclosure, or unavailability. DER applications and systems should follow an information classification process that provides a classification that is unique and separate from the other applications without any overlapping effects. The classification process should also outline how information is handled through its life cycle (from creation to consumption to disposition). All data handled must be reviewed for classification. Also, classification rules must apply to data in any format such as digital, video, fax, audio, paper, etc. Data and information should be classified by the business owner or designee.

Classification must be done in a consistent manner and following a sensitivity scheme that is agreed first. A sensitivity scheme could be based on two or more layers. Choosing the appropriate sensitivity scheme is based on the applications and business goals. Businesses (e.g. IT systems) are usually more concerned with the confidentiality and integrity of information, while power systems and control systems are more concerned with the availability and integrity of information. In general, organizations follow one of the data classification models: commercial or military. Commercial organizations use the levels of sensitivity from the highest to lowest as follows [Harris 2013]:

- Confidential
- Private
- Sensitive
- Public.

Military use the levels of sensitivity from the highest to lowest as follows [Harris 2013]:

- Top secret
- Secret
- Confidential
- Sensitive but unclassified
- Unclassified.

Although the industry uses a combination of these levels, commonly data classification levels in the commercial sector may include the following:

- For office use only
- Proprietary
- Privileged
- Private.

It is recommended that each organization should develop an information classification scheme that best fits its business and security needs. Because Smart Grid applications generate high volume of data, it is recommended to use a reasonable list of classifications that are also not too restrictive and detailed oriented and to allow a classification that avoids confusion and ambiguity.

Thus, data for DER applications and systems should be provided with a classification that is unique and separate from the others and not have any overlapping effects. Once the sensitivity scheme is decided, it is necessary to develop evaluation criteria for determining which information is assigned to each classification level. However, the classification of data is reviewed periodically, and changes may occur.

4.2.3 Information Evaluation Criteria

The evaluation criteria are based on a set of parameters. Examples of criteria parameters to determine the sensitivity of data or information may include:

- Usefulness
- Value
- Age
- Level of damage if the data were disclosed
- Level of damage if data were corrupted or modified
- Legal, regulatory, business, or contractual responsibility to protect the data
- Effects on national security
- Labeling and marking requirements
- Liability
- Operational impact

- Data dictionary review
- Access levels
- Physical security
- Storage and disposal
- Secrecy (needs for encryption)
- Priorities for security protection
- Competitive edge.

Besides data and information classification, an organization may need to classify DER applications and systems. For example, an application that processes sensitive information such as secret may need to be evaluated as secret. The processes and steps to accomplish data classification and evaluation criteria should be documented as procedures. It is important to understand that each data classification level requires specific security measures. Also, it is necessary to determine how data in these classifications is stored, maintained, transmitted, or disposed.

4.3 Infrastructure

Infrastructure is the underlying foundation or basic framework (as of a system or organization) [MW]. Infrastructure comprises physical and organizational structures needed for the operation of a society or enterprise including facilities necessary for an economy to function. It is the fundamental structure of a system or organization. The basic architecture of any system (electronic, mechanical, social, political, etc.) determines how it functions and how flexible it is to meet future requirements. Also, infrastructure is a structure that can support an industry, communications, information, etc. An infrastructure refers to the entire system of facilities, equipment, and support services that organizations need in order to function [Infrastructure].

According to [ISO 9001], section 7.1.3, the infrastructure can include buildings, equipment, utilities, and technologies (both hardware and software). An overview of information infrastructure and Smart Grid infrastructure helps to understand the context of cybersecurity for an information system.

4.3.1 Information Infrastructure

One goal of information security is the protection of information infrastructure as defined in [Audit 2005]:

> Information Security is the protection of information and information systems and encompasses all infrastructure that facilitate its use - processes, systems, services, and technology. It relates to the security of any information that is stored, processed or transmitted in electronic or similar form, and is also defined as the preservation of confidentiality, integrity and availability of information.

In IT and on the Internet, infrastructure is the physical hardware used to interconnect computers and users. Infrastructure includes the transmission media, including telephone lines, cable television lines, and satellites and antennas, and also the routers, aggregators, repeaters, and other devices that control transmission paths. Infrastructure also includes the software used to send, receive, and manage the signals that are transmitted. In some usages, infrastructure refers to interconnecting hardware and software and not to computers and other devices that are interconnected. However, to some IT users, infrastructure is viewed as everything that supports the flow and processing of information.

The term information infrastructure was introduced in the United States with the National Information Infrastructure Initiative in 1991 [NII]. In this initiative, the Internet is described as an information infrastructure shared by the users. Internet is defined as an infrastructure that supports the information society: the equipment, systems, applications, support systems, and so forth that are needed for operating in the information society. Related concepts include National Information Infrastructure, Global Information Infrastructure, and Defense Information Infrastructure (see definitions in Appendix A).

An information infrastructure is a shared, evolving, open, standardized, and heterogeneous installed base [Hanseth 2002]. The information infrastructure concept has changed the perspective from single organizations to organizational networks and from systems to infrastructures, allowing for a global and emergent perspective on information systems [Bygstad 2008]. This work regards information infrastructure as an ICT-based organizational form.

Apart from National Information Infrastructure and Global Information Infrastructure, most organizations own an information infrastructure. It is the underlying foundation, basic framework, or interconnecting structural elements that support an organization [ENISA Glossary].

Information infrastructure can support different functions such as information assurance (IA), information management, and information security management. Thus, these functions are often supported by different infrastructure subcategories: IA, information management, and information security management. A new trend on outsourcing services is cloud computing that requires its own infrastructure.

4.3.2 Information Assurance Infrastructure

The IA infrastructure is the underlying security framework that lies beyond an enterprise's defined boundary but supports its IA and IA-enabled products, its security posture, and its risk management plan. IA are measures that protect and defend information and information systems by ensuring their availability, integrity, authentication, confidentiality, and non-repudiation. These measures include providing for restoration of information systems by incorporating protection, detection, and reaction capabilities.

4.3.3 Information Management Infrastructure

The information management infrastructure provides processes and best practice for requirements analysis, planning, design, deployment, and ongoing operations management and technical support of an information system infrastructure. For example, an organization's management needs an infrastructure to support a variety of activities, including reliable communication networks to support collaboration between suppliers and customers, accurate and timely data and knowledge to gain business intelligence, and information systems to aid decision-making and support business processes.

Information systems evolve into information system infrastructures. In sum, organizations rely on a complex, interrelated information system infrastructure to effectively thrive in the ever-increasing, competitive digital world. The infrastructure management processes describe those processes within enterprise that directly relate to the information, computing, equipment, and software that is involved in providing services to users. According to [Jessup 2008], modern organizations rely heavily on their information system infrastructure that include components shown in Figure 4.1:

- Hardware
- Software
- Communications and collaboration
- Data and knowledge
- Facilities
- Human resources
- Services.

However, organizations need a secure and reliable information infrastructure, and maintaining such infrastructure is challenging, specifically for small organizations. Particularly, each component of this infrastructure has to be efficient and protected. This could be beyond their means because of the costs for maintaining and upgrading hardware and software, employing in-house experts for support, and so on. Thus, organizations big and small are turning to outside service providers for their infrastructure needs.

4.3.4 Outsourced Services

Organizations can use different services to support their infrastructure needs. Table 4.1 shows examples for different types of services that organizations may outsource to another party.

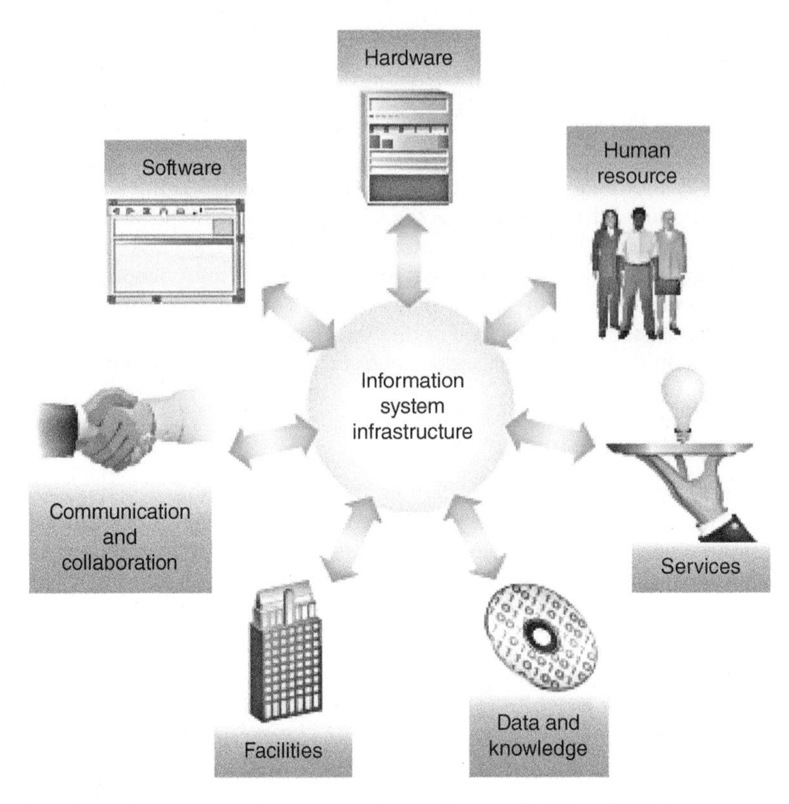

Figure 4.1 The information system infrastructure. *Source:* [Jessup 2008]. © 2008, Pearson.

Table 4.1 Examples of outsourced services.

Component	Service	Example
Hardware	Utility computing	Processing or data storage on needed basis
Software	Application service provider (ASP)	Application managed or hosted on an ASP's server (payroll, sales tax management and compliance, Web design)
Communication and collaboration	Videoconferencing	Conference rooms, meeting schedules
Data and knowledge	ASP	Data and knowledge stored on ASP by the provider
Facilities	Collocation facility	Space rented for servers in a collocation facility
Network	Network service provider	Yearly service or service per use
Back office	Customer support	Application support, Web maintenance, customer care
Projects	Construction management	New generation plant, adoption of new technologies, upgrading infrastructure

Source: Adapted from [Jessup 2008]. © 2008, Pearson.

One required solution is on-demand computing or on-demand software offered by another party. Outsourcing can help a company focus on its core processes without having to worry about supporting processes, such that outsourcing is typically limited to noncore business functions.

However, there are some noncore business functions that organizations tend to keep within their own realm [Jessup 2008], only very few organizations outsource information system security, as it is regarded as being critical for an organization's survival.

As there are pros and cons for outsourcing security [Messmer 2008], one survey concluded that while there is a market for outsourcing in some kinds of security task, the appetite for such outsourcing is not overall growing [Richardson 2008]. One argument is lack of competency and lack of customer focus. Third-party providers cannot really understand how a certain security problem impacts an organization's business [Messmer 2008]. Another argument is trust (risk of exposing confidential data, intellectual property).

Outsourcing can support a few discrete functions that include log monitoring or penetration testing [Messmer 2008]. The results of a survey show that while there is a market for outsourcing some kinds of security tasks (e.g. security testing of customer-facing Web applications), most respondents are keeping most of their operations internally based [Richardson 2008]. Another recent survey relates positive trends on increasing budgets for security operations and security program within an organization [CSI 2015]. This has helped those responsible for security get the increased resources they need to operate effectively.

An organization's information system infrastructure always needs to be secured to prevent it from inside or outside intruders. Therefore, the information security management infrastructure is vital to the mission of the organization's business.

4.3.5 Information Security Management Infrastructure

Information security management infrastructure is the combination of security technologies, procedures, and security provided to protect the information infrastructure.

Information security management infrastructure includes system components and activities that support security policy by monitoring and controlling security services and mechanisms, distributing information, and reporting security events [RFC 4949]. As argued in [Jessup 2008], no matter how organizations choose to manage their infrastructure, information system controls have to be put into place to control costs, gain and protect trust, remain competitive, or comply with internal or external governance (e.g. the Sarbanes–Oxley Act (SOX)). Such controls help ensure the reliability and security of the information. These controls may include a variety of measures, such as security policies and their implementation, access restrictions, or record keeping, to be able to trace actions for the operations and maintenance. The controls need to be applied throughout the entire information infrastructure. To be most effective, controls should be a combination of three types of controls: preventive, detective, and corrective.

Unlike traditional enterprises, most utilities do not have such infrastructures (information management infrastructure and information security management infrastructure), or it may not be feasible to expect that a small provider of a DER system could deploy one. The trend is to contract the security services from cloud providers. The trend to cloud computing is tempting to some organizations that want to reduce computing costs. It is not known if electrical industry would have preference for one model or it will embrace any of the cloud service models. Already third-party providers are providing demand and response services to utilities. Thus, a specific category of infrastructure is the cloud infrastructure.

4.3.6 Cloud Infrastructure

Cloud computing is a new way of delivering computing resources, not a new technology [ENISA 2009]. It is the result of combining load balancing techniques, virtualization, service-oriented architectures (SOA), application service providing, network emergence, and distribute computing. It is defined as a model for enabling convenient, on-demand network access to a shared pool of configurable computing resources (e.g. networks, servers, storage, applications, and services) that can be rapidly provisioned and released with minimal management effort or service provider interaction [Mell 2009]. The cloud model promotes availability and is composed of five essential characteristics, three service models, and four deployment models.

Service models include the following [NIST SP500-292]:

- Cloud software as a service (SaaS) is the capability provided to the consumer to use the provider's applications running on a cloud infrastructure.
- Cloud platform as a service (PaaS) is the capability provided to the consumer to deploy onto the cloud infrastructure consumer-created or acquired applications created using programming languages and tools supported by the provider.
- Cloud infrastructure as a service (IaaS) is the capability provided to the consumer to provision processing, storage, networks, and other fundamental computing resources where the consumer is able to deploy and run arbitrary software, which can include operating systems and applications.

Figure 4.2 is a pictorial view of service models available to a cloud consumer.

Deployment models include the following [NIST SP500-292]:

- Private cloud allows the cloud infrastructure to be operated solely for an organization; it may be managed by the organization or a third party and may exist on premise or off premise.
- Community cloud allows sharing the cloud infrastructure with several organizations and supports a specific community that has shared concerns (e.g. mission, security requirements, policy, and compliance considerations); it may be managed by the organizations or a third party and may exist on premise or off premise.
- Public cloud makes the cloud infrastructure available to the general public or a large industry group and is owned by an organization selling cloud services.
- Hybrid cloud provides the cloud infrastructure that is a composition of two or more clouds (private, community, or public) that remain unique entities but are bound together by standardized or

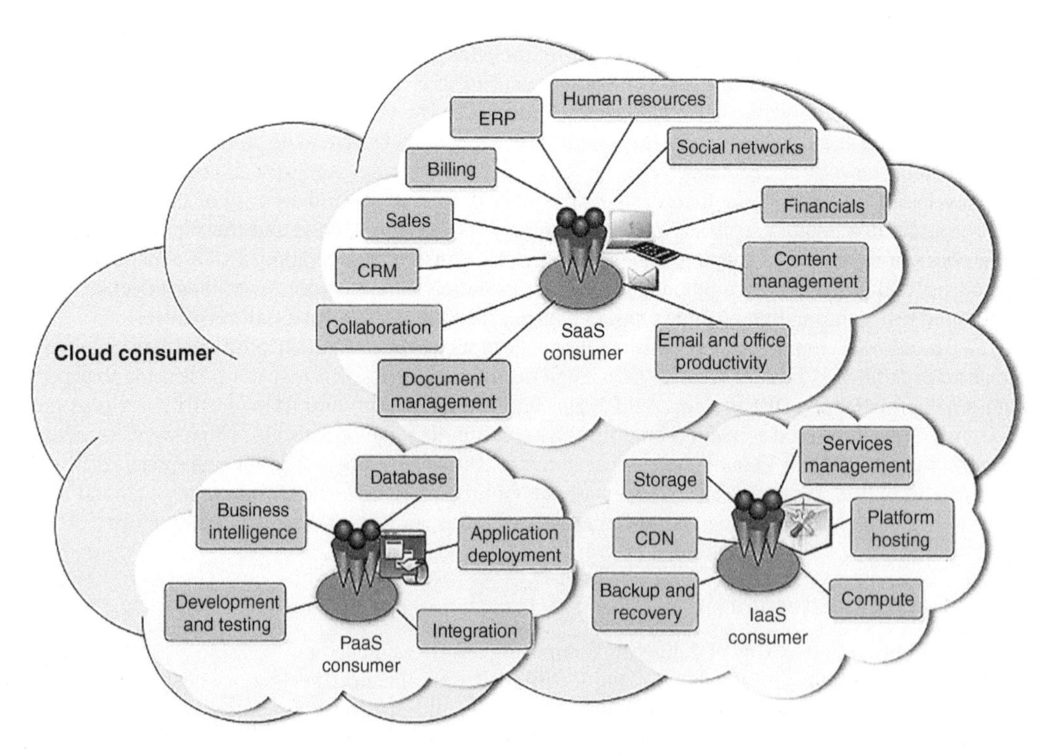

Figure 4.2 Examples of cloud services. *Source:* [NIST SP500-292]. Public Domain.

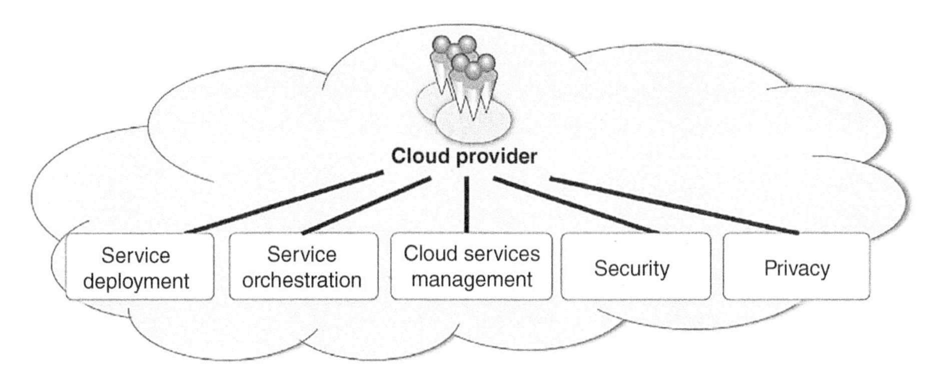

Figure 4.3 Cloud provider major activities. *Source:* [NIST SP500-292]. Public Domain.

proprietary technology that enables data and application portability (e.g. cloud bursting for load balancing between clouds).

A cloud provider's activities can be described in five major areas of service deployment, service orchestration, cloud service management, security, and privacy (see Figure 4.3).

While organizations may be offered some advantages by using cloud services, it is still controversial whether it helps or hurts security of the consumer. Though cloud computing provides a flexible solution for shared resources, software, and information, it also poses additional privacy challenges to consumers using the clouds. One aspect is that data is centralized and security resources are focused, but direct control of sensitive data is lost, and the complexity of securing dynamic and distributed environments can be overwhelming [Harris 2013]. Other challenges include the following:

- Security is a crosscutting aspect of the architecture that spans across all layers of the reference model (see [NIST SP500-292]), ranging from physical security to application security. However, security in cloud computing architecture is not solely under the responsibility of the cloud providers but also cloud consumers and other relevant actors.
- Cloud-based systems still need to address security requirements such as authentication, authorization, availability, confidentiality, identity management, integrity, audit, security monitoring, incident response, and security policy management; however there is no strict model of security features that are needed for each cloud service.
- Service models present consumers with different types of service management operations and expose different entry points into cloud systems, which in turn also create different attacking surfaces for adversaries.
- Variations of cloud deployment models have important security implication as well; a private cloud is dedicated to one consumer organization, whereas a public cloud could have unpredictable consumers coexisting with each other; therefore, workload isolation is less of a security concern in a private cloud than in a public cloud.
- Cloud provider and the cloud consumer have differing degrees of control over the computing resources in a cloud system. Security is a shared responsibility. Security controls, e.g. measures used to provide protections, need to be analyzed to determine which party is in a better position to implement.
- Privacy and compliance are not straightforward matters; it is not guaranteed that personal information is not exposed to threats.

A generic high-level conceptual model presents the NIST cloud computing reference architecture in [NIST SP500-292]. The reference architecture (see Figure 4.4) is a tool for discussing the requirements, structures, and operations of cloud computing. The NIST cloud computing reference architecture focuses on the requirements of what cloud services provide, not how to design solution and implementation. The reference architecture identifies the major actors and their activities and functions in cloud computing.

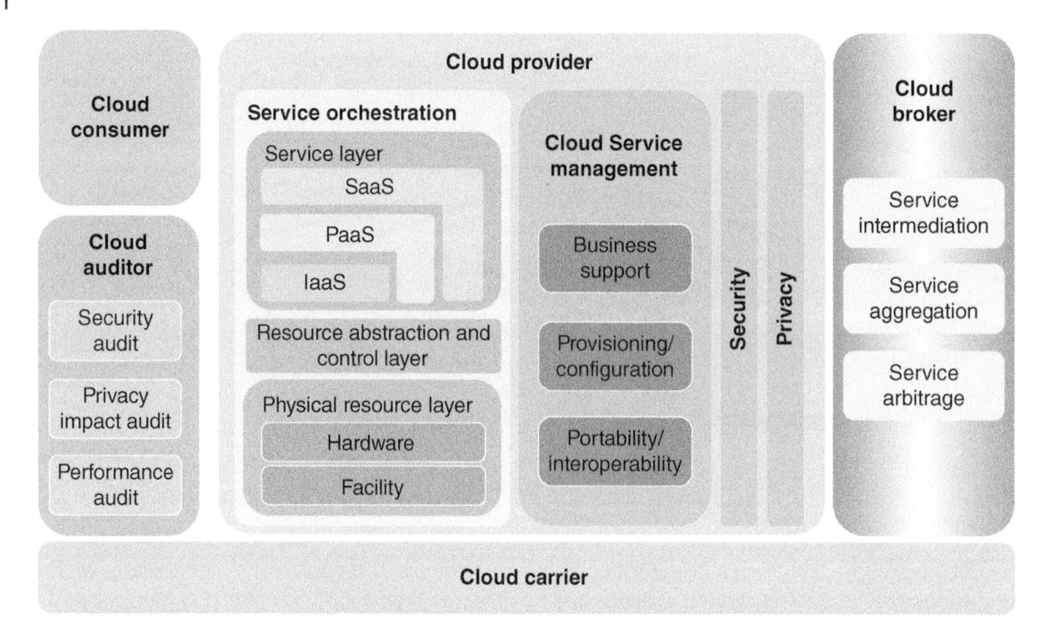

Figure 4.4 The conceptual cloud computing reference model. *Source:* [NIST SP500-292]. Public Domain.

An analysis of security concerns and solutions associated to cloud computing is described in [Gonzalez 2011]. The focus of the analysis is on classification and quantification of many security concerns that still lack a solution. While there are other authoritative references (e.g., risk assessment by ENISA [Catteddu 2009], security guidance [CSA 2011]), it is argued that there is no focus on quantifying problems related to cloud computing, encompassing from data privacy to configuration to infrastructural configuration [Gonzalez 2011]. This work provides a comparison of different guidelines [NIST SP500-292], [Catteddu 2009], [CSA 2011] including the methodology for quantifying the security problems. It shows a ranking and weight of various concerns based on published material. Also, the authors suggest a taxonomy of security architecture, security compliance, and privacy in the cloud. An updated of ENISA risk assessment presented in [Catteddu 2009] is published in this document [ENISA 2012].

As cloud computing economic model meets its promise to reduce costs and security challenges are being addressed, there is the possibility that some Smart Grid applications will be successfully running in the cloud. However, the outsourcer needs to verify the outsourcing market maturity using an outsourcing decision framework based on, core competency, asset specificity, and market maturity before signing any agreement [Zhang 2008]. Several issues both from clients and from providers should be addressed properly. In addition, technical suitability of cloud services to DER systems can be impeded by the security, privacy, and compliance issues that surround it.

4.4 Smart Grid Infrastructure

A Smart Grid uses digital technology to improve reliability, security, and efficiency (both economic and energy) of the electric system from large generation through the delivery systems to electricity consumers and a growing number of distributed generation and storage resources [DOE/OEDER 2008]. The electrical grid systems have four elements: electricity generation plants, transmission substations, distribution substations, and end users. Power bulks (or plants) generate power from various sources for distribution. Electric grid structure is strongly related to industry structure.

Therefore, the traditional electrical structure may be summarized as follows: one way energy flow from central station generators, over a transmission network, is through substations onto distribution systems

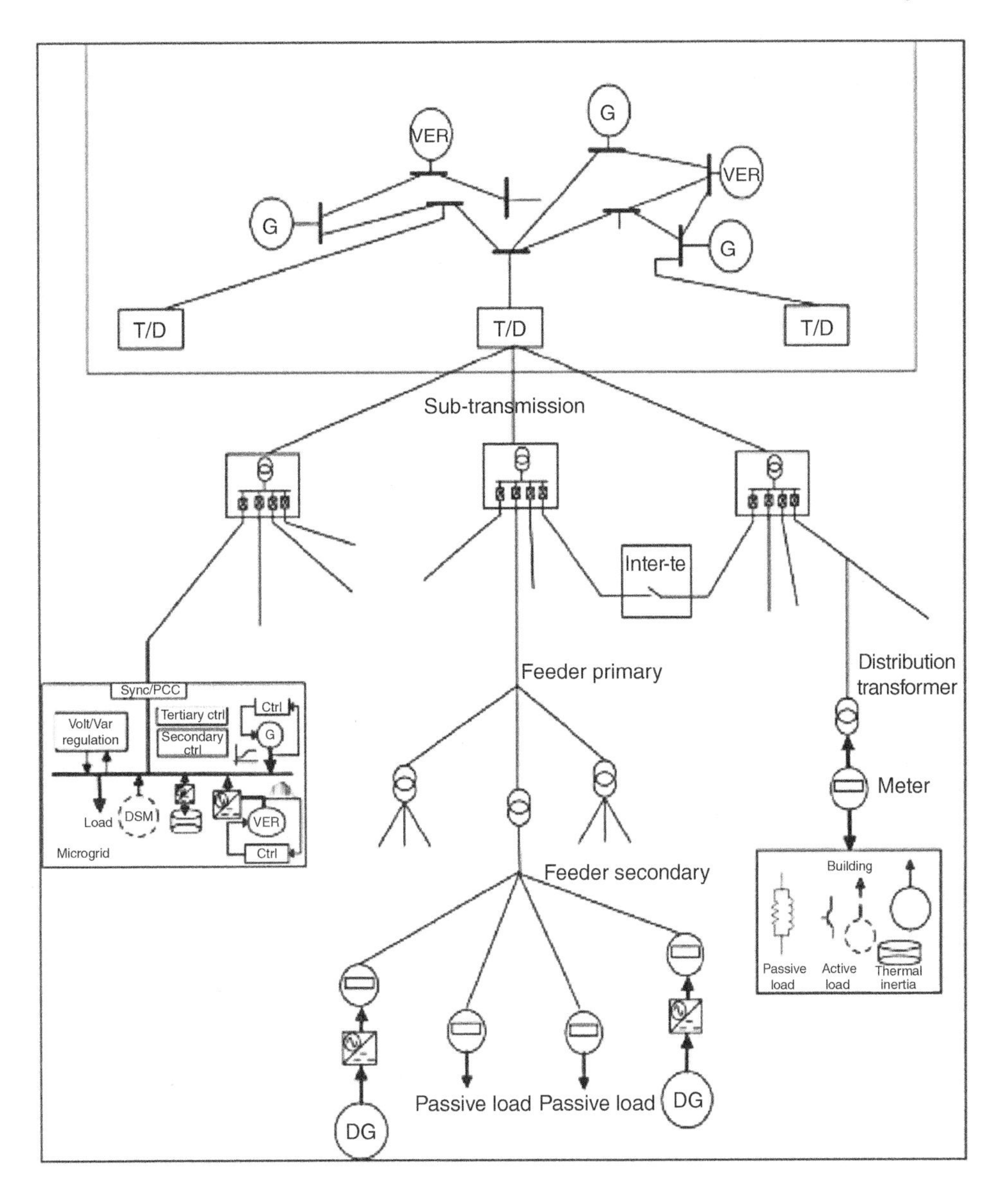

Figure 4.5 Smart Grid basic electric structure. *Source:* [PNNL 2015]. Public Domain.

over radial distribution circuits to end-use customers. Figure 4.5 provides a basic model (hierarchical view) for grid topology. Many of the recent changes to power grids have been built in a bottom-up manner, as opposed to being designed from a systems standpoint, partly due to the enormous legacy investments in infrastructure. The approach has led to decreasing the stability and, in fact, be chaotic [PNNL 2015]. As described in a report [DOE 2015], the grid is viewed as six interrelated structures and a coordination framework to understand the needs and requirements necessary to meet the performance expectations of a digital economy. The grid architecture structure types include:

- Electric infrastructure (circuit topology, generation, load composition).
- Industry structure (networks [operations, planning, markets], regulatory [federal, state, industry, other]).
- Digital infrastructure (networking, processing, persistence).

- Control structure (protection, control, synchronization).
- Convergent networks (fuels, transportation, social).

As the power approaches customers' homes, it is stepped down again to the voltage necessary for home use. Finally, home appliances access power through their electric meters.

Areas of the electric system that cover the scope of a Smart Grid include the following [DOE 2009]:

- Delivery infrastructure (e.g. transmission and distribution lines, transformers, switches).
- End-use systems and related DER (e.g. building and factory loads, distributed generation, storage, electric vehicles).
- Management of the generation and delivery infrastructure at the various levels of system coordination (e.g. transmission and distribution control centers, regional reliability coordination centers, national emergency response centers).
- Information networks themselves (e.g. remote measurement and control communication networks, inter- and intra-enterprise communications, public Internet).
- Financial and regulatory environment that fuels investment and motivates decision makers to procure, implement, and maintain all aspects of the system (e.g. stock and bond markets, government incentives, regulated or nonregulated rate of return on investment).

4.4.1 Hierarchical Structures

Based on a hierarchical structure definition, Smart Grid is viewed as a hierarchical structure [Farhangi 2010] of layers [Gao 2012]. Smart Grid applications are on the top to create electrical system/societal values. Figure 4.6 shows this structure, a pyramid view of Smart Grid infrastructure consisting of layers with security crosscutting layers from the bottom up to the top layers. In [Gao 2012], security is in another dimension and covers all layers. However, security should be provided for all layers, and it should be embedded in each layer as depicted in this figure. Each layer is defined as follows:

- At the bottom layer is the physical energy infrastructure that distributes energy and asset management framework.
- Both fundamental applications and various infrastructures constitute Smart Grid foundations [Farhangi 2010]. Smart communication infrastructures are defined on the top of the physical energy infrastructure to entire supply chain. Computing/IT is above the communication infrastructures for timely decision making. Fundamental applications layer is composed of smart meters, meter data management

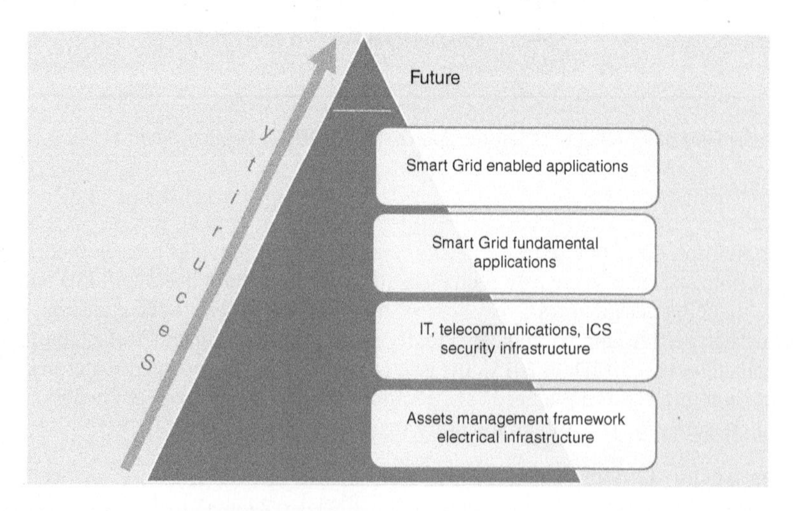

Figure 4.6 A view of Smart Grid components evolution.

system, distribution application, distribution management system, substation automation, phasor measurements, security measurements, security logging, monitoring, reporting, and control.
- Enabled applications are structured in two layers. The bottom layer includes applications such as home area network, plug-in electric vehicles, energy storage, distributed generation (e.g. renewables – solar, wind), Volt/VAR optimization applications, microgrid, and intelligent decision support systems. The above layer includes applications such as demand/response, operational efficiency, and security metrics.

As most of the issues on power systems are observed at the distribution level, the Smart Grid implementation requires a modernization of the grid to address potential issues and to have a pervasive control to prevent them [Farhangi 2010]. Information and communication technologies and a diverse infrastructure provide the foundations of the Smart Grid to support the fundamental applications.

Although there are many definitions for Smart Grid, one common understanding is that Smart Grid needs to be integrated with an information communication infrastructure in order to be smarter [GAO 2011]. Also, power system requires reliability, scalability, manageability, and extensibility but also should be interoperable, secure, and cost effective. Inadequacies in the power delivery system are manifested in the form of poor reliability, excessive occurrences of degraded power quality, vulnerability to mischief or terrorist attack, the inability to integrate renewables, and the inability to provide enhanced services to consumers [EPRI 2010].

4.4.2 Smart Grid Needs

One goal of Smart Grid is to maximize the throughput of the system and to reduce consumption of the system. To achieve this goal, Smart Grid requires an information infrastructure to meet the needs for real time, reliability, scalability, manageability, and extensibility, interoperability, security, and costs.

Therefore, the Smart Grid information infrastructure includes control systems, intelligent devices, and communication networks that keep track of electricity flowing in the grid. Closer to home, Smart Grid devices (intelligent devices) are currently being integrated into energy delivery ICS. These new devices directly control utility meters that allow energy flow into homes and track individual household energy consumption. IT is increasingly being adapted to support OT in utilities so that operating systems, computer platforms, and networks commonly used in IT are now found in some OT architectures [Hawk 2014]. In real world, the common infrastructures are shared and integrated. Further, many issues related to the old infrastructure need to be addressed. The fragmented nature of the national power grid information infrastructure complicates matters.

With the implementation of the Smart Grid, the control system infrastructure, IT infrastructure, and telecommunication infrastructure have become more important to ensure the reliability and security of the electric sector. For example, control centers have sophisticated monitoring and control systems. A control center may have SCADA and DCS systems that monitor the flow of electricity from generators through transmission and distribution lines. These electronic systems enable efficient operation and management of electric systems through the use of automated data collection and equipment control.

However, the communication systems must be carefully designed to support the goals and requirements of the power grid. It is argued that the survivability of the grid must not be implemented and assessed with power and communication in isolation; rather, a combined analysis must be conducted [Bakken 2000].

4.4.3 Cyber Infrastructure

Strategies and requirements to build a cyber infrastructure are described in [NISTIR 7628r1]. A cyber infrastructure includes electronic information and communication systems and services and the information contained in these systems and services. Information and communication systems and services are composed of all hardware and software that process, store, and communicate information, or any combination of all of these elements. Processing includes the creation, access, modification, and destruction of information. Storage includes paper, magnetic, electronic, and all other media types. Communications include sharing and distribution of information. Included in the cyber infrastructure are control centers, IA infrastructure, and information security management infrastructure. Smart Grid cyber infrastructure includes control systems, sensors, and actuators. For example, computer systems, control systems including SCADA, networks, Internet, and cyber services (e.g. managed security services) are part of cyber infrastructure [NISTIR 7298r3].

This definition is also referenced in the National Infrastructure Protection Plan (NIPP) [NIPP 2013] to ensure a common understanding with the critical infrastructure. While this definition provides only information about the components of the cyber infrastructure, the needs and challenges have to be understood too. For example, the needs for real-time grid awareness application require collection of data from sensors and dissemination through highly available, flexible, open, secure two-way communication infrastructures to any point in the grid.

The focus of the industry effort so far has been mostly on the interoperability of the communication and information model, as suggested by the National Institute of Standards and Technology (NIST) Smart Grid Interoperability standard roadmap and the International Electrotechnical Commission (IEC) documents on Smart Grid standardization. In understanding the Smart Grid from definition to deployment, the Edison Electric Institute rightly suggested that advanced controls provide the "smart" in Smart Grids [Edison]. The three value generators for the Smart Grid are application, application, and application. To enable smart applications, there is a need not only for good business logic, control, and optimization theory, but there is need for new hardware and software components that can control power flows in the network, as well as the output and the consumption of power [DOE 2015]. Thus, a carefully integrated set of Smart Grid technologies can decrease the costs and risks of integrating distributed renewables into electricity systems [IRENA 2013].

4.4.4 Smart Grid Technologies

In some terms, a wide range of communication, information management, and control technologies that contribute to the efficiency and flexibility of an electricity system's operation are called Smart Grid technologies. These technologies can be used in any of these functional categories [IRENA 2013]:

- Information collectors support data acquisition from various types of sensors that generally measure performance-related characteristics of electricity system components; examples include meters that continually measure the power and electricity output of a distributed renewable generator; sensors that track temperature, vibration, and other characteristics of a transformer; and meters that measure the electricity characteristics (voltage, current, etc.) of a distribution line.
- Information assemblers, displayers, and assessors category includes devices that accept information and display information and/or analyze information.
- Information-based controllers are devices that receive information and use it to control the behavior of other devices to achieve some goals, such as voltage stabilization.
- Energy/power resources include technologies that can generate, store, or reduce demand for electricity.

Often, Smart Grid technologies are implemented in stages, with each stage requiring a business plan for regulators to approve.

The most relevant characteristics of a Smart Grid include the following:

- Self-healing from power disturbance events.
- Enabling active participation by consumers in demand response.
- Operating resiliently against both physical and cyber attacks.
- Providing quality power that meets twenty-first-century needs.
- Accommodating all generation and storage options.
- Enabling new products, services, and markets.
- Optimizing asset utilization and operating efficiency.

These characteristics can be achieved through the application of a combination of existing and emerging technologies. Thus, the four essential technology layers and building blocks of the Smart Grid can be defined as:

- Decision intelligence.
- Communication/information.
- Sensor/actuator.
- Power conversion/transport/storage/consumption.

The smartness of the Smart Grid lies in the decision intelligence layer, which is made up of all the computer programs that run in relays, intelligent electronic devices (IEDs), substation automation systems, control centers, and enterprise back offices. These programs process the information collected from the sensors or disseminated from the communication and IT systems; they then provide control directives or support business process decisions that manifest themselves through the physical layer.

For the decision intelligence layer to work, data (information) need to be propagated from the devices connected to the grid to the controllers that process the information and transmit the control directives back to the devices. The communication and IT layer perform this task. The IT layer serves to provide responsive, secure, and reliable information dissemination to any point in the grid where the information is needed by the decision intelligence layer. In most cases, this means that data are transferred from field devices back to the utility control center, which acts as the main repository for all the utility's data.

Device-to-device (e.g. controller to-controller or IED-to-IED) communication, however, is also common, as some real-time functionality can only be achieved through inter-device communication.

The physical layer is where the energy is converted, transmitted, stored, and consumed. Solid-state technology, power electronics-based building blocks, superconducting materials, new battery technologies, and so on all provide fertile ground for innovations.

Interoperability and security are essential capabilities to assure ubiquitous communication between systems of different media and topologies and to support plug and play for devices that can be autoconfigured when they are connected to the grid, without human intervention.

For example, energy management systems monitor the energy consumption data from private residences to office buildings. These systems can determine the consumption patterns and preferences of the consumers, as well as real-time conditions (e.g. market prices, grid stress). They may use the collected information to autonomously interact with the grid to determine the charging and discharging cycles of various appliances to dynamically balance load and resources and maximize energy delivery efficiency and security in real time.

Now, more than ever, the electric power industry must play an equally active and coordinated role in defining and developing the hardware, software, standards, and protocols that are needed to realize a secure and interoperable nationwide Smart Grid.

4.5 Building an Information Infrastructure for Smart Grid

Although a final and universal Smart Grid architecture is not agreed yet, many developments are based on various architectures as provided by standards organizations, professionals of different countries, or academia. On the other hand, many challenges on the development of Smart Grid are the result of various requirements, financial needs, approaches, and engineering skills.

Smart Grid deployment must include not only technology, market and commercial considerations, environmental impact, regulatory framework, standardization usage, information and communication technology (ICT), and migration strategy but also societal requirements and governmental edicts [Nordell 2012]. However, the Smart Grid is a vision, and achieving the full benefits requires approaches to build the information infrastructure.

Thus, the path or road to Smart Grid or future grid is discussed from different perspectives and at many levels, from governments to customers. As we discuss issues about the Smart Grid in the United States, it is important to understand that other countries may have a different agenda for Smart Grid including a different power grid architecture when the modernization starts. As several Smart Grid projects are initiated in the United States and other parts of the world, we discuss in general about the infrastructure issues as published in literature. How to realize the Smart Grid infrastructure is not the intent of this section. One view is that Smart Grid is grid edge with drivers to include distributed generation, customer engagement, and grid modernization [Lacey 2013].

We describe only relevant issues that can be useful to a security engineer in understanding the scope of Smart Grid and developments for infrastructure, specifically the context and foundations for an information security program.

4.5.1 Various Perspectives

The Smart Grid is a network of networks, including power, communications, and intelligence [EPRI 2011]. That is, many networks with various traditional ownership and management boundaries are interconnected to provide end-to-end services between stakeholders and among IEDs.

Therefore, a Smart Grid infrastructure aims to support a variety of applications that may require its own infrastructure or developments and updates to existing infrastructure. Many Smart Grid applications are built on new infrastructure; some may be built upon the existing IT infrastructure requiring upgrades. IT infrastructure consists of the equipment, systems, software, and services used in common across an organization, regardless of mission/program/project. For example, an earlier assessment indicates that there is already a significant installed base of sensors at substations, but there is still limited bandwidth connecting the substation to the enterprise [EPRI 2011]. IT infrastructure also serves as the foundation upon which selected Smart Grid applications and mission/program/project-specific systems and capabilities are built. Examples of functionalities that are critical for ongoing and near-term deployments of Smart Grid technologies and services include the following [NIST SP1108r3]:

- Demand response and consumer energy efficiency is necessary for optimizing the balance of power supply and demand.
- Wide area situational awareness is to help understanding and anticipating behavior and performance of power network components and ultimately optimize the management of power network components.
- DER and DER systems enable advanced functionalities to ensure power quality and grid stability.
- Energy storage new capabilities especially, for distributed storage, benefit the entire grid, from generation to end use.
- Electric transportation enables an increasing use of renewable sources of energy that help ameliorate peak load demands.
- Network communications include development and upgrades as well as common semantic framework (e.g. agreement as to meaning) and software models for enabling effective communication and coordination across intersystem interfaces.
- Advanced metering infrastructure (AMI) enables near real-time monitoring of power usage and many other applications (e.g. dynamic pricing).
- Distribution grid management and the automation of distribution systems enable efficient and reliable operation of the overall power system.
- Cybersecurity encompasses measures to ensure the protection of the information used for the management and operation of power grid.

Although these functionalities are priority areas that have to be considered, utilities need to prepare feasible and cost-effective plans for the implementation of these applications according to their needs and jurisdictions restrictions. In addition, the foundational technologies and smart technologies for the Smart Grid need to be included. Examples of technologies include communications, sensors, distributed generation, power electronics, and controls [EPRI 2011].

Although no formal definition, smart technologies are real-time, automated, interactive technologies that optimize the physical operation of appliances and consumer devices for metering, communications concerning grid operations and status, and distribution automation (DA) [EPRI 2011]. The Smart Grid goal is simply to make the grid more intelligent (the ability to recognize and adapt to changing conditions [EPRI 2011]).

Generally, Smart Grid is a data communication network integrated with the electrical grid that collects and analyzes data captured in near real time about power transmission, distribution, and consumption. Based on these data, Smart Grid technology then provides predictive information and recommendations to utilities, their suppliers, and their customers on how best to manage power.

As argued in [Schoechle 2012], the Smart Grid in the United States evolves from, first, emerging Smart Grid using ICT as means to improving electricity reliability, to more improving efficiency, to reducing pollution, and to incorporating more renewable generation. Then, a recent grid modernization initiative calls for accelerating the path to Smart Grid developments. Grid modernization vision is a future grid that will solve the challenges of seamlessly integrating conventional and renewable sources, storage, and central and distributed generation. An information network is fundamental to support the FERC priorities and all functionalities needed for Smart Grid.

An assessment of the enterprise back-office systems revealed that these systems need additional features in order to enable the Smart Grid. Medium and large utilities need complete systems of their own. Small utilities may aggregate their needs or use service providers. Upgrades and new developments are needed to make a robust communication infrastructure in response to requirements for system operators (independent system operators (ISOs), transmission system operators (TSOs), and other independent operators) to incorporate increasing functionality [EPRI 2011]. Figure 4.7 shows the key components of the ISO infrastructure.

Briefly described in [NIST SP1108r1], an information network is a collection, or aggregation, of interconnected computers, communication devices, and other ICT. Technologies in a network exchange information and share resources.

Figure 4.8 shows a diagram of communications for Smart Grid as envisioned by IEEE [Adams 2013]. The diagram is a comprehensive view of networks organized as a communication network management layer that supports the security layer. This view shows the various networks utilized by the power system from generation to energy delivery. The Smart Grid consists of many different types of networks, not all of which are shown in the diagram. The networks include the enterprise bus that connects control center applications to markets, generators, and with each other; wide area networks that connect

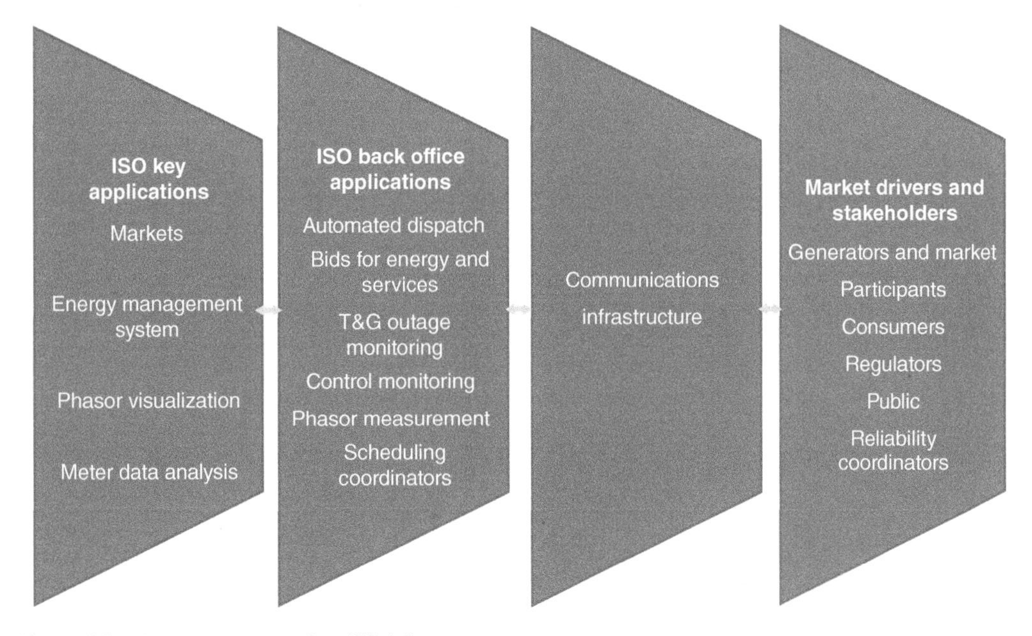

Figure 4.7 Key components of an ISO infrastructure.

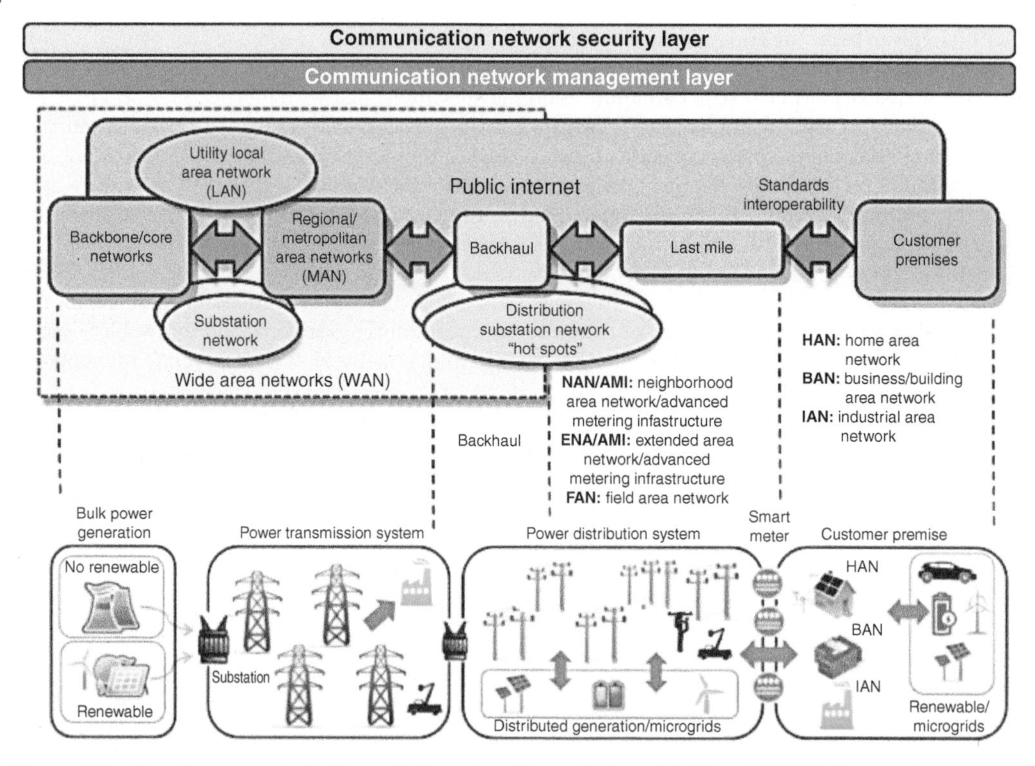

Figure 4.8 Complex networks in Smart Grid market. *Source:* [Adams 2013]. © 2013, IEEE.

geographically distant sites; field area networks that connect devices, such as IEDs that control circuit breakers and transformers; and premises networks that include customer networks as well as utility networks within the customer domain. These networks may be implemented using public (e.g. the Internet) and nonpublic networks in combination. There are instances where communications could use the public networks including Internet. Examples may include customer to third-party providers, bulk generators to grid operators, markets to grid operators, and third-party providers to utilities.

Both public and nonpublic networks require implementation and maintenance of appropriate security and access control to support the Smart Grid applications. Thus, installing the right solutions requires a change on security professionals' view that the survival of the mission depends on the ability of the network to provide continuity of service, even though degraded, in the presence of attacks, failures, or accidents.

Moreover, electricity infrastructure is interconnected with energy, information, telecommunications, transportation, and financial infrastructures, thus posing new challenges for their secure, reliable, and efficient operation. All of these infrastructures are complex networks – geographically dispersed, nonlinear, and interacting both among themselves and with their human owners, operators, and users [Amin 2012]. Figure 4.9 suggests that achieving Smart Grid goals of power system reliability and power quality requires a complex of interconnected networks to include integrated communications, compatible devices and appliances, and secure energy infrastructure.

The key challenge is to enable secure and very high confidence sensing, communications, and control of a heterogeneous, widely dispersed, yet globally interconnected system. It is even more complex and difficult to control it for optimal efficiency and maximum benefit to the ultimate consumers while still allowing all its business components to compete fairly and freely. In the electric power industry and other critical infrastructures, new ways are being sought to improve network efficiency by eliminating congestion problems without seriously diminishing reliability and security. Nevertheless, the goal of transforming the current infrastructures into self-healing energy delivery, computer, and communication networks with unprecedented robustness, reliability, efficiency, and quality for customers and society is ambitious [Amin 2012].

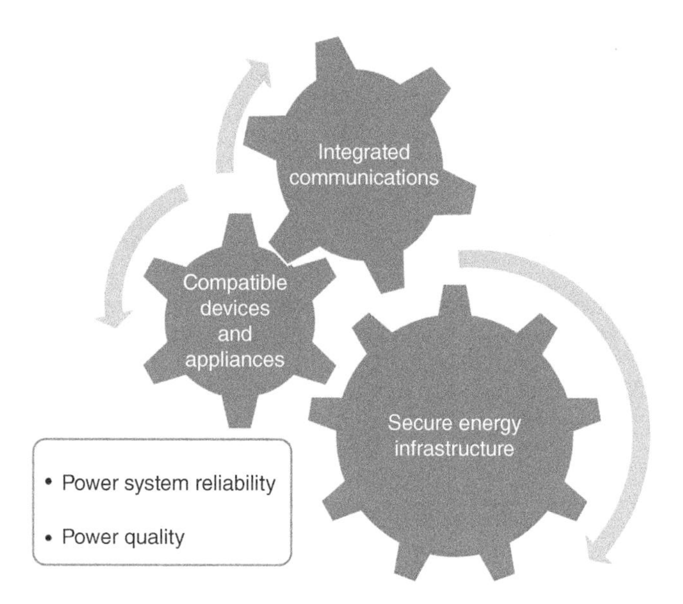

Figure 4.9 Smart Grid: a complex of interconnected components.

Several published surveys and recent research reveal novel applications and recommendations for the transition of power grid to Smart Grid. Therefore, no single approach can resolve challenges of Smart Grid implementation. A few potential challenges and relevant approaches related to Smart Grid integration [Ruth 2014] and building an information infrastructure should still be considered.

4.5.2 Challenges and Relevant Approaches

The availability of new technologies such as distributed sensors, two-way secure communication, advanced software for data management, and intelligent and autonomous controllers has opened up new opportunities for changing the energy system [Gharavi 2011]. Ultimately, the dynamic energy system is a driver for changes to information infrastructure and information security. Energy system integration (ESI) is the process of optimizing energy systems across multiple pathways and scales. Although each utility may take integrating its system with different Smart Grid functionalities, utilities rethink the distribution system to include the integration of high levels of DER.

However, strategic integration (DERs) not only requires interconnection but also plans to maximize the benefit these technologies and resources can provide to society while supporting secure and reliable electric power system operations. The modernization of power system in the United States requires [EPRI 2011]:

- Automation – At the heart of a smart power delivery system.
- Communication architecture – The foundation of the power delivery system of the future and the enabler of Smart Grid integration.
- DER and storage development and integration.
- Power electronics-based controllers and widely dispersed sensors throughout the delivery system.
- An AMI.
- End-user infrastructure.

US grid modernization vision is a future grid that will solve the challenges of seamlessly integrating conventional and renewable sources, storage, and central and distributed generation. It will deliver reliable, affordable, secure, resilient, and clean electricity to consumers where they want it, when they want it, and how they want it [Lynn 2015].

The grid integration initiative addresses challenges associated with ESI and the physical operation of the power system when these technologies are deployed at scale. Seamlessly integrating these technologies into the grid in a safe, reliable, and cost-effective manner is critical to enable deployment at scale. ESI is the process of optimizing energy systems across multiple pathways and scales.

The objective for security is minimization of disruptions to grid infrastructure caused by natural or human-induced disturbances, including cyber attacks, and mitigation of the impacts of disruptions, including economic impacts [MYPP 2014]. The Smart Grid R&D Multi-Year Program Plan (MYPP) was first published in May 2010 and had its first annual update in September 2011 [MYPP 2012].

As these MYPP plans continue to develop, system planners need to conduct a thorough analysis of all the cost-effective energy efficiency, load modifying, storage, and flexible generation resources that can avoid costlier investments in communications and information infrastructure. Therefore, cost–benefit analyses are necessary.

A sensitivity analysis as described in [Liu 2013] overviews the design of a Smart Grid to meet the demand of costs. It is needed to understand the economic indicators as optimal objectives and design variables to show the ways of economic analysis and cost savings of a Smart Grid project. The cost–benefit analysis is critical when implementing an infrastructure and cybersecurity. Also, it is required to understand the Smart Grid goals.

Several application domains such as AMI, smart home, distributed generation/microgrids, and wide area measurement and control are also key in driving the Smart Grid's development [Sollecito 2009]. Figure 4.10 shows potential DER technologies that require interconnection, communication, and information infrastructure.

However, utilities need to achieve a level of integration before they can implement any smart applications. The absence of industry-wide standards and blueprints for Smart Grid integration is compounding to the slow process of integration besides some hurdles such as technological, organizational, cultural, and business issues. The diversity of views on building the Smart Grid can only be attributed to the fact that there is more than one way to integrate a smarter grid. The ultimate Smart Grid is a vision. It is a loose integration of complementary components, subsystems, functions, and services under the pervasive control of highly intelligent management and control systems. Given the vast landscape of the Smart Grid research, different researchers may express different visions for the Smart Grid due to

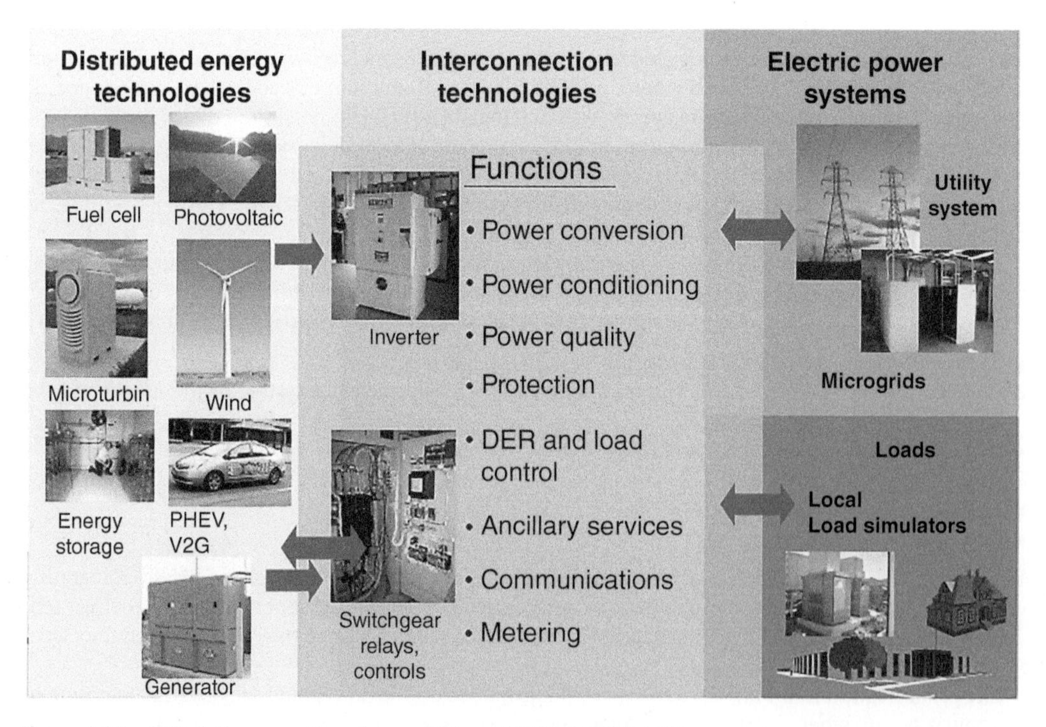

Figure 4.10 DERs interconnection. *Source:* [Adams 2013]. © 2013, IEEE.

different focuses and perspectives [Fang 2012]. Thus, different approaches can result on building the information infrastructure in different ways.

A survey of the technologies and Smart Grid projects/programs/trials reveals a trend toward three major technical systems [Fang 2012]:

- Smart infrastructure system includes energy, information, and communication infrastructure underlying support for advanced electricity generation, delivery, and consumption; advanced information metering, monitoring, and management; and advanced communication technologies.
- Smart management system is a subsystem in Smart Grid that provides advanced management and control services.
- Smart protection system is a subsystem in Smart Grid that provides advanced grid reliability analysis, failure protection, and security and privacy protection services.

Further, the smart infrastructure is divided into three subsystems: smart energy subsystem, smart information subsystem, and smart communication subsystem. Each subsystem is revised, and the scope is extended as shown in Figure 4.11. For example, energy subsystem includes microgrids and vehicle to grid/grid to vehicle (V2G/G2V); aspects of vehicle integration are discussed in [Tabari 2016]. Specific infrastructure needs are described in [Sarangi 2012].

An approach based on a holistic view is presented in [Farhangi 2014]. The author argues for the importance of developing system integration maps based on a utility's strategic Smart Grid development roadmap prior to making such large investments in their assets and infrastructure. Also, shortcomings of the smart meter strategy and proposal of a roadmap for transformation are described in [Schoechle 2012].

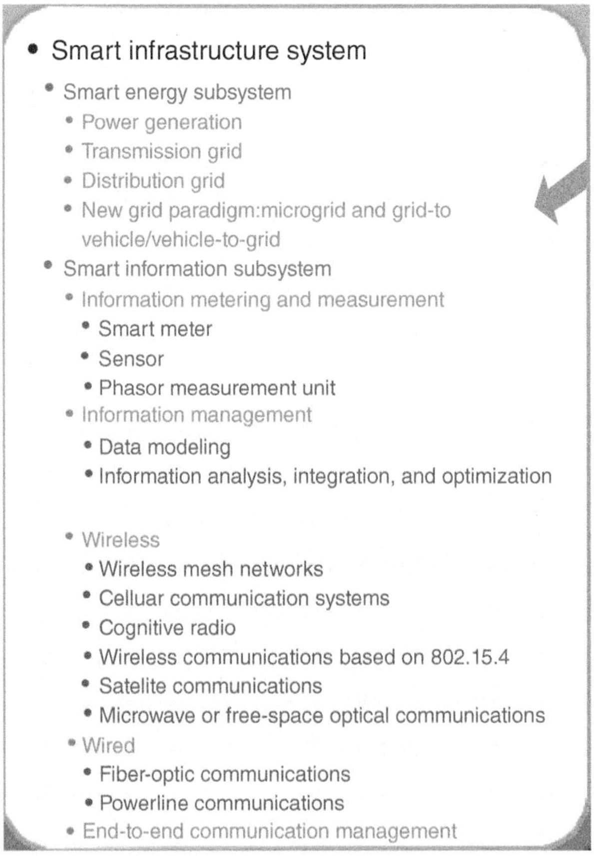

Figure 4.11 A smart infrastructure view. *Source:* [Fang 2012]. © 2012, IEEE.

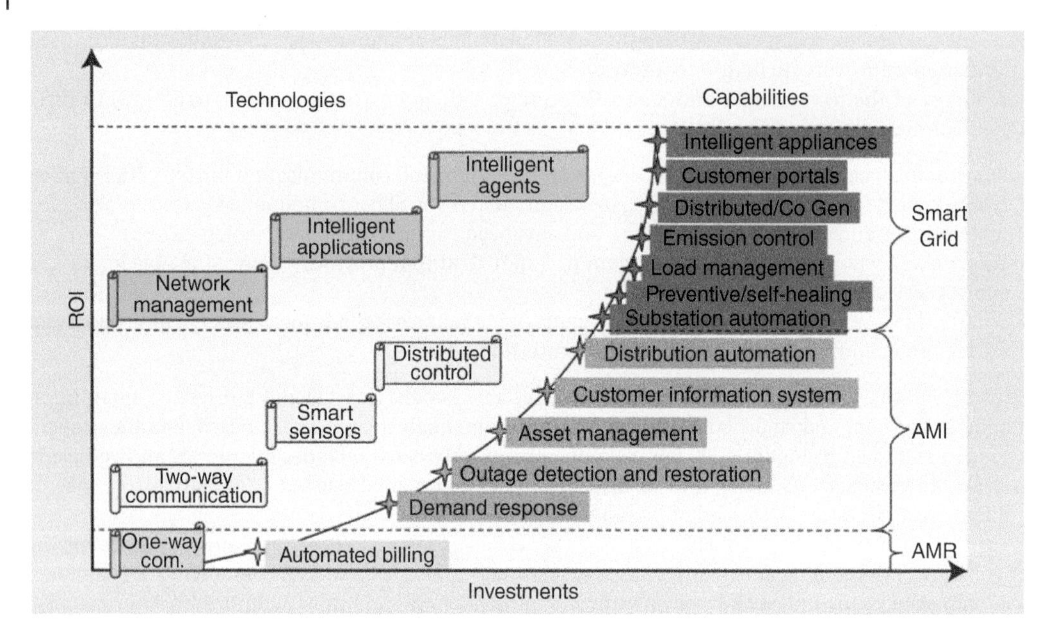

Figure 4.12 Smart Grid technologies and capabilities. *Source:* [Farhangi 2014] © 2014, IEEE.

As discussed in [Farhangi 2010], basic technologies have to be implemented to create foundations for more advanced applications. First, it needs to be understood that there is a relationship between technologies and capabilities (see Figure 4.12). For example, AMI would have to be perceived as an enabling platform for two-way communication and distributed command and control among all previously unmonitored and uncontrolled components of the distribution system. The architecture of AMI systems had to be designed to include sources of sensory, status, and alarm information to allow future developments of Smart Grid capabilities.

Also, Figure 4.12 suggests that although AMR technology proved to be initially attractive, utility companies have realized that AMR does not address the major issue they need to solve, that is, demand-side management. Instead of investing in AMR, utilities across the world moved toward AMI. AMI provides utilities with a two-way communication system to the meter, as well as the ability to modify customers' service-level parameters. Besides meeting their basic targets for load management and revenue protection, utilities realized that through AMI they not only can get instantaneous information about individual and aggregate demand, but they can also impose certain caps on consumption. So, utilities can enact various revenue models to control their costs.

Another aspect is Smart Grid integration that requires incremental developments meeting the functional and operational requirements of a gradually evolving Smart Grid system. Smart Grid integration can be broadly divided into two categories [Farhangi 2014]. One operates in local domains using global system attributes, such as demand response or outage detection, that require access to real-time local data with local analytics and local decision-making processes. The second operates over multiple domains, requiring wide area measurement system (WAMS) and an overview of the system constraints as a whole and system operational objectives, such as management of distributed energy resources, self-healing, outage prevention, and so on.

The realization of the integrated network domains emphasizes the need for a distributed command and control system. This system can use a system of intelligent agents running across multiple domains of the utility network to provide end-to-end communication and data exchange among all utility assets. This proposed approach [Farhangi 2014] could actually help utility to implement cost-effective and efficient Smart Grid capabilities. Thus, the first category is enabled through an AMI system that is designed with appropriate latency, throughput, availability, security, privacy, and resilience requirements. The second relies on an optimally integrated network of distributed systems that is designed with suitable security, scalability, and performance requirements that enables efficient distributed command and control through a multilayer, multitier, and multiagent system. Figure 4.13 shows the realization of the utility's integrated network domains based on multiagents.

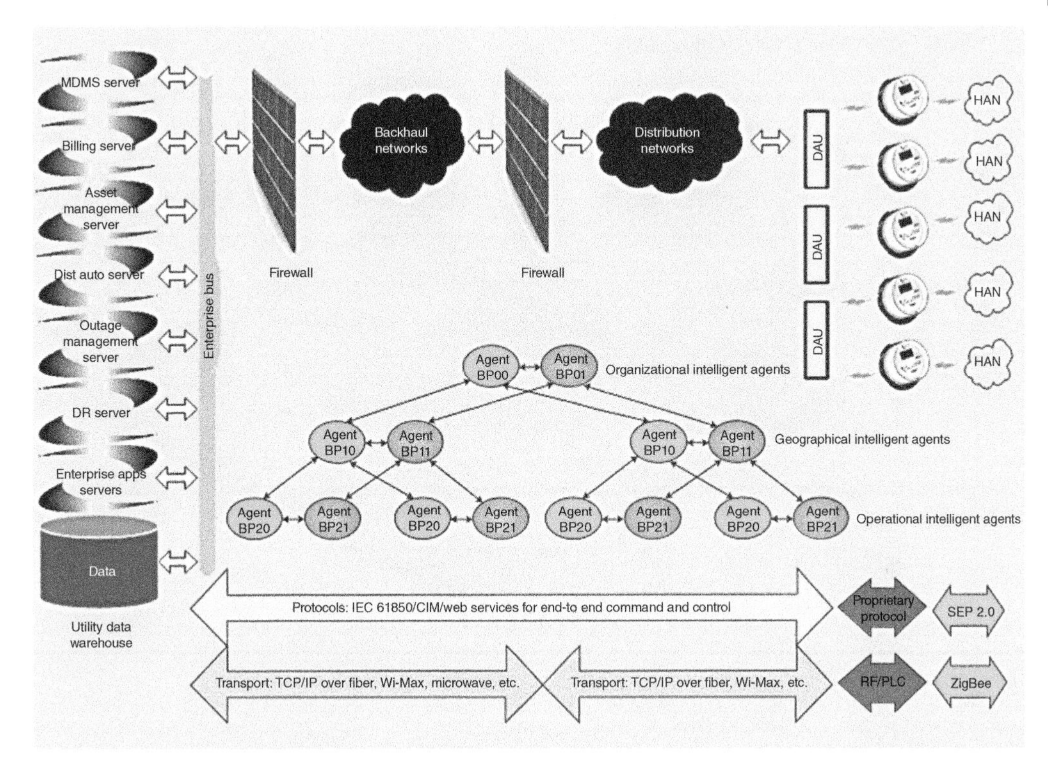

Figure 4.13 Utility integrated network. *Source:* [Farhangi 2014] © 2014, IEEE.

Another work shows an approach to integration of energy technologies into the electric power system by separating into three tiers: the customer level, the community level, and the regional level [DOE 2015]. These three tiers are interconnected and interdependent as shown in Figure 4.14.

A holistic systems approach is required to ensure that various energy technologies can be integrated in a safe, reliable, cost-effective, secure, and resilient manner. The systems approach also requires development and adoption of standards for cybersecurity and interoperability with the grid.

Another issue is the use of automation for renewable sources. Although technologies such DA are mature, the renewable technologies are the least well integrated into current DA [IRENA 2013]. Renewable forecasts can be integrated into distribution operations, allowing for optimized operations. When installing DA, the present and future communication needs should be evaluated in order to select appropriate and cost-effective communication methods.

However, utilities follow different paths for technologies integration based on the needs and issues specific to the territories they serve [GTM 2015]. There are scenarios that require utilities to update and modernize their infrastructure including distribution infrastructure. This opens up opportunities for applications that can provide utilities and grid operators with necessary intelligence across the distributed infrastructure. Therefore, common employed infrastructures should be considered.

4.5.3 Common Employed Infrastructures

Common employed infrastructures include AMI and WAMS.

4.5.3.1 Advanced Metering Infrastructure

AMI enables an entirely new range of consumer-based operational technology. Beyond the reach or control of the traditional utility, AMI is needed for implementing residential demand response and to serve as the mechanism for dynamic pricing. AMI provides customers real-time (or near-real-time) pricing of electricity, and it can help utilities achieve necessary load reductions. The accelerating

Credit: National Renewable Energy Laboratory

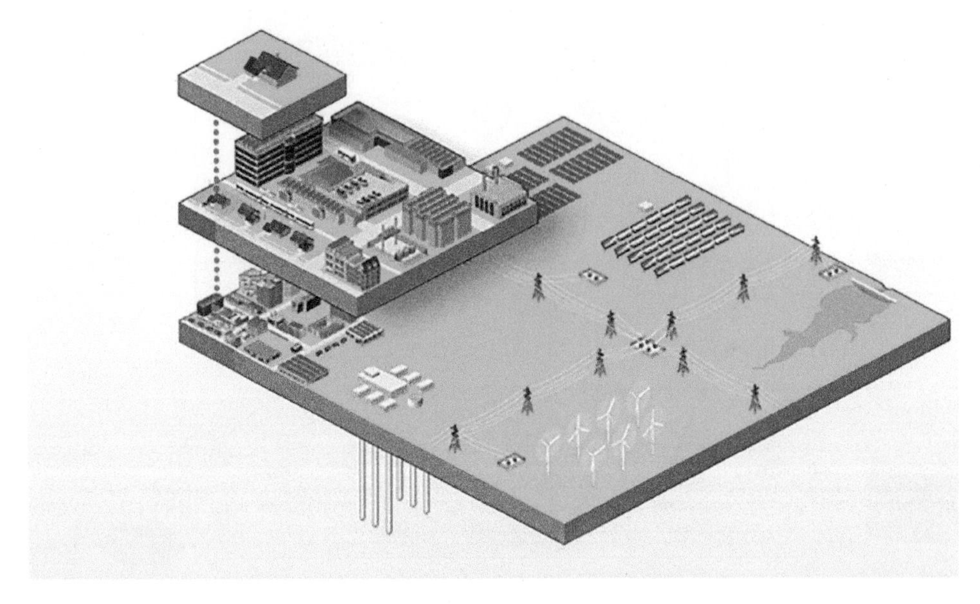

Figure 4.14 Different spatial scales of the electric power system. *Source:* [DOE 2015]. Public Domain.

deployment of AMI around the world is a big step in building a two-way communication platform for enabling demand response and other advanced distribution applications.

AMI consists of the communications hardware and software, and the associated system and data management software, that together create a two-way network between advanced meters and utility business systems, enabling collection and distribution of information to customers and other parties, such as the competitive retail supplier or the utility itself.

4.5.3.2 Wide Area Measurement System

WAMS is a communication system that monitors all phase measurements and substation equipment over a large geographical base that can use visual modeling and other techniques to provide system information to power system operations. The deployment of a WAMS is an important part of the solution to better situational awareness and advanced applications to improve planning, operations, and maintenance. The deployment of WAMS faces challenges with respect to communications and security. Designs of these system have to address requirements for real-time response, availability, security, and reliability. A comprehensive analysis of security issues for a WAMS and research efforts required to be taken are discussed in [Rihan 2013], [Bobba 2012].

4.6 IT Systems Versus Industrial Control System Infrastructure

ICS are computer-based systems that monitor and control industrial processes that exist in the physical world. An ICS consists of various combinations of control components that act together to achieve an industrial objective. ICS encompasses several types of control systems, including SCADA, DCS, programmable logic controllers (PLCs), and other types of industrial measurement and control systems.

A control system is a device, or set of devices, that manages, commands, directs, or regulates the behavior of other devices or systems [Control system, https://en.wikipedia.org/wiki/Control_system]. It is a system in which deliberate guidance or manipulation is used to achieve a prescribed value for a variable [NIST SP800-82r2]. Examples of different types of control systems that are needed for DERs and Smart Grid include [EPRI 2011]:

- AMI
- DA
- Substation automation
- Distribution feeder circuit automation
- DER and load control
- Building automation systems.

IT systems are built using traditional enterprise networks (IT infrastructure), while control systems are built on industrial control networks composed of interconnected equipment used to monitor and control physical equipment in industrial environments. These infrastructures differ quite significantly due to the specific requirements of their operation and functions supported.

4.6.1 Industrial Control Systems General Concepts

Typical control systems are a collection of control loops with sensors and actuators interacting with the physical world, human–machine interfaces (HMIs), and remote diagnostics and maintenance utilities. Figure 4.15 shows a very simple ICS configuration of components and its logic control loop (ICS operation in [NIST SP800-82r2]).

The HMI allows human operators to monitor the state of a control process and issue commands to change the control objective or manually override automatic controls in emergency situations. The control loop includes hardware such as controllers (e.g. PLCs) that interpret signals from sensors, set variables based on those signals, transmit the variables to controllers, and physically actuate components such as switches, breakers, and motors. Remote diagnostics and maintenance utilities prevent, identify, or recover from abnormal operations or failure.

In addition to the basic HMI, control loop, and remote diagnostics and maintenance utilities, there are other typical components that may be included in a control system configuration. These include control components, control network components, and SCADA systems.

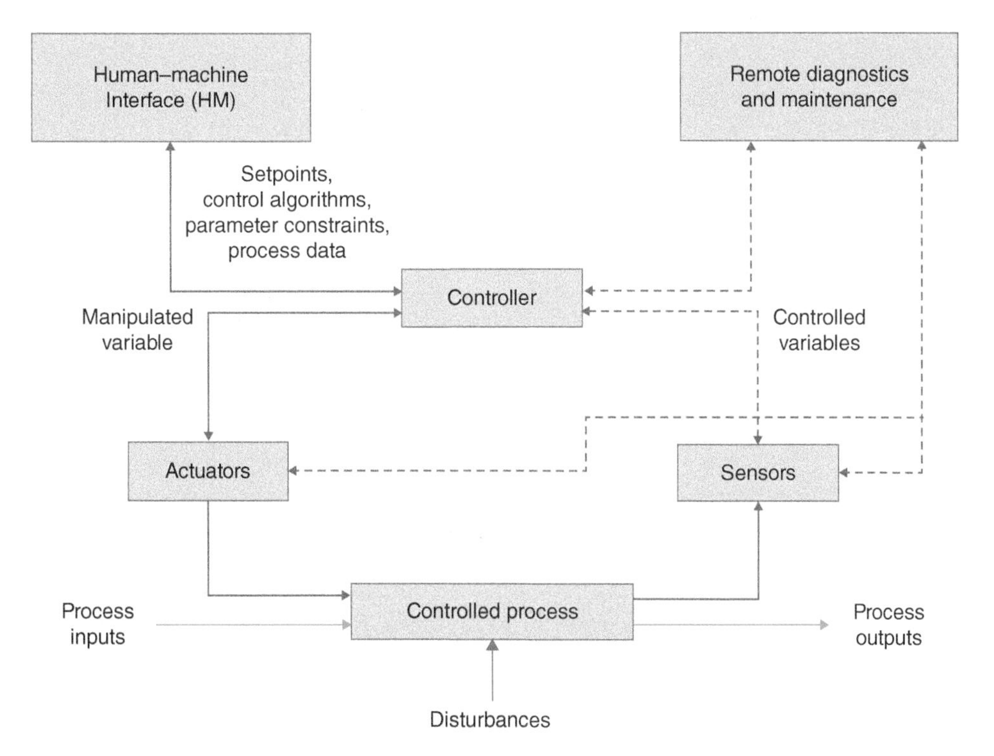

Figure 4.15 ICS operation. *Source:* [NIST SP800-82r2]. Public Domain.

4.6.1.1 Control Components

- Distributed control systems (DCS) containing multiple, geographically dispersed subsystem controllers.
- PLCs include industrial process control computers used in DCS and SCADA systems; remote PLCs are often referred to as field devices.
- SCADA are highly distributed systems controlling geographically dispersed equipment; SCADA depends on centralized data acquisition and control functions to automate critical functions; SCADA systems are implemented with a variety of hardware and software components; the electric transmission and distribution grid relies on SCADA operations.
- Control server hosts the DCS or PLC supervisory control software.
- Master terminal unit (MTU) is a control unit that acts as the master in a SCADA system.
- Remote terminal unit (RTU) is a control unit designed to support SCADA remote stations.
- IED is a smart sensor or actuator containing operational intelligence for automatic control at the local level.
- Data historian is a database storing all process information.
- Input/output (IO) server collects, buffers, and provides access to process information, and may also interface with other components.

4.6.1.2 Control Network Components

Within a control system hierarchy, there are different network components with specific characteristics for each layer. Regardless of the network topologies in use, the major components of an ICS network include:

- Fieldbus network to link sensors and other devices to a PLC or other controller; fieldbus technologies eliminate the need for point-to-point wiring between the controller and each device; the devices communicate with the fieldbus controller using a variety of protocols; the messages sent between the sensors and the controller uniquely identify each of the sensors.
- Control network to connect the supervisory control level to lower-level control modules.
- Communications router to transfer packets between networks such as connecting a LAN to a WAN or connecting MTUs and RTUs to a long-distance network medium for SCADA communication.
- Firewall to control access to devices on a network by monitoring and controlling communication packets using predefined filtering policies; firewalls are also useful in managing ICS network segregation strategies.
- Modem to convert between serial digital data and analog signal for transmission over a telephone line to allow devices to communicate; modems are often used in (i) SCADA systems to enable long-distance serial communications between MTUs and remote field devices; (ii) SCADA systems, DCS, and PLCs for gaining remote access for operational and maintenance functions such as entering commands or modifying parameters, and diagnostic purposes.
- Remote access point as a distinct device for remotely configuring control systems and accessing process data; for example, a personal digital assistant (PDA) accesses data over a LAN through a wireless access point and a laptop and modem connection to remotely access an ICS system.

4.6.2 Supervisory Control and Data Acquisition Systems (SCADA)

SCADA system is a category of control systems used to monitor or control processes such as chemical transport, water supply, power generation and distribution, and gas and oil supply. A control system is a device or set of devices to manage, command, direct, or regulate the behavior of other devices or systems. However, SCADA is not a full control system, but rather focuses on the supervisory level. Usually, SCADA systems involve a human-in-the-loop control and decision-making processes.

SCADA systems differ from DCSs, which are generally found in plant sites. While DCSs cover plant site, SCADA systems cover much larger geographic areas. Also, due to the remoteness, many of these often require the use of wireless communications. A real-world SCADA system can monitor and control hundreds to hundreds of thousands of I/O points. SCADA systems evolve rapidly and are now penetrating the market with a number of I/O channels from 100K up to near 1M I/O channels currently under development. Figure 4.16 shows an integrated SCADA architecture.

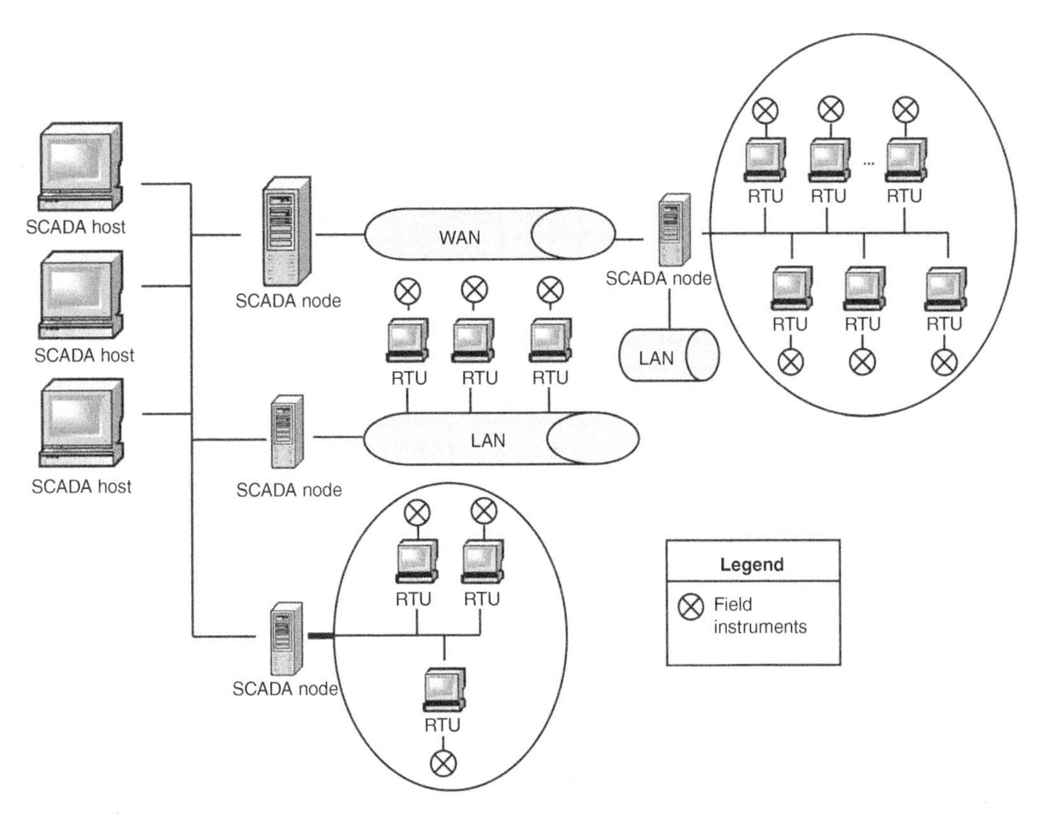

Figure 4.16 Integrated SCADA architecture. *Source:* [Hentea 2008]. © 2008, Informing Science Institute.

The architecture of a SCADA system consists of one of more MTUs that are used by engineers in a control station to monitor and control a large number of RTUs located in field or industrial plants. An MTU is a general-purpose computer running SCADA utility programs, and RTUs are generally small dedicated devices designed for rough field or industrial environment. One or more SCADA MTUs retrieve real-time analog and status data from RTUs, store, and analyze these data. MTUs automatically send control commands to the RTUs or enable the engineers to do so manually.

SCADA systems operate widely dispersed control systems and acquire system data for monitoring and control at the central server, or MTU. The data acquisition process is integral to system functions; the acquired data is critical to operational integrity. SCADA system control crucial infrastructure, including the power grid, oil and gas pipelines, railway traffic, water distribution, and waste treatment plants. Any significant disruption to a critical SCADA system could directly threaten public health and safety. SCADA may include some or all of the components defined above, plus additional specialized components.

SCADA architecture supports TCP/IP, UDP, or other IP-based communication protocols as well as strictly industrial protocols such as Modbus TCP, Modbus over TCP, or Modbus over UDP all working over private radio, cellular, or satellite networks. In complex SCADA architectures, there is a variety of both wired and wireless media and protocols involved in getting data back to the central monitoring site. This enables implementation of powerful IP-based SCADA networks over mixed cellular, satellite, and landline systems. SCADA communications can employ a diverse range of both wired (leased line, dial-up line, fiber, ADSL, cable) and wireless media (licensed radio, spread spectrum, cellular, WLAN, or satellite). The choice depends on a number of factors that characterize the existing communication infrastructure. Factors such as existing communication infrastructure, available communications at the remote sites, data rates and polling frequency, remoteness of site, installation budget, and ability to accommodate future needs all impact the final decision for SCADA architecture.

SCADA systems and DCS are often networked together. This is the case for electric power control centers and electric power generation facilities. Although the electric power generation facility operation is controlled by a DCS, the DCS must communicate with the SCADA system to coordinate production output with transmission and distribution demands.

These systems evolved from static to dynamic systems. The increased connectivity to Internet and mobile device technology has also a major impact on control systems architectures. Standardization and use of open market technologies are current requirements in control systems. Modern products are often based on component architectures using commercial off-the-shelf (COTS) product elements as units. This architecture leads to control systems that are becoming very complex software applications with the following characteristics [Sanz 2003]:

- Time critical
- Embedded
- Fault tolerant
- Distributed
- Intelligent
- Large
- Open
- Heterogeneous.

SCADA systems are exposed to the same cyberspace threats like any business system because they share the common vulnerabilities with the traditional IT systems. Also, most SCADA systems are not protected with appropriate security safeguards. The operating personnel is lacking the security training and awareness. Threats against SCADA systems are ranked high in the list of government concerns, since terrorists have threatened to attack several SCADA systems of critical infrastructure and successfully launched near-disastrous attacks. In addition, recent attacks are becoming more sophisticated, and the notion of what kind of vulnerabilities actually matter is constantly changing. For example, timing attacks are now common threats, whereas only a few years ago they were considered exotic. The threats are often poorly understood and ignored, and the vast majority of organizations lag in realizing secure infrastructures. In complexly interactive systems whose elements are tightly coupled, great accidents are inevitable. Vulnerabilities and attacks could be at different levels – software controlling or controlled device, application, storage, data access, LAN, enterprise, Internet, and communications. Therefore, a review of SCADA systems evolution allows us to better understand many security concerns.

In the past decades, there were changes in the architecture of SCADA systems. As depicted in Figure 4.17, there was an evolution from the first generation (monolithic systems connected via WAN to the RTU) to second generation (more distributed with LAN used to interconnect the components and real-time information sharing) and to third generation (connected to WAN and Internet). Similarly, DCS systems evolved too by applying the advances in computation and communications as SCADA systems. Therefore, a review of SCADA systems evolution allows us to better understand many security concerns.

As described in [Karnouskos 2011], there is a trend for next generation of SCADA systems taking advantage of other capabilities. These capabilities include integration of services and large-scale participating systems, which no longer have a single controlling/management authority, have components that are developed and evolve independently, and are using several emergent technologies in hardware and software.

These advances include networked (embedded) systems, industrial PCs, and the application of new paradigms like the SOA and system of systems (SoS). Figure 4.18 depicts the overall architecture of the next-generation SCADA system.

As described in [Karnouskos 2011], the vision for the next-generation SCADA/DCS is an information-driven network rather than a communication one, with SOA and SoS paradigms using communication-driven interaction and focusing on the information available via services, moving toward a service-driven interaction. Services can be dynamically discovered, combined, and integrated in mash-up applications. Although a communication infrastructure connecting all components is still the WAN and Internet backbone, it is much more diversified over several wired and wireless channels, providing QoS depending on the application's dynamic needs. The service-based interactions enable the integration and interoperability and lead to lower costs and rapid development of customized solutions.

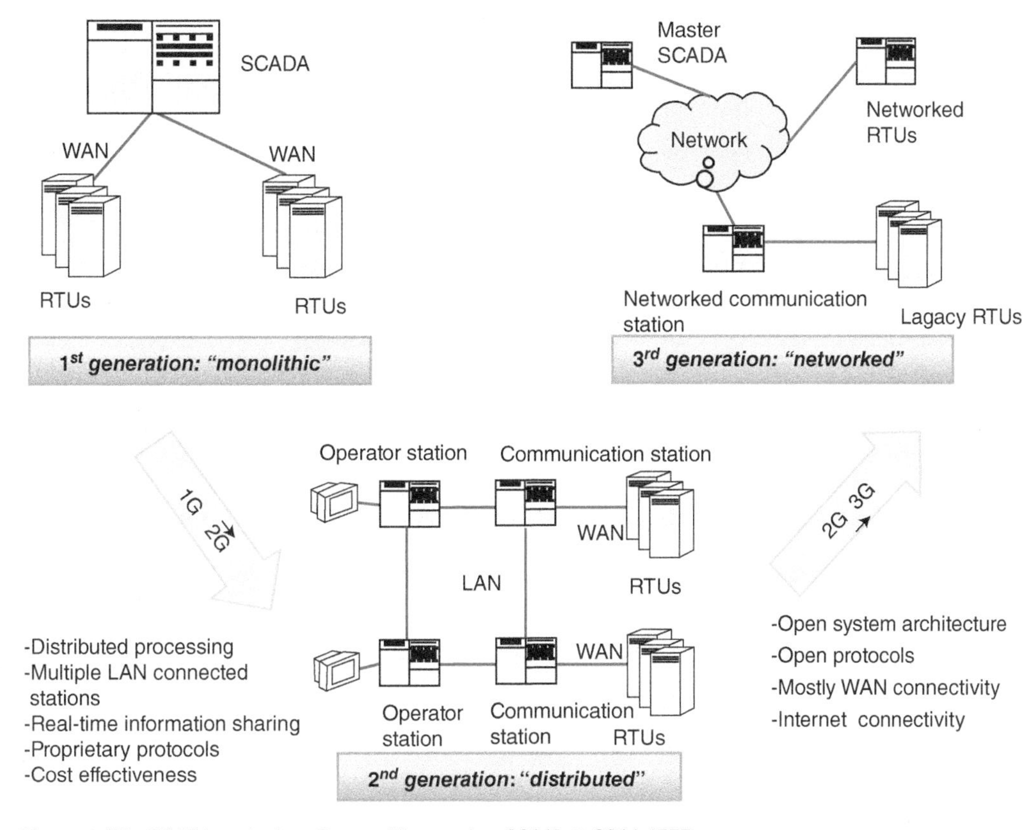

Figure 4.17 SCADA evolution. *Source:* [Karnouskos 2011]. © 2011, IEEE.

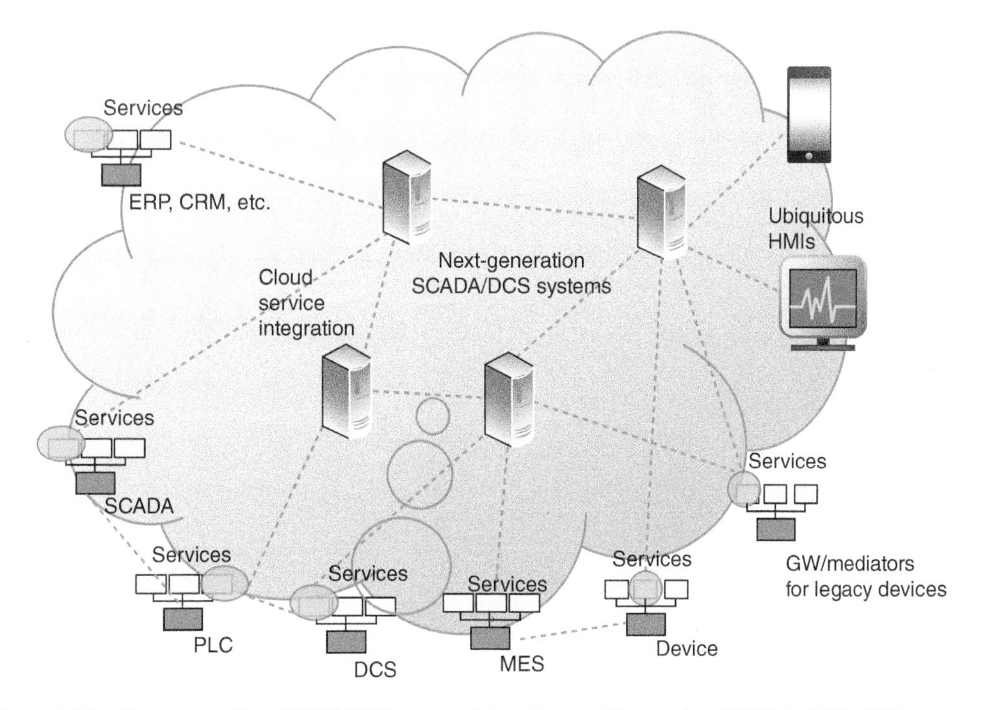

Figure 4.18 Next-generation SCADA/DCS system vision. *Source:* [Karnouskos 2011]. © 2011, IEEE.

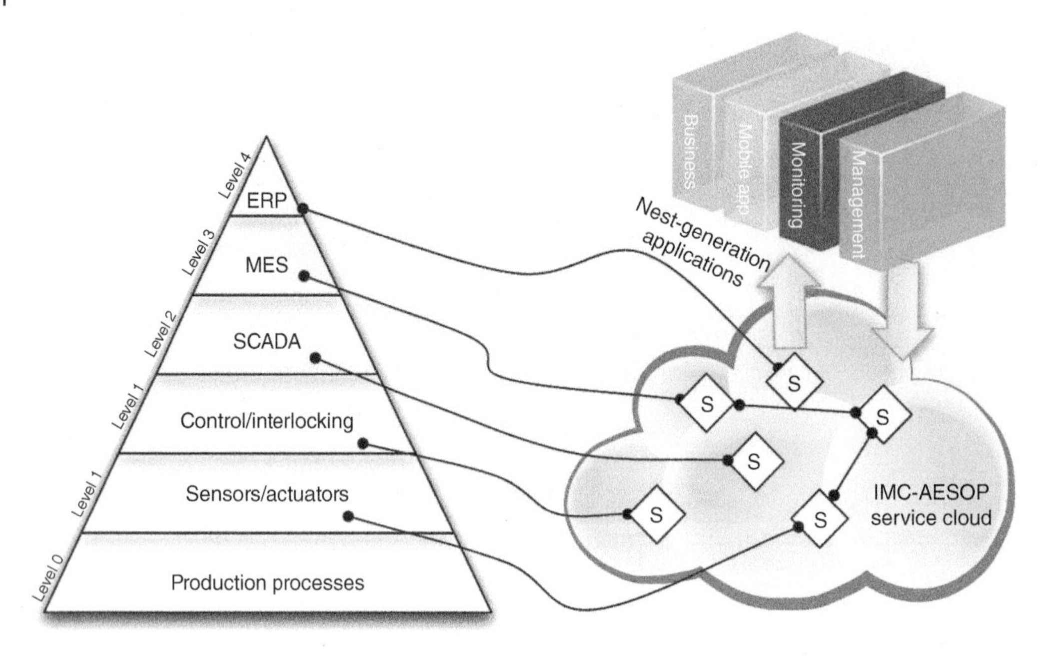

Figure 4.19 Future cloud-based industrial systems view. *Source:* [Karnouskos 2012]. © 2012, IEEE.

The industrial automation systems are becoming complex SoS that will empower a new generation of today's hardly realizable applications and services [Karnouskos 2012]. The envisioned transition to the future cloud-based industrial systems view is depicted in Figure 4.19.

The empowerment offered by modern SOA and the functionalities of each system or even device can be offered as one or more services of varying complexity, which may be hosted in the cloud and composed by other services [Karnouskos 2012]. Although the traditional hierarchical view coexists, there is now a flat information-based architecture that depends on a big variety of services exposed by the cyber–physical systems and their composition of cyber–physical services. Next-generation industrial applications can now rapidly be composed by selecting and combining the new information and capabilities offered (as services in the cloud) to realize their goals.

This work [Karnouskos 2012] proposes an SOA, which attempts to cover the basic needs for monitoring, management, data handling and integration, etc. Table 4.2 includes a list of services with a proposed prioritization of services as necessary for future systems. The authors argue that all services are essential for next-generation cloud-based collaborative automation systems.

The groups of services have been rated with high priority (+) if they constitute a critical service absolutely mandatory, with medium priority (0) if this is not a critical but nevertheless highly needed service, and lastly with low priority (−), which mainly means nice to have services that enhance functionalities but are optional.

For example, the Alarms service group contains services for alarm processing and configuration. These services support simple and complex events that are composed based on several other events coming from different parts of the system. Some of the alarms are generated in lower-level services and devices, but alarms may also be generated at other layers, e.g. in the alarm processing service using process values and limits. The alarm configuration end-processing services also support very flexible hierarchical alarm area definitions and correlations, e.g. via heuristics.

Security is a critical area especially when it comes down to enabling interactions among multiple stakeholders with various goals and access levels. The security management focuses on enforcement or execution of security measures, and policy management is about definition and management of security rules or policies. The security services are implicitly used by all architecture services.

Table 4.2 Detailed architecture services and prioritization.

Service group	Service	Priority
Alarms	Alarm configuration	+
	Alarm and event processing	+
Configuration and deployment	Configuration repository	+
	System configuration service	+
	Configuration service	+
Control	Control execution engine	+
Data management	Sensory data acquisition	+
	Actuator output	+
	Data consistency	0
	Event broker	+
	Historian	0
Data processing	Filtering	+
	Calculation engine	0
	Complex event processing service	+
Discovery	Discovery service	+
	Service registry	+
HMI	Graphics presentation	+
	Business process management and execution service	0
	Composition service	+
	Gateway	+
	Service mediator	+
	Model mapping service	+
Lifecycle management	Code repository	−
	Lifecycle management	+
Migration	Infrastructure migration solver	−
	Migration execution service	−
Mobility support	Mobile service management	0
Model	Model repository service	0
	Model management service	0
Process monitoring	Monitoring	+
Security	Security policy management	+
	Security management	+
Simulation	Constraint evaluation	0
	Simulation execution	0
	Simulation scenario manager	0
	Process simulation service	0
System diagnostic	Asset monitor	+
	Asset diagnostics management	+
Topology	Naming service	+
	Location service	+

Source: [Karnouskos 2012] © 2012, IEEE.

4.6.3 Differences and Similarities

Power grid information security and protection needs include aspects of both control operation systems and enterprise business systems. The planning of security for Smart Grid systems including DER systems requires information about differences and similarities between IT systems and control systems. A comparison of these systems highlights the following issues:

- CIA triad model has specific functional and performance requirements as well as organizational policies for determining a system security posture that is quite different. Confidentiality, integrity, and availability are the order of priority for a business IT system. However, availability, integrity, and confidentiality are often the prioritization order for control and protection systems.
- Technology capability is the choice for building these systems in a different way: control systems use process control technology (for example, a PID algorithm ensures that a predetermined set point is achieved and maintained within identified limits); IT systems are built on the criteria of allowing a device to join a network without any checks for minimum capabilities and minimum level of trust.
- Security features are designed and implemented differently: (i) control systems including SCADA networks were initially designed with little attention to security, and most SCADA systems are not protected with appropriate security safeguards; IT systems are more concerned with security during the system development, and security technologies are more mature [Hentea 2008]; (ii) security technologies are not ready to detect and isolate a cyber attack for a real-world SCADA system that monitors hundreds to hundreds of thousands of I/O points, with no security or limited features, while an IT system is more contained and supports more security features built.
- Failure propagation in control systems is greatly increased by (i) the likelihood of multiple simultaneous failures and (ii) the high speed of SCADA networks facilitating quick propagation of malicious code; although this scenario is true for IT systems, an IT device failure can often be detected and isolated faster because the management of IT devices is based on a centralized and controlled approach for the configurations of the endpoints (IT device).
- Network security and encryption application of conventional network security measures work well in IT environment, but it is not always possible to implement these controls in ICS; another real vulnerability is the PLCs that control the processes in a facility or plant; these PLCs are more exposed to unauthorized access than a SCADA system.
- Control system technologies have limited security, and if they do the vendor-supplied security capabilities are generally only enabled if the administrator is aware of the capability; control system communication protocols are absent of security functionality; considerable amount of open-source information is available regarding control system configuration and operations, escalations of privileges through code manipulation, network reconnaissance and data gathering, covert traffic analysis, and unauthorized intrusions into networks either through or around perimeter defenses.
- Safety concerns are higher in computer systems (including SCADA control systems) because failure to support applications as intended can cause harm and catastrophic damage [Dunn 2003]; in general, failures of IT systems do not affect the safety of users.
- Security is approached from the perspective of different threats: control systems are (i) more focused on accidental/inadvertent threats such as equipment failures, employee errors, and natural disasters and (ii) not focused on intentional threats (e.g. malware) as most of IT systems employ policies, procedures, and security technologies.
- Limited capabilities command the security features implementation; often limited availability of computation and communication capabilities in control systems make repurposing security mitigations commonly effective in IT domain much more challenging for the power grid operation domain [Ray 2010].
- Cost–benefit scenarios lead to challenges in favor of mitigating security risks; for example, many DER systems are not designed to employ dedicated communication infrastructure because of high costs.
- End-user representation drives the risks and choice of security controls; in general, end user for an IT system is a person, while for a control system is a computer or other intelligent control device [Weiss 2010].
- Costs are not equal, although the needs are balanced; IT systems strive to consolidate and lower the operational costs; control systems are distributed requiring remote access and additional protection costs [Weiss 2010].

- Operating environments present more differences than similarities; IT and control system operating environments often preclude the same security implementation to be feasible across both domains. As an example, in most cases, power operation systems cannot be easily restarted without adversely affecting power generation or delivery, thereby compromising high availability, reliability, and maintainability requirements. Hence use of IT restoration measures like rebooting a component is usually not acceptable owing to such adverse impacts on requirements for the control systems [Ray 2010].
- Security training and skills are not adequate; most operating personnel of control systems is lacking the security skills, training, and awareness, while IT users and personnel have more security skills and more awareness training; however, IT personnel needs more education and skills in general concepts in power engineering and control.
- Complexity is a challenge since control systems are becoming very complex software applications with the following characteristics [Sanz 2003]: time critical, embedded, fault tolerant, distributed, intelligent, large, and heterogeneous, while IT systems are open and tolerant to delay.
- Compliance is driven by different views and needs; due to their complexity, control systems have a greater need to compliance for safety, quality of service, and security of systems and data [Conference 2006], while IT systems are more compliant on security required by regulations (e.g. US federal regulations such as Health Insurance Portability and Accountability Act (HIPAA), SOX, Gramm–Leach–Bliley Act (GLBA)).
- System design principles must be applied for properly understanding the organization's protection needs and subsequently employing sound security architectural design principles and concepts through the engineering processes; in the past, control systems were developed on the principle of obscurity, while IT systems were developed more as open systems [Hentea 2008].
- Software development skills and experience necessary are not always properly identified and understood; specifically, the software-intensive system design skills for the construction of control systems are often misunderstood. In control industry, two separate groups of engineers are typically involved in the development of any nontrivial controller: control engineers and programmers. These two groups tend to have very different perspectives and working practices, and both lack the global picture needed for the task. IT system development is more under control by applying software management and practices for the verification of errors in the code, using security models, following compliance, auditing, and improving the processes [Hentea 2008].
- Programming skills are not sufficient for developing control systems; developers of control systems require skills in both power and control engineering and programming to be able to understand the broad picture and detailed control problems; programmers of IT systems are focused on business applications and have no skills or limited skills in real-time applications.
- Security personnel for the operating personnel of control systems is often lacking the security training and awareness; power and control engineers lack appropriate security education and training, while information security professionals are more and periodically trained [Hentea 2008].
- Best security practices applicability and availability differ; IT security professionals developed generally accepted best practices to protect the information infrastructure, but these practices cannot be applied directly to control systems without accounting for the different requirements of IT and control systems [Krutz 2006].
- Logging activities for IT systems take benefit of monitoring and logging activities for analysis and improvement, but many control systems lack the logging capabilities, or they are very limited [Krutz 2006].
- Latency for IT systems may not be affected by delays due to overhead of security protocols and controls, while control systems and networks cannot tolerate nondeterministic delays due to security mechanisms that require large memory capacities, locking out operators, and relatively long processing times [Krutz 2006].
- Security features priorities and assumptions drive different goals; control systems/SCADA and IT system security, while similar in function, greatly differ in terms of priority. These systems have different assumptions as to what security is and how to stay secure. Each system type has unique uptime requirements, risk-avoidance tactics, architectures, goals, and performance requirements. IT system security's first priority is typically known to protect data and help employees continue working

Table 4.3 ICS infrastructure versus IT infrastructure.

Differences/ criteria	Industrial control networks	IT infrastructure
Primary function	Control of physical equipment in industrial environment	Data processing, transfer, storage
Applicable domain	Electrical sector, manufacturing, processing, building automation, critical infrastructure, etc.	Corporate and home environments
Hierarchy	Deep, functionally separated hierarchies	Shallow, integrated hierarchies
Failure severity	High	Low
Reliability required	High	Moderate
Roundtrip times	250 μs to 10 ms	50+ ms
Determinism	High	Low
Data composition	Small messages and aperiodic traffic	Large, ad hoc traffic
Temporal consistency	Required	Not required
Operating environment	Hostile conditions, high levels of dust, heat, and vibration	Clean environments, often specifically intended for sensitive equipment
Protocols	Large number, variety of highly specialized tasks	Defined set, uniform, broad tasks
Security features	Problematic	Defined
Maturity	Constantly under development	Established
Security technologies	Not suitable everywhere; research is going on to define the appropriate technologies	More mature and broader applicability

Source: Adapted from [Galloway 2013]. © 2013, IEEE.

without being interrupted. ICS/SADA system device security, on the other hand, is known to focus on protecting the reliability of process without affecting productivity. Because of unique differences, securing ICS/SCADA systems must also be treated uniquely and approached with care [Hentea 2008], [Ray 2010].

Main differences exist in implementation, architecture, failure severity, real-time requirements, temporal consistency and event order, traffic type, data type, data size, communication protocols, and security [Galloway 2013]. Table 4.3 exhibits a summary of key differences between industrial control networks and conventional (enterprise networks, IT networks) infrastructures.

More information about these infrastructures is available in other publications such as [Hentea 2008], [Weiss 2010], [Krutz 2006], [Kadrich 2007], [NIST SP800-82r2]. Despite the functional and operational differences between industrial and enterprise networks, trends are observed toward integration.

4.7 Convergence Trends

Lately, the technology in use in industrial networks is also beginning to display a greater reliance on Ethernet and Web standards as in IT infrastructures, especially at higher levels of the network architecture. This has resulted in a situation where engineers involved in the design and maintenance of control networks must be familiar with both traditional enterprise concerns, such as network security, and traditional industrial concerns, such as determinism and response time.

The technology used in industrial networks is also more relying on Ethernet and Web technologies and standards, especially at higher levels of the network architecture. This has resulted in a networking environment that appears to be similar to conventional networks at the physical level, but which has significantly different requirements. Network topologies across different ICS implementations vary with modern systems using Internet-based IT and enterprise integration strategies. Control networks have merged with corporate networks to allow control engineers to monitor and control systems from outside of the control system network. The connection may also allow enterprise-level decision makers to obtain access to process data.

Also, this trend redefines the skills and expertise of control engineers involved in the design and maintenance of control networks. They need to be familiar with both traditional enterprise concerns such as network security and traditional industrial concerns such as determinism and response time. Similarly, security engineers and professionals need to understand the industrial control network developments and operations.

New applications are being developed that include IT functions both and control functions; examples include:

- Plug-in electric vehicle (PEV) management systems
- Building management control systems
- Electronic security systems
- Emergency management systems
- Exterior lighting control systems
- Physical access control systems
- Renewable energy geothermal systems
- Renewable energy photovoltaic systems
- Electric storage control systems.

4.8 Standards, Guidelines, and Recommendations

Standards relevant to topics and content of this chapter need to be investigated and applied where is needed. Given the trend of cloud services, the following standards could be used:

- [NIST SP500-292] describes the cloud computing reference architecture. It is a generic high-level conceptual model that is not tied to any specific vendor products, services, or reference implementation. It defines a set of actors, activities, and functions that can be used in the process of developing cloud computing architectures and relates to a companion cloud computing taxonomy. The reference architecture contains a set of views and descriptions that are the basis for discussing the characteristics, uses, and standards for cloud computing. This actor/role-based model is intended to serve the expectations of the stakeholders by allowing them to understand the overall view of roles and responsibilities in order to assess and assign risk.
- [NIST SP500-291] includes a list of standards relevant to cloud computing. This list is continuously revised and updated.

Information security, like most technical subjects, uses a complex web of terminology that is evolving. Examples of additional documents that are needed for Smart Grid cybersecurity development include glossaries of security terms (see Appendix A).

Concepts related to ICS are defined in:

- ANSI/ISA-99, Manufacturing and Control Systems Security, Part 1: Concepts, Models and Terminology
- NIST SP800-82r2, Guide to Industrial Control Systems (ICS) Security

Guidance on quality of systems and software is provided in these standards:

- [ISO/IEC 25010] defines a quality in use model and a product quality model as well as characteristics for each model. The characteristics defined by both models are relevant to all software products and

computer systems, wider systems, and services. It provides consistent terminology for specifying, measuring, and evaluating system and software product quality.
- [ISO/IEC 25012] contains a model for data quality that is complementary to the model.
- [IEC 61069-5] describes in detail the method to systematically assess the dependability of industrial process measurement and control systems. It uses the assessment methodology given in [IEC 61069-2] standard.

Guidance for IT security evaluation is provided in these standards:

- [ISO/IEC 15408] supports what is called Common Criteria for the integration and testing of many different software applications in a secure way. It provides assurance that the process of specification, implementation, and evaluation of a computer security product has been conducted in a rigorous and standard and repeatable manner at a level that is commensurate with the target environment for use.

Security guidance and requirements for compliance are defined in:

- [NERC CIP] standards are concerned with bulk electric system (BES). NERC initiated a program to help industry transition directly from the currently enforceable CIP Version 3 standards to CIP Version 5.

Given the importance of electrical energy storage (EES) in the development of Smart Grid applications, research organizations and associations (e.g., IEEE, ESA) contribute and support the development of the DER and EES technologies and standards. Issues related to standards that are necessary to be addressed for EES applications are discussed in [IEC 2014]:

- Terminology.
- Basic characteristics of EES components and systems, especially definitions and measuring methods for comparison and technical evaluation for capacity, power, discharge time, lifetime, and standard EES unit sizes.
- Communication between EES components including protocols, security, interconnection requirements such as power quality, voltage tolerances, frequency, synchronization, and metering.
- Safety for disturbances such as electrical, mechanical, chemical, etc.
- Testing, implementation, maintenance, and simulation guides.

5

Planning Security Protection

5.1 Threats and Vulnerabilities

Technological development, technical problems, extreme environmental events, adverse physical plant conditions, human frailty, and inadequacies of social, political, and economic institutions all present challenges to the smooth functioning of information systems. In order to make sound decisions about information security, management must be informed about the various threats facing the organization, people, information systems, information, and data.

5.1.1 Threats Characterization

The presence of any potential event that, if realized, can cause an undesirable impact on the organization is called a threat. A threat is any circumstance or event with the potential to adversely impact organizational operations (including mission, functions, image, or reputation), organizational assets, individuals, other organizations, or the nation through an information system via unauthorized access, destruction, disclosure, modification of information, and/or denial of service (DoS) [CNSSI 4009]. A threat is an object, person, or other entity that represents a constant danger to an asset [Whitman 2012].

A threat is a potential for violation of security, which exists when there is an entity, circumstance, capability, action, or event that could cause harm [RFC 4949]. It can create a loss of confidentiality, integrity, and availability. A threat can cause a disruption event that interrupts or prevents the correct operation of Smart Grid system services and functions.

The term threat is frequently misused with the meaning of either threat action or vulnerability; in some contexts, threat is used more narrowly to refer only to intelligent threats; in some contexts, threat is used more broadly to cover both definition and other concepts. Intelligent threat is a circumstance in which an adversary has the technical and operational ability to detect and exploit a vulnerability and also has the demonstrated, presumed, or inferred intent to do so.

Threat is a characteristic, circumstance, or condition that may manifest as an adverse event with the potential to cause harm to an asset or condition that impacts the normal operation. Threats to information systems may arise from intentional or unintentional acts and may come from internal or external sources.

Examples of threats to information security include human error or failure, deliberate acts of espionage, sabotage, vandalism, theft, software attacks, deviations in quality of service, forces of nature, hardware failures or errors, and technological obsolescence [CSI 2015], [Whitman 2012]. Some threats manifest in accidental (unintentional) occurrences, while others are intentional. For example, forces of nature or natural disasters (e.g. tornadoes, floods, etc.) are accidental threats.

The common criteria, an international standard, characterize a threat in terms of a [ISO/IEC 15408]:

- Threat agent, the originator.
- Presumed method of attack, an event that causes harm.

Building an Effective Security Program for Distributed Energy Resources and Systems: Understanding Security for Smart Grid and Distributed Energy Resources and Systems, Volume 1, First Edition. Mariana Hentea.

- Any vulnerabilities that are the foundation for the attack.
- Target, the system resource that is attacked.

A realization of a threat or threat action as a result of an accidental or an intentional act is called attack (or cyber attack) and has consequences (security violations) that are as follows:

- Unauthorized disclosure is circumstance or event whereby an entity gains access to information for which the entity is not authorized.
- Deception is circumstance or event that may result in an authorized entity receiving false data and believing it to be true.
- Disruption is circumstance or event that interrupts or prevents the correct operation of system services and functions.
- Usurpation is circumstance or event that results in control of system services or functions by an unauthorized entity.

Examples of threat agents include hackers, malicious users, non-malicious users (who sometimes make errors), computer processes and accidents, terrorists, and natural event. There are many other ways to categorize the various threats. One way to categorize is to look at those threats that come from outside of the organization versus those that are internal. Another way is to look at the various levels of sophistication of the attacks. A different way is to examine the level of organization of the various threats, from unstructured threats to highly structured threats. Also, from a hacker's (or threat agent's) motivation could be targeted (attempting to successfully penetrate or damage a particular asset or organization) or untargeted (looking for any system to compromise).

As discussed in [Kettani 2019], there is a large taxonomy of attack vectors that are paths or means by which a threat agent gains access to a computer or network for the purpose of malicious activity. Attack vectors include the human element, Web and browser attacks, Internet exposed threat, mobile application stores, malicious USB drives, etc.

Examples of numerous categories of threats are commonly listed in courses and published books by Whitman (e.g. [Whitman 2012], [Harris 2013]) include:

- Acts of human error and failure.
- Compromises to intellectual property.
- Deliberate acts of espionage or trespass, sabotage, information extortion, acts of theft, and software attacks.
- Deviations in quality of service from service providers.
- Forces of nature.
- Technical hardware or software failures and errors.
- Technological obsolescence.

However, these categories should be continuously evaluated because of evolving threats. Table 5.1 shows an updated list of categories of threats to information security. These threats are evolving threats that could have high impacts to economy and public in the coming years.

Systematic analysis of data breaches (compromise of the confidentiality attribute) for the past three years finds threat agents (actors) to be classified in these categories [Verizon 2016]:

- External threat agents originate from sources outside of the organization and its network of partners; examples include criminal groups, lone hackers, former employees, and government entities; typically, no trust or privilege is implied for external entities.
- Internal threat agents are those originating from within the organization; this category encompasses company full-time employees, independent contractors, interns, and other staff; insiders are trusted and privileged (some more than others).
- Partners include any third party sharing a business relationship with the organization; this third-party category includes suppliers, vendors, hosting providers, outsourced information technology (IT) support, etc.; some level of trust and privilege is usually implied between business partners.

As pointed by Bellovin [Bellovin 2015], the biggest security problem we face stems from one simple fact: software is often buggy. These bugs are often exploitable by attackers.

Table 5.1 Categories of evolving threats to information security.

Categories of threats	Examples
Threats to physical safety as well as periods of operational downtime	Attacks that remove connectivity or that target key individuals working from home; operational downtime caused by attacks on infrastructure, devices, or people
Threats to uninterrupted connectivity	Attacks on core Internet infrastructure, devices used in daily business and key people with access to mission-critical information (e.g. when the electricity is cut off, a major issue occurs); IT processing and recovery are impacted; corporations have backups in place for other utilities – generators, for instance, so no one can use the backups
Threats to vital critical infrastructure components	Disruptions to electricity, financial sector, supply and chain model may cause outages that could lead to cascading failures
Threats to emerging technologies (e.g. Internet of Things (IoT))	Ransomware hijacks the IoT; encrypting a victim's data and then demanding payment for the encryption key; ransomware efforts on smart devices connected to the IoT; use the devices as gateways to install ransomware on other devices and systems; disrupt business operations and automated manufacturing production lines
Threats of violence to privileged insiders	Threaten privileged insiders to give up mission-critical information assets (e.g. financial details, intellectual property and strategic plans)
Threats to information integrity	Spread lies or distorting internal information for gaining a competitive or financial advantage at the expense of targets' reputations or operational effectiveness
Threats organization's reputation	Spreading misinformation about organization's working practices or products; a single attacker could deploy hundreds of chatbots (based on AI), each spreading malicious information and rumors over social media and news sites

Physical security has a different set of threats categorized in a broad set of categories:

- Natural environmental threats (floods, earthquakes, storms and tornadoes, fires, extreme temperature conditions, chemical contamination, etc.)
- Supply system threats (power distribution outages, communications interruptions, water interruptions, gas interruptions, etc.)
- Human threats (unintentional or intentional; examples: unauthorized access – internal and external, explosions, vandalism, fraud, theft, etc.)

Current analysis similar to the survey [Kettani 2019] of annual change (see examples listed in Table 5.2) in ranking of the top fifteen or twenty threats could help security professionals and managers to improve their strategies on measures for prevention and detection of threats. The authors observe that the top fifteen threats in 2017 have been the same top fifteen threats since 2014, although some order and trending have changed.

This list of Table 5.2 shows that some of the top threats belong to same distinct threat category and the top three threats (malware, Web-based, and Web application attacks) have remained consistent since 2012.

Therefore, there is a risk, an expectation of loss expressed as the probability that a specific threat will exploit a particular vulnerability with a particular harmful result [RFC 4949].

5.1.2 Vulnerabilities Characteristics

A vulnerability is a weakness in an information system, system security procedures, internal controls, or implementation that could be exploited or triggered by a threat source [CNSSI 4009]. Vulnerability, sometimes also called a security hole, is an aspect of the system that permits attackers to mount an attack.

Table 5.2 An example of annual change in threats ranking.

Top threats	Year						
	2018	2017	2016	2015	2014	2013	2012
Malware	1	1	1	1	1	2	2
Web-based attacks	2	2	2	2	2	1	1
Web application attacks	3	3	3	3	3	3	3
Phishing	4	4	6	8	7	9	7
Denial of service	5	6	4	5	5	8	6
Spam	6	5	7	9	6	10	10
Botnets	7	8	5	4	4	5	5
Data breaches	8	11	12	11	9	12	8
Insider threat	9	9	9	7	11	14	—
Physical manipulation/theft loss	10	10	10	6	10	6	12
Information leakage	11	13	14	13	12	13	14
Cryptojacking	13	—	—	—	—	—	—
Ransomware	14	7	8	14	15	11	9
Cyber espionage	15	15	15	15	14	—	—
Exploit kits	—	14	11	10	8	4	4

Source: [Kettani 2019]. © 2019, IEEE.

A vulnerability is a flaw or weakness in a system's design, implementation, or operation and management that could be exploited to violate the system's security policy [RFC 4949]. A system can have three types of vulnerabilities:

- Vulnerabilities in design or specification.
- Vulnerabilities in implementation.
- Vulnerabilities in operation and management.

Most systems have one or more vulnerabilities, but this does not mean that the systems are too flawed to use. Not every threat results in an attack, and not every attack succeeds. Success depends on the degree of vulnerability, the strength of attacks, and the effectiveness of any countermeasures in use. If the attacks needed to exploit a vulnerability are very difficult to carry out, then the vulnerability may be tolerable. If the perceived benefit to an attacker is small, then even an easily exploited vulnerability may be tolerable. However, if the attacks are well understood and easily made, and if the vulnerable system is employed by a wide range of users, then it is likely that there will be enough motivation for someone to launch an attack. A vulnerability is a characterization of a vulnerable state that distinguishes it from all non-vulnerable states. If generic, the vulnerability may characterize many vulnerable states; if specific, it may characterize only one [Bishop 1996].

Thus, many vulnerabilities in IT products can arise through failures as described in [ISO/IEC 15408]:

- Requirements – An IT product may possess all the functions and features required and still contain vulnerabilities that render it unsuitable or ineffective with respect to security.
- Development – An IT product does not meet its specifications, and/or vulnerabilities have been introduced as a result of poor development standards or incorrect design choices.
- Operation – An IT product has been constructed correctly to a correct specification, but vulnerabilities have been introduced as a result of inadequate controls upon the operation.

Weakness is a potential vulnerability whose risk is not clear. Sometimes several weaknesses might combine to yield a full-fledged vulnerability. Bishop and Bailey [Bishop 1996] argue that a vulnerability is a characterization of a vulnerable state that distinguishes it from all non-vulnerable states. If generic, the vulnerability may characterize many vulnerable states; if specific, it may characterize only one. By definition, a cyber attack begins in a vulnerable state [Bishop 1996].

Vulnerabilities range from a flaw in a software package to an unprotected system port or an unlocked door. The presence of a vulnerability does not in itself cause harm; a vulnerability is merely a condition or set of conditions that may allow the system or activity to be harmed by an attack.

Common vulnerabilities can be grouped in a vulnerability class. Usually, classification of vulnerabilities is performed using a framework during vulnerability analysis. The frameworks are distinct by the method to describe vulnerabilities such as [Bishop 2005]:

- Techniques used to exploit them.
- System's components such as software, hardware, and interfaces.
- Nature by their nature.

The practice of identifying, classifying, remediating, and mitigating vulnerabilities is called vulnerability management. There are an incredible number of possible vulnerabilities one can find in a system. During the design and implementation of a system, formal verification and property-based testing techniques are recommended to detect vulnerabilities [Bishop 2005].

Attackers usually know well enough to carry out manually or have scripts and vulnerability scanning tools that have the capability to discover vulnerabilities in operating systems, databases, Web servers, applications, firewalls, etc. However, vulnerability scanning may have penetration features. These tools are also used by security professionals, although their use is under strict control and only used if approved by the management. Some tools attempt to exploit a vulnerability to determine the actual degree of vulnerability. When a vulnerability is uncovered, the security professional can fix it before an attacker finds it. However, penetration testing is a testing technique, not a proof technique.

Vulnerability scanning is scanning computer systems for weaknesses in configurations or patch levels in order to identify potential entry points for intruders. Vulnerability scanning alone will not protect any computer system. Vulnerability scanning does not detect:

- Legitimate users who may have inappropriate access.
- An intruder who is already in the system.

A recent study of ten popular Internet of Things (IoT) devices uncovered a total of 250 security flaws among them [Halpern 2014].

Understanding vulnerabilities and the associated attack vectors to exploit the systems is essential to building effective security mitigation strategies. Therefore, a vulnerability is a potential avenue of attack.

5.2 Attacks

Attack is an attempt to destroy, expose, alter, disable, steal, or gain unauthorized access to or make unauthorized use of an asset [ISO/IEC 27000]. Attack is an actual assault on system security; it is a method or technique used in an assault [RFC 4949]. Attack is an act that is an intentional or unintentional attempt to compromise the information and/or the systems that support it [Whitman 2012]. An attack can be done with a malicious intent, or it might be accidental. Attacks can be performed against information (in paper form or electronic form), information systems, physical assets, information infrastructures, or organizations.

There is a plethora of names for an attack such as incident, intrusion, security incident, security intrusion, event, attack event, security event, information security attack, cyber attack, compromise, security compromise, etc. These terms are used in security books (e.g. [Whitman 2012], [Harris 2013], [Anderson 2008], [Vacca 2012], [Hansche 2004], [Krutz 2006], [Krutz 2004], [Maiwald 2004]), standards such as [ISO/IEC 27000], [NIST SP800-53r4], and dictionaries such as [RFC 4949], [CNSSI 4009], although there could be slight differences in a context of an application that may apply for some terms. Understanding the terms and context is essential. For example, the term called security event is defined as an occurrence in a system that is relevant to the security of the system. However, this term covers both events that are security incidents (called also intrusion, attack, etc.) and those that are not.

5.2.1 Attack Categories

The means of an information security attack can be:

- Technical (e.g. brute force to guess a password).
- Nontechnical or social engineering (making phone calls or walking into a facility and pretending to be an employee).

There are several ways to categorize attacks. Attacks can be characterized according to intent [RFC 4949]:

- Active attack attempts to alter system resources or affect their operation.
- Passive attack attempts to learn or make use of information from a system but does not affect system resources of that system (e.g. wiretapping).

Attacks can be characterized according to point of initiation [RFC 4949]:

- Inside attack is one that is initiated by an entity inside the security perimeter (an insider of the organization), e.g. an entity that is authorized to access system resources but uses them in a way not approved by the party that granted the authorization.
- Outside attack is initiated from outside the security perimeter, by an unauthorized or illegitimate user of the system (an outsider). In the Internet, potential outside attackers range from amateur pranksters to organized criminals, international terrorists, and hostile governments.

Attacks can be characterized according to method of delivery [RFC 4949]:

- Direct attack is when the attacker addresses attacking packets to the intended target or victim.
- Indirect attack is when the attacker addresses packets to a third party, and the packets either have the address of the intended victim as their source address or indicate the intended victim in some other way; the third party responds by sending one or more attacking packets to the intended victims.

The term attack relates to some other basic security terms as shown in the following diagram (see Figure 5.1). Similar graphical representation is provided by [Ææqwerty 2020].

A system resource (either physical or logical) may have one or more vulnerabilities that can be exploited by a threat agent in a threat action. The result can potentially compromise the confidentiality, integrity, or availability of resources belonging to a system or organization.

The attack can be active when it attempts to alter system resources or affect their operation: so it compromises integrity or availability. A passive attack attempts to learn or make use of information from the system but does not affect system resources: so, it compromises data confidentiality.

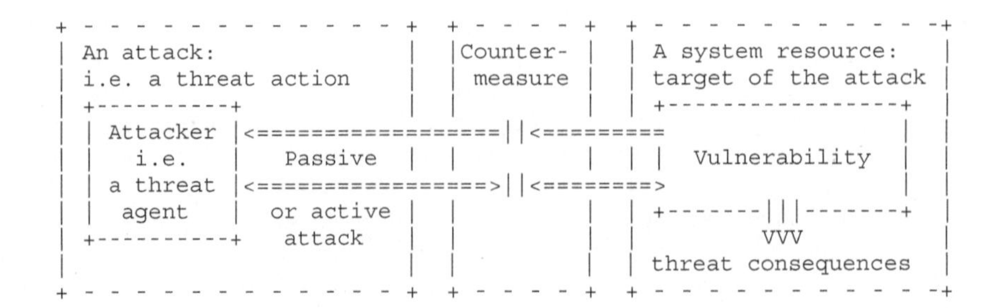

Figure 5.1 Attack types and related concepts. *Source:* [RFC 4949]. Public Domain.

Whatever the type of attack, the attacks primarily fall into four main categories:

- Access – An attempt to gain information that the attacker is not authorized to see; an access attack can be against information in storage or in transit (e.g. snooping, eavesdropping, interception, stealing the storage media or portable device).
- Modification – An attempt to modify information that an attacker is not authorized to modify; a modification attack can be against information in storage or information in transit; types of modification attacks include changes, insertion, or deletion.
- Denial of service – An attempt to deny the use of resources to legitimate users of the system, information, or capabilities; a DoS attack can be against information in storage or information in transit causing that information to be unavailable; this may be caused by the destruction of the information, or changing of the information into an unusable form, or moving the information to an inaccessible location.
  A variation of DoS is distributed denial-of-service (DDoS) attack in which a coordinated stream of requests is launched against a target from many locations at the same time.
- Repudiation – An attempt to give false information or to deny that a real event or transaction should have occurred (e.g. masquerading, denying an event).

Although no unique classification, attack methods can also be classified as methods used by untargeted threat agent (e.g. hacker) and methods used by a targeted threat agent [Maiwald 2004].

The untargeted hacker is not looking to access a specific system. Most untargeted hackers identify individual systems and attempt the exploit on one system at a time. More sophisticated hackers use the reconnaissance tools to identify many vulnerable systems and then write scripts to exploit these systems.

A targeted hacker attempts to successfully penetrate or damage a particular organization. They are motivated by a desire for something that organization has. Usually information of some type may include intellectual property, research and design projects, market information, and personnel and customer information. In some cases, the hacker chooses to do damage to a specific organization for some perceived wrong. Many of DoS attacks occur in this way. The skill level of targeted hackers tends to be higher than that for untargeted hackers. The targeted threat agent may use electronic attack methods or physical attack methods.

In general, some common security attacks target security goals of the information [Singh 2014]:

- Confidentiality – Snooping and traffic analysis are two types of attack that threaten the confidentiality of information. Snooping refers to unauthorized access to or interception of data (e.g. overhearing, eavesdropping over a communication line). Traffic analysis refers to other types of information collected by an intruder by monitoring online traffic.
- Integrity – Several kinds of attack such as modification, masquerading, replaying, and repudiation threaten the integrity of information. Modification involves corrupting transmitted data or tampering with it before reaching its destination. It happens when a perpetrator captures, modifies, steals, or deletes important information via network access or direct access using executable codes.
- Availability – The attacker can use several strategies to threaten availability. An example is interruption attack – e.g. blocking access to a service by overloading an intermediate network or network device, a network service is made degraded or unavailable for legitimate use; interruption attack intends to disrupt traffic (e.g. physically breaking a communication line). A DoS attack in which the perpetrator seeks to make a machine or network resource unavailable to its intended users by temporarily or indefinitely disrupting services of a host connected to the Internet.
- Authenticity – Fabrication creates illegitimate information, processes, communications, or other data within a system – faking data as if it were created by a legitimate and authentic party. When a known system is compromised, attackers may use fabrication techniques to gain trust, create a false trail, collect data for illicit use, or spawn malicious or extraneous processes.

The author provides a graphical representation of different states of information flow when no attacks occur (normal flow) and after attacks (interruption, interception, modification, and fabrication). Figure 5.2 shows paths through which false injection attacks (active attacks) could adversely affect a power system.

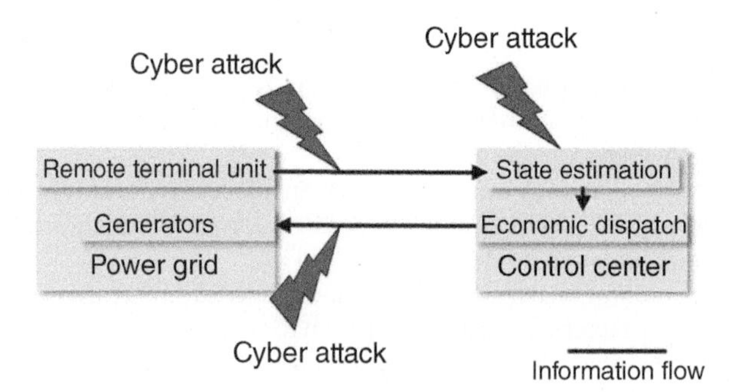

Figure 5.2 Cyber attacks on a power system. *Source:* [Xu 2020]. Licensed under CC BY-SA 4.0.

These attacks aim to compromise the readings of multiple power grid sensors and phasor measurement units in order to mislead the operation and control centers [Rahman 2012]. This work [Xu 2020] reviews the research on false injection attacks on economic dispatch, state estimation, and power system dynamic stability. For example, false injection attack can induce the generation of inappropriate control commands by directly targeting economic dispatch.

5.2.2 Reasons for Attack

Other ways to look at avenues of attack are the reasons as described in [Conklin 2004]. There are two general reasons that a specific system is attacked: either it is specifically targeted by the attacker, or it is an opportunistic target. For the first type of attack, the target is known, but the software and hardware are not known. An attack against a target of opportunity is conducted against a site that has hardware or software that is vulnerable to a specific exploit. This is an opportunity for a hacker to obtain credit card or other personal information. Targeted attacks are more difficult to realize and take more time than attacks on target of opportunity. The latter simply relies on the fact that with any piece of widely distributed software, there will almost always be somebody who has not removed a vulnerability (e.g. patched a system).

Understanding the various categories of attacks and threat's motivations is critical on deriving the adequate protection. Information security involves identifying the threats and vulnerabilities of the organization and managing them appropriately. Implementing a proper information security system is not a one-time activity. It requires a constant vigilance against new security threats that might occur.

Risk is the combination of threat and vulnerability. Threats without vulnerabilities pose no risk. Likewise, vulnerabilities without threat pose no risk [Maiwald 2004]. However, in the real world, neither of these conditions exists.

The measurement of risk is an attempt to identify the likelihood that a detrimental event will occur [Maiwald 2004]. It is difficult to estimate the value of information, but the estimation of the probability of a threat occurrence or attack is even more difficult. However, resources should be allocated to identify and evaluate the potential risk. Security represents a process of maintaining an acceptable level of risk. Risk can be measured in quantitative and qualitative terms. There are circumstances and scenarios that have to be considered in the process of measuring risk. These may include the following:

- In most cases, an organization needs to rely on its own internal information to calculate the risk and exposure factors.
- Often these values are estimated based on best guess.
- There could be various levels and types of controls that are necessary to successfully counter attacks, reduce risk, and improve the security within an organization.

The following are examples of attacks that compromise key security attributes (CIA):

- Attacks against confidentiality – e.g. interception, eavesdropping.
- Attacks against integrity – e.g. IP spoofing, sequence number attack, man-in-the-middle attack.
- Attacks against availability – e.g. DoS, traffic redirection; precursor to attack is port scanning.

Therefore, understanding the information about the agent's motivations for an attack including the vulnerabilities and potential risk to information and information systems is key to smart security [Verizon 2016].

5.3 Energy Sector: Threats, Vulnerabilities, and Attacks Overview

We need to understand relevant information about the threat agent's motivations for an attack to energy sector including the vulnerabilities and potential risk to information and information systems.

5.3.1 Threats

Examples of common and specific threats to energy sector and electrical utilities are summarized in Table 5.3 based on selected information from [Kaspersky 2014], [Hentea 2008], [Pagliery 2014].

Table 5.3 Examples of threats to energy sector and electrical utilities.

Threats	Examples
Sophisticated malware (e.g. BlackEnergy, Havex Trojan) campaign	BlackEnergy and Havex Trojan target ICS environments; specifically, BlackEnergy targets Internet-connected HMIs, which can be compromised by this campaign that has been ongoing since at least since 2011
Remote SCADA access	Exploitation of vulnerabilities and changes to SCADA controls and changes to IED functions
Collateral software	Collateral software to issue commands to IED devices; manipulation of DNP3 protocol analyzer used as man-in-the-middle attack to issue commands to IED devices
Traditional IT security technologies	Use of traditional IT security technologies (e.g. VPN with encryption) to camouflage compromised packets
	Compromised packets of the OPC (Object linking and Embedding (OLE) for Process Control) flat files are used in the distributed control system (DCS), workstations, or both to change voltages of electrical system to a remote location
Use of non-routable protocols	Compromise of non-routable protocols; frequency-hopping radio spectrum (900 MHz) can be compromised
Aurora threat	Flaw in equipment design of rotating equipment can be the threat associated with rapidly disconnecting and reconnecting a generator to the grid but out of phase – via physical or cyber intrusion of control systems conducted maliciously or unintentionally – could have serious effects on system operation and damage the equipment
Insider threat	Knowledgeable people of equipment can manipulate configurations and issue changes to SCADA control commands, sending false data and instructions, disabling alarms
Use of wireless technologies	Use of wireless technologies and two-way communication without being restricted and authorized to access the system; lack of security policy and procedures on using wireless technologies to control field equipment
Information security mismanagement	Lack of auditing and logging security activities and events, identification and authentication, and monitoring systems; unauthorized access is not detected and faults cannot be determined
Personnel skills	Lack of awareness and training – untrained personnel cannot recognize obvious faults in the control systems, functions, and control commands

(Continued)

Table 5.3 (Continued)

Threats	Examples
Operations	Lack of operations controls and policy to secure critical information used to manage an ICT infrastructure; single owner of critical information such as configurations, passwords, and other commands could refuse disclosure of the information to other parties if policy and procedures are not in place
Security controls	Lack of detection and intrusion techniques; vendors remotely perform real-time updates to ICS software and changes to configurations and installation of unauthorized software in SCADA systems without being detected
Network design	Lack of appropriate ICS network separation from corporate networks; ICS are connected to corporate networks without any controls
Tools	Lack of monitoring tools to detect unintentional operations or broadcast storm, broadcasting messages to ICS networks with limited bandwidth
Testing	Untested tools and configurations may cause interactions; these occur when security hardware is installed without being checked and tested in advance
Failure management	Inappropriate design of failures and isolation; hard drive unavailability from a single event is not isolated and impacts multiple critical applications
EMI and EMC levels	ICS devices are affected by electromagnetic interference (EMI) increased levels; if new Smart Grid equipment does not have adequate levels of electromagnetic compatibility (EMC) for the immunity may be causing increased interference among devices and danger to safety
Penetration testing	Inappropriate design of penetration testing; active penetration testing can cause failures to control system workstation
Vulnerability management	Vulnerability mismanagement can cause failure of communication networks; router failure due to worm infection in unpatched software causing loss of communications
Access controls	Backdoors can be used by malware to access several computers such that a single point of compromise may provide extended access because of preexisting trust established among interconnected resources (e.g. corporate networks connected to SCADA systems)

Source: [Kaspersky 2014], [Hentea 2008], [Pagliery 2014].

5.3.2 Vulnerabilities

Besides common IT vulnerabilities, the energy sector has to counter more vulnerabilities such as those related to control systems and SCADA systems. Dependency on technology and SCADA systems including wireless intrusion are potential risks to the security of Smart Grid and DER systems.

Table 5.4 includes a list of ten common vulnerabilities on SCADA systems based on the analysis of SCADA systems between 2003 and 2010 according to Idaho National Laboratory Report [INL 2011]. The report, prepared by Idaho National Laboratory for the Department of Energy (DOE), describes the common vulnerabilities on energy sector control systems and provides recommendations for vendors and owners of those systems to identify and reduce those risks.

Also, the report describes a standard metrics-based approach to evaluate the relative risk associated with common SCADA vulnerabilities based on the Common Vulnerability Scoring System Version 2 (CVSS) and CWE methodologies. The report provides mitigation approaches for vulnerabilities that reduce the risk. A vulnerability can result in harm to the system or its operation, especially when it is exploited by a hostile actor or it is present in conjunction with particular events or circumstances.

Table 5.4 Common SCADA vulnerabilities.

Common vulnerability	Reason for concern
Unpatched published known vulnerabilities	Most likely attack vector
Web human–machine interface (HMI) vulnerabilities	Supervisory control access
Use of vulnerable remote display protocols	Supervisory control access
Improper access control (authorization)	SCADA functionality access
Improper authentication	SCADA applications access
Buffer overflows in SCADA services	SCADA host access
SCADA data and command message manipulation and injection	Supervisory control access
SQL injection	Data historian access
Use of standard IT protocols with clear-text authentication	SCADA host access
Unprotected transport of application credentials	SCADA credentials gathering

Source: [INL 2011]. Public Domain.

One example of statistics is the relative frequency of vulnerabilities found in SCADA components categorized by the SCADA architecture functional level using the ISA99 reference model [ISA 99] (see Figure 5.3).

The reference model describes a SCADA system as a series of logical levels based on functionality. The relative assessment findings of the report describe vulnerabilities as follows [INL 2011]:

- Level 1: Local or basic control (10%)
- Level 2: Supervisory control (45%)
- Level 3: Operations management (40%)
- Level 4: Enterprise systems (5%).

The most significant vulnerabilities identified in SCADA are those that allow unauthorized control of the physical system. Compromise of the SCADA's availability and ability to function correctly may also have significant consequences. Likelihood of a successful attack must also be considered when assessing risk. Exposure to attack, attacker awareness of the vulnerability, and exploitation knowledge help assess the probability of a successful attack.

One survey report describes SCADA vulnerabilities from a different perspective [Caswell 2011]. This report provides additional information about different categories of vulnerabilities (security flaws resulting from legacy devices and software, configurations, inadequate security policies) as well as how to increase awareness and improve security controls.

A list of vulnerability classes is included in [NISTIR 7628r1]. This list was developed using information from several existing documents and Web sites, [NIST SP800-82r2], and the Open Web Application Security Project (OWASP) vulnerabilities list. The NIST document is focused on using these vulnerability classes to ensure that the security controls address the identified vulnerabilities. However, a risk management approach would provide better results, and it is highly recommended.

5.3.3 Energy Sector Attacks

Recent reports show that several utilities were attacked by various threat agents. One survey report provides statistics on incident reports of US major industry sectors chosen (financial, insurance, retail, utilities, and education) [Verizon 2016]. The analysis shows that about 73 utilities experienced security incidents and about 10 utilities confirmed data loss in 2015. Also, a comparison of malware events/week during the year of 2014 shows that utilities experienced an average of 772 events/week, much higher than financial (350 events/week) and insurance (575 events/week), and only a little bit lower than retail (801

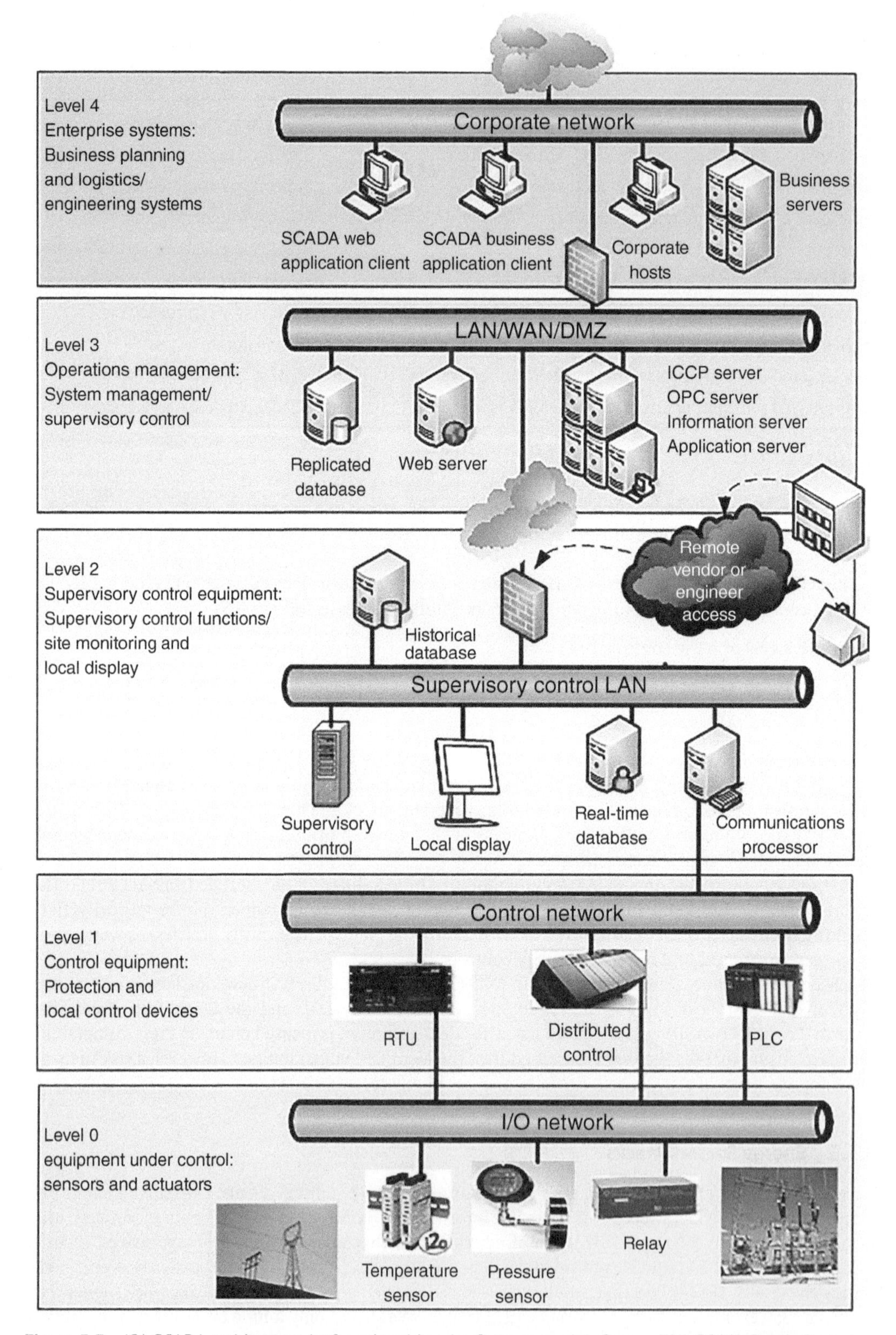

Figure 5.3 ISA SCADA architecture by functional level reference model. *Source:* [INL 2011]. Public Domain.

events/week). In addition, utilities rank higher than financial and retail sectors in regard to time to fix a vulnerability exploited by a security intrusion. This time could be up to 55 days. This situation is contrary to expectations because electric utilities are critical infrastructures and should have mechanisms to detect malware and fix vulnerabilities with a higher speed. The response time is unexpectedly higher even than retail sector. Also, the report includes an analysis of victims based on similar intrusion characteristics such as threat actors, actions, and compromised assets. The analysis shows clustering trend in the top 10 espionage plotted graph with utilities on fourth place in this ranking list.

Although there were fewer attacks reported by utilities in 2014 than in 2013, still hackers attacked the US energy grid 79 times in 2014 according to this report [Pagliery 2014].

In February 2011, the Director of National Intelligence testified that in 2010, during one year, there had been a dramatic increase in malicious cyber activity targeting US computers and networks, including a more than tripling of the volume of malicious software since 2009 [Director]. Cyber threats can be unintentional or intentional. Unintentional threats can be caused by natural disasters, human errors, software upgrades, or maintenance procedures that inadvertently disrupt systems. Intentional threats include both targeted and untargeted attacks from a variety of sources, including criminal groups, hackers, disgruntle employees, foreign nations engaged in espionage and information warfare, and terrorists. These cyber threats can exploit various vulnerabilities and adversely affect the functions of computers, software, a network, an enterprise's operations, an industry, or the Internet itself.

Throughout 2011 and 2012, it is observed an increasing trend in cyber attacks targeted at energy and pipeline infrastructure around the world. In 2012 alone, attacks against the energy sector comprised over 40% of all incidents reported to ICS-CERT [Virus 2012], [ICS-Letter]. Many of these incidents targeted information pertaining to the industrial control system (ICS)/SCADA environment, including data that could facilitate remote access and unauthorized operations. An identity theft attack was discovered too. In 2012, ICS-CERT published 332 information products, warning the ICS community about various threats and vulnerabilities that could impact control systems. The energy sector led all others again in 2014 with the most reported incidents [ICS-CERT 2015]. One recent reported incident in a power center is unauthorized access to systems due to design errors and inappropriate security controls [ICS-CERT 2016].

Several cases of confidentiality breaches and lack of access controls in DR applications by disclosing the information to competitors are reported [FERC 2012]. In one case, demand forecast and metered data information pertaining to roughly 150 market participants, covering a period of approximately 32 months, had been inadvertently disclosed to 5 market participants. The information was confidential in nature, and each market participant should only have been allowed access to its own information. Other cases point violations due to human error or computer system capabilities. In another case, an error caused a discrepancy of price displayed via a public Web site and actual price used for settlements during trading.

A recent virus infection at an electric utility in the United States resulted in downtime for the impacted systems and delayed the plant restart by approximately three weeks [ICS-Letter]. The virus infection occurred in a turbine control system that impacted approximately 10 computers on its control system network. The recovering time of three weeks points that not appropriate procedures were in place for prevention and recovery.

A test performed by a group of researchers in April 2012 revealed that many critical infrastructure assets are directly facing the Internet. Devices were having either weak, default, or nonexistent logon credential requirements, freely available such that anyone with malicious intent could locate these devices and attempt logon, leaving these systems exposed to cyber attacks [ICS-Letter]. Once accessed, these devices may be used as an entry point onto a control systems network, making their Internet facing configuration a major vulnerability to critical infrastructure. ICS-CERT evaluated approximately 7200 devices in the United States that appear to be directly related to control systems associated with critical infrastructure [ICS-Letter].

The Stuxnet worm [Falliere 2011] is a threat that was primarily written to target an ICS or set of similar systems. The ultimate goal of Stuxnet is to sabotage a facility by reprogramming programmable logic controllers (PLCs) to operate as the attackers intend them to, most likely out of their specified boundaries. Stuxnet's ability to reprogram the logic of control hardware and alter physical processes demonstrates the danger of modern cyber threats. The worm exploited vulnerabilities in the operating system (Windows). Stuxnet was discovered in July 2010, but it is confirmed to have existed at least one year prior and likely even before [Falliere 2011]. In July 2010, the malware was installed via USB media during an activity on the configuration of the nuclear enrichment facility in Iran. The success of the Stuxnet worm was an example of how nowadays sophisticated and well-resourced attackers can develop complex cyber attacks causing severe damage to power infrastructures [Leszczyna 2013].

The threats and attack vectors emerged as a sophisticated criminal ecosystem that has matured to the point that it functions much like any business – management structure, quality control, offshoring, and so on [Loveland 2012]. As pointed in this work, companies also must worry about a new type of risk – the advanced persistent threat, which is a type of hacking predominantly about stealing intellectual property and typically is associated with state-sponsored espionage. The motives go beyond financial gain, in which at risk is not only intellectual property but possibly national security. According to this work, an overview of cybersecurity landscape reveals that:

- Events such as breaches and financial losses are on the rise. Of those, risks associated with customers, partners, or suppliers are a major concern, having nearly doubled in the period of 2010–2012.
- Given the economic uncertainty, security has not been a priority; the levels of investment, awareness, and training all have declined.

Another recent survey of International Data Group for 2016 [IDG 2016] reveals the following facts about electric and energy sector:

- Operational systems of power and utilities organizations are targeted by highly skilled nation state actors, organized crime, and terrorists that attack power grids as an act of cyberwarfare.
- Advancement in incidents corresponds with a dramatic rise in theft of intellectual property as well as exploits of operational and embedded technologies.
- Compromise of operational systems more than doubled in 2015 and that exploits of embedded systems quadrupled.
- Compromise of customer records soared 62% in 2015 – despite a significant drop in the overall number of security incidents detected.

While utilities observed that the number of security incidents decreased in 2015, the employee and customer records remain the top targets of cyber attacks and continue to increase and theft of hard intellectual property tripled in 2015 [PwC 2016a], [PwC 2016b].

Physical threats to information systems fall into two broad categories: extreme environmental events and adverse physical plant conditions. Extreme environmental events include earthquake, fire, flood, electrical storms, and excessive heat and humidity. The information system may be housed in a building, in which, in addition to computers and communication lines located throughout the building, there may be dedicated computer rooms and data storage rooms. Connections for power supply and communication may lead to and from the building. Adverse physical plant conditions may arise from breach of physical security measures, power failures or surges, air-conditioning malfunction, water leaks, static electricity, and dust. An organization may be affected by lapses either directly at its premises or indirectly at a vital point outside the organization, such as power supply or telecommunication channels.

Human beings and the institutions they establish to reflect their values, whether social, economic or political, as well as the lack of such institutions, all contribute to security problems. The diversity of system users – employees, consultants, customers, competitors, or the general public – and their various levels of awareness, training, and interest compound the potential difficulties of providing security. Lack of training and follow-up about security and its importance perpetuate ignorance about proper use of information systems. Without proper training, operators and users may not be aware of the potential for harm from system misuse.

5.3.4 Smart Grid Cybersecurity Challenges

With the ongoing transition to the Smart Grid, the IT and telecommunication sectors are more directly involved. These sectors have existing cybersecurity standards to address vulnerabilities and assessment programs to identify known vulnerabilities in their systems. The same vulnerabilities need to be assessed in the context of the Smart Grid infrastructure. However, the Smart Grid could have additional vulnerabilities due to its complexity, large number of stakeholders, and highly time-sensitive operational requirements.

Summarizing the issues, these are challenges to the smooth functioning of Smart Grid and DER systems:

- Threats include technological development, technical problems, extreme environmental events, adverse physical plant conditions, human frailty, and inadequacies of social, political, and economic institutions.
- Attacks to information systems may arise from intentional or unintentional acts and may come from internal or external sources.

The Smart Grid is vulnerable to traditional cyber or computer-based attacks typically geared toward disabling devices or networks. However, the Smart Grid is also vulnerable to physical attacks where sensors can be tricked into reporting false conditions that cause the control system to react in an inappropriate manner.

Cyber–physical attacks blending both cyber and physical attack components are also a possibility. The probability of success of a cyber attack to Smart Grid control infrastructures can increase with the massive deployment of advanced automation and communication technologies relying on standardized protocols.

The migration from older legacy-type architectures to modern operating systems and platforms can force ICS to inherit many cybersecurity vulnerabilities, with some of these vulnerabilities having countermeasures that often cannot be deployed in automation systems.

Historically, the security issues have been the responsibility of the corporate IT security organization, usually governed by security policies and operating plans that protect information assets. As ICS become part of larger conjoined architectures, the main concern is providing security procedures that cover the control systems domain as well.

One of the key applications of Smart Grid is the efficient ability to manage energy loads and consumption of energy within many domains. This application allows utilities to work with customers to control and manage home energy consumption. However, the presence of both Smart Grid networks and public Internet connections at the customer site (e.g. within the home) may introduce security concerns that must be addressed. With the customer potentially having access to utility-managed information or information from a third party, safeguards are required to prevent access to the utility control systems that manage power grid operations.

As multiple DER devices are introduced in the network, the need for remote connection between them and their interconnection with the power grid introduce more vulnerabilities because of interactions between systems and vulnerabilities that belong to other systems interconnected with the DER systems in the grid.

The security of advanced metering infrastructures (AMIs) is of critical importance because of interconnection with DER systems and applications. Examples of vulnerabilities in devices and networks that connect smart meters, distributed energy resources, and energy storage systems are numerous and if exploited can result in serious damages [ICS-Letter], [McManus 2012], [Robertson 2010], [ReportLinker 2013]. AMIs are composed of smart meters, data concentrators, communication networks, and headend.

The security and accuracy of data collected via AMI are essential to ensure the proper functionality of DR applications, energy storage systems, and other DER-related systems.

Energy theft is an immense concern in smart metering [McLaughlin 2009]. Along with physical tampering, a smart meter in AMI is more vulnerable than a previous analog meter because current smart meters are equipped with software that can be compromised easily. Therefore, end-to-end security must be provided across the AMI systems, encompassing the customer end systems as well as the utility and third-party systems that are interfaced to the AMI systems.

AMI systems consist of the hardware, software, and associated system and data management applications that create a communication network between end systems at customer premises (including meters, gateways, and other equipment) and diverse business and operational systems of utilities and third parties [Faisal 2012]. An overview of AMI components and networks (highlighted with dotted lines) within the bigger context of electric power distribution, consumption, and renewable energy generation and energy storage is shown in Figure 5.4 [Faisal 2015].

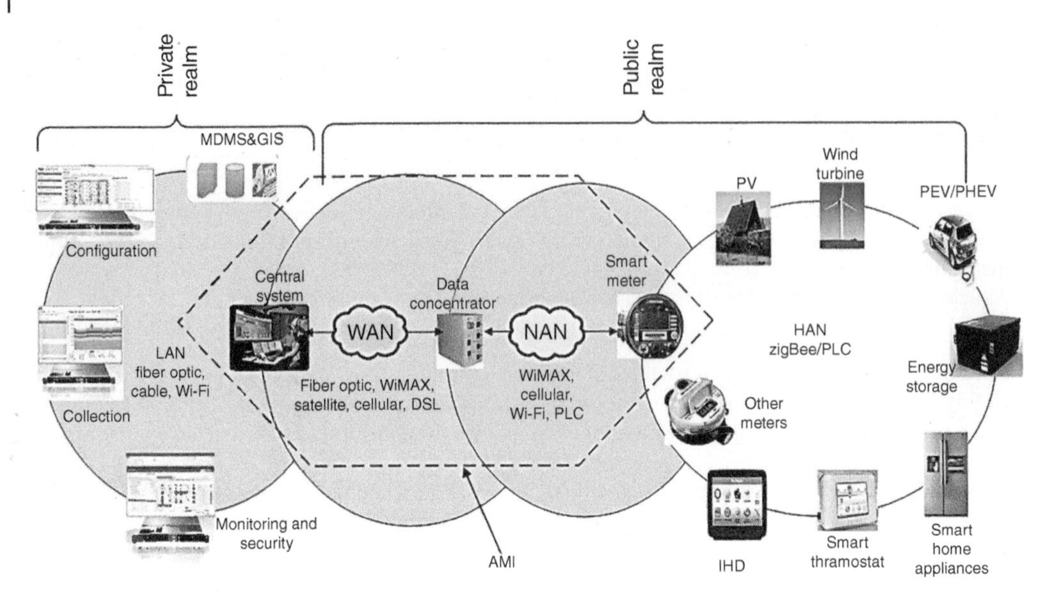

Figure 5.4 Overview of AMI components and networks. *Source:* [Faisal 2015]. © 2015, IEEE. AMI = Advanced Metering Infrastructure; DSL = Digital Subscriber Line; GIS = Geographic Information System; HAN = Home Area Network; IHD = In Home Display; LAN = Local Area Network; MDMS = Meter Data Management System; NAN = Neighborhood Area Network; PEV = Plug-in Electric Vehicle; PHEV = Plug-in Hybrid Electric Vehicle; PLC = Power Line Communication; PV = Photovoltaic.

Smart meter flaws could allow hackers to tamper with power grid [Robertson 2010]. The attacks could be stealing meters, which can be situated outside of a home, and reprogramming them, wirelessly hacking the meter from a laptop. Flaws can be attributed to the smart meter technology (even protocols that were designed recently exhibit security failures known for more than a decade) and the technologies that utilities use to manage data from meters. Another reason for security flaws is that smart meters are being developed without enough security testing and probing.

The complexity and diversity of technologies used in home networks is another security challenge because all these technologies have not implemented or have limited security features.

Vulnerabilities in home protocols such as ZigBee [Masica 2007] or G.hn [Gartner 2013] should be analyzed, and countermeasures have to be designed to avoid the exposure to cyber threats.

A major concern against the smart meters is privacy. Privacy concerns focus upon the collection of detailed energy data from customers, the accessibility of that data through the utility and, possibly, at the site of the meter, and the potential for sharing of this energy data without the knowledge or desire of customers. It is reported that consumers lack information about security risks associated with the home devices and utilities avoid any investments for security protection [GAO 2011].

As described in [UCAIUG 2008], AMI is the convergence of the power grid, the communication infrastructure, and the supporting information infrastructure. By decomposing the complex AMI system in domains, a security service model is developed for applying security requirements to implement a robust secure AMI solution. Figure 5.5 shows a view of six services, and Table 5.5 provides a description of these services.

Each AMI implementation may be specific to each utility that may choose technologies to meet their business model and have their policies and deployment environment.

With the customer potentially having access to utility-managed information or information from a third party, safeguards are required to prevent access to the utility control systems that manage power grid operations.

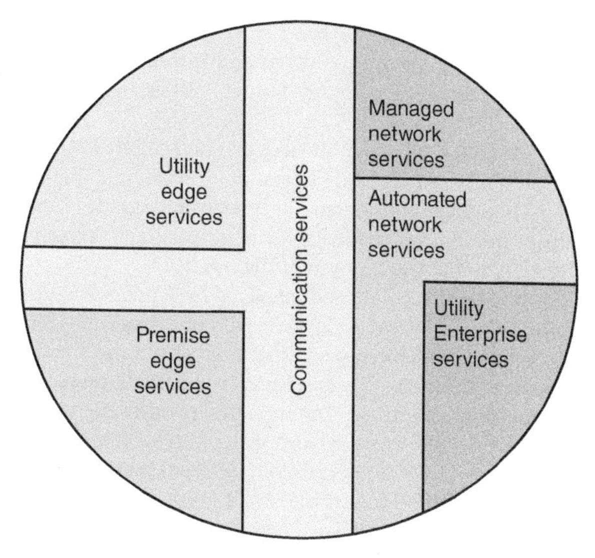

Figure 5.5 AMI security domain model. *Source:* [UCAIUG 2008].

Table 5.5 AMI security domain description.

Security domain	Description
Utility edge services	All field service applications including monitoring, measurement, and control managed by the utility
Premise edge services	All field service applications including monitoring, measurement, and control managed by the customer (customer has control to delegate to third party)
Communication services	Applications that relay, route, field aggregate, field communications aggregate, and field communications management information
Management services	Attended support services for automated and communications services (includes device management)
Automated services	Unattended data collection, data transmission, transformation, response, and staging
Business services	Core business applications (includes asset management)

Source: [UCAIUG 2008].

As more embedded computers are used in cars, thermostats, meters, dishwashers, cell phones, and many other systems that connect to Internet, security for these systems is an open question and could prove a more difficult long-term problem than security does today for desktop and enterprise [Koopman 2004]. There is a potential danger if an intruder might trick a number of thermostats into thinking that it is not a peak day, thereby increasing demand. If done on a broad enough scale, this could cause power grid failure, especially if the electricity provider has factored the ability to change set points into its plan for sizing its generating capacity. If a thermostat is connected to the Internet, an attacker could run the battery down simply by repeatedly querying the thermostat's status. A low-voltage detection circuit could disable the wireless connection before the battery died, but if the developer did not design this capability into the system, this vulnerability is critical. In battery-powered devices that use power-hungry wireless communication, too many networking conversations can run the battery down quickly [Koopman 2004]. Similarly, a washing machine can be used to launch a DoS attack to a utility or government.

While the convergence of once isolated ICS has helped organizations simplify and manage their complex environments, by connecting these networks and introducing IT components into the ICS domain, security problems arise because of many vulnerabilities in protocols and computing platforms [DHS 2011a], [ICS-Letter].

One or more vulnerabilities in IT products or devices can be exploited by threats that can attack the resources (information or devices to manage it). For example, accidental quality defects, or worse case, intentional corruption of components and systems intended to degrade, compromise, or control the system, create vulnerabilities through embedded malware, backdoors, Trojans, etc. and many supply chain vulnerabilities in SCADA and power equipment [Hawk 2014].

Although vulnerabilities should be identified and managed by vendors and utilities, recent report indicates that researchers publish vulnerabilities for financial gain or recognition [Higgins 2012].

It is necessary to address not only the individual vulnerabilities but the breadth of risks that can interfere with critical operations. For example, by sending a false control message from a computer connected to the Internet, an unauthorized intruder can manipulate traffic signals, electric power switching stations, chemical process control systems, or sewage water valves, creating major concerns to public safety and health. Also, threats need to be analyzed from different perspective.

Threats to Smart Grid can be classified into three broad groups [Yousuf 2010]: system level threats that attempt to take down the grid, attempts to steal electrical service, and attempts to compromise the confidentiality of data on the system.

All information held and processed by an organization is subject to threats of attack, error, nature (for example, flood or fire), etc. and is subject to vulnerabilities inherent in its use. Information security is generally based on information being considered as an asset that has a value requiring appropriate protection, for example, against the loss of availability, confidentiality, and integrity. Enabling accurate and complete information to be available in a timely manner to those with an authorized need is a requirement for business efficiency. Thus, organizations face a range of risks that may affect the functioning of assets and address their perceived risk exposure by implementing security controls.

Information and communication technologies (ICT) have experienced impressive evolutions and advances. As a consequence, they have enabled a number of technologies (laptops, smartphones, tablets, TV, etc.) with important computational power capabilities compared with devices of some years ago. These advances have made the development of applications that facilitate an important number of activities through the Internet such as communication, collaboration, learning activities, purchases, or remote monitoring and control possible. Although all these possibilities are advantageous to the user, they also present a number of security risks. These risks are derived from the need to establish communications between remote entities that cannot mutually verify in a physical manner who they are.

A threat (terrorist attack or natural event) realized as an attack to Smart Grid may cause a disruption (an interruption in functionality), and the system may lose functionality and enter an interim state in which one or more additional threats are encountered. Privacy implications for Smart Grid include technology deployment centers on the collection, retention, sharing, or reuse of electricity consumption information on individuals, homes, or offices. Besides threat consequences to Smart Grid functions and systems, equally important is to address potential privacy consequences. Examples of potential privacy consequences of Smart Grid systems include [EPIC Policy]:

- Identity theft.
- Determine personal behavior patterns.
- Determine specific appliances used.
- Perform real-time surveillance.
- Targeted home invasions (latchkey children, elderly, etc.).
- Activity censorship.
- Decisions and actions based upon inaccurate data.
- Profiling.
- Unwanted publicity and embarrassment.
- Public aggregated searches revealing individual behavior.

Smart Grid may also affect consumers who adopt the use of storage systems and renewable resources such as solar and wind power. The Smart Grid could make it possible to transfer excess electricity from power generating customers to others during peak periods. Examples of personally identifiable information that may be collected include details on battery charging information, e.g. amount of life remaining, date, time, location of last recharge, etc.; type of personal device; a unique item identification number as well as personalized information, e.g. user name, address, etc.; and location where the item was recharged as well as how long the device was connected to the power source. Therefore, the appropriate protection of the security of the Smart Grid and DER systems is mandatory.

In conclusion, Smart Grid cybersecurity must address not only deliberate attacks, such as from disgruntled employees, industrial espionage, and terrorists, but also inadvertent compromises of the information infrastructure due to user errors, equipment failures, and natural disasters.

5.4 Security Controls

Generally, all information held and processed by an organization is subject to threats of attack, error, nature (for example, flood or fire), etc. and is subject to vulnerabilities inherent in its use. The term information security is generally based on information being considered as an asset that has a value requiring appropriate protection, for example, at minimum, against the loss of availability, confidentiality, and integrity. Enabling accurate and complete information to be available in a timely manner to those with an authorized need is a catalyst for business efficiency.

Information security requires the application and management of appropriate controls (called also security controls, countermeasures, safeguards, measures, or security measures) that involves consideration of a wide range of threats, with the aim of ensuring sustained business success and continuity and minimizing impacts of information security incidents.

An information security incident is a single or a series of unwanted or unexpected information security events that have a significant probability of compromising business operations and threatening information security [ISO/IEC 27000]. Security events are identified occurrence of a system, service, or network state indicating a possible breach of information security policy or failure of controls, or a previously unknown situation that may be security relevant.

The controls encompass actions, devices, procedures, or techniques that meet or oppose (e.g. counters) a threat, a vulnerability, or an attack by eliminating or preventing it, by minimizing the harm it can cause, or by discovering and reporting it so that corrective action can be taken [CNSSI 4009], [RFC 4949]. Security controls – classified as management, operational, and technical controls – should be prescribed for an information system that, taken together, satisfy the specified security requirements and adequately protect the confidentiality, integrity, and availability of the system and its information [RFC 4949].

These controls need to be specified, implemented, monitored, reviewed, and improved where necessary to ensure that the specific information security and business objectives of the organization are met. Controls should ensure that risks are reduced to an acceptable level taking into account [ISO/IEC 27000]:

- Requirements and constraints of national and international legislation and regulations.
- Operational requirements and constraints.
- Cost of implementation and operation in relation to the risks being reduced and remaining proportional to the organization's requirements and constraints.
- Selection and implementation to achieve monitoring and improving the efficiency and effectiveness of information security controls to support the security requirements and organization's aims.
- Balance the investment in implementation and operation of controls against the loss likely to result from information security incidents due to security failures.

Security failures may result in direct and consequential losses. Direct losses are those to the hardware, including processors, workstations, printers, disks and tapes, and communication equipment; software, including systems and applications software for central and remote devices; documentation, including specifications, user manuals, and operating procedures; personnel, including operators, users, and managerial, technical, and support staff; and physical environment, including computer rooms, communications rooms, air conditioning, and power supply equipment. Although direct losses may account for a small percentage of total losses arising from a security failure, nonetheless, the absolute investment in developing and operating the system will usually have been significant. The system requires protection in its own right as the container and channel for the data and information.

A consequential loss may occur when an information system fails to perform as intended. Consequential losses arising from security failures may include loss of goods, other tangible assets, funds, or intellectual property; loss of valuable information; loss of competitive advantage; reduction in cash flow; loss of orders or business; loss of production efficiency, effectiveness, or safety; loss of customer or supplier goodwill; penalties from violation of statutory obligations; and public embarrassment and loss of business credibility. Consequential losses account for most of the losses arising from security lapses.

Therefore, protection against consequential loss, which, above all, means protecting the data and information, must be a top priority. The technical controls are used to mitigate the potential risk. A technical control may be a software configuration (e.g. closing an open port on the server), a hardware device, or a procedure that eliminates the vulnerability or reduces the likelihood that a threat agent will be able to exploit a vulnerability.

The reality is that no set of controls can achieve complete information security. Additional management actions should be implemented to monitor, evaluate, and improve the efficiency and effectiveness of information security controls to support the organization's business objectives.

Sometimes security measures can affect the interoperability because of reasons such as weak design, misconfigured software products, incompatible versions, inoperable standards, proprietary protocols and implementations, etc.

While openness can be an objective of interoperability, open systems could impact security. An open system is a system that continuously is interacting with the environment. However, more openness could raise security concerns.

Generally, when we talk about protection or defense, we refer to a certain number of countermeasures. Because the threat agents and the threat strengths are often not known or cannot be quantified, it can be very difficult to determine the right level of defense. Another consequence becomes that security is an objective perception. The question arises whether user's and public's perception of threat in any way conforms to the objective assessment of threat. While it is certainly true that each individual operates within its own perception, information threats need to be assessed without human biases and limitations.

An exploratory analysis of characteristics of the five most dangerous threats (hackers, worms, viruses, Trojan horses, and backdoor programs) and the five least dangerous threats (spam, piratical software, operation accidents, users' online behavior being recorded, and deviation in quality of service) was discussed and compared using a six-factor structure, which includes factors of knowledge, impact, severity, controllability, possibility, and awareness [Huang 2010]. This work determines the relationships between the factors and the perceived overall danger of threats. Also, significant effects were found in people's perception of information security related to computer experience and types of loss.

5.4.1 Security Controls Categories

Security controls can vary in nature, but the use of standards, books, and best practices could help on understanding the capabilities and limitations of each control. Although a list of controls in different areas is provided in known standards (e.g. [ISO/IEC 27001]), we need to understand that such a list is not exhaustive. Basically, there are more criteria to classify security controls. Published standards and more guidance are available in [Harris 2013], [Whitman 2012], [Maiwald 2004], [Conklin 2004].

Controls can be classified based on different goals to be achieved such as preventative, detective, and reactive controls. The control types can be blended for comprehensive protection at reasonable cost.

The controls include management, operational, and technical controls prescribed for an information system that, taken together, satisfy the specified security requirements and adequately protect the confidentiality, integrity, and availability of the system and its information [RFC 4949]. A summary of the most important categories of security controls is provided in this standard [NIST SP800-53r4]:

- Technical controls traditionally include products and processes (such as firewalls, antivirus software, intrusion detection, and encryption techniques) that focus mainly on protecting an organization's ICTs and the information flowing across and stored in them.

- Operational controls include enforcement mechanisms and methods of correcting operational deficiencies that various threats could exploit; physical access controls, backup capabilities, and protection from environmental hazards are examples of operational controls.
- Management controls, such as usage policies, employee training, and business continuity planning, target information security's nontechnical areas.

Table 5.6 shows specific security controls for each category of the control groups (families) that are recommended in [NIST SP800-53r4]. This standard was superseded by [NIST SP800-53r5] in September of 2020. This revision includes two additional control families: PM - Program Management and SR - Supply Chain Risk Management.

The document [NIST SP800-53Ar4] provides recommendations on assessing security and privacy controls. More information on specific topics for each category is provided in this guide [NIST SP800-12]. The following topics are described in this guide:

- Management controls – Policy, IT security program management, risk management, and life cycle security.
- Operational controls – Personnel and user issues, contingency planning, incident handling, awareness and training, computer support and operations, and physical and environmental security issues.
- Technical controls – Identification and authentication, logical access controls, audit trails, and cryptography.

Another criterion is grouping the controls in the categories of administrative, logical, and physical. These three areas are often further divided into several different subsets of classifications for management and implementation purposes. Controls can be grouped to associate similar types together. As they are used for protection of information, their effectiveness is measured as the ability to reduce the probability of a threat to exploit a specific vulnerability. These controls cover a diverse range of activities

Table 5.6 Control families.

Control family acronym	Control family	Type
AC	Access control	Technical
AT	Awareness and training	Operational
AU	Audit and accountability	Technical
CA	Certification, accreditation, and security assessments	Management
CM	Configuration management	Operational
CP	Contingency planning	Operational
IA	Identification and authentication	Technical
IR	Incident response	Operational
MA	Maintenance	Operational
MP	Media protection	Operational
PE	Physical and environmental protection	Operational
PL	Planning	Management
PS	Personnel security	Operational
RA	Risk assessment	Management
SA	System and services acquisition	Management
SC	System and communications protection	Technical
SI	System and information integrity	Operational

Source: [NIST SP800-53r4]. Public Domain.

from policy and procedure development to granular technical controls for automatic incident response mechanisms. Organizations that choose to develop a granular and structured approach to controls group similar types of controls together for better management and efficiency. Different security controls are implemented based on the level of protection that management and security team have determined to be needed for an asset.

Administrative (also called procedural or management) controls form the framework for running the business and managing people. Laws and regulations created by government or industry bodies are also a type of administrative control. Smart Grid applications and DER systems need to follow electrical policies, procedures, standards, and guidelines as provided by DOE, FERC, NERC, NIST, etc. They consist of approved written policies, procedures, standards, and guidelines.

Administrative controls are of paramount importance. They form the basis for the selection and implementation of logical and physical controls. Often, logical and physical controls are manifestations of administrative controls.

Logical controls (also called technical) use software and data to monitor and control accesses to information and computing systems. These controls include access controls, detection controls, and communications controls.

Access controls are a collection of mechanisms that work together to protect the assets of the enterprise. Examples of threats related to access controls include:

- Malicious software (e.g. viruses, worms, Trojan horses, logic bombs, spyware/adware, botnets, backdoor/trap door, rootkits, etc.)
- Password crackers
- Spoofing/masquerading
- Sniffers
- Eavesdropping
- Shoulder surfing
- Tapping
- Object reuse
- Data remanence
- Unauthorized targeted data mining
- Dumpster diving
- Theft
- Intruders
- DOS
- DDoS
- Social engineering.

The access control categories are used for preventative, detective, corrective, compensating, deterrent, and recovery.

Physical security includes a different set of threats, vulnerabilities, and risks (e.g. health and safety). Therefore, security controls for physical security protection (called physical controls) should comprise safety and security mechanisms. Physical controls monitor and control the environment of the workplace and computing facilities. They also monitor and control access to and from such facilities. Examples may include doors, locks, heating and air conditioning, smoke and fire alarms, fire suppression systems, cameras, barricades, fencing, security guards, cable locks, etc. Separating the network and workplace into functional areas is also a physical control.

Physical and environmental controls are a subset of measures that physically protect the facilities of an organization by implementing physical access controls, conditioning power lines, providing backup power, and establishing reuse and data disposal policies. Physical security controls protect people, data, equipment, systems, facilities, and many other assets.

In addition to devices mentioned, physical security mechanisms include site design and layout, building, facility, computer rooms, data center, windows, doors, fences, parking lots, environmental components, emergency response readiness, power control, fire protection, HVAC, guards, dogs, etc. Physical security mechanisms protect people, data, equipment, systems, facilities, and many other assets.

The international standard [ISO/IEC 27002] describes controls for each domain of 14 domain identified in the standard. The threats therefore give rise to risks to the assets, based on the likelihood of a threat that may occur and the impact on the assets when the security incident occurred. Subsequently countermeasures are imposed to reduce the risks to assets:

IT countermeasures (such as firewalls and smart cards) and non-IT countermeasures (such as guards and procedures). Every control that is used in information security provides at least one of these principles: confidentiality, integrity, and availability.

Safeguarding assets of interest is the responsibility of owners who place value on those assets.

Actual or presumed threat agents may also place value on the assets and seek to abuse assets in a manner contrary to the interests of the owner. The owners of the assets perceive such threats as potential for impairment of the assets such that the value of the assets to the owners would be reduced. Security-specific impairment commonly includes, but is not limited to, loss of asset confidentiality, loss of asset integrity, and loss of asset availability.

It is critical to understand all the possible ways that these principles can be provided and circumvented. Control selection should follow and should be based on the risk assessment and risk management techniques that are discussed further in the book.

5.4.2 Common Security Controls

A control is a risk mitigation that takes the form of either technical, administrative, or physical actions in response to one or more specific events. Commercial standards, such as the Payment Card Industry Standards, address the same scope of controls but are not as granular in detail, allowing greater latitude in their implementation by the voluntary participants of that organization. However, common controls that can be implemented for an organization include the following:

- Strict and granular access controls for all levels of sensitive data and programs
- Restricted access to critical data
- Auditing and monitoring
- Separation of duties
- Auditing and monitoring
- Periodic reviews
- Proper disposal
- Backup and recovery procedures
- Secure information flow channels
- Others.

5.4.3 Applying Security Controls to Smart Grid

In the power industry, the focus has been on implementation of equipment that could improve power system reliability. Until recently, communications and IT equipment were typically seen as supporting power system reliability. However, these sectors are becoming more critical to the reliability of the power system. In addition, safety and reliability are of paramount importance in electric power systems. Any cybersecurity measures in these systems must not impede safe, reliable power system operations.

In today's world, the security focus of the electric industry has expanded to include withstanding disturbances caused by man-made physical or cyber attacks. It is critical, as cybersecurity measures are considered, that these measures do not inadvertently impair system reliability. This places a constraint on the acceptable complexity and computational intensiveness of the cybersecurity measures to be adopted [Nordell 2012]. As defined in [ISA 62443-1-1], security refers to measures taken to protect a system, where measures can be controls related to physical security (controlling physical access to computing assets) or logical security (capability to login to a given system and application).

Achieving security in any kind of networked system, application, or service requires the provision of different security mechanisms at the different levels composing a system. Thus, it is required to understand the threats, vulnerabilities, and means to protect the information.

5.5 Security Training and Skills

The operational and risk differences between ICS and IT systems create the need for increased sophistication in applying cybersecurity and operational strategies. A cross-functional team of control engineers, control system operators, and IT security professionals has to collaborate and work closely to understand the possible implications of the installation, operation, and maintenance of security solutions in conjunction with control system operation. IT professionals working with ICS need to understand the reliability impacts of information security technologies before deployment. Some of the OSs and applications running on ICS may not operate correctly with commercial off-the-shelf (COTS) IT cybersecurity solutions because of specialized ICS environment architectures.

Development of Smart Grid and DER applications require skilled security professionals that understand the security requirements for both IT and control systems including energy-related applications. Skills are a combination of ability, knowledge, and experience that enable a person to do something well. Many of those individuals responsible for auditing, installing, or operating ICS are aware of the need for cybersecurity, yet the training opportunities are dependent on the budgets. In many electric companies, management is not spending and allocating money in security education, training, and awareness. Traditionally, security training has been a low priority for the management. Lately, information security becomes more visible in organizations, so the values of security education, training, and awareness are recognized.

5.5.1 Education, Training, and Awareness

Recently, more universities offer degree programs in information security education. However, these programs are mostly in computer science, information systems, or IT departments and provide mostly skills for an information security specialist, a beginner level at the graduation.

Given the complexity of cybersecurity problems, there is a need for people educated and trained with a multitude of skills such as technical knowledge and management approaches. For example, more advanced positions (e.g. security systems engineering, security designer, security architect) demand continuous education via training courses or more advanced degree courses offered in a Master of Science or PhD program.

Further, the role of a systems security engineer is to participate in a multidisciplinary systems engineering team, applying fundamental systems security understanding, skills, expertise, and experience to develop a system that satisfies stakeholder mission/business requirements. Similarly, security verification activities may require specialized tools and the application of certain techniques that require specialized expertise and skills. Security personnel that are responsible to operate and to interact with the protection-related portions of the system need the capabilities, skills, and experience necessary to service and maintain the system.

However, information security education is not offered by programs in engineering education. Process control engineers and power engineers need also education and more training to understand and manage the security requirements of Smart Grid and challenges of new technologies. It is argued that degree programs or specializations in information security should be encouraged to be established in engineering and engineering technology schools in the United States and around the globe [Hentea 2007], [Hentea 2006].

Some employers accept certification along with work experience in lieu of a degree or specialization in information security. Certification is typically offered through industry groups and associations, training institutions, and product vendors as validation of an individual's demonstrated understanding of an area of expertise.

Although professional certification in information security may help to quickly train workforce with basic skills, it is demonstrated that the lack of higher skills in information security has put the industry in a stagnant mode because more research and progress has to be made in order to ensure the resilience of Smart Grid in face of current and more sophisticated cyber threats in the future.

Guidance, at a higher strategic level, on how to build an IT security awareness and training program is provided in the standard [NIST SP800-50]. At a lower tactical level, an approach to role-based IT security training is described in [NIST SP800-16r1]. The approach uses the concept of continuum learning; learning starts with awareness, builds to training, and evolves into education. Figure 5.6 shows the conceptual relationship between awareness, training, and education, called IT security learning continuum in this standard [NIST SP800-50]. Learning starts with awareness, builds to training, and evolves into education. The learning levels are defined in [NIST SP800-50] as follows:

- Awareness presentations are intended to allow individuals to recognize IT security concerns and respond accordingly. The purpose of awareness presentations is simply to focus attention on security. Awareness is not training.
- Training strives to produce relevant and needed security skills and competencies by practitioners of functional specialties other than IT security (e.g. management, systems design and development, acquisition, auditing). The most significant difference between training and awareness is that training seeks to teach skills, which allow a person to perform a specific function, while awareness seeks to focus an individual's attention on an issue or set of issues.
- Education integrates all security skills and competencies of the various functional specialties into a common body of knowledge and strives to produce IT security specialists and professionals capable of vision and proactive response. An example of education is a degree program at a college or university. Some people take a course or several courses to develop or enhance their skills in a specific discipline. This is training as opposed to education.

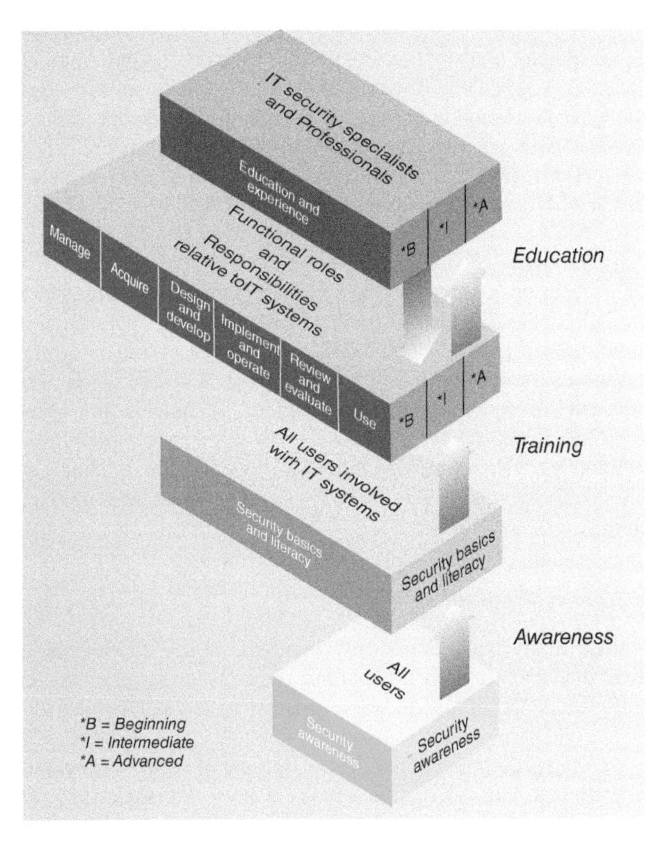

Figure 5.6 The IT security learning continuum. *Source:* [NIST SP800-50]. Public Domain.

The purpose of computer security awareness, training, and education (SETA) program is to enhance security by [NIST SP800-50]:

- Improving awareness of the need to protect system resources.
- Developing skills and knowledge so computer users can perform their jobs more securely.
- Building in-depth knowledge, as needed, to design, implements, or operate security programs for organizations and systems.

Therefore, these definitions clarify also the differences between awareness and training, while the commonality of these concepts is focus on security:

- In awareness activities, the learner is the recipient of information, whereas the learner in a training environment has a more active role.
- Awareness relies on reaching broad audiences with attractive packaging techniques.
- Training is more formal, having a goal of building knowledge and skills to facilitate the job performance. A companion training document could be [NIST SP800-12].

Also, there is a distinction between training and education. The following clarify differences:

- An example of education is a degree program at a college or university.
- Taking a course or several courses to develop or enhance skills in a particular discipline is training as opposed to education.
- Certificate programs are more characteristic of training than education. Many colleges and universities offer certificate programs, wherein a student may take two, six, or eight classes, for example, in a related discipline, and is awarded a certificate upon completion. Often, these certificate programs are conducted as a joint effort between schools and software or hardware vendors.

Therefore, those responsible for security training need to assess both types of programs (education and training) and decide which one is better addressing needs for workforce, industry, and whole economy. In addition to these learning levels, the job functions require specific knowledge, skills, and competencies.

One can find information on security training requirements for federal IT employees in [NIST SP800-16r1]. Two publications, [NIST SP800-50] and [NIST SP800-16r1], are complementary: [NIST SP800-50] guide works at a higher strategic level, discussing how to build an IT security awareness and training program, while [NIST SP800-16r1] guidance is at a lower tactical level, describing an approach to role-based IT security training.

While these documents focus on security of IT-based environments, the training for ICS is also supported in the United States by CERT [ICS-CERT] and national laboratories (e.g. Idaho National Laboratory).

According to [NISTIR 7628], [NIST SP800-53r4], a developer of the information system, system component, or information system service is required to provide organization defined training on the correct use and operation of the implemented security functions, controls, and/or mechanisms. Further, a guide to ICS security [NIST SP800-82r2] provides awareness and training requirements: all users of an information system should have been provided with basic information system security awareness and training materials before authorization to access the system is granted; personnel training must be monitored and documented.

5.5.2 Security Awareness Program

Organizations can reduce many types of security incidents if the employees know how to access resources, why controls are in place, and understand the consequences of not using these concepts properly. Security awareness is part of an education program called SETA (security education, training, and awareness).

Security awareness training is designed to reduce the number of security incidents that occur through a lack of employee security awareness. An active security awareness program is probably the single most effective method to counter potential social engineering attacks [Conklin 2004], [NIST SP800-12]. Social engineering refers to manipulation of people and use of tricks into performing actions or divulging confidential information.

Security awareness training should be comprehensive, tailored to specific groups, and organization-wide. The extent of training may vary depending on the organization's environment and the level of threat. The goal is that each employee understands the importance of security to the company as a whole and to each individual. It is important that employees understand not only how security works in their environments but also why it is important. Users of information systems are required to understand their basic responsibilities for the information security. Users include employees, contractors, and third parties [NISTIR 7628]. More important to know is that security awareness should happen periodically and continually. Obviously, the program requires allocation of resources in terms of time and money. Special attention should be given consumers that are connected to Smart Grid (utilities) via smart meters or renewables.

Different elements of awareness education can be employed to be interesting, effective, and low costs [Conklin 2004], [Anderson 2008], [Whitman 2012]. In [DOE 2012], it is suggested that a utility can develop an internal Web site with relevant information on threats, vulnerabilities, and best practices to be shared with the personnel. Personnel should be informed how to identify suspicious behavior, avoid spam or spear phishing, and recognize social engineering attacks to avoid providing information about the utility to potential adversaries.

In addition, a security awareness program should target audiences such as management, staff, and technical employees such that each group understands particular responsibilities, liabilities, and expectations. Each group needs to know how to report suspicious activity and how to handle these situations. Security awareness training is a type of control, and it requires monitoring and evaluation for effectiveness.

DOE supports cybersecurity awareness and training (CSAT) program, a comprehensive approach to training and awareness, to all levels of departmental personnel to be informed of their cybersecurity responsibilities and possess the skills appropriate for their functional roles to adequately protect DOE information and information systems.

This control applies to external and internal (in-house) developers. Training of personnel is an essential element to ensure the effectiveness of security controls implemented within organizational information systems. Training options include classroom-style training, Web-based/computer-based training, and hands-on training. Organizations can also request sufficient training materials from developers to conduct in-house training or offer self-training to organizational personnel. Organizations determine the type of training necessary and may require different types of training for different security functions, controls, or mechanisms.

5.6 Planning for Security and Privacy

Efficient and effective security can be achieved if an organization devotes resources for information security and privacy planning. Information security planning includes planning for developing the organizational plan for information security, information security implementation, and contingency planning [Whitman 2012]. Planning for security and privacy usually involves many interrelated groups and organizational processes. The planning may include responsibilities for the capital planning and investment and allocation of resources to provide oversight for the information security-related aspects. The planning results are documented in a formal plan. This plan provides a set of management controls that are typically implemented at the organization level and not directed at individual organizational information systems.

5.6.1 Plan Structure

Because information security planning depends on the same planning process for organizational plan, this commonly includes the common components such as mission, vision, values, and strategy. So the information security department or division needs to define these components. For example, a security plan for an organization developing Smart Grid should include components such as mission, vision, strategy, and other specific objectives:

- Mission should include a concise statement about the importance of availability, integrity, and confidentiality (including other goals) of information for supporting the reliability of the Smart Grid. Reliability remains a fundamental principle of grid modernization efforts, but in today's world, reliability requires cybersecurity [Hawk 2014].
- Vision component should include the ambitious goal of achieving a resilient Smart Grid such as surviving a cyber incident while sustaining critical energy delivery functions.
- Strategy should include all levels strategic, tactical, and operational planning:
  - At the strategic level, the plan may include goals such as unified security program (includes both IT and OT technologies) based on risk management and proactive methods, security- and privacy-driven Smart Grid architecture, and security culture instead of compliance.
  - At the tactical level, the plan may include plans for architecting cybersecurity into the solution, beginning at the concept stage; use of risk-balanced methodologies based on advanced techniques and standards; business impact analysis for disaster recovery; and improving security awareness, training, and education.
  - Operational planning should include the development of unified coordination activities for the integration of OT and IT procedures, continuous security and privacy monitoring, improving communication between OT and IT teams, developing progress reports, etc.

As with security strategies, it is necessary to define a plan for proactive and reactive security planning. The proactive plan is developed to protect assets by preventing attacks and employee mistakes. The reactive plan is a contingency plan to implement when proactive plans have failed [Benson 2014].

Planning for information security implementation may include processes, actions, and methods on approaches such as bottom-up (initiated by administrators and technicians) or top-down (management support and a formal development strategy referred to as the system development life cycle).

Contingency planning includes plans to prepare for, detect, react to, and recover from events that threaten the security of information resources and assets, both human and natural. It includes incident response plan, disaster recovery plan, and business continuity plan.

This plan, called information security program plan, is a formal document. It could also provide an overview of the following issues [NIST SP800-53r4]:

- The requirements for the security program and a description of the security program management controls and common controls in place or planned for meeting those requirements.
- The identification and assignment of roles, responsibilities, management commitment, coordination among organizational entities, and compliance.

The development of security plans for information systems is also addressed in [NIST SP800-100]. A security plan provides an overview of the security requirements for an information system and describes the security controls in place or planned for meeting those requirements.

5.6.2 Security Team

Information security is a field where technical and nontechnical skills are required. A project team should include individuals who are experienced in one or multiple requirements of both the technical and nontechnical areas [Whitman 2012]. Many of the same skills needed to manage and implement security are needed to design it. Security professionals should be educated in all aspects of information security from both technical and nontechnical areas. It takes a wide range of professionals to support a diverse security program. A security program is a framework made up of many entities: logical, administrative, and physical protection mechanisms, procedures, business processes, and people that work together to provide a protection level for an environment [Harris 2013]. All levels of management should understand how roles with security-related responsibilities are identified within their organization.

The security team functions are grouped into three categories [Schwartz 2001]:

- Define – Writing policies, standards, and guidelines; performing risk assessment, analysis, design security products, and architecture; in general, these are people with a broad knowledge of security and business.
- Build – Install and test security solutions; these are technical people that deploy security controls.
- Operate – Maintaining and monitoring the security operations; administrating the security tools; these are people who do a specific task.

Although this grouping could provide a balance of job functions, organizations tend to use one person for several functions regardless of skills.

Government and some organizations may be structured in different layers of responsibilities from upper management to functional management to operational management and staff. Every layer has its own insight into what type of role security plays within an organization to ensure the agreed security level.

However, most private businesses and smaller organizations do not have these layers. In electricity subsector, it is common that cybersecurity responsibilities are not restricted to traditional IT roles; for example, some operations engineers may have cybersecurity responsibilities [DOE 2012]. This strategy for cybersecurity is not ensuring adequate security for Smart Grid. The personnel, either IT personnel or operations engineers, should have dedicated responsibilities to security and be provided with adequate education and training. IT security skills do not apply well to Smart Grid, and operations engineers need to have skills in security.

Also, companies developing DER systems need to assign responsibilities to a security team that ensures the products are designed and implemented to support the appropriate security features needed by the users. Likely, the DER service providers may need to have their own security team or outsource the work to an organization specialized in security services.

A typical organization like a utility may need to have a number of individuals with information security responsibilities. Most of the job functions fit into one of the following information security roles [Whitman 2012]:

- Chief information security officer (CISO)
- Chief privacy officer (CPO)
- Security managers
- Security administrators and analysts
- Security technicians
- Security staffers
- Security consultants
- Security officers and investigators
- Security personnel.

The roles may be assigned to just a few people or to a large security team. Other roles include board of directors, data owner, data custodian, system owner, application owner, supervisor (user manager), security administrator, change control analyst, internal auditor, etc. [Harris 2013], [IT Handbook]. Although each layer of management is important to the overall security of an organization, many other specific roles must be defined. For example, security engineer and security systems engineer are job functions that should be part of security team for any Smart Grid or DER system. More information about the functions, roles, and skills is provided in [NIST SP800-16r1]. Guidance and information regarding roles, function areas, competencies, and other requirements are provided in this document.

In order to implement information security for Smart Grid and DER systems, security team must include specific skills in electrical engineering, power engineering, control engineering, computer engineering, environmental engineering, and maybe other specialized engineering skills. These skills have to be augmented with competencies in information security. Implementing security of control systems requires a wide range of skills. For example, the guide [NIST SP800-16r1] defines the following knowledge (security-specific units) for roles in technology research and development:

- Digital forensics
- Software

- Compliance
- Cryptography and encryption
- IT Systems and operations
- Information systems
- Network and telecommunication security
- Architecture
- Modeling and simulation
- Physical and environmental security
- Procurement
- Security risk management
- Systems and applications security
- Emerging technologies.

Other needed skills that are needed include systems engineering responsibilities and knowledge units such as:

- Identity management/privacy
- Incident management
- Information assurance.

Identifying skills for specific roles implied by security professionals, the [NIST SP800-16r1] guidance is a good start. For example, tasks oriented on physical security of Smart Grid – analysis of threats and vulnerabilities and design of security controls – have to be performed by a team of individuals with various backgrounds from management, construction, environment, electronics, electrical, power, control, information security, mechanical, safety, health, etc.

5.7 Legal and Ethical Issues

Smart Grid policy is organized in Europe as Smart Grids European Technology Platform. The European SmartGrids Technology Platform for the Electricity Networks of the Future was started by the European Commission Directorate-General for Research in 2005 [EC]. Its aim is to formulate and promote a vision for the development of Europe's electricity networks looking toward 2020 and beyond [EC 2006]. Among stakeholders of Smart Grid in Europe are regulators that support the European market for energy and related services with a stable and clear regulatory framework, with well-established and harmonized rules across Europe. Regulatory frameworks should have aligned incentives that secure a grid with increasingly open access and a clear investment remuneration system and keep transmission and distribution costs as low as possible.

In the United States, Smart Grid policy is described in the Energy Independence and Security Act (EISA) of 2007, originally named the Clean Energy Act of 2007 [EISA 2007]. The EISA aims to [EPA]:

- Move the United States toward greater energy independence and security.
- Increase the production of clean renewable fuels.
- Protect consumers.
- Increase the efficiency of products, buildings, and vehicles.
- Promote research on and deploy greenhouse gas capture and storage options.
- Improve the energy performance of the federal government.
- Increase US energy security, develop renewable fuel production, and improve vehicle fuel economy.

In the United States, DOE is working with federal government, the Department of Homeland Security (DHS), and NIST to address Smart Grid issues such as security and interoperability as documented in [FERC 2009].

With the implementation of Smart Grid, there are privacy and ethical concerns such as the following:

- Utilities and service providers are not transparent with the use of data collected about the customers.
- Companies and individuals act dishonestly and unethically to increase the electricity price.

- Smart Grid technology may enable some people to get control of the power supply market.
- Smart Grid may also affect consumers who adopt the use of solar and wind power; there is fear that utilities could make it possible to transfer excess electricity from power-generating users to others during peak periods.
- Communication between utilities and the meters at residential homes and businesses increases the chance of someone gaining control over the power supply of a single building or an entire neighborhood.
- Granularity of control of electricity use could be down to the appliances or devices located within a home or office.
- Marketers may view Smart Grid systems as another opportunity to learn more about consumers and how they use the items they purchase.
- Examples of the personally identifiable information that may be collected include details on battery charging information, e.g. amount of life remaining, date, time, location of last recharge, etc.; type of personal device; a unique item identification number as well as personalized information, e.g. user name, address, etc.; and location where the item was recharged as well as how long the device was connected to the power source.

As energy and telecommunication services are increasingly intersecting, there are ICT-specific policy implications that are of specific relevance to ICT policy makers, telecommunications, and utilities regulators [OECD 2012]:

- Converging energy and telecommunication services
  The open-access provisions allowing smart meter service providers and utilities access to data capacity over telecommunication networks are reason for converging energy and telecommunication services. Policy makers and regulatory authorities have started consultations on the communications needs for the Smart Grid (e.g. by DOE) and integrating energy policy objectives into national broadband plans (e.g. by FCC). Increased service bundling might require enhanced coordination between regulatory authorities. Thus, the coordination could in some instances be challenging given that the regulatory domains of energy and telecommunications are traditionally located with separate institutions and industries.
- IT and utilities issues
  IT innovation and value chain disruption require policy makers to reinforce the alignment involving utilities, IT firms, and other stakeholders. A number of ICT companies are providing entirely new services or services formerly reserved to utilities only. It is therefore important for incumbent stakeholders to define and formulate requirements and for newcomers to be able to engage in cross-industry dialogue. IT companies need to listen carefully to electricity sector requirements while at the same time leveraging the innovation that IT and the Internet provide.
- Health impacts
  The health concerns are related to wireless technologies from potential increases in electromagnetic radiation from smart meters. While the whole evidence of health impacts is not identified, customers have been reluctant to allow installation of smart meters in many jurisdictions. One option might be to allow customers to selectively turn off wireless data transmission, e.g. at night, or potentially default to wired infrastructures when available. Therefore, regulators have to mediate these options with customers.

Among other issues, the need for coordination of cybersecurity with the legal and regulatory system is a fundamental problem that has to be addressed. It raises questions about the role of government, the relationship between the public and private sectors, the balance between privacy and public safety, and the definition of security [Wulf 2001].

In order to assure the smooth connection of DERs to grids, regulatory frameworks need to be investigated besides technical requirements. Regulators need to work actively on the public aspects and to create the incentives to encourage consumers and private actors to play their part.

The rapid development of new technologies and installations such as electrical energy storage in the near future by consumers demand specific regulations for safety and environment to be enforced to protect the public and users.

In the United States, reporting cyber incidents is facilitated by the Internet Crime Complaint Center [IC3] associated with several private and public organizations on investigating and referring the cases to law enforcement. Policies of business ethics and conduct are in the agenda of several groups. For example, FERC is enforcing several conduct rules for the electrical sector (e.g. FERC Order 717, Standards of Conduct for Transmission Providers [FERC 2008]), while other groups such as Electronic Privacy Information Center (EPIC) advocate privacy protection [EPIC Policy].

The following sections introduce key standards relevant to information security and development of Smart Grid and DER systems.

5.8 Standards, Guidelines, and Recommendations

Besides existing Smart Grid and cybersecurity standards, the development of standards for energy storage safety and environment compatibility is recommended.

Several issues in establishing Smart Grid standards include [Gilbert 2011]:

- Complexity of Smart Grid.
- Accommodating multiple communication layers.
- Smart Grid evolving at different rates.
- Unaligned stakeholder interests.
- New security and privacy challenges.
- Balancing certainty with flexibility to enable innovation.

Competing priorities for adopting standards include:

- Quality and performance of standards in meeting Smart Grid objectives.
- Costs resulting from standards.
- Timing of standards in meeting Smart Grid objectives.
- Institutional endorsement and market position in the competing priorities.

The OPEN METER initiative of Europe has the goal to specify a comprehensive set of open and public standards for AMI, supporting electricity, gas, water, and heat metering, based on the agreement of all the relevant stakeholders in this area, and taking into account the real conditions of the utility networks so as to allow for full implementation [OPEN METER].

There are several guidelines, documents, standards, and legislation that can provide guidance on the physical security controls and planning. Examples include the President's Commission on Critical Infrastructure Protection [PCIP], Crime Prevention Through Environmental Design [CPTED], International CPTED Association [ICA], Federal Information Processing Standards (FIPS) publication series [FIPS], and regulations issued by DHS, DOE, Labor Department, Occupational Safety and Health Administration [OSHA Laws], Environmental Protection Agency [EPA], etc. FIPS publications are standards issued by NIST after approval by the Secretary of Commerce pursuant to the Federal Information Security Management Act (FISMA).

References Part 2

[Ackoff 1989] Ackoff, R.L. (1989). From data to wisdom. *Journal of Applied Systems Analysis, 15*, 3–9.

[Adams 2013] Adams, Jr, W.C. (2013). IEEE standards association smart grid strategic focus. (Presentation). *Wireless World Research Forum Meeting*, Vancouver, Canada, (October 2013). https://sgc2013.ieee-smartgridcomm.org/sites/ieee-smartgridcomm.org/files/WilbertAdams.pdf

[Ææqwerty 2020] Ææqwerty. (2020). *Passive vs active attack.* CC BY-SA 4.0. https://upload.wikimedia.org/wikipedia/commons/6/6b/Passive_vs_active_attack_.png

[AHIMA] AHIMA. (2012). *Pocket Glossary of Health Information Management and Technology* (3rd Edition). AHIMA Press.

[Amin 2012] Amin, S.M., Giacomoni, A.M. (2012). Smart grid—safe, secure, self-healing. *IEEE Power & Energy Magazine, 10*(1), 33–41. https://doi.org/10.1109/MPE.2011.943112

[Anderson 2008] Anderson, R.J. (2008). *Security Engineering: A Guide to Building Dependable Distributed Systems* (2nd Edition). Wiley.

[ASQ Glossary] ASQ. (n.d.). *Quality Glossary*. http://asq.org/glossary/. American Society for Quality.

[Audit 2005] ANAO. (2006). *IT Security Management Audit* (Report No. 23 2005–2006). https://www.anao.gov.au/sites/default/files/ANAO_Report_2005-2006_23.pdf

[Bakken 2000] Bakken, D., Bose, A., Bhowmik, S. (2000). Survivability and status dissemination in combined electric power and computer communications networks. *Proceedings of the IEEE Third Information Survivability Workshop (ISW-2000)*. Boston, Massachusetts, USA, (October 2000). IEEE.

[Bellovin 2015] Bellovin, S.M. (2015). What a real cybersecurity bill should address. *IEEE Security & Privacy, 13*(3), 92. https://doi.org//10.1109/MSP.2015.52

[Benson 2014] Benson, C. (2014). *Security Planning*. Microsoft. https://technet.microsoft.com/en-us/library

[Bernstein 2009] Bernstein, J.H. (2009). The Data-Information-Knowledge-Wisdom Hierarchy and its Antithesis. *Proceedings of the 2nd North American Symposium on Knowledge Organization* (Vol 2, pp. 68–75), Syracuse University, Syracuse, New York, USA, (June 2009).

[Bishop 1996] Bishop, M., Bailey, D. (1996). *A Critical Analysis of Vulnerability Taxonomies* (CSE-96-11). http://nob.cs.ucdavis.edu/bishop/notes/1996-cse-11/1996-cse-11.pdf

[Bishop 2005] Bishop, M. (2005). *Introduction to Computer Security*. Addison-Wesley Professional.

[Bobba 2012] Bobba, R.B., Dagle, EJ., Heine, E., Khurana, H., Sanders, W.H., Sauer, P., Yardley, T. (2012). Enhancing grid measurements: wide area measurement systems, NASPInet, and security. *IEEE Power & Energy Magazine, 10*(1), 67–73.

[Borlase 2017] Borlase, S. (2017). *Smart Grids Advanced Technologies and Solutions* (2nd Edition, edited by Stuart Borlase). CRC Press.

Building an Effective Security Program for Distributed Energy Resources and Systems: Understanding Security for Smart Grid and Distributed Energy Resources and Systems, Volume 1, First Edition. Mariana Hentea.

[Buckman] Buckman, J.F. *Center for Leadearship in Quality*. Carlson School of Management University of Minnesota Minneapolis, USA.

[Bundschuh 2008] Bundschuh, M., Dekkers, C. (2008). Software measurement and metrics: fundamentals. *The IT Measurement Compendium Estimating and Benchmarking Success with Functional Size Measurement* (pp. 179–206). Springer. https://doi.org/10.1007/978-3-540-68188-5_7

[Bygstad 2008] Bygstad, B. (2008). Information infrastructure as organization: a critical realist view. *ICIS 2008 Proceedings of the International Conference on Information Systems*, Paper 190, Paris, France, (December 2008). https://aisel.aisnet.org/icis2008/190

[Caswell 2011] Caswell, J. (2011). *Survey of Industrial Control Systems Security* [Project Report]. http://www.cse.wustl.edu/~jain/cse571-11/ftp/ics/index.html

[Catteddu 2009] Catteddu, D. and Hogben, G. (2009). *Cloud Computing Benefits, Risks and Recommendations for Information Security* (Version 1, rev. A). European Network and Information Security Agency. *Replaced by (Haeberlen 2012).

[CC] Commoncriteriaportal. (n.d.). *The Common Criteria for Information Technology Security Evaluation V3.1 R4* (September 2012). http://www.commoncriteriaportal.org/cc/

[Cherdantseva 2012] Cherdantseva, Y., Hilton, J. (2012). *The Evolution of Information Security Goals from the 1960s to Today*. http://users.cs.cf.ac.uk/Y.V.Cherdantseva/LectureEvolutionInfoSecGOALS.pdf

[Cherdantseva 2013a] Cherdantseva Y., Hilton J. (2013a). Information security and information assurance. The discussion about the meaning, scope and goals. In F. Almeida, I. Portela (Eds.), *Organizational, Legal, and Technological Dimensions of Information System Administrator* (pp. 546–555). IGI Global. https://doi.org/10.4018/978-1-4666-4526-4.ch010

[Cherdantseva 2013b] Cherdantseva, Y., Hilton, J. (2013b). A reference model of information assurance & security. *IEEE Proceedings of 2013 Eighth International Conference on Availability, Reliability and Security (ARES)* (pp. 546–555). Regensburg, Germany, (September 2013). https://doi.org/10.1109/ARES.2013.72

[Cleveland 2013] Cleveland, F. (2013). *List of Cybersecurity for Smart Grid and Standards Guidelines*. IEC TC57 WG15. http://iectc57.ucaiug.org/wg15public/Public%20Documents/List%20of%20Smart%20Grid%20Standards%20with%20Cybersecurity.pdf

[Cleveland 2016] Cleveland, F. (2016). *IEC 62351 Security Standards for the Power System Information Infrastructure*. IEC TC57 WG15. http://iectc57.ucaiug.org/wg15public/Public%20Documents/White%20Paper%20on%20Security%20Standards%20in%20IEC%20TC57.pdf

[CNSSI 4009] CNSSI. (2010). *Committee on National Security Systems (CNSS) Information Assurance (IA) Glossary*. Committee on National Security Systems (CNSS) Instruction No. 4009. *Revised in 2015.

[Conference 2006] IEEE. (2006). Conference reports four focused forums. *IEEE Control Systems Magazine*, *26*(4), 93–98.

[Comer 2006] Comer, D. (2006). *Internetworking with TCP/IP - Principles, Protocols, and Architectures* (5th Edition). Prentice Hall.

[Conklin 2004] Conklin, Wm. A., Williams, D., White, G.B., Cothren, C. (2004). *Principles of Computer Security Security+ and Beyond*. McGraw Hill Technology Education.

[CPTED] Wikipedia. (n.d.). *Crime Prevention Through Environmental Design*. https://en.wikipedia.org/wiki/Crime_prevention_through_environmental_design

[CSA 2011] CSA. (2011). *Security Guidance for Critical Areas of Focus in Cloud Computing V3.0*. Cloud SecurityAlliance.https://cloudsecurityalliance.org/artifacts/security-guidance-for-critical-areas-of-focus-in-cloud-computing-v3/

[CSI 2015] CSI. (2015). State of Cybersecurity: Implications for 2015 An ISACA and RSA Conference Survey [Paper presentation]. *RSA Conference 2015*, San Francisco, California, USA, (April 2015). https://www.slideshare.net/robertestroud/isaca-rsa-csx-presentation-from-the-rsa-2015-conference

[CYBRARY Models] Cybrary. (n.d.). *Information Security Models*. https://www.cybrary.it/study-guides/cissp/information-security-models/

[DAMA Dictionary] DAMA. (2011). *The Data Management Association International (DAMA) Dictionary* (1e).

[DAMA Framework] DAMA. (2008). *The Data Management Association International (DAMA-DMBOK) Functional Framework* (Version 3.02). https://dama.org/sites/default/files/download/DAMA-DMBOK_Functional_Framework_v3_02_20080910.pdf

[Dhillon 2007] Dhillon, G. (2007). *Principles of Information Systems Security: Texts and Cases*. Wiley.

[DHS 2011a] DHS. (2011a). *Common Cybersecurity Vulnerabilities in Industrial Control Systems*. U.S. Department of Homeland Security. Control Systems Security Program. https://us-cert.cisa.gov/sites/default/files/recommended_practices/DHS_Common_Cybersecurity_Vulnerabilities_ICS_2010.pdf

[DHS 2011b] DHS-Catalog. (2011b). *Catalog of Control Systems Security: Recommendations for Standards Developers*. U.S. Department of Homeland Security. Control Systems Security Program. https://us-cert.cisa.gov/sites/default/files/documents/CatalogofRecommendationsVer7.pdf

[Director] FAS. (n.d). *Statement for the Record on the Worldwide Threat Assessment of the U.S. Intelligence Community for the Senate Select Committee on Intelligence* (February 16, 2011). Office of the Director of National Intelligence. https://fas.org/irp/congress/2011_hr/021611clapper.pdf

[DOE/OEDER 2008] Energy. (2008). *The Smart Grid: An Introduction*. U.S. Department of Energy Office of Electricity Delivery and Energy Reliability. Litos Strategic Communication. http://energy.gov/sites/prod/files/oeprod/DocumentsandMedia/DOE_SG_Book_Single_Pages%281%29.pdf

[DOE 2009] DOE. (2009, July). *Smart Grid System Report*. U.S. Department of Energy. http://energy.gov/sites/prod/files/2009%20Smart%20Grid%20System%20Report.pdf

[DOE 2012] DOE. (2012). *Electricity Subsector Cybersecurity Capability Maturity MODEL (ES-C2M2) Version 1.0* (May 2012). U.S. Department of Energy. *Replaced by Version 1.1 February 2014. https://www.energy.gov/sites/prod/files/2014/02/f7/ES-C2M2-v1-1-Feb2014.pdf

[DOE 2015] DOE. (2015). *Quadrennial Technology Review 2015 Chapter 3: Enabling Modernization of the Electric Power System Technology Assessments*. U.S. Department of Energy. http://www.energy.gov/sites/prod/files/2015/09/f26/QTR2015-3D-Flexible-and-Distributed-Energy_0.pdf

[Dunn 2003] Dunn, W.R. (2003). Designing safety-critical computer systems. *IEEE Computer, 36*(11), 40–46.

[Durai 2012] Durai, A., Varakantam, V. (2012). *Building Smart Grid Core Networks* (October 2012). https://resourcecenter.smartgrid.ieee.org/common/content-list.tag.html/arvind_durai

[EC 2006] EC. (2006). *European SmartGrids Technology Platform Vision and Strategy for Europe's Electricity Networks of the Future* (EUR 22040). European Commission. Directorate General for Research Information Communication Unit. https://ec.europa.eu/research/energy/pdf/smartgrids_en.pdf

[EC] EC. (n.d.). *European Technology Platform (ETP) SmartGrids*. European Commission. https://www.edsoforsmartgrids.eu/policy/eu-steering-initiatives/smart-grids-european-technology-platform/

[Edison] Wikia. (n.d.). *EEI Principles for Cybersecurity and Critical Infrastructure Protection*. Edison Electric Institute. IT Law Wiki. https://itlaw.wikia.org/wiki/EEI_Principles_for_Cybersecurity_and_Critical_Infrastructure_Protection

[EISA 2007] EISA. (2007). *U.S. Energy Independence and Security Act of 2007, Public Law No: 110-140 Title XIII, Sec. 1301*. http://www.gpo.gov/fdsys/pkg/BILLS-110hr6enr/pdf/BILLS-110hr6enr.pdf

[Ellison 1997] Ellison, R.J., Fisher, D.A., Linger, R.C., Lipson, H.F., Longstaff, T., Mead, N.R. (1997). *Survivable Network Systems: An Emerging Discipline* (Technical Report CMU/SEI-97-TR-013). Software Engineering Institute, Carnegie Mellon University, Pittsburgh, USA.

[English 2009] English, Larry P. (2009*). Information Quality Applied: Best Practices for Improving Business Information, Processes and Systems.* Wiley.

[ENISA 2009] ENISA. (2009). *Cloud computing: benefits, risks and recommendations for information security* [News]. ENISA Media. https://www.enisa.europa.eu/media/news-items/cloud-computing-speech

[ENISA 2012] ENISA. (2012). *Smart Grid Security Related Standards Guidelines and Regulatory Documents (Annex IV).* European Network and Information Security Agency (ENISA). https://www.enisa.europa.eu/topics/critical-information-infrastructures-and-services/smart-grids/smart-grids-and-smart-metering/smart-grid-security-related-standards-guidelines-and-regulatory-documents

[ENISA Glossary] ENISA. (2015). *Glossary*. European Union Agency for Cybersecurity. https://www.enisa.europa.eu/topics/threat-risk-management/risk-management/current-risk/risk-management-inventory/glossary

[EPA] EPA. (n.d.). *Summary of the Energy Independence and Security Act Public Law 110-140 (2007).* United States Environmental Protection Agency. https://www.epa.gov/laws-regulations/summary-energy-independence-and-security-act

[EPIC] EPIC. (n.d.). *The Smart Grid and Privacy Concerning Privacy and Smart Grid Technology. Electronic Privacy Information Center (EPIC).* https://epic.org/privacy/smartgrid/smartgrid.html

[EPIC Policy] EPIC. (2011). *Smart Grid Policy Summit.* Electronic Privacy Information Center (EPIC). https://epic.org/2011/04/smart-grid-policy-summit-1.html#

[EPRI 2010] EPRI. (2010). *Smart Grid Technologies Report* (1020415). https://www.epri.com/research/products/000000000001020415

[EPRI 2011] EPRI. (2011). *Estimating the Costs and Benefits of the Smart Grid a Preliminary Estimate of the Investment Requirements and the Resultant Benefits of a Fully Functioning Smart Grid Technical Report* (1022519). https://www.epri.com/research/products/000000000001022519.

[Faisal 2012] Faisal, M.A., Aung, Z., Williams, J.R., Sanchez, A. (2012). Securing advanced metering infrastructure using intrusion detection system with data stream mining. In M. Chau, G.A. Wang, W.T. Yue, H. Chen (Eds.), *Intelligence and Security Informatics (PAISI) (PAISI 2012, Lecture Notes in Computer Science* (Vol 7299, pp. 96–111). Springer. https://doi.org/10.1007/978-3-642-30428-6_8

[Faisal 2015] Faisal, M.A., Aung, Z., Williams, J.R., Sanchez, A. (2015). Data-Stream-Based Intrusion Detection System for Advanced Metering Infrastructure in Smart Grid: A Feasibility Study. *IEEE Systems Journal, 9*(1), 31–44. https://doi.org/10.1109/JSYST.2013.2294120

[Falliere 2011] Falliere, N., Murchu, L.O. and Chien, E. (2011, February 11). Symantec Response W32. Stuxnet Dossier (Version 1.4) [Blog]. *Wired.* http://www.wired.com/images_blogs/threatlevel/2011/02/Symantec-Stuxnet-Update-Feb-2011.pdf

[Fang 2012] Fang, X., Misra, S., Xue, G., Yang, D. (2012). Smart grid - the new and improved power grid: a survey. *IEEE Communications Surveys and Tutorials, 14*(4), 944–980. https://doi.org/10.1109/SURV.2011.101911.00087

[Farhangi 2010] Farhangi, H. (2010). The path of the smart grid. *IEEE Power & Energy Magazine,* 1(1), 18–28. https://doi.org/ 10.1109/MPE.2009.934876

[Farhangi 2014] Farhangi, H. (2014). A road map to integration: Perspectives on Smart Grid Development. *IEEE Power & Energy Magazine, 12*(3), 52–66. https://doi.org/10.1109/MPE.2014.2301515

[FERC 2008] FERC. (2008). *Standards of Conduct for Transmission Providers, Federal Energy Regulatory Commission 131 FERC DocketNo.RM07-1-002;OrderNo.717-C*. https://www.ferc.gov/whats-new/comm-meet/2008/101608/M-1.pdf

[FERC 2009] FERC. (2009). *Smart Grid Policy 128 FERC Docket No. PL09-4-000, 16 July 2009*. Federal Energy Regulatory Commission. https://www.ferc.gov/sites/default/files/2020-08/E-3-Fercs-Smart-Grid-Policy.pdf

[FERC 2012] FERC. (2012). *Assessment of Demand Response and Advanced Metering* [Report]. Federal Energy Regulatory Commission. https://www.ferc.gov/sites/default/files/2020-04/12-20-12-demand-response.pdf

[FIPS] NIST. (n.d.). *Federal Information Processing Standards (FIPS) Publications*. NIST Computer Security Resource Center. http://csrc.nist.gov/publications/PubsFIPS.html

[Galloway 2013] Galloway, B., Hancke, G.P. (2013). Introduction to Industrial Control Networks. *IEEE Communications Surveys & Tutorials*, *15*(2), 860–880. https://doi.org/10.1109/SURV.2012.071812.00124

[GAO 2011] GAO. (2011). *Critical Infrastructure protection cybersecurity guidance is available, but more can be done to promote its use* (GAO-11-117). United States Government Accountability Office. http://www.gao.gov/assets/590/587529.pdf

[Gao 2012] Gao, J., Xiaoa, Y., Liua, J., Liang, W., Chen, C.L.P. (2012). A survey of communication/networking in smart grids. *Future Generation Computer Systems*, *28*(2), 391–404. https://doi.org/10.1016/j.future.2011.04.014

[Gartner 2013] Firstbrook, F.P., Girard, J., MacDonald, N. (2013). *Magic Quadrant for Endpoint Protection Platforms* (G00239869). Gartner Information Technology Research. https://www.gartner.com/en/documents/2292216/magic-quadrant-for-endpoint-protection-platforms

[Gharavi 2011] Gharavi, H., Ghafurian, R. (2011). Smart grid: the electric energy system of the future. *Proceedings of the IEEE* (Vol 99, No 6, pp. 917–921).

[Giarratano 2005] Giarratano, J.C., Riley, G.D. (2005). *Expert Systems: Principles and Programming* (4th Edition). Thomson Course Technology.

[Gilbert 2011] Gilbert, E.I., Violette, D.M., Rogers, B. (2011*). Paths to Smart Grid Interoperability A Smart Grid Policy Center* [White Paper]. https://www.smartgrid.gov/files/documents/Paths_Smart_Grid_Interoperability.pdf

[Gonzalez 2011] Gonzalez, N.M., Miers, C., Redigolo, F.F., Carvalho, T.C.M.B., Simplicio, M., Naslund, M., Pourzandi, M. (2011). A quantitative analysis of current security concerns and solutions for cloud computing. *2011 IEEE 3rd International Conference on Cloud Computing Technology and Science (CloudComp)* (pp. 231—238), Athens, Greece, (November 2011). https://doi.org/10.1109/CloudCom.2011.39

[GS1] GS1. (n.d.). *Data Quality Framework*. The Global Language of Business. http://www.gs1.org/gdsn/dqf/data_quality_framework

[GTM 2015] GTM. (2015). *Evolution of the Grid Edge: Pathways to Transformation* [White paper]. Greentech Media Inc. cconcernedcitizens.org/wp-content/uploads/2015/02/Evolution-Grid-Edge-Ecosystem-Whitepaper.pdf

[Hale 1998] Hale, J., and Bose, A. (1998). Information survivability in the electric utility industry. *1998 Information Survivability Workshop*, Orlando, Florida, USA, (October 1998).

[Halpern 2014] Halpern, S. (2014, November 20). *The Creepy New Wave of the Internet* [News]. The New York Review. http://www.nybooks.com/articles/archives/2014/nov/20/creepy-new-wave-internet

[Hansche 2004] Hansche, S., Berti, J., Hare, C. (2004). *Official (ISC)2 Guide to The CISSP Exam*. Auerbach Publications.

[Hanseth 2002] Hanseth, O. (2002). *From systems and tools to networks and infrastructures - from design to cultivation. Towards a theory of ICT solutions and its design methodology implications.* http://heim.ifi.uio.no/~oleha/Publications/ib_ISR_3rd_resubm2.html

[Harris 2013] Harris, S. (2013). *All in One CISSP Exam Guide* (6th Edition). McGraw Hill Education.

[Hawk 2014] Hawk, C., Kaushiva, A. (2014). Cybersecurity and the smarter grid. *The Electricity Journal*, *27*(8), 84–95. https://doi.org/10.1016/j.tej.2014.08.008

[Hentea 2006] Hentea, M. (2006). Information Security Awareness through a Research Agenda. *60 Minute Presentation at SecureIT 2006 Conference*, Anaheim, California, USA, (March 2006).

[Hentea 2007] Hentea, M., Dhillon, H. (2007). New competencies for control engineers to meet the market demands in control systems. *Proceedings of ICEE (International Conference in Engineering Education) 2007 Conference*, Coimbra, Portugal, (September 2007).

[Hentea 2008] Hentea, M. (2008). Improving security for SCADA control systems. *Interdisciplinary Journal of Information, Knowledge, and Management*, *3*, 73–86. https://doi.org/10.28945/3185

[Higgins 2012] Higgins, K.J. (2012, July 19). *Smart grid researcher releases open source meter-hacking tool* [News]. Dark Reading. https://www.darkreading.com/vulnerabilities---threats/smart-grid-researcher-releases-open-source-meter-hacking-tool/d/d-id/1138052

[HP PPM] PPMetrics. (n.d.). *Adoption and Maturity Management Add-on for HP PPMTM A Holistic Approach to PPM*. PPMetrics. http://www.ppmetrics.com/wp-content/uploads/2017/08/AMM_Data_Quality_WP.pdf

[Huang 2010] Huang, D-L., Rau, P-L.P., Salvendy, G. (2010). Perception of information security. *Behaviour & Information Technology*, *29*(3), 221–232. https://doi.org/10.1080/01449290701679361

[IC3] IC3. (n.d.). *Federal Bureau of Investigation*. Internet Crime Complaint Center. https://www.ic3.gov/default.aspx

[ICA] CPTED. (n.d.). *A Brief History*. International Crime Prevention Through Environmental Design (CPTED) Association. https://cpted.net/A-Brief-History

[ICS-CERT] ICS-CERT. (n.d.). *Training Available Through CISA*. Cybersecurity & Infrastructure Security Agency. https://us-cert.cisa.gov/ics/Training-Available-Through-ICS-CERT

[ICS-CERT 2015] ICS-CERT. (2015). *Monitor September 2014–February 2015 (MM201502)*. Cybersecurity & Infrastructure Security Agency. https://us-cert.cisa.gov/ics/monitors/ICS-MM201502

[ICS-CERT 2016] ICS-CERT. (2016). *Monitor January–February 2016 (MM201602)*. Cybersecurity & Infrastructure Security Agency. https://us-cert.cisa.gov/ics/monitors/ICS-MM201602

[ICS-Letter] ICS-CERT. (n.d.). *ICS Monitor Newsletters*. Cybersecurity & Infrastructure Security Agency. https://us-cert.cisa.gov/ics/monitors

[IDG 2016] IDG. (2016). *Global economic crime survey 2016: US results adjusting the lens on economic crime*. The Global State of Information. https://www.pwc.com/us/en/services/forensics/library/economic-crime-survey-us-supplement.html

[IEC 2014] IEC. (2014). *Electrical Energy Storage* [White Paper]. International Electrotechnical Commission. http://www.iec.ch/whitepaper/pdf/iecWP-energystorage-LR-en.pdf

[IEC 61069-2] *IEC 61069-2:2016 Industrial-process measurement, control and automation Evaluation of system properties for the purpose of system assessment - Part 2: Assessment methodology.*

[IEC 61069-5] *IEC 61069-5*:2016 *Industrial-process measurement and control – Evaluation of system properties for the purpose of system assessment - Part 5: Assessment of system dependability.*

[IEC 62443-2-1] *IEC 62443-2-1 Edition 1.0 2010-11 Industrial communication networks – Network and system security – Part 2-1: Establishing an industrial automation and control system security program.*

[IEC/TC56] IEC. (n.d.). *IEC/TC56 Dependability*. International Electrotechnical Commission. https://tc56.iec.ch/

[IEEE 1061] *IEEE 1061:1998 Standard for a Software Quality Metrics Methodology.*

[IEEE 15026-1] *ISO/IEC TR 15026-1*:2010 *Systems and Software Engineering – Systems and Software Assurance – Part 1: Concepts and Vocabulary*. *Replaced by ISO/IEC TR 15026-1:2013.

[IEV 191-02-5] Electropedia. (n.d.). *Dependability and quality of service IEV ref 191-02-5*. http://www.electropedia.org/iev/iev.nsf/display?openform&ievref=191-02-05

[IIC 2015] IIC. (2015). *Industrial Internet Vocabulary* (Version 1.0). Industrial Internet Consortium. http://www.iiconsortium.org/IIRA.htm

[INCOSE SEH] INCOSE. (2015). *Systems Engineering Handbook V4* (March 2015). International Council on Systems Engineering. http://www.incose.org/ProductsPublications/sehandbook

[Infrastructure] Wikipedia. (n.d.). *Infrastructure*. https://en.wikipedia.org/wiki/Infrastructure

[INL 2011] INL. (2011). *The Vulnerability Analysis of Energy Delivery Control Systems* (INL/EXT-10-18381). Idaho National Laboratory. http://energy.gov/sites/prod/files/Vulnerability%20Analysis%20of%20Energy%20Delivery%20Control%20Systems%202011.pdf

[IQ/DQ Glossary] IQINT. (n.d.). *IQ/DQ Glossary*. The International Association for Information and Data Quality (IAIDQ). ** IAIDQ has changed name to IQ International*. https://www.iqint.org

[IRENA 2013] IRENA. (2013). *Smart grids and renewables A guide for effective deployment.* International Renewable Energy Agency (IRENA). https://www.irena.org/DocumentDownloads/Publications/smart_grids.pdf

[ISA 99] ISA. (n.d.). *ISA99 Standards, International Society for Automation (ISA), Security for Industrial Automation and Control Systems*. * In 2010, the standards were renumbered to be the ANSI/ISA-62443 series

[ISA 62443-1-1] *ANSI/ISA 62443-1-1 (99.01.01)*-2007 *Security for Industrial Automation and Control Systems Part 1: Terminology, Concepts, and Models formerly, ANSI/ISA-99, Manufacturing and Control Systems Security, Part 1: Concepts, Models and Terminology.*

[ISACA Glossary] ISACA. (n.d.). *Glossary*. http://www.isaca.org/Pages/Glossary.aspx?tid=466&char=I

[ISO 27k] ISO. (n.d.). *About the ISO27k Standards*. International Standards Organization. https://www.iso27001security.com/html/iso27000.html

[ISO 5725-1] *ISO 5725-1:1994 Accuracy (trueness and precision) of measurement methods and results - Part 1: General principles and definitions.*

[ISO 9000] *ISO 9000:2015 Quality Management Systems – Fundamentals and Vocabulary.*

[ISO 9001] *ISO 9001:2015 Quality Management Systems – Requirements.*

[ISO 84024] *ISO 8402*:1994. *Quality Management and Quality Assurance – Vocabulary*. * Replaced by [ISO 9000] in 2015.

[ISO Guide 73] *ISO Guide 73*:2009 *Risk Management – Vocabulary.*

[ISO TR 15443-1] *ISO/IEC TR 15443-1:2012 Information technology -- Security Techniques - Security Assurance Framework -- Part 1: Introduction and Concepts.*

[ISO/IEC 2382] *ISO/IEC 2382:2015 Information Technology – Vocabulary.*

[ISO/IEC 9126] *ISO/IEC 9126*:2001 *Software Engineering — Product Quality*. *Replaced by [ISO/IEC 25010] in 2011.

[ISO/IEC 15288] *ISO/IEC 15288:2008 Systems and Software Engineering -- System Life Cycle Processes, February 2008.*

[ISO/IEC 15408] *ISO/IEC 15408 Information Technology – Security Techniques – Evaluation Criteria for IT Security, Parts 1–3.*

[ISO/IEC 15939] *ISO/IEC 15939*:2007 *Systems and Software Engineering – Measurement Process.*

[ISO/IEC 25010] *ISO/IEC 25010:2011 Systems and software engineering - Systems and software Quality Requirements and Evaluation (SQuaRE) - System and software quality models.*

[ISO/IEC 25012] *ISO/IEC 25012:2008 Software engineering – Software product Quality Requirements and Evaluation (SQuaRE) – Data quality model.*

[ISO/IEC 27000] *ISO/IEC 27000:2018 Information technology — Security techniques — Information security management systems - Overview and vocabulary* (Fifth edition).

[ISO/IEC 27001] *ISO/IEC 27001:2013 Information technology — Security techniques — Information security management systems — Requirements.*

[ISO/IEC 27002] *ISO/IEC 27002:2013 Information Technology — Security Techniques — Code of Practice for Information Security Controls.*

[IT Handbook] FFIEC. (2016). *FFIEC Information Technology Examination Handbook Information Security* (September 2016). I.B. Responsibility and Accountability. https://ithandbook.ffiec.gov/it-booklets/information-security/i-governance-of-the-information-security-program/ib-responsibility-and-accountability.aspx

[IUT-T 2008] *ITU-T X.1205 (04/2008) Overview of Cybersecurity.* International Telecommunications Union for the Telecommunication Standardization Sector. https://www.itu.int/rec/T-REC-X.1205-200804-I

[Jessup 2008] Jessup, L., Valacich, J. (2008). *Information Systems Today: Managing in the Digital World* (3rd Edition). Prentice Hall.

[JohnManuel 2009] JohnManuel. (2009). *The Information Security triad: CIA Second version.* CC BY-SA 3.0. https://en.wikipedia.org/wiki/Information_security#/media/File:CIAJMK1209.png

[Juran 1999] Juran, J.M., Godfrey, B.A. (1999). *Juran's Quality Handbook* (5th Edition, p. 2.2). McGraw-Hill.

[Kabay 2000] Kabay, M.E. (2000). *The Parkerian Hexad.* http://www.mekabay.com/overviews/hexad.pptx

[Kadrich 2007] Kadrich, M.S. (2007). *Endpoint Security.* Addison-Wesley.

[Karnouskos 2011] Karnouskos, S., Colombo, A.W. (2011). Architecting the next generation of service-based SCADA/DCS system of systems. *IECON 2011 - 37th Annual Conference of the IEEE Industrial Electronics Society* (pp. 359–364), Melbourne, VIC, Australia, (November 2011). https://doi.org/10.1109/IECON.2011.6119279

[Karnouskos 2012] Karnouskos, S., Colombo, A.W., Bangemann, T., Manninen, K., Camp, R., Tilly, M. Stluka, P., Jammes, F., Delsing, J., Eliasson, J. (2012). A SOA-based architecture for empowering future collaborative cloud-based industrial automation. *IECON 2012 - 38th Annual Conference on IEEE Industrial Electronics Society* (pp. 5766–5772), Montreal, QC, Canada, (October 2012). https://doi.org/10.1109/IECON.2012.6389042

[Kaspersky 2014] Kaspersky Lab. (2014). *Cyber threats to ICS Systems You don't have to be a target to become a victim industrial security.* Industrial Security. https://media.kaspersky.com/en/business-security/critical-infrastructure-protection/Cyber_A4_Leaflet_eng_web.pdf

[Kettani 2019] Kettani, H., Wainwright, P. (2019). On the top threats to cyber systems. *2019 IEEE 2nd International Conference on Information and Computer Technologies (ICICT)* (pp. 175–179), Kahului, HI, USA, (March 2019). https://doi.org/10.1109/INFOCT.2019.8711324

[Knight 2000] Knight, J.C., Sullivan, K.J. (2000). *On the definition of Survivability.* http://citeseerx.ist.psu.edu/viewdoc/download?doi=10.1.1.626.3985&rep=rep1&type=pdf

[Knight 2003] Knight, J.C., Strunk, E.A., Sullivan, K.J. (2003). Towards a rigorous definition of information system survivability. *Proceedings of the 3rd DARPA Information Survivability Conference and Exposition (DISCEX)* (Vol 1, pp. 78–89), Washington, DC, USA, (April 2003). https://doi.org/10.1109/DISCEX.2003.1194874

[Koopman 2004] Koopman, P. (2004). Embedded system security. *Computer, 37*(7), 95–99.

[Krutz 2004] Krutz, R.L., Vines R.D. (2004). *The CISSP Prep Guide* (2nd edition). Wiley.

[Krutz 2006] Krutz, R.L. (2006). *Securing SCADA Systems.* Wiley.

[Lacey 2013] Lacey, S. (2013, October). *The grid edge: A New way to define the changing power sector.* Greentech Media Inc. http://www.greentechmedia.com/articles/read/the-grid-edge-a-new-way-to-define-the-changing-power-sector

[Leszczyna 2013] Leszczyna, R., Fovino, I.N. (2013). Evaluating security and resilience of critical networked infrastructures after Stuxnet. In P. Theron, S. Bologna (Eds.), *Critical Information Infrastructure Protection and Resilience in the ICT Sector* (pp. 242–256). IGI-Global. https:/doi.org/10.4018/978-1-4666-2964-6.ch012

[Liu 2013] Liu, G., Rasul, M. G., Amanullah, M. T. O., Khan, M. M. K. (2013). Economy of smart grid. In A.B.M.S. Ali (Ed.), *Smart Grids Opportunities, Developments, and Trends* (pp. 215–228). Springer-Verlag. https://doi.org/10.1007/978-1-4471-5210-1_10

[Longlivetheux] Longlivetheux. (2015). *DIKW Pyramid: Data, Information, Knowledge, and Wisdom.* CC SA-4.0. https://commons.wikimedia.org/wiki/File:DIKW_Pyramid.svg

[Loveland 2012] Loveland, G., Lobel, M. (2012). Cybersecurity the new business priority. *PwC perspectives on current business issues and trends* (15), 24—33. www.pwc.com.

[Lynn 2015] Lynn, K. (2015). *Grid modernization initiative crosscut discussion* (July 8). Office of Energy Efficiency and Renewable Energy. The National Renewable Energy Laboratory. https://www.nrel.gov/esif/assets/pdfs/agct_day2_lynn.pdf

[Maconachy 2001] Maconachy, W.V., Schou, C.D., Ragsdale, D., and Welch, D. (2001). A model for information assurance: an integrated approach. *Proceedings of the 2001 IEEE Workshop on Information Assurance and Security Military Academy* (pp. 306–310), West Point, New York, USA, (June 2001).

[Maiwald 2004] Maiwald, E. (2004). *Fundamentals of Network Security.* McGraw-Hill/Technology Education.

[Masica 2007] Masica, K. (2007). *Recommended Practices Guide for Securing ZigBee Wireless Networks in Process Control System Environments.* https://www.energy.gov/oe/downloads/recommended-practices-guide-securing-zigbee-wireless-networks-process-control-system

[McCumber 1991] McCumber, J. (1991). Information Systems Security: a comprehensive model. *Proceedings of the 14th National Computer Security Conference.* National Institute of Standards and Technology, Baltimore, Maryland, USA, (October 1991).

[McCumber] Wikipedia. (2006). *McCumber Cube.* CC BY-SA 3.0. https://en.wikipedia.org/wiki/File:Mccumber.jpg

[McLaughlin 2009] McLaughlin, S., Podkuiko, D., McDaniel, P. (2010). Energy Theft in the Advanced Metering Infrastructure. In E. Rome, R. Bloomfield (Eds.), *Critical Information Infrastructures Security. CRITIS 2009. Lecture Notes in Computer Science* (Vol. 6027, pp. 176–187). Springer. https://doi.org/10.1007/978-3-642-14379-3_15

[McManus 2012] McManus, M. (2012, November 13). *Maxim smart-meter SoC combines metrology, security, and communication* [News]. DIGITIMES. http://www.digitimes.com/news/a20130121VL205.html

[Mell 2009] Mell, P., Grance, T. (2009, October 7). *The NIST Definition of Cloud Computing* [Technical Report 15]. National Institute of Standards and Technology. http://www.nist.gov/itl/cloud/upload/cloud-def-v15.pdf

[Messmer 2008] Messmer, E. (2008, March 24). Outsourcing security presents pros & cons. *Network World 25*(12), 1.

[MYPP 2012] Energy. (2012). *Smart grid R&D multi-year program plan (2010–2014) September 2012 update*. Office of Electricity. U.S. Department of Energy. http://energy.gov/oe/downloads/smart-grid-rd-multi-year-program-plan-2010-2014-september-2012-update

[MYPP 2014] EERE. (2014). *Grid integration multi-year program plan* (February 2014). Office of Energy Efficiency and Renewable Energy. U. S. Department of Energy. http://iiesi.org/assets/pdfs/iiesi_lynn.pdf

[NERC CIP] NERC. (n.d.). *CIP Standards*. North American Electric Reliability Corporation. http://www.nerc.com/pa/Stand/Pages/CIPStandards.aspx

[NII] NII. (n.d.). *High Performance Computing Act of 1991 PubL 102–194, 9 December 1991*. National Information Infrastructure (NII).

[NIPP 2013] DHS. (2013). National Infrastructure Protection Plan (NIPP): 2013 Partnering for Critical Infrastructure Security and Resilience. U.S. Department of Homeland Security. https://www.cisa.gov/publication/nipp-2013-partnering-critical-infrastructure-security-and-resilience

[NIST SP1108r1] *NIST Special Publication (SP) 1108r1 NIST Framework and Roadmap for Smart Grid Interoperability Standards* (Release 1.0, January 2010). https://doi.org/10.6028/NIST.sp.1108

[NIST SP1108r3] *NIST Special Publication (SP) 1108r3 NIST Framework and Roadmap for Smart Grid Interoperability Standards* (Release 3.0, September 2014). http://dx.doi.org/10.6028/NIST.SP.1108r3

[NISTIR 7628] *NIST Interagency Report (NISTIR) 7628 Guidelines for Smart Grid Cyber Security* (August 2010).

[NISTIR 7628r1] *NIST Interagency Report (NISTIR) 7628r1 Guidelines for Smart Grid Cybersecurity* (Revision 1, September 2014). http://dx.doi.org/10.6028/NIST.IR.7628r1

[NISTIR 7298r3] *NIST Interagency Report (NISTIR) 7298r3 Glossary of Key Information Security Terms* (Revision 3, July 2019). https://doi.org/10.6028/NIST.IR.7298r3

[NIST SP500-291] *NIST Special Publication (SP) 500-291 NIST Cloud Computing Standards Roadmap* (Version 2.0, July 2013). *Road Map Working Group*. http://dx.doi.org/10.6028/NIST.SP.500-291r2

[NIST SP500-292] *NIST Special Publication (SP) 500-292 Cloud Computing Reference Architecture Recommendations of the National Institute of Standards and Technology* (September 2011). https://doi.org/10.6028/NIST.SP.500-292

[NIST SP800-12] *NIST Special Publication (SP) 800-12r1 An Introduction to Computer Security* (Revision 1, June 2017). https://doi.org/10.6028/NIST.SP.800-12r1

[NIST SP800-14] *NIST Special Publication (SP) 800-14 Generally Accepted Principles and Practices for Securing Information Technology Systems*. *Replaced by other standards.

[NIST SP800-16r1] *NIST Special Publication (SP) 800-16r1. A Role-Based Model for Federal Information Technology/Cybersecurity Training*, (Revision 1, 3rd draft, March 2014).

[NIST SP800-30r1] *NIST Special Publication (SP) 800-30r1 Guide for Conducting Risk Assessments Revision 1*, (September 2012). Computer Security Division Information Technology Laboratory.

[NIST SP800-50] *NIST Special Publication (SP) 800-50 Building an Information Technology Security Awareness and Training Program* (October 2003). Computer Security Division Information Technology Laboratory.

[NIST SP800-53Ar4] *NIST Special Publications (SP) 800-53Ar4 Assessing Security and Privacy Controls in Federal Information Systems and Organizations: Building Effective Assessment Plans Revision 4*, (December 2014). Computer Security Division Information Technology Laboratory. https://dx.doi.org/10.6028/NIST.SP.800-53Ar4

[NIST SP800-53r4] *NIST Special Publication (SP) 800-53 Security and Privacy Controls for Federal Information Systems and Organizations Revision 4*, (April 2013). Computer Security Division Information Technology Laboratory. http://dx.doi.org/10.6028/NIST.SP.800-53r4 * Superseded by [NIST SP800-53r5] in September 2020; Withdrawal Date September 23, 2021.

[NIST SP800-53r5] *NIST Special Publication (SP) 800-53 Security and Privacy Controls for Information Systems and Organizations Revision 5*, (September 2020). Joint Task Force. https://doi.org/10.6028/NIST.SP.800-53r5

[NIST SP800-82r2] *NIST Special Publication (SP) 800-82r2 Guide to Industrial Control Systems (ICS) Security Supervisory Control and Data Acquisition (SCADA) systems, Distributed Control Systems (DCS), and other control system configurations such as Programmable Logic Controllers (PLC) Revision 2*, (May 2015). http://dx.doi.org/10.6028/NIST.SP.800-82r2

[NIST SP800-100] *NIST Special Publication (SP) 800-100 Information Security Handbook Recommendations of the National Institute of Standards and Technology* (October 2006). Computer Security Division Information Technology Laboratory.

[NIST SP800-160] *NIST Special Publication (SP) 800-160 Systems Security Engineering Considerations for a Multidisciplinary Approach in the Engineering of Trustworthy Secure Systems Volume 1*, (November 2016, updates of March 2018). https://doi.org/10.6028/NIST.SP.800-160v1

[NIST Wiki] NIST. (n.d.). *SGIP NIST Smart grid collaboration site*. TWiki. https://collaborate.nist.gov/twiki-sggrid/bin/view/SmartGrid/WebHome

[Nordell 2012] Nordell, D.E. (2012). Terms of protection: The Many Faces of Smart Grid Security. *IEEE Power & Energy Magazine*, *10*(1), 18–23. https://doi.org/10.1109/MPE.2011.943194

[NSTISSI 4011] NSTISS. (1994). *NSTISSI No. 4011 National Training Standard for Information Systems Security (Infosec) Professionals*. NATIONAL SECURITY TELECOMMUNICATIONS AND INFORMATION SYSTEMS SECURITY. https://www.linuxglobal.com/wp-content/uploads/2012/02/nstissi_4011.pdf

[OASIS 2006] OASIS. (2006). *OASIS reference model for service oriented architecture* 1.0 (soa-rm-cs). Organization for the Advancement of Structured Information Standards (OASIS). https://www.oasis-open.org/committees/download.php/19679/soa-rm-cs.pdf

[OECD 1992] OECD. (1992). *Recommendation of the Council Concerning Guidelines for the Security of Information Systems*. Organisation for Economic Co-operation and Development. https://www.oecd.org/internet/ieconomy/oecdguidelinesforthesecurityofinformationsystems1992.htm

[OECD 2002] OECD. (2002). *OECD Guidelines for the Security of Information Systems and Networks: Towards A Culture of Security*. Organisation for Economic Co-operation and Development. http://www.oecd.org/sti/ieconomy/oecdguidelinesforthesecurityofinformationsystemsandnetworkstowardsacultureofsecurity.htm

[OECD 2012] OECD. (2012, January 10). *ICT Applications for the Smart Grid: Opportunities and Policy Implications. OECD Digital Economy Papers* (No. 190) (DSTI/ICCP/REG(2011)12/FINAL). OECD Publishing. http://dx.doi.org/10.1787/5k9h2q8v9bln-en

[OECD 2015] OECD. (2015*). Digital Security Risk Management for Economic and Social Prosperity: OECD Recommendation and Companion Document*. OECD Publishing. http://dx.doi.org/10.1787/9789264245471-en

[OECD Glossary] OECD. (n.d.). *Statistics portal. Glossary of statistical terms*. http://stats.oecd.org/glossary/detail.asp?ID=3108

[OPEN METER] OpenEI. (n.d.). *Open meter (Smart Grid Project)*. https://openei.org/wiki/Open_meter_(Smart_Grid_Project)

[OSHA Laws] OSHA. (n.d.). *Law and regulations*. Occupational Safety and Health Administration. https://www.osha.gov/laws-regs

[Pagliery 2014] Pagliery, J. (2014, December 29). *Hackers attacked the U.S. energy grid 79 times this year* [News]. *CNN Money*. http://money.cnn.com/2014/11/18/technology/security/energy-grid-hack

[Parker 2002] Parker, D.B. (2002). Toward a New Framework for Information Security. In Boswort, S., Kabay, M.E. (Eds.), *The Computer Security Handbook* (4th Edition). Wiley.

[Paton 1999] Paton, S.M. (1999). *A century of quality: an interview with quality legend Joseph M. Juran. QualityView.* http://www.qualitydigest.com/feb99/html/body_juran.html

[PC Encyclopedia] PCMAG. (n.d). *PC Encyclopedia. API.* http://www.pcmag.com/encyclopedia/term/37856/api

[PCC] Necula, G. (1997). Proof-Carrying Code. *Proceedings of the 24th SIGPLAN-SIGACT Symposium on Principles of Programming Languages (POPL'97)* (pp. 106–119), Paris, France, (January 1997).

[PCIP] DHS. (n.d.). *President's Commission on Critical Infrastructure Protection Executive Order 13010: Critical Infrastructure Protection, 61 Fed. Reg. 37347* (July 15, 1996). https://www.hsdl.org/?abstract&did=1613

[Perrin 2008] Perrin, C. (2008, June 30). The CIA Triad [Blog]. *TechRepublic.* http://www.techrepublic.com/blog/security/the-cia-triad/488

[PNNL 2015] PNNL. (2015). *Grid Architecture Final* (PNNL-24044). Pacific National Northwest Laboratory. https://gridarchitecture.pnnl.gov/media/white-papers/GridArchitecture2final.pdf

[PwC 2016a] PwC. (2016a). *Global economic crime survey 2016: US results adjusting the lens on economic crime* [Report]. http://www.pwc.com/us/en/forensic-services/economic-crime-survey-us-supplement.html

[PwC 2016b] PwC. (2016b). *Turnaround and transformation in cybersecurity: Power and utilities Key findings from The Global State of Information Security® Survey 2016. Cybersecurity, Privacy, Forensics.* https://www.pwc.com/us/en/services/consulting/cybersecurity.html

[Rahman 2012] Rahman, M.A., Mohsenian-Rad, H. (2012). False data injection attacks with incomplete information against smart power grids. 2012 IEEE Global Communications Conference (GLOBECOM) (pp. 3153-3158), Anaheim, California, USA, (December 2012). https://doi.org/10.1109/GLOCOM.2012.6503599

[Ray 2010] Ray, P.D., Harnoor, R., Hentea, M. (2010). Smart power grid security: a unified risk management approach. In D.A. Pritchard, L.D. Sanson (Eds.), *44th Annual 2010 IEEE International Carnahan Conference on Security Technology Proceedings* (pp. 276–285), San Jose, California, USA, (October 2010). https://doi.org/10.1109/CCST.2010.5678681

[ReportLinker 2013] ReportLinker. (2013, January 22). *Global smart home network equipment market 2012-2016* [Press Release]. The Business Journals PR NewsWire. htps://www.slideshare.net/ReportLinker?utm_campaign=profiletracking&utm_medium=sssite&utm_source=ssslideview

[RFC 4949] *IETF Request for Comments (RFC) 4949 Internet Security Glossary Version 2* (August 2007).

[Richardson 2008] Richardson, R. (2008). *2008 CSI computer crime & security survey.* Computer Security Institute. http://www.kwell.net/doc/FBI2008.pdf

[Rihan 2013] Rihan, M., Ahmad, M., Beg, M. (2013). Vulnerability analysis of wide area measurement system in the smart grid. *Smart Grid and Renewable Energy, 4*(6A), 1–7.

[Roadmap 2006] DOE. (2006). *Roadmap to secure control systems in the energy sector executive summary* (January 2006). http://energy.gov/sites/prod/files/oeprod/DocumentsandMedia/2-_exec_sum.pdf

[Roadmap 2011] DOE. (2011). *Roadmap to achieve energy delivery systems cybersecurity (September 2011).* Energy Sector Control System Working Group. http://energy.gov/sites/prod/files/Energy%20Delivery%20Systems%20Cybersecurity%20Roadmap_finalweb.pdf

[Robertson 2010] Robertson, J. (2010, March 26). *'Smart' meters have security holes* [News]. NBCNEWS. http://www.nbcnews.com/id/36055667

[Roebuck 2011] Roebuck, K. (2011). *Data Quality: High-impact Strategies - What You Need to Know: Definitions, Adoptions, Impact, Benefits, Maturity, Vendors.* Emereo Publishing.

[Ruth 2014] Ruth, M. F., Kroposki, B. (2014). Energy systems integration: an evolving energy paradigm. *The Electricity Journal, 27*(6), 36–47. https://doi.org/10.1016/j.tej.2014.06.001

[Sanz 2003] Sanz, R., Arzen, K.E. (2003). Trends in software and control. *IEEE Control Systems Magazine, 23*(3), 12–15.

[Sarangi 2012] Sarangi, S. R., Dutta, P., Jalan, K. (2012). IT infrastructure for providing energy-as-a-service to electric vehicles. *IEEE Transactions on Smart Grid, 3*(2), 594–604. https://doi.org.10.1109/TSG.2011.2175953

[SGIP SGCC] NIST. (n.d.). *Smart Grid Smart Grid Interoperability Panel.* https://www.nist.gov/programs-projects/smart-grid-national-coordination/smart-grid-interoperability-panel-sgip

[Schoechle 2012] Schoechle, T. (2012). *Getting smarter about the smart grid* [White paper]. National Institute of Science Law & Public Policy. https://gettingsmarteraboutthesmartgrid.org/pdf/Smart%20Grid%20Report%203-15-13.pdf

[Schwartz 2001] Schwartz, E., Erwin, D., Weafer, V., Briney, A. (2001). Roundtable: InfoSec staffing help wanted! *Information Security Magazine Online* (April 2001).

[Singh 2014] Singh, A., Vaish, A., Keserwani, P. K. (2014). Information security: Components and Techniques. *International Journal of Advanced Research in Computer Science and Software Engineering, 4*(1), 1072–1077.

[Sollecito 2009] Sollecito, L. (2009). Smart grid: the road ahead. *Protection & Control Journal, 8th Edition*, 15–19.

[Sommerville 2004] Sommerville, J. (2004). *Software Engineering* (7th Edition). Pearson Addison Wesley.

[Sullivan 1999] Sullivan, K.J., Knight, J.C., Du, X., Geist, S. (1999). Information survivability control systems. *Proceedings of the 1999 International Conference on Software Engineering, ICSE'99* (pp. 184–192), Los Angeles, CA, USA, (May 1999).

[Tabari 2016] Tabari, M., Yazdani, A. (2016). An energy management strategy for a DC distribution system for power system integration of plug-in electric vehicles. *IEEE Transactions on Smart Grid, 7*(2), 659–668. https://doi.org/10.1109/TSG.2015.2424323

[Talbot 2009] Talbot, J., Jakeman, M. (2009). *Security Risk Management Body of Knowledge.* Wiley.

[Turban 2007] Turban, E., Aronson, J.E., Liang, T-P., Sharda, R. (2007). *Decision Support and Business Intelligence Systems* (8th Edition). Pearson Education.

[UCAIUG 2008] UCAIUG. (2008). *AMI system requirements V1.01* [UCAIUG: AMI-SEC-ASAP]. UCA International User Group. http://energy.gov/sites/prod/files/oeprod/DocumentsandMedia/14-AMI_System_Security_Requirements_updated.pdf

[Uludag 2016] Uludag, S., Lui, K., Ren, W., Nahrstedt, K. (2016). Secure and scalable data collection with time minimization in the smart grid. *IEEE Transactions on Smart Grid, 7*(1), 43–54. https://doi.org/ 10.1109/TSG.2015.2404534

[Vacca 2012] Vacca, J.R. (2012). *Computer and Information Security Handbook* (2nd Edition). Morgan Kaufmann. https://doi.org/10.1016/C2011-0-07051-5

[Virus 2012] ICS-CERT. (2012). Malware infections in the control environment. Monitor (ICS-MM201212) October-December. https://us-cert.cisa.gov/sites/default/files/Monitors/ICS-CERT_Monitor_Oct-Dec2012.pdf

[Verizon 2016] Verizon. (2016*). Data breach digest scenarios from the field.* VERIS. http://veriscommunity.net

[Wang 1996] Wang, R., Strong, D. (1996). Beyond accuracy: what data quality means to data consumers. *Journal of Management Information Systems, 12*(4), 5–34.

[Weiss 2010] Weiss, J. (2010). *Protecting Industrial Control Systems from Electronic Threats.* Momentum Press.

[Whitman 2012] Whitman, M.E., Mattord, H.J. (2012). *Principles of Information Security* (4th Edition). Thomson Course Technology.

[Whitman 2014] Whitman, M.E., Mattord, H.J. (2014). *Management of Information Security* (4th Edition). Cengage Learning.

[Wikilubina] Wikilubina. (2014). *Reference Model of Information Assurance and Security (RMIAS).* CC BY-SA 3.0. Wikimedia. https://commons.wikimedia.org/wiki/File:A_Reference_Model_of_Information_Assurance_and_Security_(RMIAS).png

[Wiki Audit] Wikipedia. (n.d.). *Information Technology Security Audit.* https://en.wikipedia.org/wiki/Information_technology_security_audit

[Wiki Safety] Wikipedia. (n.d.). *Safety.* Wikipedia. http://en.wikipedia.org/wiki/Safety

[Wulf 2001] Wulf, W.A. (2001, October 10). *Cyber security: Beyond the Maginot line. Statement before the House Science Committee U.S House of Representatives* [Speech]. National Academy of Engineering. http://www.nae.edu/News/SpeechesandRemarks/CyberSecurityBeyondtheMaginotLine.aspx

[Xu 2020] Xu, Y. (2020). A review of cyber security risks of power systems: from static to dynamic false data attacks. *Protection and Control of Modern Power Systems* 5, Article: 19. https://doi.org/10.1186/s41601-020-00164-w

[Yousuf 2010] Yousuf, S. (2010, October 1). *Smart grid security: Critical success factors.* http://www.cio.com.au

[Zhang 2008] Zhang, H., Liu, J., Yan, Z. (2008). A research of outsourcing decision-making based on outsourcing market maturity. *Proceedings of The International Symposium on Electronic Commerce and Security* (pp. 629–632), Guangzhou, China, (August 2008).

Part III

Security of Critical Infrastructure

6

Critical Infrastructure

6.1 Introduction

An infrastructure is the framework of interdependent networks and systems comprising identifiable industries, institutions (including people and procedures), and distribution capabilities that provide a reliable flow of products and services essential to the defense and economic security of the United States, the smooth functioning of government at all levels, and society as a whole. Consistent with the definition in the Homeland Security Act, infrastructure includes physical, cyber, and/or human elements [SPP Glossary].

6.1.1 Critical Infrastructure

Critical infrastructure is the body of systems, networks, and assets that are so essential that their continued operation is required to ensure the security of a given nation, its economy, and the public's health and/or safety. An infrastructure is considered to be critical when it meets a standard of importance for the national interest such as the goods or services it provides that are essential to national way of life.

In most OECD countries' definitions, the word critical refers to infrastructure that provides an essential support for economic and social well-being, for public safety, and for the functioning of key government responsibilities. For example, Canada's definition of criticality involves serious impact on the health, safety, security, or economic well-being of Canadians or the effective functioning of governments in Canada [Canada]. Germany refers to significant disruptions to public order or other dramatic consequences. The Netherlands' critical infrastructure policy refers to infrastructure whose disruption would cause major social disturbance, tremendous loss of life, and economic damage. Thus, the word critical refers to infrastructure that, if disabled or destroyed, would result in catastrophic and far-reaching damage.

One common characteristic of critical infrastructures is that they are complex in their structure, which makes the issue of dependencies and common cause failure an important topic to society. There is not a commonly accepted definition of critical infrastructure and other related terms, but all definitions emphasize the contributing role of a critical infrastructure to the society or the debilitating effect in the case of disruption [CIPRNET]. Examples of other terms include disruption, dependency, and common cause failure. Examples of definitions include the following:

- Disruption is a circumstance or event that interrupts or prevents the correct operation of system services and functions [RFC 4949]. However, this term can have a national definition. It is being defined by each country according to their own criteria. For example, Canada defines disruption as a disturbance that compromises the availability, delivery, and/or integrity of services of an organization [CIPRNET].

Building an Effective Security Program for Distributed Energy Resources and Systems: Understanding Security for Smart Grid and Distributed Energy Resources and Systems, Volume 1, First Edition. Mariana Hentea.

- A dependency is defined as:
  - The relationship between two critical infrastructure products or services in which one product or service is required for the generation of the other product or service [CIP Glossary]
  - A unidirectional relationship of two infrastructures through which the state of the depending infrastructure is influenced by or is correlated to the state of the other [Rinaldi 2001]
  - The one directional reliance of an asset, system, network, or collection thereof – within or across sectors – on an input, interaction, or other requirement from other sources in order to function properly [NIPP 2009].
- Common cause failure is the failure of two or more structures, systems, or components due to a single event or cause [Safety Glossary].

The definitions of infrastructure used in official descriptions of critical infrastructure tend to be broad. It refers to physical infrastructure and includes intangible assets and/or to production or communication networks. Australia, for example, refers to physical facilities, supply chains, information technologies, and communication networks. Canada refers to physical and information technology (IT) facilities, networks, services, and assets. The United Kingdom refers to assets, services, and systems. Most governments adopt a broad sectoral perspective on critical infrastructure – they include sectors that account for substantial portions of national income and employment. So, critical infrastructure is described as physical or intangible assets whose destruction or disruption would seriously undermine public safety, social order, and the fulfillment of key government responsibilities [OECD 2008a].

Related to the term critical infrastructure in Europe is the term critical infrastructure systems (CIS), which includes the public, private, and governmental infrastructure assets and interdependent cyber and physical networks.

The European Council Directive 2008/114/EC [EU 2008] defines critical infrastructure an asset, system, or part thereof located in member states that is essential for the maintenance of vital societal functions, health, safety, security, economic or social well-being of people, the disruption or destruction of which would have a significant impact in a member state as a result of the failure to maintain those functions. More definitions can be found in [CIP Glossary], [EUCOM 2005].

Table 6.1 exhibits definitions for critical infrastructure as used by different countries. The definition of ECI refers to the doctrine and programs created to identify and protect critical infrastructure. More definitions for other countries may be found at [CA], [CIPedia].

Table 6.1 Examples of critical infrastructure definitions.

Country	Critical infrastructure definition
Australia	Those physical facilities, supply chains, information technologies, and communication networks that, if destroyed, degraded, or rendered unavailable for an extended period, would significantly impact on the social or economic well-being of the nation or affect Australia's ability to conduct national defense and ensure national security [AU]
European Union	European critical infrastructure or ECI means critical infrastructure located in EU states, the disruption or destruction of which would have a significant impact on at least two EU states [EU]
United Kingdom	Critical infrastructure (or critical national infrastructure (CNI) in the United Kingdom) is a term used by governments to describe assets that are essential for the functioning of a society and economy – the infrastructure [UK]
United States	Critical infrastructure as those systems and assets, whether physical or virtual, so vital to the United States that the incapacity or destruction of such systems and assets would have a debilitating impact on security, national economic security, national public health or safety, or any combination of those matters [US]

Generally, the critical infrastructure includes telecommunication, transportation, energy, banking, finance, water supply, emergency services, government services, agriculture, and other fundamental systems and services that are critical to the security, economic prosperity, and social well-being of the public. Identification of the critical infrastructure may differ in any of these countries [Maloor 2012].

In the United States, critical infrastructure is the backbone of nation's economy, security, and health [DHS 2015]. It provides the essential services that underpin American society [DHS 2013]. It is described in [CNSSI 4009], [SPP Glossary], [DHS CIS]:

> Systems and assets, whether physical or virtual, so vital to the United States that the incapacity or destruction of such systems and assets would have a debilitating impact on security, national economic security, national public health or safety, or any combination of those matters [CNSSI 4009].
>
> System and assets, whether physical or virtual, so vital to the U.S. that the incapacity or destruction of such systems and assets would have a debilitating impact on security, national economic security, national public health or safety, environment, or any combination of those matters, across any Federal, State, regional, territorial, or local jurisdiction [SPP Glossary].
>
> The physical and cyber systems and assets so vital to the United States that their incapacity or destruction would have a debilitating impact on our physical or economic security or public health or safety [DHS CIS].

Although these definitions are almost the same, there are some specific elements that are included for a different perspective. Thus, many definitions are national definitions (established by a country), although academic definitions are emerging and standard organizations (e.g. IETF, ITU-T) provide common definitions of concepts and related terms: key resources, critical infrastructure, and critical information infrastructure. In the United States, critical infrastructure is classified as a national concern [DHS 2003] because of its scope and its importance to the nation.

6.1.2 Critical Information Infrastructure

Related to critical infrastructure concept are key resources and critical information infrastructure. As defined in the Homeland Security Act of the United States, key resources are publicly or privately controlled resources essential to the minimal operations of the economy and government [SPP Glossary]. Critical information infrastructure is a fundamental component in the design and operation of all forms of traditional critical infrastructure (e.g. electricity grids, transportation systems, water supply, etc.). Figure 6.1 depicts a perspective on the relationships between these terms.

The term critical resources is one of most expansive of all the terms; these resources include those assets within the sphere of critical infrastructure and critical information infrastructure. They have been defined by some national governments to include natural and environmental resources such agriculture, energy, freshwater, rainforests, etc.

With emergence of a global information society, the term critical Internet resources is considered by many as related to critical information infrastructure in the Internet era. As a result, some countries evolved their critical information infrastructure definitions to include the Internet concept [Maloor 2012].

Information that is not customarily in the public domain and is related to the security of critical infrastructure or protected systems is called critical infrastructure information (CII). CII consists of records and information concerning any of the following:

- Actual, potential, or threatened interference with, attack on, compromise of, or incapacitation of critical infrastructure or – protected systems by either physical or computer-based attack or other similar conduct (including the misuse of or unauthorized access to all types of communications and data transmission systems) that violates federal, state, or local law; harms the interstate commerce of the United States; or threatens public health or safety.
- The ability of any critical infrastructure or protected system to resist such interference, compromise, or incapacitation – including any planned or past assessment, projection, or estimate of the

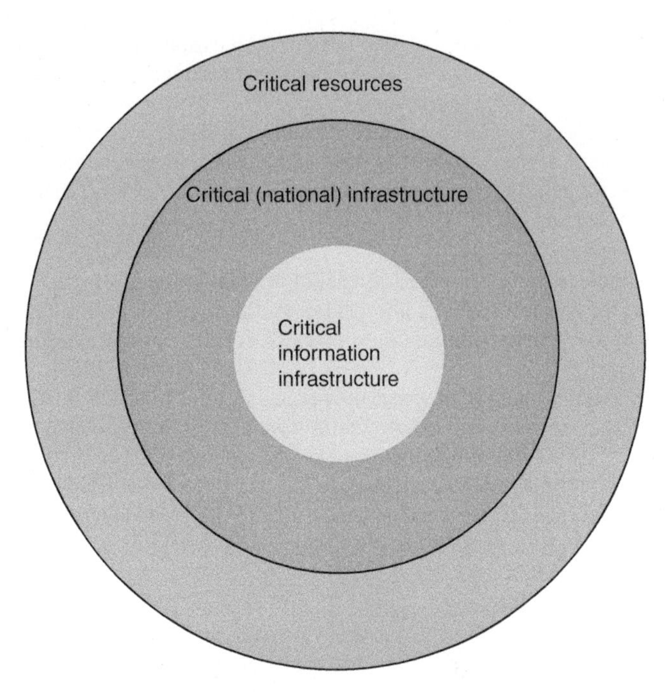

Figure 6.1 Critical infrastructure related terms.

vulnerability of critical infrastructure or a protected system, including security testing, risk evaluation thereto, risk management planning, or risk audit.

- Any planned or past operational problem or solution regarding critical infrastructure or protected systems, including – repair, recovery, insurance, or continuity, to the extent that it is related to such interference, compromise, or incapacitation.

Supporting the CII is the critical information infrastructure. An information infrastructure is defined by Hanseth [Hanseth 2002] as a shared, evolving, open, standardized, and heterogeneous installed base.

Critical information infrastructure includes those systems that are so vital to a nation that their incapacity or destruction would have a debilitating effect on national security, the economy, or public health and safety [RFC 4949]. Critical information infrastructures should be understood as referring to those interconnected information systems and networks, the disruption or destruction of which would have serious impact on the health, safety, security, or economic well-being of citizens or on the effective functioning of government or the economy [OECD 2008c].

Some members of OECD, like Australia, Canada, Netherlands, the United Kingdom, and the United States, describe critical information infrastructure as the information systems (software, hardware, and data) and services that support one or more critical infrastructure(s), the disruption or outage of which causes severe damage to the functioning of that dependent critical infrastructure(s) [OECD 2008c].

The United States relies on critical information infrastructure for government operations, a vibrant economy, and the health and safety of its citizens. Critical infrastructure and key resources (CIKR) may be directly exposed to the event themselves or indirectly exposed as a result of the dependencies and interdependencies among CIKR [NIPP 2009]. Malicious actors can and do conduct attacks against critical cyber infrastructure on an ongoing basis. The national strategy to secure CII identifies the responsibilities of the various CIKR partners with a role in securing cyberspace. Both public and private sector owners and operators actively manage the risk to their operations through monitoring and mitigation activities designed to prevent daily incidents from becoming significant disruptions. Risk is defined as the potential for an unwanted outcome resulting from an incident, event, or occurrence, as determined by its likelihood and the associated consequences [SPP Glossary].

Key terms and concepts related to the cyber dimension for critical information infrastructure are defined in Presidential Policy Directive 21 [PPD-21], [SPP Glossary]. Many of these terms are different than common cybersecurity terms used in private businesses.

Therefore, in order to follow the security recommendations, it is necessary to understand the referenced security terms. For example, cybersecurity is the prevention of damage to, unauthorized use of, or exploitation of, and, if needed, the restoration of electronic information and communications system and the information contained therein to ensure confidentiality, integrity, and availability. It includes protection and restoration, when needed, of information networks and wireline, wireless, satellite, public safety answering points, and 911 communication systems and control systems [NIPP 2009].

6.2 Associated Industries with Critical Infrastructure

Critical infrastructures are those physical and cyber-based systems essential to the minimum operations of the economy and government. They include, but are not limited to, telecommunications, energy, banking and finance, transportation, water systems, and emergency services, both governmental and private. Most commonly associated with the critical infrastructure term are facilities for electrical power grid, utilities (gas, oil, and water), telecommunication, water supply, agriculture, public health, transportation, food, and financial services.

6.2.1 US Critical Sectors

According to [PPD-21], the following 16 sectors are considered critical infrastructures in the United States (critical infrastructure sectors):

- Chemical
- Commercial facilities
- Communications
- Critical manufacturing
- Dams
- Defense industrial base
- Emergency services
- Energy
- Financial services
- Food and agriculture
- Government facilities
- Healthcare and public health
- IT
- Nuclear reactors, materials, and waste
- Transportation systems
- Water and wastewater systems.

National monuments and icons sector and postal and shipping sector are not anymore considered critical sectors as previously defined in [NIPP 2009] of the United States. The [PPD-21] directive designates an associated federal sector-specific agency (SSA) for each sector. In some cases, co-SSAs are designated where those departments share the roles and responsibilities of the SSA. The Department of Homeland Security (DHS) critical infrastructure program provides more information about sectors and their associated critical functions and value chains [DHS CIS].

6.2.2 Other Countries

While other countries may have a different categorization of the critical sectors, critical infrastructure protection (CIP) is the universal goal for each country and government.

6.3 Critical Infrastructure Components

Critical infrastructures are composed of physical, personal, and cyber components. It includes distributed networks, varied organizational structures and operating models (including multinational ownership), interdependent functions and systems in both the physical space and cyberspace, and governance constructs that involve multilevel authorities, responsibilities, and regulations [PPD-21].

Critical infrastructures have physical or mechanical processes controlled electronically by systems, usually called supervisory control and data acquisition (SCADA) or process control systems (PCSs), composed of computers interconnected by networks. SCADA systems, distributed control systems (DCS), and other smaller control system configurations including skid-mounted programmable logic controllers (PLCs) are often found in the industrial sectors and critical infrastructures. These are also known under a general term, industrial control system (ICS). A control system is a device or set of devices to manage, command, direct, or regulate the behavior of other devices or systems. ICSs are typically used in industries such as electrical, water, oil and gas, and chemical including experimental and research facilities such as nuclear fusion laboratories. The reliable operation of modern infrastructures depends on computerized control of SCADA systems.

Critical infrastructures are supported by public and private owners and operators and other supporting entities that play a role in securing the nation's critical infrastructure. Each sector performs critical functions that are supported by IT, ICSs, and, in many cases, both IT and ICS.

ICSs are computer-based systems used to monitor and control physical processes. They are usually composed of a set of networked devices such as sensors, actuators, controllers, and communication devices.

Control systems and networks are essential to monitoring and controlling many critical infrastructure assets (e.g. electric power distribution, water treatment, and transportation management) and industrial plants (e.g. those used for manufacturing chemicals, pharmaceuticals, and food products). Most of these infrastructures are safety critical – an attack can impact public health, the environment, and the economy and even lead to the loss of human life [Huang 2009].

Control systems are becoming more complex and interdependent and, therefore, more vulnerable. The increased risk of computer attacks has led to numerous investigations of control system security. Generally, industrial control devices are designed with more consideration for hardening against environmental and physical threats including extensive forensics for physical parameters. However, they have inadequate security against Internet-based threats and few cyber-related forensics. ICS devices typically use non-hardened networking stacks, common operating systems (DOS, Windows NT/2000, and Linux), and applications that are seldom patched after their initial deployment. As a result, such systems can easily fall prey to malware (viruses, worms, and Trojans).

Critical information infrastructure, also called cyber infrastructure in the United States that exists in most, if not all, sectors, includes business systems, control systems, access control systems, and warning and alert systems. Critical information infrastructure is diverse and complex because of many dependencies and relationships among information infrastructures of critical sectors.

Critical information infrastructure may be described as referring to one or more of the following components [OECD 2008c]:

- Information components supporting the critical infrastructure.
- Information infrastructures supporting essential components of government business.
- Information infrastructures essential to the national economy.

Therefore, the critical information infrastructure should be [NIPP 2009]:

- Robust enough to withstand attacks without incurring catastrophic damage.
- Resilient enough to sustain nationally critical operations.
- Responsive enough to recover from attacks in a timely manner.

A trustworthy system for critical information infrastructure is characterized by attributes such as security, reliability, resilience, and QoS, ensuring user privacy and providing usable and trusted tools that support user's security management.

6.4 Energy Sector

The energy sector includes assets related to three key energy resources: electric power, petroleum, and natural gas. Energy assets and critical infrastructure components are owned by private, federal, state, and local entities, as well as by some types of energy consumers, such as large industries and financial institutions (often for backup power purposes). More than 80% of the US energy infrastructure is owned by the private sector supporting supplying fuels to the transportation industry, electricity to households and businesses, and other sources of energy that are integral to growth and production across the nation.

How dependent society is on its infrastructure, the following events need to be considered. In May 1998, PanAmSat's Galaxy IV satellite's onboard controller malfunctioned, disrupting service to an estimated 80–90% of the nation's pagers, causing problems for hospitals trying to reach doctors on call, emergency workers, and people trying to use their credit cards at gas pumps, and many others [Zuckerman 1998], [Felps 1999]. A software bug in the control software caused the electricity blackout in August 2003, the world's second most widespread blackout in history [Blackout 2003]. The electricity blackout in the United States and Canada left without electricity about 55 million people in the United States and Canada together. Also, it interrupted the services of railroad transportation, airports, financial markets, and many more economies. It took about a week to restore the service in some areas. It illustrated the interdependencies between electricity and other elements of the energy market such as oil refining and pipelines, as well as communications, drinking water supplies, etc.

A healthy energy infrastructure is one of the defining characteristics of a modern global economy. Any prolonged interruption of the supply of basic energy – electricity, petroleum, or natural gas – would do considerable harm to the US economy and the American people.

Cybersecurity is the prevention of damage to, unauthorized use of, or exploitation of, and, if needed, the restoration of electronic information and communication systems and the information contained therein to ensure confidentiality, integrity, and availability. It includes protection and restoration, when needed, of information networks and wireline, wireless, satellite, public safety answering points, and 911 communication systems and control systems [SPP Glossary].

On 30 June 2006, the US DHS announced completion of the National Infrastructure Protection Plan (NIPP), a comprehensive risk management framework that defines CIP roles and responsibilities for all levels of government, private industry, and other sector partners. The NIPP builds on the principles of the President's National Strategy for Homeland Security and strategies for the protection of CIKR. The NIPP was reissued in January 2009, and CIKR protection responsibilities are assigned to select federal agencies called SSAs.

The US Department of Energy (DOE) has been designated the Energy SSA. In this role, it has closely collaborated with dozens of government and industry partners to rewrite and revise the 2007 Energy Sector-Specific Plan (SSP). The DOE also conducted formal review and comment periods for the draft 2010 Energy SSP.

CIKR protection and resilience are not new concepts to energy sector asset owners and operators. The electricity and oil and natural gas subsectors have faced challenges from both natural and man-made events well before 11 September 2001. Since that time, the energy sector has made significant progress in developing plans to protect the energy CIKR and to prepare for restoration and recovery in response to terrorist attacks or natural disasters. More recently, potential threats from cyber penetration have created new concerns and industry responses. Through the Energy SSP process, government and industry have established unprecedented cooperation and close partnership to develop and implement a national effort that brings together all levels of government, industry, and international partners. This updated 2010 Energy SSP is reflection of that partnership and the achievements of the sector over the last three years.

6.4.1 Electrical Subsector

The US electricity subsector includes power generation, transmission, conversion, and distribution. The electric power industry comprises electricity generation (AC power), electric power transmission, and ultimately electricity distribution to an electricity meter located at the premises of the end user of the electric power. The electricity then moves through the wiring system of the end user until it reaches the load.

The electricity is produced by combusting coal (primarily transported by rail), in nuclear power plants, by combusting natural gas, and by hydroelectric plants, oil, and renewable sources such as solar, wind, geothermal, and others.

The power grid is a system of power plants, substations, power lines, and control centers that are more specifically described as interconnected generators, transformers, transmission lines, buses, circuit breakers, reclosers, protective relays, switches, voltage control devices, distribution lines, and computers. Each component or collection of components (e.g. the varied subcomponents of a substation) plays its operability role through accepting inputs and yielding outputs. A substation's input is voltage at one level, and its output is voltage at another level. Thus, some physical assets rely on nonphysical services, for example, the transmission of voice and data packets integral to communication networks. Supporting components are integral to each of the primary infrastructure components. Figure 6.2 shows typical elements for power generation, transmission, conversion, and distribution.

Electrical power infrastructure refers to power grid that covers very large areas of the United States that includes three main power grids, called interconnections (see Figure 6.3):

- Western Interconnection
- Eastern Interconnection
- Texas Interconnection.

There is a substantial degree of interconnection within these grids, and computer networks play an important role in managing grid operation and the production of electrical power. Each grid contains a main transmission system (trunk) to which hundreds of distribution systems, local and regional power companies and load centers, are connected. The pooled power is referred to as system power, which is tapped and distributed throughout the grid region.

Many economic, geographic, and historical practicalities complicate the interoperability of the US power grid because the United States and Canada are served not by a single grid as much as by four interlocking regional grids, namely, the Eastern, Western, Texas, and Quebec systems, as depicted in Figure 6.4.

The electric power industry is diverse in its ownership, geography, and asset type. Electricity system facilities are dispersed throughout the North American continent [Interconnections]. Although most assets are privately owned, no single organization represents the interests of the entire sector. NERC,

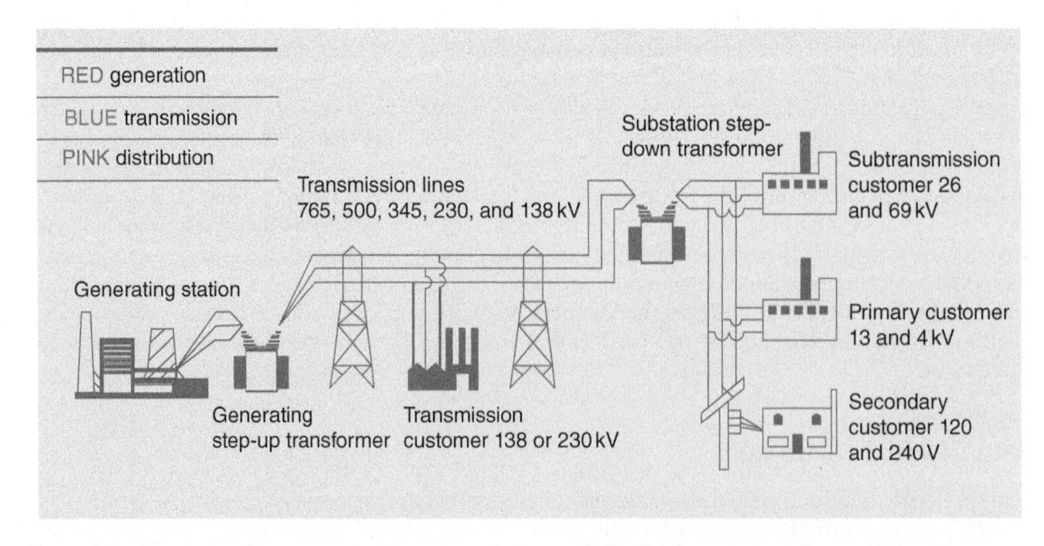

Figure 6.2 Elements of power generation, transmission, and distribution systems. *Source:* Federal energy regulatory commission 2016. Public Domain.

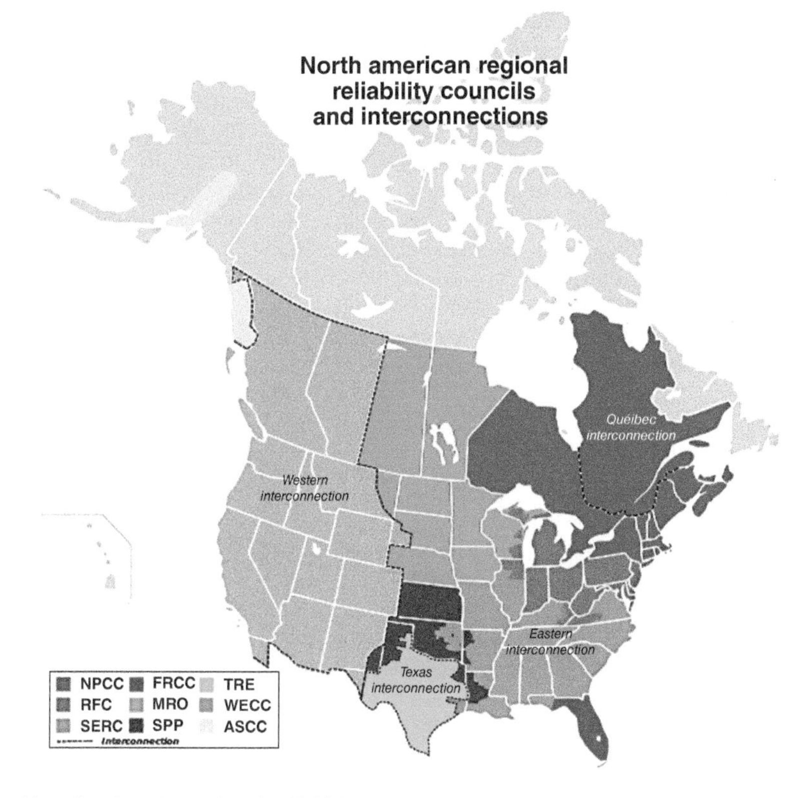

Figure 6.3 Map showing the regional reliability councils and interconnections in North America. *Source:* [Bouchecl]. Licensed under CC BY-SA 3.0.

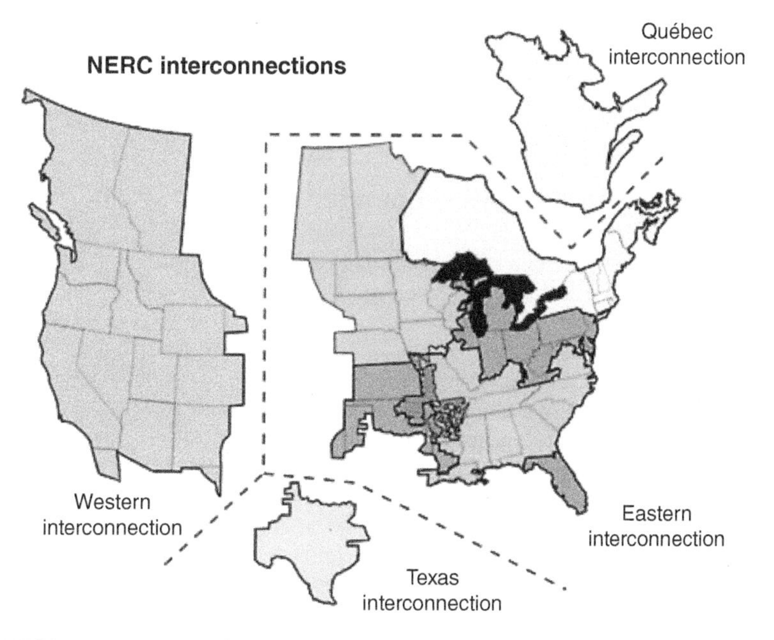

Figure 6.4 NERC interconnections. *Source:* NERC Interconnections, U.S Department of Energy. Public Domain.

through its eight regional reliability councils, provides a platform for ensuring reliable, adequate, and secure supplies of electricity through coordination with the many asset owners.

Significantly, while these regional grids are technically connected to a small degree, they are not in fact integrated in terms of the ability to share large amounts of power. As a result, in the event of a prolonged disruption to any of the single discrete segments of the North American power grid today, it would be both very desirable and essentially technically impossible to have the remaining grids share their power across and into the affected grid.

The traditional electrical grids are generally used to carry power from a few central generators to a large number of users or customers. In contrast, the new emerging Smart Grid uses a two-way flow of electricity and information to create an automated and distributed advanced energy delivery network.

As demand for energy accelerates, governments, power producers, distributors, equipment providers, and other stakeholders are working together to modernize power infrastructure and to add intelligence, communications, and decentralized control to become a Smart Grid.

6.4.2 Smart Grid Infrastructure

A Smart Grid is defined as modernized electrical grid that uses information and communication technology (ICT), remote control, automation, and computer processing to gather and act on information such as information about the behaviors of suppliers and consumers, in an automated fashion to improve the efficiency, reliability, economics, and sustainability of the production and distribution of electricity. Besides electrical infrastructure, we need to consider information infrastructures within the electrical sector that are critical to support markets and power quality and distribution. Figure 6.5 depicts components of Smart Grid infrastructure (power networks, distributed energy resources (DER), consumer networks, etc.).

Smart Grid generally refers to a class of technology people are using to bring utility electricity delivery systems into the twenty-first century. These systems are made possible by two-way communication technology and computer processing that has been used for decades in other industries. They are beginning to be used on electricity networks, from the power plants and wind farms all the way to the consumers of electricity in homes and businesses.

Automation and sensing devices and communication networks make the Smart Grid possible. Devices on the electric utility network equipped with sensors to gather data include power meters, voltage sensors, fault detectors, etc., plus two-way digital communication between the device in the field and the utility's network operations center. Then, automation technology lets the utility adjust and control each individual device or millions of devices from a central location.

6.5 Critical Infrastructure Interdependencies

Interdependency is a mutually reliant relationship between entities (objects, individuals, or groups). The degree of interdependency does not need to be equal in both directions [SPP Glossary]. Historically, many critical national infrastructures were physically and logically separate systems that had little interdependence [Goertzel 2007].

During the last half of the twentieth century, technical innovations and developments in digital information and telecommunications dramatically increased interdependencies among the nation's critical infrastructures including collaborating governments. As shown in Figure 6.6, each infrastructure depends on other infrastructures to function successfully. Disruptions in a single infrastructure can generate disturbances within other infrastructures and over long distances, and the pattern of interconnections can extend or amplify the effects of a disruption. An incident in one infrastructure can directly and indirectly affect other infrastructures through cascading and escalating failures.

As digital information became a predominant part of each critical national infrastructure, the cyber components evolved to a connected complex system, and the relationships between infrastructures increased the dependence of one infrastructure on other infrastructures. The US critical infrastructures

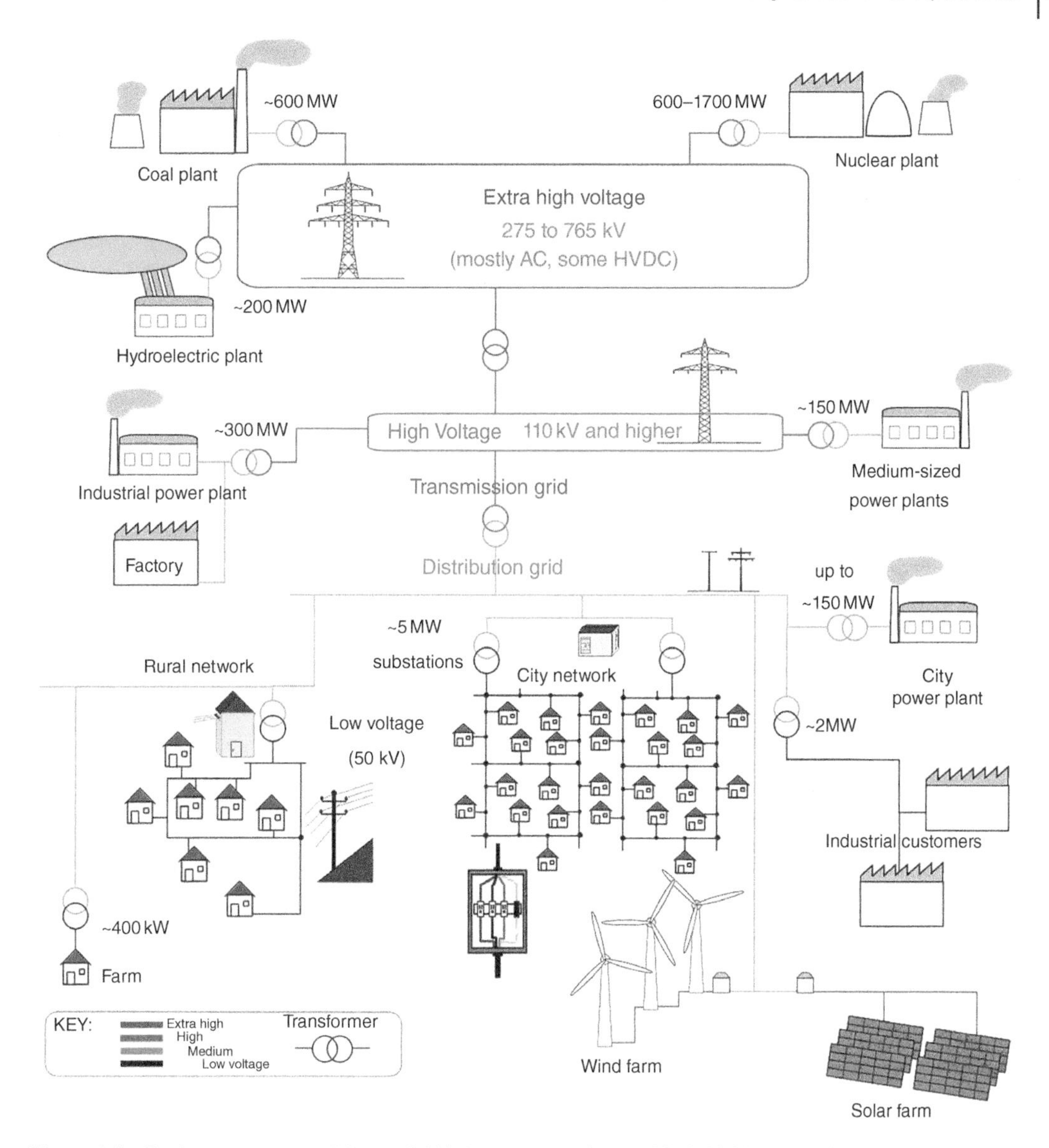

Figure 6.5 Basic components of Smart Grid infrastructure. *Source:* [GAO 2011]. Public Domain.

are highly interconnected and mutually dependent in complex ways, both physically and through a host of ICT (so-called cyber-based systems) [Rinaldi 2001] (see more information in Appendix C).

All businesses are dependent on critical inforation infrastructure in order to conduct their day-to-day operations. By acknowledging this dependency and becoming educated on critical infrastructure, businesses are better prepared to protect their facilities and develop a more resilient environment.

Disruption of a particular gas pipeline or storage facility could impact the ability of numerous power generation assets to function because of lack of fuel, which could in turn affect key water treatment facilities, telecommunication facilities, transportation facilities, or other critical infrastructures.

The US critical infrastructure is often referred to as a system of systems (SoS) because of the interdependencies that exist between its various industrial sectors as well as interconnections between business partners.

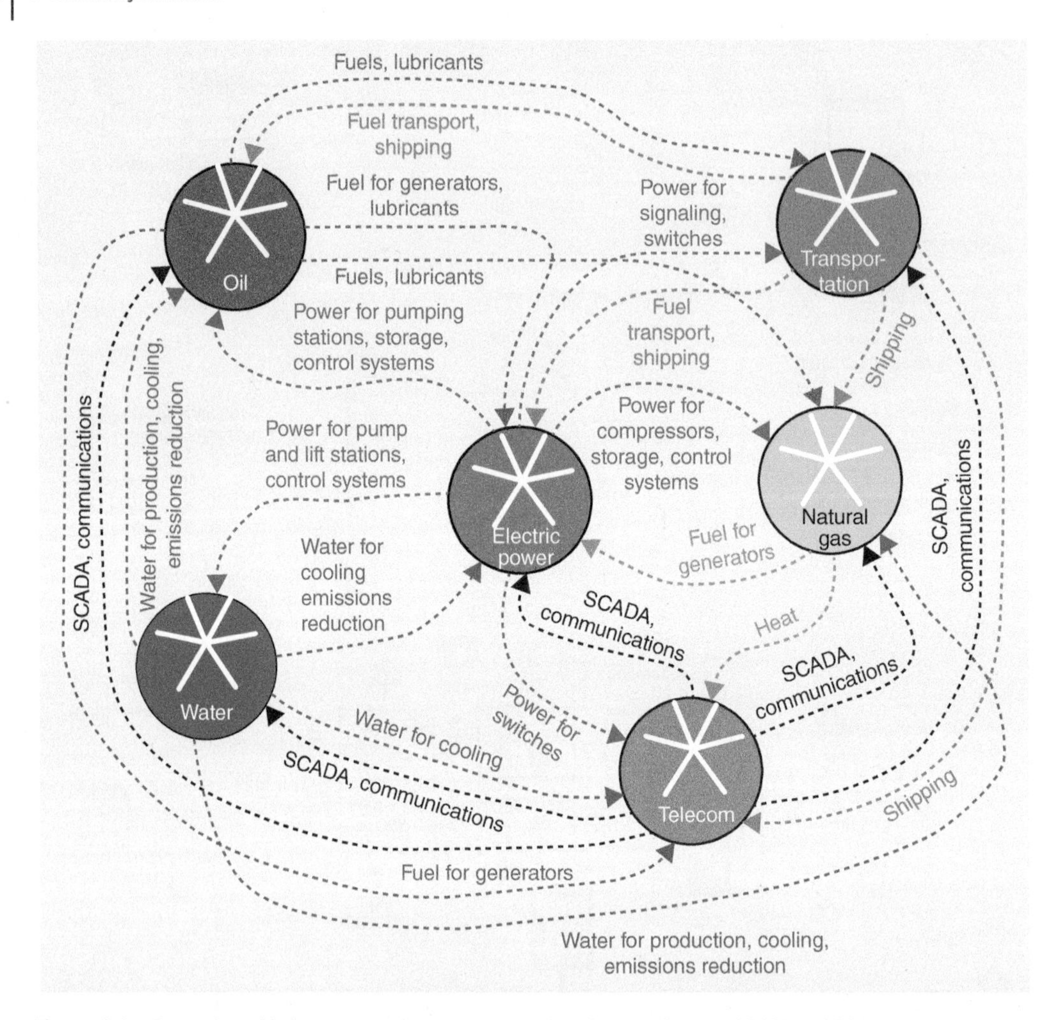

Figure 6.6 Examples of infrastructure interdependencies. *Source:* [Rinaldi 2001]. © 2001, IEEE.

As shown in the diagram, critical infrastructures are highly interconnected and mutually dependent in complex ways, both physically and through a host of ICT. Therefore, there is an infrastructure dependency that is either a link or a connection between two products or services, through which the state of one influences or correlates to the state of the other.

When examining the more general case of multiple infrastructures connected as a system of systems, interdependencies need to be considered. Infrastructures are frequently connected at multiple points through a wide variety of mechanisms, such that a bidirectional relationship exists between the states of any given pair of infrastructures; that is, infrastructure i depends on j through some links, and j likewise depends on i through other links.

Jonkeren [Jonkeren 2013] recognizes that ICT systems are part of a SoS where induced disruption of an infrastructure's operations can occur even if it pertains to another system and/or the disruption initiates in another sector, infrastructure, etc. This reveals not only interdependency but also possible domino or cascading effects, e.g. the disruption of a system, sector, or organization can lead to a cascading disruption across (and within) systems, sectors, and organizations.

Interdependency is a bidirectional relationship between two infrastructures through which the state of each infrastructure influences or is correlated to the state of the other. More generally, two infrastructures are interdependent when each is dependent on the other (e.g. telecommunication and energy sector). Telecommunication infrastructure needs power to function, and power infrastructure needs telecommunications to carry information to manage and distribute power.

6.5.1 Interdependency Dimensions

Identifying, understanding, and analyzing such interdependencies pose significant challenges, which are magnified by the breadth and complexity of critical national infrastructure [Rinaldi 2001]. A basic discussion of the concept and importance of interdependencies is presented in a framework called dimensions. The authors identify six dimensions of infrastructure interdependencies: infrastructure environment, coupling, response behavior, failure types, infrastructure characteristics, and state of operation. Analyzing infrastructure in these terms yields new insights into infrastructure interdependencies. They also identify four types of interdependencies: physical, cyber, logical, and geographical. A broad range of interrelated factors and system conditions are represented and described in terms of these six dimensions, as depicted in Figure 6.7.

Comprising the environment – technical, economic, business, social/political, legal/regulatory, public policy, health and safety, and security concerns – these factors and other challenges affect infrastructure operations [Rinaldi 2001]:

- The environment influences normal system operations, emergency operations during disruptions and periods of high stress, and repair and recovery operations.

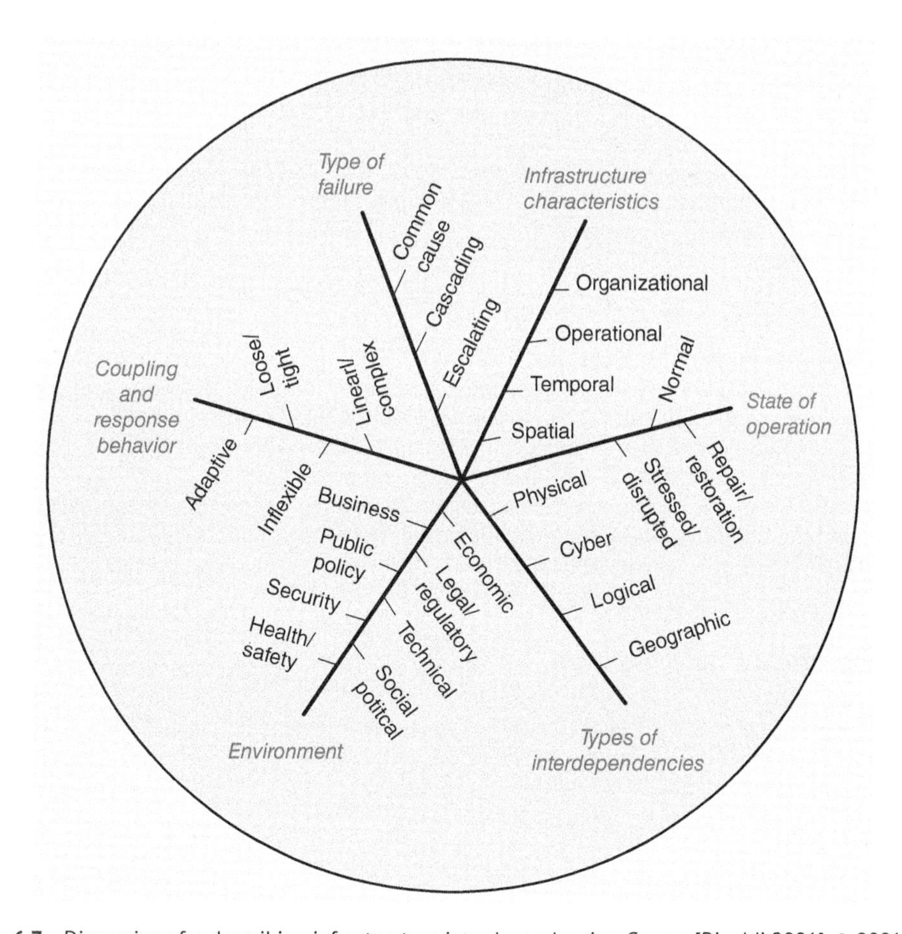

Figure 6.7 Dimensions for describing infrastructure interdependencies. *Source:* [Rinaldi 2001]. © 2001, IEEE.

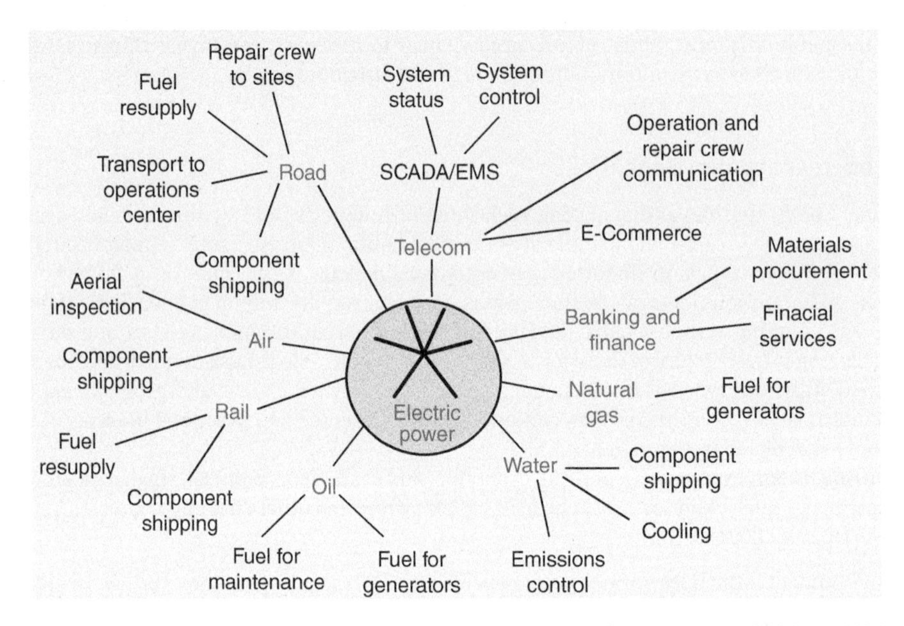

Figure 6.8 Examples of electric power infrastructure dependencies. *Source:* [Rinaldi 2001]. © 2001, IEEE.

- The degree to which the infrastructures are coupled, or linked, strongly influences their operational characteristics.
- Some linkages are loose and thus relatively flexible, whereas others are tight, leaving little or no flexibility for the system to respond to changing conditions or failures that can exacerbate problems or cascade from one infrastructure to another; these linkages can be physical, cyber, related to geographic location, or logical in nature.

6.5.2 Dependencies

If we consider a specific, individual connection between two infrastructures, such as the electricity used to power a telecommunication switch, the relationship is usually unidirectional; that is, infrastructure i depends on j through the link, but j does not depend on i through the same link. For example, at the higher level, one can say that food infrastructure depends on transport infrastructure to carry food to consumers. Figure 6.8 illustrates the dependency concept for power grid.

As depicted in Figure 6.8, electric power is the supported infrastructure, and natural gas, oil, transportation, telecommunications, water, and banking and finance are supporting infrastructures. Although not shown, emergency and government services are also supporting infrastructures.

6.6 Electrical Power System

Electrical power system (EPS) is where the majority of economic activities depend upon as well as the operation of many other infrastructures. In real world, an EPS is one of the most critical infrastructures with its protection being a priority worldwide [Milis 2012].

6.6.1 Electrical Power System Components

The EPS consists of several and heterogeneous components, all connected through complex electrical networks/grids. The current trend is private and public electrical utilities operate in interconnected power grids, thus increasing the reliability of the whole system but also increasing the complexity of the EPS.

More, the EPS is an interdependent infrastructure. A fault in a transformer at a local utility can take a cloud infrastructure down, causing inability the EPS operator to respond to EPS failure via a computing system. The reliance of the EPS operator to external communication networks obviously creates a cyber interdependency, which needs to be carefully engineered to avoid future contingencies. Another dependence is contractual agreements between EPS operator and ISP provider or between ISP provider and carrier network provider. These are some type of logical interdependencies that are harder to identify and harder to protect from. Contractual agreements between network provider and carrier may prohibit disclosure of a temporary network congestion by the network provider to EPS operator. Thus, network congestion can cause service degradation or communication disruption to the EPS operator.

However, the EPS operators would like to know about these issues to proactively enable alternative communication means for isolated remote terminal units (RTUs) (e.g. via GPRS/3G data links) or other backup means. The simple scenario described reveals several interdependencies between electrical sector and telecommunication sector. The fact is that in real EPS grids with hundreds of buses, the complexity of the interconnected ICT networks and their dependencies are daunting.

The advent of the Smart Grid is amplifying these interdependencies, further increasing the need for a resilient ICT infrastructure. An efficient combination of IEDs, PLCs, and RTUs to monitor and communicate could lead to a good level of EPS substation automation in order to improve customer service. The functions of IEDs (protection, monitoring, control, metering, and communication) are very important to the reliability of EPS. There is a strong need to have adequate backup in case communication is lost, e.g. as a result of a disturbance. The communication can be power line carrier, microwave, leased phone lines, dedicated fiber, and even combination of these.

In the power industry, the focus has been on implementation of equipment that could improve power system reliability. Until recently, communications and IT equipment were typically seen as supporting power system reliability. However, these sectors are becoming more critical to the reliability of the power system. In addition, safety and reliability are of paramount importance in electric power systems. Any cybersecurity measures in these systems must not impede safe, reliable power system operations.

Traditionally, cybersecurity for IT focuses on the protection required to ensure the confidentiality, integrity, and availability of the electronic information communication systems. Cybersecurity needs to be appropriately applied to the combined power system and IT communication system domains to maintain the reliability of the Smart Grid and privacy of consumer information. Cybersecurity in the Smart Grid must include a balance of both power and cyber system technologies and processes in IT and power system operations and governance. Poorly applied practices from one domain that are applied into another may degrade reliability.

6.6.2 Electrical Power System Evolution and Challenges

However, the optimization of EPS operations and resources comes with the cost of increased risk from cyber attacks, so the need to protection. To ensure the effective protection of the EPS, the following three elements are considered fundamental: measurements (data defining the state of the system), data processing, and control. These elements are tightly interconnected with the ICT infrastructure to provide the necessary two-way communication between EPS elements.

Smart Grid calls for computers, communication, sensing, and control technologies to operate in parallel toward enhancing the reliability, minimizing the costs, and facilitating the interconnection of new sources in the EPS. The mobilization of the above ICT-enabled technologies needs to happen across broad temporal, geographical, and industry scales, so as to close loops where they have never been closed before.

According to IEEE the Smart Grid is seen as a large and complex system where different domains are expanded into three foundational layers [P2030 2011]:

- Power and energy layer
- Communication layer
- IT/computer layer.

The communication and IT/computer layers are enabling infrastructure platforms of the power and energy layer that makes the grid smarter. Therefore, the IEEE viewpoint is representative of the awareness around the interdependent nature of the EPS and the ICT infrastructure.

Also, Smart Grid evolves to include devices that were not previously considered, such as distributed energy sources, storage, electric vehicles, and appliances. Such devices comprise heterogeneous systems that serve as integrated components of the emerging EPS and that have different characteristics, requirements for security, requirements or fault detection, protection, and metering. Thus, the need to achieve real control over the operation of EPS drives EPS to evolve to an SoS.

An SoS is defined as a collaborative set of systems in which component systems have these characteristics:

- Fulfill valid purposes in their own right and continue to operate to fulfill those purposes if disconnected from the overall system.
- Being managed in part for their own purposes rather than the purposes of the whole.

However, the challenge is to build this type of an SoS system that allows:

- Components to be added, replaced, or modified individually in case of malfunctions, without affecting the remainder of the system.
- Components to be distributable.
- Interfaces for use in replacing components.

Therefore, problems (created by accidental or malicious intervention) can be detected if these components and their interactions are known.

Understanding these relationships is crucial for efficient and effective CIP. In many situations, the relationships and the effect of cascading failures are not well understood because the controls for limiting the effect of failure to one infrastructure are limited. Often one infrastructure uses another as part of its underlying technology. For example, the telecommunication infrastructure relies on the power grid for electricity. It is possible to limit cascade effects by understanding the relationships and compensating for them, taking steps to limit the damage that can cascade from one infrastructure to the other, but in a limited way. However, limiting factors contribute to the overall health of the infrastructures.

Another aspect is the interconnection of old SCADA systems to the Internet, networks, or telephone lines, and the every more intensive use of computer networks and wireless systems has risen the fact that potential terrorists, bored hackers, unhappy employees, and smart kids around the world could access the controls of power systems and hurt countries in one of the key factors of their economies, producing billionaire losses in almost all the sectors of their economies [Watts 2003].

Therefore, security measures and practices that the electric industry must implement in order to reduce these risks are economically desirable. One main need is to reduce at a minimum the threat of a cascading failure of the electric power system and the electric market. This requires important changes in the way the control systems can isolate the failures. Security measures based on detecting and thwarting intrusion only recently became commercially available for SCADA systems [Fairley 2016].

As a result of advances in IT and the necessity of improved efficiency, however, these infrastructures have become increasingly automated and interlinked. The critical infrastructure is characterized by interdependencies (physical, cyber, geographic, and logical) and complexity (collections of interacting components). Cyber interdependencies are a result of the pervasive computerization and automation of infrastructures [Rinaldi 2001]. The critical infrastructure disruptions can directly and indirectly affect other infrastructures, impact large geographic regions, and send ripples throughout the national and global economy. For example, under normal operating conditions, the electric power infrastructure requires fuels (natural gas and petroleum), transportation, water, banking and finance, telecommunication, and SCADA systems for monitoring and control.

The analysis of such interdependencies is a challenge magnified by the complexity of US critical national infrastructures [Rinaldi 2001]. Further complicating this challenge is the range of interrelated factors and system conditions called six dimensions, as depicted in Figure 6.7.

Interdependent infrastructures also display a wide range of spatial, temporal, operational, and organizational characteristics, which can affect the capability to adapt to changing system conditions; more information about interdependencies is provided in [Rinaldi 2001]. For example, this work describes topological interdependencies as follows:

- Physical interdependency arises from a physical linkage between the inputs and outputs of two components; an output of one infrastructure is required as input to another infrastructure for it to operate.
- Cyber interdependency when the state of an infrastructure depends on information transmitted through the information infrastructure; computerization and automation of modern infrastructures with the use of SCADA systems and wide-area monitoring systems.
- Geographical interdependency when a local environmental event or local infrastructure event can create state changes in other infrastructures; it implies close spatial proximity of components of different infrastructures.
- Logical interdependency when the state of an infrastructure depends on the state of another via a mechanism that is not a physical, cyber, or geographic connection. For example, various policies, laws, or regulatory regimes can give rise to logical linkage between two or more infrastructures.

The heavy reliance on pipelines to distribute products across the nation highlights the interdependencies between the energy and transportation systems sector. Without a stable energy supply, health and welfare are threatened, and the US economy cannot function. [PPD-21] identifies the energy sector as uniquely critical because it provides an enabling function across all critical infrastructure sectors. The reliance of virtually all industries on electric power and fuels means that all sectors have some dependence on the energy sector.

The energy infrastructure provides essential fuel to all of the other critical infrastructures and in turn depends on the nation's transportation, IT, communications, finance, and government infrastructures. Over time cyber/IT dependencies have increased. For example, electricity and natural gas suppliers rely heavily on data collection systems to ensure accurate billing. Energy control systems and the ICT on which they rely play a key role in the North American energy infrastructure. They are essential in monitoring and controlling the production and distribution of energy. They create the highly reliable and flexible energy infrastructure in the United States [SPP 2010].

While this increased reliance on interlinked capabilities helps make the economy and nation more efficient and perhaps stronger, it also makes the country more vulnerable to disruption and attack. This interdependent and interrelated infrastructure is more vulnerable to physical and cyber disruptions because it has become a complex system. In the past, an incident that would have been an isolated failure can now cause widespread disruption because of cascading effects. As an example, capabilities within the information and communication sector have enabled the United States to reshape its government and business processes while becoming increasingly software driven. It is argued that one catastrophic failure in this sector has the potential on bringing down multiple systems including air traffic control, emergency services, banking, trains, electrical power, and dam control.

Telecommunication services are another national level network. The telecom backbone that supports the Internet and voice communications is composed of a number of large networks. An incident that disrupts the services provided by several of these large networks could disrupt communication traffic but cannot take down all the communication services. The presence of multiple overlapping connections means that there is no single point of failure. The use of satellites in communication services also introduces a degree of redundancy and assurance to keep minimum services.

In the case of the power grid blackout, many critical facilities have installed backup power generation. For example, many critical electronic components of the telecommunication network are on battery backup to prevent disruption resulting from short-term power failures. However, in a situation where the power is limited or not available for each device, the functionality of telecommunication networks will be limited, impacting users and other infrastructures relying on the telecommunication network.

6.6.3 Needs

According to [PPD-21], the US critical infrastructure provides the essential services that underpin American society. Critical infrastructure must be secure and able to withstand and rapidly recover from all hazards.

Proactive and coordinated efforts are necessary to strengthen and maintain secure, functioning, and resilient critical infrastructure – including assets, networks, and systems – that are vital to public confidence and the nation's safety, prosperity, and well-being. The policy establishes the national policy on critical infrastructure security and resilience. It describes that critical infrastructure:

- Provides the essential services that underpin American society.
- Requires proactive and coordinated efforts to strengthen and maintain security, functions, and resilience.
- Implies that all assets, networks, and systems part of critical infrastructure should be secure.
- Must be secure and able to withstand and rapidly recover from all hazards defined as a threat or an incident, natural or man-made, that warrants action to protect life, property, the environment, and public health or safety and to minimize disruptions of government, social, or economic activities.

Also, the directive [PPD-21] mandates that critical infrastructure owners and operators should manage risks to their individual operations and assets and to determine effective strategies to make them more secure and resilient. The directive provides a list of updated cyber terms, guidelines to be followed, and some deadlines. Examples of definitions include:

- Security or secure refers to reducing the risk to critical infrastructure by physical means or defense cyber measures to intrusions, attacks, or the effects of natural or man-made disasters.
- Resilience as the ability to prepare for and adapt to changing conditions and withstand and recover rapidly from disruptions. Resilience includes the ability to withstand and recover from deliberate attacks, accidents, or naturally occurring threats or incidents.
- Hazards term includes natural disasters, cyber incidents, industrial accidents, pandemics, acts of terrorism, sabotage, and destructive criminal activity targeting critical infrastructure.
- National essential functions are that subset of government functions that are necessary to lead and sustain the nation during a catastrophic emergency.
- Primary mission essential functions include those government functions that must be performed in order to support or implement the performance of the national essential functions before, during, and in the aftermath of an emergency.

However, many issues of this directive [PPD-21] are controversial as argued in [Stiennon 2013] or remain to be resolved. Stiennon argues that a risk management approach may not work for all kinds of hazards (e.g. natural disasters). From the stated goal of applying a risk management approach to all hazards, establishing two national critical infrastructures (one for physical and one for virtual (cyber)) and centralized information collection and dissemination, there are many controversial issues, and they do not fit well in a security framework. Also, by defining two independent centers, there is a risk of no collaboration and sharing information to detect and prevent threats. Further, a centralized information collection could be a burden on critical information infrastructure.

Submitting Freedom of Information Act Request, EPIC raises privacy concerns. It evaluates the existing public–private partnerships as well as offering recommendations for improving the effectiveness of the partnerships [EPIC]. A security approach based on privacy and user security could be incentives to engage the public to proactively detect threats and defend CIIs when security compromises might occur. Cybersecurity levels are defined in terms of consequences for the nation rather than consequences for responsible entities.

6.7 Recent Threats and Vulnerabilities

The components of the critical infrastructures are increasingly vulnerable to a dangerous mix of traditional and nontraditional types of threats. Traditional and nontraditional threats include equipment failures, human error, weather and natural causes, physical attacks, and cyber attacks. Other possible threats include extortion, terrorism, and government-sponsored attacks.

In the United States, the Internet Crime Complaint Center, a joint operations consisting of the Federal Bureau of Investigation (FBI) and the National White Collar Crime Center, categorizes the cybercrime spectrum to include intellectual property (IP) rights matters, computer intrusions (hacking), economic espionage (theft of trade secrets), online extortion, international money laundering, identity theft, and a growing list of Internet-facilitated crimes [IC3].

A growing international concern for end users is mass-marketing fraud, which exploits mass communication techniques (such as email, instant messaging, bulk mailing, and social networking sites) to target victims for financial profit. Using the Internet, criminals target many more potential victims [Whitty 2015].

Attacks against critical infrastructures can have direct or indirect effects. Direct effects would be stoppage or disruption of the functions of critical infrastructures or key assets through direct attack on a critical part, system, or function. Indirect effects of an attack are the disruption and problems that result from a reaction to attacks on other critical infrastructure.

Attacks on one critical infrastructure have effects to other infrastructure. If the transportation infrastructure is damaged, infrastructures like food and health, emergency services, and other infrastructures can be affected. For example, a rural electric utility may be vulnerable and a target for different threats. The utility may find itself inundated with attacks from customers protesting increases in service fees. It may be targeted by hacktivists who do not condone its methods of power generation. Or it may fall prey to foreign intelligence operatives attempting to exploit a weak link in the nation's power grid infrastructure [Radware 2015].

These effects can escalate to major catastrophes as called in [Jackson 2013]. These catastrophes may be caused by natural phenomena such as earthquakes or hurricanes. Or they may be terrorist attacks, or they may be caused by operator error or design error. One important aspect of these effects is a recurring pattern among major catastrophes that is the loss of communications, command and control, and other capabilities that are intended to enable rescue and recovery efforts [Jackson 2013]. These disruptions occur in a plurality of events when one disruption gives rise to another disruption. A disruption is an interruption in functionality caused by a threat. The consequences of disruption are multiples, and one is the economic effects on neighboring countries and also at national level. This criterion is an attempt, defined in the European Directive (2008/114/EC), to get all European CI actors or operators involved in the assessment of their infrastructures [Jonkeren 2013].

Each year new threats are discovered or old threats become more sophisticated. According to the Information Security Forum (ISF) predictions for the year 2014, the global threats include BYOD (bring your own device) trends in the workplace, data privacy in the cloud, brand reputational damage, privacy and regulation, cybercrime, reputational damage, and continued expansion of ever present technology [ISF 2013].

Also, several reports published by the Industrial Control Systems Cyber Emergency Response Team (ICS-CERT) reveal vulnerabilities in software including control system anomalies in operations at water and wastewater treatment facilities, manufacturing factories, medical devices, and utilities [ICS-CERT 2014b].

6.7.1 Reported Cyber Attacks

Several sectors of industry are targeted by cyber incidents. Breaches are on the rise in control systems [SANS 2014], [Harp 2015]. In the last six months of 2012, the average number of targeted attacks observed per day was 87 (with 14 in the energy sector). In the first six months of 2013, the average number decreased to 60 targeted attacks per day (5 in the energy sector). Multiple attacks are observed against financial services, public sector, and IT service organizations. At the top of the list was the government and public sector – quantitatively the most attacked sector. Then, going down on the list, there were manufacturing, financial services, IT systems, energy, and others.

Several hundred machines including critical medical equipment in an undisclosed number of US hospitals were infected with the Conficker [Forum 2009].

Cyberespionage campaigns and sabotage attacks are becoming increasingly common, with countless threat actors attempting to gain access in some of the best protected organizations as described in this survey [Wueest 2014]. Also, the survey reports roughly five targeted attacks per day being mounted on organizations of the energy sector. These attacks have become increasingly sophisticated, more frequent, more persistent, and intensive.

Another report [ICS-CERT 2015b] provides incident response statistics for the fiscal year by sector (see Figure 6.9). It shows that manufacturing and energy were the target of an increased number of security incidents compared to 2014. In 2015, ICS-CERT identifies advanced malware (ransomware) growth that obstructs victims' access to their device, often through the encryption of files with a key accessible only to the attacker [ICS-CERT 2015a]. The malware then displays a message containing demands for the release of the system or files, which is typically payment via some form of cryptocurrency or untraceable prepaid card. The victim is given a period of time, as defined by the attacker, to comply with the ransom demands or face loss of system or files. Attackers use more advanced mechanisms of persistence and secrecy by implanting malware at the firmware level in USB devices, the firmware of hard drives, or the basic input/output system (BIOS) of PCs. This creates new and difficult challenges for system security because of the trust models that were designed into these devices' standards.

As anti-malware research continues to advance, some of these inherent issues have been addressed, but they remain a high concern that has not yet been solved.

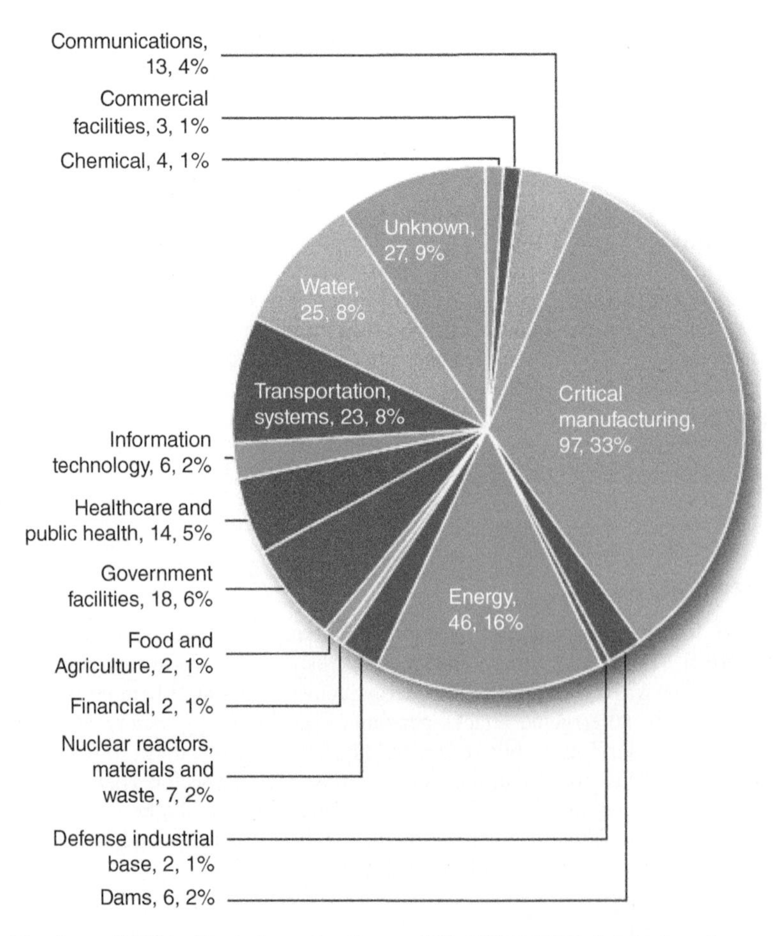

Figure 6.9 Fiscal year 2015 incidents by sector. *Source:* [ICS-CERT 2015b]. Public domain.

Attacks against the energy sector often follow the same pattern. It can be broken down in different phases of attack. [Wueest 2014] describes phases of targeted attacks to include reconnaissance, incursion, discovery, capture, and exfiltration. Attackers modify their behavior and exceptions from the norms. This is possible especially if the target company has special circumstances or security measures in place. While there is no exact number of steps that a cyber attack may take, the typical steps involved in a cyber intrusion are depicted in Figure 6.10. Security professionals argue that these may be between four and seven steps.

Four types of security attacks remain top of the list [Radware 2015]:

- Cybercrime – Criminal attacks are typically motivated by money; large in number and present in virtually every country around the globe, these groups range in skill level from basic to advanced; mass-marketing fraud is a serious, complex, and often organized crime [Whitty 2015].
- Hacktivism – Hacktivists are primarily motivated not by money but rather by a desire to protest or seek revenge against an entity. As with criminals, there are a large number of hacktivist groups. However, most of these groups have basic IT skills, while some individuals possess advanced skills and motivate a potentially larger set of followers.
- Espionage – These attacks are aimed at acquiring secrets that support national security and gaining economic benefit or both. A growing number of countries have the ability to use cyber attacks for espionage – and a larger array of groups is being supported or tolerated with such activities.
- War (cyber) – This type of attack is arguably most nefarious type of attack. Those involved in these attacks are motivated by a desire to destroy, degrade, or deny. A growing number of countries have the ability to use this form of politics by other means. Further, non-state actors are determined to undertake cyber attacks as a form of war.

A total of 295 incidents involving critical infrastructure in the United States were reported to the ICS-CERT in the fiscal year 2015, compared with 245 in the previous year [Kovacs 2016]. The energy sector, which in 2014 accounted for 32% of critical infrastructure incidents, reported only 46 incidents in 2015, which represents 16% of the total. Incidents were also reported in sectors such as water (25), transportation systems (23), government facilities (18), healthcare (14), and communications (13).

Often, these critical infrastructures are dependent on their human operators, whose actions are supported, reinforced, or carried out using computers and networks. This human element reduces the risk of

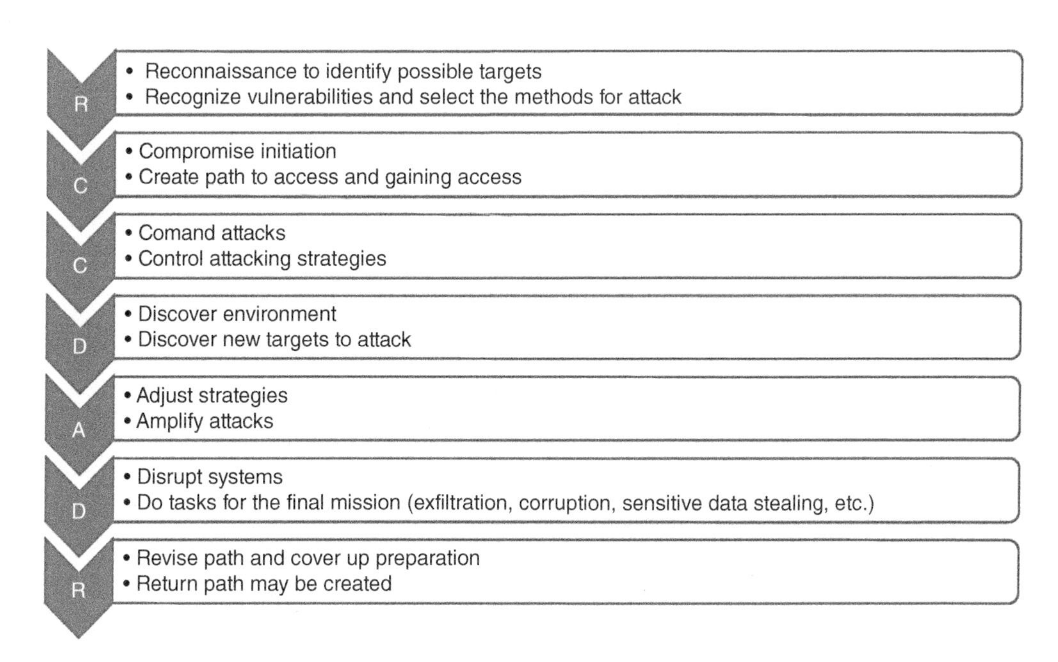

Figure 6.10 Example of typical phases of targeted attacks.

cyber attack to critical infrastructures. For example, during the 2003 Slammer worm attack, computers of police departments in Washington State had a slow response to the point of uselessness as the worm spread and implemented its instructions [CAIDA 2003]. The departments compensated by using paper notes to record calls, allowing 911 services to continue uninterrupted. It is argued if the critical emergency service was affected, although computers were vulnerable and affected.

Thus, if a cyber attack damages one company in a critical sector but leaves its competitors operational, it limits the overall risk to critical national functions. For many critical infrastructures, while computer networks are vulnerable to attack, the critical infrastructures they support are not equally vulnerable. While the dependence on computer networks continues to grow, many critical functions remain insulated from cyber attack or capable of continuing to operate even when computer networks are degraded.

In the United States and around the globe, surveyed organizations report the rise of cybercrime from 44% in 2014 to 54% in 2015. Main concerns include several security incidents including economic crime. The US Director of National Intelligence has ranked cybercrime as the top national security threat – higher than that of terrorism, espionage, and weapons of mass destruction [DNI 2014]. As described in this survey [Radware 2015], the year of 2014 was a watershed year for the security industry with cyber attacks reaching a tipping point in terms of quantity, length, complexity, and targets. Threats have expanded to a broader range of industries, organizational sizes, and technology deployments. This means that the overall number of cyber attacks as well as the frequency and intensity of these attacks has increased in 2014. The attack volumes, complexities, and frequency have increased year over year.

The Ring of Fire (grouping of results in the report) reflects five risk levels, with organizations closer to the red center more likely to experience DoS/DDoS and other cyber attacks and to experience them at a higher frequency. As companies move toward the center of the Ring of Fire, susceptibility to cyber attack grows, creating a security gap.

The critical infrastructure sector has emerged as an especially vulnerable attack vector, due to SCADA aging facilities and the prevalence of bolt-on half measures and Band-Aids in lieu of comprehensive security systems [TrendMicro 2014]. More information about the most brutal and worst cyber attacks in 2014 and 2015 is provided in these works [McGregor 2014], [Szoldra 2015]. Two noteworthy critical infrastructure attacks of 2014 are described in this survey [TrendMicro 2015]:

- A Russian hacker group called Energetic Bear caused significant disruption for companies in the United States in the energy sector; other targets include governments and critical infrastructures.
- An attack launched against a steel plant in Germany resulted in massive damage to the plant; the compromise caused frequent failures of individual control components and the various systems and ultimately led to the operators being unable to adequately regulate and promptly shut down a blast furnace.

Also, the annual security survey [TrendMicro 2015] reports that numerous organizations worldwide, including banks and public utilities, lost millions of customer records and credentials to attackers in 2014. In 2014, the America region was affected by different malware, each with their distinct characteristics that could aid would-be attackers in their exploits. This report provides an analysis of cybercrime activity and trends that take place in the Americas within the critical infrastructure sector; more specifically it provides highlights and analyses of January 2015 survey responses from member states to questions around cybersecurity, attacks, preparedness, and critical infrastructure. The survey covers issues such as:

- Experience with various incidents and percentage of organizations that experienced attempts to have information deleted or destroyed by organization type: according to the survey results, the government and energy sectors are the top two industries that experience destructive attacks by threat, followed by communications and finance and banking.
- Types of cyber attack methods used against organization: the majority of regions surveyed indicated their ICS/SCADA equipment is being targeted by hackers, which indicate a broad amount of activity by threat actors. Most regions are dealing with phishing attacks against their organizations. This may also indicate threat actors' initial attempts to penetrate an organization in an effort to move laterally across other systems, such as their ICS/SCADA devices.

- Perception of preparedness for cyber incidents: most countries feel somewhat prepared for a cyber incident, but the survey results suggest that the efforts to improve preparedness may be more difficult than it appears.

The collected data during the survey shows a specific increase in critical infrastructure attacks (43%) with an alarming response of 31% that are unsure if they have been attacked. A major challenge today is the sophistication of attacks (76% say they are getting more sophisticated), which are difficult to detect. Also, the survey identifies issues and trends:

- There is still a lack of proactive partnership between governments and private organizations in the western hemisphere; in the absence of a formal public–private partnership, these cybercriminals will thrive.
- The most significant trend is the use of malware to compromise SCADA systems, including human–machine interface (HMI), historians, and other connected devices; the malware has manifested itself in two major ways: malware disguised as valid SCADA applications and malware used to scan and identify specific SCADA protocols, the purpose is likely intelligence gathering for industrial espionage or future targeting for an attack.
- Each DER could become a potential entry point for a cyber attack; the cyber network that controls and optimizes the physical power network also amplifies the scale, speed, and complexity of an attack; it can also quicken and broaden failure propagation on the physical network, making blackout mitigation much more difficult.
- Research is focused on mitigating potential threats: monotonicity properties of grid failures caused by possible cyber attacks can be utilized to develop load shedding strategies; security constraints can be included in optimal power flow (OPF) formulas to maintain network efficiency; ubiquitous load-side frequency control may be utilized to maintain network stability through distributed algorithms that adjust for potential cyberattacks. OPF is the problem of minimizing certain cost functions, such as power loss over a network, the fuel cost of electricity generation, or security constraints. OPF is fundamental in power system operations and planning as it underlies numerous applications.

According to this survey [PwC 2016a], organizations could expect to be hit by an increase of all types of crime in the following two years: cybercrime (49%), asset misappropriation (36%), IP infringement (25%), and bribery and corruption (14%), among others. Top systems challenges include complexity of upgrading systems (28%) and data quality (26%). Many systems are still hampered by legacy monitoring systems that are proving to be burdensome and expensive to tune and maintain. Cyber attackers are classified in groups by different surveys, with similar groups or light differences. Also, this report identifies cyber attackers as:

- Nation-states – Threats include espionage and cyber warfare; victims include government agencies, infrastructure, energy, and IP-rich organizations.
- Insiders – Organization's employees and trusted third parties with access to sensitive data who are not directly under the organization's control.
- Organized crime syndicates – Threats include theft of financial or personally identifiable information (sometimes with the collusion of insiders); victims include financial institutions, retailers, medical, and hospitality companies.
- Hacktivists – Threats include politically focused service disruptions or reputational damage; victims include high-profile organizations, governments, or even individuals.
- Terrorists – Still a relatively nascent threat, threats include disruption and cyber warfare; victims include government agencies, infrastructure, and energy.

Other key findings of the global state of information security are reported in a survey of many industries such as energy, power, telecommunications, manufacturing, technology, media, public sector, government, pharmaceutical, retail, etc. [PwC 2016b]; there is a 38% increase in detected security incidents in 2015; cyberattacks continue to escalate in frequency, severity, and impact; prevention and detection methods have proved largely ineffective against increasingly adept assaults; and many organizations do not know what to do, or do not have the resources to combat highly skilled and aggressive cybercriminals. Damage as a result of cybercrime include:

- Reputational damage
- Actual financial loss
- Legal, investment, and/or enforcement costs
- IP theft, including theft of data
- Regulatory risks
- Service disruption
- Theft or loss of personally identifiable information.

Although financial losses due to monetizable cybercrime cannot be ignored, they rarely pose an existential threat to companies [PwC 2016a]. According to this survey, the more critical economic crime facing organizations is that of international cyberespionage: the theft of critical IP – trade secrets, product information, negotiating strategies, and the like. Such breaches are called extinction level events because damages could extend to billions of dollars and include the destruction of an entire line of business, a company, or even a larger economic ecosystem. Not only are these kinds of attacks difficult to detect, but they may not even be on an organization's threats list.

6.7.2 ICS/SCADA Incidents and Challenges

Advances in IT and the necessity of improved efficiency to use Internet have created new vulnerabilities and exposures to cyber attacks. As automation continues to evolve and becomes more important worldwide, the use of ICS/SCADA systems is going to become even more prevalent. ICSs are devices, systems, networks, and controls used to operate and/or automate industrial processes. These devices are often found in nearly any industry, from the vehicle manufacturing and transportation segment to the energy and water treatment segment. SCADA networks are systems and/or networks that communicate with ICS to provide data to operators for supervisory purposes as well as control capabilities for process management.

In 2015, ICS-CERT responded to a significant number of incidents involving improperly configured infrastructure where ICS networks were connected to corporate networks and even directly to the Internet [Kovacs 2016].

Most critical infrastructures nowadays are controlled by SCADA systems. There is a continuous and growing concern about the security and safety of the SCADA and control systems in terms of vulnerabilities, lack of protection, and awareness, although these were reported for more than decade [Byres 2005], [Byres 2006], [Hentea 2008]. The threats to these systems are a continuous concern. These systems were the target of several cyber attacks (intentional and unintentional). An individual or an entity with malicious intent might disrupt the operation of the system by blocking, delaying the flow of information through the control networks. They can also make unauthorized changes to programmed instructions in the PLCs, RTUs, and DCS controllers. This may result in malfunctioning of an infrastructure.

These control systems are used to be based on proprietary solutions, which provided some form of security, essentially by obscurity. However, critical infrastructure companies are now using standard hardware and software. The common defense, the hackers do not know our systems, is no longer true. Controllers are industrial PCs running off-the-shelf operating systems (such as Windows, Linux, or Linux), the networks are often wired (or even wireless) Ethernet, and control and supervision protocols are normally encapsulated on top of UDP-TCP/IP protocols. Furthermore, the technologies being used, specifically the software that often is not high quality, and older versions plagued with high-level vulnerabilities.

Software bugs (e.g. buffer overflow) and other vulnerabilities in ICS devices manufactured by different vendors can expose utilities, nuclear plants, and other critical operations to cyber attacks [Fisher 2014]. This analysis suggests that the actors likely use automated tools to discover and compromise vulnerable systems. Also, ongoing sophisticated malware is exploiting the vulnerabilities of protocols [ICS-CERT 2014a], [ICS-CERT 2014b]. The ICS-CERT biannual reports provide not only incidents but also summarize common problems across several sectors of critical infrastructure that include different categories of vulnerabilities such as flaws in design of systems, lack of adequate security controls, third-party software installers for critical infrastructure, and lack of monitoring and operation procedures including mitigation strategies.

Organizations are encouraged to report incidents, check ICS systems for vulnerabilities, be aware of ICS-CERT alerts and warnings, and use the ICS-CERT services to assess their own systems and infrastructures. A test performed by a group of researchers in April 2012 revealed that many critical infrastructure assets are directly facing the Internet: devices having either weak, default, or nonexistent logon credential requirements, freely available such that anyone with malicious intent could locate these devices and attempt logon, leaving these systems exposed to cyber attacks [ICS-CERT 2012]. According to this report, ICS-CERT evaluated approximately 7200 devices in the United States that appear to be directly related to control systems associated with critical infrastructure. Once accessed, these devices may be used as an entry point onto a control systems network, making their Internet facing configuration a major vulnerability to critical infrastructure. Several incidents and penetration attacks are reported in [Caswell 2011] for different sectors of critical infrastructure.

6.7.2.1 Stuxnet Exploitation

Stuxnet is a sophisticated virus that uses seven vulnerabilities to spread and infect its targets. This is the first known autonomous threat to target and sabotage ICSs to such an extent. Stuxnet's payload focused on PLCs, which are used to control different industrial components. The target of the Stuxnet operation was the uranium enrichment facility in Iran. Seizing control of the automation system, the worm was able to reconfigure the centrifuge drive controllers, causing the equipment to slowly destroy itself. The sabotage payload disrupted and partially destroyed the cascaded high-frequency gas centrifuges. This malware had a degree of sophistication with no precedent and many infection and destruction capabilities as described in this report [Wueest 2014]. The impacts of Stuxnet infection can be summarized as follows:

- Stuxnet had a specific target, but like all attacks, cyber or conventional, there were collateral damages.
- Several companies in the United States had PLCs that were reconfigured by Stuxnet.
- A lot of labor charges were incurred and shutdowns occurred in many systems.
- Hackers and criminals discovered that SCADA/ICS products are attractive targets; these systems soon became targets of choice for public security disclosures.

6.7.2.2 Exposure to Post-Stuxnet Malware in Rise

Following the media coverage of Stuxnet in July 2010, the fabrication of viruses, worms, and all sorts of malware increased. The Conficker worm appeared in 2008 and revealed in 2009 as the largest mutation. In 2011, reports show that about four million IP addresses still attempting to connect to a Conficker update server on a daily basis [Mohan 2011]. Also, the Havex malware is focused on critical infrastructure dating back to 2011, and possibly earlier. The campaign involves multiple intrusion vectors including phishing emails and redirects to compromised Web sites and software update installers on at least four industrial software vendor Web sites. This attack methodology is commonly referred to as a watering hole attack [ICS-CERT 2014b]. Also, the presence on network scanner tools for ICS requires monitoring and incident reporting.

The continuous debate in the information security world regarding the occurrence of the ICS/SCADA system-related incidents and attacks motivated researchers and security agencies to prove the security exposure of these systems. A study of vulnerabilities for ICS identified the following categories [DHS 2011]:

- Credential management
- Weak firewall rules
- Network design weaknesses
- Lack of formal documentation.

ICS/SCADA systems have been the talk of the security community for the past years after the media coverage of the Stuxnet, followed by other cyber attacks (some variations of Stuxnet), the most sophisticated cyber attack that targeted the ICS devices. Based on DHS report about warnings of vulnerabilities, the US ICS-CERT released 104 security advisories for SCADA/ICS products with obvious vulnerabilities from 39 different vendors in 2011. Prior to Stuxnet, only five SCADA vulnerabilities had ever been reported in vulnerability databases.

While Stuxnet was created for political reasons, the opportunities for corporate exploitation were soon realized by governments and criminals alike. In February 2011, a new attack, Night Dragon, against energy industry was released. This malware, based on the techniques from Stuxnet, was stealing sensitive data such as oil field bids and SCADA operations data from energy and petrochemical companies. Then, other malware like Duqu and Nitro attacks with different targets and capabilities were released [Byres 2012], [Wueest 2014].

In the year 2012, several new exploit tools were publicly released that specifically were targeting PLCs, the building blocks of many ICSs. There were exploits targeting PLCs from specific vendors [ICS-CERT Alerts]. The increase in the number of cyber attacks was about 52% in 2012 compared with the previous year [ICS-CERT 2012], and US power, water, and nuclear systems were targeted by cybercriminals [Goldman 2013].

Compared with 2011, in 2012 the number of vulnerabilities increased; there were about 171 unique vulnerabilities in ICS products from 55 different vendors. According to these reports, the attacks targeted natural gas pipeline and chemical industries, but the energy sector was the most targeted sector with 41% of attacks, followed by the water industry. However, those attacks were reported. Often companies choose not to report incidents, and the majority of cyber attacks go undiscovered, according to industry researchers. These new threats defeat the myth of air gap principle, used by software manufacturers on engineering the ICS software.

6.7.2.3 Inappropriate Design and Lack of Management

Another view of broad categories of ICS/SCADA vulnerabilities includes inappropriate planning and design, operations and management negligence, and safety concerns and health of equipment.

In 2013, many ICSs were still vulnerable to viruses such as Conficker, despite the fact that the corresponding security update is available since 2009 [Kaspersky 2014] report. The report includes analysis of the main causes of incidents in industrial networks that shows the following categories:

- Software error 23%
- Malware attacks 35%
- Operator error 11%
- SCADA component failures 19%
- Other 12%.

Also, the report shows the sources from which malicious code penetrates industrial networks to include:

- Mobile devices 4%
- Wi-Fi 5%
- Internet connections 9%
- Outside contractors 10%
- Via corporate networks 35%
- HMI interface 8%
- Via remote access 26%.

Another report [Goldman 2013] released by researchers from a security advocacy group documents more nearly 500,000 devices across the United States appeared to tap into key control systems and being directly reachable through the Internet because of either weak, default, or nonexistent logon credential requirements. ICS-CERT evaluated approximately 7200 devices in the United States that appear to be directly related to control systems associated with critical infrastructure. Similar tests were done in Europe and revealed several vulnerabilities in home automation systems and smart meters.

Besides vulnerabilities and exposures, another big concern is geographical location of ICS devices. Most critical infrastructures are geographically concentrated, which means that the physical location of critical infrastructures and assets are in sufficient proximity to each other, and vulnerable to disruption of the same, or successive regional events [Robles 2008]. As described in this work, examples of these infrastructures include:

- Transportation – About 33% of US waterborne container shipments pass through the ports of Los Angeles and Long Beach of California.
- Transportation – About 37% of US freight railcars pass through Illinois.
- Chemical industry and hazardous materials (chlorine) – About 38% of US chlorine production is located in coastal Louisiana.
- Public health and healthcare – About 25% of US pharmaceuticals are manufactured in San Juan Metropolitan Area, Puerto Rico.
- Energy – About 43% of US oil refineries are located along the Texas and Louisiana coasts.

Another issue with many devices used in critical infrastructure is the location visibility and exposure to information and documentation via Google Maps and Google Earth applications or other Internet search engines. Researchers from many countries discover more zero-day vulnerabilities and methods to unauthorized access that are discussed in published papers and professional conferences. All these materials are easily accessible to cybercriminals. Figure 6.11 shows a manufacturer ICS security architecture published on the Web, so visible to criminals.

Using published documentation, several vulnerabilities are disclosed by a researcher for SCADA systems used in oil and gas facilities, water management systems, and manufacturing. The researcher published tests that demonstrate successful exploitation of buffer overflow vulnerabilities, denial-of-service

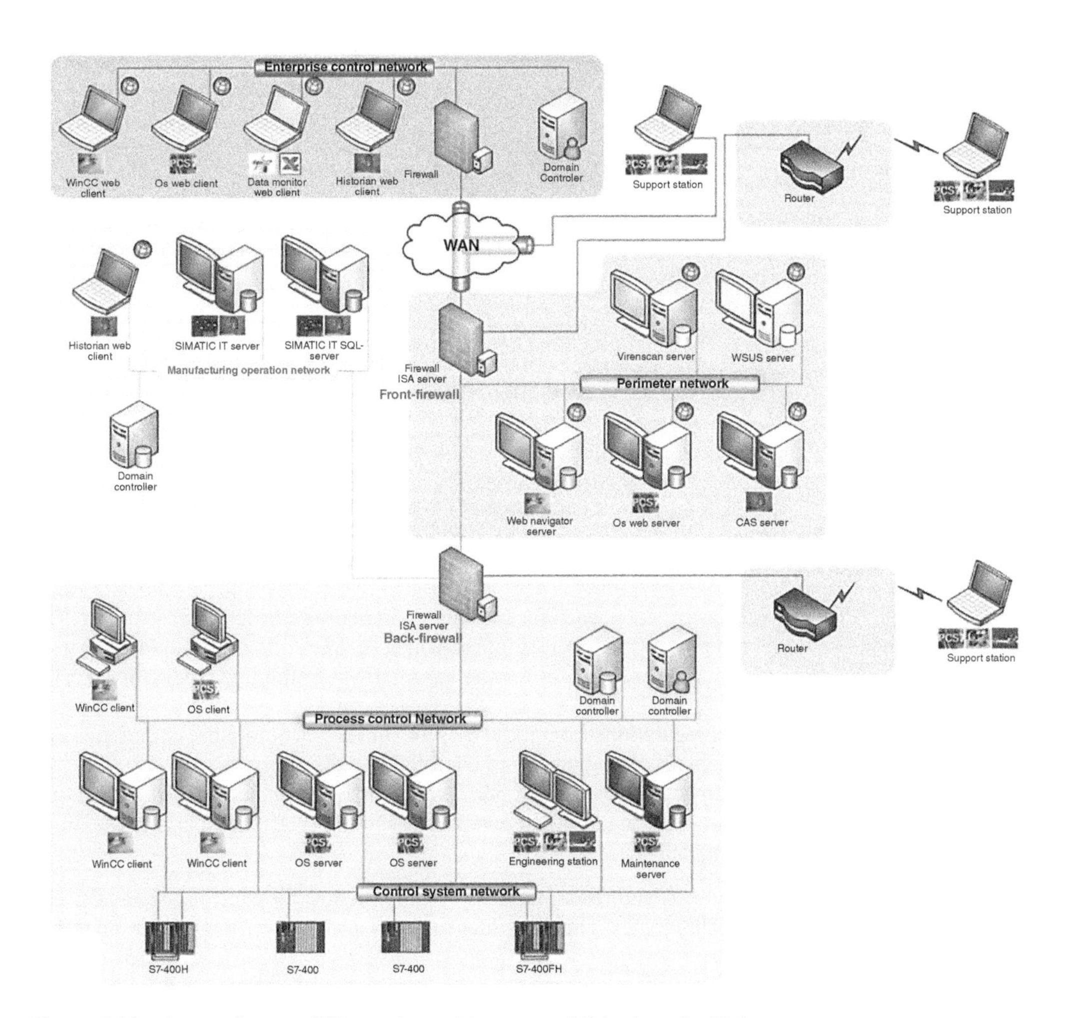

Figure 6.11 A manufacturer ICS security architecture published on the Web.

attacks, foreign file insertion onto systems, altered data displayed to operators monitoring system operations, and enablement of remote execution for malicious code [Zetter 2011]. In addition, the air gap myth – believed to be a good architecture strategy for the protection of ICS – is not real because there is no such separation between business systems and control systems (see more information in [Tofino]).

Search engines used to identify and directly access controllers and industrial software applications are freely available via Web search engines. Coupled with the ease of obtaining documentation for common ICS protocols, security exploits can be designed fairly quickly [ICS-CERT 2011]. As highlighted in the report, intruders are able to breach ICS on an all too frequent basis. Given that many ICS are largely unsecured, a continued rise in the rate of new cyber threats to critical infrastructure is quite likely.

6.7.2.4 Safety

Besides security concerns, the computer systems including ICS and SCADA control systems raise the issue of safety causing harm and catastrophic damage when they fail to support applications as intended [Dunn 2003]. In January 2003, Slammer worm infected the safety monitoring systems at the Davis–Besse nuclear plant in the United States (see more about Slammer worm in [Moore 2003]). In 2003, two hackers gained access to control technology for the US Amundsen–Scott South Pole Station that ran life-support technology for scientists. This attack disabled the safety monitoring system for nearly five hours [Poulsen 2003]. The infamous breach of SCADA for Maroochy water system in Australia [Gellman 2002] plagued the wastewater system for two months. This caused a leak of hundreds of thousands of gallons of putrid sludge into parks, rivers, and private properties as a result of which marine life died, the creek water turned black, and the stench was unbearable for residents.

6.7.3 Equipment Failure

Often equipment failure or malfunctioning may create disruptions. At a generating station near Toronto in 2002, a backup reactor shutdown system that had been operating for weeks, in what appeared to be working order, was actually incapable of catching a dangerous rise in radiation, owing to an incorrectly calibrated neutron detector [Betts 2006].

6.8 Standards, Guidelines, and Recommendations

In the United States, several recommendations and guidelines for protection of critical infrastructures were published. There are specific requirements for the standards that have to be counted in the security management for critical infrastructures:

- Control system security standards must differ from existing general IT security standards because their mission and goals differ.
- Specific methods for risk assessment are needed for each component of the critical infrastructure. For example, assessing risks for the power grid involves impacts on both safety and financial losses, safety as a top priority, compared with assessing risks for the banking where the impacts involve financial losses.
- Consolidation of standards and regulations from industries and government to more comprehensive standards that avoid ambiguity, goals discrepancy, and compliance differences.
- Emphasis on a novel architecture and protocols that preserve legacy systems and on a new device that provides incremental protection, assuring different levels of resilience to different parts of the infrastructure, according to their criticality.

A broad array of standards, guidelines, and recommendations are provided for control systems devices and technologies including IT technologies. The following includes examples of most common documents that contain security topics:

- International Organization for Standardization (ISO) 31000:2009.
- NIST Special Publication (SP) 800-39, Managing Information Security Risk: Organization, Mission, and Information System View, March 2011.

- Electricity Subsector Cybersecurity Risk Management Process, DOE/OE-0003, May 2012.
- FIPS 199, Standards for Security Categorization of Federal Information and Information Systems.
- ANSI/ISA-99.00.01-2007, Security for Industrial Automation and Control Systems: Concepts, Terminology and Models, International Society of Automation (ISA).
- ANSI/ISA-99.02.01-2009, Security for Industrial Automation and Control Systems: Establishing an Industrial Automation and Control Systems Security Program, ISA.
- Control Objectives for Information and Related Technology (COBIT).
- Council on Cyber Security (CCS) Top 20 Critical Security Controls (CSC).
- ANSI/ISA-62443-2-1 (99.02.01)-2009, Security for Industrial Automation and Control Systems: Establishing an Industrial Automation and Control Systems Security Program.
- ANSI/ISA-62443-3-3 (99.03.03)-2013, Security for Industrial Automation and Control Systems: System Security Requirements and Security Levels.
- ISO/IEC 27001, Information technology – Security techniques – Information security management systems – Requirements.
- ISO/IEC 27005, Information technology – Security techniques – Information security risk management.
- NIST Special Publication (SP) 800-53 Rev. 4, Security and Privacy Controls for Federal Information Systems and Organizations, April 2013 (including updates as of 15 January 2014).
- NIST Roadmap for Improving Critical Infrastructure Cybersecurity, 12 February 2014.
- Other examples of documents include policy regulations and documents dealing with CIIs and security matters in the United States such as [PPD-8], [PPD-21], [EO 13636].
- One key directive is the Presidential Decision Directive 63 [PDD-63] that established the framework to protect the critical infrastructure. Ensuring the resilience of critical information infrastructures (CII) require protection that may involve coordination beyond national borders.

7

Critical Infrastructure Protection

7.1 Critical Infrastructure Attacks and Challenges

The power grid consists of geographically dispersed production sites that distribute power through different voltage level stations (from higher to lower voltage) until energy eventually flows to consumers. Both the production and distribution sites are typically controlled by supervisory control and data acquisition (SCADA) systems, which are remotely connected to supervision centers and to the corporate networks (intranets) of the companies managing the infrastructures. The intranets are linked to the Internet to facilitate, for example, communication with power regulators and end clients. These links can be exposed to cyber attacks. Operators access SCADA systems remotely for maintenance operations, and sometimes equipment suppliers keep links to the systems through modems.

7.1.1 Power Grid

It has been reported that an Internet worm entered a nuclear plant's supervision systems through a supplier's modem, which almost caused a disaster. An unprotected modem line to a system that controlled a high-voltage power transmission line is reported in [Shipley 2001]. Thus, allowing open access to the control systems is the most dangerous state from security perspective.

Control systems in power industry play an increasing role in critical infrastructure performance and protection. But the current designs lack capabilities of being highly autonomous and flexible systems. Many modern control systems are just digital versions of the analog architectures they replaced [Rieger 2013]. Although these control systems provided a reliable means to establish central monitoring and have eased integration of feedback/supervisory controls, they have only a limited ability to recognize infrastructure degradation wherever it may occur and optimize a corporate response. This results in a heavily dependency on human interaction. In addition, the ability to network distributed components is producing additional interdependencies, resulting in even greater system rigidity or brittleness, which in turn increases the likelihood of cascading failures. A cascading failure occurs when a disruption in one infrastructure causes the failure of a component in a second infrastructure, which subsequently causes a disruption in the second infrastructure.

The Smart Grid delivers electricity from suppliers to consumers using two-way digital technology to communicate with end loads and appliances at industrial, commercial, and residential premises to save energy, reduce capital and operational cost by improving efficiency, and increase reliability and transparency.

Also, the Smart Grid includes control systems, intelligent devices, and communication networks that keep track of electricity flowing in the grid. Since the emergence of Internet and World Wide Web technologies, these systems were integrated with the business and information technology (IT) systems and became more exposed to cyber threats. The smart power grid infrastructure is characterized by interdependencies (physical, cyber, geographical, and logical) and complexity (collections of interacting

Building an Effective Security Program for Distributed Energy Resources and Systems: Understanding Security for Smart Grid and Distributed Energy Resources and Systems, Volume 1, First Edition. Mariana Hentea.

components). Cyber interdependencies are a result of the pervasive computerization and automation of infrastructures. The Smart Grid disruptions can directly and indirectly affect other infrastructures, impact large geographic regions, and send ripples throughout the national and global economy.

An attack from any area of the Smart Grid can propagate to another area either adjacent or interdependent infrastructures employed in the generation, transmission, and distribution of power. However, the problem quickly could become intractable due to the different areas in which interdependencies manifest themselves if not appropriate protections are not implemented to prevent cascading effects. Industrial control systems (ICSs) and IT systems are highly interconnected and interdependent both within power industry and with external infrastructures and economic sectors. The interdependencies are manifested at different levels. Assessment must take into account the interdependencies at all levels. Examples of critical infrastructures interdependencies are included in Appendix C.

In complex interactive systems like Smart Grid whose elements are tightly coupled, likelihood of targeted attack and failures from erroneous operations and natural disasters and accidents are quite high. Vulnerabilities and attacks could be at different levels – software controlling or controlled device, application, storage, data access, local area network (LAN), enterprise, and private communication links as well as public PSTN and Internet-based communications.

In several of these cases, the relationships are not well understood. A natural extension of the cascade effect is the effect of multiple, coordinated, sustained attacks on several infrastructures simultaneously. The results of the cascade effect could include the following [Goertzel 2007]:

- Poor coordination of infrastructures relying on the Internet.
- Infrastructures using the Internet as the underlying technology for an operational intranet between remote locations will lose connections.

Infrastructures supporting both an operational network and Internet connections may expose control of the operational network to attackers, possibly resulting in collapse of the infrastructure.

Recent reports show a variety of threats, attacks, and vulnerabilities in the software and hardware used in power industry [PwC 2016b], [Verizon 2016], [Radware 2015]. There are kinds of domino effects yielded by particularly potent attacks on the information systems of the energy infrastructure. As utility assets and systems are increasingly interconnected and generate more data, these systems and information assets are more at risk [PwC 2016b]. This survey indicates that cybersecurity concerns include theft of customer records (234% increase in 2015) and theft of intellectual property (tripled in 2015); compromise of customer records soared 62% in 2015 despite a significant drop in the overall number of security incidents detected; security compromises of operational systems more than doubled in 2015; and exploits of embedded systems quadrupled. The estimated total financial losses as a result of all security incidents almost doubled over the year before.

Attacks on DER systems, specifically the cyber attacks on renewables, can result in impacts on energy infrastructure leading to cascading blackouts [Krancer 2015]. Also, the attacks on oil and gas sectors (e.g. single-point assets such as refineries, storage terminals, cyber systems) can disrupt the supply of oil and gas or even create an environmental disaster. Cyber system is defined as any combination of facilities, equipment, personnel, procedures, and communications integrated to provide cyber services. Examples include business systems, control systems, and access control systems [SPP Glossary].

7.1.2 Attacks on Information Technology and Telecommunications

Communications and IT play an increasingly important role both within and between national critical infrastructures [Johnson 2013]. As we rely on food, energy, and transportation systems, we also rely on information infrastructures. Similarly, a range of "Smart Grid" initiatives depend upon computational infrastructures to coordinate the supply and demand of renewable and conventional power sources.

The benefits that are provided by telecommunications and IT also create new vulnerabilities; for instance, it is increasingly difficult for national critical infrastructures to recover and reorganize their service provision in the aftermath of computational failures.

However, [Hanseth 2002] argues that current and future IT solutions are integrating numbers of systems across organizational and geographical borders and in many respects are significantly different from traditional information systems. This change in the nature of IT is reflected in public discourses about technology where the term IT has been replaced by ICT to reflect the so-called convergence between information and communication technologies. This convergence process is an extension and enhancement of change processes related to the nature of information systems. From the times when organizations developed and implemented their first systems, the number and types of systems in use have increased. The current solutions support communication, collaboration, and information exchange between any units (people, organizations, information systems) globally. In parallel, as the number of systems grows, so does their integration. Thus, the security risks are increasing, and the occurrence of security incidents plagues all economy sectors, governments, and public.

In 2015, technology companies detected twice as many incidents over the year before, according to [PwC 2016b]. After a significant decline in 2014, average information security spending soared 51% in 2015.

In 2015, telecoms reported a 45% rise in detected information security incidents over the year before, according to [PwC 2016b]. The Radware's 2014 survey [Radware 2015] found that Internet not only has it increased as a point of failure, but the Internet now is being the number one failure point. Also, hackers are trying their way through every protocol to determine whether and how to use it for the next big reflective attack. The result is that reflective attacks represent 2014s single largest DDoS. Multi-attack vector campaigns have become so commonplace that to have a campaign with a single-attack vector is far more exotic.

What changed in 2014 is attack duration, which has increased, and extra-large attacks have become common. In 2014, one in seven attacks was larger than 10 Gbps; the intensity can go up to 50 Gbps. Such attacks challenge the victim's Internet pipe as well as the overall health of the Internet service provider (ISP).

Attackers are combining multiple techniques in a single attack – enabling them to bypass defense lines, exploit server vulnerabilities, and strain server-side resources. Such attacks include anonymization and masquerading, fragmentation, encryption, dynamic parameters, evasion and encoding, parameter pollution, and extensive functionality abuse.

There were about 52 insider cyber attacks reported in IT and telecommunications between 1996 and 2002 [Kowalski 2008]. Of the 52 incidents, 24 involved solely sabotage, 11 involved solely theft of intellectual property, 8 involved solely fraud, 6 involved both sabotage and theft of intellectual property, and 3 involved both fraud and theft of intellectual property. Detecting and controlling insider activities is a challenge because malicious insiders have knowledge, capabilities, and authorized access to information systems [Omar 2016]. Organizations affected by insider activity in the IT and telecommunication sector included organizations such as:

- Internet Service Providers.
- Companies conducting e-business.
- Software, hardware, network, and telecommunication equipment manufacturers and suppliers.
- Newspapers.
- Companies that provide IT and telecommunication-related technical consulting services.

The motives were revenge (in just over half the cases), financial gain, theft of the information/property, and sabotage to the organization. Other key findings of the report include the following:

- Insiders had authorized access to the systems/networks at the time of the incidents.
- Insiders used relatively sophisticated tools or methods for their illicit activities, including scripts or programs, autonomous agents, toolkits, probing, scanning, flooding, spoofing, compromising computer accounts, or creating unauthorized backdoor accounts.
- Insiders exploited systemic vulnerabilities in applications, processes, and/or procedures.
- Insiders committed their illicit activities from within the workplace or remotely in nearly equal numbers, and incidents took place during and outside normal working hours.

Several reports of 2014 describe more issues related to hacking critical infrastructures, ranging from small dish satellite systems and very small aperture terminals (VSATs) to taking full control of ICSs [Storm 2014], [Storm 2013], [Wueest 2014], [Kirk 2014].

Several vulnerabilities were identified in the devices related to satellite communications. Satellite communications play a vital role in the global telecommunications system. The vulnerabilities include backdoors, hardcoded credentials, undocumented and/or insecure protocols, and weak encryption algorithms, design flaws, and features in the devices that pose security risks [Santamarta 2013].

Also, security engineers need to assess the vulnerabilities of VSATs that can be a target for cyber attacks. VSATs are dish-based computer systems that provide narrowband and broadband Internet access to remote locations or transmit point-of-sale credit card transactions, SCADA, and other narrowband data. A typical VSAT system consists of user computer or laptop, VSAT system, satellite, hub station, internet infrastructure including servers.

Reports (e.g. [IntelCrawler], [Paganini 2014]) discuss the vulnerabilities and concerns related to the security of over 2.9 million active VSAT terminals that are installed in the world. Many of these devices with exposure to Internet (telnet access, weak passwords, default passwords, easy unauthorized access from remote location) are used by various industries and critical infrastructures or governments of different countries. They are being used by several sectors. Defense sector transmits government and classified information. Financial organizations like banks transmit sensitive data. Industrial sectors such as energy transmit data from power grid substations, or oil and gas use VSAT terminals to transmit climate data or data from oil rigs. Over 10000 of those devices are identified as open and targets to cyber attacks [Paganini 2014].

7.1.3 Attacks in Manufacturing

Already, manufacturers report that security compromises Internet of Things (IoT) technologies like operational systems and embedded devices more than doubled in 2015, according to [PwC 2016b]. Various analysis reports for 2015 [ICS-CERT 2015b], [Kovacs 2016] show one-third of the incidents impacted the critical manufacturing sector in 2015, which in 2014 were accounted for 27% of incidents. The increase was the result of a spear-phishing campaign launched by an advanced persistent threat (APT) actor against organizations in critical manufacturing and other sectors. The attacker, believed to be the threat group known as APT3, exploited a zero-day vulnerability in Adobe Flash Player in its operations. In 2014, the same actor launched a reconnaissance operation that used social engineering tactics to trick the employees of the targeted organizations into handing over valuable information.

7.1.4 Defense

The various categories of attackers that military may be faced with in the cyber arena are identified as follows [DCSINT 2006]:

- Hackers – Computer users who spend a lot of time on or with computers and work hard to find vulnerabilities in IT systems; some hackers known as white hat hackers or black hat hackers.
- Hacktivists – Combinations of hackers and activists; usually, they have a political motive for their activities and identify that motivation by their actions, such as defacing opponents' Web sites with counter-information or disinformation; they may be an unrelated activity or a supporting piece of a terrorist campaign.
- Computer criminals – Individuals that exploit computer systems, primarily for financial gain; computer extortion is a form of this type of crime.
- Industrial espionage – Industrial spies could be government sponsored or affiliated from commercial organizations or private individuals; they look for proprietary information on financial or contractual issues or acquire classified information on sensitive research and development efforts.
- Insiders – Employees with authorized access to a system can conduct an attack; the identification of these insiders with criminal intent is difficult; they may be disgruntled employees working alone, or they may be excellent workers in concert with other terrorists to use their access to help compromise the system.
- Consultants/contractors – The practice by many organizations to use outside contractors to develop software systems is sometimes a risk to security of an organization.

- Terrorists – Individuals or groups that may be able to use the Internet as a direct instrument to cause casualties, either alone or in conjunction with a physical attack.

Intruders look for vulnerabilities in software, design, implementations, and operations. For example, flaws and ad hoc procedures provided by different agencies exposed NASA systems to serious intrusions. NASA reported that in 2011 hackers stole employee credentials and gained access to mission-critical projects in 13 major network breaches that could compromise US national security [Reuters 2012].

7.2 The Internet as a Critical Infrastructure

When considering damaging effects on critical national infrastructures, we must examine the information infrastructure itself and how it can be affected by a sustained attack on the Internet. A sustained denial-of-service attack against the Internet would disconnect a large portion of the information infrastructure and probably bring down almost the entire infrastructure.

The cascading effect caused by single points of failure has the potential to pose severe consequences. Examples of factors contributing to the cascade effect of on a successful cyber attack could be the following:

- The increasingly important role played by the Internet in the National Information Infrastructure.
- Increased reliance on the Internet as the transport of the information for other networks.
- Infrastructure component – Other critical infrastructures use the Internet to a greater or lesser degree to exchange business, administrative, developmental, and research information between remote sites.

It is argued that the Internet is the single large infrastructure that could be attacked with cyber weapons. However, Internet is a shared global network. An attack against it will affect both target and attacker. There are different opinions on the Internet capabilities for sustaining its functions against a cyber attack or during an attack. Some argue that the Internet is very robust. Its design and architecture emphasize survivability. The Internet could deal with disruption by automatically rerouting to ensure that a message would arrive despite the complete destruction of key nodes from the network. The Internet addressing system, which is critical to the operations of the system, is multilayered and decentralized and can continue to operate, possible with some degradation of service, even if updating the routing tables that provide the addressing function is interrupted for several days.

A DDoS attack against the 13 domain name system (DNS) root servers that govern Internet addresses was launched by unknown parties in October 2002. The attack forced 8 of the 13 servers offline and did not noticeably degrade Internet performance. Still the system functioned as designed, demonstrating overall robustness in the face of a concerted, synchronized attack against all 13 root servers [Vixie 2002]. It was unnoticed by most of the public, but had it been continued for a longer period (and if the perpetrators remained undetected), there could have been a significant slowdown in traffic. In response to these attacks of 2002, the DNS system has been strengthened by dispersing the root servers to different locations and by using new software and routing techniques. The new redundancy measure makes shutting down the DNS system a difficult task for an attacker. Then, a subsequent DDoS attack occurred in February 2007, but only a few servers were affected for a limited time. Although no one person or group is in charge of the servers or of coordinating their operators, a global effort among security professionals and engineers across the globe stopped the attack, and it had a very limited impact on actual Internet users [ICANN].

7.3 Critical Infrastructure Protection

Critical infrastructure protection (CIP) is defined as [DCSINT 2006]:

> Actions taken to prevent, remediate, or mitigate the risks resulting from vulnerabilities of critical infrastructure assets. Depending on the risk, these actions could include changes in tactics, techniques, or procedures; adding redundancy; selection of another asset; isolation or hardening; guarding, etc.

In Australia, critical infrastructure organizations are urged to embrace best practice-based approach for information security protection. It is recognized that the approach to information security may vary between organizations due to differences in objectives and resources. Given the various challenges to information security (e.g. convergence, emerging technologies, deperimeterization), there is an underlying set of requirements based on seven basic principles of information security that all organizations must follow in order to ensure the security of their information assets. These basic principles that are identified from other sources are described in [TISN 2007] as follows:

1) Information security is integral to enterprise strategy.
2) Information security impacts on the entire organization.
3) Enterprise risk management defines information security requirements.
4) Information security accountabilities should be defined and acknowledged.
5) Information security must consider internal and external stakeholders.
6) Information security requires understanding and commitment.
7) Information security requires continual improvement.

In order to protect the critical infrastructures against cyber threats, the US President issued several policies.

7.3.1 Policies, Laws, and Regulations

In May 1998, the Presidential Decision Directive on Critical Infrastructure Protection was designed to defend the US nation's critical infrastructure from physical and cyber intrusions [PDD-63]. This directive calls for a national effort to assure the security of the vulnerable and interconnected infrastructure of the United States, most notably telecommunications. It stresses the critical importance of cooperation between the government and the private sector because the critical infrastructure of the United States is primarily owned and operated by the private sector.

In October 2001, the US Executive Order 13231, Critical Infrastructure Protection in the Information Age [EO 13231], created the President's Critical Infrastructure Protection Board (PCIPB). The PCIPB's core mission is to secure cyberspace.

The National Strategy to Secure Cyberspace [CERT 2003] identifies five national priorities to achieve this ambitious goal. These are as follows:

1) A national cyberspace security response system.
2) A national cyberspace security threat and vulnerability reduction program.
3) A national cyberspace security awareness and training program.
4) Securing governments' cyberspace.
5) National security and international cyberspace security cooperation.

These five priorities serve to prevent, deter, and protect against attacks. In addition, they also create a process for minimizing the damage and recovering from attacks that do occur. Cybersecurity and personal privacy need not be opposing goals. Cyberspace security programs must strengthen, not weaken, such protections. The federal government needs to continue to regularly meet with privacy advocates to discuss cybersecurity and the implementation of this [CERT 2003] strategy.

The most recent two policies of February 2013 include Executive Order (EO) on Improving Critical Infrastructure Cybersecurity [EO 13636] and Presidential Policy Directive (PPD) on Critical Infrastructure Security and Resilience [PPD-21]. These policies reinforce the need for holistic thinking about security and risk management. Implementation of the EO and PPD require action toward system and network security and resiliency and also enhance the efficiency and effectiveness of the US government's work to secure critical infrastructure and make it more resilient. These policies established new overall goals for protecting critical infrastructure from both physical threats and cyber threats. The Executive Order 13636 directs the Executive Branch to [DHS FACT]:

- Develop a technology-neutral voluntary cybersecurity framework.
- Promote and incentivize the adoption of cybersecurity practices.

- Increase the volume, timeliness, and quality of cyber threat information sharing.
- Incorporate strong privacy and civil liberties protections into every initiative to secure our critical infrastructure.
- Explore the use of existing regulation to promote cybersecurity.

The Presidential Policy Directive-21 (PPD-21) replaces Homeland Security Presidential Directive 7 and directs the Executive Branch to [DHS FACT]:

- Develop a situational awareness capability that addresses both physical and cyber aspects of how infrastructure is functioning in near real time.
- Understand the cascading consequences of infrastructure failures.
- Evaluate and mature the public–private partnership.
- Update the National Infrastructure Protection Plan (NIPP).
- Develop comprehensive research and development plan.

PPD-21 assigns a federal agency, known as a Sector-Specific Agency (SSA), to lead a collaborative process for critical infrastructure security within each of the 16 critical infrastructure sectors. Each SSA is responsible for developing and implementing a sector-specific plan (SSP), which details the application of the NIPP concepts to the unique characteristics and conditions of their sector.

PPD-21 also made some adjustments to sector designations: National Monuments and Icons was designated as a subsector of Government Facilities; Postal and Shipping was designated as a subsector of Transportation; Banking and Finance was renamed Financial Services; and Drinking Water and Water Treatment was renamed Water and Waste Water Systems. Table 7.1 shows the current list of sectors and their lead agencies. Aspects related to organization and coordination challenges among sectors are discussed in [Moteff 2015].

Together the policies establish an ambitious set of tasks for an array of federal government agencies to carry out over the next three years. The Department of Homeland Security (DHS) is working with industry to increase the sharing of actionable threat information and warnings between the private sector and the US government and to spread industry-led cybersecurity standards and best practices to the most vulnerable critical infrastructure companies and assets.

Meeting the requirements of PPD-21, the plan [NIPP 2013] outlines how government and private security and resilience outcomes. The plan was developed through a collaborative process involving

Table 7.1 Current lead agency assignments.

Department/agency	Sector/subsector
Agriculture	Agriculture, food, meat/poultry, all other
Health and Human Services	Public health and healthcare
Treasury	Financial services (formerly banking and finance)
Environmental Protection Agency (EPA)	Water and waste water systems (formerly drinking water and water treatment systems)
Defense	Defense industrial base
Energy	Electric power, oil and gas
Homeland Security	Transportation systems (includes postal and shipping), information technology, communications, commercial nuclear reactors, materials, and waste, chemical, emergency services, dams, commercial facilities, government facilities (includes national monuments and icons), critical manufacturing

Source: Adapted from [Moteff 2015].

stakeholders from all 16 critical infrastructure sectors, all 50 states, and from all levels of government and industry. It provides a clear call to action to leverage partnerships, innovate for risk management, and focus on outcomes.

This plan [NIPP 2013] represents an evolution from concepts introduced in the initial version of the NIPP released in 2006 and revised in 2009. The National Plan is streamlined and adaptable to the current risk, policy, and strategic environments. It provides the foundation for an integrated and collaborative approach to achieve the vision of a nation in which physical and cyber critical infrastructure remain secure and resilient, with vulnerabilities reduced, consequences minimized, threats identified and disrupted, and response and recovery hastened. SSP are periodically being updated to align with the NIPP 2013.

Therefore, the terrorist attacks of 11 September 2001 and the subsequent anthrax attacks demonstrated the need to reexamine protections in light of the terrorist threat as part of an overall CIP policy [Moteff 2015]. The author provides a historical background and tracks the evolution of such an overall policy and its implementation in the United States.

Similar directives in Europe include [EU 2008], [EU 2009] protecting Europe from large-scale cyber attacks and disruptions: enhancing preparedness, security, and resilience and adoption of network and information security policies [Italy 2011].

According to Council Directive [EU 2008], a priority is identification and designation of European critical infrastructures and the assessment of the need to improve their protection. Critical infrastructure system protection should be based on an all-hazards approach, recognizing the threat from terrorism as a priority. Several threats need to be considered, including, among others, natural disasters, accidental damage, equipment failure, human error, and terrorist attacks.

Development of policies for the protection of critical information infrastructures (CII) varies even among countries that are members of OECD. Regardless of difficulties or recommendations sponsored by a coordinating agency such as OECD, the following issues have to be considered by governments when implementing national policies for the protection of the CII and cybersecurity programs [OECD 2008b]:

- A national strategy.
- Legal foundations.
- Incident response capability.
- Industry–government partnerships.
- A culture of security.
- Information sharing mechanisms.
- Risk management approach.

In the United States, the DHS has the mission to safeguard and secure cyberspace. The department has the lead for the federal government for securing civilian government computer systems and works with industry and state, local, tribal, and territorial governments to secure critical infrastructure and information systems. The DHS is focused on to:

- Analyze and reduce cyber threats and vulnerabilities.
- Distribute threat warnings.
- Coordinate the response to cyber incidents to ensure that nation's computers, networks, and cyber systems remain safe.

7.3.2 Protection Issues

The CIP problem is the security of the interconnections among infrastructure providers, regulators, operators, and others. For nations, securing cyberspace is a matter of survival, and US presidents have made it a national priority. Cybersecurity is such a high priority because the critical infrastructures are not all peers. The invisibility of cyberspace makes it an insidious national threat. The reality of critical infrastructures being owned in large measure by private entities is a challenge for the public and private sectors to collaborate on new interaction models and rules of engagement that should provide the needed protection.

The continuous growth of cybersecurity threats and attacks including the increasing sophistication of the malware is impacting the security of critical infrastructure, ICSs, and SCADA control systems. Since SCADA systems involve a human in the loop control and decision-making processes, they are more vulnerable to threats. Among the fastest-growing needs in security is better protection for ICSs, including SCADA [Landau 2008].

Therefore, several issues have been acknowledged for almost two decades by researchers and industry experts. For example, a strategy to deal with cyber attacks against the nation's critical infrastructure requires first understanding the full nature of the threat [Hentea 2008]. It is imperative to develop depth defense and proactive solutions to improve the security of SCADA control systems to ensure the future of control systems and survivability of critical infrastructure. However, as this argument said, what the future brings depends on two factors: available technology and societal concern [Bell 1999]. Perrow said that the public is unaware of our basic (US) vulnerabilities in the chemical industry and electric power industry including nuclear plants [Perrow 2006]. Recognizing that the threats are real and the challenges are complex, meaningful action has to be taken to avoid catastrophes [Radware 2015]. Putting controls and measures in place to ensure the cybersecurity of the energy infrastructure should be a task of paramount importance [Krancer 2015].

Basically, aspects of a multilateral cooperation in the protection of critical national infrastructure may include the following [Maloor 2012]:

- Nations consider protection of their critical infrastructure as closely linked to the protection of their national sovereignty and have a variety of national legislations in place to safeguard this infrastructure.
- General agreement that the protection of critical national infrastructure requires multilateral cooperation.

The failure of these critical bilateral contract resources could significantly disrupt the operation of the Internet. For example, in *Canada–US Action Plan for Critical Infrastructure*, the complexity and interconnectedness of Canada–US critical infrastructure requires that the Canada–US Action Plan be implemented using organizational structures and partnerships committed to sharing and protecting information and managing risks, while Australia defines critical information resources as a shared responsibility across governments and the owners and operators of critical infrastructure.

Numerous characteristics of the nation's energy infrastructure, including the wide diversity of owners and operators and the variety of energy supply alternatives and delivery mechanisms, make protecting it a challenge. Energy infrastructure assets and systems are geographically dispersed. Millions of miles of electricity lines and oil and natural gas pipelines and many other types of assets exist in all 50 states and territories. In many cases these assets and systems are interdependent. In addition, the energy sector is subject to regulation in various forms.

Therefore, the protection of the electric power system (EPS) is defined as the effort to ensure its reliable operation within an environment with effects of disturbances, failures, and events that put the system at risk. More specifically, it is necessary to ensure public safety, equipment protection, and quality of service by limiting the extent and duration of service interruption, as well as by minimizing the damage to the involved system components. Examples of EPS components that need to be protected include:

- Human personnel.
- EPS equipment (transmission/distribution lines, generators, transformers, motors, busbars).
- Customer-owned equipment (smart meters, DERs, etc.).
- Operations (power delivery, stability, power quality).
- Services (financial, transportation, telecommunication, and other infrastructure services).

Therefore, the multiple challenges for protection of critical infrastructures could be addressed by using defined cybersecurity frameworks.

7.4 Information Security Frameworks

An information security framework is a series of documented processes that are used to define policies and procedures around the implementation and ongoing management of information security controls in an enterprise environment. The following is an overview of most known security frameworks.

7.4.1 NIST Cybersecurity Framework

In response to [EO 13636] for improving critical infrastructure cybersecurity, the National Institute of Standards and Technology (NIST) released [NIST 2014] document that describes a framework as the basis to provide a common language that organizations can use to assess and manage cybersecurity risk. In this document, cybersecurity is defined as the process of protecting information by preventing, detecting, and responding to attacks.

The framework recommends risk management processes that enable organizations to inform and prioritize decisions regarding cybersecurity based on business needs, without additional regulatory requirements. It enables organizations – regardless of sector, size, degree of cybersecurity risk, or cybersecurity sophistication – to apply the principles and effective practices of risk management to improve the security and resilience of critical infrastructure. The framework is designed to complement, and not replace or limit, an organization's risk management process and cybersecurity program. Each sector and individual organization can use the framework in a tailored manner to address its cybersecurity objectives.

The framework is designed to and has been used to:

- Self-assessment, gap analysis, budget and resourcing decisions.
- Standardizing communication between business units.
- Harmonize security operations with audit.
- Communicate requirements with partners and suppliers.
- Describe applicability of products and services.
- Identify opportunities for new or revised standards.
- Categorize college course catalogs as a part of cybersecurity certifications.
- Categorize and organize requests for proposal responses.

The framework also supports:

- Consistent dialogue, both within and among countries.
- Common platform on which to innovate and identify market opportunities where tools and capabilities may not exist today.

The framework is a risk-based approach to managing cybersecurity risk and is composed of three parts:

- Core provides a set of activities to achieve specific cybersecurity outcomes and references examples of guidance to achieve those outcomes. The core comprises four elements: functions, categories, subcategories, and informative references.
- Implementation tiers provide context on how an organization views cybersecurity risk and the processes in place to manage that risk; tiers reflect a progression from informal, reactive responses to approaches that are agile and risk informed.
- Profiles can be used to identify opportunities for improving cybersecurity posture by comparing a current profile (the as is state) with a target profile (the to be state). Profiles can be used to conduct self-assessments and communicate within an organization or between organizations; the profile can be characterized as the alignment of standards, guidelines, and practices to the framework core in a particular implementation scenario.

Each framework component reinforces the connection between business drivers and cybersecurity activities.

The framework core elements work together as follows:

- Functions are identify, protect, detect, respond, and recover; they organize basic cybersecurity activities at their highest level; they aid an organization in expressing its management of cybersecurity risk by organizing information, enabling risk management decisions, addressing threats, and improving by learning from previous activities. Functions also align with existing methodologies for incident management and help show the impact of investments in cybersecurity. For example, investments in planning and exercises support timely response and recovery actions, resulting in reduced impact to the delivery of services.

- Categories are the subdivisions of a function into groups of cybersecurity outcomes closely tied to programmatic needs and particular activities. Examples of categories include asset management, access control, and detection processes.
- Subcategories further divide a category into specific outcomes of technical and/or management activities. They provide a set of results that, while not exhaustive, help support achievement of the outcomes in each category. Examples of subcategories include external information systems are catalogued, data at rest is protected, and notifications from detection systems are investigated.
- Informative references are specific sections of standards, guidelines, and practices common among critical infrastructure sectors that illustrate a method to achieve the outcomes associated with each subcategory. The informative references presented in the framework core are illustrative and not exhaustive. They are based upon cross-sector guidance most frequently referenced during the framework development process.

The five framework core functions are defined below:

- Identify is organizational understanding to manage cybersecurity risk to systems, assets, data, and capabilities.
- Protect includes development and implementation of the appropriate safeguards to ensure delivery of critical infrastructure services.
- Detect includes development and implementation of the appropriate activities to identify the occurrence of a cybersecurity event.
- Respond includes development and implementation of the appropriate activities to take action regarding a detected cybersecurity event.
- Recover is development and implementation of the appropriate activities to maintain plans for resilience and to restore any capabilities or services that were impaired due to a cybersecurity event.

Although the framework is not mandatory, it is emerging as an important tool for technologists to communicate with organizational leaders on managing cyber risks. One report lists organizations representing manufacturing, financial, and pharmaceutical sectors that are committed to use the framework or committed for establishing support for information sharing [FACT SHEET].

More guidance to assist the energy sector organizations is provided in [DOE 2015]. This guide outlines a general approach to framework implementation, followed by an example of a tool-specific approach to implementing the framework. The tool selected for this example is the Department of Energy (DOE)- and industry-developed Cybersecurity Capability Maturity Model (C2M2) [DOE 2014a]. The model is used to assess an organization's cybersecurity capabilities and prioritize their actions and investments to improve the security.

However, there are currently other variants of the C2M2: the Electricity Subsector Cybersecurity Capability Maturity Model (ES-C2M2) [DOE 2012] and Oil and Natural Gas Cybersecurity Capability Maturity Model (ONG-C2M2) [DOE 2014b] contain guidance and examples pertinent to those subsectors. In addition, the mapping may result in outputs that require more interpretation at the sector level. Therefore, the more general C2M2 [DOE 2014a] can be used by organizations regardless of their sector. It is possible that an organization that performs C2M2 practices mapped to a specific framework tier may determine that some C2M2 practices do not satisfy the tier characteristics to a degree required by that organization. Organizations utilizing this mapping should therefore review it and ensure that it aligns with their needs. Users may find also guidance in this document [DOE 2014c].

NIST still calls request for information such as experience with the cybersecurity framework or questions focused on awareness and roadmap areas. Among several valuable responses and comments, there we can mention a few important suggestions and questions as included in [NIST 2016]:

- NIST and other federal agencies could drive additional use of the framework and best practices by providing more real-world examples of the application and use of the framework, demonstrating the business value proposition of the framework and developing incentives for its use. Just as importantly,

the government can best promote cyber best practices by avoiding prescriptive regulatory regimes. However, there are questions about what constitutes a best practice and how to evaluate it.
- One of the steps the US government can take to increase the sharing of best practices is to promote alignment of federal information security practices with the framework core. A majority of information security vendors service both the public and private sectors. Aligning Federal Information Security Management Act requirements with the framework subcategories and mapping these requirements to other global standards referenced in the framework enable more vendors to compete in the public and private sector information security marketplace, driving further innovation and improving security capabilities.
- Another question is what could be the best way to align Federal Information Security Management Act requirements with framework (e.g. mapping [NIST SP800-53r4] security controls, mapping FISMA language, restructuring subcategories or categories).
- While the future success of the framework will depend in large part on the extent to which individual enterprises share their experiences and learn from the experiences of others, the question of automated indicator mechanism can be leveraged for sharing is waiting for a response.

As many issues were suggested and questioned, further improvements to the framework are expected by organizations to fully adopt the approaches of this framework. However, education and awareness remain major barriers to improved cybersecurity for small businesses, as noted in the analysis [NIST 2016]. Also, the most important impact of this framework is that it has raised awareness among senior management in the oil and natural gas industry and highlighted the importance of cybersecurity in protecting critical infrastructure. In the face of implementing a global digital economy, the suggestion of implementing an international framework for cybersecurity requires to address some challenges.

As further discussed in this report [Moteff 2015], one issue is regulation. The degree to which some of the security protection activities are mandated varies across sectors such that in some cases (nuclear plants), sectors are quite regulated. On the other hand, there are sectors such as information and telecommunication, oil and gas, and commercial (e.g. malls and office buildings) where similar activities (e.g. vulnerability assessments, etc.) are encouraged but not mandated. The author argues that it has proven difficult to pass additional regulations although the security community, the President's administration, industry, and Congress have debated the need to regulate more comprehensively the cybersecurity of critical infrastructure assets. There is a debate about these needs. Some in the security community suggest that strategic national needs are market externalities that require regulation to encourage more owner/operators (in particular, those who may not be at the forefront in cybersecurity capabilities or practices) to take the type of action that the security community considers necessary. Industry groups are concerned about the costs and benefits and the potential for duplicative reporting requirements associated with additional regulations.

In the United States, as a general statement of policy, owners and operators of critical infrastructure are to work with the federal government on a voluntary basis. Sharing information with the federal government about vulnerability assessments, risk assessments, and the taking of additional protective actions is meant to be voluntary.

The framework is to form the basis for a Voluntary Critical Infrastructure Cybersecurity Program that would encourage critical infrastructure owners and operators to improve the security of their information networks. Also, those agencies that have regulatory authority over certain critical infrastructure owner and operators are to consider using or modifying the framework in any regulatory action.

Although this framework is designed specifically for companies that are part of the US critical infrastructure, many other organizations in the private and public sectors are using and gaining value from the approach. The Presidential Executive Order [EO 13800] of 2017 requires federal agencies to use it, but the cybersecurity framework remains voluntary for industry. Twenty-one states are using it, and it has been reported an increase in the use and adaptation of the framework internationally.

7.4.2 NIST Updated Cybersecurity Framework

A new version of the [NIST 2014] document is the [NIST 2018] document that describes an updated NIST Cybersecurity Framework, published in 2018.

The new document maintains the same structure of the framework (core, implementation tiers, profile). The core structure comprises the same four elements as described in the previous version to include functions, categories, subcategories, and informative references (see Figure 7.1). It provides a set of activities to achieve specific cybersecurity outcomes and reference examples of guidance to achieve those outcomes.

The framework uses risk management processes to enable organizations to inform and prioritize decisions regarding cybersecurity. It supports recurring risk assessments and validation of business drivers to help organizations select target states for cybersecurity activities that reflect desired outcomes. Thus, the framework gives organizations the ability to dynamically select and direct improvement in cybersecurity risk management for the IT and ICS environments.

The framework is adaptive to provide a flexible and risk-based implementation that can be used with a broad array of cybersecurity risk management processes. Examples of cybersecurity risk management processes include [ISO 31000], [ISO/IEC 27005], [DOE 2012], [NIST SP800-39].

The framework provides a common organizing structure for multiple approaches to cybersecurity by assembling standards, guidelines, and practices that are working effectively today. Because it references globally recognized standards for cybersecurity, the framework can serve as a model for international cooperation on strengthening cybersecurity in critical infrastructure as well as other sectors and communities.

7.4.2.1 Examples of Enhancements

This new document refines, clarifies, and enhances Version 1.0 of NIST Cybersecurity Framework, which was issued in February 2014. For example, the updated document incorporates:

- Refinements to better account for authentication, authorization, and identity proofing.
- A new section on self-assessment to explain how the framework can be used by organizations to understand and assess their cybersecurity risk, including the use of measurements.
- Expanded guidelines on using the framework for cyber supply chain risk management (SCRM) purposes.
- An expanded section, communicating cybersecurity requirements with stakeholders, that helps users better understand SCRM processes.
- A new section, buying decisions, that highlights use of the framework in understanding risk associated with commercial off-the-shelf products and services.

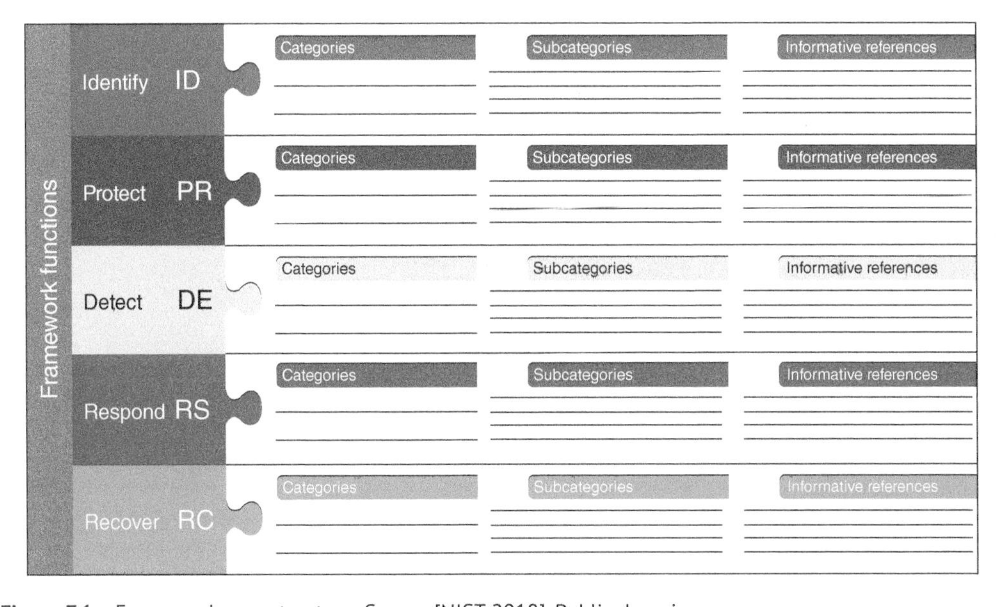

Figure 7.1 Framework core structure. *Source:* [NIST 2018]. Public domain.

- Additional cyber SCRM criteria that were added to the implementation tiers and a SCRM category, including multiple subcategories, that has been added to the framework core

7.4.2.2 Communicating Cybersecurity Requirements with Stakeholders

The framework provides a common language to communicate requirements among interdependent stakeholders responsible for the delivery of essential critical infrastructure products and services. Examples include the following:

- An organization may use a target profile to express cybersecurity risk management requirements to an external service provider (e.g. a cloud provider to which it is exporting data).
- An organization may express its cybersecurity state through a current profile to report results or to compare with acquisition requirements.
- A critical infrastructure owner/operator, having identified an external partner on whom that infrastructure depends, may use a target profile to convey required categories and subcategories.
- A critical infrastructure sector may establish a target profile that can be used among its constituents as an initial baseline profile to build their tailored target profiles.
- An organization can better manage cybersecurity risk among stakeholders by assessing their position in the critical infrastructure and the broader digital economy using implementation tiers.

Communication is especially important among stakeholders up and down supply chains. Supply chains are complex, globally distributed, and interconnected sets of resources and processes between multiple levels of organizations. Supply chains begin with the sourcing of products and services and extend from the design, development, manufacturing, processing, handling, and delivery of products and services to the end user. Given these complex and interconnected relationships, SCRM is a critical organizational function. On the other hand, cyber SCRM is the set of activities necessary to manage cybersecurity risk associated with external parties.

7.4.2.3 Cyber Supply Chain Risk Management

More specifically, cyber SCRM addresses both the cybersecurity effect an organization has on external parties and the cybersecurity effect external parties have on an organization. A primary objective of cyber SCRM is to identify, assess, and mitigate products and services that may contain potentially malicious functionality, are counterfeit, or are vulnerable due to poor manufacturing and development practices within the cyber supply chain.

Cyber SCRM activities may include activities such as:

- Determining cybersecurity requirements for suppliers.
- Enacting cybersecurity requirements through formal agreement (e.g. contracts).
- Communicating to suppliers how those cybersecurity requirements will be verified and validated.
- Verifying that cybersecurity requirements are met through a variety of assessment methodologies.
- Governing and managing the above activities.

As depicted in Figure 7.2, cyber SCRM encompasses technology suppliers and buyers, as well as non-technology suppliers and buyers, where technology is minimally composed of IT, ICSs, cyber–physical systems (CPS), and connected devices more generally, including the IoT. However, the diagram shows an organization at a single point in time. Through the normal course of business operations, most organizations will be both an upstream supplier and downstream buyer in relation to other organizations or end users.

The parties included in the diagram comprise an organization's cybersecurity ecosystem. These relationships highlight the crucial role of cyber SCRM in addressing cybersecurity risk in critical infrastructure and the broader digital economy. These relationships, the products, and services they provide and the risks they present should be identified and factored into the protective and detective capabilities of organizations, as well as their response and recovery protocols.

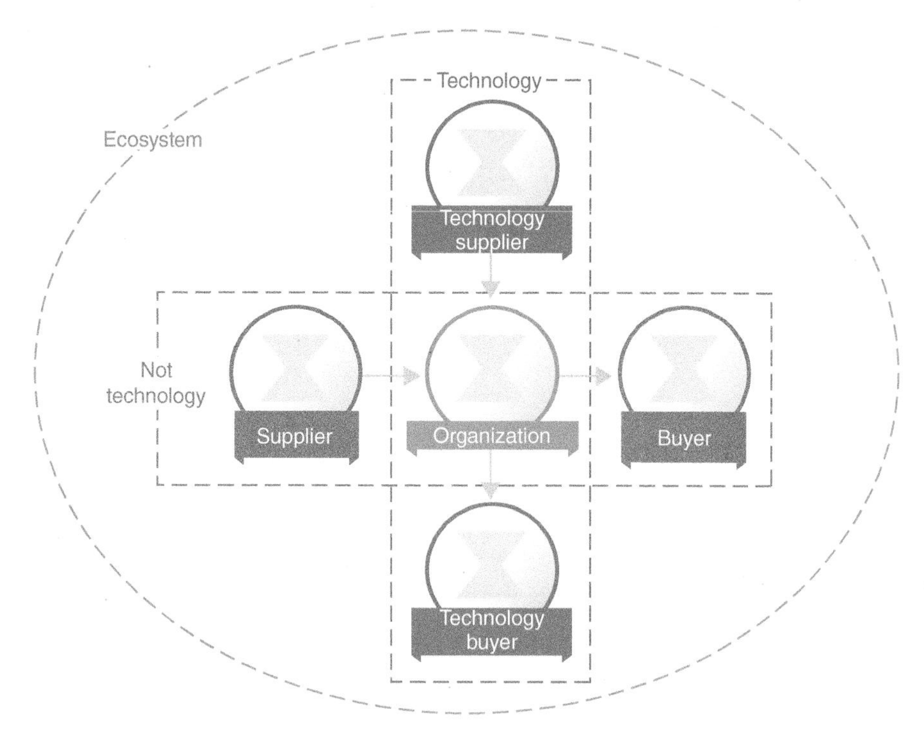

Figure 7.2 Cyber supply chain relationships. *Source:* [NIST 2018]. Public domain.

7.4.3 Generic Framework

Critical Information Infrastructure Protection (CIIP) is therefore generally acknowledged to be an indispensable component of national security policy. While some countries (in particular, the Western European and North American states) have established programs to cover all the different facets of CIIP, ranging from reducing vulnerabilities and fighting computer crime to defense against cyberterrorism, some countries may lack the resources and programs to protect their critical infrastructure. Also, due to their complexity and country context, these CIIP models are not necessarily applicable to other countries and therefore not suitable for the majority of countries in the world. A generic framework in order to help these countries to determine their response to the challenges of CIIP is described in [Suter 2007]. The model was also presented at World Summit on the Information Society (WSIS), Geneva 2003 – Tunis 2005, organized by the United Nations. Figure 7.3 illustrates the four pillars of CIIP. Essential tasks called the four pillars of CIIP model are:

- Prevention – The main function of prevention is to ensure that companies operating critical infrastructures are prepared to cope with incidents. Prevention and early warning cannot be approached on a purely technical level – potential dangers have to be weighed up constantly in a trade-off against risk situations.
- Detection – It is crucial that new threats are discovered as quickly as possible. In order to recognize emerging threats on a timely basis, the CIIP unit depends on a broad national and international network. In close collaboration with technical experts from Computer Emergency and Response Teams (CERTs), the CIIP unit should identify new technical forms of attacks as soon as possible.
- Reaction – Includes the identification and correction of the causes of a disruption. Initially, the CIIP unit should provide technical help and support to the targeted company. However, the CIIP unit cannot take on the management of incident response for these companies. The activities of the CIIP unit should complement, but not replace, the efforts of companies. Instead, the CIIP unit usually provides

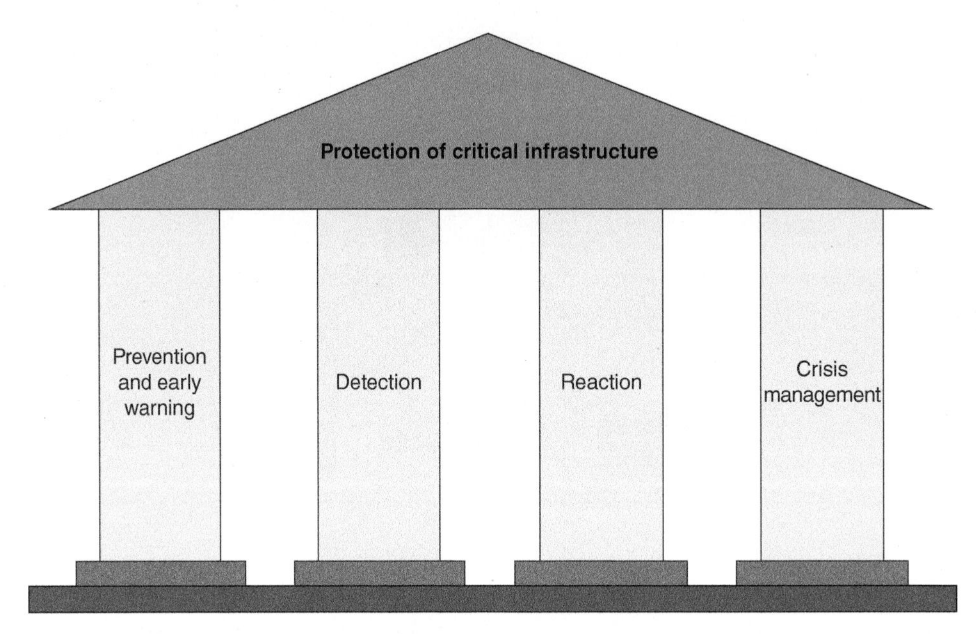

Figure 7.3 The four pillars of CIIP. *Source:* [Suter 2007], Research Paper sponsored by ITU-T, August 2007, Zurich, Switzerland.

advice and guidance on how to tackle an incident, rather than offering complete solutions. Therefore, incident response must start as quickly as possible.

- Crisis management – In case of a national crisis, the CIIP unit must be able to offer advice directly to the organizations and government. Within the administration, the CIIP unit should act as the center of competence for all questions related to information security. The CIIP unit should cooperate with various partners within the government. Hence, crisis management needs to be rehearsed regularly. All crucial actors must be familiar with their responsibilities, duties, and risks in times of crisis. Also, the CIIP unit should raise awareness of the various existing interdependencies. Operators of CI are dependent on each other in many respects. For instance, energy supply is crucial for the communication sector and vice versa. Since companies may tend to focus on their own business, they may often lack awareness of these interdependencies. The government in turn may also neglect the fact that it is dependent on the functioning of these critical infrastructures. A key task for the CIIP unit is to reinforce understanding of these dependencies among different actors, for example, through workshops and exercises. CIIP can only succeed if all stakeholders work together, with the same goals.

The cooperation model requires different specialized competencies, implying a large and complex organization. However, the CIIP unit can be streamlined without loss of capacity by establishing cooperation among different partners who are best qualified to cope each with one of the tasks. If each partner concentrates on their competency, existing know-how can be applied most efficiently, saving costs and manpower. According to the tasks of each of the four pillars, CIIP requires different organizational, technical, and analytical competencies. The CIIP unit should ideally include three partners:

- A governmental agency, providing strategic leadership and supervision (the head of the CIIP unit).
- An analysis center with strong linkages to the intelligence community (the situation center).
- A technical center of expertise, usually consisting of staff members of a national CERT (the CERT team).

Thus, the wide range of threats makes the protection of critical infrastructures a highly complex and interdisciplinary research domain. Several challenges need to be understood.

Recognizing the need for CIIP is a key priority in the countries of the European Union. This document [ENISA 2015] describes different approaches to follow on the governance of critical information infrastructures. It also includes examples of plans implemented in different countries to protect their national critical infrastructure. The information included in the document aims to help member states that were still working in designing and implementing their national strategy and includes priorities and models followed. Also, the document aims to provide assistance to the private sector to better understand their role in the implementation of the provisions of the national strategy. However, the approach each country takes on the topic is diverse and according to their national requirements and legislation.

Therefore, the evolution toward a more holistic and comprehensive critical infrastructure protection and resilience paradigm requires the integration of such elements as all-hazards, dependencies/interdependency analysis, stakeholder integration, information sharing, and cyber analysis. This evolution also incorporates regional considerations of protection and resilience that build strong information sharing processes and public–private partnerships.

However, there are challenges that need to be addressed to facilitate and support the ongoing conceptual and operational evolution [Moteff 2015].

7.5 NIST Privacy Framework

While some organizations have a robust grasp of privacy risk management, a common understanding of many aspects of this topic is still not widespread. To promote broader understanding, NIST recently published a preliminary draft [NIST 2019] that describes NIST Privacy Framework: A Tool for Improving Privacy through Enterprise Risk Management. It is intended for voluntary use to help organizations:

- Better identify, assess, manage, and communicate privacy risks when designing or deploying systems, products, and services.
- Foster the development of innovative approaches to protecting individuals' privacy.
- Increase trust in systems, products, and services.

This document covers concepts and considerations that organizations may use to develop, improve, or communicate about privacy risk management. Also, the document provides additional guidance on key privacy risk management practices.

The privacy framework can drive better privacy engineering and help organizations protect individuals' privacy by:

- Building customer trust by supporting ethical decision making in product and service design or deployment that optimizes beneficial uses of data while minimizing adverse consequences for individuals' privacy and society as a whole.
- Fulfilling current compliance obligations, as well as future proofing products and services, to meet these obligations in a changing technological and policy environment.
- Facilitating communication about privacy practices with customers, assessors, and regulators.

While managing cybersecurity risk contributes to managing privacy risk, it is not sufficient, as privacy risks can also arise outside the scope of cybersecurity risks. Figure 7.4 illustrates how NIST considers the overlap and differences between cybersecurity and privacy risks. The intersection of cybersecurity risks and privacy risks is the place where privacy breach occurs.

The NIST approach to privacy risk is to consider potential problems individuals could experience arising from system, product, or service operations with data, whether in digital or non-digital form, through a complete life cycle from data collection through disposal.

The privacy framework describes these data operations in the singular as a data action and collectively as data processing. The problems individuals can experience as a result of data processing can be expressed in various ways, but NIST describes them as ranging from dignity-type effects such as embarrassment or stigmas to more tangible harms such as discrimination, economic loss, or physical harm. Problems can arise as unintended consequences from data processing that organizations conduct to meet their mission or business objectives.

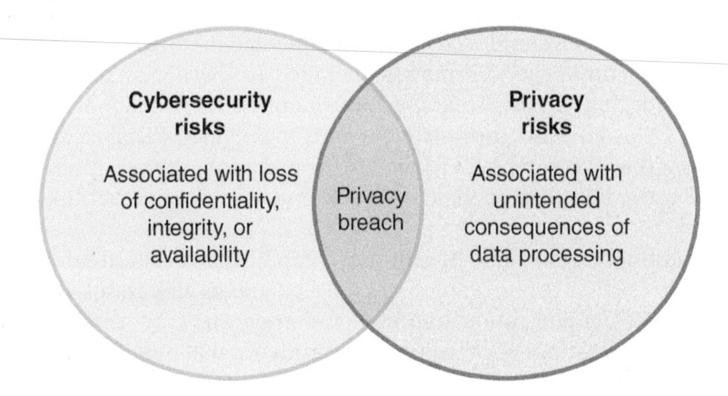

Figure 7.4 Cybersecurity and privacy risk relationship.

An example might be the concerns that certain communities have about the installation of smart meters as part of the Smart Grid. The ability of these meters to collect, record, and distribute highly granular information about household electrical use could provide insight into people's behavior inside their home. The meters were operating as intended, but the data processing could lead to unintended consequences that people might feel surveilled. However, these problems also can arise from privacy breaches where there is a loss of confidentiality, integrity, or availability at some point in the data processing, such as data theft by external attackers or the unauthorized access or use of data by employees who exceed their authorized privileges.

Also, Figure 7.4 shows privacy breach as the overlap between a loss of confidentiality, integrity, or availability and unintended consequences of data processing for mission or business objectives. Once an organization can identify the likelihood of any given problem arising from the data processing, which the privacy framework refers to as a problematic data action, it can assess the impact should the problematic data action occur. This impact assessment is where privacy risk and organizational risk intersect. Individuals, whether singly or in groups (including at a societal level), experience the direct impact of problems. As a result of the problems individuals experience, an organization may experience impacts such as noncompliance costs, customer abandonment of products and services, or harm to its external brand reputation or internal culture. These organizational impacts can be drivers for informed decision making about resource allocation to strengthen privacy programs and to help organizations bring privacy risk into parity with other risks they are managing at the enterprise level.

The privacy framework follows the structure of the cybersecurity framework. Like the cybersecurity framework, the privacy framework is composed of three parts: core, profiles, and implementation tiers. Each component reinforces privacy risk management through the connection between business and mission drivers and privacy protection activities. It describes the role of privacy risk management and privacy risk assessment as well as the relationship between these activities. The core comprises three elements, namely, functions, categories, and subcategories, depicted in Figure 7.5.

Functions aid an organization in expressing its management of privacy risk by understanding and managing data processing, enabling risk management decisions, determining how to interact with individuals, and improving by learning from previous activities. There are five functions: Identify-P, Govern-P, Control-P, Communicate-P, and Protect-P. The first four can be used to manage privacy risks arising from data processing, while Protect-P can help organizations manage privacy risks associated with privacy breaches. Protect-P is not the only way to manage privacy risks associated with privacy breaches. For example, organizations may use the cybersecurity framework functions in conjunction with the privacy framework to collectively address privacy and cybersecurity risks. Profiles can be used to describe the current state or the desired target state of specific privacy activities.

However, these problems also can arise from privacy breaches where there is a loss of confidentiality, integrity, or availability at some point in the data processing, such as data theft by external attackers or the unauthorized access or use of data by employees who exceed their authorized privileges.

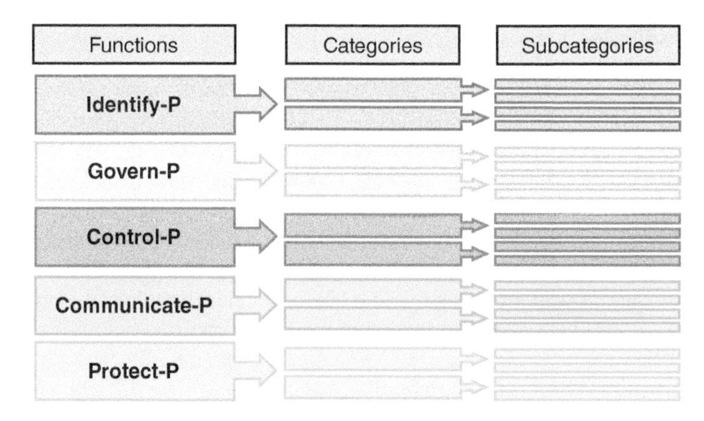

Figure 7.5 Privacy framework core structure. *Source:* [NIST 2018]. Public domain.

Effective privacy risk management requires an organization to understand its business or mission environment; its legal environment; its enterprise risk tolerance; the privacy risks caused by its systems, products, or services; and its role or relationship to other organizations in the ecosystem.

7.6 Addressing Security of Control Systems

There is a growing concern about the security of the SCADA control systems in terms of vulnerabilities, lack of protection, and awareness. The threat to the control systems is very real, as evidenced by several incidents of actual attacks [Constantin 2012]. Although the process control industry seems to understand that the threat of attack exists and must be addressed, it does not seem to have a shared view of how big this risk really is, which has led to a wide spectrum of proposed responses ranging from ignoring the topic to total panic. Addressing these vulnerabilities and threats requires several approaches from best security practices to advanced methods such as those described further in this book.

7.6.1 Challenges

Other problems and concerns with CIP include the following:

- Monoculture systems running the same software for business, government, and information systems increasingly control the critical infrastructures. This puts the entire networked systems at risk because they share the same vulnerabilities and hacker's exploits [Lala 2009].
- Control systems manage physical processes that can lead to safety and environment hazards if they are intentionally or unintentionally disturbed.
- Security as an operational task is often overlooked in the discussions about security in process control systems (PCSs); therefore, it is crucial how a certain level of security is maintained during operation.
- Unsuccessful central control and response by a designated security organization to provide timely information to critical infrastructure owners and government departments about threats, actual attacks, and recovery techniques.
- Ensuring the security of interconnections among infrastructure providers, regulators, operators, and others [Bessani 2008], there are a few networks that are national in scope and interconnect thousands of entities in ways that make them mutually dependent. However, these networks – finance, telecommunications, and electrical power – are among the most critical for national security and economic health, and their interconnectedness, national scope, and criticality may make them more attractive targets for cyber attack.

- Difficulty of estimating the actual cost of a cyber attack for planning of CIP; estimates of damage from cyber attacks generally overestimate or underestimate damages; there is also considerable variation in the methodologies, which are often not made public.
- Security of critical infrastructure systems is challenged by two of the most interesting domains, ICSs and emergency management, which are both characterized by severe requirements for both security and safety. Meeting needs at the intersection of security and safety can be especially challenging, particularly when standardized interfaces are also required [Landau 2008].
- Quantification of attack probabilities from different threat agents is a difficult task especially in the absence of large-scale data correlating prior attacks and vulnerability paths (static and dynamic) exploited by them. Similarly, quantifying the role of a particular threat or an identified vulnerability in a given scenario in the overall risk assessment of the IT and control systems is also extremely challenging [Ray 2010].
- Integrating security into ICS/SCADA domain that was previously closed is a challenge for ICSs. ICSs have a unique architecture that comprises both embedded (programmable logic controllers) and conventional computing devices and applications. ICSs were originally closed-loop systems with low security exposure. More recently, factors such as market pressure (requiring that control systems be connected to planning systems such as enterprise requirements planning) and deregulation (necessitating a need for energy generators or distributors to give energy traders access to SCADA systems) have altered these systems' connectivity topology, making them more widely accessible and thus increasing their security exposure.
- Maintaining updates to all critical sector plans and ensuring smooth and efficient coordination among different agencies responsible for plans can be a hurdle. [PPD-21] assigns a federal agency, known as a SSA, to lead a collaborative process for CIP within each of the 16 critical infrastructure sectors.
- Several agencies, tasks, and actions planned for all critical sectors including implementation of NIST framework have to be coordinated. Examples of actions, tasks, and plans are summarized in [Cedarbaum 2013].
- Although each agency is responsible for developing and implementing a plan, there could be details such as the application of the NIPP concepts to the unique characteristics and conditions of their sector that are not identified and specified. In addition, SSPs should be updated periodically to align with the most current NIPP. Achieving an effective implementation of the NIPP and SSP requires integrated and effective public–private partnerships, as well as communication and coordination at all levels. Cyberspace does not respect national boundaries, so government and industry task forces must address the pressing need for cybersecurity collaborations across jurisdictions such as the Multi-State Information Sharing and Analysis Center (www.msisac.org), with public–private partnerships (such as proposals from the Internet Security Alliance) and internationally. Cybersecurity is a global priority and shared responsibility, which will motivate the development of more comprehensive definitions and models [Agresti 2010].
- Management and responsibilities are not well defined. There are fears that the frequency and severity of critical infrastructure incidents will increase in the future and emergency management require improvement in government relationships with the private sector because 85% of the nation's critical infrastructure are in the hands of private sector and clear response and recovery capabilities are absent within government [Givens 2011]. Although efforts are under way, there is no unified national capability to protect the interrelated aspects of the country's infrastructure. One reason for this is that a good understanding of the interrelationships does not exist. There is also no consensus on how the elements of the infrastructure mesh together or how each element functions and affects the others. Securing national infrastructure depends on understanding the relationships among its elements. There is an unrealistic believe in the myth "My government has the solution and will protect me" [Hathaway 2009].
- Security costs are not planned and budgeted. The management of the private sector is divided on the opinion about spending on security. Still many managers of the private sector view security as a complete cost, while others completely understand the value to their particular business of investing in information security [CERT 2007]. Operated and maintained by process control engineers, these traditionally dedicated, stand-alone control systems are modernized or replaced at only every 10 or more

years [Palmer 2008]. However, reliance only on private sector to support security is still questionable for many security professionals.

- Although two CI security frameworks are available, NIPP framework of 2013 and NIST framework of 2014, the organizations may not know which framework to choose. The adoption is lacking incentives and direction on choosing a framework that is appropriate for an organization. In addition, NIST document does not provide any guidance on privacy.
- Technical challenges and the policy challenges on globalization undoubtedly generate a quite complex security and privacy standard landscape that affects critical infrastructures and control systems.
- Another issue is the supply chain security for major sectors such as DoD and NASA as they try to make sure that hardware, software, and other components have not been tampered with by other nations. This has proved challenging because so many parts come from overseas and American companies often contract for programming work abroad.
- Consequently, some of the biggest challenges in making control systems more secure relate to human behavior and organizational processes. Most people perceive security as a purchasing task and that buying a secure control system will solve everything. This attitude disregards the fact that technical security controls (e.g. a virus checker, intrusion detection system, or firewall) installed and configured in one day could become unsecure next day if the plant owner does not allocate enough effort and money into maintaining such mechanisms. Different activities must address overall security throughout a system's entire life cycle. Security in system development is the vendor's responsibility, secure installation and upgrades belong to the system integrator, and the plant owner must maintain secure plant operations, which constitute the longest part of the life cycle. Technology alone cannot address security, or as Schneier [Schneier 2000] argues, security is a process, not a product.
- Lack of trained professionals to handle the multitude aspects of critical infrastructures is a serious concern. Awareness, training, and education programs need to provide government officials and critical infrastructure owners and operators with the knowledge and skills needed to implement critical infrastructure security and resilience activities.
- If cloud computing services are used, they do face some specific risks, such as their staff to potentially compromise large quantities of sensitive data. Cloud infrastructures tend to concentrate data and resources, presenting an attractive target to attackers. They are globally distributed, meaning that confidential data may be held across a number of jurisdictions. However, through replication of systems and more robust and scalable operational security, they may achieve a level of security that would be beyond most smaller-scale enterprises [ENISA 2012]. Also, less attention so far has been paid to the impact of catastrophic events on cloud services. Without careful resilience planning, customers risk a loss of processing capacity and of essential data. If these customers are organizations that are part of critical sectors, the consequences of compromised data or losing data could affect not only the organization but also other interconnected organizations and critical sectors.
- Interdependencies among critical infrastructures have to be investigated to prevent propagation and cascading of failures between infrastructures. The understanding of interdependencies is still immature and requires the coordinated involvement of several different disciplines [Milis 2012].
- Modeling the interdependencies among interlinked infrastructures and assessing their impacts on the ability of each system to provide resilient and secure services are of high importance and require robust analysis. Specifically, the interdependency analysis for the EPS requires to take steps to mitigate any identified vulnerabilities and protect the system's operation from any internal or external threat. Since information exchange between infrastructure operators is limited, several physical and geographical interdependencies go unnoticed until a disruptive event occurs [Milis 2012].
- The complexity of the EPS drastically increases due to the amplified impact of the EPS to other critical infrastructures as well as due to the increased impact of other infrastructures to the operation of the EPS.
- The scale of complexity involved in present and future power system architectures is significantly greater than in the past. Electric power systems evolved over years from local independent entities toward large interconnected networks monitored and controlled by sophisticated ICT technologies, which eventually will be transformed into Smart Grids where also distributed energy sources, storage, electric vehicles, and appliances will be active components of the system [Asprou 2012].

- The nature and scale of interdependency between these two critical systems, EPS and ICT, require analysis and robust engineering design to avoid disturbance on communications and cascading effects on the power system's monitoring process as result of a communication node failure, which is responsible for transferring phasor measurement unit (PMU) data to the control center.
- Complexity is another issue. CIP is harder to address than ICT protection because of these infrastructures' interconnection complexity, which can lead to various kinds of problems. For example, the power grid includes geographically dispersed production sites that distribute power through different voltage level stations (from higher to lower voltage) until energy eventually flows into customers' buildings. CII are formed by facilities, such as power transformation substations or corporate offices, modeled as collections of LANs, which are linked by a wide-area network (WAN), such as dedicated phone lines or the Internet, and modeled as a WAN. Also, the socio-technical origins of ICT and information infrastructures need to be investigated [Bygstad 2008].
- Many infrastructure systems (e.g. power, transportation, and telecommunications) are complex adaptive systems, that is, their collective, systemic behavior is emergent (e.g. it follows patterns that result, yet are not analytically predictable from, dynamic, nonlinear, spatiotemporal interactions among a large number of components or subsystems [Coveney 1995]); capabilities of components and decision rules change over time in response to interactions with other components and external interventions [Gell-Mann 1994].
- A complex adaptive system is greater than the sum of its parts, so the system can only be described at levels higher than the components [Heller 2001].
- Human loop has to be accounted. The production and distribution sites are typically controlled by SCADA systems, which are remotely connected to supervision centers and to the corporate networks (intranets) of the companies that are managing the infrastructures. The intranets are linked to the Internet to facilitate, for example, communication with power regulators and end clients. These links create a path for external attackers. Operators access SCADA systems remotely for maintenance operations, and sometimes equipment suppliers keep links to the systems through modems.
- Legacy systems continue to function. Commonly, critical infrastructures feature numerous legacy subsystems and non-computational components, such as controllers, sensors, and actuators, which cannot be modified for operational or other reasons, for several years to come. In addition, an organization's main concern is keeping its critical infrastructure working at the expected level of service.
- Security measures are missing or incomplete. Lack of protection and increasing sophistication of cyber threats require several security measures to be employed to protect critical infrastructure. Technology alone cannot address security; security is a process. Consequently, some of the biggest challenges in making control systems more secure relate to human behavior and organizational processes.
- Conflict of operation has to be avoided. Security engineers and architects need to understand that security mechanisms that stand in the way of power grid operation are unacceptable.
- Policy management requires continuous refining. Differences in both policy semantics and enforcement strategies across multiple platforms and application domains require building a protection policy management framework flexible enough to handle enforcement mechanism for adaptation to a variety of systems.
- Risk management in critical infrastructures is much more complicated than in other domains of application because of specific factors [Bialas 2016]:
  - Unprecedented critical infrastructure complexity, even when compared with very large business organizations or technical facilities
  - Continuous evolution and enhancement of critical infrastructures
  - Mutual interrelations between different infrastructures (interdependencies)
  - Problem diversity and to many other issues, like complex system architectures, complex interactions, behavioral aspects, reliability theory, vulnerability analysis, resilience, emerging behavior, knowledge of architecture and functioning principles of complex systems that are fuzzy and the data incomplete, different abstraction levels applied to manage CIs and cross-sectoral relations, high-impact and low-probability events that may occur, and increased needs for communication and coordination among the CI operators

- New frameworks are needed for understanding systems of infrastructure systems as a basis for modeling the complex behaviors of individual infrastructure systems as well as coupled systems [Heller 2001].
- Infrastructure systems that were engineered to facilitate the competitive flow of people, goods, energy, and information have expanded far beyond their original design specifications, so they need to be reengineered to serve their original purposes under new conditions, such as globalization, deregulation, telecommunication intensity, and increased customer requirements [Heller 2001].
- Concerns remain that the private sector is less well prepared against commercial and state espionage to an extent that could damage the long-term national competitiveness of advanced economies [OECD 2011].
- There is a significant danger that the public–private partnership remains a description of an aspiration rather than a well-worked-out set of formal relationships and understandings. The ownership of the critical national infrastructure of OECD member countries is partly public and partly private. For a wide variety of catastrophes, the two elements will need to work together to achieve adequate levels of protection and ability to recover [OECD 2011]. The public trust in the public–private partnerships to secure critical infrastructure can be achieved with effective privacy safeguards and civil society involvement [OECD 2011].
- Mass adoption of cyber attack laws, including nationalistic rules, is foreseen [OECD 2011]. As government faces an increasingly dissatisfied, frustrated constituency, and growing threats around state-sponsored espionage, legislators may begin the process of writing laws on cyber attacks. Such laws will likely aim to dictate network traffic flows, security levels at critical infrastructure companies, and acceptable data processing rules. They could also provide guidelines on what constitutes acceptable Internet behavior.
- Efficient management and processes may be needed for developing international partnerships, promoting public–private partnerships, and improving information sharing and associated education and training.
- Private operators have incentives to maintain continuity of service to their customers, but without some government intervention, they may not be willing to commit resources to protecting such wider interests of society as public confidence promoted by the general availability of shelter, electricity and gas, and telecommunications [OECD 2011].

7.6.2 Terrorism Challenges

The elements of the critical infrastructure themselves are also considered possible targets of terrorism. Traditionally, critical infrastructure has been lucrative target for anyone wanting to attack another country. Now, because the infrastructure has become a national lifeline, terrorists can achieve high economic and political value by attacking elements of it. Disrupting or even disabling the critical infrastructure may reduce the ability to defend the nation, erode public confidence in critical services, and reduce economic strength. Additionally, well-chosen terrorist attacks can become easier and less costly than traditional warfare because of the interdependence of infrastructure elements.

Computers and the connectivity they brought to societies and businesses while increasing productivity and creativity also increased the vulnerability to attacks by criminals and terrorists. Definitions of terrorism-related elated terms are defined in [DCSINT 2006]:

- A terrorist is an individual who uses violence, terror, and intimidation to achieve a result.
- Terrorism is the calculated use of violence or threat of violence to inculcate fear; intended to coerce or to intimidate governments or societies in the pursuit of goals that are generally political, religious, or ideological.
- Cyber terrorism is a criminal act perpetrated by the use of computers and telecommunications capabilities, resulting in violence, destruction and/or disruption of services to create fear by causing confusion and uncertainty within a given population, with the goal of influencing a government or population to conform to a particular political, social, or ideological agenda.

Several studies examining the cyber threat have shown that critical infrastructures are potential targets of cyber terrorists. It is argued that the most critical infrastructures include [DCSINT 2006]:

- Energy systems
- Emergency services
- Telecommunication
- Banking and finance
- Transportation
- Water system.

Almost every economic and social function is based in some way on the sourcing of energy, telecommunication services, transportation, etc. An attack to these infrastructures would bring devastating effects on the economy and in the people's life [Watts 2003].

Terrorists could use some installations of the power system to attack civil infrastructure, for example, terrorists could couple an electromagnetic pulse through the grid to damage computer or telecommunication infrastructure. These infrastructure elements can become easier targets where there is a low probability of detection. As described in [Verizon 2016], external threat is one category of threats employed by actors that are capable of organized crime and terrorism. Three kinds of attacks could be initiated against Smart Grid include [Amin 2002]:

- Attacks upon the power system – The target is the electric infrastructure; for example, terrorists could attack simultaneously two substations or key transmission towers in order to cause a blackout in a big area of the grid; other example could be an attack to the electric market.
- Attacks by the power system – Terrorists could use some installations of the power system to attack the population, for example, using power plant cooling towers to disperse chemical or biological agents.
- Attacks through the power system.

Terrorists are employing easy-to-use encryption programs (easy downloads from the Internet) so they are able to communicate in a secure environment. Using other technique such as steganography, they hide instructions, plans, and pictures for their attacks in pictures and posted comments in chat rooms. Also, they could use system warfare that is a technique or method that identifies critical system components and then attacks them to degrade or destroy the use or importance of the overall system. The enemy targets single points of failure to cripple larger systems. Examples of systems that might be targeted by system warfare are logistics, command and control, medical evacuation, commerce, and transportation [DCSINT 2006].

In addition, the DHS (www.dhs.gov) alerts include information about the attack methods used by intruders. Although this information is well received, the companies need to focus on getting a better security posture by refining their strategies. Instead of just fixing holes discovered by intruders, the companies need to focus on improving the protection and removing vulnerabilities.

In 2016, it is observed an increase in incidents attributed to technically proficient threat actors like foreign nation states, organized crime, and terrorists [PwC 2016b]. This survey shows that the advancement in incidents corresponds with an increase in theft of intellectual property as well as exploits of operational and embedded technologies.

Recently, several groups are responsible for targeted attacks aimed at organizations around the globe. Different types of malware have been used in numerous targeted attacks aimed at government organizations and critical infrastructure organizations in the United States [Kovacs 2016]. A threat group has been using the Russia-linked BlackEnergy malware family in attacks aimed at news media and electrical power organizations in Ukraine. A Ukrainian power company blamed some recent power outages on outsiders who remotely tampered with automatic control systems. The attack of 23 December 2015 includes many vectors: malware destroys data needed to operate equipment; synchronized, remote operation of substation breakers causes blackout; control room backup power supplies are remotely disconnected; phone are jammed. The consequences of the attacks on electrical sector in Ukraine include unplugged 225,000 people, damage to ICT systems, removal of the Windows event logs, and destruction of files (documents, images, databases, configuration files, industrial system sabotage, making the operating system unbootable, to name a few).

The attack on Ukraine's power grid started before December 2015 with extensive reconnaissance of distribution utilities' networks and theft of credentials for accessing SCADA systems; as researchers reveal, it was sustained and multipronged [Fairley 2016].

Reported in [Verizon 2016] are incidents carried by external factors that affect safety and health of people. Water utility detects incidents that manipulated the PLCs that managed the amount of chemicals used to treat the water to make it safe to drink, as well as affecting the water flow rate, causing disruptions with water distribution.

Besides loss of life, the terrorist attacks of 11 September 2001 disrupted the services of a number of critical infrastructures (including telecommunications, the Internet, financial markets, and air transportation). In some cases, protections already in place (like off-site storage of data, mirror capacity, etc.) allowed for relatively quick reconstitution of services. In other cases, service was disrupted for much longer periods of time [Moteff 2015].

In May 2013, the US DHS warned a rising tide of attacks aimed at sabotaging processes at energy companies. In the first half year of 2013, media (e.g. *Washington Post*, *New York Times*) covered intensively alerts sent by DHS at the beginning of May 2013 about potential cyber attacks against big companies in the United States [Nakashima 2013], [Sanger 2013]. DHS alerts included also specific measures that could be taken to prevent disruptive attacks. DHS warned that most attacks, especially those coming from a foreign country, have been attempts to obtain confidential information, steal trade secrets, and gain competitive advantage. On the other hand, the new attacks seek to destroy data or to manipulate industrial machinery and take over or shut down the networks that deliver energy or run industrial processes in different industries. Sometimes, alerts could be false. For example, a water pump failure in Illinois reported in 2011 was not a cyber attack [Nakashima 2011].

Terrorist groups use cyber capabilities to assist them in planning and conducting their operations and also to create destruction and turmoil by attacking critical infrastructures. For example, the terrorists of April 2013 attack in Boston used the Internet for finding a home bomb-making recipe. Although many people believe terrorists only operate in the world of physical violence, many terrorist groups have well-educated people and modern computer equipment to compete in cyberspace. As terrorists gain experience and technology, cyber attacks on all infrastructures become an increasing concern.

Campaigns of malware, known as Desert Falcon, are using multiple technical and social engineering methods to infect even well-protected entities like governments, banks, and top media in several Arab countries and Israel in 2013 and 2014 [Mimoso 2015].

Researchers are interested to know who are the actors targeting the ICS/SCADA systems facing the Internet and what methods they use. As described in [Wilhoit 2013], a honeypot architecture emulates PLCs. All three honeypots were Internet facing and using different static IP addresses in different subnets scattered throughout the United States. One honeypot imitates the activities of the real systems it mimics, that of a water pressure station. The results of the study show that in a very short time after the deployment, about 18 hours, there were signs of attacks on one of the three honeypots. As reported, the findings concerning the deployments proved disturbing [Wilhoit 2013]. Based on the data collected in the study for a period of 28 days, the following statistics are reported:

- A total of 39 attacks were initiated.
- 12 attacks were unique, targeted.
- 13 attacks were repeated by several of the same actors over a period of several days and could be considered targeted and/or automated.
- All attacks were preceded by port scans performed by the same IP address or an IP address in the same /27 network bits.
- Attackers were from 14 different countries.
- China accounted for the majority of the attack attempts at 35%, followed by the United States at 19% and Laos at 12%.
- Repeat offenders from countries such as Laos and China attempted intrusions at dedicated times during a 24 hour period, and they tried to discover new vulnerabilities and exploit known vulnerabilities.
- Attacks on the honeypot servers were carried with various malware, some new and encrypted.
- Reconnaissance attempts were initiated from China, Russia, and the United States.

Cyber attacks targeting drone systems caused damage to important functions while in international missions [Schachtman 2011]. Electricity grid in the United States was penetrated by spies from other countries (e.g. China, Russia, and others) in April 2009 [Gorman 2009].

In May 2007, Estonia's infrastructure was the target of a cyber attack [Estonia 2007]. The cyber attack impaired the country's network with data traffic, clogging it and rendering major services unusable. People were not able to access financial utilities, communication, and data services for several hours and some for days together. Several attacks against Georgia Web sites and government occurred in 2008 in Georgia before and during Georgia war with Russia [Georgia 2008].

However, it is argued that the complexity of successfully carrying out a cyber attack against national infrastructures like telecommunications or the electrical grid, combined with a lower probability of success than a physical attack, may make it unattractive to terrorists. Terrorists want to have impact on people to run in terror past mangled bodies in the street. In theory, the idea of a cyber attack against telecommunication systems in coordination with a physical attack is attractive, as it could compound damage and terror, but coordinating two simultaneous attacks adds a degree of complexity that may overwhelm a terrorist planning capabilities while increasing the chances of detection. The same constraints do not apply to a nation state attacker. In the event of a conflict, however, a nation state opponent is likely to use cyber weapons to attempt to disrupt the large US national networks.

7.7 Emerging Technologies and Impacts

Technological changes continue to disrupt how organizations compete and create value in ways that often alter operating models. For example, some of today's most significant business trends such as the explosion of data analytics, the digitization of business functions, and a blending of service offerings across industries have expanded the use of technologies and data, which is creating more risk than ever before [PwC 2016b]. Analytics and big data emerged as themes underscoring the growing importance of increased security intelligence.

Thus, the emerging technologies pose a dilemma [BoozAllen 2011]. New technologies are constantly emerging in the market with promises to improve efficiency and increase a business potential. On the other hand, cyber threats continuously evolve increasing the risks incurred by adopting new and current technologies into the mainstream of businesses. Successfully establishing a solid cyber environment depends on the ability to leverage new technologies and better conduct a business while simultaneously mitigating the risks threatening to compromise mission.

It is perceived that the probability of success of a cyber attack to Smart Grid control infrastructures will increase with the massive deployment of advanced automation and communication technologies relying on standardized protocols [Dondossola 2013]. Another survey reports that there are serious privacy and security risks associated with connected homes and vehicles, however, since providers will amass and store an unprecedented amount of information about consumer activity and create points of access into home and car networks that did not exist five years ago [PwC 2016b].

Other trends, believed to be incredibly disruptive to information security, include the continued migration to cloud (and the accompanying dissolution of enterprise IT), the rise in the IoT, and the move toward the software-defined network (SDN) [Radware 2015]. This report argues that controlling employee endpoint devices with security hardware and software is no longer feasible economically, technically, or politically. IoT brings an end to controlled endpoints and introduces new threats. Also, it increases the attack surface, increases the sophistication of the attack itself, and complicates mitigation requirements.

ICT advancements allow governmental and industrial sectors to develop complete new infrastructures and infrastructure services, the so-called next-generation infrastructures (NGI) [Luiijf 2013]. The author predicts threats and cybersecurity failures alike for the envisioned NGI such as smart (energy) grids, smart road transport infrastructure, smart cities, and e-health unless fundamental changes in the approach to security of ICT-based and ICT-controlled infrastructures are taken.

7.7.1 Control Systems Open to Internet

Process control system characteristics and system architectures have changed drastically since the Internet became the worldwide network connecting public, governments, and organizations.

In the past, control systems were mostly isolated or connected to other systems only through specialized proprietary communication mechanisms and protocols. In addition, it was not customary to directly exchange information between the process control network and business networks. However, needs to make fast and cost-effective decisions, which in turn require accurate and up-to-date information about both the plant and process status, pushed for more communications between control and business systems including Internet. Controllers are industrial PCs running off-the-shelf operating systems (such as Windows or Linux), the networks are often wired (or even wireless) Ethernet, and control and supervision protocols are normally encapsulated on top of UDP/IP or TCP/IP protocol stack. Modern safety-critical systems use not only increasing numbers of microcomputers and microprocessors but also dedicated hardware to process the growing amounts of data needed to control the systems and monitor their status.

IT and Internet have a huge impact on power grid, making it increasingly possible for utilities to balance power generation and load demands more efficiently and do it in a quick and intelligent manner [Grose 2009]. Building a Smart Grid is all about using IT to control load so that it better matches generation.

Currently, plant owners in manufacturing can exchange vital information by interconnecting PCSs with each other and also with enterprise and office systems. These connections are increasingly based on open and standardized tools, commercial off-the-shelf solutions, Internet protocols, and the Web. Although these new avenues help companies meet market requirements and stay competitive, they also introduce IT security as a new challenge for PCSs [Brandle 2008]. Among several questions that are rising, the most relevant include the following [Palmer 2008]:

- How can these critical real-time control systems continue to work uninterrupted when introduced to the wild and lawless world of the Internet?
- How are these legacy systems going to keep pace with the rapidly changing environment of modern IT security?
- Can the IT security industry evolve its products to be effective on these real-time systems, where the emphasis is often on resiliency and availability first and confidentiality second?

7.7.2 Wireless and Mobile

During the past two decades, wireless cellular industry evolved on adding mobility to access Internet and information *anywhere* and *anytime*. The convergence of wireless and Internet marked a new era in the telecommunications industry. The market offers Internet-enabled home appliances and security systems, and some hospitals use wireless IP networks for patient care equipment. Cars could have indirect Internet connections – via a firewall or two – to safety-critical control systems. There have already been proposals for using wireless roadside transmitters to send real-time speed limit changes to engine control computers.

Controllers are industrial PCs running off-the-shelf operating systems (such as Windows or Linux); the networks are often wireless. Smart Grid is using digital IT – the Internet, sensors, controls, and wireless devices – to make it more efficient, reliable, and better able to accommodate renewable energy sources, particularly solar and wind. Renewable producers are intermittent and scattered producers of power based on various sources such as wind or sun. These producers and utility service providers use wireless and mobile networks to support communications of market, controllers, and sensors data.

Although Smart Grids push for using broadband power line (BPL) in rural connectivity and data communications, not all Smart Grids are using it. Some, such as the vast grid just proposed for some cities, use wireless ties rather than BPL technology [Schneider 2009]. A broad assessment of many wireless technologies that may be used for Smart Grid applications is provided in [NISTIR 7761]. In addition, wireless technology continues to evolve. The simulation and modeling studies include many technologies anything from radio, satellite, Bluetooth, Wi-Fi, and WiMAX. The wireless technologies chosen in the study encompass different capabilities, cost, and ability to meet different requirements for advanced power system applications. However, a complete technology assessment with respect to capacity and coverage including technology attributes such as latency and security must also be taken into account for a full Smart Grid suitability assessment. Although wireless technologies hold many promises for the future, their use is not without limitations.

Also, control systems continuously need to collect tens of thousands of measurements about the process and then either take immediate action through actuators such as valves and pumps or present the information along with alarms to the human operators overseeing the process. The communication of measurements and commands is often based on wireless networks. These complex, fault-tolerant, real-time computer systems are typically called automation systems, distributed control systems, SCADA systems, or PCSs.

It is argued that Wi-Fi technology is not ready for critical applications, mainly because of its intrinsic security and robustness problems, but next-generation wireless networks are expected to incorporate modern security features that will enable Wi-Fi become a fundamental building block for critical applications [Aime 2007]. In addition, Wi-Fi technology promises to provide extensions and changes to maintain its supremacy among the various wireless technologies.

7.7.3 Internet of Things and Internet of Everything

Also, technologies such as IoT and Internet of Everything (IoE) are emerging. Recently market studies acknowledge different devices enabled with Wi-Fi IoE technology [Wi-Fi 2014]:

- Thermostats.
- Lighting.
- Home security, monitoring, and control.
- Appliances: refrigerators and air conditioners.
- Automotive.
- Wearable devices and health/fitness devices.

Also, the study reveals that home security systems, lighting, thermostats, cars, irrigation systems, personal health devices, and appliances topped the list of areas for which 25% or more of respondents stated that Wi-Fi connectivity would be a useful feature. Although consumer interest in this technology continues to grow, the use of the smart devices and suitability of this technology and other wireless technologies for critical infrastructures have to meet security and interoperability requirements before they become mainstream technologies (see more in [Davis-Felner 2014]). More information about the future of connectivity is provided in [Wi-Fi 2018].

7.7.4 Web Technologies

Control systems, PCSs, ICSs, distributed control systems, and other similarly named systems control the processes that manage the components of our critical infrastructures, such as the electric power grid, factories and refineries, oil and gas pipelines, and water infrastructures. Since the emergence of Internet and World Wide Web technologies, these systems were integrated with the business systems and became more exposed to the cyber threats. People using these systems, however, want to enjoy increased functionality that recent applications, email, and Web browsing offer – things they can easily get by connecting directly to the Internet. Therefore, appropriate security requirements and designs have to provide adequate security measures to avoid security intrusions.

7.7.5 Embedded Systems

From cars to cell phones, video equipment to MP3 players, and dishwashers to home thermostats, embedded computers increasingly penetrated societies. But security of these devices is an open question and could prove a more difficult long-term problem than security does today for desktop and enterprise computing [Koopman 2004]. Security issues are nothing new for embedded systems. However, as more embedded systems are connected to the Internet, the potential damages from such vulnerabilities scale up dramatically. Because embedded systems can effect changes in the physical world, the consequences of exploiting their security vulnerabilities can go beyond mere annoyance to significant societal disruption. Attackers that break into a computer and get complete control of it can do anything they want with the attached sensors and actuators – send commands to traffic lights, shut down power stations, and so on.

7.7.6 Cloud Computing

While organizations are excited by the opportunities to reduce the capital costs and the chance to divest themselves of infrastructure management and focus on core competencies, a new paradigm, cloud computing, transforms the way IT is consumed and managed. Although cloud computing has promising benefits such as improved cost efficiencies, accelerated innovation, faster time to market, and the ability to scale applications on demand [Leighton 2009], there are still security and privacy issues that pose significant challenges [Sen 2016]. These challenges need to be addressed before a ubiquitous adoption may happen and reversal of consequences is not easy or possible.

7.8 Standards, Guidelines, and Recommendations

In the United States, cyber-specific authorities include various federal strategies, directives, policies, and regulations that provide the basis for federal actions and activities associated with implementing the cyber-specific aspects of the NIPP (e.g. [NIPP 2009] and most current updated plan). The following is a brief overview of the agencies.

7.8.1 Department of Homeland Security (DHS)

The DHS is the principal federal agency to lead, integrate, and coordinate the implementation of efforts to protect critical infrastructures and key resources [GAO 2011]. Some regulated entities may also be required to meet regulations or mandatory requirements that address cybersecurity problems or face enforcement actions [GAO 2011]. For example, IT, communications, water critical infrastructure sectors, and oil and natural gas subsector of the energy sector are not subject to direct federal cybersecurity-related regulation. Electrical sector is one of regulated entities that may be required to meet regulations or mandatory requirements that address cybersecurity problems or face enforcement actions [GAO 2011].

The National Strategy to Secure Cyberspace [DHS 2003] is an attempt to engage and empower Americans to secure portions of cyberspace. Securing SCADA systems became a national priority because many industries require the many monitoring and control capabilities that SCADA offers. SCADA systems are a key component of critical infrastructures like energy sector, manufacturing, food processing, telecommunications, and IT system of communications.

The National Strategy to Secure Cyberspace [US-CERT] is part of DHS overall effort to protect the nation [DHS Cyber]. It is an implementing component of the National Strategy for Homeland Security and is complemented by a National Strategy for the Physical Protection of Critical Infrastructures and Key Assets [DHS 2003] document. This document identifies a set of national goals and objectives and outlines the guiding principles that underpin efforts to secure the infrastructures and assets vital to the national security, governance, public health and safety, economy, and public confidence. It establishes a foundation for building and fostering the cooperative environment in which government, industry, and private citizens can carry out their respective protection responsibilities more effectively and efficiently. Cyberspace is composed of hundreds of thousands of interconnected computers, servers, routers, switches, and fiber-optic cables that allow our critical infrastructures to work [DHS Cyber].

To help the private sector learn about and adopt the NIST Cybersecurity Framework [NIST 2014], DHS created and launched the Critical Infrastructure Cyber Community (C3) Voluntary Program [DHS C3], which gives critical infrastructure owners and operators access to services and cybersecurity experts in the DHS. Through this program, DHS assists stakeholders with understanding use of the NIST

Cybersecurity Framework and other cyber risk management efforts and supports development of general and sector-specific guidance for framework implementation. This program is a good example of information sharing with confidentiality, privacy, and civil liberties protections built into its structure. Through sharing cyber threat information with its partners, DHS enables the detection, prevention, and mitigation of threats. One strategy for security risk mitigation is transferring the risk to an insurance company.

This work [Rosson 2019] reports that collaboration between the DHS Cybersecurity and Infrastructure Security Agency (CISA) and public sector partners has revealed a lack of cyber incident data combined with the unpredictability of cyber attacks that have contributed to a shortfall in first-party cyber insurance protection in the critical infrastructure community. The authors propose an extended insurance framework to assess the security risk profiles of the US power enterprise. The framework considers industry provided reliability indicators, estimated loss ratios, and various insurance features to recommend an optimal insurance package that minimizes risk to both the insurance offeror and insured party.

Further, the authors [Rosson 2019] argue that the cycle of adoption of effective cyber insurance in critical infrastructure has the potential to lead not only to a more secure power industry but also to improve cybersecurity across all industrial control system regulated sectors of critical infrastructure. The framework's risk assessment capability can help power companies in finding the right mix of risk transfer and risk mitigation strategies. Potentially, this strategy could help on preventing catastrophic failure of one system from impacting other interconnected systems.

7.8.2 Federal Communications Commission (FCC)

The Federal Communications Commission's (FCC) role in cybersecurity is to strengthen the protection of critical communications infrastructure, to assist in maintaining the reliability of networks during disasters, to aid in swift recovery after, and to ensure that first responders have access to effective communication services.

7.8.3 National Institute of Standards and Technology (NIST)

NIST plays an important role in the National Strategy to Secure Cyberspace. NIST published several standards for cybersecurity and continues to publish new standards as well as to revise the old standards and update to new constraints and developments for the protection of critical infrastructure. Examples of standards include Federal Information Processing Standards (FIPS) signed/approved by the Secretary of Commerce (FISMA made FIPS mandatory for federal organizations), Special Publications (SPs) providing guidance to federal organizations on IT security since 1990 (not mandatory for use), and NIST Interagency Reports (NISTIRs) (describe research of a technical nature to a specialized audience of interest including projects such as Smart Grid and computer/cyber/information security and privacy-related topics). For example, NIST document [NISTIR 7628r1] provides guidance to organizations that are addressing cybersecurity for the Smart Grid (e.g. utilities, regulators, equipment manufacturers and vendors, retail service providers, and electricity and financial market traders). This report is based on what is known at the current time about the Smart Grid and cybersecurity including technologies and their use in power systems as well as the risk environment in which those technologies operate.

Also, NIST supports the National Vulnerability Database (NVD) and Common Vulnerability Scoring System Support (CVSS) v2 [CVSS]. A new version, CVSS v3, is released and planned for deployment.

CVSS provides an open framework for communicating the characteristics and impacts of IT vulnerabilities. CVSS is a standard measurement system for industries, organizations, and governments that need accurate and consistent vulnerability impact scores. The NVD provides CVSS scores for almost all known vulnerabilities.

The NVD contains content (and pointers to scanning products) for performing configuration checking of systems implementing the checklists using the Security Content Automation Protocol (SCAP) validated tools.

Since [NIPP 2013], [NIST 2014] documents recommend a framework that complement and do not replace an organization's risk management process and cybersecurity program, an organization can use its current processes and leverage the frameworks to identify opportunities to strengthen and communicate its management of cybersecurity risk while aligning with industry practices. Alternatively, an organization without an existing cybersecurity program can use a framework as a reference to establish one.

Since the frameworks are not industry specific, the common taxonomy of standards, guidelines, and practices that are provided also is not country specific. Organizations outside the United States may also use the frameworks that can contribute to developing a common language for international cooperation on critical infrastructure cybersecurity.

7.8.4 North American Electric Reliability Corporation (NERC)

The North American Electric Reliability Corporation (NERC) is a nonprofit corporation dedicated to ensuring that the bulk electric system (BES) in North America is reliable, adequate, and secure. NERC maintains comprehensive reliability standards that define requirements for planning and operating the collective bulk power system (BPS). Examples include the Critical Infrastructure Protection (CIP) Cyber Security Standards, which are engineered to ensure the protection of cyber assets that are critical to the reliability of North America's BES. NERC's standards for governing critical infrastructure apply to entities that materially impact the reliability of the BES. These entities include owners, operators, and users of any portion of the system. Pursuant to the Energy Policy Act of 2005, NERC works with electric industry, regional entities, and state and federal agencies to develop reliability and cybersecurity standards that apply across the North American grid, including parts of Canada and Mexico. NERC standards are subject to FERC review and approval and are periodically updated as potential threats evolve.

On 20 March 2014, FERC approved the revised definition of BES, which includes core criteria with various enumerated inclusions and exclusions. As a result of the application of these BES definition provisions, all elements and facilities necessary for the reliable operation and planning of the interconnected BPS are being included as BES elements. Entities apply the definition of BES, including the respective inclusions and exclusions, to their asset inventory effective 1 July 2014. A BPS is a large interconnected electrical system made up of generation and transmission facilities and their control systems. A BPS does not include facilities used in the local distribution of electric energy. If a BPS is disrupted, the effects are felt in more than one location.

NERC CIP Version 5 was released on 22 November 2013. It categorizes systems based on their impact to BES cyber assets, helping organizations identify risks to their infrastructure and prioritize mitigating efforts. Information about the schedules and compliance is available via Web site [NERC CIP].

These standards are defined based on terms and concepts defined in the dictionary; they have different status such as NERC approved, availability, and enforcement [NERC Glossary].

Any security professional should get familiar with the terms and concepts used in the Version 5 CIP cybersecurity standards. Many terms have specific meaning in the context of these standards, may have a temporary definition, change definition, and/or do not correspond with the known security terms such as the following:

- The terms program and plan are sometimes used in place of documented processes where it makes sense and is commonly understood. For example, documented processes describing a response are typically referred to as plans (e.g. incident response plans and recovery plans). Likewise, a security plan can describe an approach involving multiple procedures to address a broad subject matter.
- Similarly, the term program may refer to the organization's overall implementation of its policies, plans, and procedures involving a subject matter. Examples in the standards include the personnel risk assessment program and the personnel training program. The full implementation of the CIP cybersecurity standards could also be referred to as a program. However, the terms program and plan do not imply any additional requirements beyond what is stated in the standards.

- Cyber assets, BES cyber assets, critical cyber assets, and critical BES assets have different meanings and may have different security requirements (see definitions and other related definitions in Appendix B).

7.8.5 Federal Energy Regulatory Commission

The Federal Energy Regulatory Commission (FERC) is an independent agency that regulates the interstate transmission of electricity, natural gas, and oil. FERC also reviews NERC proposals for standards and oversees the development of mandatory reliability and security standards. The commission monitors and directs NERC to ensure compliance with the approved mandatory standards by the users, owners, and operators of the BPS. As of now, electricity grid and nuclear generation are the only critical infrastructure sectors with mandatory and enforceable cybersecurity standards.

On 21 January 2016, FERC approved advancements to Critical Infrastructure Protection (CIP) Reliability Standards that address cybersecurity of the BES. The NERC CIP Version 5 standards identify and categorize BES cyber structures based on whether such structures have a low, medium, or high impact on the reliable operation and set specific requirements for each category, with which categorized entities must comply. To ensure that the electricity grid – a vast and complex system of transmission and communication networks – can withstand both natural events and cyber and physical attacks, the Energy Policy Act of 2005 subjects the electric power sector to mandatory cybersecurity standards under FERC jurisdiction.

The tiered impact rating methodology of CIP version 5 standards would bring all cyber assets that could impact BES facilities into the scope of the CIP standards. The action reflects the dynamic cybersecurity environment, which is moving toward proactive efforts for flexible and timely response to threats rather than basic compliance. While mandatory standards provide protection against known threats, electric utility sector and government agencies are increasingly coordinating their activities to maintain reliability against new and evolving threats [ENERKNOL 2016].

7.8.6 DOE Critical Infrastructure Guidance

On 8 January 2015, the DOE released guidance [DOE 2015] to help the energy sector establish or align existing cybersecurity risk management programs to meet the objectives of the NIST Cybersecurity Framework released [NIST 2014]. The voluntary cybersecurity framework consists of standards, guidelines, and practices to promote the protection of critical infrastructure through collaboration between industry and government. In developing this guidance, the DOE collaborated with private sector stakeholders through the Electricity Subsector Coordinating Council and the Oil and Natural Gas Subsector Coordinating Council. The Department also coordinated with other SSA representatives and interested government stakeholders.

7.8.7 US-CERT

The US Computer Emergency Readiness Team (US-CERT) supports activities that include:

- Providing cybersecurity protection to federal civilian executive branch agencies through intrusion detection and prevention capabilities.
- Developing timely and actionable information for distribution to federal departments and agencies; state, local, tribal, and territorial (SLTT) governments; critical infrastructure owners and operators; private industry; and international organizations.
- Responding to incidents and analyzing data about emerging cyber threats.
- Collaborating with foreign governments and international entities to enhance the nation's cybersecurity posture.

US-CERT supports information for:

- ICS owners, operators, and vendors.
- Government users including resources for information sharing and collaboration among government agencies.

- Home and business users including information for system administrators and technical users about latest threats.

The periodic ICS-CERT *Monitor Newsletter* offers a means of promoting preparedness, information sharing, and collaboration with the 16 critical infrastructure sectors. The publication highlights recent activities and information products affecting ICSs and provides a look ahead at upcoming ICS-related events. Also, yearly reports provide valuable information to users and manufacturers about threats and security measures.

ICS-CERT conducts on-site cybersecurity assessments of control systems to help strengthen the cybersecurity posture of critical infrastructure owners and operators and of ICS manufacturers. For example, ICS-CERT worked with an ICS asset owner following a report of possible intrusion activity targeting the entity's network. The asset owner operates in the power and water sectors, providing both power and water to their local community [ICS-CERT 2016a].

Recently, ICS-CERT conducted unclassified in-person briefings and online webinars for asset owners and federal, state, local, tribal, and territorial government representatives to increase awareness of the threat and provide additional information related to attacks against Ukrainian power infrastructure in December 2015. The briefing sessions provided details about the events surrounding the attack, techniques used by the threat actors, and strategies for mitigating risks and improving the cyber defensive posture of an organization. The ICS-CERT, the US DHS, the Federal Bureau of Investigation (FBI), the DOE, and other federal agencies have been actively working with the government of Ukraine to understand these criminal activities of December 2015 [ICS-CERT 2016b].

References Part 3

[Agresti 2010] Agresti, W. (2010). The Four Forces Shaping Cybersecurity. *Computer, 43*(2), 101–104. https://doi.org/10.1109/MC.2010.53

[Aime 2007] Aime, M.D., Calandriello, G., Lioy, A. (2007). Dependability in wireless networks: Can We Rely on WiFi?. *IEEE Security & Privacy*, *5*(1), 23–29. https://doi.org/10.1109/MSP.2007.4

[Amin 2002] Amin, M. (2002). Security challenges for the electricity infrastructure. *Computer, 35*(4), supl8-supl10. https://doi.org/10.1109/MC.2002.1012423

[Asprou 2012] Asprou, M., Hadjiantonis, A.M., Ciornei, I., Milis, G. (2012). On the complexities of interdependent infrastructures for wide area monitoring systems. *2012 Complexity in Engineering (COMPENG). Proceedings* (pp. 1–6), Aachen, Germany, (June 2012). https://doi.org/10.1109/CompEng.2012.6242951

[AU] AustralianGovernment. (n.d.). *Critical Infrastructure Centre.* https://www.homeaffairs.gov.au/nat-security/files/cic-factsheet-what-is-critical-infrastructure-centre.pdf

[Bell 1999] Bell, T.E., Dooling, D., Fouke, J. (1999). Threshold of the new millennium. *IEEE Spectrum, 36*(10), 59–64.

[Bessani 2008] Bessani, A., Sousa, P., Correia, M., Neves, N.F., Verissimo, P. (2008). The crutial way of critical infrastructure protection. *IEEE Security & Privacy*, *6*(6), 44–50. https://doi.org/10.1109/MSP.2008.158

[Betts 2006] Betts, B. (2006). Smart sensors New standard could save lives and money. *IEEE Spectrum, 43*(4), 50–53.

[Bialas 2016] Bialas, A. (2016). Risk management in critical infrastructure – foundation for its sustainable work. *Sustainability*, *8*(3), 240. https://doi.org/10.3390/su8030240

[Blackout 2003] Wikipedia. (n.d.). *Northeast electricity blackout* (2003, August 14). https://en.wikipedia.org/wiki/Northeast_blackout_of_2003

[BoozAllen 2011] Federal. (2011, October 11). *The cybersecurity dilemma: incentives to drive change rather than reactive response* [News]. Federal News Network. https://federalnewsnetwork.com/technology-main/2011/10/the-cybersecurity-dilemma-incentives-to-drive-change-rather-than-reactive-response

[Bouchecl] Bouchecl. (2009). *Map showing the regional reliability councils and interconnections in North America.* CC BY 3.0. https://commons.wikimedia.org/wiki/File:NERC-map-en.svg

[Brandle 2008] Brändle, M., Naedelle, M. (2008). Security for process control systems an overview. *IEEE Security & Privacy*, *6*(6), 24–29. https://doi.org/10.1109/MSP.2008.150

Building an Effective Security Program for Distributed Energy Resources and Systems: Understanding Security for Smart Grid and Distributed Energy Resources and Systems, Volume 1, First Edition. Mariana Hentea.

[Bygstad 2008] Bygstad, B. (2008). Information infrastructure as organization: a critical realist view. *ICIS 2008 Proceedings of the International Conference on Information Systems*, Paper 190, Paris, France, (December 2008). https://aisel.aisnet.org/icis2008/190

[Byres 2005] Byres, E.J., Franz, M. (2005). Finding the security holes before the hackers do. *ISA Technical Conference, Instrumentation Systems and Automation Society*, Chicago, IL, USA, (October 2005). Instrumentation Systems and Automation Society.

[Byres 2006] Byres, E.J., Hoffman, D., Kube, N. (2006). On shaky ground – a study of security vulnerabilities in control protocols. *5th American Nuclear Society International Topical Meeting on Nuclear Plant Instrumentation, Controls, and Human Machine Interface Technology*, Albuquerque, New Mexico, USA, (November 2006). American Nuclear Society.

[Byres 2012] Byres, E. (2012). SCADA security 2012 crystal ball [Blog]. *Tofino*. https://www.tofinosecurity.com/blog/scada-security-2012-crystal-ball.

[CA] PublicSafety. (n.d.). *Public safety Canada. Critical infrastructure*. https://www.publicsafety.gc.ca/cnt/ntnl-scrt/crtcl-nfrstrctr/index-en.aspx

[CAIDA 2003] CAIDA. (2003). *The spread of the Sapphire/Slammer worm*. https://www.caida.org/publications/papers/2003/sapphire/sapphire.html

[Canada] Wiki. (n.d.) *Disruption, National definition, Canada*. https://websites.fraunhofer.de/CIPedia/index.php/Critical_Infrastructure#Canada

[Caswell 2011] Caswell, J. (2011). Survey of *industrial control systems security* [Project Report]. http://www.cse.wustl.edu/~jain/cse571-11/ftp/ics/index.html

[Cedarbaum 2013] Cedarbaum, J.G., Schloss, L. (2013, April 22). *Implementation of the cybersecurity executive order and presidential policy directive: timetable and processes*. http://www.wilmerhale.com

[CERT 2003] US-CERT. (2003, February). *The National strategy to secure cyberspace*. https://www.us-cert.gov/sites/default/files/publications/cyberspace_strategy.pdf

[CERT 2007] US-CERT. (2007, October). *Business resilience: A more compelling argument for information security transcript* [Podcast]. Podcast Series. https://resources.sei.cmu.edu/library/asset-view.cfm?assetid=34714

[CIP Glossary] CIPRNet. (n.d.). *CIPRNet CIP Glossary*. Critical Infrastructure Preparedness and Resilience Research Network. https://ciprnet.eu/services/glossary/

[CIPedia] CIPedia. (n.d.). *Critical Infrastructure*. https://websites.fraunhofer.de/CIPedia/index.php/Critical_Infrastructure

[CIPRNET] CIPRNet. (n.d.). *Critical infrastructure preparedness and resilience research network*. https://www.ciprnet.eu/summary.html

[CNSSI 4009] CNSSI. (2010). *Committee on National Security Systems (CNSS) Information Assurance (IA) Glossary*. Committee on National Security Systems (CNSS) Instruction No. 4009. *Revised in 2015.

[Constantin 2012] Constantin, L. (2012, December 21). *Poor SCADA security will keep attackers and researchers busy in 2013* [News]. Computerworld. http://www.computerworld.com/article/2494135/malware-vulnerabilities/poor-scada-security-will-keep-attackers-and-researchers-busy-in-2013.html

[Coveney 1995] Coveney, P., Highfield, R. (1995). *Frontiers of Complexity: The Search for Order in A Chaotic World*. Random House.

[CVSS] NIST. (n.d.). *The common vulnerability scoring system (CVSS)*. https://nvd.nist.gov/cvss.cfm

[Davis-Felner 2014] Davis-Felner, K. (2014). *Five important things to know about the Wi-Fi® Internet of Everything* [White paper]. Wi-Fi Alliance. https://www.wi-fi.org/beacon/kelly-davis-felner/five-important-things-to-know-about-the-wi-fi-internet-of-everything

[DCSINT 2006] DCSINT. (2006, August 10). *Handbook No. 1.02 Critical Infrastructure Threats and Terrorism*. https://fas.org/irp/threat/terrorism/sup2.pdf

[DHS 2003] DHS. (2003, February). *The National strategy for the physical protection of critical infrastructures and key assets.* https://www.dhs.gov/xlibrary/assets/Physical_Strategy.pdf

[DHS 2011] DHS. (2011, May). *Common cybersecurity vulnerabilities in industrial control systems.* https://us-cert.cisa.gov/sites/default/files/recommended_practices/DHS_Common_Cybersecurity_Vulnerabilities_ICS_2010.pdf

[DHS 2013] DHS. (2013, January 8). *What is critical infrastructure?* *see update CRITICAL INFRASTRUCTURE SECTORS. https://www.cisa.gov/critical-infrastructure-sectors

[DHS 2015] DHS. (2015). *Critical 5 Role of Critical Infrastructure in National Prosperity Shared Narrative* (October 2015). https://www.cisa.gov/sites/default/files/publications/critical-five-shared-narrative-ci-national-prosperity-2015-508.pdf

[DHS Cyber] DHS. (n.d.). *National strategy to secure cyberspace.* https://www.dhs.gov/national-strategy-secure-cyberspace

[DHS C3] DHS. (n.d.). *Critical infrastructure cyber community C^3 voluntary program.* https://www.dhs.gov/ccubedvp

[DHS CIS] DHS. (n.d.). *Critical infrastructure sectors.* http://www.dhs.gov/critical-infrastructure-sectors

[DHS FACT] DHS. (n.d.). *Background.* https://www.cisa.gov/sites/default/files/publications/eo-13636-ppd-21-fact-sheet-508.pdf

[DNI 2014] DNI. (2014, January 29). *Worldwide threat assessment of the US Intelligence Committee.* https://www.dni.gov/files/documents/Intelligence%20Reports/2014%20WWTA%20%20SFR_SSCI_29_Jan.pdf

[DOE 2012] DOE. (2012, May 31). *The electricity subsector cybersecurity capability maturity model (ES-C2M2) Version 1.0.* * see also Version 1.1. https://www.energy.gov/sites/prod/files/Electricity%20Subsector%20Cybersecurity%20Capabilities%20Maturity%20Model%20%28ES-C2M2%29%20-%20May%202012.pdf

[DOE 2014a] DOE. (2014a, February). *U.S. Department of energy cybersecurity capability maturity model.* http://energy.gov/oe/cybersecurity-capability-maturity-model-c2m2-program/cybersecurity-capability-maturity-model-c2m2

[DOE 2014b] DOE. (2014b, February). *U.S. Department of energy oil and natural gas subsector cybersecurity capability maturity model.* https://www.energy.gov/oe/downloads/oil-and-natural-gas-subsector-cybersecurity-capability-maturity-model-february-2014

[DOE 2014c] DOE. (2014c, February). *U.S. Department of energy cybersecurity capability maturity model facilitator guide.* http://energy.gov/oe/downloads/cybersecurity-capability-maturity-model-facilitator-guide-february-2014

[DOE 2015] DOE. (2015, January). *Energy sector cybersecurity framework implementation guidance.* http://www.energy.gov/oe/downloads/energy-sector-cybersecurity-framework-implementation-guidance

[Dondossola 2013] Dondossola, G., Garrone, F., Szanto, J. (2013). Cyber risks in energy grid ICT infrastructures. In P. Théron, S. Bologna (Eds.), *Critical Information Infrastructure Protection and Resilience in the ICT Sector* (pp. 198–219). IGI Global. https://doi.org/10.4018/978-1-4666-2964-6.ch010

[Dunn 2003] Dunn, W.R. (2003). Designing safety-critical computer systems. *IEEE Computer, 36*(11), 40–46.

[ENERKNOL 2016] ENERKNOL. (2016, February 22). *FERC's revised critical infrastructure protection demands active vigilance.* https://www.powersystemsdesign.com/articles/fercs-revised-critical-infrastructure- protection-demands-active-vigilance/8/9855

[ENISA 2012] ENISA. (2012). *Cloud computing benefits, risks and recommendations for information security* (Version 2, rev. B). European Network and Information Security Agency. https://resilience.enisa.europa.eu/cloud-security-and-resilience/publications/cloud-computing-benefits-risks-and-recommendations-for-information-security

[ENISA 2015] ENISA. (2015, July). *Critical information infrastructures protection approaches in EU final document* (Version 1). TLP: Green. https://resilience.enisa.europa.eu/enisas-ncss-project/CIIPApproachesNCSS.pdf

[EO 13231] WhiteHouse. (2001, October 16). *Executive Order (EO) 13231 Critical Infrastructure Protection in the Information Age.* Federal Register The Daily Journal of United States Government. https://www.federalregister.gov/documents/2001/10/18/01-26509/critical-infrastructure-protection-in-the-information-age

[EO 13636] WhiteHouse. (2013, February 12). *Executive Order (EO) 13636 Improving Critical Infrastructure Cybersecurity.* https://www.whitehouse.gov/the-press-office/2013/02/12/executive-order-improving-critical-infrastructure-cybersecurity

[EO 13800] WhiteHouse. (2017, May 11). *Executive Order (EO) 13800 Strengthening the Cybersecurity of Federal Networks and Critical Infrastructure.* https://www.whitehouse.gov/presidential-actions/presidential-executive-order-strengthening-cybersecurity-federal-networks-critical-infrastructure/

[EPIC] EPIC. (2013, February 12). *PPD-21* [News]. EPIC. https://epic.org/foia/dhs/ppd-21.html

[Estonia 2007] Trainor, I. (2007, May 16). *Russia accused of unleashing cyberwar to disable Estonia* [World News]. The Guardian. http://www.theguardian.com/world/2007/may/17/topstories3.russia

[EU] EuropeanCommission. (n.d.). *European critical infrastructure.* https://ec.europa.eu/home-affairs/tags/critical-infrastructure_en

[EU 2008] EU. (2008, December 8). *Cyber security – European Directive EPCIP EU Directive 2008/114/EC.* Commission of the European Communities. https://eur-lex.europa.eu/LexUriServ/LexUriServ.do?uri=OJ:L:2008:345:0075:0082:EN:PDF

[EU 2009] Europa. (2009). *Operator Security Plan CIIP EU Communication COM(2009)149.* Commission of the European Communities. https://eur-lex.europa.eu/LexUriServ/LexUriServ.do?uri=COM:2009:0149:FIN:EN:PDF

[EUCOM 2005] Europa. (2005, November 17). *Green Paper on a European Programme for Critical Infrastructure Protection* (COM (2005) 576 Final). Commission of the European Communities. https://eur-lex.europa.eu/LexUriServ/LexUriServ.do?uri=COM:2005:0576:FIN:EN:PDF

[FACT SHEET] WhiteHouse. (2015, February 13). *FACT SHEET: White House Summit on Cybersecurity and Consumer Protection.* https://www.whitehouse.gov/the-press-office/2015/02/13/fact-sheet-white-house-summit-cybersecurity-and-consumer-protection

[Fairley 2016] Fairley, P. (2016). Cybersecurity at U.S. utilities due for an upgrade: Tech to detect intrusions into industrial control systems will be mandatory. *IEEE Spectrum, 53*(5), 11–13. https://doi.org/10.1109/MSPEC.2016.7459104

[Felps 1999] Felps, B. (1999, May 17). *Whiskers' caused satellite failure: galaxy IV outage blamed on interstellar phenomenon* [News]. WirelessWeek. https://web.archive.org/web/20090303130808/http://www.wirelessweek.com/article.aspx?id=104090

[Fisher 2014] Fisher, D. (2014, December 1). *Researcher Releases Database of Known Good ICS and SCADA Files* [News]. Threatpost. http://threatpost.com/researcher-releases-database-of-known-good-ics-and-scada-files/109652

[Forum 2009] Electricityforum. (2009, April 9). *Conficker infected critical hospital equipment* [News]. Electricity Forum. http://www.electricityforum.com/news/apr09/Confickerinfectedcriticalequipment.html

[GAO 2011] GAO. (2011). *Critical Infrastructure protection cybersecurity guidance is available, but more can be done to promote its use* (GAO-12-92). United States Government Accountability Office. http://www.gao.gov/assets/590/587529.pdf

[Gellman 2002] Gellman, B. (2002, June 27). *Cyber-attacks by Al Qaeda feared terrorists at threshold of using Internet as tool of bloodshed, experts say* [News]. Washington Post (Page AD1).

[Gell-Mann 1994] Gell-Mann, M. (1994). Complex adaptive systems. In G.A. Cowan, D. Pines, D. Meltzer (Eds.), *Complexity: Metaphors, Models and Reality* (pp. 17–45). Addison-Wesley.

[Georgia 2008] Kirk, J. (2008, July 21). *Georgia president's web site falls under DDOS attack* [News]. Computerworld. http://www.computerworld.com/article/2534930/networking/georgia-president-s-web-site-falls-under-ddos-attack.html

[Givens 2011] Givens, A. (2011, May 27). *Deepwater horizon oil spill is an ominous sign for critical infrastructure's future* [News]. Emergency Management. http://www.emergencymgmt.com/disaster/Deepwater-Horizon-Oil-Spill-Critical-Infrastructure-052711.html

[Goertzel 2007] Goertzel, K.M. et al. (2007). Software Security Assurance: State-of-the-Art Report (SOAR). *Information Assurance Technology Analysis Center/Data and Analysis Center for Software.*

[Goldman 2013] Goldman, D. (2013, January 9). *Hacker hits on U.S. power and nuclear targets spiked in 2012.* CNN Business The Cybercrime Economy. https://money.cnn.com/2013/01/09/technology/security/infrastructure-cyberattacks/

[Gorman 2009] Gorman, S. (2009, April 4). *Electricity grid in U.S. penetrated by spies* [News]. Wall Street Journal. Technology. http://online.wsj.com/article/SB123914805204099085.html

[Grose 2009] Grose, T.K. (2009). The cyber grid. *ASSE Prism, 19*(2), 26–31.

[Hanseth 2002] Hanseth, O. (2002). *From systems and tools to networks and infrastructures – from design to cultivation Towards a theory of ICT solutions and its design methodology implications.* http://heim.ifi.uio.no/~oleha/Publications/ib_ISR_3rd_resubm2.html

[Harp 2015] Harp, D., Gregory-Brown, B. (2015). *The state of security in control systems today a SANS survey* (June 2015). https://www.sans.org/webcasts/state-security-control-systems-today-survey-webcast-99817

[Hathaway 2009] Hathaway, M. (2009, December 21). *Five myths about cybersecurity.* Belfer Center for Science and International Affairs. https://www.belfercenter.org/publication/five-myths-about-cybersecurity

[Heller 2001] Heller, M. (2001). Interdependencies in civil infrastructure systems. *Frontiers in Engineering, 31*(4), 47–55.

[Hentea 2008] Hentea, M. (2008). Improving security for SCADA control systems. *Interdisciplinary Journal of Information, Knowledge, and Management, 3,* 73–86. https://doi.org/10.28945/3185

[Huang 2009] Huang, Y.-L., Cárdenas, A.A., Amin, S., Lin, Z.-S., Tsai, H.-Y., Sastrya, S. (2009). Understanding the physical and economic consequences of attacks against control systems. *International Journal of Critical Infrastructure Protection, 2*(2), 69–134.

[IC3] IC3. *Internet crime complaint center.* https://www.ic3.gov/about/default.aspx

[ICANN] ICANN. *Internet corporation for assigned names and numbers.* https://www.icann.org/

[ICS-CERT 2011] ICS-CERT. (2011). *ICS-CERT monthly monitor October 2011.* https://us-cert.cisa.gov/sites/default/files/Monitors/ICS-CERT_Monitor_Oct2011.pdf

[ICS-CERT 2012] ICS-CERT. (2012). *Monitor April* (ICS-MM201204). https://ics-cert.us-cert.gov/monitors/ICS-MM201204

[ICS-CERT 2014a] ICS-CERT. (2014a). *Monitor January—April* (ICS-MM201404). https://ics-cert.us-cert.gov/monitors/ICS-MM201404

[ICS-CERT 2014b] ICS-CERT. (2014b). *Monitor May—August* (ICS-MM201408). https://ics-cert.us-cert.gov/monitors/ICS-MM201408

[ICS-CERT 2015a] ICS-CERT. (2015a). *Monitor September—October* (ICS-MM201510). https://ics-cert.us-cert.gov/sites/default/files/Monitors/ICS-CERT_Monitor_Sep-Oct2015.pdf

[ICS-CERT 2015b] ICS-CERT. (2015b). *Monitor November—December* (ICS-MM201512). https://ics-cert.us-cert.gov/sites/default/files/Monitors/ICS-CERT_Monitor_Nov-Dec2015_S508C.pdf

[ICS-CERT 2016a] ICS-CERT. (2016a). *Monitor January—February* (ICS-MM201602). https://ics-cert.us-cert.gov/monitors/ICS-MM201602

[ICS-CERT 2016b] ICS-CERT. (2016b). *Monitor March—April* (ICS-MM201604). https://ics-cert.us-cert.gov/monitors/ICS-MM201604

[ICS-CERT Alerts] ICS-CERT. (2012). *ICS-CERT Alerts*. https://ics-cert.us-cert.gov/alerts

[ISO 31000] *ISO 31000:2009, Risk management – Principles and guidelines.*

[IntelCrawler] IntelCrawler. (n.d.). *Cyber threat intelligence*. https://www.crunchbase.com/organization/intelcrawler

[Interconnections] DOE. (n.d.). *Interconnections. Learn more about interconnections.* http://energy.gov/oe/information-center/recovery-act/recovery-act-interconnection-transmission-planning/learn-more

[ISF 2013] ISF. (2013). *Information Security Forum*. https://www.securityforum.org/

[ISO/IEC 27005] *ISO/IEC 27005:2018 Information Technology – Security Techniques – Information Security Risk Management* (third edition).

[Italy 2011] Legislative Decree n. 61 from 11 April 2011 following the Directive 2008/114/EC. *Italian Official Journal, 4* (May 2011, n. 102).

[Jackson 2013] Jackson, S. (2013). Resilience principles for the ICT sector. In P. Théron, S. Bologna (Eds.), *Critical Information Infrastructure Protection and Resilience in the ICT Sector* (pp. 36–49). IGI Global. https://doi.org/10.4018/978-1-4666-2964-6.ch002

[Johnson 2013] Johnson, C.W. (2013). The telecoms inclusion principle: the missing link between critical infrastructure protection and critical information infrastructure protection. In P. Théron, S. Bologna (Eds.), *Critical Information Infrastructure Protection and Resilience in the ICT Sector* (pp. 277–303). IGI Global. https://doi.or/10.4018/978-1-4666-2964-6.ch014

[Jonkeren 2013] Jonkeren, O., Ward, D. (2013). Modelling economic consequences of ICT infrastructure failure in support of critical infrastructure protection policies. In P. Théron, S. Bologna (Eds.), *Critical Information Infrastructure Protection and Resilience in the ICT Sector* (pp. 115–138). IGI Global. https://doi.org/10.4018/978-1-4666-2964-6.ch006

[Kaspersky 2014] Kaspersky. (2014). *Cyber threats to ICS Systems You don't have to be a target to become a victim industrial security*. Industrial Security. https://media.kaspersky.com/en/business-security/critical-infrastructure-protection/Cyber_A4_Leaflet_eng_web.pdf

[Kirk 2014] Kirk, J. (2014, January 7). *Satellite links for remote networks may pose soft target for attackers* [News]. Computerworld. https://www.computerworld.com/article/2487472/satellite-links-for-remote-networks-may-pose-soft-target-for-attackers.html

[Koopman 2004] Koopman, P. (2004). Embedded system security. *Computer, 37*(7), 95–99.

[Kovacs 2016] Kovacs, E. (2016, January 20). *Critical infrastructure incidents increased in 2015: ICS-CERT* [News]. SecurityWeek. https://www.securityweek.com/critical-infrastructure-incidents-increased-2015-ics-cert

[Kowalski 2008] Kowalski, E.F., Cappelli, D.M., Moore, A.P. (2008). *Insider threat study: illicit cyber activity in the information technology and telecommunications sector joint SEI and U.S. secret service report* (January 2008).

[Krancer 2015] Krancer, M.L. (2015, November 4). *The biggest cybersecurity threat: the energy sector.* [News]. Forbes/Energy. http://www.forbes.com/sites/michaelkrancer/2015/11/04/the-biggest-cybersecurity-threat-the-energy-sector/#4f8d501260ba

[Lala 2009] Lala, J.H., Schneider, F.B. (2009). IT monoculture security risks and defenses. *IEEE Security & Privacy*, *7*(1), 12–13. https://doi.org/10.1109/MSP.2009.11

[Landau 2008] Landau, S. (2008). Security and privacy landscape in emerging technologies. *IEEE Security & Privacy*, *6*(4), 74–77. https://doi.org./10.1109/MSP.2008.95

[Leighton 2009] Leighton, T. (2009). *Akamai and cloud computing: a perspective from the edge of the cloud* [White Paper]. Akamai Technologies. http://docshare01.docshare.tips/files/4393/43936871.pdf

[Luiijf 2013] Luiijf, E. (2013). Next generation information-based infrastructures: new dependencies and threats. In P. Théron, S. Bologna (Eds.), *Critical Information Infrastructure Protection and Resilience in the ICT Sector* (pp. 304–317). IGI Global. https://doi.org/10.4018/978-1-4666-2964-6.ch015

[Maloor 2012] Maloor, P. (2012). Protection of critical national infrastructure. [Paper presentation]. *Joint APT-ITU Workshop on the International Telecommunications Regulations (ITRs)*, Bangkok, Thailand, (February 2012).

[McGregor 2014] McGregor, J. (2014, July 28). *The top 5 most brutal cyber attacks of 2014 so far* [News]. Forbes. https://www.forbes.com/sites/jaymcgregor/2014/07/28/the-top-5-most-brutal-cyber-attacks-of-2014-so-far/

[Milis 2012] Milis, G.M., Kyriakides, E., Hadjiantonis, A.M. (2012). Electrical power systems protection and interdependencies with ICT. In A.M. Hadjiantonis, B. Stiller (Eds.), *Telecommunications Economics. Lecture Notes in Computer Science* (Vol *7216*, pp. 216–228). Springer. https://doi.org/10.1007/978-3-642-30382-1_27

[Mimoso 2015] Mimoso, M. (2015, February 17). *First Arabic Cyberespionage Operation Uncovered* [News]. Threatpost. http://threatpost.com/first-arabic-cyberespionage-operation-uncovered/111068

[Mohan 2011] Mohan, R. (2011, April 6). *Two Years after the Conficker Worm, Are We Still at Risk?* [News]. SecurityWeek. http://www.securityweek.com/two-years-after-conficker-worm-are-we-still-risk

[Moore 2003] Moore, D., Paxson, V., Savage, S., Shannon, C., Staniford, S., Weaver, N. (2003). Inside the slammer worm. *IEEE Security & Privacy*, *1*(4), 33–39. https://doi.org/10.1109/MSECP.2003.1219056

[Moteff 2015] Moteff, J.D. (2015, June 10). *Critical infrastructures: background, policy, and implementation* (RL30153). Congressional Research Service. https://www.fas.org/sgp/crs/homesec/RL30153.pdf

[Nakashima 2011] Nakashima, E. (2011, November 25). *Water-pump failure in Illinois wasn't cyberattack after all* [News]. Washington Post. http://www.washingtonpost.com/world/national-security/water- pump-failure-in-illinois-wasnt-cyberattack-after-all/2011/11/25/gIQACgTewN_story.html

[Nakashima 2013] Nakashima, E. (2013, February 10). *U.S. said to be target of massive cyber-espionage campaign* [News]. *Washington Post*. https://www.washingtonpost.com/world/national-security/us-said- to-be-target-of-massive-cyber-espionage-campaign/2013/02/10/7b4687d8-6fc1-11e2-aa58-243de81040ba_story.html

[NERC CIP] NERC. (n.d.). *Critical infrastructure protection committee* (CIPC). CIP Implementation. https://www.nerc.com/comm/CIPC/Pages/default.aspx

[NERC Glossary] NERC. (n.d.) *Glossary of Terms Used in NERC Reliability Standards*. http://www.nerc.com/files/Glossary_of_Terms.pdf

[NIPP 2009] DHS. (2009). *National infrastructure protection plan (2009): partnering to enhance protection and resiliency*. U.S. Department of Homeland Security. https://www.cisa.gov/publication/nipp-2009-partnering-enhance-protection-resiliency

[NIPP 2013] DHS. (2013). *National infrastructure protection plan (NIPP): 2013 partnering for critical infrastructure security and resilience*. U.S. Department of Homeland Security. https://www.cisa.gov/publication/nipp-2013-partnering-critical-infrastructure-security-and-resilience

[NIST 2014] *NIST Framework for Improving Critical Infrastructure Cybersecurity Version 1.0* (February 2014). https://www.nist.gov/system/files/documents/cyberframework/cybersecurity-framework-021214.pdf

[NIST 2016] NIST. (2016). *Analysis of cybersecurity framework RFI responses* (March 2016). http://www.nist.gov/cyberframework/upload/RFI3_Response_Analysis_final.pdf

[NIST 2018] *NIST Framework for Improving Critical Infrastructure Cybersecurity Version 1.1* (April 2018). https://nvlpubs.nist.gov/nistpubs/CSWP/NIST.CSWP.04162018.pdf

[NIST 2019] *NIST Privacy Framework A Tool for Improving Privacy through Enterprise Risk Management Preliminary draft* (September 6, 2019).

[NISTIR 7628r1] *NIST Interagency Report (NISTIR) 7628r1 Guidelines for Smart Grid Cybersecurity* (Revision 1, September 2014). http://dx.doi.org/10.6028/NIST.IR.7628r1

[NISTIR 7761] *NIST Interagency Report (NISTIR) 7761 Guidelines for Assessing Wireless Standards for Smart Grid Applications* (Revision 1, February 2011). http://collaborate.nist.gov/twiki-sggrid/bin/view/SmartGrid/SGIPCosSIFNISTIR7761

[NIST SP800-39] *NIST Special Publication (SP) 800-39 Managing Information Security Risk Organization, Mission, and Information System View* (March 2011). https://nvlpubs.nist.gov/nistpubs/Legacy/SP/nistspecialpublication800-39.pdf

[NIST SP800-53r4] *NIST Special Publication (SP) 800-53 Security and Privacy Controls for Federal Information Systems and Organizations Revision 4*, (April 2013). Computer Security Division Information Technology Laboratory. http://dx.doi.org/10.6028/NIST.SP.800-53r4 * Superseded by [NIST SP800-53r5] in September 2020; Withdrawal Date September 23, 2021.

[OECD 2008a] OECD. (2008a). *Protection of 'Critical Infrastructure' and the Role of Investment Policies Relating to National Security* (May 2008). https://www.oecd.org/daf/inv/investment-policy/40700392.pdf

[OECD 2008b] OECD. (2008b). *Development of Policies for Protection of Critical Information Infrastructures*, Seoul, Korea, (June 2008). http://www.oecd.org/internet/ieconomy/40761118.pdf

[OECD 2008c] OECD. (2008c). *OECD Recommendation of the Council on the Protection of Critical Information Infrastructures*. http://www.oecd.org/sti/40825404.pdf

[OECD 2011] OECD. (2011). *Reducing Systemic Cybersecurity Risk*. Organisation for Economic Co-operation and Development. http://www.oecd.org/dataoecd/57/44/46889922.pdf

[Omar 2016] Omar, M. (2016). Insider Threats: Detecting and Controlling Malicious Insiders. In M. Dawson, M. Omar (Eds.), *New Threats and Countermeasures in Digital Crime and Cyber Terrorism* (pp. 162–172). IGI Global. http://doi:10.4018/978-1-4666-8345-7.ch009

[P2030 2011] *IEEE Std 2030-2011-IEEE Guide for Smart Grid Interoperability of Energy Technology and Information Technology Operation with the Electric Power System (EPS), End-Use Applications, and Loads.*

[Paganini 2014] Paganini, P. (2014, January 9). *VSAT Terminals Are Opened for targeted Cyber Attacks* [News]. Security Affairs. https://securityaffairs.co/wordpress/21049/hacking/vsat-terminals-opened-targeted-cyber-attacks.html

[Palmer 2008] Palmer, C.C., Trellue, R. (2008). Process control system security bootstrapping a legacy. *IEEE Security & Privacy*, *6*(6), 22–23. https://doi.org/10.1109/MSP.2008.148

[Perrow 2006] Perrow, C. (2006). Shrink the targets. IEEE Spectrum, *43*(9), 46–49. https://doi.org/10.1109/MSPEC.2006.1688258

[Poulsen 2003] Poulsen, K. (2003, August 19). *Slammer worm crashed Ohio nuke plant network* [News]. SecurityFocus. https://www.securityfocus.com/news/6767

[PPD-8] WhiteHouse. (2011, March 30). *The White House Presidential Policy Directive/PPD-8 – National Preparedness*. http://www.dhs.gov/xlibrary/assets/presidential-policy-directive-8-national-preparedness.pdf

[PPD-21] WhiteHouse. (2013, February 12). *The White House Presidential Policy Directive/PPD-21 – Critical Infrastructure Security and Resilience*. http://www.whitehouse.gov/the-press-office/2013/02/12/presidential-policy-directive-critical-infrastructure-security-and-resil

[PDD-63] WhiteHouse, (1998, May 22). *The White House Presidential Decision Directive/PDD- 63 Critical Infrastructure Protection*. https://clinton.presidentiallibraries.us/items/show/12762

[PwC 2016a] PwC. (2016a). *Global economic crime survey 2016: US results adjusting the lens on economic crime* [Report]. http://www.pwc.com/us/en/forensic-services/economic-crime-survey-us-supplement.html

[PwC 2016b] PwC. (2016b). *Turnaround and transformation in cybersecurity: Power and utilities Key findings from The Global State of Information Security® Survey 2016. Cybersecurity, Privacy, Forensics* [Report]. https://www.pwc.com/us/en/services/consulting/cybersecurity.html

[Radware 2015] Radware. (2015). *Global Application & Network Security Report 2014-2015* [Report]. http://docs.media.bitpipe.com/io_12x/io_122872/item_1121486/Radware_ERT_Report_2014-2015.pdf

[Ray 2010] Ray, P.D., Harnoor, R., Hentea, M. (2010). Smart power grid security: a unified risk management approach. In D.A. Pritchard, L.D. Sanson (Eds.), *44th Annual 2010 IEEE International Carnahan Conference on Security Technology Proceedings* (pp. 276–285), San Jose, California, USA, (October 2010). https://doi.org.10.1109/CCST.2010.5678681

[Reuters 2012] Reuters. (2012, March 3). *NASA says was hacked 13 times last year* [News]. Reuters. http://www.reuters.com/article/us-nasa-cyberattack-idUSTRE8211G320120303

[RFC 4949] *IETF Request for Comments (RFC) 4949 Internet Security Glossary* (Version 2, August 2007).

[Rieger 2013] Rieger, C. (2013, November 13). *A multi-agent approach to smart grid control architecture* [Newsletter Article]. https://resourcecenter.smartgrid.ieee.org/publications/newsletters/SGNL0140.html

[Rinaldi 2001] Rinaldi, S.M., Peerenboom, J.M., Kelly, T.K. (2001). Identifying, understanding and analyzing critical infrastructure interdependencies. *IEEE Control Systems Magazine*, *21*(6), 11–25. https://doi.org/10.1109/37.969131

[Robles 2008] Robles, R.J, Choi, M-K., Cho, E-S., Kim, S-S., Park, G-C., Lee, J-H. (2008). Common threats and vulnerabilities of critical infrastructures. *International Journal of Control and Automation*, *1*(1), 17–22.

[Rosson 2019] Rosson, J., Rice, M., Lopez, J., Foss, D. (2019). Incentivizing cyber security investment in the power sector using an extended cyber insurance framework. *Homeland Security Affairs 15* (Article 2, May 2019). https://www.hsaj.org/articles/15082

[Safety Glossary] IAEA. (2007). *IAEA Safety Glossary*. International Atomic Energy Agency. http://www-pub.iaea.org/mtcd/publications/pdf/pub1290_web.pdf

[Sanger 2013] Sanger, D.E., Perlroth, N. (2013, July 13). *Nations buying as hackers sell flaws in computer code* [News]. New York Times. http://www.nytimes.com/2013/07/14/world/europe/nations-buying-as-hackers-sell-computer-flaws.html?_r=0

[SANS 2014] SANS. (2014). *Breaches on the rise in control systems: A SANS survey*. https://ics.sans.org/media/sans-ics-security-survey-2014.pdf

[Santamarta 2013] Santamarta, R. (2013). *A Wake-up Call for SATCOM Security* [White paper]. IOActive. http://www.ioactive.com/pdfs/IOActive_SATCOM_Security_WhitePaper.pdf

[Schachtman 2011] Schachtman, N. (2011, October 7). *Exclusive: Computer Virus Hits U.S. Drone Fleet* [News]. Wired. http://www.wired.com/dangerroom/2011/10/virus-hits-drone-fleet/

[Schneider 2009] Schneider, D. (2009). Is this the moment for broadband over power lines? *IEEE Spectrum*, *46*(7), 17. https://doi.org/10.1109/MSPEC.2009.5109438

[Schneier 2000] Schneier, B. (2000). The process of security. *Information Security Magazine*, (April 2000). https://www.schneier.com/essays/archives/2000/04/the_process_of_secur.html

[Sen 2016] Sen, J. (2016). *Security and privacy issues in cloud computing*. Innovation Labs, Tata Consultancy Services Ltd. https://arxiv.org/ftp/arxiv/papers/1303/1303.4814.pdf

[Shipley 2001] Shipley, P., Garfinkel, S.L. (2001). *An analysis of dial-up modems and vulnerabilities*. https://www.researchgate.net/publication/239563561_An_Analysis_of_Dial-Up_Modems_and_Vulnerabilities

[SPP 2010] DOE. (2010). *Energy Sector Specific Plan 2010 An Annex to the National Infrastructure Protection Plan*. http://energy.gov/sites/prod/files/oeprod/DocumentsandMedia/Energy_SSP_2010.pdf

[SPP Glossary] DHS. (2010). *Appendix 1: Glossary of Key Terms of SPP* (2010) *Energy Sector Specific Plan 2010 An Annex to the National Infrastructure Protection Plan*. http://www.dhs.gov/xlibrary/assets/nipp-ssp-energy-2010.pdf

[Stiennon 2013] Stiennon, R. (2013, February 14). *PPD 21: extreme risk management gone bad* [News]. Forbes. http://www.forbes.com/sites/richardstiennon/2013/02/14/ppd-21-extreme-risk-management-gone-bad/2/#7c1fd508577d

[Storm 2013] Storm, D. (2013, July 2). *Brute-force cyberattacks against critical infrastructure, energy industry, intensify* [Opinion]. *Computerworld*. http://www.computerworld.com/article/2473941/cybercrime-hacking/brute-force-cyberattacks-against-critical-infrastructure--energy-industry--intens.html

[Storm 2014] Storm, D. (2014, January 15). Hackers exploit SCADA holes to take full control of critical infrastructure [Opinion]. *Computerworld*. http://www.computerworld.com/article/2475789/cybercrime-hacking/hackers-exploit-scada-holes-to-take-full-control-of-critical-infrastructure.html

[Suter 2007] Suter, M. (2007). *A generic national framework for critical information infrastructure protection (CIIP)*. https://www.itu.int/ITU-D/cyb/cybersecurity/docs/generic-national-framework-for-ciip.pdf

[Szoldra 2015] Szoldra, P. (2015, December 29). *The 9 worst cyber attacks of 2015* [News]. TechInsider. https://www.businessinsider.com/cyberattacks-2015-12

[TISN 2007] TISN. (2007). *Secure your information: information security principles for enterprise architecture 2007 report*. https://static.aminer.org/pdf/PDF/000/305/486/application_of_security_principles_to_integration_of_enterprise_system_and.pdf

[Tofino] Byres, E. (2012, July 5). #1 ICS and SCADA Security Myth: Protection by Air Gap [Blog]. *Tofinosecurity*. https://www.tofinosecurity.com/blog/1-ics-and-scada-security-myth-protection-air-gap

[TrendMicro 2014] TrendMicro. (2014, July 23). Breaking down old and new threats to critical infrastructure [Blog]. *TrendMicro*. https://blog.trendmicro.com/breaking-down-old-and-new-threats-to-critical-infrastructure

[TrendMicro 2015] TrendMicro. (2015). *Report on Cybersecurity and Critical Infrastructure in the Americas, Organization of American States (OAS)* [Report]. https://www.sites.oas.org/cyber/Documents/2015%20-%20OAS%20Trend%20Micro%20Report%20on%20Cybersecurity%20and%20CIP%20in%20the%20Americas.pdf

[UK] Wikipedia. (n.d.). *Critical Infrastructure. United Kingdom*. https://en.wikipedia.org/wiki/Critical_infrastructure#United_Kingdom

[US] Wikipedia. (n.d.). *Critical Infrastructure. United States*. https://en.wikipedia.org/wiki/Critical_infrastructure#United_States

[US-CERT] US-CERT. (2003). *The National strategy to secure cyberspace February 2003*. https://www.us-cert.gov/sites/default/files/publications/cyberspace_strategy.pdf

[Verizon 2016] Verizon. (2016). *Data breach digest scenarios from the field*. VERIS. http://veriscommunity.net

[Vixie 2002] Vixie, P., Sneeringer, G., Schleifer, M. (2002). *Events of 21-Oct-2002* (ISC/UMD/Cogent, November 24, 2002). https://web.archive.org/web/20110302164416/http://www.isc.org/f-root-denial-of-service-21-oct-2002

[Watts 2003] Watts, D. (2003). Security & vulnerability in electric power systems. *NAPS 2003, 35th North American Power Symposium* (pp. 559—566), University of Missouri-Rolla, Rolla, Missouri, (October 2003).

[Whitty 2015] Whitty, M.T. (2015). Mass-marketing fraud: a growing concern. *IEEE Security and Privacy, 13*(4), 84–87. https://doi.org/10.1109/MSP.2015.85

[Wi-Fi 2014] Wi-Fi Alliance. (2014). *Connect your life: Wi-Fi® and the Internet of Everything*. [White paper]. Wi-Fi Alliance. http://www.wi-fi.org/system/files/wp_Wi-Fi_Internet_of_Things_Vision_20140110.pdf

[Wi-Fi 2018] Wi-Fi. (2018). *Next generation Wi-Fi®: The future of connectivity*. [White paper]. Wi-Fi Alliance. https://www.wi-fi.org/download.php?file=/sites/default/files/private/Next_generation_Wi-Fi_White_Paper_20181218.pdf

[Wilhoit 2013] Wilhoit, K. (2013). *The SCADA that didn't cry wolf who's really attacking your ICS equipment? (Part 2)* [Research paper]. https://media.blackhat.com/us-13/US-13-Wilhoit-The-SCADA-That-Didnt-Cry-Wolf-Whos-Really-Attacking-Your-ICS-Devices-Slides.pdf

[Wueest 2014] Wueest, C. (2014). *Security Response Targeted Attacks Against the Energy Sector Version 1.0* [White Paper]. http://www.symantec.com/content/en/us/enterprise/media/security_response/whitepapers/targeted_attacks_against_the_energy_sector.pdf

[Zetter 2011] Zetter, K. (2011, March 22). *Attack Code for SCADA Vulnerabilities Released Online* [News]. Wired. http://www.wired.com/threatlevel/2011/03/scada-vulnerabilities/

[Zuckerman 1998] Zuckerman, L. (1998, May 21). *Satellite failure is rare, and therefore unsettling* [News]. The New York Times. http://www.nytimes.com/1998/05/21/business/satellite-failure-is-rare-and-therefore-unsettling.html

Part IV

The Characteristics of Smart Grid and DER Systems

8

Smart Power Grid

8.1 Electric Power Grid

A broad definition of electric power system is the system that deals with generation, transmission, and distribution of electrical energy. Electric power system is a complex assemblage of equipment and circuits for generating, transmitting, transforming, and distributing electrical energy [Encyclopedia], [Free Dictionary]. Separate systems are interconnected, and higher levels of bulk power system (BPS) reliability are attained through properly coordinated interconnections among separate systems.

Electric power is generated in generating stations and then carried to consumers via transmission system. Power delivered by transmission circuits is stepped down to lower voltages in facilities called substations. Distribution system takes power from a bulk power substation to customers' switches for industrial and residential use.

The electric power grid refers to the electricity infrastructure that lies between the generation sources and the consumer (e.g. transmission and distribution or electricity delivery). Figure 8.1 depicts a hierarchical view of electrical power grid system.

8.1.1 Power System Services

The services provided within a power system are classified in terms of destination, commercial feature, and the entity that provides them. From the commercial point of view, the power system services can be classified in two categories [Venkata 2013]:

- System services are those services provided by the system operator and transmission operators using some functions or means designed to monitor and control the continuous operation of the power system necessary to support the electrical energy supply to consumers while ensuring the secure conditions at the lowest costs.
- Ancillary services are those services procured by the system operator from the network users, producers, and consumers by means of which the system operator is carrying out its functions to coordinate the transport of the electrical energy from producers to competitive markets and the providers are financially remunerated. Examples of services include frequency control, voltage control, active power replacement in case of blackouts, and blackout start capability. The ancillary services are system support services necessary for the operation of a transmission or distribution system. These services are required to provide system reliability and power quality. They are provided by generators and system operators.

8.1.2 Power System Operations

Traditionally, managing power distribution networks had consisted mainly of performing planned switching and restoring systems after unplanned outages using manual paper-based methods. The early

Building an Effective Security Program for Distributed Energy Resources and Systems: Understanding Security for Smart Grid and Distributed Energy Resources and Systems, Volume 1, First Edition. Mariana Hentea.

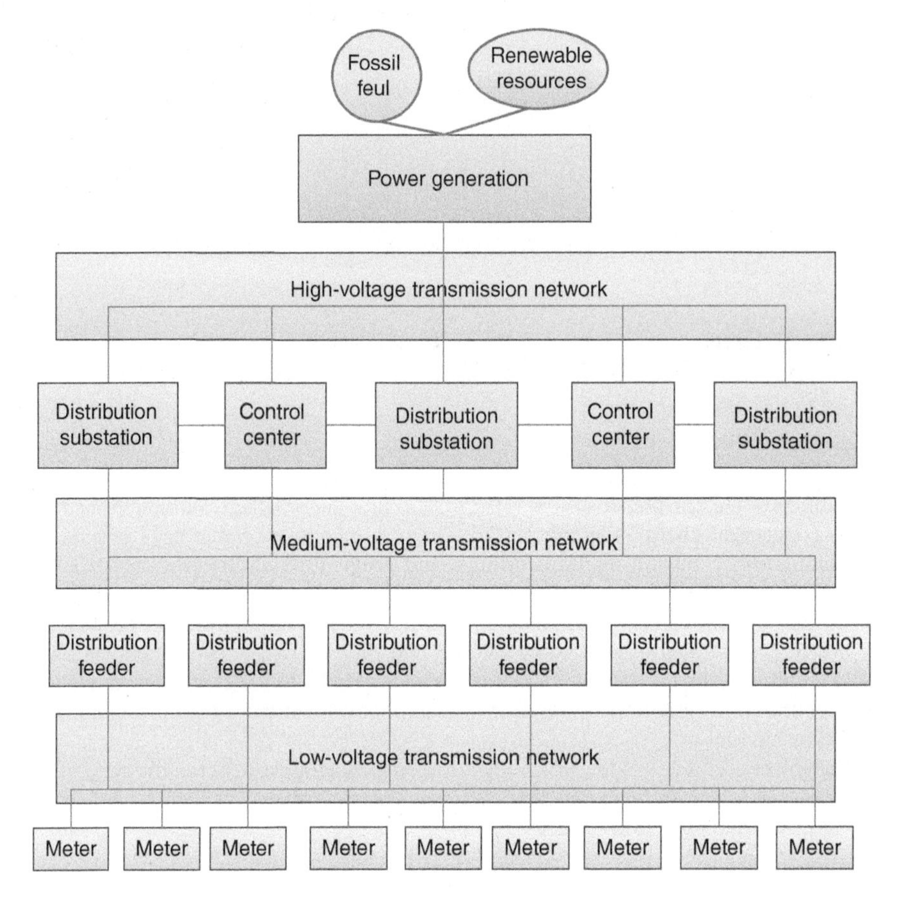

Figure 8.1 Electrical grid system overview. *Source:* [Gao 2012]. © 2012, Elsevier.

power systems were simple with isolated generation and load. The monitoring and control of a power system from a centralized control center became desirable early in the development of electric power systems. Then, these grew into separate grids that were then interconnected.

The operation and control of the generation, transmission, and distribution grid is quite complex because this large system has to operate in synchronism and because many different organizations are responsible for different portions of the grid. In North America and Europe, many public and private electric power companies are interconnected, often across national boundaries. Thus, many organizations have to coordinate to operate the grid, and this coordination can take many forms, from a loose agreement of operational principles to a strong pooling arrangement of operating together. Power system operations can be classified in three stages:

- Operations planning that includes optimal scheduling of generation resources to meet anticipated demand in the next few hours, weeks, or months.
- Real-time control to respond to actual demand of electricity and any unforeseen contingencies (equipment outages) including maintaining the security of the system so that a contingency cannot disrupt a power supply.
- After-the-fact accounting is the tracking of purchases and sales of energy between organizations so that billing can be generated.

Electrical utilities are facilities that play a key role in the electric power system. As electrical utilities became more interconnected and evolved into complex networks, the control center became the operations headquarters for each utility. This system is referred to as energy management system (EMS). Figure 8.2 depicts

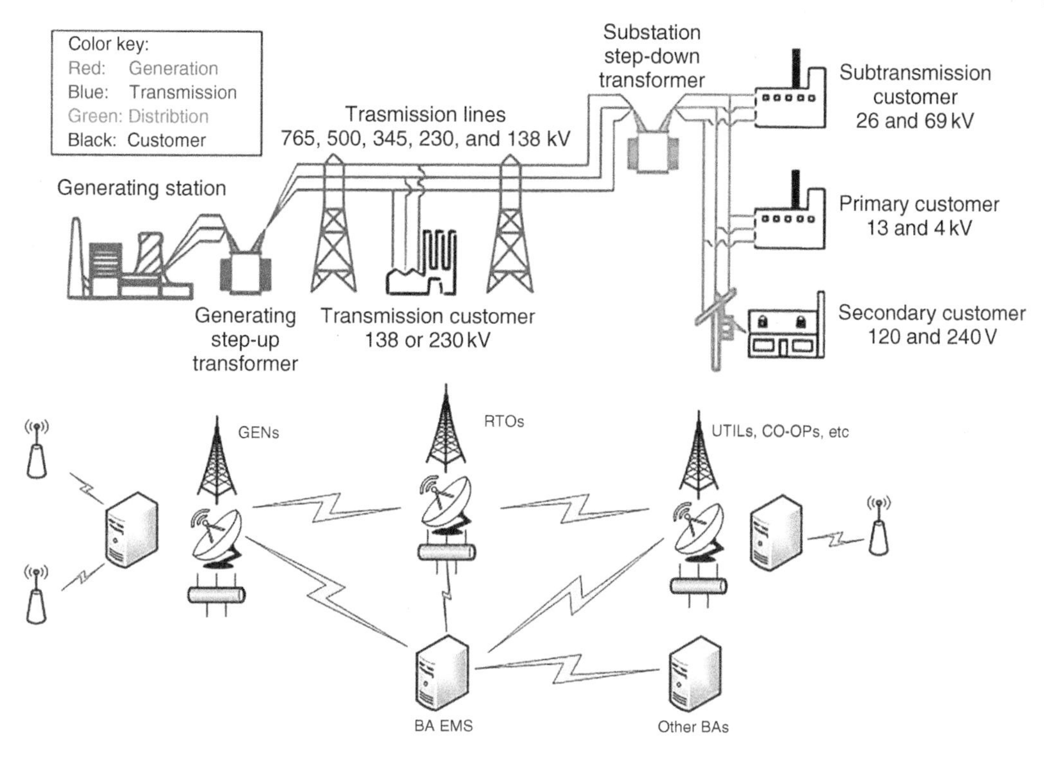

Figure 8.2 Delineation of power grid systems. *Source:* [DOE 2016]. Public Domain.

two view layers: a view of power grid (top) and communications (low) necessary among generators, regional transmission organizations, utilities, etc. Computers and networks are used to manage, monitor, protect, and control the continuous real-time delivery of electrical power.

Operational technology (OT) computers and networks for energy delivery systems allow operators to maintain situational awareness, perform economic dispatch of energy resources, plan for contingencies, and balance generation with load in real time. These capabilities are often provided by EMS that resides in a utility control center and performs state estimation, contingency analysis, and automatic generation control (AGC).

The EMS performs functions such as state estimation, contingency analysis, and automatic generation control. It receives data from a SCADA system that acquires power system operating measurements periodically (every two to five seconds) from specialized devices in substations. The EMS state estimator uses this data to estimate the operational state of the power grid every few minutes. This information provides operators with the situational awareness to make informed decisions.

Smart Grid network communication refers to a variety of public and private communication networks, both wired and wireless, that will can be used for Smart Grid domains and subdomains. The increasing penetration of distributed energy resources and electrical vehicles brings new challenges and requires the evolution of the electricity distribution grids to enable their full utilization and effective automation.

Therefore, hybrid network architectures based on wireless and wired networks are promising solutions to Smart Grid communication infrastructures due to the balanced trade-off between investments and benefits and meeting the critical requirements of the Smart Grid applications [Zhang 2018].

Many utilities are leveraging existing communication infrastructure from telecommunications companies to provide connectivity between generation plants and control centers, between substations and control centers (particularly SCADA), and increasingly between pole top AMI collectors and AMI headend systems, and pole top distribution automation equipment and distribution management systems.

As data is communicated among different entities, new capabilities for enhanced grid include reliability, resiliency, and efficiency. However, reliability requires cybersecurity. This work [Hawk 2014] discusses different proposed solutions for the Smart Grid cybersecurity with the vision of surviving a cyber attack while sustaining critical energy delivery functions.

8.1.3 Energy Management System Overview

The EMS receives data from a supervisory control and data acquisition (SCADA) system that acquires power system operating measurements every two to five seconds from specialized devices in substations. The EMS state estimator uses SCADA data conveyed from other utility's control centers and the laws of physics to estimate the operational state of the power grid every few minutes. This information provides operators with the situational awareness to make informed decisions, such as optimized power flow for economic and efficient generation dispatch. State estimators also detect and reject corrupted data from malfunctioning sensors. New methods, such as security-oriented cyber–physical state estimation (SCPSE), are being developed to detect data that have been maliciously compromised with the intent to misrepresent grid operations [Zonouz 2012]. SCPSE fuses uncertain information from different types of distributed sensors, such as power system meters and cyber-side intrusion detectors, to detect the malicious activities within the cyber–physical system (CPS).

The EMS performs real-time contingency analysis to anticipate grid instabilities that might result from a major grid component failure, such as the loss of a generator or transmission line. This analysis shows how power grid operating conditions could evolve in response to the loss of particular components at that moment and supports planning to ensure that grid operating limits would not be violated if such a contingency were to occur. Automated remedial action schemes (RASs) or special protection systems (SPS) ensure the grid remains stable even if a major component is unexpectedly lost. From a cybersecurity perspective, physical consequences of malicious commands can be modeled as contingencies to assess risk and develop mitigations well in advance.

The AGC allows a balancing authority to adjust generation to meet power demand in real time as load connects to and disconnects from the grid. Protection and control devices, such as intelligent electronic devices (IEDs) with embedded operating systems, are used at the generation, transmission, and increasingly the distribution levels. These devices measure and automatically react to grid operating conditions within milliseconds, a few cycles at 60 Hz, to prevent equipment from exceeding safe operating limits and keep the grid stable.

Information technology (IT) is increasingly being adapted to support OT in utilities so that operating systems, computer platforms, and networks commonly used in IT are now found in some OT architectures. Segmented communication paths are architected to provide business IT systems with secure access to selected OT data, only when needed. The increasing use of IT computers and networks in OT architectures brings the need to protect these systems against malware developed to attack IT systems.

Cybersecurity protections are imperative and must not interfere with energy delivery functions. Cybersecurity for the Smart Grid is bringing together two communities that until recently have spoken different languages. IT speaks the language of computers and networks that support utility business administrative processes. OT speaks the language of electronic devices with embedded operating systems streamlined to support energy delivery functions and operational networks. Utility IT and OT differ in important ways, making cybersecurity protections that are appropriate for one often inappropriate, and even potentially damaging, for the other. However, each has benefits that can be gained from the other.

8.1.4 Electrical Utilities Evolution

The electrical utilities are at different points due to changes in generation, transmission, and distribution. The transformation is based on the introduction of different systems as described in [Abi-Samra 2014]:

- EMS is a system of computer-aided tools used by operators of electric utility grids to monitor, control, and optimize the performance of the generation and/or transmission system. The monitor and control functions are known as SCADA; the optimization packages are often referred to as advanced applications.

EMS are also often commonly used by individual commercial entities to monitor, measure, and control their electrical building loads for centrally control devices like HVAC units and lighting systems across multiple locations, such as retail, grocery, and restaurant sites. EMS can also provide metering, submetering, and monitoring functions that allow facility and building managers to gather data and insight that allows them to make more informed decisions about energy activities across their sites.

- Automation since 1970s but more since 1990s when utilities came under pressure to improve the reliability and quality of delivered power. More effective management and control, including SCADA, outage management system (OMS), and distribution management system (DMS), along with supporting technologies such as geographic information systems (GIS).
- Energy management includes planning and operation of energy-related production and consumption units. Objectives are resource conservation, climate protection, and cost savings, while the users have permanent access to the energy they need. It is connected closely to environmental management, production management, logistics, and other established business functions. Energy management is the proactive, organized, and systematic coordination of procurement, conversion, distribution, and use of energy to meet the requirements, taking into account environmental and economic objectives.

However, there are challenges on this transformation. The main challenges include the following:

- Deregulation unleashed unprecedented energy trading across regional power grids, presenting power flow scenarios and uncertainties the system was not designed to handle.
- The increasing penetration of renewable energy in the system further increases the uncertainty in supply and at the same time adds stress to the existing infrastructure due to the remoteness of the geographic locations where the power is generated.
- Our digital society depends on and demands a power supply of high quality and high availability.
- The threat of terrorist attacks on either the physical or the cyber assets of the power grid introduces further uncertainty.
- There is an acute need to achieve sustainable growth and minimize environmental impact via energy conservation, for example, by switching to green and renewable energy sources. This objective can be achieved by increasing energy efficiency, reducing peak demand, and maximizing the use of renewable energy.

The national governments' reaction is that Smart Grid technology is the answer to these challenges.

8.2 Smart Grid: What Is It?

The establishment of the European Technology Platform (ETP) for the Electricity Networks of the Future was for the first time suggested by the industrial stakeholders and the research community at the first International Conference on the Integration of Renewable Energy Sources and Distributed Energy Resources, which was held in December 2004 [ETP 2006]. The European Commission Directorate General for Research with the support of an existing research cluster developed the initial concept and guiding principles of the ETP SmartGrids (SG). The ETP SG began its work in 2005 to formulate and promote a vision for the development of European electricity networks looking toward 2020 and beyond. In April 2006, the ETP SG presented the document for SmartGrids and its vision and strategy for Europe's electricity networks of the future.

The Smart Grid represents a full suite of current and proposed responses to the challenges of electricity supply, demand, and environment. Because of the diverse range of factors, there are numerous competing taxonomies and no agreement on a universal definition. In general, Smart Grid refers to modernizing the electricity grid. The following includes examples of such definitions.

8.2.1 Definitions

Based on the descriptions included in the Energy Independence and Security Act of 2007 [EISA 2007], [NETL 2007], one definition in the United States is provided by Electric Power Research Institute (EPRI) [EPRI 2009]:

> Smart Grid refers to a modernization of the electricity delivery system so it monitors, protects and automatically optimizes the operation of its interconnected elements from the central and distributed generator through the high-voltage transmission network and the distribution system, to industrial users and building automation systems, to energy storage installations and to end-use consumers and their thermostats, electric vehicles, appliances and other household devices.

The American National Institute of Standards and Technology (NIST) defines Smart Grid in a more comprehensive way [NIST SmartGrid]:

> A modernized grid that enables bidirectional flows of energy and uses two-way communication and control capabilities that will lead to an array of new functionalities and applications.

The US Department of Energy (DOE) defines Smart Grid as follows:

> Smart Grid is the electric delivery network, from electrical generation to end-use customer, integrated with the latest advances in digital and information technology to improve electric system reliability, security, and efficiency [SPP Dictionary] and uses a growing number of distributed generation and storage resources [DOE/OEDER 2008].

There are international definitions such as the European Technology Platform for the Electricity Networks of the Future, also called ETP SmartGrids [EC 2011]:

> A Smart Grid is an electricity network that can intelligently integrate the actions of all users connected to it—generators, consumers—and those that do both—in order to efficiently deliver sustainable, economic and secure electricity supplies.

Also, there are many definitions promoted by academia and industry experts such as [Uslar 2013]:

> Smart Grid (an intelligent energy supply system) comprises the networking and control of intelligent generators, storage facilities, loads and network operating equipment in power transmission and distribution networks with the aid of Information and Communication Technologies (ICT). The objective is to ensure sustainable and environmentally sound power supply by means of transparent, energy and cost-efficient, safe and reliable system operation.

A definition that blends both functions and components is proposed by the International Energy Agency [IEA 2011]:

> A Smart Grid is an electricity network that uses digital and other advanced technologies to monitor and manage the transport of electricity from all generation sources to meet the varying electricity demands of end-users. Smart Grids coordinate the needs and capabilities of all generators, grid operators, end-users and electricity market stakeholders to operate all parts of the system as efficiently as possible, minimizing costs and environmental impacts while maximizing system reliability, resilience and stability.

The concept of Smart Grids shall enable the generation and consumption of electrical power to become more efficient and sustainable to meet the challenges of climate change and reduce the dependence on fossil fuels and nuclear power [Uslar 2013]. To achieve this, the coordination between distributed generation assets, storages, and the consumers shall be realized by using information and communication technologies (ICT).

A common definition adheres to this idea by defining Smart Grid as an electricity network that can intelligently integrate the actions of all the users connected to it – generators, consumers, and those that do both – in order to efficiently deliver sustainable economic and secure electricity supply [EC 2011].

Turning to the components, Smart Grids are typically described as electricity systems, complemented by communication networks, monitoring and control systems, smart devices, and end-user interfaces [OECD 2009], [OECD 2010]. The smartness of the Smart Grid lies in the decision intelligence layer, all the computer programs that run in relays, IEDs, substation automation systems, control centers, and enterprise back offices [OECD 2012].

A common element to most definitions for Smart Grid is the application of digital processing and communications to the power grid, making data flow and information management central to the Smart Grid. Various capabilities result from the deeply integrated use of digital technology with power grids, and integration of the new grid information flows into utility processes and systems is one of the key issues in the design of Smart Grids. The most important characteristics include a two-way flow of electricity and information to create an automated, widely distributed energy delivery network. It incorporates into the grid the benefits of distributed computing and communications to deliver real-time information and enable the near-instantaneous balance of supply and demand at the device level [EPRI 2009]. Electric utilities now find themselves making transformations such as improvement of infrastructure, addition of the digital layer, and business process transformation. Much of the modernization work that has been going on in electric grid, especially substation and distribution automation, is now included in the general concept of the Smart Grid, but additional capabilities are evolving as well.

The advanced power grid relates to widespread integration of renewables along with large-scale storage; widespread deployment of grid sensors and secure cyber-based communication within the grid; climate change through the reduction of energy waste in homes, businesses, and factories; and the accommodation of millions of electric vehicles (EVs) through innovative approaches to battery charging [NIST SP1108r3]. In this document, Smart Grids are viewed from the perspective of CPS, hybridized systems that combine computer-based communication, control, and command with physical equipment to yield improved performance, reliability, resiliency, and user and producer awareness.

Another related term is the grid edge, which is another way to define the changing power sector, coined by GTM Research in 2013 [Lacey 2013]. Grid edge is about the transformation of the electricity sector, evolution of markets, and integration of technologies. It is defined as follows [Ishimuro 2015]:

> The grid edge comprises the technologies, solutions and business models advancing the transition toward a decentralized, distributed and transactive electric grid.

The taxonomy diagram attempts to define the various technology layers and individual market segments that will shape the power grid to Smart Grid (called next-generation power grid). The top plane of the diagram outlines some opportunities that will be present in this new landscape.

Since grid edge includes DER systems, it is important to understand challenges and opportunities at the grid edge as identified in [Thompson 2013]:

- Grid modernization largely focused on technologies and architectures to support a growing distributed energy system.
- Distribution automation evolving to support variable generation sources and the challenges they pose related to voltage, frequency, forecasting, and overall grid reliability.
- Grid communication networks (the networked grid) will become even more important within the distribution grid, and reaching behind the meter, to support new requirements and applications.
- Grid edge analytics increase is one of the most important segments of the soft grid.
- Utility business models and regulatory models are evolving substantially given the trends at the grid edge.
- Energy efficiency technologies and programs will evolve to support smart connected buildings and advanced demand response (DR) programs.

8.2.2 Vision of the Future Smart Grid

The advanced and future power grid relates to a number of key scientific and technological areas. These include power quality, reliability, and resilience. As stated by Edison Electric Institute [EEI], the vision is that a smarter grid will enable utilities to:

- Empower customers to control and optimize their energy usage.
- Rely on greater amounts of distributed generation (wind, solar, etc.).

- Use electricity as a fuel for vehicles (PEVs).
- Enhance the reliability and efficiency of the power grid.
- Provide the framework and foundation for future economic growth.

Figure 8.3 depicts the Smart Grid concept, an interconnected network of power flow and data flow sharing intelligently information among generation, transmission, distribution, operation, and consumers; an integrated grid with renewable energy sources, such as solar and wind to augment the power supplied by fossil fuels; and a monitoring system to ensure energy flowing efficiently and economically.

Consumers can use the grid in different ways. More consumers will become prosumers – both consumers and producers of energy. Power will flow both ways, and other ancillary services may also be provided by these new prosumers. In systems where consumers deploy significant DERs, the distribution grid is to be the enabling platform. However, key success factors for Smart Grid include relevant characteristics, key technology areas, metrics, and performance.

Many aspects of the electric system touched by the Smart Grid applications are illustrated in Figure 8.4.

Some aspect of the electricity system touches every person in the nation. The stakeholder landscape for a Smart Grid is complex (see Figure 8.5), and the lines of distinction are not always crisp, as corporations and other organizations can take on the characteristics and responsibilities of multiple functions.

8.2.3 Tomorrow's Utility

A vision of tomorrow's utility (Utility X.0) includes more renewable generation, storage, distributed generation, and EVs and will need to manage such widely distributed systems, more participants, and two-way flows of electricity and information [Abi-Samra 2014]. New application emerging for utility enterprise, called demand response management system (DRMS), provides operational support in an automated and integrated way that supports the utility to control, operate, and monitor remote assets and a flexible DR solution. Also, the development and installation of new EMS will both facilitate and help system operators manage these complexities

Table 8.1 shows a comparison of features for the utility of the future (Utility X.0) and current utility. Since these systems overlap to a degree, operators often need multiple graphical user interfaces (GUIs) to monitor, control, and optimize networks. Many utilities have installed systems from different vendors, or employed systems that were user customized, limiting the ability of the various systems to easily interact. Thus, many issues have to be considered for these systems included in this table. In addition, it is recognized that the Smart Grid developments facilitate solutions for addressing EMS issues, energy trade, consumer, and automation solutions.

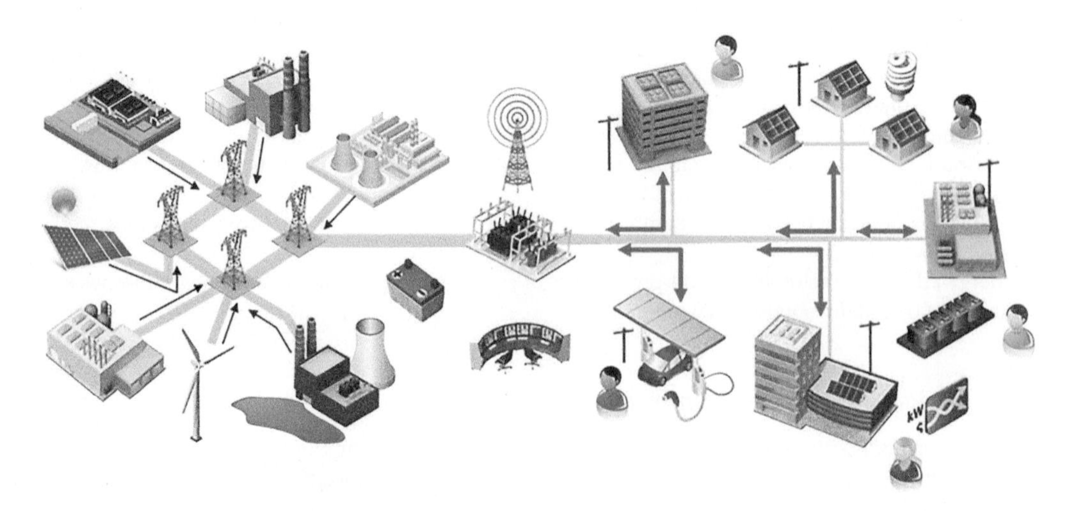

Figure 8.3 Smart Grid integrated view concept. *Source:* [Cleanpng SGP-I].

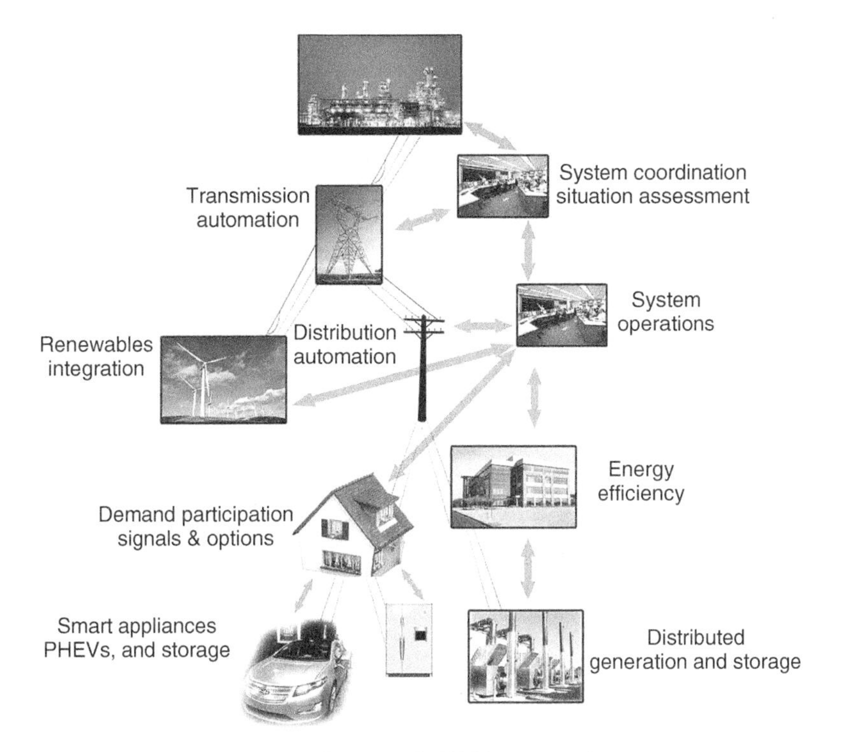

Figure 8.4 Scope of Smart Grid concerns. *Source:* [DOE 2009a]. Public Domain.

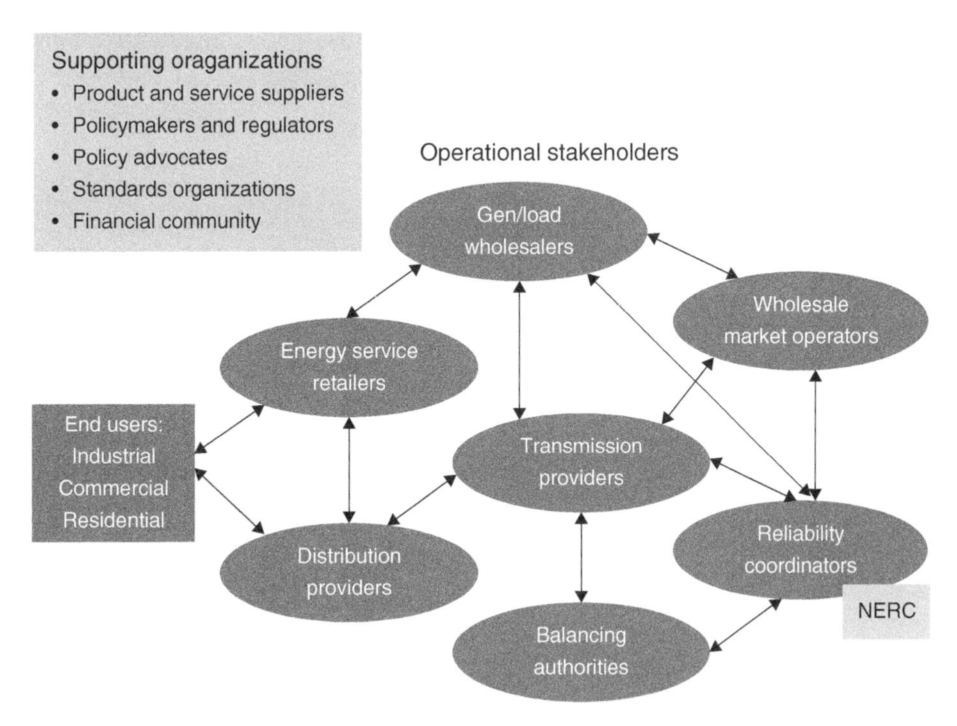

Figure 8.5 Stakeholder landscape. *Source:* [DOE 2009a]. Public Domain.

Table 8.1 Comparison of features for the current utility and future utility.

Features	Present	Utility X.0
Supply and resources	Heavily dependent on central generation plants; some distributed generation; some storage and renewables (such as solar energy and biofuels); and demand-side management	Microgrids; virtual power plants; high renewable penetration; measures to optimize system efficiency; more storage for renewables and peak load/spinning reserve reduction; and load management before and behind the meter
Component hardening	Enhanced wood pole test/treat programs; strengthened critical poles and upgrades; pole attachment and loading audits; and sealing and waterproofing equipment in substations	Resilient overhead conductor designs; steel, concrete, and composite poles; physical barriers in and around substations; and raising critical components in flood-vulnerable substations
System resiliency	Plans for critical load infrastructures; adequate backups; mobile command centers; central working team; use of social media; and better placement of lateral fusing	Advanced weather prediction systems; protecting of critical circuits; microgrids for critical loads; hydrophobic and nanoparticle coatings of conductors; feeder automation and reconfiguration; and proactive placement of repair crews
Substation management	Some distributed monitoring and control; some substation data collected over DNP3/IP protocols; and some vendor interoperability	Fiber-connected systems; IEC 61850-based messaging; remote engineering access; and modular substation designs
Condition monitoring	Time-based maintenance/manual inspection; local condition monitoring; some sensors; and limited equipment condition monitoring	Online advanced sensors; advanced condition monitoring; and centralized software tools with condition monitoring analytics
Integration of systems	Some movement away from proprietary hardware and software; some LAN-connected substation devices; and some vendor interoperability	Service-oriented architecture; open communication protocols; and wide-scale substation device interoperability
Data management	Some use of spreadsheets; distributed and paper-based systems; grid data historians; and limited use of common information model (CIM) and meter data management	Data portals that deliver flexibility to all users across the organization and real-time and offline predictive analytics for management of grids, customer data, and assets
Grid management	Some automated reactions to critical grid disturbances; SCADA extended to all distribution systems; and software tools (DMS and EMS)	Holistic and central view of the grid; decentralized and hybrid control; wide-scale usage of phasor measurement units (PMUs); extensive wide area monitoring; and high levels of situational awareness and self-healing

Source: [Abi-Samra 2014]. Text used based on courtesy of the author.

8.2.4 EMS Upgrades

EMS has to be upgraded to meet the requirements of the power grid, which is evolving at an exponential pace into a highly interconnected, complex, and interactive network of power systems, telecommunications, Internet, and electronic commerce applications. Virtually every element of the power system will incorporate sensors, communications, and computational ability.

Since EMS is critical element to grid reliability, it is desired to improve the EMS to include new technologies (e.g. exploit the phasor measurement units (PMUs) being installed across the grid) and advanced computational and communication techniques and the introduction of more closed-loop controls requiring less human intervention [EAC 2012]. The EMS provides the system operator with the necessary information to operate the grid from day to day, hour to hour, and minute to minute. In terms of system condition, the EMS (often referred to as real-time monitoring and control) provides near-real-time information by using data coupled with input from sensors to estimate the condition of the grid 20–30 seconds after the fact. However, faster response time on the order of subseconds is expected. Power distribution microgrids are key component for monitoring and control capabilities that can identify and isolate faults, so customers and businesses no longer have to accept power outages as a way of life. Besides these issues, security requirements for the information, communications, and existing cybersecurity controls have to be included in the EMS.

8.2.5 Electricity Trade

The move toward more competitive electricity markets requires a much more sophisticated infrastructure to support a myriad of informational, financial, and physical transactions between the members of the electricity value chain that supplement or replace the vertically integrated utility. One perspective on defining electricity system is to include electricity commodity and trading. So, electricity system is defined as the collection of all systems and actors involved in electricity production, transport, delivery, and trading [Kok 2010]. The electricity system is considered as two subsystems: the physical subsystem, centered around the production, transmission, and distribution of electricity, and the commodity subsystem, in which the energy product is traded. Electricity is an example of a flow commodity: physical flows that are continuous, for example, the commodity is (virtually) infinitely divisible. The continuous nature of flow commodities makes their infrastructure behavior fundamentally different from infrastructures transporting discrete objects such as cars or data packets. These subsystems are highly interrelated. One subsystem is for commodity trade in which the energy product is traded (commodity subsystem). Physical subsystem is centered around the production, transmission, and distribution of electricity. This special aspect of electricity has consequences on the developed economies (electricity networks and electricity markets). One important consequence of this perspective is the separation between commodity trades and network operations, both performed in two separate subsystems with limited, yet crucial, interaction. The financial flows that result from the electricity trade are referred to as the commodity transaction to distinguish it from transactions related to the physical electricity flows.

The physical subsystem consists of all hardware that physically produces and transports electricity to customers, as well as all equipment that uses the electricity. The structure of the physical subsystem is determined by the nature of the components that make up the electricity supply system: generators (large power producers and distributed generation operators), transmission network (TSO), distribution networks (DSOs), and loads (consumers). The commodity subsystem is defined as the actors involved in the production, trade, or consumption of electricity in supporting activities or their regulation and mutual relations. The commodity subsystem controls the physical subsystem, but it is constrained by this system in the same time.

Many difficult issues must be addressed, including technology challenges, the evolution of the regulatory model, a rethinking of how electricity is priced and electric infrastructure investments are recovered, and who can provide value-added services. To successfully develop sustainable solutions to these tough issues, a collaborative process that includes all stakeholders will be essential [GridWise 2014].

8.2.6 Trading Capabilities

The Smart Grid is the platform for enabling consumer choice, providing for the public good, and enabling future innovation. It can reliably integrate all energy sources and respond to wide fluctuations in supply from central and distributed generation [GridWise 2014].

Smart Grids, especially on distribution level, will have a significant impact on the energy landscape. They facilitate the integration of distributed energy resources (DERs) such as renewable energy sources and the adaptation of demand patterns, forming a central part of a sustainable, decentralized, participatory, secure, and safe energy system of the future, which helps to empower the consumer and corresponds to local needs [CEDEC].

For larger DERs, the utility often sends pricing information, allowing the DER automation system to perform make/buy decisions on a real-time basis. When utility power is cheaper than internally generated power, the DER automation system can reduce power output. When utility power is more expensive than internally generated power, the DER can run at full capacity.

DR programs support electricity market and reliability of electricity system. Emerging software systems have created a wave of new capital investment and technology advancement in support of these new DR market dynamics. For example, software solutions for building a DR business network are envisioned in [Vojdani 2006]. A DR business network enables many entities such as the independent system operator (ISO), DR service providers and aggregators, distribution utilities, metering agents, settlement agents, billing agents, and customers to collaborate in real time to efficiently execute DR business processes.

DERs participate in markets to some extent today and will participate to a greater extent as the Smart Grid becomes more interactive. Smart Grid facilitates new market key actors in operating and trading energy such as trading agency, market operators, market management, DER aggregation, trading, wholesale, etc. Also, Smart Grid markets may also include several technologies related to ICT, transmission and distribution, energy storage systems, renewables, electric vehicles, etc. Desired frameworks include game theory, optimization, and simulation that enable benefits such as profit for energy trading, improved power system efficiency, and reduction of emissions. A commercial trading platform is illustrated in [COMSAR].

Figure 8.6 shows a generic market platform for coordinating various energy agents representing power suppliers, customers, and prosumers. These energy agents implement a generic bidding

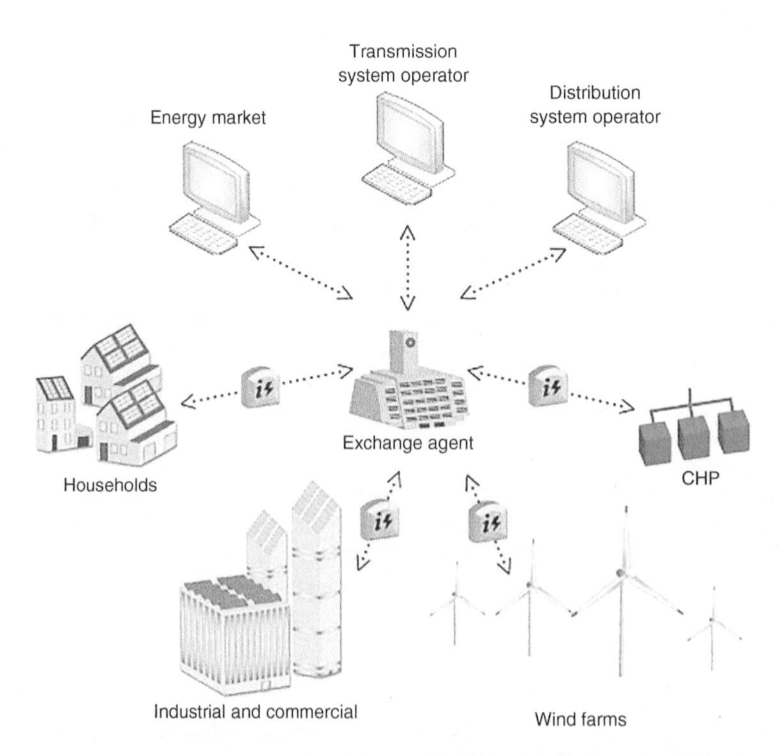

Figure 8.6 View of market exchange structure. *Source:* [ETP 2006]. Licensed under CC BY-SA 4.0.

strategy that can be governed by local policies. These policies represent consumer's preferences or constraints for the devices controlled by the agent. Efficient coordination between the agents is realized through a market mechanism that incentivizes the agents to reveal their policies truthfully to the market. The market platform supports various market structures ranging from a single local energy exchange to a hierarchical energy market structure. A hierarchical structure for the electricity market can facilitate the coordination of energy markets in distribution and transmission networks.

By adding sophisticated ICT and intelligent devices, various Smart Grid applications support smart transactions, peak load reductions, efficient balancing mechanisms, etc. Smart transactions are realized via smart interfaces consisting of power meters measuring bidirectional power flows and controllers monitored and controlled by EMS.

The DR application is making this linkage between the utility back office and its customers and requires a platform capable of adapting to ever-changing business needs while remaining focused on providing the infrastructure for both command and control as well as customer empowerment [Vos 2009]. So, the emerging DRMS technology enables utility to provide a DR solution such as using DR resources in a more intelligent and efficient way by planning and executing load shed at grid locations where the utility has more benefit. Also, DRMS help utilities to leverage investments in Smart Grid technology.

Another promising technology for improving electricity trading is the power distribution microgrid systems that have the objectives to meet the energy demands of the customers and to minimize the cost of the external energy imported to the system. Customers can receive dynamic pricing information via the Internet, for example, that enables them to control their electricity use and spending in the most user-friendly and beneficial way. Also, they can participate in the marketplace by selling power back to the grid through local power generation and plug-in EVs as the technology advances. These capabilities help in maximizing entrepreneurial innovation and private investment opportunities by providing consumers choice in a free retail service market. Energy management strategies for a microgrid system are described in [Wang 2013]. The objectives are to meet the energy demands of the customers and to minimize the cost of the external energy imported to the system.

However, all these solutions to be cost effective require fast, reliable, automated, and secure communications between multiple players in the DR domain in real time [Newmann 2007]. Also, the Smart Grid transformation and the introduction of assets that could enable a suite of benefits could create issues and concerns for the power systems and characteristics that must be addressed. Therefore, DER integration and the effects of DER on wholesale market behavior have to be considered. Examples of impacts include visibility of wholesale operations and retail-level interactions [Masiello 2014]. Many other technical issues are discussed in this work. In addition, DERs interact with applications from different domains, so needs for security are increasing concerns.

The trend toward renewable, decentralized, and highly fluctuating energy suppliers (e.g. photovoltaic, wind power, CHP) introduces new requirements for communications and security to ensure the market transactions.

DR communications cover interactions between wholesale markets and retail utilities and aggregators, as well as between these entities and the end load customers who reduce demand in response to grid reliability or price signals. Communications for markets domain interactions must be reliable. They must be traceable and auditable. They must support e-commerce standards for integrity and non-repudiation. As the percentage of energy supplied by small DER increases, the allowed latency in communications with these resources must be reduced.

8.3 Smart Grid Characteristics

Characteristics are defined as prominent attributes, behaviors, or features that help distinguish the grid as Smart Grid. Sometimes, people confuse the Smart Grid with smart meters and advanced metering infrastructure (AMI) or with interoperability grid.

8.3.1 Relevant Characteristics

Thus, the most relevant or principal characteristics of a Smart Grid include [NETL 2007], [NETL 2008]:

- Self-healing from power disturbance events.
- Enabling active participation by consumers in DR.
- Operating resiliently against both physical and cyber attacks.
- Providing quality power that meets twenty-first-century needs.
- Accommodating all generation and storage options.
- Enabling new products, services, and markets.
- Optimizing asset utilization and operating efficiency.

These characteristics can be achieved through the application of a combination of existing and emerging technologies. These new features require introducing the most intelligence to the existing power grid.

One feature of the Smart Grid is to ensure the balance of generation and demand in the presence of time-varying and stochastic (random) generation and DR. This goal is a departure from the deterministic paradigm of today's power grid.

Figure 8.7 shows the concept of Smart Grid from the perspective of integrated solutions for power (traditional power plants and renewables – solar, power). Commercial solutions being implemented are increasing. However, Smart Grids are developed based on current power grid systems and evolve in time to meet new trends of clean environment, climate changes, and customer and operational requirements. A comparison of a selected set of functions for the US legacy power grid and Smart Grid is provided in Table 8.2. Various research documents and newsletters about Smart Grid developments and activities in the United States and Europe are available via OPEN ACCESS at this Web site [OpenEI].

Figure 8.7 A smart control center with integrated power solutions. *Source:* chombosan/Getty Images.

Table 8.2 Legacy and Smart Grid functionality comparison.

Legacy grid	Smart Grid
Consumers have limited information and opportunity for participation with power system, unless under direct utility control	Informed, involved, and active consumers – demand response and distributed energy resources
Dominated by central generation – many obstacles exist for distributed energy resources interconnection and operation	Many distributed energy resources with plug-and-play convenience; responsive load to enhance grid reliability, enabling high penetration of renewables
Limited wholesale markets, not well integrated – limited opportunities for consumers	Mature, well-integrated wholesale markets; growth of new electricity markets for consumers; product interoperability
Focus on outages and primarily manual restoration – slow response to power quality issues, addressed case by case	Power quality is a priority with a variety of quality/price options – rapid resolution of issues
Limited integration of operational data with asset management – business process silos limit sharing	Greatly expanded data acquisition of grid parameters – focus on prevention, minimizing impact to consumer
Responds to prevent further damage – focus is on protecting assets following a fault	Automatically detects and responds to problems – focus on prevention, minimizing impact to consumers, and automated restoration
Vulnerable to inadvertent mistakes, equipment failures, malicious acts of terror, and natural disasters	Resilient to inadvertent and deliberate attacks and natural disasters with rapid coping and restoration capabilities

Source: [DOE 2010], [Ton 2010].

Another comparison of the characteristics for today's grid and Smart Grid is shown in Table 8.3. Considering that identification and optimization as the most important problem for the network operation, maintenance, and restoration planning, this work [Gudzius 2011] recommends a list of developments in the area of electric power network regimes analysis models, control algorithms, and technologies.

8.3.2 Electrical Infrastructure Evolution

The Smart Grid is an industry transformation that is shifting the business and operational models used by utilities from a fragmented approach to a more cohesive one. The changes are grouped in stages (waves) of enhancing the infrastructure as described in [Vos 2009]:

1) Acquisition and deployment of advanced metering communication systems for two-way connectivity to energy consumers, communication between the utility and the home.
2) Using the connectivity established in wave 1, utilities collect and manage massive amounts of consumption and performance data; starting the installation of systems that utilize this information for DERs, operations, DR, and customer empowerment.
3) Integration of command and control infrastructure with the transmission and distribution control infrastructure for advanced Smart Grid control applications (e.g. DER systems); DRMS becomes the critical infrastructure component – linking the utility back office to its customers.

A Smart Grid allows the power industry to observe and control parts of the system at higher resolution in time and space. This enables better services for customers to obtain cheaper, greener, less intrusive, more reliable, and higher-quality power from the grid. The legacy grid did not allow for real-time information to be relayed from the grid, so one of the main purposes of the Smart Grid is allowing real-time information to be received and sent from and to various parts of the grid to make operation as efficient and seamless as possible. It allows to manage logistics of the grid and view consequences that arise from

Table 8.3 Today's grid comparison with Smart Grid.

Characteristic	Today's grid	Smart Grid
Collaboration with energy consumers	Homogeneous customers; no collaboration	Active collaboration; consumers are well informed; involved managing their energy dependence, consumption, and resources
Compatibility of energy generation and storage	Domination of main generation sources	Distributed generation; attention paid to renewable energy
Products, services and markets	Limited and poorly integrated wholesale market, limited consumer involvement	Mature; well-integrated wholesale market with potential growth of new electricity markets
Energy quality	Focused on the outage management; low-level responsibility for the energy quality	Energy quality is an essential factor in relation to the price
Expenses and operations	Low degree of integration with operations information	Optimized operations and efficiency; full integration with information of network parameters
Reaction to disturbances	The aim is to prevent system from more damage; focused on protection of the resources, which are needed to restore the system	Automatic detection of problems and acting reasonably; focused on prevention and minimum impact to the customer
Behavior during an event of natural or other disaster	High degree of vulnerability; delayed response and restoration	Flexible and quick restoration

Source: [Gudzius 2011]. Licensed under CC BY-SA 4.0.

its operation on a time scale with high resolution, from high-frequency switching devices on a microsecond scale to wind and solar output variations on a minute scale to the future effects of the carbon emissions generated by power production on a decade scale.

In the United States, the DOE modernization plan goal is a modernized grid that will therefore balance these characteristics [DOE MYPP-2015]:

- Resilient – Quick recovery from any situation or power outage.
- Reliable – Improves power quality and fewer power outages.
- Secure – Increases protection to our critical infrastructure.
- Affordable – Maintains reasonable costs to consumers.
- Flexible – Responds to the variability and uncertainty of conditions at one or more time scales, including a range of energy futures.
- Sustainable – Facilitates broader deployment of clean generation and efficient end-use technologies.

A study of Smart Grids in Europe [DeepResource 2012] reports an increase of smart meter penetration in some countries: Europe about 240 million, India about 130 million, Brazil about 64 million, and the United States about 60 million.

The global utility Smart Grid IT systems market is expected to grow from millions in 2013 to more than two billion in 2020. Smart Grid IT systems are segmented by category such as software purchases and upgrades, software maintenance fees, services, applications, and software as a service (SaaS) [Martin 2014].

IT systems are used for many applications. Currently, the most demanded need is building energy controls. Commercial buildings are increasingly employing alternative strategies to draw less energy from the grid, either through efficiency gains from fewer AC/DC conversions or through reducing the need for inverters by using renewable or fuel cell power sources. Direct control (DC)-based systems enable stable, high-quality power distribution to critical services in buildings including security, lighting, and environmental control systems. Besides power equipment, these systems require powered and biometric door locks, security cameras and sensors, networking devices such as wireless access points, and actuators and controls for heating, ventilation, and air conditioning that communicate and exchange information.

Plug and play is an approach to connecting resources in the future electric power system. In computing, a plug-and-play device or computer bus is one with a specification that facilitates the discovery of a hardware component in a system without the need for physical device configuration or user intervention in resolving resource conflicts. Similar to a computer system, plug and play in a power system implies that a resource can be placed at any point on the electrical system without reengineering the controls.

The plug-and-play concept eliminates the need for relatively costly wiring and installation by a professional electrician; it can significantly reduce the balance of systems costs of purchasing and installing residential and small business DER systems. The advantage of plug-and-play systems (e.g. PV systems) is the ability to start drawing renewable energy from the PV panels by simply plugging in an extension cord such that the applications including markets for these self-contained systems are expanding [Zhang 2013]. Already, many commercial products are portable and remotely managed. For example, residential PV panel kits include all the wiring, inverters, charge controllers, and mounting hardware needed to quickly and easily install the system.

The vision of the future is to build supergrids (also called super grids). One vision is to interconnect regional electricity networks to form a globe-spanning supergrid [Gellings 2015]. Use of renewables is the key to this concept. The fact is that energy from the sun can be stored and wisely used by consumers around the globe. It is claimed that within six hours deserts receive more energy from the sun than humankind consumes within a year [DESERTEC 2015]. This is the first proposed project by a German-led consortium in 2009. The DESERTEC project aims to harvest solar power in the Mediterranean and other deserts of the world and use HVDC to transmit the electricity to population centers. Similar to DESERTEC project, other projects are on the way [Gellings 2015].

The European SuperSmart Grid is a proposal of linking renewable energy projects like DESERTEC and Medgrid across North Africa, the Middle East, and Europe and could serve as the backbone for the hypothetical super Smart Grid. The Medgrid project vision is for developing 20 GW of solar power generation in North Africa, of which 5 GW would be exported to Europe. The Medgrid electricity network would become the backbone of the European SuperSmart Grid. Figure 8.8 depicts the proposed network of the European SuperSmart Grid.

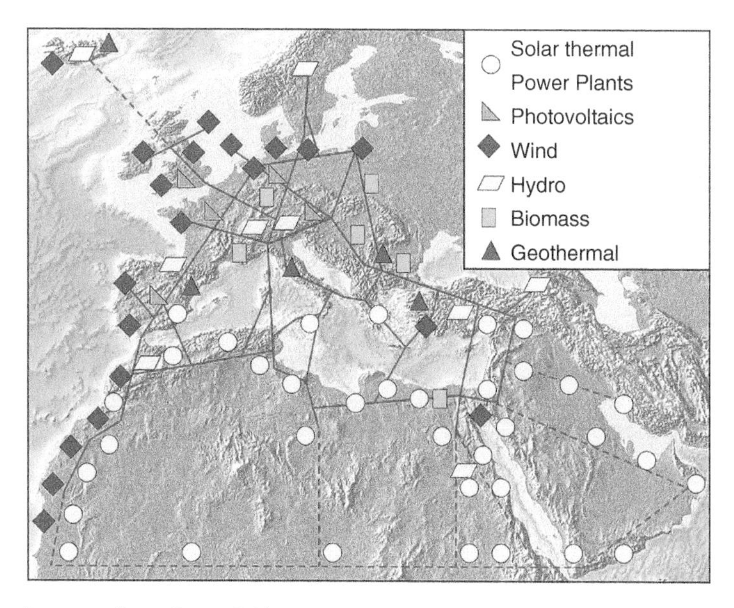

Figure 8.8 The European SuperSmart Grid conceptual plan. *Source:* Trans-Mediterranean Renewable Energy Cooperation, https://commons.wikimedia.org/wiki/File:TREC-Map-en.jpg. Licensed under CC BY-SA 2.5.

In the United States, the Unified National Smart Grid is a proposed wide area grid that is a national interconnected network relying on a high-capacity backbone of electric power transmission lines linking all local electrical networks that have been upgraded to Smart Grids.

The electric system of the future will include both central and distributed generation sources with a mix of dispatchable (e.g. controllable) and non-dispatchable resources. While central generation will continue to play a major role, there will also be other generation supply options, and a key component will be the grid, the network that serves as the backbone of the electric power system [GridWise 2014]. In addition, DERs and DR will play a more significant role, but it requires an ability of the grid to reliably integrate these new generation options; basically it will require to be more flexible and adaptable.

The analysis of systems is facilitated by using a conceptual model (see definition in Appendix D). The following section provides a brief introduction of different conceptual models and architectures for the Smart Grid.

8.4 Smart Grid Conceptual Models

The Smart Grid is a complex system of systems for which a common understanding of its major building blocks and how they interrelate must be broadly shared. A conceptual architectural reference model facilitates this shared view.

8.4.1 NIST Conceptual Model

The NIST developed the Smart Grid conceptual model, which provides a high-level framework for the Smart Grid [NIST SP1108r1]. The diagram is a simplified model of the multiple and complex systems of Smart Grid. This model provides a means to analyze use cases, to identify interfaces for which interoperability standards are needed, and to facilitate development of a cybersecurity strategy. A revised conceptual diagram that includes DERs is provided in [NIST SP1108r3]. Evolved from the conceptual model defined in [NIST SP1108r1], the generation domain includes renewables (in 2010 generation refers to bulk generation). Since 2009, NIST has worked cooperatively with industry to develop and refine this framework.

The conceptual model is not only a tool for identifying actors and possible communication paths in the Smart Grid but also a useful way for identifying potential intra- and inter-domain interactions and potential applications and capabilities enabled by these interactions.

The model is a set of views (diagrams) and descriptions that are the basis for discussing the characteristics, uses, behavior, interfaces, requirements, and standards of the Smart Grid. The model is not a design diagram or final architecture; it provides a high-level and overarching perspective; it is a tool to help in analysis. It is a tool for discussing the structure and operation of the power system. Also, the model helps stakeholders understand the building blocks of an end-to-end Smart Grid system, from generation to (and from) customers, and explores the inter- and intra-relation between domains.

8.4.2 IEEE Model

An interactive model of the NIST model with most current references for each domain is provided at [IEEE Smart Grid]. IEEE views the Smart Grid as a large system of systems where each NIST domain is expanded into three Smart Grid foundational layers [P2030 2011]:

- Power and energy layer.
- Communication layer.
- IT/computer layer.

The communication and IT/computer are enabling infrastructure platforms of the power and energy layer that make the grid smarter.

8.4.3 European Conceptual Model

The European Commission's Mandate 490 (EU-M490) for Smart Grid with the European Telecommunications Standards Institute (ETSI), European Committee for Standardization (Comité Européen de Normalisation (CEN)), and the European Committee for Electrotechnical Standardization (CENELEC) completed their first version of the Smart Grid Reference Architecture in November 2012.

The European conceptual model is an evolution of the NIST model in order to take into account some specific requirements of the European context that NIST model did not address [CENELEC 2012].

As pointed in this document, the major one requirement is the integration of DERs. The second one is the flexibility concept that brings consumption, production, and storage together in a flexibility entity. The EU conceptual model is a top layer model that acts as a bridge between the underlying models in the different viewpoints of the reference architecture.

A reference architecture describes the structure of a system with its element types and their structures, as well as their interaction types, among each other and with their environment. Through abstraction from individual details, a reference architecture is universally valid within a specific domain. Further architectures with the same functional requirements can be constructed based on the reference architecture [ISO/IEC 42010].

Figure 8.9 defines the scope of PAN European Energy Exchange System and application area of a microgrid architecture. The application area of the hierarchical mesh cell architectures (microgrids) includes the customer, distribution, and DER domains.

The conceptual model is the basis for defining the Smart Grid Architecture Model (SGAM) framework (see Figure 8.10). The interoperability categories introduced by the GridWise Architecture Council [Grid Interoperability] represent a widely accepted methodology to describe requirements to achieve interoperability between systems or components. The individual categories of GWAC stack are divided among the three drivers: technical, informational, and organizational. These interoperability

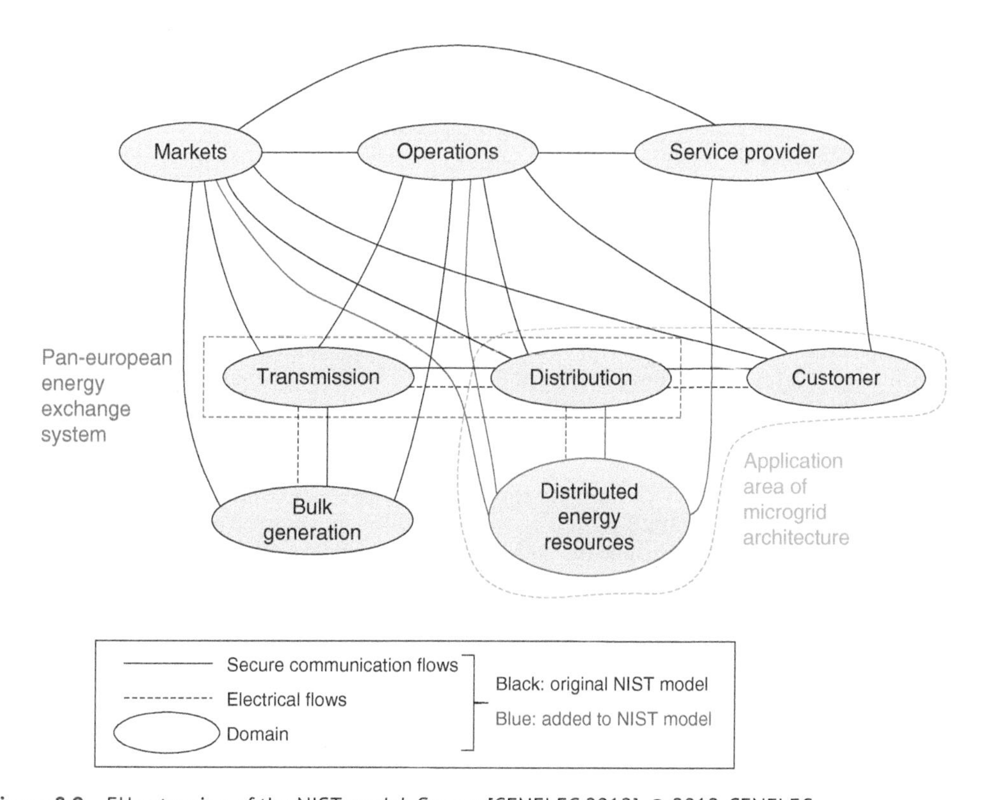

Figure 8.9 EU extension of the NIST model. *Source:* [CENELEC 2012]. © 2012, CENELEC.

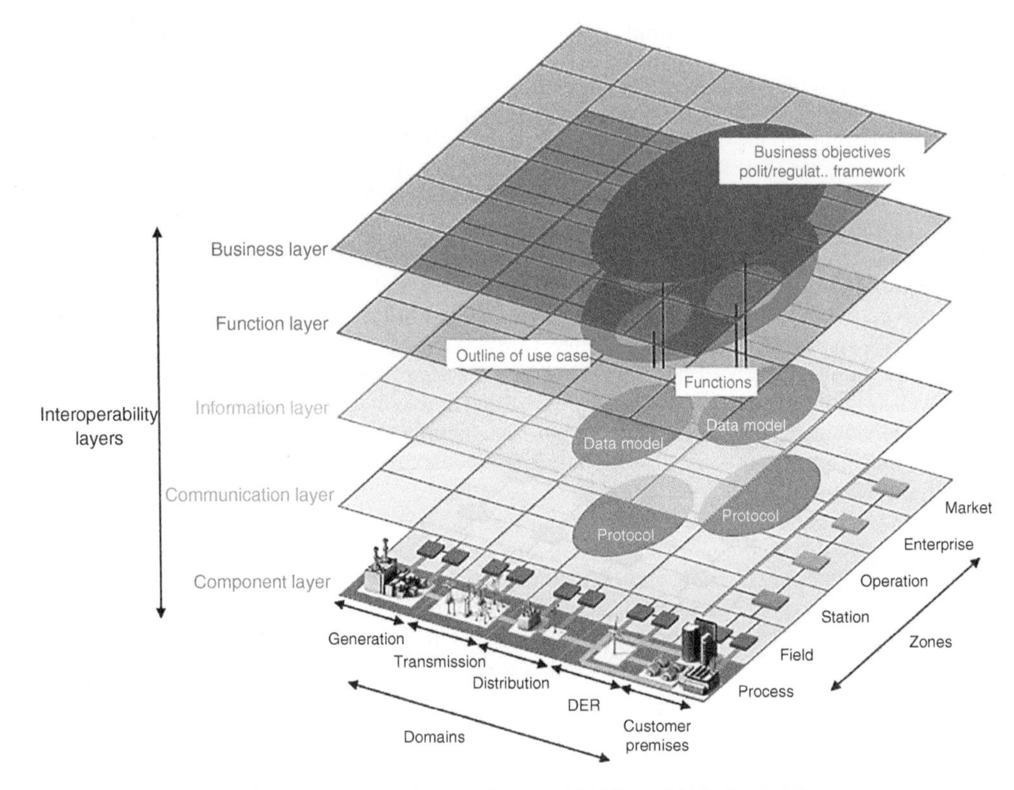

Figure 8.10 European SGAM framework. *Source:* [CENELEC 2012]. © 2012, CENELEC.

categories underline the definition of interoperability. Interoperability is seen as the key enabler of Smart Grid. Hence, for the realization of an interoperable function, all categories have to be covered by means of standards or specifications.

This framework spans three dimensions, namely, domain, interoperability (layer), and zone defined as follows:

- Interoperability refers to the ability of two or more devices from the same vendor, or different vendors, to exchange information and use that information for correct cooperation [IEC 61850].
- SGAM interoperability layer allows a clear presentation and simple handling of the architecture model; the interoperability categories described in the GridWise Architecture model are aggregated in SGAM into five abstract interoperability layers: business, function, information, communication, and component.
- SGAM Smart Grid plane is defined from the application to the Smart Grid conceptual model of the principle of separating the electrical process viewpoint (partitioning into the physical domains of the electrical energy conversion chain) and the information management viewpoint (partitioning into the hierarchical zones [or levels] for the management of the electrical process [IEC 62357], [IEC 62264]).
- SGAM domain of the Smart Grid plane covers the complete electrical energy conversion chain, partitioned into five domains: bulk generation, transmission, distribution, DER, and customer premises.
- SGAM zone of the Smart Grid plane represents the hierarchical levels of power system management, partitioned into six zones: process, field, station, operation, enterprise, and market [IEC 62357].

So, the SGAM framework is merging the dimension of five interoperability layers (business, function, information, communication, and component) with the two dimensions of the Smart Grid plane, e.g. zones (representing the hierarchical levels of power system management: process, field, station, operation, enterprise, and market) and domains (covering the complete electrical energy conversion chain: bulk generation, transmission, distribution, DER, and customer premises).

Also, crosscutting issues are topics that need to be considered and agreed on when achieving interoperability [Grid Interoperability]. These topics may affect several or all categories to some extent. Typical crosscutting issues are cybersecurity, engineering, configuration, energy efficiency, performance, and others.

The following viewpoints have been selected as the most appropriate to represent the different aspects of Smart Grid systems with associated architectures [CENELEC 2012]:

- Business architecture is addressed from a methodology point of view in order to ensure that whatever market or business models are selected, the correct business services and underlying architectures are developed in a consistent and coherent way.
- Functional architecture provides a metamodel to describe functional architectures and gives an architectural overview of typical functional groups of Smart Grids (intended to support the high-level services).
- Information architecture addresses the notions of data modeling and interfaces and how they are applicable in the SGAM model; it introduces the concept of logical interfaces, which is aimed at simplifying the development of interface specifications especially in case of multiple actors with relationships across domains.
- Communication architecture deals with communication aspects of the Smart Grid, considering generic Smart Grid use cases to derive requirements and to consider their adequacy to existing communication standards in order to identify communication standard gaps.

The SGAM framework aims at offering a support for the design of Smart Grids with an architectural approach allowing for a representation of interoperability viewpoints in a technology neutral manner, both for current implementation of the electrical grid and for future implementations of the Smart Grid.

8.5 Power and Smart Devices

Smart equipment refers to all field equipment that is computer based or microprocessor based, including controllers, remote terminal units (RTUs), and IEDs; it includes the actual power equipment, such as switches, capacitor banks, or breakers, equipment inside homes, buildings, and industrial facilities; the embedded computing equipment must be robust to handle future applications for many years without being replaced [EPRI 2011]. Also, end-user equipment is no longer the dumb device, but an interactive and intelligent node on the Smart Grid.

Current deployments of Smart Grid technologies include a large array of hardware devices and software solutions for:

- Customer – Displays, Internet portals, direct load controls, programmable thermostats, EV chargers, smart appliances, smart meters, and smart sensors.
- Distribution – Automated switches, automated capacitors, automated voltage regulators, equipment monitoring, energy storage, equipment health monitors, weather forecasting, energy storage systems, AMI, and intelligent universal transformers.
- Transmission – Wide area monitoring, synchrophasor technology, phasor data concentrators, dynamic line rating, and energy storage systems.
- Control – IEDs.

The following sections provide a brief introduction of key devices and security features.

8.5.1 Smart Meters

The term smart meter often refers to an electricity meter, but it also may mean a device measuring natural gas or water consumption. A smart meter is an electronic device that records consumption of electric energy in intervals of an hour or less and communicates that information at least daily back to the utility for monitoring and billing. It is an embedded system that contains a microcontroller with nonvolatile and volatile memory, analog/digital ports, timers, real-time clock, and serial communication facilities.

Smart meters usually involve real-time or near-real-time sensors that may also notify the utility of a power outage and power quality monitoring or allow the utility to remotely switch electricity service on or off. Smart meters enable two-way communication between the meter and the central system. These additional features are distinct and more than simple automated meter reading (AMR). They are similar in many respects to AMI meters.

Smart meters are part of the AMI and provide a communication path extending from generation plants to electrical outlets (smart socket) and other Smart Grid-enabled devices that can shut down during times of peak demand, if the customer chooses this option.

8.5.2 Intelligent Electronic Devices

IEDs encompass a wide array of microprocessor-based controllers of power system equipment, such as circuit breakers, transformers, and capacitor banks. IEDs receive data from sensors and power equipment and can issue control commands, such as tripping circuit breakers if they sense voltage, current, or frequency anomalies, or raise/lower voltage levels in order to maintain the desired level.

Common types of IEDs include protective relaying devices, load tap changer controllers, circuit breaker controllers, capacitor bank switches, recloser controllers, voltage regulators, network protectors, relays, etc. A typical IED today can perform 5–12 protection functions and 5–8 control functions, including controls for separate devices, an auto-reclose function, self-monitoring function and communication functions, etc. It can do this without compromising security of protection – the primary function of IEDs. IEDs used in power system automation include RTU, meter, digital fault recorder, PLC, protective relay, and controlling output or communication devices (see more information in Appendix D).

A new generation of IEDs is rapidly being deployed throughout the power system to support a higher level of intelligence to enable distributed data acquisition and decentralized decision making [Denton 2014]. They are equipped with advanced technologies that make two-way digital communication possible where each device on the network is equipped with sensing capabilities to gather important data for wide situational awareness of the grid.

Also, these devices can be remotely controlled and adjusted at the node level as changes and disturbances on the grid occur. Additionally, these IEDs not only communicate with SCADA systems but among each other, enabling distributed intelligence to be applied to achieve faster self-healing methodologies and fault location/identification. With all IEDs connected to the substation concentrator, it also becomes possible to have a remote maintenance connection to most of IEDs [Borlase 2013].

8.5.3 Phasor Measurement Units

PMUs or synchrophasors are devices that measure the electrical waves (magnitude and phase angle of the sine waves) on an electricity grid using a common time source for synchronization, a global positioning system (GPS) radio clock. Time synchronization allows synchronized real-time measurements of multiple remote measurement points on the grid. Synchronization of the voltage frequency and phase allows electrical systems to function in their intended manner without significant loss of performance or life. A PMU can be a dedicated device, or the PMU function can be incorporated into a protective relay or other device. PMUs are considered as one of the most important measuring devices in the future of power systems.

Synchrophasors placed in different locations collect measurements that can be compared in real time. These comparisons can be used to assess system conditions such as frequency, phase, and voltage changes. The phase angle measurements can indicate shifts in system (grid) stability. The phasor data is collected either on-site or at centralized locations using phasor data concentrator technologies, although the cost of installation of communication links is an issue as described in [Liu 2001].

Wide area measurement systems based on PMUs are an integral part of the Smart Grid. The wide area monitoring, control, and protection extend the time resolution. Traditional SCADA and EMS have a measurement update rate of several minutes or minutes and are not suitable to monitor and control the power system. Better control requires time resolution to sub-second time scale in order to react to dynamic power instabilities. This can be achieved with PMUs located at different locations. The PMUs support data acquisition with high enough sampling rate and at the same time instant.

However, there are various concerns related to the security of these systems that have been identified in [Rihan 2013].

8.5.4 Intelligent Universal Transformers

The intelligent universal transformer (IUT) is a first-generation power electronic replacement of conventional distribution transformers. Conventional transformer limitations include poor energy conversion efficiency at partial loads, use liquid dielectrics that can result in costly spill cleanups, and provide only one function – stepping voltage, no real-time voltage regulation, no monitoring capabilities, etc. The new concept of IUT includes an interface point for integration of distributed resources, from storage to plug-in hybrid electric vehicles. New transformer designs allow add-on intelligence to enhance power quality compatibility between source and load and can isolate a disturbance from either source or load [Arindam 2009].

It also incorporates command and control functions for system integration, local management, and islanding. The controller interfaces with DMS, EMS, and DR systems to optimize overall grid performance and improve reliability.

8.6 Examples of Key Technologies and Solutions

It is recognized that technology is critical to the Smart Grid, but in isolation it is not transformational [Borlase 2013]. The author argues that in order to fully achieve the benefits of Smart Grid, it requires a strong partnership between utilities, government, regulators, and the public. More information about Smart Grid drivers, benefits, challenges, and success factors is available in [Borlase 2013]. The authors discuss a wide range of technologies that can enable the Smart Grid including brief overviews of Smart Grid efforts in the United States and around the world. The following sections provide an overview of key technologies and common networking solutions for the Smart Grid applications.

8.6.1 Communication Networks

The American Recovery and Reinvestment Act of 2009 (ARRA 2009) included funding to modernize infrastructure and accelerate the deployment by electric utilities of foundational Smart Grid technologies [EPRI 2011]:

- Communications
- Computational ability
- Distributed generation
- Power electronics and controls
- Energy storage
- Home and building automation systems
- Efficient appliances and devices
- Sensors.

Communication networks refer to the media and to communication protocols have to preserve interoperability and security of the systems. Smart Grid is the inclusion of many things, but the key to achieving these potential benefits of Smart Grid is to successful buildup of communication network that can support all identified functionalities such as:

- AMI
- DR
- EVs
- Wide area situational awareness (WASA)
- DER and storage
- Distribution grid management.

A Smart Grid is one that incorporates ICT into every aspect of electricity generation, delivery, and consumption in order to minimize environmental impact, enhance markets, improve reliability and service, and reduce costs and improve efficiency [EPRI SmartGrid]. The Smart Grid is a network of networks, including power, communications, and intelligence [EPRI 2011]. That is, many networks with various traditional ownership and management boundaries are interconnected to provide end-to-end services between stakeholders and in and among IEDs. While communication architecture is considered the foundation of the power delivery system of the future and the enabler of Smart Grid integration, automation is the heart of a smart power delivery system [EPRI 2011]. These facilitate any of the following:

- DER and storage development and integration.
- Power electronic-based controllers and widely dispersed sensors throughout the delivery system.
- An AMI.
- End-user infrastructure.

As pointed in this definition [Cisco 2010], Smart Grid is a data communication network integrated with the electrical grid that collects and analyzes data captured in near real time about power transmission, distribution, and consumption. Based on this data, Smart Grid technologies can be used to provide predictive information and recommendations to utilities, their suppliers, and their customers on how to manage power.

From one perspective view, Smart Grid is a network of networks, including power, communications, and intelligence. That is, many networks with various traditional ownership and management boundaries are interconnected to provide end-to-end services between stakeholders and among IEDs. Thus, any domain application could communicate with any other domain application via the information network, subject to the necessary network access restrictions and quality of service requirements. The applications in each domain are the end points of the network. For example, an application in the Customer domain could be a smart meter at the customer premise; an application in the Transmission domain could be a PMU on a transmission line or in a Distribution domain at a substation; an application in the Operation domain could be a computer or display system at the operation center. Each of these applications has a physical communication link with the network.

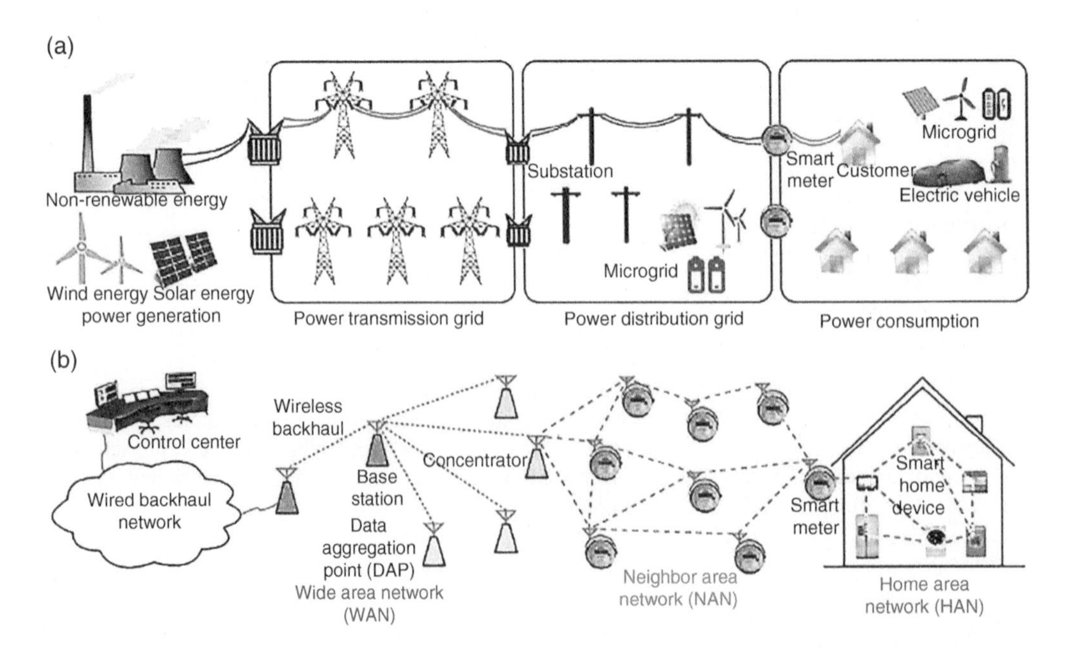

Figure 8.11 The overall layered architecture of Smart Grid. *Source:* [Ho 2013b]. © 2013, IEEE.

Figure 8.11 shows the overall layered architecture of Smart Grid described in [Ho 2013b]. The Smart Grid is the electricity delivery system (from point of generation to point of consumption) integrated with communications and information technology for enhanced grid operations, customer services, and environmental benefits.

Another proposed solution for the Smart Grid is a hierarchical architecture. This is divided into four layers as follows [Shi 2012]:

- Application layer provides Smart Grid applications for customers and utilities and includes power transmission, distribution, and customer applications.
- Communications layer supports a two-way reliable, efficient, and secure information exchange for the application layer and includes wide area network (WAN) networks (core, metro, and backhaul networks), neighborhood area network and field area network (NAN/FAN), and premise network (e.g. the home area network (HAN), the building area network (BAN), and the industrial area network (IAN)).
- Power control networks (e.g. SCADA) enable functions for monitoring, control, and management in the grid.
- Power system layer is the bottom layer where the electric power flows via power generations, transmission, and distribution systems.

The communication network is the fundamental infrastructure to provide end-to-end information exchange between and among stakeholders as well as smart devices in the Smart Grid. The communication network is a complicated network of networks interconnecting a large number of heterogeneous devices and systems with various ownerships and management boundaries. However, understanding the network architecture is essential for the construction of communication networks and building security protection for these networks. For example, a network infrastructure based on a multitier architecture may not dynamically adapt the various business models and services of the system [Kim 2015] (see Figure 8.12):

Figure 8.12a shows an example of a three-tier hierarchical Smart Grid network that is organized by WANs, NANs, and HANs. Figure 8.12b shows the layer-based information and power flow model of the Smart Grid. In these diagrams, all objectives between the power facility and end-user system, such as the power generator and substation and AMI, are connected to the wired power grid infrastructure. The information data from each objective is exchanged via the data communication infrastructure, which can be established using both wired and wireless communication protocols.

However, the hierarchical system architecture is not unique [Shi 2012]. Several models of conceptual Smart Grid are being proposed, and innovative solutions continue to emerge (see more in [Gao 2012], [Kim 2015]).

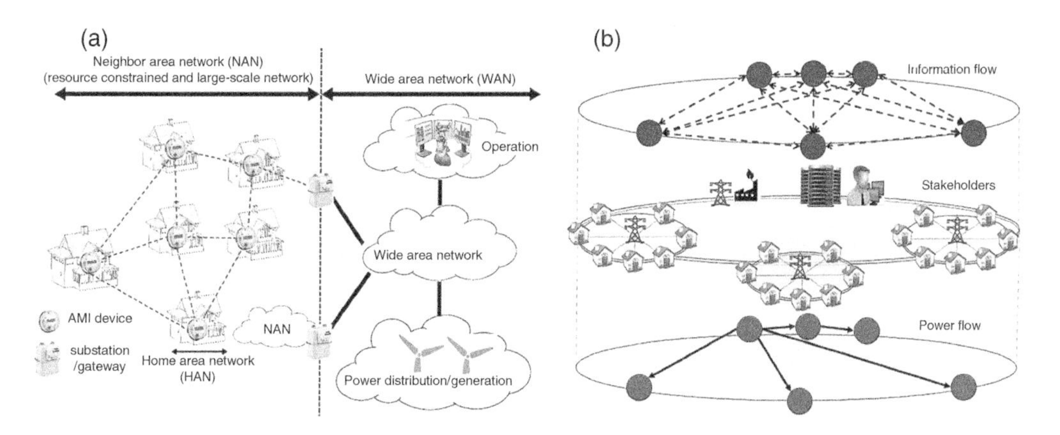

Figure 8.12 View of the multitier Smart Grid infrastructure. (a) Multitier-based Smart Grid network architecture; (b) abstraction of the information, stakeholders, and power flow. *Source:* [Kim 2015]. Licensed under CC BY-SA 4.0.

This work [Gao 2012] argues that a utility's Smart Grid integration map must support the realization of a distributed network for the utility's integrated network domains. Sometimes, a utility network will have two domain networks at the transmission level and distribution level. The authors emphasize the need for a distributed command and control system (using a system of intelligent agents) running across multiple domains of the utility network and providing end-to-end communication and data exchange among all utility assets. The suggested distributed architecture (see Figure 8.13) is composed of different components within the communication system; it does not specify any communication technology that can be employed.

However, several technology and architecture solutions are proposed by researchers worldwide. All technologies have direct impact on cybersecurity and management of the information, so understanding the security requirements of Smart Grid applications is crucial in the design of solutions for communication networks.

8.6.2 Integrated Communications

One key technology to make the modern grid dynamic and interactive for two-way real-time information and power exchange is integrated communications (IC) [NETL 2007], [NETL 2009]. Besides so many definitions for Smart Grid, some definitions have one common element, which is the integration of electrical networks with information communication infrastructure to make it more reliable and secure.

Integrated communications need to be based on high speed, various types of technologies, and medium. Integrated communications are based on advanced control methods as foundational technology that enables communication between other key technologies (sensing and measurement, advanced components, advanced controls, decision support) as depicted in Figure 8.14.

8.6.3 Sensor Networks

Another key technology includes sensor networks on the low-voltage (LV) side of the distribution system. Although such sensory data on the LV side (such as those from PMUs) have not yet been established as a critical requirement, one should assume that should become a necessity [Farhangi 2014]. Although the chosen AMI infrastructure could be the primary means of supporting such real-time data,

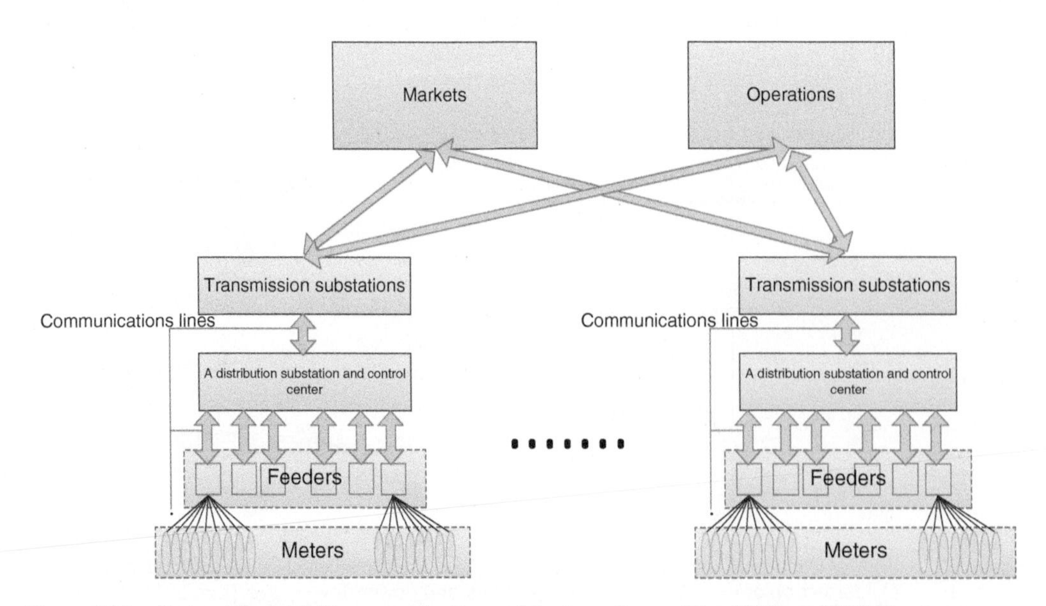

Figure 8.13 Abstract Smart Grid communication architecture. *Source:* [Gao 2012]. © 2012, Elsevier.

a utility's Smart Grid integration map must include provisions for supporting additional data networks including chosen architecture for the implementation of sensor networks.

Figure 8.15 is an illustration of the sensor system architecture with interconnecting boundary between cyber and object domain. A proposed sensor system solution for Smart Grid is discussed in [EPRI 2011]. Although there is already a significant installed base of sensors at substations that is needed for EMS and SCADA systems, the limited bandwidth connecting the substation to the enterprise does not satisfy the needs of more advanced applications including cybersecurity features for Smart Grid, so there is a need for new communication infrastructure.

A recent trend is to take benefits from mergers between virtual networking and physical actuation to reliably perform all conventional and complex sensing and communication tasks. Oil and gas pipeline monitoring provides a novel example of the benefits of CPS, providing a reliable remote monitoring platform to leverage environment, strategic, and economic benefits.

Figure 8.15 depicts the architecture of cyber–physical sensor network for infrastructure monitoring described in [Ali 2015]. This infrastructure is used for pipeline monitoring based on sensor networks and functionalities distributed among the nodes. The nodes at the lowest level are the sensing nodes that sense the parameters from the environment. Relay nodes closer to sensing nodes collect sensed data and pass it on to data dissemination nodes, which finally transmit the data over long-distance communication links to the control center.

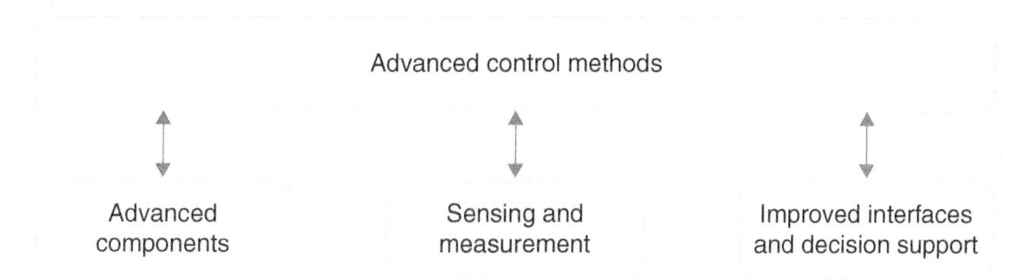

Figure 8.14 A system view of advanced control methods.

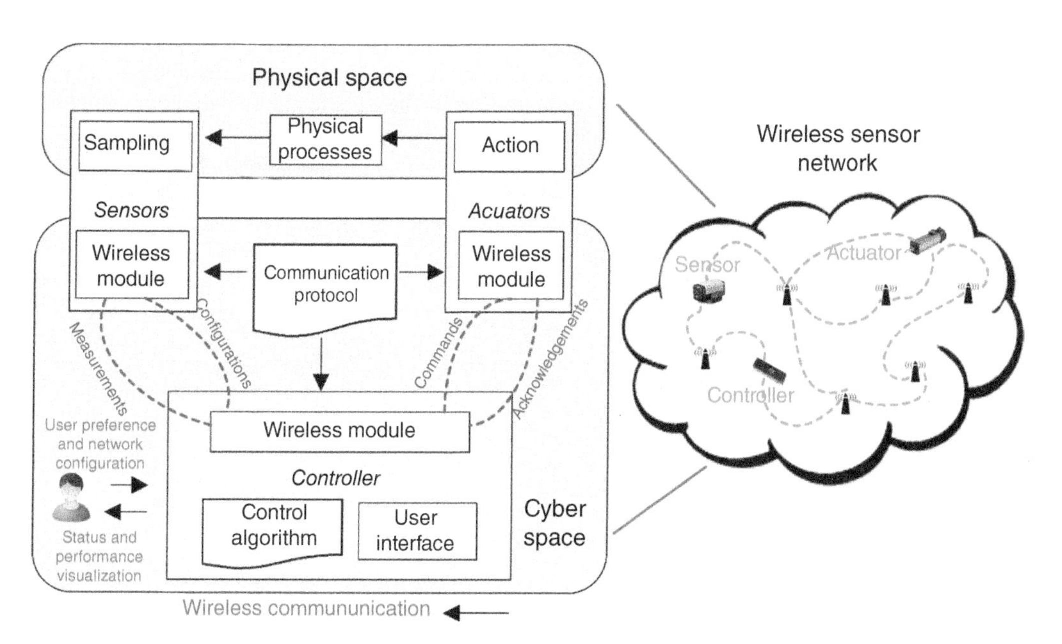

Figure 8.15 Architecture of sensor network for infrastructure monitoring. *Source:* [Ali 2015]. Licensed under CC BY-SA 4.0.

This work [Ali 2015] evaluates the applications and technical requirements for seamlessly integrating CPS with sensor networks from a reliability perspective and review the strategies for communicating information between remote monitoring sites and the widely deployed sensor nodes. The authors suggest that functionalities of nodes on each level can be made distinctively diverse such as to make the data collecting system more intelligent.

However, there are security concerns related to this cyber–physical sensor network infrastructure that are summarized in [Ali 2015]. The major challenges and security-related problems arise due to the decentralized nature and heterogeneity of protocol and hardware used to interconnect devices. The physical interactions between devices over an untrusted environment are exposing potential vulnerabilities. Therefore, all such concerns need to be addressed with thorough understanding, risk assessment, and a defense mechanism. Security goals need to be defined, and security controls should be applied in relation to system complexity. Defensive mechanisms for CPS normally must ensure continuous interaction between physical devices to highlight a clear gap between the role of attacker and defender. These interactions could be complex enough depending upon the system distribution and environmental reliability.

8.6.4 Infrastructure for Transmission and Substations

The new distributed infrastructure for smart substations supports the higher level of information monitoring, analysis, and control required for Smart Grid operations, as well as the communication infrastructure to support full integration of upstream and downstream operations [EPRI 2011]. As reported in this work, there are needs for capabilities installed at smart substations in the United States such as:

- WAN interface to receive and respond to data from an extensive array of transmission line sensors, dynamic thermal circuit ratings, and strategically placed PMUs.
- Integration of variable power flows from renewable energy systems in real time; maintain a historical record or have access to a historical record of equipment performance.
- Predictive maintenance facilitated by combining real-time monitoring of equipment and historical records.

PMUs provide system operators with feedback about the state of the power system with much higher accuracy than the conventional SCADA systems, which typically take observations every four seconds. Because PMUs provide more precise data at a much faster rate, they provide a much more accurate assessment of operating conditions and limits in real time. The availability of near-real-time data enables a system to make better decisions with potential flow on benefits to the consumer [Tariq 2013]. PMU samples and uploads the synchronized real-time data through the communication system called wide area measurement system (WAMS) (see a generic architecture in [Law 2014]).

The new infrastructure allows the coordination of the flow of intelligence from critical equipment, such as self-diagnosing transformers, with downstream operations, and be able to differentiate normal faults from security breaches. Since there is already a significant installed base of sensors at substations, but there is still limited bandwidth connecting the substation to the enterprise, the smart substation will build upon the existing platform. Historically, the communication channel to the substation was justified as part of the installation of the EMS and SCADA systems. However, a key consideration for the future is that these legacy systems have limited bandwidth [EPRI 2011]. Thus, improvements to existing communications and IT infrastructure for transmission and substations are planned [NIST SP1108r3].

8.6.5 Wireless Technologies

While wireless networks are one option contending for the deployment of Smart Grid applications, they are more considered as a cost-effective solution for some applications or considered for hybrid solutions that include wireless and wireline networks. There are different types of wireless technologies that have different availability, time sensitivity, and security characteristics that may constrain the applications.

Wireless technologies can be used across the Smart Grid system including generation plants, transmission systems, substations, distribution systems, and customer premises communications [Fernandez 2012]. This work provides a survey of wireless technologies that can be used for normal operation or as a backup option within the Smart Grid network. The authors argue that there is no single technology appropriate for all Smart Grid applications. Smart Grid communication solutions need to be designed based on the analysis of existing solutions and specific requirements such as user density, propagation characteristics, response time, data throughput, latency, jitter, etc.

Another analysis (more detailed) of requirements and performances of wireless technologies for the Smart Grid applications is provided in [NIST 2012], [DOE Comm]. Guidelines for assessing wireless Standards for Smart Grid applications are provided in [NISTIR 7761r1]. Promising wireless communication technologies that could be used for the implementation of the Smart Grid communication network are discussed in [Ho 2014].

Figure 8.16 depicts a proposed solution based on wireless technologies for the Smart Grid networks (WAN, NAN/FAN, customer premises network). A proof of concept is documented in [Ho 2014] with detail analysis and comparison of different wireless technologies including analysis of simulation results for the integrated proposed solution.

The transformation of a network from conventional modes of utility operation to an integrated network based on the Smart Grid architecture framework is the theme of many publications and research projects. More guidelines on planning networks with special emphasis on wireless broadband networks are included in [Budka 2014]. The authors propose a detailed roadmap for electric power utilities to migrate from existing multiple disparate networks to an integrated network.

8.6.6 Advanced Metering Infrastructure

Government agencies and utilities are turning toward AMI systems as part of larger Smart Grid initiatives. AMI extends current advanced meter reading (AMR) technology by providing two-way meter communications, allowing commands to be sent toward the home for multiple purposes. Wireless technologies are critical elements of the NANs, aggregating a mesh configuration of up to thousands of meters for backhaul to the utility's IT headquarters.

AMI involves two-way communications with smart meters, customer and operational databases, and various EMS. AMI is a network connecting all key enterprise systems including meter data management, customer care, auto-DR system, and energy management. The goal is to provide a highly secure,

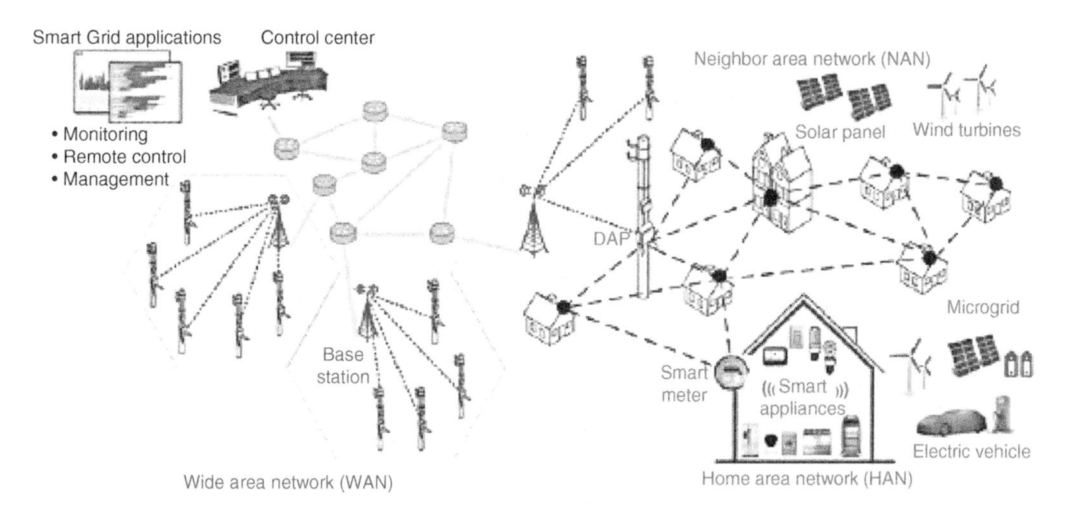

Figure 8.16 Wireless solution for Smart Grid networks. *Source:* [Ho 2014]. © 2014, Springer.

resilient, and flexible technology upgrade to the core business components that involve functions for data collection and transfer requiring security and privacy protection including the following:

- Smart meters installed at consumer site are the main component of AMI.
- Communication system that is redundant and self-healing and includes related hardware and software systems to communicate between smart meters, substation and distribution automation equipment, customer EMS, and headend software applications/meter data management systems.
- Meter data management system capable of storing and organizing data, allowing for advanced analysis and processing, and interfacing AMI headends with a range of other enterprise software applications.

An overview of AMI components and networks within the bigger context of electric power distribution, consumption, and renewable energy generation and energy storage is presented in [NIST SP1108r1].

8.7 Networking Challenges

Networking requirements for a utility network are different in many aspects compared with those for common data networks in an enterprise or for a network service provider used for data services offered to its customers. Therefore, there are many challenges that need to be understood and addressed by security engineer and control engineer during planning, design, and implementation of networks for the Smart Grid.

A major roadblock to harnessing the power of DERs and DR assets has been the lack of efficient, affordable, and secure communication paths. These devices – such as digital thermostats and water heaters – are typically sold through third-party retail outlets directly to the end users, meaning they operate on a wide diversity of operating systems and communication platforms.

8.7.1 Architecture

One of the biggest challenges of networking is the architecture of industrial networks that generally have a much deeper architecture than commercial networks. Figure 8.17 depicts a side-by-side view of a typical network architecture for the commercial network of an enterprise and a typical network architecture industrial networks. While a commercial network may consist of branch or office local area networks (LANs) connected by a backbone network or WAN, industrial networks tend to have a hierarchy three or four levels deep. Generally, the connection of instruments to controllers may happen at one level, the interconnection of controllers at the next, the human–machine interface (HMI) may be situated above that, with a final network for data collection and external communication sitting at the top.

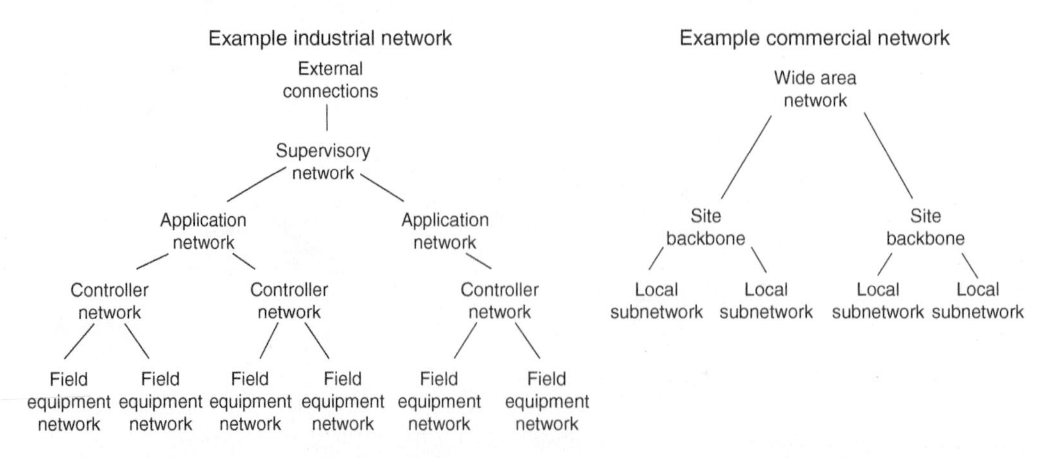

Figure 8.17 Difference in network architecture. *Source:* [Galloway 2013]. © 2013, IEEE.

8.7.2 Protocols

Different protocols and/or physical media often are used in each level, requiring gateway devices to facilitate communication. Although improvements to industrial networking protocols and technology have resulted in some flattening of typical industrial hierarchies (e.g. by combining higher layers), often the network architecture is not flattened as much as is possible in order to retain correlation to the functional hierarchy of the controlled equipment [Galloway 2013].

Also, the authors argue that power islands within a power-generating utility need to retain independent control networks in order to retain a logical separation between units at both mechanical and control level.

8.7.3 Constraints

In addition, other networking aspects to deal include the following:

- Failure severity of an industrial network system has a much more severe impact than that of commercial systems. Effects of failure of an industrial network can include damage to equipment, production loss, environmental damage, loss of cybersecurity and reputation, and even loss of life [Galloway 2013].
- Real-time requirements include response time and delays. Response time for the industrial control networks is expected to be in the range of 250 μs to 1 ms, although less stringent processes may only require response times of 1–10 ms. Delays in information delivery can severely impact the performance of control loops, especially in the case of closed-loop systems. The response time for commercial networks can be in the range of tens of hundreds of milliseconds or more commonly rather seconds. Higher levels of the hierarchy of an automation network tend to have progressively lower time requirements and at the highest levels begin to resemble commercial networks [Galloway 2013].
- Deterministic and predictable transmission of data used in the lowest levels of an industrial network such that it must be possible to predict when a reply to a transmission will be received. Also, the latency of a signal must be bounded and have a low variance referred as jitter. Low jitter is required because variance in time has a negative effect on control loops. The derivative and integral portions of a control loop are affected by time variation, and digital signal processing methods such as fast Fourier transforms require fixed intervals between sampled data. Commercial networks are as a whole not affected by jitter as severely as industrial networks are, although some exceptions exist, such as in Voice over Internet Protocols, which require low jitter to transport speech. While voice over Internet is implemented on standard networks by discarding data with a high jitter, such a solution is not appropriate for industrial use, and determinism must be built into industrial network protocols [Galloway 2013].
- Higher-capacity home networking equipment will be needed to accommodate the analysis of the energy consumption patterns of individual consumers and their equipment and the large data rate demand due to increase in data rates (meter readings will be taken at a greater frequency compared with the legacy grid); existing home networks are normally capable of handling only a limited number of nodes, while the Smart Grid may introduce up to 30 nodes, with a commensurate increase in the complexity of both network connectivity and inter-device communications.
- Strategies for either alleviating or minimizing interference of Smart Grid devices (mostly wireless and increased density); specifically, new effective inter-HAN interference mitigation schemes are needed; in densely populated urban areas, the coverage of neighboring HAN will often overlap so generating interference between the HANs.
- Common communication interface needs to be established, which is independent of the manufacturers and works on all communication media [Tariq 2013].
- Enable communication and data flow between a variety of diverse networks; for example, sensor networks communicate differently to cellular networks, so the internetwork interface has to be designed to ensure that the data can be seamlessly transferred from the consumer to the central controller over multiple heterogeneous networks [Tariq 2013].

- Security and privacy measures have to be robust and implemented to avoid the risk of large-scale damage, malicious data eavesdropping, and wormholes [Tariq 2013].
- Control of distributed generation needs to be addressed. The increase in use of residential PV, wind, and storage requires utilities to investigate the integration of this production capacity into their distribution systems and provide for control of distributed generation. There are several avenues available for the two-way communication between utilities to these renewable resources. One solution to communicate with customer-owned generation is using the AMI network.
- An increasing number of IT solutions have emerged in the areas of DERs to meet the Smart Grid needs, although there are challenges to acquire, integrate, and maintain these systems. IT vendors are working to become more interoperable in terms of their solutions and to also develop creative purchasing models, such as managed services. The potential benefits are real and increasingly measurable in terms of grid efficiency, reliability, and financial viability.
- With the new hardware and associated or embedded software, the concerns for security and privacy are mounting, so detailed analysis of vulnerabilities and threats has to be considered by engineers and security professionals for all devices, applications, and users. Consumers need to be provided with simple and easy-to-use systems that do not create a venue for successful malicious behavior. Therefore, the security of DER systems is critical to ensure no security incident to DERs has impact on the energy landscape.
- The plug-and-play market is growing faster than regulators' acceptance to implement this technology for DERs. The plug-and-play concept eliminates the need for relatively costly wiring and installation by a professional electrician; it can significantly reduce the balance of systems costs of purchasing and installing residential and small business DER systems.
- Many advantages of plug-and-play technology are needed, but this requires heavy reliance on reliable software and networks including highly secure and automated services. In most all instances, a DER automation system must be connected to utility control and communication systems.
- Integration of DERs into the utility generation, transmission, and integration system presents numerous challenges. Integration of DERs is a quite complex subject requiring the consideration of many power grid and DER variables. Modern automation systems are one of the best methods for dealing with these challenges and successfully integrating DERs into utility systems.
- The management of a distribution grid with high levels of DERs is one of the central issues confronting distribution utilities so utilities need a new set of tools to manage the technical, operational, and economic ramifications of having a high penetration of interconnected DERs. Key within this tool set is the need to provide situational awareness and to account for and mitigate the impact of DERs on grid reliability including capability to capture the economic values of DERs for distribution and BPS and for retail customers.
- The coordinated management of distributed and renewable generation, energy storage, and DERs can deliver significant value to a utility. However, in the absence of effective tools to manage high levels of DERs, the grid may experience voltage variations, overloads, phase load imbalances, frequency issues (involving microgrids and islanding), and other variations from utility operating standards.

8.8 Standards, Guidelines, and Recommendations

Although the list of standards related to the topics covered in this chapter is not complete, this section provides a list of the most common standards used for different purposes besides security specific. The security engineer and security professional have to understand, investigate, and apply a suite of standards and protocols that support functionality of DER and Smart Grid including cybersecurity.

The evidence of the essential role of standards is growing for technologies such as smart meters, microgrids, electrical battery storage systems, etc. It is reported in [Congressional], Congress that the ongoing deployment of smart meters as an area in need of widely accepted standards. The US investment in smart meters is predicted to be on the order of at least $40 billion to million new smart meters over a five-year period.

Standards are critical to enabling interoperable systems and components. Mature, robust standards are the foundation of mass markets for the millions of components that have a role in the future Smart Grid. Standards enable innovation where thousands of companies may construct individual components. Standards also enable consistency in systems management and maintenance over the life cycles of components.

8.8.1 Smart Grid Interoperability

Criteria for Smart Grid interoperability standards are discussed in [NIST SP1108r3]. European Standards are under the responsibility of the European Standardization Organizations (ESOs) that include the following:

- The European Committee for Standardisation (CEN).
- The European Committee for Electrotechnical Standardisation (CENELEC).
- The European Telecommunications Standards Institute (ETSI).

In March 2011, the European Commission and EFTA issued the Smart Grid Mandate M/490, which was accepted by the three ESOs (CEN, CENELEC, and ETSI) in June 2011. This mandate highlights the following key points: the need for speedy action and the need to accommodate a huge number of stakeholders and to work in a context where many activities are international.

M/490 requests CEN, CENELEC, and ETSI to develop a framework to enable ESOs to perform continuous standard enhancement and development in the Smart Grid field. These reports and additional information on the Smart Grid standardization activities are available on this Web site [CENELEC].

As summarized in a February 2013 report from the White House's National Science and Technology Council [NSTC 2013], interoperability standards make markets more efficient, help open new international markets to US manufacturers, and reduce the costs of providing reliable safe power to US households and businesses.

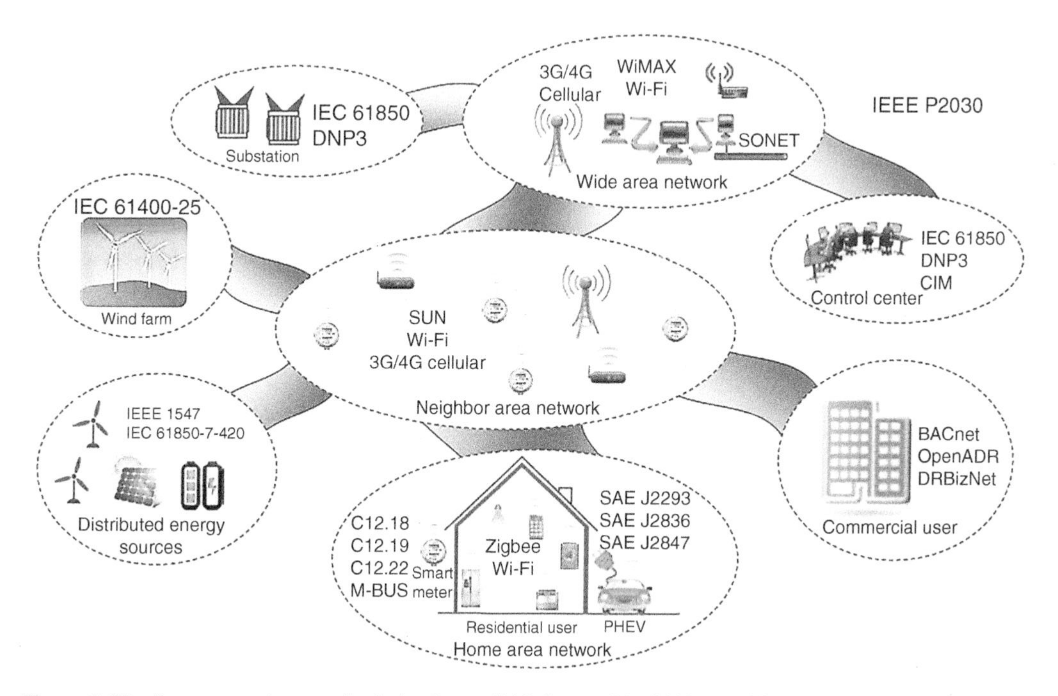

Figure 8.18 Representative standards for Smart Grid. *Source:* [Ho 2013a]. © 2013, Elsevier.

Improved standards and conformance programs, transparent market requirements, and ongoing research and development for hardware and software advances in power, communications, and information technologies are paramount for successfully accelerating the evolution of the future grid.

However, technical standards alone are not immediately effective unto themselves. Effectiveness involves partnerships and acceptance among a broad consensus and coherent implementation across the transmission and distribution infrastructure, physical boundaries, utilities, system integrators, and customers understanding and using those standards.

8.8.2 Representative Standards

Figure 8.18 depicts a summary of representative standards for Smart Grid. The standards depicted in the diagram are directly considered as relevant to the Smart Grid functions and interoperability. Information related to IT and Smart Grids is available for European countries at [CORDIS].

This is an example of the overall architecture of the Smart Grid communication network as an integration of HAN, NAN, and WAN including protocols that can be used in Smart Grid networks. Each type of network has its own set of protocols that support a large variety of Smart Grid applications. Many standards applying for the Smart Grid are sponsored by national and international organizations. As illustrated in this diagram, many technologies and protocols are used to support the communication of the information. Standards are continuously revised or new standards are developed to support the interoperability, QoS, and cybersecurity by different organizations and forums.

9

Power System Characteristics

9.1 Analysis of Power Systems

As described in [GridWise 2014], the future grid will serve as a very dynamic two-way power flow system of the future, ensuring that the system remains stable and resilient. Therefore, changing conditions make it imperative that the grid and grid operations evolve. The evolution has significant implications for reliability, stability, interconnectivity, resilience, security, and other aspects such as transmission and distribution operations, consumer choice, etc. In addition, the impacts of distributed energy resources (DERs) on the grid's physical stability and other characteristics require attention before any system is designed.

9.1.1 Analysis of Basic Characteristics

The general characteristics that have to be accounted in the analysis of power systems may include [Venkata 2013], [Kundur 2004]:

- Most complex system ever devised by man.
- Highly nonlinear system that operates in a constantly changing environment; loads, generator outputs, and key operating parameters change continually.
- Constrained dynamical systems, as their state trajectories are restricted to a particular subset in the state space.
- Nonautonomous and time varying, interacting with its (typically unmodeled) environment; examples include load variations and network topology changes due to switching in substations.
- High-order multivariable processes whose dynamic response is influenced by a wide array of devices with different characteristics and response rates.

A power system can be characterized as having multiple states, or modes, during which specific operational and control actions and reactions are taking place. These modes can be described as [Amin 2008]:

- Normal – e.g. economic dispatch, load frequency control, maintenance, and forecasting.
- Disturbance – e.g. faults, instability, and load shedding.
- Restorative – e.g. rescheduling, resynchronization, and load restoration.

In addition to these spatial, energy, and operational levels, power systems are also multi-scale in the time domain, from nanoseconds to decades. A time hierarchy of power systems is shown in Appendix B. Reliability and stability are the most monitored characteristics in power systems. The concept of reliability is also used in networking. In industrial and systems engineering, reliability is accompanied by statistical and probabilistic approaches that characterize system performance after predicted and unpredicted failures. Besides these two concepts of stability and reliability that are well monitored and analyzed in power systems, other characteristics include interconnectivity, security, robustness, and resilience.

Building an Effective Security Program for Distributed Energy Resources and Systems: Understanding Security for Smart Grid and Distributed Energy Resources and Systems, Volume 1, First Edition. Mariana Hentea.

An emerging trend is the concept of power system cyber–physical resilience. The following sections provide an overview of these characteristics.

9.1.2 Stability

Power system stability is similar to the stability of any dynamic system and has fundamental mathematical underpinnings. Although precise definitions of stability can be found in the literature dealing with the rigorous mathematical theory of stability of dynamic systems, a physically motivated definition of power system stability is provided in [Kundur 2004]:

> Power system stability is the ability of an electric power system, for a given initial operating condition, to regain a state of operating equilibrium after being subjected to a physical disturbance, with most system variables bounded so that practically the entire system remains intact.

Stability of a power system refers to the continuance of intact operation following a disturbance. It depends on the operating condition and the nature of the physical disturbance. Generally, stability is the system's ability to tolerate small perturbations. The small perturbations often come from uncertainties in measurements and system models [Arghandeh 2016].

Power system must be able to adjust to the changing conditions due to disturbances and must be able to survive numerous disturbances from small and large or of a severe nature, such as a short circuit on a transmission line or loss of a large generator. Small and large disturbances can cause short- and long-term stability problems.

Power system stability problem gets more pronounced in case of interconnection of large power networks. The interconnected power systems undergo various electric and magnetic interactions due to random behavior of the consumers and sometimes of the power plants, which lead to power, voltage, or frequency oscillations. When subjected to a disturbance, the stability of the system depends on the initial operating condition as well as the nature of the disturbance [Venkata 2013], [Kundur 2004].

The analysis of stability problem based on a practitioners' approach is motivated toward the need for a more comprehensive treatment of stability theory for power systems [Kundur 2004].

Because of power system characteristics (high dimensionality, complexity, large number of variables, etc.), simplifying assumptions are made in order to allow the analysis of specific types of problems with satisfactorily accuracy.

In the analysis of stability problem, it is necessary to identify the key factors that contribute to instability and provide methods of improving stable operation. This task is facilitated by classification of stability into appropriate categories based on factors such as:

- Physical nature of the resulting mode of instability as indicated by the main system variable in which instability can be observed.
- Size of the disturbance considered, which influences the method of calculation and prediction of stability.
- Devices, processes, and the time span that must be taken into consideration in order to assess stability.

A clear understanding of different types of instability and how they are interrelated is essential for the satisfactory design and operation of power system.

9.1.3 Partial Stability

Classification of the power system stability by focusing mainly on only one variable (e.g. voltage, frequency, or rotor angle) is a meaningful practice; such approach is also known as partial stability. Figure 9.1 shows categories of stability:

- Rotor angle stability – Ability of synchronous machines of an interconnected power system to remain in synchronism after being subjected to a disturbance.
- Voltage stability – Ability of a power system to maintain steady voltages at all buses in the system after being subjected to a disturbance from a given initial operating condition.

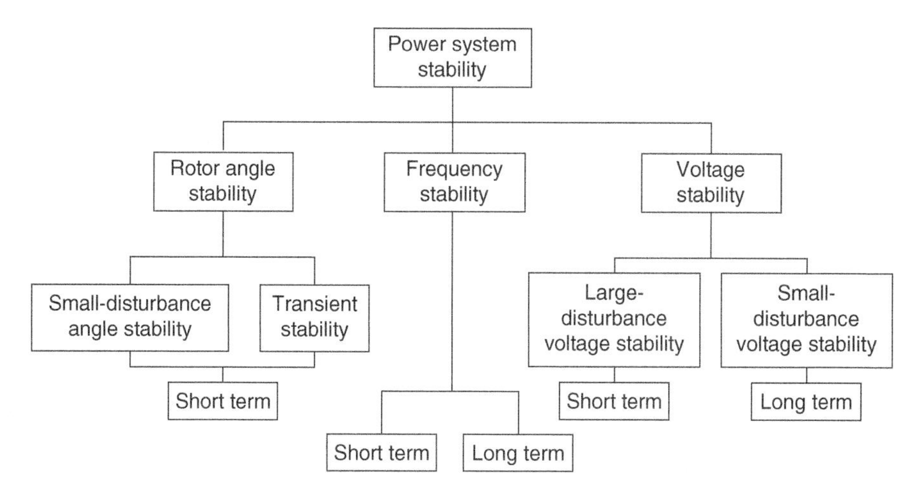

Figure 9.1 Classification of power system stability. *Source:* [Kundur 2004]. © 2004, IEEE.

Table 9.1 Stability study time frames.

Stability type	Short term	Long term
Rotor angle	3–5 seconds following a severe disturbance; 10–20 seconds following a small disturbance	
Voltage	1–3 seconds	Tens of minutes
Frequency	A few seconds	Tens of seconds to several minutes

- Frequency stability – Ability of a power system to maintain steady frequency following a severe system upset, resulting in a significant imbalance between is needed.

Following a disturbance, stability studies aim to devise remedies. However, these studies need to be conducted within a time frame following a disturbance. Control systems and security applications should be designed to support these studies within the time frames. Based on the analysis results of studies discussed in (see [Kundur 2004], [Ouyang 2012]), a summary of stability time frames is provided in Table 9.1.

The needs for analysis studies to devise remedies and actions mandate requirements for control systems applications, communications, and information security controls. For example, the time frames for stability studies and response time constraints should be criteria when designing the security access controls to avoid delay and support the availability of the information when it is needed.

One concern is the integration of DERs with the power grid that could cause technical problems in the areas of voltage stability, protection, control, and security. It is common knowledge that each DER has its own integration issues. For example, solar PV and wind power both have intermittent and unpredictable generation, so they create many stability issues for voltage and frequency. A large scale of distributed generation (DG) deployment may affect grid-wide functions such as frequency control and allocation of reserves. Similarly, the power grid outages can cause problems to DER functions. The simulations presented in this paper [Banu 2014] show that the short-circuit faults in the power grid have disturbing effects on optimal operation of grid-connected PV systems. The impact of grid faults on the PV system performance depends on the grid fault type, and some faults have a higher impact on performance both at the PCC and inside the PV system.

A cybersecurity attack could create instability problems. Therefore, understanding the concepts of stability and impacts is critical to properly mitigate the risks and ensure adequate protection to information and systems. [IEEE PES-TR22] provides useful guidance on the stability problems.

9.2 Analysis of Impacts

While Smart Grid is focusing on incorporating renewables and increasing energy efficiency, different regions emphasize slightly different aspects of environmental impact. For example, North America leads renewable energy integration and consumption shifting. Therefore, there is a need to understand the impacts of DERs and interconnectivity scenarios.

9.2.1 DER Impacts

There are two aspects of DERs (known collectively as combined heat and power, wind and photovoltaic (PV) systems, demand response, energy storage, vehicle-to-grid systems, and microgrids) integration with system operations and energy markets. The impacts include grid's physical stability and price responsive on wholesale market behavior [Masiello 2014]. These are two very different issues, but both have implications for wholesale operations visibility and retail-level interactions.

For example, the intermittent nature of wind and solar resources can create voltage stability problems between the transmission system and provider. In response, utilities and regulators are increasingly mandating strict interconnection requirements. Renewable energy producers must condition the power produced in order to interconnect with the power grid and not interfere with the grid's overall performance. Solutions for steady-state voltage regulations are well-documented practices. However, it is important to investigate the reliability impacts of these resources during power system analysis, especially in the planning process. The uncertainties that DG brings to the system need to be considered in power system analysis [Dong 2010].

Maintaining the power systems in a secure operating state assumes that some electrical variables should be maintained within admissible limits. Control systems are designed to monitor the variables and regulate the values including sending notifications to operators. The control systems (see open-loop and closed-loop control definitions in Appendix B) attempt to restore the power system to steady conditions and balance load with power generation supply. The various types of stability problems of a power system cannot be properly understood and effectively dealt with by treating them as a single problem.

In addition, control systems need to have the capability to monitor cybersecurity events and detect if the disturbance is caused by a cybersecurity intrusion. Cybersecurity intrusion is a security event that involves a security violation of the confidentiality, integrity, or availability of an information system; or the information the system processes, stores, or transmits; or that constitutes a violation or imminent threat of violation of security policies, security procedures, or acceptable use policies, where an information system can be a control system [CNSSI 4009], [RFC 4949].

9.2.2 Interconnectivity

One of the great engineering achievements of the last century has been the evolution of large synchronous alternating current (AC) power grids, in which all the interconnected systems maintain the same precise electrical frequency. Most of the electric utilities in the contiguous United States and a large part of Canada now operate as members of power pools, and these pools in turn are interconnected into one gigantic power grid known as the North American Power Systems Interconnection. Their participation in a power pool affords a higher level of service reliability and important economic advantages. Minimizing the likelihood that an interconnection will lead to such problems as voltage collapse, dynamic and transient instability, or cascading outages due to propagated disturbances requires careful planning and well-coordinated operation.

The North American power system is composed of four synchronous systems, namely, the Eastern, Western, Texas, and Quebec interconnections. The Eastern Interconnection by itself has been called the largest machine in the world, consisting of thousands of generators, millions of kilometers of transmission and distribution lines, and more than a billion different electrical loads. Despite this complexity, the network operates in synchronism as a single system. Similarly, Western European Interconnection reaches from the United Kingdom and Scandinavia to Italy and Greece, embracing along the way much of Eastern Europe (for example, Poland, Hungary, Slovakia, and the Czech Republic).

Other synchronous interconnections among countries are expanding in Central and South America, North and sub-Saharan Africa, and the Middle East. The greatest benefits of interconnection are usually derived from synchronous AC operation, but this operating mode can also entail greater reliability risks. In any synchronous network, disturbances in one location are quickly felt in other locations. After interconnecting, a system that used to be isolated from disturbances in a neighboring system could become vulnerable to those disturbances.

At the same time that synchronous AC networks have reached the continental scale, the use of high-voltage direct current (HVDC) interconnections is also rapidly expanding as a result of technical progress over the last two decades. HVDC permits the asynchronous interconnection of networks that operate at different frequencies, or are otherwise incompatible, allowing them to exchange power without requiring the tight coordination of a synchronous network. HVDC has other advantages as well, especially for transmitting large amounts of power over very long distances.

There are number of technical rationales for grid interconnections, many of which have economic components. Interconnections obviously entail the expense of constructing and operating transmission lines and substations or, in the case of HVDC, converter stations. Interconnections also entail other costs, technical complexities, and risks.

Power system interconnection requires a high degree of technical compatibility and operational coordination, which grows in cost and complexity with the scale and inherent differences of the systems involved. For example, when systems are interconnected, even if they are otherwise fully compatible, fault currents (the current that flows during a short circuit) generally increase, requiring the installation of higher-capacity circuit breakers to maintain safety and reliability.

However, all these changes needed by interconnection require extensive planning studies, computer modeling, and exchange of data between the interconnected systems. Fundamentals of both AC and DC interconnections are discussed in the [UN 2006]. Technical planning of a grid interconnection should be coordinated with economic, organizational, legal, and political aspects of a potential interconnection project from the outset of project consideration. Several basic technical issues must be addressed early in the planning process for a grid interconnection. Other options based on technologies – FACTS and HVDC – are being considered as alternatives or complements to traditional transmission upgrades. Simulation software is an essential tool for planning and operating an interconnection. For modeling to be effective, however, extensive technical data must first be gathered and shared between systems, and personnel must be trained.

Another aspect is the interconnection of old SCADA systems to Internet, networks, or telephone lines, and the every more intensive use of computer networks and wireless systems have risen the potential threats to interconnections and power systems. Technical planning of a grid interconnection should be coordinated with economic, organizational, legal, and political aspects of a potential interconnection [UN 2006], [Meslier 2000].

Generating solar polar at equator in Africa and transporting to Europe or other countries can be achieved by means of integrated power grids across regional and national borders. This can ensure clean air, long-term efficiency, and sustainability adequately tackled [Smrcka 2010]. Data transparency and requirements for planning exchange of data between the owners/operators of the systems to be interconnected is essential from the outset of an interconnection project.

9.3 Reliability

Reliability of a power system refers to the probability of its satisfactory operation over the long run. It denotes the ability to supply adequate electric service on a nearly continuous basis, with few interruptions over an extended time period. It is defined as the probability that power system will perform its intended function satisfactorily (from the customer point of view) during its estimated life under specified environmental and operating conditions [Encyclopedia].

Power grids are complex systems whose security and reliability depend on the collaboration among all the concerning entities, including peer operators, government authorities, etc. [Tong 2013]. Therefore, specific definitions are used for different aspects:

- Reliability is the ability of a component or system to perform required functions under stated conditions for a stated period of time [IEEE Std 493].
- The North American Electric Reliability Council (NERC) defines reliability for different components (bulk power system or operations; see Appendix B).

9.3.1 Reliable System Characteristics

Power systems have largely operated without smart technology for decades. In fact, many power systems operate at 99.999% reliability at the bulk transmission level [EPRI 2011]. As long as reliability levels have been maintained (the lights were still on) and costs were low (rates have been essentially flat for decades), the norms of reliability were satisfactory. As discussed in this report, the reliability at the distribution is at different values when it is compared with the bulk and transmission systems. Factors such as greater complexity, exposure, and geographic reach of the distribution system result in inherently lower reliability, reduced power quality, and greater vulnerability to disruptions of any kind. Using a reliability measure of average total duration of the interruptions experienced by a customer in a year, over 90% of the minutes lost by consumers are attributable to distribution events. In 2004, EPRI estimated that a fully automated distribution system could improve reliability levels by 40%. Many factors contribute to the reliability characteristic of power system as discussed in [Osborn 2001], [Brown 2002]. Digital-based technologies can improve the current value of today's power system.

Therefore, reliability definitions are emerging to reflect new trends and requirements such that one defines reliability in [Brown 2002], [Arghandeh 2016], [Osborn 2001]:

> Reliability is the ability of the power system to deliver electricity to customers with acceptable quality and in the amount desired while maintaining grid functionality even when failures occur.

Characteristics of a reliable electric system include [GridWise]:

- Meeting end-use customer demands consistently and effectively.
- Failing rarely.
- Redundant controls and assets enabling the system to keep operating despite the loss of that asset when one portion of the system fails (whether one specific asset, such as a power plant or transmission line or a geographic region).
- Sufficient resilience that one asset or region can lose service without causing other portions of the system to fail.
- Graceful failure, in the sense that the failure is slow enough for affected pieces to protect themselves from damage (whether those pieces are power plants shutting down or traffic lights and hospitals switching to backup power quickly enough to protect public safety).
- Quick restoration of service once the system has failed so that critical services and societal transactions are not affected for a long period of time.

However, advances in Smart Grid technologies are not a substitute for good maintenance practices, inspection, and vegetation management [EPRI 2011].

9.3.2 Addressing Reliability

Reliability can be addressed by considering two basic functional aspects of the power systems:

- Adequacy – Ability of the power system to supply the aggregate electric power and energy requirements of the customer at all times, taking into account scheduled and unscheduled outages of system components.

- Security – Ability of the power system to withstand sudden disturbances such as electric short circuits or non-anticipated loss of system components.

Security of a power system refers to the degree of risk in its ability to survive imminent disturbances (contingencies) without interruption of customer service. It relates to robustness of the system to imminent disturbances and, hence, depends on the system operating condition as well as the contingent probability of disturbances.

However, meeting reliability objectives in modern grids is becoming increasingly more challenging due to various factors that include renewable resources (solar, wind, biofuels, geothermal), load management, demand response, electrical transportation, and storage devices [Moslehi 2010].

Addressing the reliability engineering design of a Smart Grid architecture requires an understanding what reliability means in the power industry as well as the ICT industry in order to effectively communicate reliability requirements and capabilities. This involves considering various aspects and designs to include requirements and metrics for each aspect [P2030 2011]:

- Reliable delivery of safe electrical power.
- Reliable data.
- Reliable communications.
- Reliable operation of Smart Grid equipment.
- Contingency mode of operation.

9.3.3 Evaluating Reliability

Power engineers typically think of reliability in terms of electric power reliability – a measure of the continuous safe delivery of electric power to customers. Often, regulatory agencies evaluate the electric power reliability provided by a utility. Examples of measures include reliability indices such as system average interruption duration index (SAIDI) and system average interruption frequency index (SAIFI) [IEEE Std 493].

When considering the reliability of data delivery to support Smart Grid applications, a different concept of data reliability needs to be included. Examples of data reliability characteristics include [P2030 2011]:

- Priority.
- Availability.
- Level of assurance.
- Ability to withstand high-impact low-frequency events.

Therefore, one set of measures can describe electric power reliability, while a different set of measures can describe data reliability and communications. By the end, the safety and reliability of the power system should prevail in the event of a disruption of the ICT infrastructure. In addition, when evaluating the reliability and performance of information and computing technology in power systems, it is important to consider the effects of real-world environments.

Examples of factors that have the potential to degrade the reliability include [P2030 2011]:

- Weather (e.g. tornados and lightning).
- Electromagnetic interference (EM) in normal environments, as well as natural and man-made events; broad categories of EMC to be considered include:
  - Commonly occurring EMC events such as electrostatic discharges, fast transients, and power line disturbances
  - Radio-frequency (RF) interference from various kinds of wireless transmitters
  - Coexistence of wireless communications systems so that wireless communications can be incorporated beneficially into the Smart Grid
  - High-level EM disturbances, both from intentional criminal and terrorist acts and from naturally occurring events such as lightning surges and geomagnetic storms
- Security issues (e.g. cybersecurity and physical security) such as:
  - Accidents (e.g. dig-ins)
  - Geography (e.g. distance, terrain, topography)
  - Other major events affecting large areas (e.g. earthquakes, hurricanes, tsunami, forest fires).

The Smart Grid has the potential to improve electrical distribution reliability [P2030 2011]. Accordingly, the ability of a Smart Grid applications to withstand lost or delayed data has to be implemented, and workaround of these issues should be provided. It is important for power engineers to communicate reliability needs clearly to avoid unneeded ICT costs as well as excessive electric power interruptions. Also, reliability is a quantifiable measure that can be used in the control and management of an information system.

9.3.4 ICT Reliability Issues

Reliability provides an early warning about the quality of the information system and identifies the areas where there are problems. Also, it can be used to compare various information systems. Information technology (IT) contributes to improvements on the calculation of the reliability index, also called as SAIFI.

This index is often used by regulatory agencies to evaluate the electric power reliability provided by a utility. This encompasses the performance of distribution, transmission, and generation. A typical goal of any utility is to keep the values of SAIFI as small as practical. The accuracy of this index is improved by using measurements provided by different applications. For example, weather-related failures including all of the failure and repair data are used to calculate electrical power reliability metrics. In addition, electrical distribution reliability is improved by using various sensory and control data delivered through ICT-based networks.

However, some applications data paths require extremely high levels of reliability (e.g. fault clearing and autonomous restoration of power), while most applications should be able to withstand relatively long and/or frequent interruptions without negatively impacting electric power reliability (e.g. meter reading). Therefore, the reliability required for a given data flow is a basic requirement that needs to be set based on the power system requirements.

The process for developing specific criteria for a data flow is an iterative and collaborative process. The power engineers and ICT professionals need to work together to determine the optimum implementation of interfaces and data links to support the data flows.

The reliability metrics to be used and the ranges of possible values must be understood by all in order for this process to work effectively. Power engineers need to understand and translate between the electric power reliability requirements they are used to and the data reliability characteristics that ICT systems use. The power reliability inputs can be structured using concepts typical of the power industry: duration of failure and frequency of failures.

In the context of the Smart Grid, information data loss does not necessarily result in power outages. Thus, it may be useful for electric power engineers to think in terms of values that parallel traditional terminology.

Reasonable consideration needs to be made for possible causes of data interruption because of such factors when determining the reliability and performance of planned data links. Some of the data reliability characteristics that should be considered include priority, availability, level of assurance, and capability to withstand high-impact events.

Another concept to be understood is contingency mode of operation in which power is safely delivered to customers in the event of a failure of the communication aspects of the system. This concept is important for events that impact local power delivery reliability as well as for high-impact and low-probability regional events. But, perhaps most importantly, the existence of this contingency mode for some DER systems helps in mitigating the need to spend exorbitant amounts of money on extremely reliable interfaces. During the process for defining data flow requirements, appropriate consideration should be given to the system's ability to perform its functions through certain failures in contingency mode.

Therefore, information reliability can be characterized using qualitative and quantitative metrics. Reliability is the ability of a component or system to perform required functions under stated conditions for a stated period of time [IEEE Std 493]. Different technologies and applications of missions often establish specific reliability indices definitions or establish qualitative definitions of reliability attributes. Examples of qualitative terms include [P2030 2011]:

- Informative – Not necessarily important for operations.
- Important – Failure of information transfer may result in loss of revenue, damage to public image, or loss of technical information.
- Critical – Failure of information transfer may result in compromised safety or damage to equipment.

In addition to ICT issues, DER impacts to reliability have to be understood and addressed accordingly.

9.3.5 DER Impacts

Reliability of some types of DG resources such as renewables is a concern for the transmission and distribution network service providers. Although the renewable resources may have adverse effects on reliability, as renewables grow over the long run, increased penetration of demand response, storage, devices, and PEVs will complement the conventional remedies [Moslehi 2010]. The authors argue that improving reliability is realized by the following:

- Demand response can serve as an ancillary resource to help reliability.
- Storage devices make the net demand profile flatter and alleviate congestion in both transmission and distribution; most battery storage devices can respond in sub-second time scales, so they can be used as enablers of fast controls in a Smart Grid.
- Various forms of stored energy to fully harvest the potential of renewable resources; integrated, grid-tied storage is helping to make solar energy a dependable, on-demand power that is deliverable with controllable power ramp rate sources [Scaini 2012].
- Selecting the right energy storage system is key to optimizing price per performance, maximizing energy production, and minimizing grid impact for utilities [Scaini 2012].
- Implementing modern communication and information technologies to enable an infrastructure that provides grid-wide coordinated monitoring and control capabilities and fail-proof and nearly instantaneous bidirectional communications among all devices ranging from individual loads to the control centers including all important equipment at the distribution and transmission levels and utilization of modern cybersecurity measures.

It is predicted that by 2016, there will be a new distributed solar PV installation every 83 seconds in the United States, and GTM Research is forecasting another doubling over the next 2.5 years [Thompson 2013].

As demands for renewable deployment increase, integration requirements increase for all levels of bulk power system, distribution system, and customer system. Numerous technologies can be used for the integration of DERs (see examples in Appendix B). Utilities argue that inverters play a larger role, but there are concerns. For example, there is a need for inverters to be smarter to incorporate a range of automated control capabilities that help smooth out potential grid fluctuations [Berdner 2014]. In addition, design documents for Smart Grid applications need to provide reliability requirements for different components and functions including security controls.

9.4 Resiliency

As with many other definitions, resilience has different definitions and meanings to different entities. Resilience is defined as the ability to resist, absorb, recover from, or successfully adapt to adversity or a change in conditions (see more information in Appendix D). It also means the ability to prepare for and recover rapidly from disruptions [SGAC 2014].

Resilience can apply to individual components or to systems. As applied to power grid, it is needed to also include the context of energy security and clarify that resilience is measured in terms of robustness, resourcefulness, and rapid recovery [SPP Dictionary]. Resiliency absolutely impacts grid reliability. A more resilient grid is a more reliable grid.

The energy sector envisions a robust, resilient energy infrastructure. Energy infrastructure resilience, including electrical infrastructure, is defined as the ability to reduce the magnitude and/or duration of disruptive events [SPP Dictionary]. Resiliency aspects of the Smart Grid include self-healing from power disturbance events, operating resiliently against physical and cyber threats. It resists attacks on both the physical infrastructure (substations, poles, transformers, etc.) and the cyber structure (markets, systems, software, communications, etc.).

9.4.1 Increasing Resiliency

Different approaches are explored to increase the resilience of power grid. One approach to resilience is assessment of needs and technologies to be used (e.g. advanced sensor networks and metering infrastructure), new architectures that may be enabled, or improvement of existing large-scale generation, transmission, and distribution components that serve as its backbone. Sensors, cameras, automated switches, and intelligence are built into the infrastructure to observe, react, and alert when threats are recognized within the system. The system is resilient and incorporates self-healing technologies to resist and react to diverse threats. Constant monitoring and self-testing can be conducted against the system to detect threats and control threats.

Other approaches are described by goals identified in [SPP 2010]. Examples of key goals include:

- Information sharing and communication – Establish robust situational awareness within the energy sector through timely, reliable, and secure information exchange among trusted public and private sector partners.
- Physical security and cybersecurity – Use sound risk management principles to implement physical and cyber measures that enhance preparedness, security, and resilience.
- Public confidence – Strengthen partner and public confidence in the sector's ability to manage risk and implement effective security, reliability, and recovery efforts.

In the United States, DHS suggests that the primary goal of resilient electric grid is to develop and demonstrate an inherently fault current limiting, high-temperature, superconducting cable for increased electric grid resiliency [DHS Initiative]. The second goal is recovery transformer to enable recovery within days instead of months (or years). However, the resilience of an infrastructure or enterprise depends on its ability to anticipate, absorb, adapt to, and/or rapidly recover from a disruptive event.

Another approach is exploring Smart Grid resilience at three stages (initial failure, damage propagation, recovery process) to determine practical parameters that can be extracted to characterize resilience features. Then, resilience metrics are determined under different hazard types. Based on these metrics, the effects of different resilience improvement measures can be quantified and provide insight and direction for efficient implementation of the Smart Grid. Figure 9.2 depicts typical performance response curve of an infrastructure system following a disruptive event.

As described in [Ouyang 2012], the performance response process of an infrastructure system following a disruptive event is divided into three stages, reflecting system-resistant, absorptive, and restorative capacities. These capacities together determine system-level resilience in a quantitative fashion. Figure 9.3 shows the time sequence of more specific events (e.g. detection time, decision time) that can be used to characterize the performance of a dynamic system after a disruptive event. [Elsafdi 2019] discusses also the concept of dynamic resilience and methodology to enhance the dynamic resilience.

Other approaches used by utilities to improve outage management are documented in [DOE 2014]. Utilities have plans to continue investments in Smart Grid technologies, tools, and techniques to improve storm response capabilities and reliability, build resilience, lower costs, and improve customer services. The key technologies deployed by utilities include:

- Distribution automation, including automated feeder switching (AFS) and fault location, isolation, and service restoration (FLISR).
- Integrating advanced metering infrastructure (AMI) capabilities with outage management systems.

In addition, researchers and laboratories advocate new approaches and technologies in academic published papers and research projects. Resilience is defined as the capacity of a control system to maintain state awareness and to proactively maintain a safe level of operational normalcy in response to

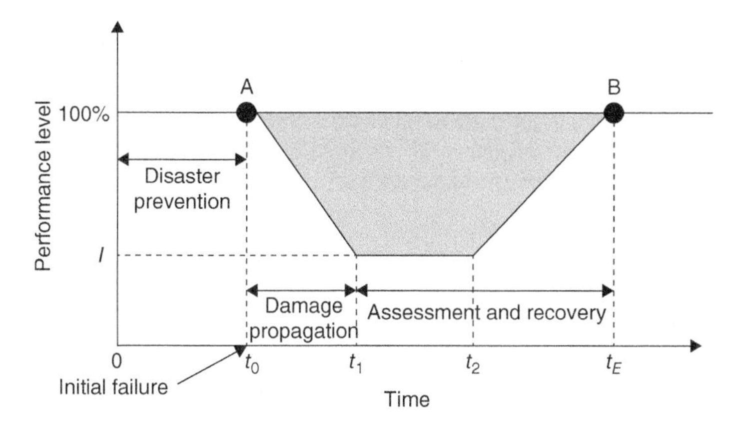

Figure 9.2 Typical performance response curve following a disruptive event. *Source:* [Ouyang 2012]. © 2012, Elsevier.

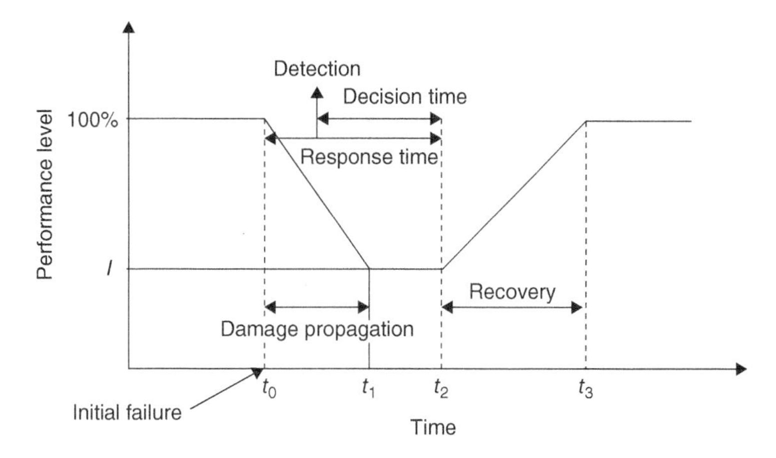

Figure 9.3 Overall performance response after a disruptive event. *Source:* [Ouyang 2012]. © 2012, Elsevier.

anomalies, including threats of a malicious and unexpected nature [Rieger 2016]. The authors' vision is to identify ways to cope with failure and attack besides adaptive and transformative systems. Adaptive systems include components designed to function in more than one role, allowing self-modification and leading to emergent properties that counterbalance anomalies while preserving function. Transformable systems have the capacity to reconstitute into fundamentally new systems when external forces render an existing system untenable. Ideally, both adaption and transformability are integrated in resilient systems.

State awareness enables the transformation from reactive to proactive control of power generation, transmission, and distribution systems. It also provides defense in depth against malicious actors and actions and support to overlapping and sometimes competing regulatory agencies. State of awareness can be achieved when resiliency is designed into control systems – when resilience is the primary design objective.

Researchers argue that current resilience research is mainly focused in two areas: organizations and IT. However, a resilient control system design should be based on multidisciplinary aspects that need to include [Rieger 2016]:

- Human systems
- Complex control networks
- Cyber awareness
- Data fusion.

Practical metrics that define how systems operate at an acceptable level of normalcy despite disturbances or threats are suggested in [Rieger 2014]. Metrics are applied to system integrity to establish performance and proper operation including impact and effects on the process that roll into the business case.

A practical approach is taken to integrate the cognitive, cyber–physical aspects that should be considered when defining solutions for resilience. A proposed definition for power system cyber–physical resilience is described in [Arghandeh 2016] as follows:

> Power system cyber-physical resilience is the system's ability to maintain continuous electricity flow to customers given a certain load prioritization scheme. A resilient power system responds to cyber-physical disturbances in real-time or semi real-time, avoiding service interruptions. A resilient power system alters its structure, loads, and resources in an agile way.

Power system cyber–physical resilience centers around the system's ability to recognize, adapt to, and absorb disturbances in a timely manner. It includes understanding the system's boundary conditions and their changes during disturbances [Hollnagel 2015]. As emphasized in this work, one goal in resilience engineering is to model and predict the short-term and long-term effects of change and line management decisions on resilience and thus on risk (see more at Web site Resilience Engineering, https://www.resilience-engineering-association.org/blog/2019/11/09/what-is-resilience-engineering/).

9.4.2 DER Opportunities

It should be noticed that risk-based protective programs and resilience are complementary elements of the comprehensive risk management strategy pursued under the Energy Sector Specific Plan (SSP) (the most recent is SSP of 2015 – an update of 2010 SSP). Distribution grids that have interconnected microgrids and other DER assets are more resilient to disruptions resulting from natural or human causes because they can reduce the numbers of homes and business affected by an outage. Also, the concept of distribution as a distributed multidimensional power flow network is of growing importance. Machine-to-machine (M2M) technologies add intelligence and enable communications between grid devices to create new levels of operational awareness. Increasingly efficient and cost-effective micro-generation and energy storage products offer alternatives to existing generation strategies.

Microgrids contain generation assets, typically renewables, but can include cogeneration and traditional energy sources too. The electricity generated by these assets is consumed within the microgrid, or it can be sold back to the larger utility grid through interconnection agreements.

Generation and energy storage assets can be distributed across the grid, along with the intelligent devices and software to securely manage these assets. Also, new software applications and services to manage the increasing numbers of devices that create, monitor, and control electricity across the supply chain aim to improve the resiliency of the systems. As the proportion of intermittent and variable power generation increases, the ability to accurately forecast the generation capacity – along with loads – is becoming essential. Grid resiliency through sensible cybersecurity and DER strategies can improve grid reliability. However, poorly integrated DG and electric vehicles (EVs) could reduce grid reliability, and mitigation solutions are required. That means to have a technology gap assessment and modernize the infrastructure. Build and rebuild smart is offered as an alternate definition for resilience [SGAC 2014]. The report retains the need for new regulatory policies and metrics that encourage utilities and consumers – both residential and business – to invest in DER and microgrid solutions to build resiliency as well as reliability into the Smart Grid.

The Smart Grid could become a network of integrated microgrids that are internally regulated whereby power from different sources is fine-tuned to local supply and demand by consumers of different types [Wolsink 2012]. Microgrids support a flexible and efficient electric grid by enabling the integration of growing deployments of DERs such as intermittent renewable energy resources. In [Wolsink 2012], the author argues that critical to the development of decentralized renewable power generation is the need to optimize the microgrid business model within the community on which the microgrid is based. Therefore, there could be many business model resulting in different types of controls and different components for the microgrid architecture.

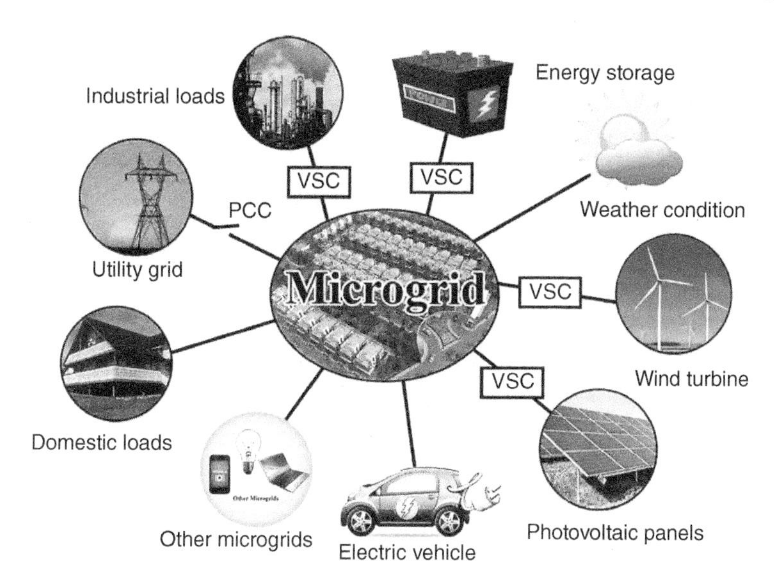

Figure 9.4 An illustration of islanded microgrid. *Source:* [Badal 2019]. Licensed under CC BY 4.0.

Figure 9.4 shows the concept of Smart Grid vision as a network of integrated DERs. The microgrid is a small network of power system with DG units connected in parallel. This figure illustrates a microgrid in islanded mode connected to the main grid and in the same time can be connected to other microgrids. A microgrid is a group of interconnected loads and DERs within clearly defined electrical boundaries that acts as a single controllable entity with respect to the grid. The integration challenges of renewable energy sources and the control methods for microgrids are described in [Badal 2019]. The increasing number of DG units becomes a common challenge to maintain the operation of a microgrid. Different types of control – centralized or decentralized – that are based on the functions of controllers, computational algorithms, and communication of the information introduce a need for cybersecurity protection. Thus, cybersecurity is another challenge that has to be considered in the design of microgrids as well in the integration of microgrids. Specific security measures such as authentication or physical security need to be considered for the protection of the information or a device.

9.5 Addressing Various Issues

Several issues and concerns need to be addressed. The most important include cybersecurity and resiliency of DERs. In the cybersecurity area, there are two key concepts for the Smart Grid: resiliency and cyber–physical. Resiliency implies that the power system critical infrastructure is designed not only to prevent malicious cyber attacks and inadvertent failures but also to cope with and recover from such attacks and failures.

9.5.1 Addressing Cybersecurity

Specific issues related to cybersecurity include providing secure communications with adaptive protection mechanisms, implementing a balanced security with operations, and enabling trust in the system. Security efforts should not only focus on smart meters but also on substation automation, microgrids, SCADA, telecommunication networks, DERs, etc. Cybersecurity features are required in the automated systems. Automation addresses DER issues by supporting capabilities that allow utility to:

- Monitor DER power output.
- Request power from DER.
- Shut off power from DER and prevent back feed.
- Send pricing signals to DER.
- Send synchronization and other power system information to DER.

One of the main functions of the automation system is monitoring of DER power output. To balance load, utilities must monitor demand and generation in real time. Local DER automation systems provide information to the utility concerning the exact amount of power being produced. Power is typically measured by components such as current transformers and potential transformers, both of which are inputs to the automation system.

The automated system requires the implementation of many functions for various scenarios. To improve energy management, utilities need to monitor DER generation condition such as power transitions and deviation conditions. Another function of the automation system is to process requests from the utility. In times of peak demand, a utility might request the DER to run at full capacity and may offer pricing incentives to do so. When work needs to be done on local distribution systems, the utility can request a shutdown of the DER. In cases of safety-related shutdowns, the local automation system typically provides a fail-safe method to verify complete shutdown of equipment and correct status of all interposing switchgear. Also, these decisions have to be quick and accurate. All these functions need to include specific security features (e.g. access, availability, and integrity are critical) that support adequate protection in any situation.

Resilience can be affected by the interdependencies of other industries, such as water, fuel, and telecommunication. Therefore, greater sharing of information between industries and sectors can be helpful in emergency situations and when disruptions occur.

The key to achieving resiliency of Smart Grid is to pursue designs and implementations of electrical, communications, and information infrastructures. Partial implementation only in one infrastructure will not guarantee the benefits of Smart Grid resiliency [Amin 2001]. Although mature cybersecurity technologies and solutions can be easily deployed, the change for people and processes is hard. Utilities must incorporate the appropriate skills, policies, and practices into their daily operations to maintain a secure and thus more resilient grid.

Some countries have started moving toward network schemes that provide some intelligence to the grid, isolating the failures before they destabilize the whole system; intelligent adaptive islanding is considered to be a good next step in electric power industry. Following this step, every node in the power network of the future will be awake, responsive, adaptive, price-smart, eco-sensitive, real time, flexible, humming, and interconnected with everything else [Silberman 2001].

9.5.2 Cyber–Physical System

Cyber–physical implies that the power system consists of both cyber and physical assets that are tightly intertwined and can be used in combination to improve the resiliency of the power system infrastructure [EPRI 2013]. This report describes five levels of DER system architectures, and level 1 (the lowest level) describes the autonomous cyber–physical DER systems (see Figure 9.5). DER systems consist of

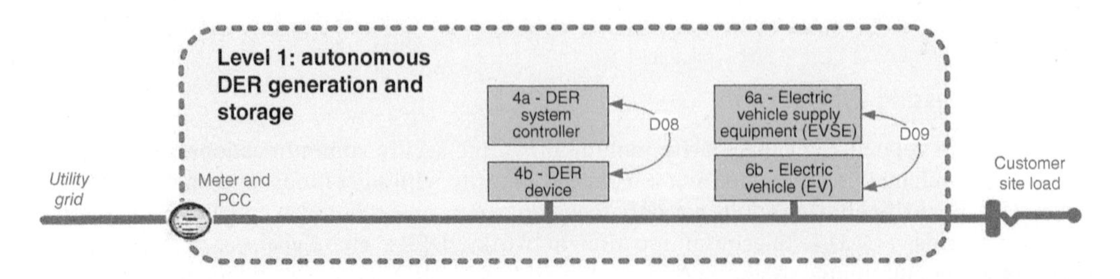

Figure 9.5 Autonomous DER systems at smaller customer and utility sites. *Source:* [EPRI 2013]. © 2013, Electric Power Research Institute, Inc.

physical/electrical components (e.g. the electric generator and storage components) and cyber components (e.g. DER controllers), thus making them cyber–physical systems. They are generally designed to operate autonomously according to settings in the DER controller. While it is argued that DER systems can impact the resiliency of the power system, at the same time, the power system cyber–physical capabilities can be used to increase resiliency, but only if appropriate cybersecurity measures are combined with these capabilities [EPRI 2013]. Therefore, greater assurances of safety, security, scalability, and reliability are necessary.

The DER equipment is connected to the local electric power system (EPS) (shown as solid red lines in the diagram) as are customer loads if they exist. This local EPS is connected to the utility's area EPS through a circuit breaker and meter at the point of common coupling (PCC). The logical interface D08 supports interactions between DER system controllers and their DER devices, while logical interface D09 supports battery charging/discharging (V2G) interactions between the electric vehicle supply equipment (EVSE) and the EV.

The interfaces D08 and D09 are documented in [NISTIR 7628]. Most DER systems are supplied as complete units. The controllers are usually located within a short distance of the physical DER devices, with any communications between them limited, point to point, and generally using proprietary communication protocols provided by the DER manufacturer. In the diagram, these communication channels are shown as curved red arrows. Some DER systems include a simple human–machine interface (HMI) to provide status information and may be used during maintenance (connection to HMI may be local or remote). The only external communications between the utility and these DER systems are the meter readings, typically measured at the PCC between the local EPS and the area EPS.

These individual DER systems operate autonomously most of the time, with the cyber controllers closely monitoring the physical devices and responding to the local electrical conditions by controlling the physical DER systems according to prespecified settings. The controllers can also respond to customer preferences and to HMI commands to change these settings or to react to emergency situations. Therefore, the HMI is the local access for viewing the status and measurements of the DER system and for changing the settings of the autonomous actions of the DER system. In the same time, HMI can be the point of unauthorized access and malware distribution.

9.5.3 Cyber–Physical Resilience

The authors of this work [Arghandeh 2016] argue that:

- There is not a clear and universally accepted definition of cyber–physical resilience in the electric grid community.
- Current literature on power system resilience presents many conflicting and vague descriptions; specifically, the definitions of electric grid resilience in different publications do not always converge.
- Definitions of resilience have evolved and expanded over the years.
- The breadth of and number of definitions for resilience have increased significantly over the last decade, making it difficult to find a universal understanding of the term resilience for power grid systems (Appendix D includes different definitions for resilience in different disciplines).
- Definitions of resilience in other disciplines can help in the definition of resilience in power systems because the electric grid is a socio-ecological system with different spatial, temporal, and organizational parameters that are affected by policy, economy, and society.
- Different disciplines use different definitions that share similar concepts from different perspectives; in resilience operations, response time and service availability are key.
- Power system community needs a tailored resilience definition that includes physical and cyber network characteristics and service outage consequences.
- Resilience is especially critical immediately following an event that challenges system performance and functionality.
- Events that are considered in resilience design are given various names by different authors or various disciplines; a generic term, disturbing events, is suggested; these terms describe consequences of rapid changes both in the environment and in system operation that are caused by system/component failures, attacks, and natural disasters.

- The resilience of a system presented with an unexpected set of disturbances is the system's ability to reduce the magnitude and duration of the disruption; a resilient system downgrades its functionality and alters its structure in an agile way.

In this work [Arghandeh 2016], the authors suggest a practical definition of cyber–physical resilience for power transmission and distribution networks. This paper aims to illuminate the value of resilient distribution network operation and how it goes beyond reliability and stability. The concept of a power system cyber–physical resilience centers around maintaining system states at a stable level in the presence of disturbances. Resilience is a multidimensional property of the electric grid, but it requires managing disturbances originating from physical component failures, cyber component malfunctions, and human attacks [Arghandeh 2016].

Dynamic system components like loads and DG force sudden changes in system behavior. Resilient power systems know how to reroute electricity to customers using alternative paths and alternative local sources during natural disasters.

In addition, due to the responsibility for system management, stability, and integrity at all times, it appears inevitable that distribution system operators (DSOs) need to be in charge of the data management in a Smart Grid environment. As regulated noncommercial parties, DSOs also are well placed to ensure a level-playing field for all market parties relying on the data for their business while being a safeguard of consumer data privacy and security [CEDEC 2012].

Further, cyber resilience has to be considered in Smart Grid implementation. As discussed in [Davis 2014], cyber resilience covers the ability to keep operating during a detected attack or incident, to keep operating under the assumption that an undetected compromise has occurred, to operate with reduced capability or capacity, and to provide graceful degradation and recovery during and after an incident.

Focus should be on building resilience on the Smart Grid systems. This requires the application of standard IT and information security techniques such as backup, testing of recovery and business continuity procedures, use of hot/warm/cold sites, alternative service provisioning, and incident response and management.

Also, IT should be treated as a commodity, so design choices to build or purchase resilience in systems and infrastructure should be adopted. Organizations must be involved in testing, response, setting the minimum capability required, and stating tolerance of failure during operations. Resilience is both a technical and a business responsibility [Davis 2014].

Similarly, infrastructure resilience should be understood. This is defined as the ability to reduce the magnitude and/or duration of disruptive events. The effectiveness of a resilient infrastructure or enterprise depends upon its ability to anticipate, absorb, adapt to, and/or rapidly recover from a potentially disruptive event [NIAC 2010].

9.5.4 Related Characteristics, Relationships, Differences, and Similarities

Reliability and stability are two common concepts in power system operation, and both concepts are related to robustness and resilience. Understanding resilience in the electricity sector is one topic discussed in the [NIAC 2010] report. According to this report, the predominant risk management concept within the electricity sector is reliability as the ability to meet the electricity needs of end-use customers, even when events reduce the amount of available electricity – in other words, keeping the lights on.

Commonly called the world's largest and most complicated machine, the North American electric grid, which covers the United States, Canada, and a small portion of Baja California Mexico, operates at 99.9% reliability, a feat that requires advanced monitoring and control technology and trained operators working in concert 24/7/365. System interconnection and close cooperation among utilities, power producers, and transmission operators enable the grid to withstand equipment failures and disruptive events while keeping the lights on [NIAC 2010].

Resilience in the electric sector is often described in many ways. Some would talk about resilience as the ability to ride through events and bring back facilities after an event. Others may describe as an element of the overall electric system design: the capacity of a large interconnected grid to absorb shocks. Specific definitions of resilience, however, are less important than fundamental concepts of resilience that are defined in this framework [NIAC 2010]. The NIAC resilience construct includes robustness, resourcefulness, rapid recovery, and adaptability.

The terms robustness and resilience are sometimes used interchangeably in some disciplines like social systems and organizational systems [Arghandeh 2016]. The authors argue that the term resilience is similar to the term robustness, but these are two different and sometimes mutually exclusive properties. In infrastructure systems like power systems, the terms robustness and resilience are more distinct due to power systems' structure function centering around conductor lines delivering electric power to a certain area within specific voltage and frequency ranges.

Robustness is defined as the ability of a system to cope with a given set of disturbances and maintain its functionality. Resilience is taking one step forward while taking quick actions to maintain system functionality.

Robustness and resilience belong to two different design philosophies; robustness is embedded in the system's design and is concerned with strength, whereas resilience is integrated into the system's operational components like control systems and is concerned with flexibility. In this work [Arghandeh 2016], the authors argue on robustness versus resilience by identifying issues (included also in Table 9.2) such as:

Table 9.2 Robustness vs. resilience in power systems.

Criteria	Robustness	Resilience
Application*	Network hardening*	Network flexibility*
Enterprise focus*	Utility assets*	Utility services*
Value proposition*	Design*	Operations*
Security approach*	Passive*	Active*
Network preference*	Isolated*	Interdependent*
Network coupling*	Loose*	Tight*
System configuration	Fixed	Flexible, adaptable
Power system restoration performance	Prescribed restoration procedures	Automated and coordinated restoration procedures based on analysis of system operations during restorations, such as load variations, transmission line reconnections, and loop closures
Restoration approach	Analysis of vulnerable and critical buses; voltage stability analysis	Planned optimization based on load redistribution and avoiding cascade failures; quickly recover from natural threats or cyber attack or from high-impact, low-frequency events
Metrics approach	Simple metrics, complex metrics	Resilience metrics based on the performance of power systems, as opposed to relying on attributes of power systems; specified in the context of low-probability, high-consequence potential disruptions
Events considered	High-probability, low-consequence hazards	Low-probability, high-consequence hazards
Binary or continuous	Operationally, the system is robust or not (0, 1); confidence is unspecified	Resilience is considered a continuum; confidence is specified

*Source: [Arghandeh 2016]. © 2016, Elsevier.

- Robustness in the enterprise world is more focused on asset utilization, whereas resilience centers around service quality.
- Robustness is embedded in the system architecture design; resilience is more concerned with system operation.
- Robustness can be a passive approach for system security; distribution pole hardening and putting cables underground are examples of passive system security enhancement; resilience, on the other hand, is an active approach with real-time reactions to disturbances; resilience can mean a set of real-time switching and islanding actions; resilience can involve explicitly partitioning the grid into different subnetworks (microgrids).
- Robust electric grid networks try to maintain system functionality by damping perturbations; resilient networks, on the other hand, rely on interdependencies to withstand perturbations. Therefore, multiple couplings between network components are crucial in resilient systems. Table 9.2 compares robustness and resilience against different criteria (adapted from [Arghandeh 2016]).

By comparing reliability, security, and stability, there are differences that have to be understood. There is a relationship among these characteristics as described in [Kundur 2004]:

- A system is reliable if the power system is secure most of the time.
- A system is secure if the system is stable but must also be secure against other contingencies that would not be classified as stability problems, e.g. damage to equipment such as an explosive failure of a cable and fall of transmission towers due to ice loading or sabotage. Maintaining the power system in a secure operating state assumes that some electrical variables should be maintained within admissible limits [Venkata 2013].
- A system is stable if power system variables are within admissible stability limits that depend on the power reserves, on the existence and performance of some regulation and control systems, and on the existence of advanced devices (FACTS, HVDC links, etc.) and the voltage limit, which can be influenced by the performance of the automatic voltage regulations. The performance of control systems may be impacted by cybersecurity incidents (intrusions or interactions due to cybersecurity controls).
- Security and stability are time-varying attributes that can be evaluated by studying the performance of the power system under a particular set of conditions. Reliability, on the other hand, is a function of the time-average performance of the power system; it can only be judged by consideration of the system's behavior over an appreciable period of time.

DER devices are connected with different systems in the Smart Grid. These systems include microgrid system, distribution system, and synchrophasor system. These systems along with DER help in providing high-quality energy with increased efficiency and reliability where consumer can produce and manage their energy usage. Distribution system helps in connecting these independent generation units with the main power grid.

Emerging standards (e.g. [IEEE 1547]) are defining the requirements that DER technologies must meet in order to safely and reliably interconnect. These standards ensure that new technologies do not jeopardize the safety or reliability of the electric power system. Besides these characteristics, power system functionality is dependent on interoperability capability. More information about DER interconnections is provided in [EPRI 2013]. Also, this report includes DER use case categorizations by configuration and different use case scenarios. The understanding of DER interconnections, uses (e.g. scenarios, configurations), technologies, environment, and management issues are aspects that have to be understood when planning for security measures.

9.6 Power System Interoperability

Interoperability is the ability of diverse computing systems to cooperate in solving computational problems or the ability of making systems to work together (interoperate). Although no unique definition, from a software engineering and IT perspective, one defines interoperability as the ability of two or more systems or components to exchange information and to use the information that has been exchanged [IEEE Std 610]. From another perspective, interoperability is the ability of software and hardware on multiple machines from multiple vendors to communicate meaningfully [Comer 2001].

This term describes the goal of internetworking, namely, to define an abstract, hardware-independent networking environment that makes it possible to build distributed communications.

9.6.1 Interoperability Dimensions

Within the electricity system, interoperability means the seamless end-to-end connectivity of hardware and software from the customers' appliances all the way through the transmission and distribution system to the power source, enhancing the coordination of energy flows with real-time flows of information and analysis [GridWise 2009].

However, interoperability has many dimensions, including physical and communication interconnections to informational interoperability (content, semantics, meaning, and format) and organizational interoperability (covering matters such as transactions structures, contracts, regulation, and policy). Interoperability is often achieved and institutionalized with support from formal technical standards and implementation testing. The events of the 14 August 2003 blackout in the United States illustrate several different ways in which information flows were ineffective, reflecting the severe consequences of the lack of interoperability between organizations and power systems.

Technical interoperability covers the physical and communications connections between and among devices or systems (e.g. power plugs and USB ports). Informational interoperability covers the content, meaning, and format for data and instruction flows (such as the accepted meanings for human and computer languages). Organizational interoperability covers the relationships between organizations and individuals and their parts of the broad system, including business and legal relationships (such as contracts, intellectual property rules, or regulations).

The downside of interoperability is that it enables an increase in system size and therefore can increase system complexity. When interoperability works properly, new devices and assets can be added to the system and work effectively (after sufficient interoperability specification and testing). As a system of systems becomes larger, more geographically dispersed, and more complex, there are more things that can go wrong and more ways that system components can interact in unexpected ways.

A more recent and broad definition takes into account other factors such as social, cultural, and political factors that impact system-to-system performance as promoted by Network Centric Operations Industry Consortium [NCOIC 2014]. Figure 9.6 shows the QuadTrangle™ diagram of interoperability

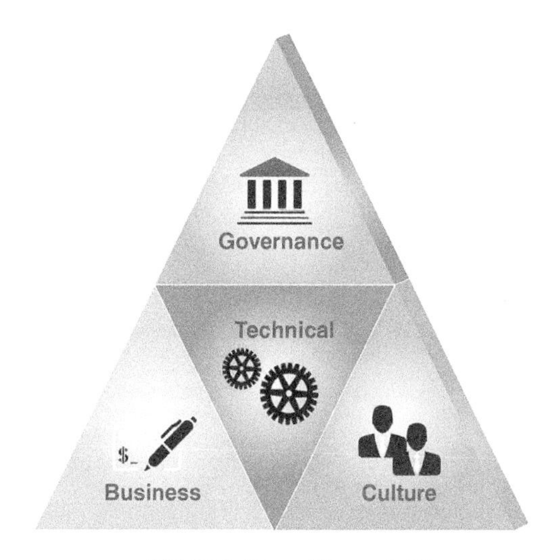

Figure 9.6 NCOIC QuadTrangle. *Source:* [QuadTrangle]. Licensed under CC BY 4.0.

with four interdependent areas to be considered when working to create a reliable and trusted interoperable environment. These areas include technical, governance, culture, and business.

The direct benefit of interoperability is that it enables large systems to become larger and more complex; the indirect consequence of interoperability may be that larger complex systems can fail in complex and unpredictable ways.

9.6.2 Smart Grid Interoperability

Under the Energy Independence and Security Act of 2007 (EISA), the National Institute of Standards and Technology (NIST) is assigned the primary responsibility to coordinate development of a framework that includes protocols and model standards for information management to achieve interoperability of Smart Grid devices and systems [EISA 2007].

EISA, which designates development of a Smart Grid as a national policy goal, specifies that the interoperability framework should be flexible, uniform, and technology neutral. The law also instructs that the framework should accommodate traditional centralized generation and distribution resources while also facilitating incorporation of new innovative Smart Grid technologies, such as DERs.

However, the concepts of interoperability for Smart Grid are also based on work related to distributed process integration and interoperation across the economic spectrum. The definition of interoperability for Smart Grid is more than exchange of information between two or more systems. It also includes organizational, technical, and informational aspects including quality attributes for the information (shared meaning and actionable) and quality attributes for the service (reliability, fidelity, security) [Grid Interoperability].

In enabling the interoperability, standards are critical. Standards are the foundation of mass markets for the millions of components that have a role in the Smart Grid. Interoperability among devices and systems of Smart Grid requires open, broadly accepted standards. Thus, open standards facilitate interoperability and data exchange among different products or services and are intended for widespread adoption. Also, open interoperability standards are widely considered to be the most effective way to:

- Reduce IT capital costs and to operate and upgrade information infrastructures.
- Reduce security risk and improved security management.

Therefore, it is needed to establish protocols and standards for the Smart Grid domains including smart sensors on distribution lines, smart meters in homes, and widely dispersed sources of renewable energy sources and DER systems. Without standards, there is the potential for technologies developed or implemented with sizable public and private investments to become obsolete prematurely or to be implemented without measures necessary to ensure security [NISTIR 7628], [NIST SP1108r1]. Most important benefits of interoperability include reliability enhancement and enhanced security of the electricity system by reducing vulnerability to physical and cyber attack, increasing the system's ability to resist and recover from an attack, and speed service restoration.

9.6.3 Interoperability Framework

Interoperability is characterized by different dimensions grouped in categories:

- Technical interoperability covers the physical and communications connections between and among devices or systems (e.g. power plugs and USB ports).
- Informational interoperability covers the content, meaning, and format for data and instruction flow such as the accepted meanings for human and computer languages.
- Organizational interoperability covers the relationships between organizations and individuals and their parts of the broad system, including business and legal relationships such as contracts, intellectual property rules, or regulations.

Each of these dimensions affects the degree to which interoperability can enhance grid reliability or, in the absence of interoperability, compromise grid reliability and security. Applications that communicate with other applications require an appropriate set of protocol layers. Each layer must be considered and accommodated within every communication path and often across multiple lower layer protocols. Communication layers represent another dimension of complexity, and appropriate standardization by layer is critical to achieving interoperability.

Figures 9.7 depicts the GridWise Architecture Council (GWAC) stack with the interoperability levels (organizational, informational, and technical), each level described by interoperability categories. This conceptual model is helpful to explain the importance of organizational alignment in addition to technical and informational interface specifications for Smart Grid devices and systems. The interoperability context-setting framework [Gridwise 2008] identifies eight interoperability categories that are relevant to the mission of systems integration and interoperation in the electrical end-use, generation, transmission, and distribution industries. The major aspects of interoperability are defined into these categories: technical, informational, and organizational. The organizational categories emphasize the pragmatic aspects (policy and business drivers for interaction) of interoperation. The informational categories emphasize the semantic aspects of interoperation (focus on what information is being exchanged and its meaning). The technical categories emphasize the syntax or format of the information. Effective systems interoperation requires that resources, at all interoperability levels, can be unambiguously identified by all automation components that need to interact. A resource refers to an instance of an information modeling concept, such as a generator, refrigerator, or building owner.

Crosscutting issues are areas that need to be addressed and agreed upon to achieve interoperation [GridWise 2008]. They usually are relevant to more than one interoperability category of the framework. Cybersecurity is depicted as a crosscutting issue.

Another interoperability framework is promoted by IEEE organization. Figure 9.8 is a graphical illustration of interoperability (based on IEEE definition, see IEEE 2030 standards) focus areas for the electric power, communications, and information technologies that constitute the technological heart of the Smart Grid. This framework is not considering the organizational aspects.

While each focus area requires its own design and implementation, cross-cutting issues such as cybersecurity require a unified approach for any focus area.

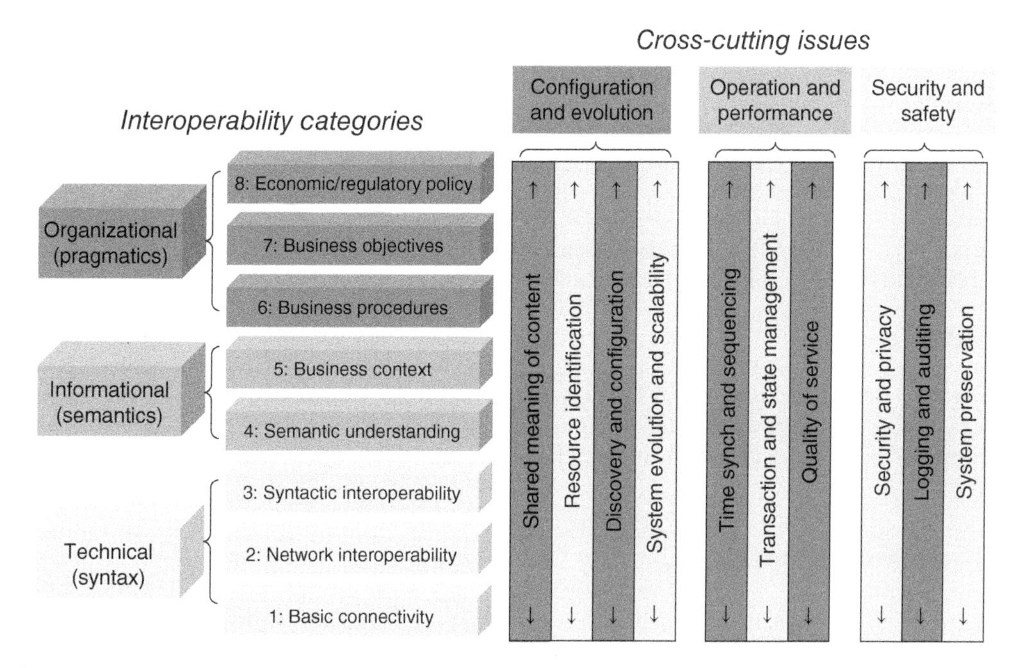

Figure 9.7 GWAC interoperability context setting framework, aka GWAC stack. *Source:* [Grid Interoperability]. Public Domain.

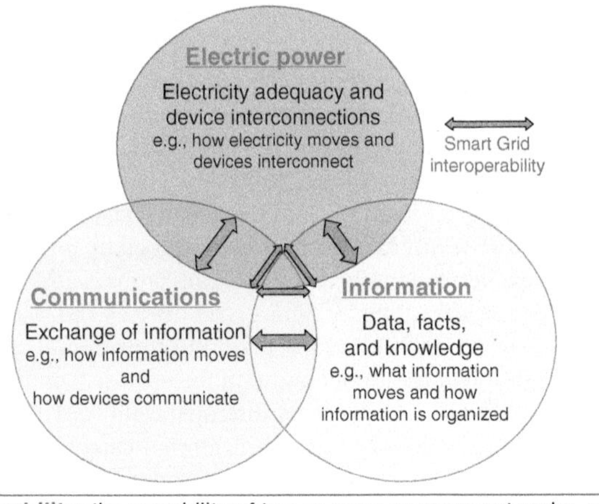

Interoperability: the capability of two or more or more networks, systems, devices, applications, or components to externally exchange and readily use information securely and effectively. (Std 2030)

Figure 9.8 Smart Grid interoperability: the integration of power, communications, and information technologies. *Source:* [NREL 2014]. Public domain.

9.6.4 Addressing Crosscutting Issues

As shown in Figure 9.7, cybersecurity is one of several crosscutting issues. When cybersecurity is applied to the information exchange standards, it is described as profiles of technologies and procedures that can include both power system methods (e.g. redundant equipment, analysis of power system data, and validation of power system states), policies, and information security technology measures (e.g. encryption, role-based access control, and intrusion detection).

Generally, cybersecurity is viewed as a stack of different security technologies and communication standards and procedures, woven together to meet the security requirements of a particular implementation of a stack of policy, procedural, and communication standards, all designed to provide specific services.

Correlating cybersecurity with specific information exchange standards, including functional requirements standards, object modeling standards, and communication standards, is needed although it is very complex. In addition, open standards could have different definitions by standards organizations, meanings, and uses in different contexts and countries, since no common definition is prevalent at this time (see Appendix A). In addition, open standards do not always guarantee interoperability since an open standard can be provided by different manufacturers with different interpretation of specifications and multiple implementations.

While the power generation and delivery process seems reasonably straightforward, in practice, many economic, geographic, and historical practicalities complicate the interoperability of national power grid in the United States, as manifested in the fact that the United States and Canada are served not by a single grid as much as by four interlocking regional grids, the Eastern, Western, Texas, and Quebec systems. The following section includes examples of interoperability challenges.

9.7 Smart Grid Interoperability Challenges

In power systems, the challenge is how to bridge a diverse set of organizations, systems, perspectives, goals, and frames of reference so that operating teams can address real-world events. The focus of the industry effort so far has been mostly on the interoperability of the communication and information model, as suggested by the NIST Smart Grid Interoperability standard roadmap and the International Electrotechnical Commission (IEC) documents on Smart Grid standardization. Therefore, other specific interoperability aspects include the following:

- Interoperability and standard interfaces between systems as well as interfaces with DER networks and systems [EPRI 2009].
- Interoperability between business and control systems is crucial to the implementation of the Smart Grid where the killer application is being able to have a controllable power and information interface [Grose 2009].
- Smart Grid is perceived as a system of systems where complex and secure communications and networks have to interwork for supporting decentralization of the power generation. Controlling load and power generation to achieve efficiencies requires two-way communications and secures information exchange.

Significant interoperability challenges toward a modern grid include [OECD 2012]:

- Information and communications asymmetries across the electricity sector value chain remain an important issue to be resolved and with them the need for effective and reliable communication channels. The electricity sector's line of command and operations in cases where electricity demand risks peaking (e.g. extremely hot or cold days) remain to a large degree patchy and mediated. On the other end, electricity consumers, in particular residential consumers, have little effective means of obtaining information about the current state of electricity production, its availability, cost, and environmental impacts. Utilities have relatively little information about disaggregated electricity consumption patterns below the distribution system level, and losses of electricity along transmission and distribution lines are not always accounted for systematically.
- Consumer acceptance, engagement, and protection need to be addressed when designing products and services. For example, various survey results show that consumers are concerned about privacy issues and costs related to smart meters. There is substantial debate among utilities, consumer associations, and policy makers about the initial costs.
- Converging IT and operational technologies (OT) require IT firms that provide value-added services in the Smart Grid to have a more detailed understanding of operational processes. This refers to services targeting utilities (e.g. distribution grid management), consumers (e.g. energy consumption optimization), or both (e.g. operating virtual power plants). Also, a tighter integration of IT into operational processes in the electricity, transport, and building sectors requires the alignment of research and policy agendas.
- Sometimes security control measures can affect the interoperability, because of reasons such as weak design, misconfigured software products, incompatible versions, inoperable standards, proprietary protocols and implementations, etc.
- Although openness can be an objective of interoperability, open systems could impact security. An open system is a system that continuously is interacting with the environment, and more openness could raise security concerns.
- Connectivity and communication channels need to be available across the economy to all electricity users to maximize the potential benefits of Smart Grids. Ensuring communication channels that are available universally across the economy is a key goal of policy makers, and there are significant potential synergies that could be exploited between communication and electrical distribution companies (e.g. utility pole or duct sharing).

9.8 Standards, Guidelines, and Recommendations

The overall Smart Grid system is based on many standards that often are lacking widely acceptance. Different markets may adopt a distinct suite of standards. The following sections describe documents such as recommendations, guidelines, and standards applicable to Smart Grid, information, IT systems, control systems, and utilities.

9.8.1 ISO/IEC Standards

A set of IT standards that are published as free (some are more current, also some old standards) can be found at [ISO IT]. These standards are made freely available for standardization purposes, although they are protected by copyright. The standards apply to many areas that support the Smart Grid developments.

The IEC provides a mapping tool via Web for the Smart Grid [IEC Map]. The map provides an architectural view of components and relationships between components of the Smart Grid including various standards. A user guide helps in finding the standards and needed information. There are multiple paths to finding a needed standard. Also, the map includes crosscutting issues such as security. Examples of the recommended standards for cybersecurity include:

- IEC 61400-25 – Communications for monitoring and control of wind power plants.
- IEC 61850 – Communication networks and systems for power utility automation.
- IEC 61850-90-5 – Protocol for transmitting digital state and time synchronized power measurements (synchrophasor measurements) over wide area networks and systems.
- IEC 62056-5-3 – Electricity metering data exchange.
- IEC 62351 – Handle security based on different security objectives such as access control, prevention of eavesdropping, prevention of playback and spoofing, and intrusion detection.
- IEC 62443 series – Formerly ISA-99, a series of standards and technical reports for implementing electronically secure industrial automation and control systems (IACS), to be used by end users and manufacturers of control systems.
- ISO/IEC 15118 – Vehicle to grid communication interface.
- ISO/IEC 27001 – Information technology, security techniques, information security management systems.
- ISO/IEC 27002 – Code of practice for information security controls.

9.8.2 IEEE Standards

The US Federal Energy Policy Act of 2005 calls for state commissions to consider certain standards for electric utilities. Under Section 1254 of the Act, i nterconnection services shall be offered based upon the standards developed by the Institute of Electrical and Electronics Engineers: IEEE 1547 Standard for Interconnecting Distributed Resources with Electric Power Systems, as they may be amended from time to time [Policy 2005]. Examples of some areas that are covered by IEEE Standards Organization include:

- Interconnection
- Interoperability
- Networking and communications (including the home)
- Cybersecurity
- Substations automation
- Distribution automation
- Renewables
- AMI
- Power quality and energy efficiency
- EVs.

A pictorial view of IEEE standards used for DER interconnection within a residential building is provided in [IEEE Consumer]. The numerous standards for enabling connectivity are grouped as follows [IEEE Consumer]:

- Home networking standards
- Smart metering standards
- Home devices standards (e.g., IEEE 1547 series, Distributed energy interconnection, Solar, Wind, Storage, etc.)
- Smart Grid into home devices standards
- 3D video standards
- Mobile video standards
- Electric vehicle standards.

In North America, protective devices are generally referred to by standard device numbers. There are nearly 100 ANSI standard device numbers describing features of different protective devices. Examples include devices that monitor phase differences and protect generators from loss of synchronization with the grid; protect transmission lines from exceeding rated current carrying capacity; monitor over- or under-frequency conditions and over- or undervoltage conditions; implement compensating voltage control when needed; or, in an emergency, prioritize load shedding to protect grid stability. Capabilities are in place today, and continue to improve, that secure these devices from cyber exploitation by adversaries seeking to misuse them, causing them to disrupt, instead of protect, power flow.

The standard [IEEE Std C37-2] provides definition and application of function numbers and acronyms for devices and functions used in electrical substations and generating plants and in installations of power utilization and conversion apparatus.

Figure 9.9 is a pictorial view of IEEE standards used for DER interconnection within a residential building. Figure 9.10 depicts a stack configuration of protocols used for Smart Grid networks as suggested in [Ho 2014].

The IEEE 1547 and IEEE 2030 series of standards provide information for DER interconnection and interoperability with the electricity grid.

The IEEE 1547 standard is a series of standards developed concerning DER interconnection. DERs include distributed generators and energy storage systems. The standard focuses on the technical specifications for, and testing of, the interconnection and not on the types of DER technologies – it is technology neutral. The standard provides requirements relevant to the performance, operation, testing, safety considerations, and maintenance of the interconnection. It includes general requirements, responses to abnormal conditions, power quality, islanding, and test specifications and requirements for design, production, installation evaluation, commissioning, and periodic tests. It provides local, state, and federal regulators and policy makers a technical basis for promoting transparency, openness, and fairness in implementing DER interconnecting to the grid.

Figure 9.11 is a schematic diagram of the IEEE 1547 series of standards (Standard P1547.5 is withdrawn since 2011].

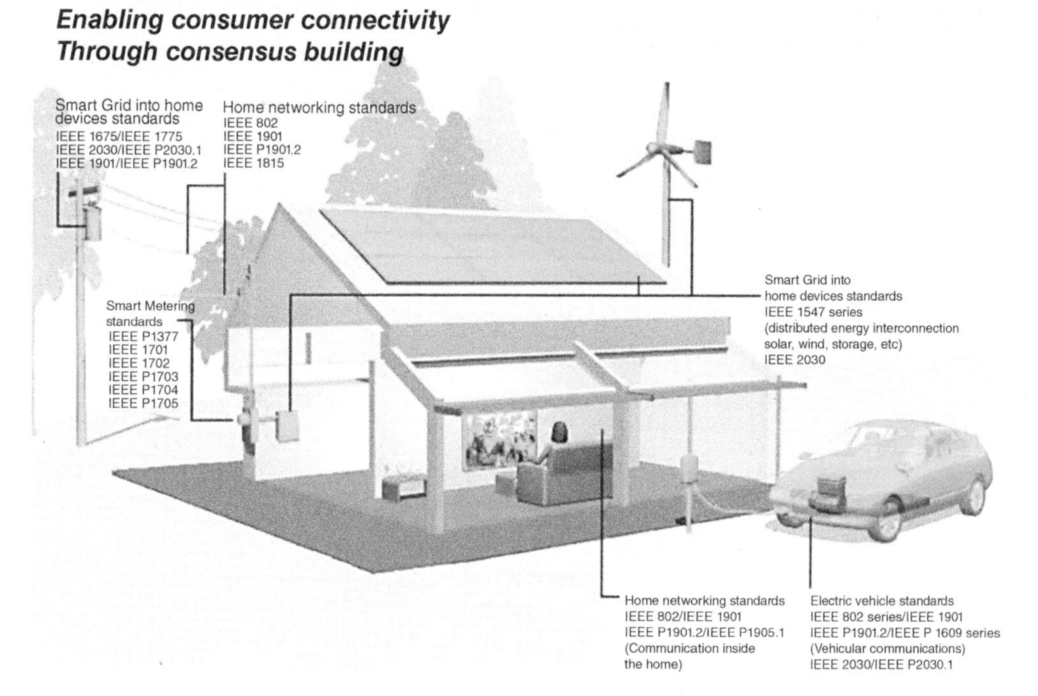

Figure 9.9 DER interconnections based on IEEE standards. *Source:* [Adams 2013]. © 2013, IEEE.

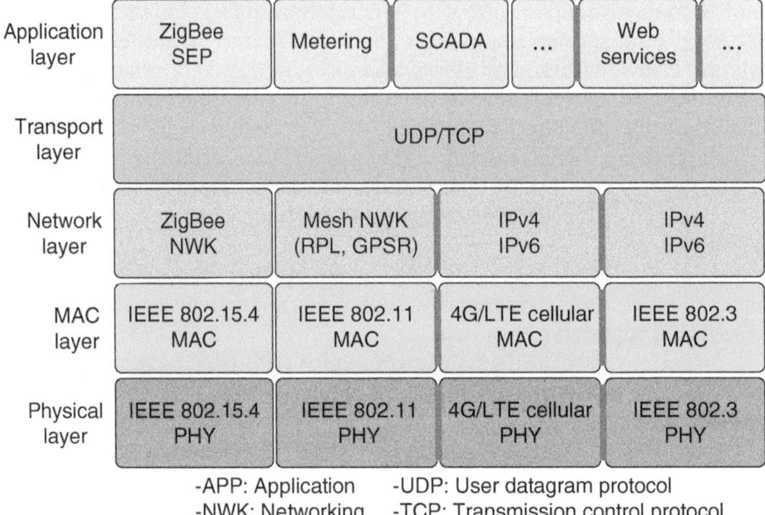

Figure 9.10 View of wireless protocols per layer. *Source:* [Ho 2014]. © 2014, Springer Nature.

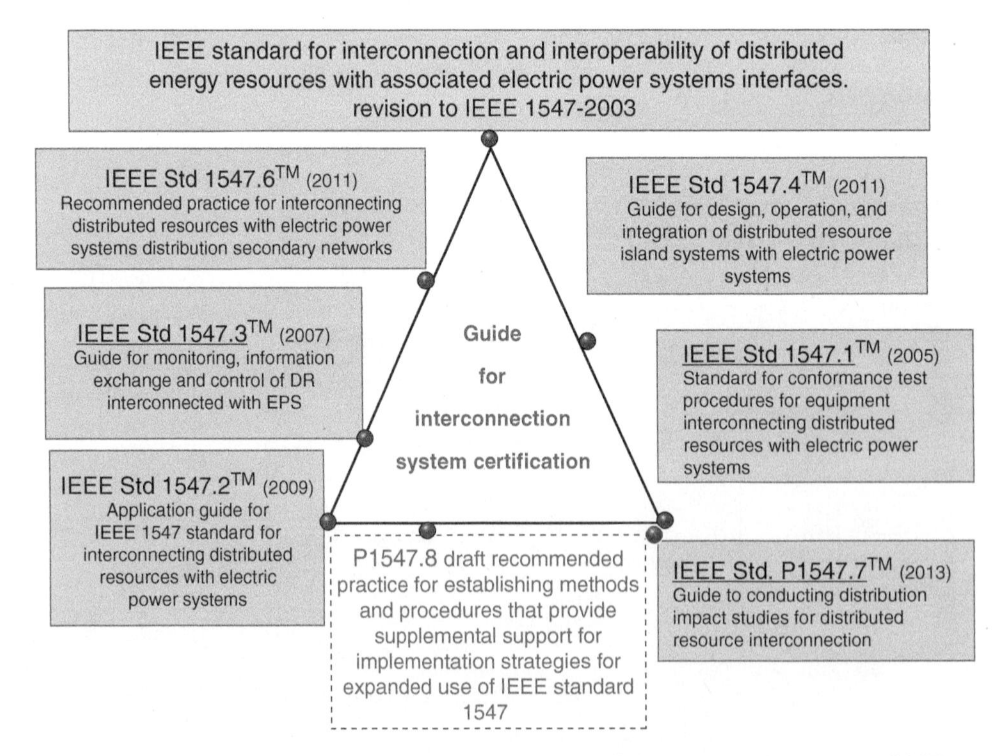

Figure 9.11 The IEEE 1547 series of interconnection standards. *Source:* Adapted from [Adams 2013]. © 2013, IEEE.

The IEEE 2030 series include:

- IEEE 2030 Guide for Smart Grid Interoperability of Energy Technology and Information Technology Operation with the Electric Power System (EPS), End-Use Applications, and Loads (September 2011).
- IEEE P2030.1 Draft Guide for Electric-Sourced Transportation Infrastructure.
- IEEE 2030.2 Guide for the Interoperability of Energy Storage Systems Integrated with the Electric Power Infrastructure (June 2015).
- IEEE P2030.3 Draft Standard for Test Procedures for Electric Energy Storage Equipment and Systems for Electric Power Systems Applications.

The IEEE 1547 and 2030 interconnection and interoperability technical standards continue to evolve as foundational documents helping accelerate the realization of the future grid (see also [Adams 2013], [Ho 2013a]).

The IEEE 1547 standard was first published in 2003, reaffirmed in 2008, and as of August 2016, the working group is focused to complete a full revision of the IEEE 1547 standard before 2018 [IEEE 1547]. The IEEE Standards Association maintains a Web site with the information about the suite of IEEE 1547 standards (see http://grouper.ieee.org/groups/scc21/1547_series/1547_series_index.html).

Examples of standards supporting specific Smart Grid topics include:

- Smart Grid networks: IEEE 1901™ (Standard for Broadband over Power Line Networks: Medium Access Control and Physical Layer Specifications), IEEE P1901.2™ (Standard for Low-Frequency (less than 500 kHz) Narrowband Power Line Communications for Smart Grid Applications), and IEEE 1905.1™ (Standard for a Convergent Digital Home Network for Heterogeneous Technologies).
- Security features: IEEE 1686 (Standard for Substation Intelligent Electronic Devices (IED) Cyber Security Capabilities), IEEE P37.240 (Standard for Cyber Security Requirements for Substation Automation, Protection and Control Systems), IEEE 1711 (Cryptographic Protocol for Cyber Security of Substation Serial Links), IEEE P1711.3 (Standard for Secure SCADA Communications Protocol (SSCP)), and IEEE 1402 (Standard for Physical Security of Electric Power Substations).

10

Distributed Energy Systems

10.1 Introduction

Classic electricity grids are based on large central power stations connected to high-voltage transmission systems. In turn, supply power is connected to medium- and low-voltage local distribution systems. They are designed for a one-way power flow – from the power stations, via the transmission and distribution (T&D) systems, to the final customer. They are unable to connect large-scale distributed energy generations. Moreover, there is little or no consumer participation and no end-to-end communications.

10.1.1 Distributed Energy

Distributed energy systems include a diverse array of distributed generation, energy storage and energy monitoring, and control solutions. Distributed generation is an approach [VirginiaTech 2007] that employs distributed energy technologies to produce electricity close to the end users of power. Distributed energy consists of a range of smaller-scale and modular devices designed to provide electricity, and sometimes also thermal energy, in locations close to consumers. They include fossil and renewable energy technologies, energy storage devices, and combined heat and power (CHP) systems [DOE Distributed].

Distributed generation, also called distributed energy, on-site generation (OSG) [EON Generation], or district/decentralized energy, is generated or stored by a variety of small, grid-connected devices referred to as distributed energy resources (DERs) or distributed energy resource systems (DERS) [Wiki DG].

Conventional power stations are centralized and often require electricity to be transmitted over long distances. They are referred as classic electricity paradigm [VirginiaTech 2007]. By contrast, in a power system composed of DER, much smaller amounts of energy are produced by numerous small units. These units can be stand alone or integrated into the electricity grid.

10.1.2 Distributed Energy Systems

Distributed energy systems include a diverse array of distributed generation, energy storage and energy monitoring, and control solutions. DERS are decentralized, modular, and more flexible technologies that are located close to the load they serve. DERS are also referred as distributed generation electricity paradigm [VirginiaTech 2007]. Figure 10.1 is an illustration of both paradigms.

In the current paradigm, the power is delivered through a passive distribution infrastructure to consumers. In this tomorrow model, a significant proportion of power generation is distributed generation (DG) with distributed storage. Power can even flow from DG into the distribution network and from distribution to transmission networks.

Building an Effective Security Program for Distributed Energy Resources and Systems: Understanding Security for Smart Grid and Distributed Energy Resources and Systems, Volume 1, First Edition. Mariana Hentea.

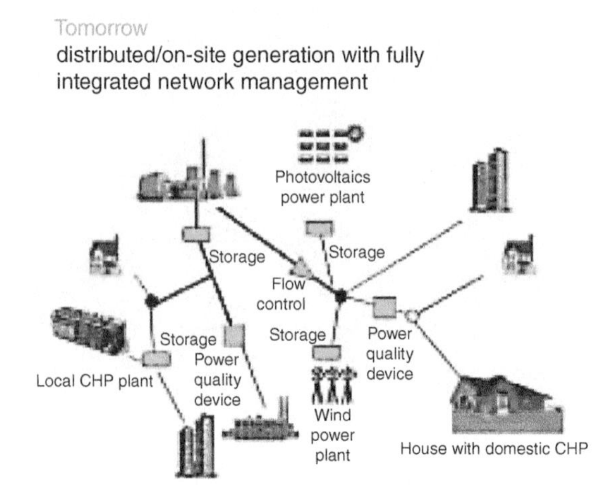

Figure 10.1 Comparison of generation sources. *Source:* [EC 2003]. Licensed under CC BY 4.0.

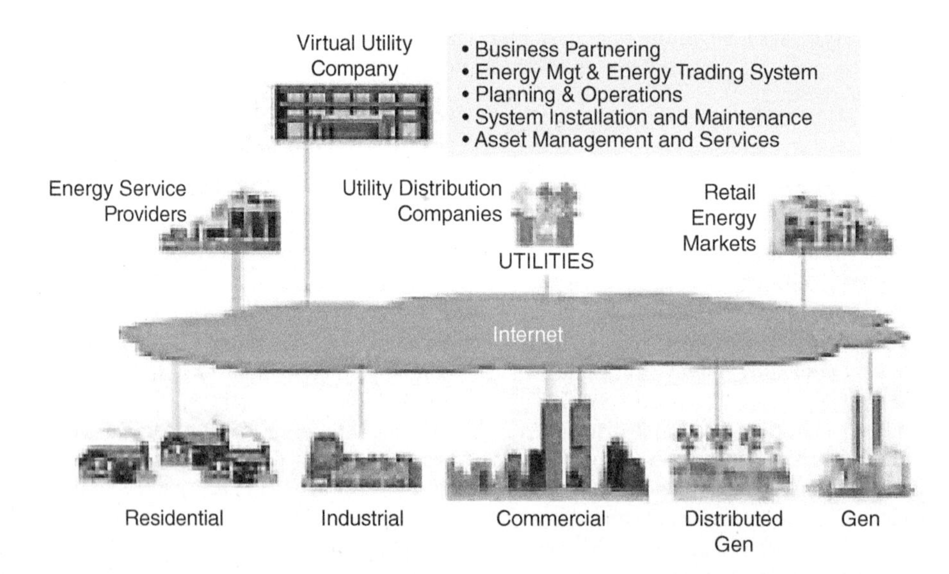

Figure 10.2 The virtual utility based on the internet model. *Source:* [EC 2003]. Licensed under CC BY 4.0.

The architecture of the future electricity systems is a continuous research and innovative projects. Models for future architecture of electricity systems are needed because of increased levels of distributed generation penetration and the distribution network can no longer be considered as a passive appendage to the transmission network [EC DG]. EC envisioned three conceptual models to include microgrids, active networks, and Internet model, all of which could have application depending on geographical constraints and market evolution. Figure 10.2 shows a representation of the Internet model.

In this model, the flow of information uses the concept of distributed control where each node, Web host computer, email server, or router acts autonomously under a global protocol. In the analogous electricity system, every supply point, consumer, and switching facility corresponds to a node [EC DG].

The virtual utility (VU) can be defined as a new model of energy infrastructure that consists of integrating different kinds of distributed generation utilities in an energy (electricity and heat) generation network controlled by a central energy management system (EMS).

As shown in Figure 10.2, a VU manages energy system and market-compatible solutions based on renewable energy sources, distributed generation, and information technology (IT). The benefits of the VU are the optimization of the utilization yield of the whole network, the high reliability of the electricity production, the complete control of the network for achieving the main aim of the EMS, the high velocity for assuming quick changes in the demand of the system, and high integration of renewable energy sources, plus the advantages of the DERs.

This study [Bryant 2018] reveals that energy utilities and energy equivalent utilities are aware of the need for innovation and new value propositions to deal with the changing energy market landscape and the need to adapt to the impact of increased variable renewable energy resources. Specifically, the utilities have to employ business models adjusted to capture value for and from their customers. Therefore, the authors propose solutions for business model adaptation to five typologies (traditional utility, green utility, cooperative utility, prosumer utility, and prosumer facilitator).

As prosumer and prosumer typologies emerged, there we should understand the security challenges associated with these new business models. We discuss briefly only the prosumer and prosumer facilitator models.

A utility operating using the prosumer utility typology achieves customer energy provision through a different set of business activities. It offers customers the prospect of green, local, self-produced electricity while seeking to maximize customers' ability to utilize their own (owned/leased) self-generation assets (e.g. rooftop solar PV). This involves the sale or leasing of distributed renewable generation equipment (solar PV, battery storage) to customers in order to build the distributed renewable energy generation capacity base that the prosumer utility subsequently seeks to access. Alongside the sale and leasing of distributed energy capacity to households, the prosumer utility deploys peer-to-peer (P2P) trading software and distributed generation control processes in order to allow for the development of virtual power plants (VPPs). These VPPs allow the prosumer utility to sell any excess generation from customers' rooftop solar PV systems to other customers, generating payments for the owner of the system and for the utility. This combination of customer-owned assets and the utility's partial control over the solar PV and battery storage assets of these customers allows the prosumer utility to link local prosumers with other customers locally to further develop their customer base (as shown in the structure of Figure 10.3). The model structure includes the key components and flow of the information and energy to overcome the variability of the energy.

The key factors that have to be considered in the security risk assessment include customer segment (residential), the operating platform (P2P/VPP platform and trading software), communication services, gateway access, signing-up prosumers into network, provision of kit (solar, energy storage, etc.), key partners (community members, prosumer customers, grid operator, utility, energy market regulator) and communication paths, users, and prosumer resources assets (DERs, smart meters, controllers, invertors, power cables, sensors, etc.).

The prosumer facilitator energy utility typology represents the most distinctive value proposition of the five typologies described in this work [Bryant 2018]. The focus is on helping customers reduce their dependence (partially or entirely) on grid supplied energy while reducing their cost of energy in the process. This business model primarily deals with the sale and leasing of rooftop solar PV systems (often combined with battery storage) to residential and small commercial customers.

The revenue is derived either from an upfront payment for the system and its installation or from monthly leasing payments, in addition to maintenance contracts designed to ensure the longevity and optimal performance of the system (as depicted in Figure 10.4).

Similarly, the business model includes customer segment focused on residential and commercial customers. Although the business activities are similar to the prosumer model, there are distinct activities that need to be consider in the risk assessment. First, there are other key players due to additional maintenance and provision program, network of equipment installers, third-party retailers, technology experts, equipment manufacturers, marketing and sale services, maintenance contracts, etc.

Compared with the prosumer model, the structure of prosumer facilitator has distinct components and a different energy and implicit information flow that have to be considered in the design of security controls.

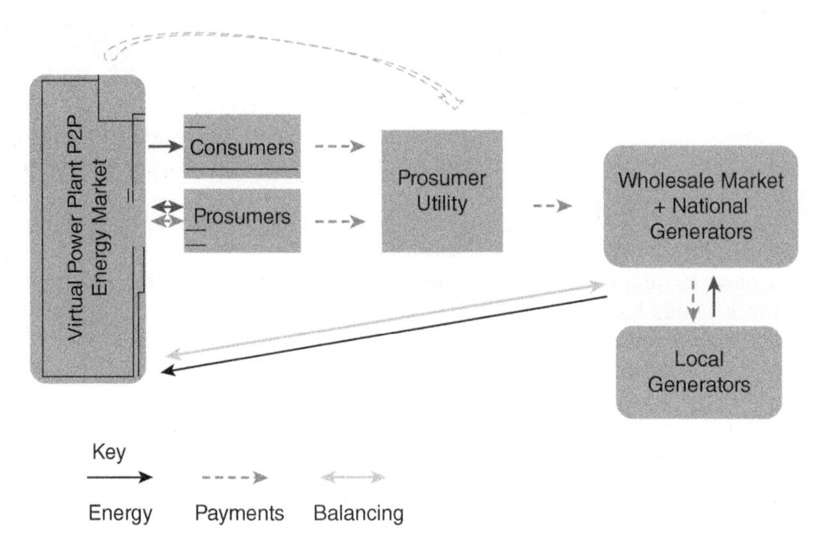

Figure 10.3 The prosumer utility business model. *Source:* [Bryant 2018]. © 2018, Elsevier.

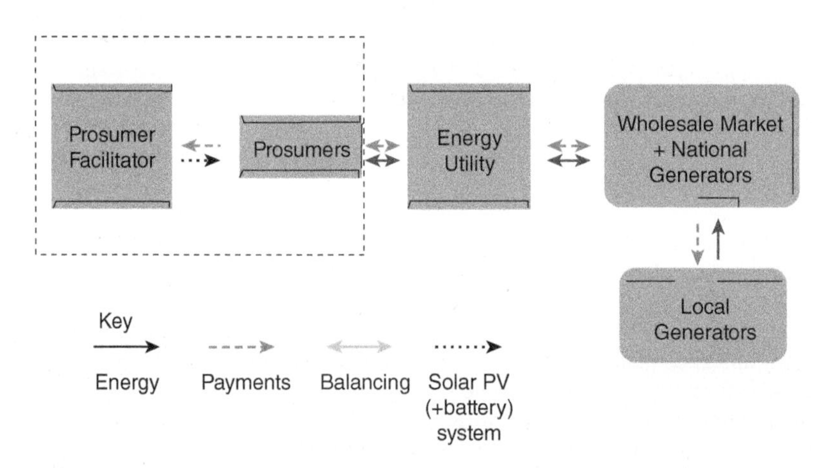

Figure 10.4 The prosumer facilitator business model. *Source:* [Bryant 2018]. © 2018, Elsevier.

10.2 Integrating Distributed Energy Resources

Traditional energy networks need to be made fit to integrate the DERs, such as renewable generation, storage and electric vehicles (EVs), and balance supply and demand at all times. Bottlenecks can be avoided by extending the network to the (new) peak demand or by managing the (peak) demand through the utilization of information and communication technologies (ICT) to manage the flexibility in the system. For every (local) part of the network, this choice must be made by the distribution system operator. By adding innovative ICT to parts of the networks, the general management of the system is facilitated.

As we discussed earlier, DERs are dispatchable energy generation and storage technologies, typically up to 10 of MWs or more in size, which are interconnected to the distribution grid to provide electric capacity and/or energy to a customer or a group of customers and potentially export the excess to the grid for economical purposes. DERs can generate electric power, as opposed to curtailable or interruptible loads that can just reduce electric loading on the grid. DER devices are small-scale power generation units that can provide energy according to the consumer demand.

DERs include any distributed generation, renewable energy resources, energy storage, and demand response (DR) connected at the distribution or end user. These resources may be under local control or under the control of a distribution management system or load aggregator [Taylor 2012]. DERs may be subdivided into:

- Distributed generation (DG) that can be classified as:
  - Renewable generation, e.g. small wind power systems, solar photovoltaic (PV), fuel cells, biofuel generators and digesters, nuclear, hydro, and other generation resources using renewable fuel
  - Nonrenewable generation, e.g. microturbines, small combustion turbines, diesel, natural gas and dual-fueled engines, etc.
- Distributed storage that can be classified as electric storage, mechanical, electrochemical, chemical thermal, etc. (one classification is shown in Table 4.5).
- Plug-in electric vehicles (PEV) and hybrid vehicles (PEV/PHEV) may also be considered as DER; the EVs may be used to supply stored electric energy back to the grid.

Behind-the-meter DER may be bundled with regular load and managed alongside the DR resources – such as a residential rooftop PV solar panel. But often, DER is treated separately in part due to its control capabilities. In addition to a regular retail tariff, behind-the-meter DER may be subject to net metering or feed-in tariff, where excess generation can be exported to the grid at an established or a dynamic price.

Similar to DR resources, DER assets can be registered and enrolled into a DR program. Furthermore, DER assets are typically required to meet additional technical requirements and certification for grid interconnection. Depending on the size of a DER and its export capabilities, submetering and telemetry capabilities may be required to monitor the impact of the DER operation on the distribution grid reliability and power quality. Also, renewable resources may receive renewable energy credits (RECs) and may also qualify as a must-run resource, e.g. wind power in some regions. These resources need to be incorporated into pricing and control signals associated with DER operation.

Distributed PV, smart inverters, distributed storage, EMS, new load types (e.g. plug-in electric vehicles (PEVs), building energy load during a DR event, EVs, etc.), and the management and control of these technologies become critical. Therefore, opportunities for growth will present themselves for new third-party energy service providers and microgrids.

10.2.1 Energy Storage Technologies

Energy storage technologies are analyzed under the criteria such as operation and advantages, applications, cost, disadvantages, and future potential applications [Connolly 2010]. The potential technologies include:

- Pumped hydroelectric energy storage (PHES)
- Underground pumped hydroelectric energy storage (UPHES)
- Compressed air energy storage (CAES)
- Battery energy storage (BES), which include lead–acid (LA), nickel–cadmium (NiCd), and sodium–sulfur (NaS)
- Flow battery energy storage (FBES), which include vanadium–redox (VR), polysulfide-bromide (PSB), and zinc–bromine (ZnBr)
- Flywheel energy storage (FES)
- Supercapacitor energy storage (SCES)
- Supermagnetic energy storage (SMES)
- Hydrogen energy storage system (HESS)
- Thermal energy storage (TES), which include air-conditioning thermal energy storage (ACTES) and thermal energy storage system (TESS)
- Electric vehicles.

Energy storage can be utilized for a broad range of applications. However, the type of technology that is suitable for some applications is primarily defined by their potential power and storage capacities that can be obtained. Based on the size of power and storage capacity that they can achieve, the following categories have been defined in [Connolly 2010]:

- Devices with large power (>50 MW) and storage (>100 MWh) capacities (e.g. PHES, UPHES, CAES).
- Devices with medium power (1–50 MW) and storage capacities (5–100 MWh) (e.g. BES, FBES).
- Devices with small power (<10 MW) and storage capacities (<10 MWh) (e.g. FES, SCES, SMES).

The HESS, TESS, and EVs have unique characteristics as these are energy systems; for example, they require a number of different technologies that can be controlled differently. From another perspective, the security has to be designed accordingly to include these differences.

10.2.2 Electric Vehicles

Electric vehicles can feed directly from the power grid while stationary, at individual homes, or at common recharging points, such as car parks or recharging stations. By implementing EVs, it is possible to make large-scale BES economical and increase system flexibility (by introducing the large-scale energy storage). Consequently, similar to the HESS and the TESS, EVs also provide a method of integrating existing energy systems more effectively. EVs can be classified under three primary categories:

- Battery electric vehicles (BEV) – Plugged into the electric grid and act as additional load.
- Smart electric vehicles (SEV) – Have the potential to communicate with the grid; for example, at times of high wind production, it is ideal to begin charging EVs to avoid ramping centralized production; at times of low wind production, charging vehicles should be avoided if possible until a later stage.
- Vehicle to grid (V2G) – Operate in the same way as SEVs; in addition, they have the capability to supply power back to the grid, increasing the level of flexibility within the system.

All three types of EVs could be used to improve renewable penetrations feasible on a conventional grid, in case of wind, with each advancement in technology increasing the wind penetrations feasible from approximately 30 to 65% (from BEV to V2G).

10.2.3 Distributed Energy Resource Systems

Distributed energy resource systems (DERS) typically use renewable energy sources, communication technologies, and interfaces to connect to utilities and consumers. A DER is not limited to the generation of electricity but may also include a device to store distributed energy. DERS enable collection of energy from many sources to improve security of supply. By means of an interface, DERS can be managed and coordinated within Smart Grid.

DERS are small-scale power generation or storage technologies that are used to provide an alternative to or an enhancement of the traditional electric power system. Societal, policy, environmental, and technological changes are increasing the adoption rate of distributed resources, and Smart Grid technologies can enhance the value of these systems.

While conventional power plants are centralized and often require electricity to be transmitted over long distances, DERS are decentralized, modular, and more flexible technologies that are located close to the load they serve and have smaller capacities (e.g. 10 megawatts (MW) or less).

DER consist of generation and/or electric storage systems that are interconnected with distribution systems, including devices that reside on a customer premise, behind the meter. A grid-connected device for electricity storage can also be classified as a DERS and is often called a distributed energy storage system (DESS). Distributed generation and storage enables collection of energy from many sources and may lower environmental impacts and improve security of supply.

DERS utilize a wide range of generation and storage technologies such as renewable energy, CHP generators, fixed battery storage, and EVs with bidirectional chargers. DERS can be used for local generation/storage, can participate in capacity and ancillary service markets, and/or can be aggregated as VPPs. VPP is an ICT structure that integrates different types of distributed energy sources, flexible consumers, and energy storage with each other and with other market segments in real time through a Smart Grid.

10.2.4 Electrical Energy Storage Systems

Energy storage means of storing energy, directly or indirectly. The most common bulk energy storage technology used today is pumped hydroelectric storage technology. New storage capabilities – especially for distributed storage – would benefit the entire grid, from generation to end use. Energy storage can be utilized for a broad range of applications. However, the type of technology, which is suitable for some applications, is primarily defined by their potential power and storage capacities that can be obtained. Therefore, to provide a fair comparison between the various energy storage technologies, they have been grouped together based on the size of power and storage capacity that they can achieve.

Besides categories defined in [Connolly 2010], storage systems are classified based on energy source. Figure 10.5 illustrates a classification of energy storage systems based on energy source (mechanical, electrochemical, electrical). DESS applications include several types of battery, pumped hydro, compressed air, and thermal energy storage.

There are two major emerging market needs for electric energy storage (EES) as a key technology: to utilize more renewable energy and less fossil fuel and the future Smart Grid. New trends in EES applications include renewable energy, energy storage systems, Smart Grids, smart microgrids, smart houses, and EVs.

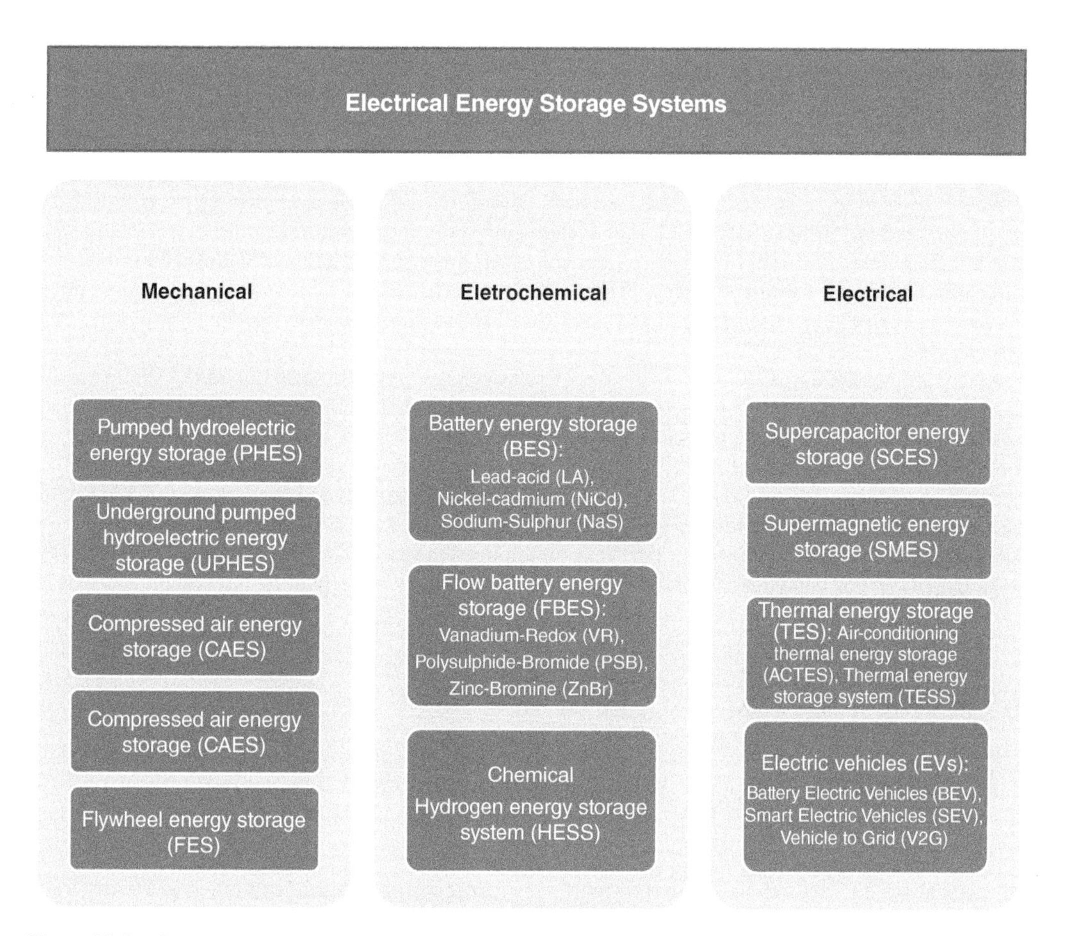

Figure 10.5 Classification of electrical energy storage systems.

10.2.4.1 Renewable Energy Generation

In order to solve global environmental problems, renewable energies such as solar and wind will be widely used. This means that the future energy supply will be influenced by fluctuating renewable energy sources – electricity production will follow weather conditions, and the surplus and deficit in energy need to be balanced. One of the main functions of energy storage, to match the supply and demand of energy (called time shifting), is essential for large- and small-scale applications.

A further option is so-called demand-side management (DSM), where users are encouraged to shift their consumption of electricity toward periods when surplus energy from renewables is available.

With the increasing number of installed PV systems, the low-voltage grid is reaching its performance limits. Therefore, self-consumption of power will become an important option for private households with PV facilities, especially as the price of electricity increases.

10.2.4.2 Energy Storage Systems

Energy storage systems could be a more promising solution for the integration of intermittent renewable energy than individual technologies. Energy storage technologies will most likely improve the penetrations of renewable energy on the electricity network, but often disregard the heat and transport sectors. One issue is uncertainties surrounding the costs and potential of energy storage systems because of their relative promise as stand-alone technologies.

An energy storage system is composed of three primary components [Connolly 2010]:

- Storage medium (SM) – It ranges from mechanical, chemical, and electrical potential energy.
- Power conversion system (PCS) – It converts convert from alternating current (AC) to direct current (DC) and vice versa for all storage devices except mechanical storage devices such as PHES and CAES.
- Balance of plant (BOP) – It includes all the devices that are used to house the equipment, communications, monitoring, and control, control the environment of the storage facility, provide the electrical connection between the PCS and the power grid.

An energy storage system requires monitoring and control of the operations. This capability is provided by a communication infrastructure that supports communications, data management, control, and security protection. Although energy storage systems may be communication agnostic and incorporate open system communication architecture [Steeley 2011], there should be information about the communication assets (hardware and software) owned by an enterprise. Examples of communication infrastructure assets include networks, gateways, software, applications, security products, protocols, and object models.

While it is essential to have local management for the safe and reliable operation of the storage facilities, it is equally important to have a coordinated control with other components in the grid when grid-wide applications are desired. Figure 10.6 depicts an energy storage system block diagram and its components.

Examples of energy storage applications include end-use applications, load management, T&D stabilization, emergency power backup, renewable energy integration, and DSM. The functionality and technical suitability of storage technologies to different applications are described in [Connolly 2010], [IEC 2011]. One aspect of any energy storage system is the management and control of storage systems as well as security.

10.2.5 Virtual Power Plant

Technical and social innovations are the key reasons for establishing and developing the VPP. Based on system theory, a VPP belongs to four systems, and it is shaped by one [Ropuszyoska-Surma]. These systems are identified as technical, legal, economic, and social. The most significant changes to such systems are summarized as follows [Ropuszyoska-Surma]:

- Technical systems are connected to information and telecommunication technologies, Smart Grids, smart metering, renewable resources, distributed storage, and electric cars to ensure energy security.

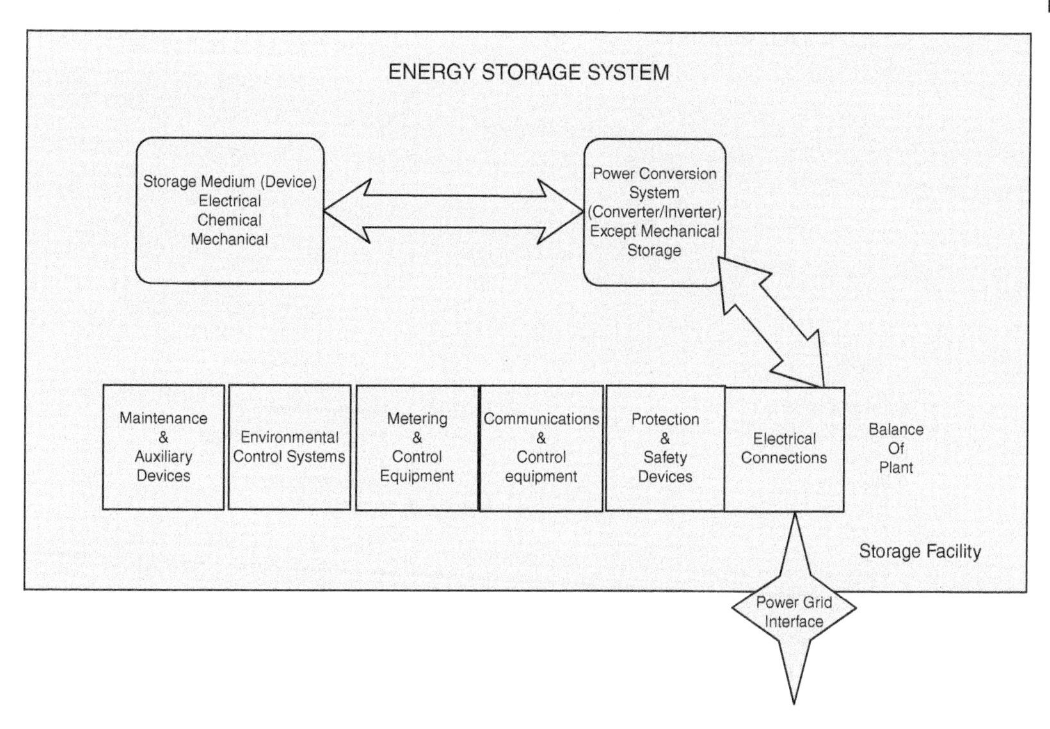

Figure 10.6 Energy storage system block diagram.

- Legal system with new regulations is focusing on the support system of renewable installations, limitation of greenhouse gas emissions, priority for the renewable, and legal regulation of prosumer.
- Economic system with elements like liberalization of energy sector, including electricity whole markets and establishment of the power markets, economic tools belonging to DSM and DR, mainly tariff system, high energy prices, economic trend to regionalization of energy market and creating local energy market connected with autonomous energy areas (called energy islands) base on local renewable resources, establishing prosumers, and economic cost connected with blackouts.
- Social system changes occur in behavior of energy consumers, higher ecological awareness of society, and establishment of network cooperation, based on local energy sources such as energy cooperatives and energy clusters to keep energy security.

As shown in Figure 10.7, there are connections among these systems; they should be taken into consideration together for the VPP characterization. Specifically, threats to VPP have to be analyzed not only from the perspective of one system, although the tendency is to focus only on technical system because the impulses for VPP establishment are technical innovations.

Therefore, risk assessment should consider more analysis paths to include dependencies among these systems that can be exploited by threat actors, creating new forms of malware causing wider disruption and impacts to Smart Grid.

10.3 DER Applications and Security

DERs are used in several core applications. Examples of core applications for Smart Grid include Volt and Var control; fault detection, isolation, and restoration (FDIR); demand response management (DRM); and DER integration and management [Feng 2012].

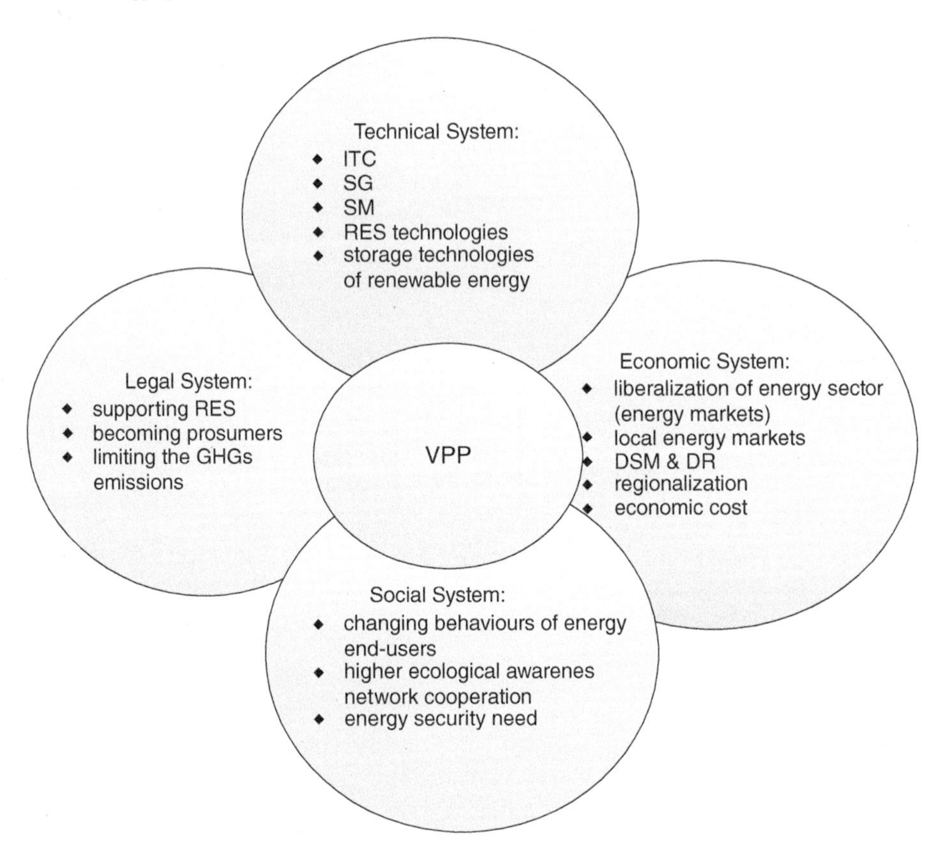

Figure 10.7 The reasons and conditions of VPP. *Source:* [Ropuszyoska-Surma]. Licensed under CC BY 4.0.

A typical solution for these applications includes power assets, communication infrastructures, control decision computing platforms, applications for control decisions, supervisory control and data acquisition (SCADA) system or distributed management system (DMS), and supporting applications (network modeling, load forecasting, state estimation, power flows, information model, etc.). In some applications, communications and automation of field devices (IEDs) and software schemes play a critical role in reliability of Smart Grid (one requirement for customer satisfaction is restoration time to be achieved within 1 to 5 minutes).

DER integration and management is applicable to both distribution and transmission systems [Feng 2012]. At the distribution system level, the DERs are small-scale wind generators, community energy storages, commercial and residential PV cells, microturbines, fuel cells, and PEV. At the transmission system level, DERs are the mega wind farms, solar farms, geothermal power plants, etc. DER and DER applications manage the dispatch of DR and DERs to reduce the power demand during peak hours and/or shift some of the demand to hours of the day when the demand is low, thus improving the load factor (utilization efficiency) of T&D assets.

Integrating any of these new applications requires the addition of new control, monitoring, and protection mechanisms. In addition, the successful implementation of automation and optimization of distributed resource systems requires to implement security requirements and maintain a security process [Abi-Samra 2013]. Examples of entities that include DERs and interfaces to DERs are illustrated in Table 10.1.

10.3.1 Energy Storage Applications

Energy storage applications differ from other DER options, such as distributed generation or energy efficiency, and have unique characteristics such as [Rastler 2010]:

- No typical operating profile or load shape that can be applied prospectively.
- Limited energy resources with a narrow band of dispatch and operation.
- Support participation in multiple wholesale markets.
- Provide several benefits simultaneously to the wholesale power system, electric distribution companies, and customers.
- Deficient in monetizing multiple stakeholder benefits.

Table 10.1 Entities including and interfacing DERs.

Entity	Description	Comments
Bulk generation	Conventional or renewable generation source that is connected to the electrical transmission system	Typical generation types include nuclear, fossil fuels, hydro, and various renewable resources
Bulk storage	Electric storage that is connected to the electrical transmission system	Bulk storage technologies include pumped hydro, compressed air, geothermal, superconducting magnetic energy storage, and batteries
Customer distributed energy resource	Customer DER includes demand response, generation, and energy storage located on and connected to the customer electrical system. Customer generation or storage that is connected to the transmission system is considered in the bulk generation domain	Customer DER may include PEVs
Customer point(s) of interface	The customer point(s) of interface is a common point for the on-premises devices that may communicate with the entities outside of the customer domain	Can be a physical interface box, meter, EMS, generator controller, controllable load devices, or directly connected. A single customer may have more than one point of interface
Customer substation	Electrical transmission or distribution substation located at customer facility that converts power to distribution voltage levels, which are then distributed within the customer site. Contains infrastructure necessary to control, monitor, and protect the electrical distribution system. Facility may include transformers, bus work, circuit breakers, capacitor banks, etc.	Customer substations can be at transmission or distribution voltage levels, depending on the size of the facility
Distribution sensors and measurement devices	Distribution sensors and measurement devices located on the distribution system but not within a distribution substation	Distribution sensors and measurement devices are typically a device associated with conversion equipment such as current transducers or voltage transducers. Examples of these items include meters, oscillographs, temperature sensors, etc.
Distribution distributed energy resources	Distribution DERs include demand response, generation, and energy storage located on and connected to the distribution system	Generation or storage connected to the transmission system is considered in the bulk generation domain

Source: [NISTIR 7628r1]. Public Domain.

Examples of applications that benefit of energy storage systems include [EPRI 2011]:

- End user: Energy management, reliability, DESS.
- Distribution, transmission, utility system: T&D investment deferral, renewable integration, renewable smoothing, T&D system support.
- Utility system, ISO: Energy arbitrage, system capacity, ancillary services.

Due to the unique characteristics of the various techniques available, there are a wide range of applications for energy storage systems that are identified in [Connolly 2010] as follows:

- End-use applications
- Emergency backup
- T&D stabilization
- Transmission upgrade deferral
- Load management
- Renewable energy integration
- Demand side management.

The highest value applications from a regional or total resource cost perspective are identified in [Rastler 2010] as follows:

- Wholesale services with regulation (15 minutes).
- Commercial and industrial power quality and reliability.
- Stationary and transportable systems for grid support and T&D deferral.
- End-use customer applications such as commercial, industrial, or home energy management.

There is a distinction between storage technologies classified as those that are best suited for power applications and those best suited to energy applications. Power applications require high power output, usually for relatively short periods of time (a few seconds to a few minutes). Examples of technologies include capacitors, SMES, and flywheels. Energy applications are uses of storage requiring relatively large amounts of energy, often for discharge durations of many minutes to hours. Examples of technologies include CAES, pumped hydro, thermal energy storage, and most battery types.

Another distinction is capacity applications versus energy applications. Capacity applications are those involving storage used to defer or to reduce the need for other equipment (e.g. reduce the need for generation or T&D equipment). Capacity applications tend to require relatively limited amounts of energy discharge throughout the year.

Energy applications involve storing a significant amount of electric energy to offset the need to purchase or to generate the energy when needed. Typically, they require a relatively significant amount of energy to be stored and discharged throughout the year.

However, there are many application synergies. Storage systems can be used for multiple applications if the capabilities of technologies satisfy the requirements of applications. For example, large storage used for load following may be especially complementary to other applications if charging and discharging for the other applications can be coordinated with charging and discharging to provide load following.

Load following could have good synergies with renewables capacity firming, electric energy time shift, and possibly electric supply reserve capacity applications. If storage is distributed, then that same storage could also be used for most of the distributed applications and for voltage support.

Issues in application synergies are technical and/or operational conflicts because storage systems do not have features or performance characteristics needed to serve multiple applications (e.g. storage that cannot tolerate many deep discharges; it could be well suited for T&D deferral, but the same storage system is not suitable for energy time shift).

Operational conflicts involve competing needs for a storage plant's power output and/or stored energy. For example, when storage is providing power for distribution upgrade deferral, it cannot be called upon to provide backup power for electric service reliability.

Other challenges may occur because many technologies, concepts and applications, systems, and infrastructures have interactions or they are competing for resources to satisfy a need.

10.3.2 Microgrid

A microgrid is a localized grouping of electricity generation, energy storage, and loads that normally operates connected to a traditional centralized grid (*macrogrid*). This single point of common coupling with the macrogrid can be disconnected. The microgrid can then function autonomously. Small power stations of 5–10 MW are used to serve the microgrids. Generation and loads in a microgrid are usually interconnected at low voltage. Microgrids generate power locally to reduce dependence on long-distance transmission lines and cut transmission losses.

Microgrids were proposed in the wake of India's blackouts in July 2012 [India Blackout]. Small microgrids cover an area of 30–50 km radius. According to [GTM 2016], microgrid operational capacity in the United States is expected to exceed 3.7 gigawatts (GW) by 2020, which is a 30% increase from previous forecasts.

Generation resources can include fuel cells, wind, solar, or other energy sources. The multiple dispersed generation sources and ability to isolate the microgrid from a larger network would provide highly reliable electric power. Produced heat from generation sources such as microturbines could be used for local process heating or space heating, allowing flexible trade-off between the needs for heat and electric power.

Microgrid is a small-scale power supply for a community. It can also be connected to the utility grid to avoid power outages. Microgrids can be developed, owned, and operated by communities, utilities, or corporations.

Microgrid is very similar to DER, as both can be used as stand-alone power generation units that can operate independently. Microgrids are much smaller version of the main grid that can generate, distribute, and regulate the flow of electricity to consumers. The consumers can participate in the electricity market, but on a smaller level. Microgrid components include:

- Distributed generation
- Loads
- Immediate storage
- Controller
- Point of common coupling.

Microgrids form the building blocks of the future power system. Three largest microgrid market segments include commercial and industrial, educational campuses, and military installations.

Forecast is that islanded microgrids will provide most of the short-term market growth; however dynamic islanding will provide the biggest opportunity in the medium term. This standard [IEEE 1547] covers the design and integration of microgrids into electrical power systems.

Figure 10.8 shows a diagram of a microgrid layers. The lower layer is the physical structure of the microgrid, including buildings, generators, solar panel generators, and one wireless access point (AP).

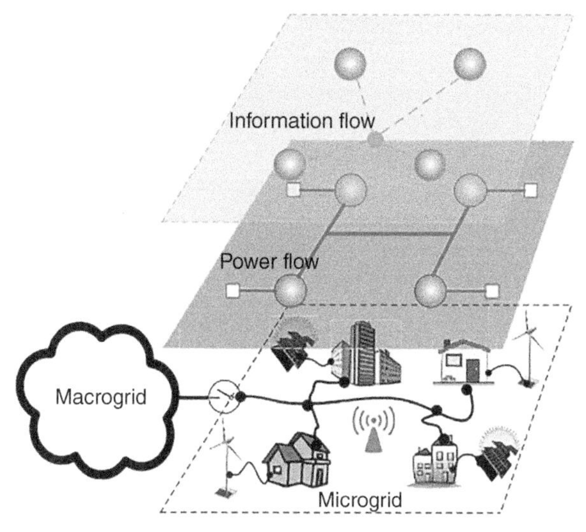

Figure 10.8 Another microgrid view. *Source:* [Fang 2012]. © 2012, IEEE.

The buildings and generators exchange power via power lines. They also exchange information via AP. The blue top layer shows the information flow within the microgrid and the red (middle) layer shows the power flow.

More commonly, microgrids are generally facilities with generation, such as large university campuses and military bases, hospitals, or communities. One rapidly increasing role of microgrids is the provision of services and benefits from the management of multiple DERs, such as electric storage, rooftop solar PV, and electric vehicles, in conjunction with building loads. However, there are challenges in the development of more complex controllers for microgrids and coordinating nested and networked microgrids with each other and with other DERs [INL 2016]. Microgrids can exist in multiple configurations: independently, networked along a feeder, or nested within another.

Microgrids serve loads with critical processes that depend on uninterrupted energy to operate. Local generation is used as a backup in case of a loss of the centralized utility source. In microgrids, DER devices are mainly used as the source of energy generation. Normal electricity market adopts a mechanism where utility sells the electricity to the consumer. Offers given by the generators are based on the energy demand that is either known or forecasted. Microgrid is a small power generation unit where the demands can change instantly. It is very difficult to forecast the demand in advance, so the supply-driven demand mechanism is used for microgrid operations.

However, it was always assumed that all those DER devices should belong to same utility, to have communication among them without any conflict. So, managing these customer-owned DER devices in microgrid, a market-based approach is proposed for electricity trading at local level within an intelligent distributed agent power microgrid.

Wholesale electricity market uses similar approach based on agent technology and Web services. It ensures the portability and interoperability among various devices and systems. Different operating modes require appropriate communication networks and adequate cybersecurity protection.

In wholesale electricity market, SCADA systems are used for the control and communication of any operations in power systems. In microgrids, local controllers and computer-based applications are used for monitoring, control, and decision making. From the point of view of the grid operator, a connected microgrid is controlled as if it were one entity.

10.4 Smart Grid Security Goals

For decades, power system operations have been managing the reliability of the power grid in which power availability has been a major requirement, with information security as a secondary but increasingly critical requirement.

10.4.1 Cybersecurity

Traditionally, cybersecurity for IT focuses on the protection of information and information systems from unauthorized access, use, disclosure, disruption, modification, or destruction in order to provide confidentiality, integrity, and availability. In the electricity sector, the historical focus has been on implementation of equipment that could improve electricity system reliability. Communications and IT equipment were formerly viewed as just supporting electricity system reliability. However, both the communications and IT sectors are becoming more critical to the reliability of the electricity system.

Cybersecurity for Smart Grid requires to address the combined IT, ICS, and communication systems and their integration with physical equipment and resources in order to maintain the reliability and the security of the Smart Grid and to protect the privacy of consumers [NIST SP1108r3]. As described in this third release of the document, measures are required to ensure the confidentiality, integrity, and availability of the electronic information communication systems and the control systems necessary for the management, operation, and protection of the Smart Grid's energy, IT, and telecommunication infrastructures.

In release 3.0 of NIST Roadmap for Smart Grid [NIST SP1108r3], Smart Grids are viewed from the perspective of cyber–physical systems (CPS) that are hybridized systems that combine computer-based communication, control, and command with physical equipment to yield improved performance, reliability, resiliency, and user and producer awareness. Cybersecurity is being expanded to address the following: combined power systems, IT and communication systems required to maintain the reliability of the Smart Grid, physical security of all components, reduced impact of coordinated cyber–physical attacks, and privacy of consumers.

CPS are increasingly being utilized in critical infrastructures and other settings. However, CPS have many unique characteristics, including the need for real-time response and extremely high availability, predictability, and reliability, which impact cybersecurity decisions, including in the DER environment [EPRI 2013]. In the United States, the federal government promotes development and deployment of a secure cyber–physical electric power grid. In addition to the efforts focused on Smart Grid and electricity subsector cybersecurity, there has also been additional attention to the cybersecurity of all the critical infrastructure sectors [NIST 2014]. This document describes a framework that includes a risk-based approach that enables an organization to gauge resource estimates (e.g. staffing, funding) to achieve cybersecurity goals in a cost-effective, prioritized manner.

Although focused on accidental/inadvertent security problems, such as equipment failures, employee errors, and natural disasters, existing power system management technologies can be expanded to provide additional security measures for Smart Grid and the integrated DERs. The objectives include:

- Ensuring adequate protection of information and systems such that every aspect of the Smart Grid infrastructure is secure; cybersecurity technologies and compliance with standards alone are not enough to achieve secure operations without policies, continuous risk assessment and management, and training.
- Security strategies should include addressing safety of people, prevention of cyber attacks, and reducing damage from cyber attacks.

Also, it is important to understand the functionality of each DER device and its role in the Smart Grid functionality, the domains, and enterprise in order to design the security to meet the business objectives of stakeholders. For example, the energy storage assets can operate in ways that resemble production, transmission, and/or distribution. Moreover, it may be possible for some energy storage assets to provide some combination of production, transmission, and distribution services simultaneously [FERC 2009b], [FERC 2012].

Advanced grid interactive DER functionalities, enabled by smart inverter interconnection equipment, are becoming increasingly available (and required in some jurisdictions) to ensure power quality and grid stability while simultaneously meeting the safety requirements of the distribution system.

Advanced DER functionalities also enable new grid architectures incorporating microgrids that can separate from the grid when power is disrupted and can interact in cooperation with grid operations to form a more adaptive resilient power system.

In addition, privacy of information is a significant concern in a widely interconnected system of systems that is the basis of the Smart Grid. Confidentiality and privacy of customer information has not been a priority for utilities, while financial and health institutions were required to comply with privacy standards such as Gramm–Leach–Bliley Act of 1999 or Health Insurance Portability and Accountability Act of 1996 (HIPAA).

10.4.2 Reliability and Security

For the Smart Grid to realize its potential, there is a need to understand how coupling between communication and power networks will affect the reliability, security, resiliency, and robustness of these networks. When information networks are supporting the power grid, we are really coupling many different communication networks to many different T&D networks [Hines 2014]. As pointed in this work, understanding the full implications of this multifaceted coupling is a significant research challenge for some issues. The relationships among key characteristics such as reliability, availability, cybersecurity, and resilience have to be understood for each implemented system or application. There could be different coupling and behavior scenarios that influence the metrics for reliability and availability.

One key issue is that reliability remains a fundamental principle of grid modernization efforts, but in today's world, reliability requires cybersecurity. Meanwhile, cybersecurity solutions for critical energy infrastructure are imperative for reliable energy delivery. Cybersecurity for the power grid must be carefully engineered to not interfere with energy delivery functions.

As discussed at the beginning of the chapter, reliability and security are important attributes for a system that have to be addressed in the context of information, system, infrastructure, network, Smart Grid, or system of systems. Security includes attributes such as availability, integrity, and confidentiality. Availability is one of the foundational components of security. However, availability is also an attribute of a resource such as network or system.

Availability in the context of this section means that necessary components of a network or system are operable and accessible when a user requires them. Accessibility means the user can use the network resources when needed. If the application is not available to a user or the network itself is unavailable, no authorized users will have access to information. So, one measure is to ensure that computing resources are available.

Often the network availability is a function of the reliability of its components. Reliability is a measure of the frequency of a resource (network) or component (hardware or software) failure. A network failure is an event that prohibits users from using the network as they normally do. One way to evaluate the reliability of a resource is to compute mean time between failure (MTBF) and mean time to repair (MTTR):

> MTBF is the average amount of time a given component can be expected to operate before failing.
> MTTR is the average time between the failure of a device or service and its repair (time to repair).

Both measures are important in determining the frequency of failure and the time required to return a network or a resource to successful operation.

Three factors influence the resource availability: operational considerations, MTBF, and MTTR.

Availability (A) can be defined by a probability function that can be simplified as suggested in Equation (10.1) [Stamper 2003]:

$$A = \text{MTBF}/\left(\text{MTBF} + \text{MTTR}\right) \tag{10.1}$$

If several components must be linked together to make the system available to the user, then system availability is given by the product of the availabilities of the components as suggested in Equation (10.2):

$$As = Ac_1 \times Ac_2 \times \cdots \times Ac_n, \tag{10.2}$$

n = number of components.

Reliability function $R(t)$ is the probability that the system will not fail during a given period, which can be defined by Equation (10.3) as

$$R(t) = e^{-bt}, \tag{10.3}$$

where b = 1/MTBF (time units for t and MTBF should be the same).

Like availability, reliability of a network or system is computed as the product of the reliability of its components (see Equation (10.4)):

$$Rs = Rc_1 \times Rc_2 \times Rc_3 \cdots Rc_n, \tag{10.4}$$

where n = number of components.

Thus, availability implies reliability. On other hand, cybersecurity posture is dependent on availability of systems and information.

These metrics that include mean time to fail (MTTF), MTTR, and MTBF are used to represent or calculate the reliability and availability [Mekkanen 2012]. As described in this work, availability is usually estimated based on two separated metrics for two events (failure and repair), defined as time to failure and time to repair, which also can be considered as reliability representing components. In addition, determining the availability of the system is dependent on the interconnections between the parts of the system for which two conditions can be considered. First, series interconnection or a failure in one part causes a failure to another part. Second, parallel interconnection or a failure in one part leads to other parts operating over the fielded one in the system. A method for calculating reliability and availability metrics is described in this work. Improving a system reliability and availability may require redundancy (e.g. in the network). On the other hand, by introducing redundancy, the system complexity and cost increases. Similarly, cybersecurity improves if reliability and availability are higher, but it can decrease with a higher complexity of the system. Often, security solutions do not scale well. Therefore, solutions have to be more practical and careful based on each implementation.

There are many facets and requirements for the relationship between reliability and security attributes (specifically, availability). For example, the most important to consider include the following:

- Availability and reliability are key performance metrics of a computing infrastructure that impact security operations. Availability and reliability are indicative of failure frequency and service duration. A system can have low reliability but have high availability.
- Internet and global e-business application requirements demand that companies increasingly implement computing infrastructures specifically designed for at least 99.999% availability. High availability of the environment, at least 99.999, is the equivalent of less than 5.3 minutes of downtime a year.
- 100% availability represents when fully operable and 0% represents the unavailability (failure) case. The failure case might be considered as component failure when it is not working as intended or some application IEDs are no longer reachable due to the communication network.
- SCADA networks are key component of power grid infrastructure and critical infrastructures. Different from the reliability that focuses on a period when the system is free of failures, the availability is concerned with a point in time at which the system does not stay at the failed state. It can be viewed as the percentage of time the system is available to perform its expected task. If the computing infrastructure is not available, it is possible that a security service using that infrastructure will not work. In response to these needs, government and SCADA owners need to address increased reliability and support for high availability.
- For systems or resources that cannot be down, solutions are available to have the information available. Examples include redundancy, fault-tolerant technologies, service-level agreements, and robust operational procedures. However a balance has to be made between the value of information and the cost of keeping the information available.
- The reliability of a component or a system is the probability that the component or the system will be operating correctly without failure in a time interval. The correlation between reliability and security is a recurring theme; it has been shown that investment in software quality reduces the incidence of computer security problems, regardless of whcther security was a target of the quality program or not.

For example, availability of communication network is an integral requirement for a PMU-based measurement system. The effect of communication failure on a network installed with PMUs has been analyzed in [Rihan 2013]. It has been shown that the number of PMUs that are required to maintain observability increases significantly as the number of locations with communication failure increases and it may even restrict complete observability. Therefore, availability of a robust communication infrastructure should be an integral consideration of a PMU placement methodology. The measurements from PMUs are time sensitive, and they must reach the point of use within about two seconds. Late arriving data is either discarded or passed on to data store. Therefore, security measures should perform within the allocated constraints and should not introduce any time delay.

As the power grid increasingly uses modern computational platforms, field devices, and communication networks, it gains access to new and higher-resolution data. New ways of measuring, analyzing, and communicating data support new capabilities for enhanced grid reliability, resiliency, and efficiency.

Nowadays, a potential additional source of unavailability of electric power distribution systems is related to the fast introduction of complex information systems supporting technical and commercial activity of the several actors active in electricity marketplace. Most important ones are:

- The widespread and increasing use of SCADA systems as part of the electric power distribution systems. Normally, SCADA systems make use of commercial off-the-shelf (COTS) hardware and software and provide connections to other company networks. The reliability of these components for highly available systems, such electric power distribution systems, is an issue.

- The rapidly proliferating industry-wide information systems, some of which mandated by regulatory bodies to facilitate competition, are based on open system architectures, centralized operations, increased communications over public telecommunication networks, and remote maintenance. The usability of this type of architecture for highly critical systems, such as electric power distribution systems and telecommunications network, is an issue.
- Taking advantage of the speed, efficiency, and effectiveness of computers and digital communications, electric power distribution systems are increasingly connected to information and communication networks, including public telecommunication network (PTN) and Internet, in a typical configuration of largely distributed and open systems of systems, multi-jurisdictional, and unbounded. Rapid increasing complexity, increasing software driven, remote managed and maintained PTN increase the possibility for unpredictable threat and exposure to cyber attacks. Vulnerability of PTN and recovery policy from external attacks and natural disasters is an issue. All such aspects contribute to form highly complex systems difficult to design and defend.
- New properties that are a consequence of the interaction between components, subsystems, and systems (emergent properties) have to be understood, represented, and faced.
- Survivability, the availability of critical functions and services after accidental events or intrusions, is an emergent property of large-scale electric power distribution and telecommunication systems. Survivability depends on the characteristics of security, integrity, reliability, and performance [Bologna 2000].

10.4.3 DER Security Challenges

Solutions for the Smart Grid focus on Internet-based, decentralized grid that accommodates DERs. This openness may allow several entry and observation points that could be used for launching cyber attacks. The protection of power grid is more difficult because it requires investments for more security, new technologies, innovation, education, and regulations. Concepts such as confidentiality and privacy, concepts historically not relevant to the power industry, are being introduced to become a goal for control engineers. Security challenges for the electrical sector are numerous, and examples include the following:

- Scalability and policy issues impede mass deployments that must integrate a vast number of smart devices and systems in order to realize Smart Grid capabilities.
- Integration of the many individual Smart Grid technologies is the largest challenge in development and deployment of Smart Grid codes [IEA 2013].
- Interoperability is a key element of technology development and has to be put into practice through technical standards and grid codes [IEA 2013].
- Electricity industry does not have an effective mechanism for sharing information on cybersecurity, no security built in in Smart Grid systems and components, lacking consistent cybersecurity awareness, missing security metrics [GAO 2011].
- Providing transparency because small delay in providing data can cause inconvenience not only to market participants but also for the whole market.
- Potentially increasing security vulnerabilities due to continuous move of automated systems toward more open architecture; for example, the use of Window as a base to the SCADA/EMS systems has been increasing, but it is not clear that these operating systems can achieve the reliability of other less commercial alternatives.
- Increasing risks with respect to usability (users are impacted by excessive security controls or lack of controls) that can negatively impact or even paralyze a system to the point that the system will fail to meet its intended operational goals.
- Need for efficient communications and network reliability planning including prioritizing and assigning budgets to those networks within Smart Grid infrastructure that are most mission-critical and whose interruption can affect service delivery.

- Effective coordination of several activities that must take place when a disruption due to a cyber threat occurs.
- Understanding how and to what degree the systems are interdependent. It is necessary to identify interdependencies and coupling at various levels such as physical, logical, information, economic, etc.; for example, by understanding the logical coupling in a system, an attacker could potentially disrupt the operation of a critical system by manipulating measurement or other data used to make system control decisions.
- Usage of technologies with known vulnerabilities and devices with no adequate security features built in.
- Limitations of current security technologies to satisfy stringent real-time requirements.
- Control system technologies have limited security, and if they do the vendor-supplied security capabilities are generally only enabled if the administrator is aware of the capability.
- No planning and organization of budgets dedicated to security problems.
- Utilities are facing infrastructure investment needs to support issues such as climate change, energy independence, and cybersecurity.
- Inconsistent support to increase security education and cybersecurity awareness.
- The electricity industry does not have an effective mechanism for sharing information on cybersecurity and other issues [GAO 2011].
- Lack of cybersecurity awareness, lack of security features built into Smart Grid systems, and lack of metrics to measure cybersecurity [GAO 2011].

The need for improving the critical infrastructure is voiced recently by many organizations and forums. Although best practices are useful and can be used for some activities in the management of a security program, there is a need to lay down a foundation for security planning. Risk assessment is the first step to follow on establishing a meaningful security strategy. These issues are discussed in another chapter.

10.5 Security Governance in Energy Industry

A global survey was conducted in 2012 on security governance, specifically how boards of directors and senior management are governing the security of their organizations' information, applications, and networks [Westby 2012]. The survey respondents included 75% participants from critical infrastructure companies and represented. The survey reveals aspects of the security pasture for the energy sector and other industries (financial sector, healthcare, manufacturing, IT, and telecommunications):

- Boards still are not undertaking key oversight activities related to cyber risks, such as reviewing budgets, security program assessments, and top-level policies; assigning roles and responsibilities for privacy and security; and receiving regular reports on breaches and IT risks.
- Utilities are one of the least prepared organizations when it comes to risk management.
- Utilities/energy sector and the industrial sector came in last in numerous areas – surprising is that these companies are part of critical infrastructure.
- All industry sectors surveyed are not properly assigning privacy responsibilities.
- Energy/utilities and IT/telecom respondents indicated that their organizations never (0%) rely upon insurance brokers to provide outside risk expertise, while the industrials sector relies upon them 100%.

Changes in the cybersecurity posture are reported in a recent survey [PWC 2016]. This survey identifies power and utilities companies that have combined their IT, cybersecurity, and physical security functions into the same organization, while others have developed a more robust governance structure. Also, the survey finds that many more companies will address cybersecurity issues in a more integrated and holistic way in 2016.

There is a growing acceptance that cybersecurity is an enterprise-wide priority including increased education and a move toward rethinking cyber risk governance. Increasingly, power and utilities organizations are sharing cybersecurity intelligence with external partners to better identify and respond to risks. Specifically, companies may be upgrading and integrating their cybersecurity, physical security, corporate, and control system environments to incorporate security and safety of data and personnel in order to provide greater protection from operational and reputational risks.

10.5.1 Security Governance Overview

Governance, as defined by the IT Governance Institute (ITGI), is the set of responsibilities and practices exercised by the board and executive management with the goal of providing strategic direction [ITGI 2003].

Governance specifies the accountability framework and provides assignment of responsibility all in an effort to manage risk [NIST SP800-100]. Governance ensures that security strategies are aligned with business objectives and consistent with regulations.

It is important that information security activities to be integrated into the corporate governance structure, because information is a vital resource for organizations [Pironti 2006]. Corporate governance involves a set of relationships between a company's management, its board, its shareholders, and other stakeholders [OECD 2004].

Information security governance is the set of responsibilities and practices exercised by the board and executive management with the goal of providing strategic direction, ensuring that objectives are achieved, ascertaining that risk is managed appropriately, and verifying that the enterprise's resources are used responsibly [ISACA Glossary]. Information security governance (or IT security governance) is the system by which an organization directs and controls information security. IT security governance should not be confused with IT security management. IT security management is concerned with making decisions to mitigate risks; governance determines who is authorized to make decisions.

10.5.2 Information Governance

Often, information security governance is interchanged with information governance that is related, but it is different. Information governance is the specification of decision rights and an accountability framework to encourage desirable behavior in the valuation, creation, storage, use, archival, and deletion of information. It includes the processes, roles, standards, and metrics that ensure the effective and efficient use of information in enabling an organization to achieve its goals [Logan 2010]. Several information governance solutions are touted by businesses because the information governance challenges facing the enterprise today is of strategic significance. One goal is turning information from a potential liability into a trusted strategic asset. Accomplishing this requires a governance program geared to proactively manage information and ensure its quality, security, and trustworthiness as the basis for making effective decisions.

10.5.3 EAC Recommendations

The Electricity Advisory Committee (EAC) recommends that DOE promote the establishment of enterprise security governance as a corporate norm in the energy sector in the near term [EAC 2014]. The document outlines recommendations such as the following:

- Responsibility for physical and cybersecurity belongs to the entire organization: it begins with senior leadership and extends throughout the enterprise to foster a more security-aware culture.
- Organizations are required to meet objectives: adequately safeguard critical energy infrastructure, satisfy regulatory requirements, and protect customer data and corporate reputations.
- Achieving objectives: it mandates full support and leadership of senior executives and a new approach to enterprise security governance.

10.5.4 Establishing Information Security Governance

With the increasing complexity of IT infrastructures in organizations and the increasing information warfare on the Internet, many organizations find that traditional approaches to information security management are no longer working. Over the past few years, information security governance and information security culture have become popular aspects of organizational information security.

Governance frameworks used in the IT governance processes and structures include COBIT [COBIT 5], ITIL [ITIL 2011], [ISO/IEC 27000] and [ISO 9000]. ITIL and ISO 27000 are common frameworks in use.

One progressive step is the growing recognition of department managers to accept responsibility for their data and its protection. Shifting the role of the ISO from compliance to offering assistance realizes the concept of security as a service. Although CERT references several useful documents at [CERT Governance], some references are outdated and not anymore aligned with the dynamics of organizations and emerging security trends. Other guidelines include best practices [ISACA 2006] and guidelines for implementing information security governance [Infosecurity 2014], which can complement the standards and frameworks.

What needs to be governed, defining an information security program, is a key characteristic of effective security governance. Activities of an information security program directly support/trace to an institutional risk management plan. In other words, the information security program is targeted to managing institutional risk [Allen 2007].

Effective security must take into account the dynamically changing risk environment within which most organizations are expected to survive and thrive [Caralli 2004a]. This work describes the responsibilities of management and security leaders. Leaders must shift their point of view from an IT-based, security-centric technology solution perspective to an enterprise-based risk management, organizational continuity, and resilience perspective. This requires moving beyond ad hoc and reactive approaches to security to preventive approaches that are process centered, strategic, and adaptive. The chief security officer (CSO) (called also chief information security officer (CISO)) must be able to draw upon the capabilities of the entire organization so that they can be deployed to address a problem requiring an enterprise-wide solution set. This also means being able to achieve security in a way that is sustainable – systematic, documented, repeatable, optimized, and adequate with respect to the organization's strategic drivers [Caralli 2004a].

Managing security refers to the process of developing, implementing, and monitoring an organization's security strategy, goals, and activities. The effectiveness of the security strategy depends on how well it is aligned with and supports the organization's business drivers: mission, business strategy, and critical success factors. Critical success factor method based on theories and experience in applying it to enterprise security management is described in [Caralli 2004b]. The enterprise security management is defined as a management and process-oriented view of security as a business process that is pervasive across and dependent on the enterprise.

Outcomes of effective information security governance should include [ISACA 2006]:

- Strategic alignment of information security with institutional objectives.
- Risk management – identify, manage, and mitigate risks.
- Resource management.
- Performance measurement – defining, reporting, and using information security governance metrics.
- Value delivery by optimizing information security investment.

Other characteristics of security governance effectiveness include [CERT Governance]:

- Leaders' accountability.
- Defined roles, responsibilities, and segregation of duties.
- Enforced policy.
- Adequate committed resources.
- Aware and trained staff.

Although standards provide general guidelines, it is necessary to adjust information security to the latest developments in security research.

10.5.5 Governance for Building Security In

More importantly, in a dynamic environment, organizations need to implement decentralized decision making to ensure the necessary flexibility and adaptability of their security posture. In such an environment of decentralized decision making, it becomes extremely important to implement the right security governance structures and practices to ensure that consistently good decisions are being made.

To support a dynamic flexible security posture based on decentralized decision making, one approach is an enterprise-wide security governance where security objectives and strategies are the main focus instead of security risks and controls.

To realize the information security governance, organizations need also to adopt an enterprise software security that is about how software resists attack, not how well the organization protects the environment in which the software is deployed [Steven 2006]. The author describes an enterprise-wide software framework to support built-in software security based on governance principles. The first step on establishing this framework is assessing the organization's current software development and security strengths and weaknesses. This applies to software built in-house, outsourced, purchased off the shelf, or integrated as part of a vendor solution. The framework should possess a who, what, when structure – that is, the framework should describe what activities each role is responsible for and at what point the activity should be conducted, and a vision for software security should be supported by organization's executives.

10.6 What Kind of Threats and Vulnerabilities?

Electric companies face both physical and computer disasters. Physical disasters include the destruction of generators or control, transmission, distribution centers, and operating and control facilities. Threats to IT, operations technology, and communication infrastructure assets are numerous and vary.

10.6.1 Threats

The threats may include malicious actors, malware (e.g. viruses and worms), accidents, and weather emergencies [DOE 2012].

In addition, disgruntled employees and subverted SCADA software manufactured by third party are considered as the biggest threats to electric companies [Hale 1998]. Consolidation and restructuring brought by impending deregulation may leave a lot of dissatisfied employees. An angry employee with an intimate knowledge of a control center's SCADA software is more capable of causing serious damage than an external hacker.

Also, the compliance to CIP standards does not guarantee a risk-free industry of energy. Intrusions into ICS systems have reportedly occurred with CIP mandatory standards in place [Behr 2014], [Campbell 2015]. Examples of threats and impacts to Smart Grid and DERs may include:

- Reputation loss – Attacks or accidents that destroy trust in Smart Grid services, including their technical and economic integrity.
- Business attack – Theft of money or services or falsifying business records.
- Gaming the system – Ability to collect, delay, modify, or delete information to gain an unfair competitive advantage (e.g. in energy markets).
- Safety – Attack on safety of the grid and its personnel or users.
- Assets – Damaging physical assets of the grid or assets of its users.
- Short-term denial or disruption of service.
- Long-term denial or disruption of service (including significant physical damage to the grid).
- Privacy violations.
- Hijacking and control of neighbor's equipment.
- Physical and logical tampering.
- Subverting situational awareness so that operators take fatal actions that disrupt the system.
- Cause automated system to waste resources on false alarms.
- Hijacking utility's services.
- Use of Smart Grid services or communication mechanisms to attack end users' residential or industrial networks (e.g. allowing end users to compromise other end users' networked systems).
- Authorization violation when an authorized peer attempts to perform actions/functions for which the peer is not authorized.

- Eavesdropping when the communication packets are being monitored by an intruder impacting the confidentiality of sensitive information or privacy of personal data.
- Information leakage such as disclosure of information to an unauthorized entity impacting the confidentiality of sensitive information.
- Interception or altering the communication packets by an intruder; the information in the packets can be then modified and forwarded to the original destination application, posing a threat to data integrity.
- Masquerade, typically referred to as spoofing; an intruder attempts to gain system access by attempting to pretend to be a different entity; this threat poses a severe control and data confidentiality risk.
- Replay when a communication packet that has been obtained through eavesdropping is retransmitted onto the network at a later time; if the captured packet contains control commands, this threat can have severe consequences.

Several cyber incidents are periodically reported in the United States by ICS-CERT, news media, and researchers.

10.6.2 Reported Cyber Incidents

This section is an overview of the recent cyber incidents targeting electricity sector and energy sector for the period of 2012–2016. More information about past incidents can be found in [ICS-CERT 2012a].

Threats to electrical substations were reported in many countries. Computer attacks could be used to manifest brownouts and blackouts, but their goal also can be disruption of mundane day-to-day operations or destruction of company data. Disrupting electrical service via computer requires an extensive knowledge of the power grid as well as familiarity of control center software. Crashing control center computing systems is unlikely to create blackouts but would adversely affect the operation of the power grid. Man-made production of brownouts and blackouts affecting large regions requires a sophisticated plan for attack. Such attacks require extensive knowledge of the power grid. This knowledge is often publicly available.

Attacks mounted on trading computers and computers in the business and engineering divisions of utility companies could wreak havoc on day-to-day operations. While such attacks can be launched from a network by hackers, the most serious threats to the power grid are likely to come from the inside by disgruntled employees or from subverted software used by electric companies but developed by third-party vendors.

It is reported that public utility companies are targets of frequent cyber incidents. The malware infections are reported in [ICS-CERT 2012b]. Compared with businesses in any other industry, the energy companies had the highest targeted malware attacks over a six-month period in 2012 [Korosec 2013]. The worldwide power loss including theft is estimated at approximately 200 billion dollars annually [Grundvig 2013].

A utility's control software was compromised when a sophisticated threat actor gained unauthorized access to its control system network via remote access [ICS-CERT 2014a].

IoT devices used in household devices were part of the first IoT botnet discovered in December 2013 by security professionals of Proofpoint [Dignan 2014]. Home routers and appliances were used to send emails to targets.

Researchers monitored energy sector for cyber attacks during one-year period, from July 2012 to June 2013 [Wueest 2014], [Wueest Blog], [Storm 2013], [Storm 2014]. They observed an average of 74 target attacks per day globally [Wueest 2014]. In addition, researchers are reporting that traditional energy utility companies are particularly concerned about scenarios created by the likes of Stuxnet (W32. Stuxnet) or Disttrack/Shamoon that can sabotage industrial facilities. In addition, there are concerns that intruders try to steal intellectual property on new technology, like wind or solar power generators or gas field exploration charts and data theft. Information stolen could be used in the future to perform more disruptive actions. As reported, the motivations and origins of attacks can vary considerably. However, the analysis report [Wueest 2014] provides statistics of other attacks. Spear phishing is, along with watering hole attacks, one of the most common attack vectors used to attack companies. In the first half of 2013, the energy sector was the fifth most targeted sector worldwide, experiencing 7.6% of all cyber attacks. A summary for the security of the energy sector includes categories of threat agents, threat agents' motivations, methods, targets, impacts to organizations and energy sector, and a forecast for the future security.

Also, studies show that threat agents are focusing on control systems and operating with varying motivations and intentions [Hull 2012].

10.6.3 Vulnerabilities

A set of possible vulnerabilities are identified in management, operational, and technical categories [NISTIR 7628]. In the same time, many new vulnerabilities could be discovered with the technological advances and human behavior.

Other concerns include security design by untrained engineers, reliance on nonstandard techniques and unproven algorithms, and security through obscurity. These practices are the key cause of many vulnerabilities. In addition, the fragmented nature of the national power grid information infrastructure complicates matters. A brief overview of vulnerabilities and threats for Smart Grid and DERS include:

- SCADA systems – The software control itself poses potential dangers to the electric utility companies. Mission-critical SCADA software is produced by third-party vendors. Subverted software that contains Trojan horses could render host computers and networks incapacitated. In addition, SCADA software vendors are less likely to have the same level of security awareness as are control centers, even though their software must be trusted to perform reliably and securely.
- Fragmentation – The federated nature of the electric utility industry complicates survivability issues. Control centers often share sensitive information to help them regulate the regional T&D of power. Electric companies rely on the accuracy of this information to make local adjustments to their operational units. The need for cooperation between independent distributed enterprises potentially exposes communication links to external threats, man-made and natural. Moreover, attacks can be distributed and therefore less easily recognized if units do not communicate ostensibly isolated intrusions to each other.
- Systems of systems – Increasing the complexity of the grid (adding new DERs and renewable integration) could introduce vulnerabilities and increase exposure to potential attackers and unintentional errors; interconnected networks can introduce common vulnerabilities, large number of stakeholders, and highly time-sensitive operational requirements.
- Network expansion – Increased number of entry points and paths for information communication offer potential adversaries to exploit increasing vulnerabilities to communication disruptions, and introduction of malicious software could result in denial of service (DoS) or compromise the integrity of software and systems.

The migration from older legacy-type architectures to modern operating systems and platforms can force industrial control systems to inherit many cybersecurity vulnerabilities, with some of these vulnerabilities having countermeasures that often cannot be deployed in automation systems.

Historically, proprietary and intricate information exchange architecture ensured "security through obscurity" provided secure means for data sharing, data acquisition, P2P data exchange, and other business operations between well-separated corporate business system and control system domains. Security threats for such systems were mostly limited to physical access to the system. The power grid control information system is evolving from isolated clusters of computers running stand-alone applications on a proprietary platform to a highly interconnected and interdependent system of local and wide area information and communication systems. Consequently, it is being exposed to new and emergent vulnerabilities and risks, very different in size, scope, likelihood, and frequency of occurrence than what traditional system analysis would suggest [Ray 2010].

Proliferation of control systems such as SCADA and ICS devices (PLCs, DCS controllers, IEDs and RTUs, etc.) became targets to cyber attacks. These systems and devices were primarily designed for reliability and real-time I/O. Many ICS devices crash if they receive malformed network traffic or even high loads of correctly formed data. Many systems use Windows PCs in these networks that often run for months without security or antivirus updates and are susceptible to outdated malware. Even without a direct connection to the Internet, modern control systems are accessed by numerous external sources that expose the control system to many threat agents as shown in Figure 10.9. Potential vulnerabilities and exposures to cyber attack include:

- Remote maintenance/diagnostics connections.
- Historian and MES servers shared with business users.
- Remote access modems.
- Serial connections.
- Wireless systems.
- Mobile laptops.
- USB devices.
- Data files (e.g. PDF or PLC project files).

Vulnerabilities might allow an attacker to penetrate a network, gain access to control software, and alter load conditions to destabilize the grid in unpredictable ways.

Among many of the higher-ranking vulnerabilities were certain DNP3 implementations of the DNP3 protocol in various master–slave station products [Byres 2013] affecting not only power industry. The DNP3 vulnerabilities, if exploited, could cause a DoS condition against the DNP3 master over either TCP/IP or serial communication paths, impeding the DNP3 master from communicating to field devices until remediated. The most recent vulnerabilities update was reported by ICS-CERT in April 2014 [ICS-CERT 2014b]. Also, this site includes a comprehensive list of public DNP3 implementation vulnerabilities.

The predicted huge amounts of IoT-connected things and the numerous vulnerabilities are big concerns [Dignan 2013]. Vendors and enterprises have numerous hurdles ahead such as a lack of standards, scalability, and a young application ecosystem.

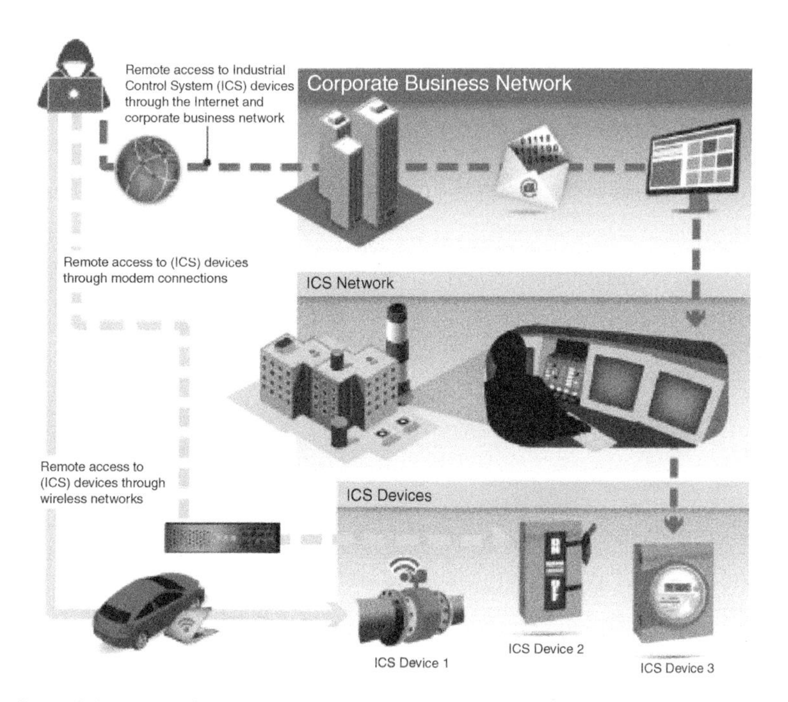

Figure 10.9 Potential accesses into a control system. *Source:* [GAO 2019]. Public Domain.

Security of the supply chain for the Smart Grid systems is a significant procurement concern because many components are obtained from many sources and vendors internationally. These sources may be considered targets of opportunity to compromise or counterfeit Smart Grid components [Opstal 2012].

Vulnerabilities occur in emerging CPS found in building automation as a result of contradictory requirements between the safety/real-time properties and the security needs of the system [Sun 2014].

The vulnerability of electric power distribution systems has been traditionally associated with the physical equipment, such as substations, generation facilities, and transmission lines. The migration from older legacy-type architectures to modern operating systems and platforms can force industrial control systems to inherit many cybersecurity vulnerabilities, with some of these vulnerabilities having countermeasures that often cannot be deployed in automation systems.

10.6.4 ICS Reported Vulnerabilities

As reported in [ICS-CERT 2014a], the 2013 year review shows that ICS-CERT received 181 vulnerability reports from researchers and ICS vendors throughout the year. Of those, 177 were determined to be true vulnerabilities that involved coordination, testing, and analysis across 52 vendors. The majority of these or 87% were exploitable remotely, while the other 13% required local access to exploit the vulnerabilities.

A fundamental recommendation for mitigating remotely exploitable vulnerabilities is to minimize network exposure and configure ICSs behind firewalls so they are not directly accessible and exploitable from the Internet. Equally important is patching and updating ICS devices as soon as practically possible, understanding that patches and upgrades must be properly tested by each asset owner/operator before being implemented in operational environments. The chart of Figure 10.10 depicts the different types of vulnerabilities reported and coordinated in 2013.

Energy theft is an immense concern in smart metering [McLaughlin 2010]. Smart meter flaws could allow hackers to tamper with power grid [Robertson 2010]. Generally, utility companies do not have the means to detect power theft or equipment failure [Grundvig 2013]. The attacks could be stealing meters – which can be situated outside of a home – and reprogramming them, wirelessly hacking the meter from a laptop. Flaws can be attributed to the smart meter technology (even protocols that were designed recently exhibit security failures known for more than a decade) and the technologies that

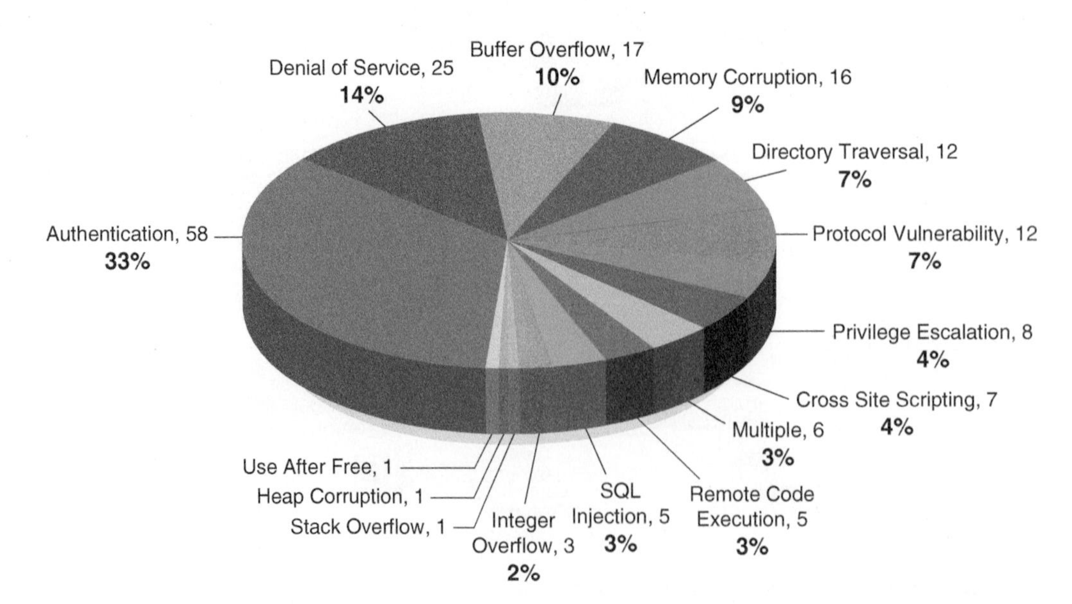

Figure 10.10 Different types of vulnerabilities reported and coordinated in 2013. *Source:* [ICS-CERT 2014a]. Public Domain.

utilities use to manage data from meters. Another reason for security flaws is that smart meters are being developed without enough security testing and probing.

The complexity and diversity of technologies used in home networks is another security challenge because all these technologies have not implemented or have limited security features. Vulnerabilities in home protocols such as ZigBee [Masica 2007] or G.hn [Gartner 2013] have to be analyzed, and countermeasures have to be designed to avoid the exposure to cyber threats. The total number of different vulnerabilities increased from 2011 to 2012, but buffer overflows still remained as the most common vulnerability type [ICS-CERT 2012a].

10.6.5 Addressing Privacy Issues

The Smart Grid brings new challenges and privacy issues, which require regulatory frameworks and adequate protection. Generally, privacy concerns are related to the collection and use of energy consumption data. These issues exist, unrelated to the Smart Grid, but Smart Grid aspects fundamentally change their impact.

Although it is uncertain how existing legal and regulatory frameworks may or may not apply to energy usage data collected, stored, and transmitted by Smart Grid technologies, it is clear that the new uses of existing data may require additional study and public input to adapt to current laws or to shape new laws and regulations. Also, organizations need to promote policies and activities to assess potential privacy concerns.

Therefore, there is a wide range of privacy concerns to address within the Smart Grid. These may impact the implementation of Smart Grid systems or their effectiveness. For example, a lack of consumer confidence in the security and privacy of their energy consumption data may result in a lack of consumer acceptance and participation, if not outright litigation [NISTIR 7628r1]. Privacy threats, vulnerabilities, and many other related issues are presented in this document.

10.7 Examples of Smart Grid Applications

In the United States, investments for the modernization of the power grid that supported several projects achieved many benefits that are presented in [Bossart 2013]. The analysis of the results provides information about the projects, investment costs, benefits, and development challenges for various applications and technologies. It also shows improvements in reliability indices, operational, and energy efficiencies due to new applications. The examples of Smart Grid projects and applications include [Bossart 2013]:

- Conservation voltage reduction.
- Improving power factors.
- Distribution and substation automation.
- Feeder reconfiguration (voltage and frequency control, Volt/VAR balance, automated load balancing).
- DR and consumer behavior.
- Outage management.
- Remote services (e.g. reading, connection).
- Condition-based maintenance.
- Transmission real-time situational awareness.
- Renewable integration.
- Distributed resources management.
- Energy storage systems.

The examples of technologies include [Bossart 2013]:

- Smart meters.
- Two-way communications (e.g. advanced metering infrastructure [AMI]).
- Automated capacitors.

- Smart sensors, switches, reclosers.
- Phasor measurement units.
- Cybersecurity.
- Weather forecasting.
- Equipment health monitors.
- Smart appliances.
- Home energy displays and networks.

Although successes are one aspect of the analysis, the report points out that there are still challenges to be addressed. These include short- and long-term vision, not fully developed concepts, shifts in technologies and business models, and technical regulatory and cultural challenges. In addition, the report includes an overview of several features of the future power grid modernization. Thus, several developments are on the way toward a Smart Grid. The following sections discusses the status of these developments.

10.7.1 Smart Grid Expectations

The Smart Grid is expected to control the demand side as well as the generation side so that the overall power system can be more efficiently and rationally operated. The Smart Grid includes many technologies such as IT and communications, control technologies, renewables, and electric energy storage.

Currently, dispatching of power and network control is typically conducted by centralized facilities, and there is little or no consumer participation. The future distribution system will become more active and will have to accommodate bidirectional power flows and an increasing transmission of information. Some of the electricity generated by large conventional plants will be displaced by the integration of renewable energy sources. An increasing number of PVs, biomass, and onshore wind generators will feed into the medium- and low-voltage grid.

Penetration of renewable energy requires more frequency control capability in the power system. EES can be used to enhance the capability through the control of charging and discharging from network operators so that the imbalance between power consumption and generation is lessened. Also, DERS play a key role on the modernization of the power grid.

DERS utilize a wide range of generation and storage technologies such as renewable energy, CHP generators, fixed battery storage, and EVs with bidirectional chargers. DERS can be used for local generation/storage, can participate in capacity and ancillary service markets, and/or can be aggregated as VPPs. Advanced grid interactive DER functionalities, enabled by smart inverter interconnection equipment, are becoming increasingly available (and required in some jurisdictions) to ensure power quality and grid stability while simultaneously meeting the safety requirements of the distribution system. Advanced DER functionalities also enable new grid architectures incorporating microgrids that can separate from the grid when power is disrupted and can interact in cooperation with grid operations to form a more adaptive resilient power system.

Future Smart Grids will not only have to integrate distributed renewable energy sources but will also have to integrate ICT for management and control. Currently, ICT integration is done by installing smart meters, which opens a wide area of new applications [Sobe 2013].

The Smart Grid is expected to control the demand side as well as the generation side so that the overall power system can be more efficiently and rationally operated. Thus, Smart Grid uses two-way flows of electricity and information to create an automated and distributed advanced energy delivery network utilizing modern information technologies, and thus it is capable of delivering power in more efficient ways and responding to widely ranging conditions and events. For example, once a medium-voltage transformer failure event occurs in the distribution grid, the Smart Grid may automatically change the power flow and recover the power delivery service immediately.

A further option is called demand-side management, involving Smart Grids and residential consumers creating what is called the smart home. With intelligent consumption management and economic incentives, consumers can be encouraged to shift their energy buying toward periods when surplus power is available. Users may accomplish this shift by changing when they need electricity, by buying and storing electricity for later use when they do not need it, or both. Future efforts target the increase of manageability and efficiency by dividing the Smart Grid into microgrids that consist of energy consumers and producers at a small scale and being able to manage themselves (see Figure 10.11). It shows the concept of generating electricity with the Smart Grid. The electricity is generated from various distributed sources, the power flow is bidirectional, and dispatching of power and network control is with consumer participation.

While various communication technologies are available for Smart Grids, there are communication and security challenges when implementing applications. For example, applications are grouped by their data rate and coverage range required for their successful deployment.

Potential secure and modern communication protocols including faster and more robust control devices and embedded intelligent devices (IEDs) for the entire grid from substation and feeder to customer resources could significantly strengthen the system security, reliability, and robustness. To provide system reliability, robustness, and availability simultaneously with appropriate installation costs and adequate security, a hybrid communication technology combined with wired and wireless solutions can be a balanced choice, but not an ideal solution as suggested in [Tsampasis 2016].

10.7.2 Demand Response Management Systems (DRMS)

Demand response is becoming an ever-increasing asset in the Smart Grid ecosystem. While these assets are critical to the ability of customers to make choices to potentially reduce their energy consumption, they are vital to the operations of many electric utilities for peak load management, economic control, and distribution system operations and optimization.

Making this linkage between the utility back office to its customers involves more than protocols, standards, and common application programming interfaces (APIs). It requires a platform capable of adapting to ever-changing business needs while remaining focused on providing the infrastructure for both command and control as well as customer empowerment [Vos 2009]. It is recognized that California's electricity blackouts of 2001 could have been easily avoided, if a small level of electricity consumption were reduced in real time by a demand response management system (DRMS) [Neumann 2006]. DRMS features have to support intelligent energy management, predictive load, dynamic pricing

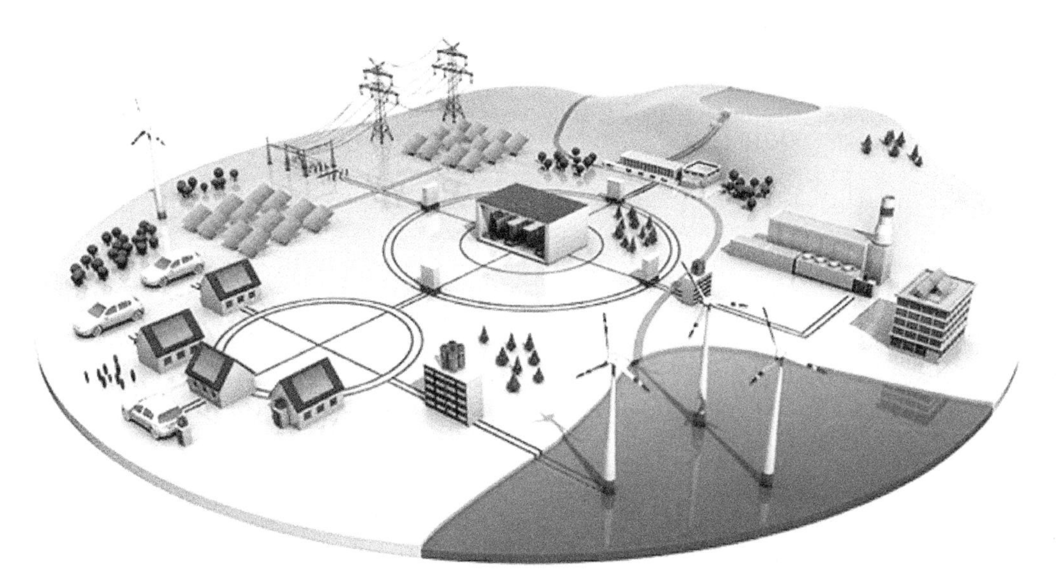

Figure 10.11 A Smart Grid view with integrated microgrids. *Source:* [Tsampasis 2016]. Licensed under CC BY 4.0.

controls, flexible accommodation of next-generation upgrades such as variable pricing, renewable energy management, and EV charging.

Security measures have to balance the protection of utilities business and customers' privacy. Unfortunately, the touted enforcement of cybersecurity standards by NERC do not cover something that is beyond the control room. Therefore, protection of communication networks must extend into customers' homes and provide secure access to systems, secure transactions, integrity and confidentiality of information, and customers' privacy.

10.7.3 Distribution Automation

Distribution automation (DA) is a family of technologies including sensors, processors, communication networks, and switches that can perform a number of distribution *system* functions [DA].

DA is used to improve reliability, service quality, and operational efficiency including automatic switching, reactive power compensation coordination, and other feeder operations/control. DA involves the integration of SCADA systems, advanced distribution sensors, IEDs, and two-way communication systems to optimize system performance.

Substation automation, when combined with automated switches, reclosers, and capacitors, enables full Smart Grid functionality. It includes building intelligence into the distribution substations, the metering infrastructure, and distribution feeder circuits and components that link these two essential parts of the grid. This allows automating switches on the distribution system, automatic reconfiguration, automating protection systems and adapting them to facilitate reconfiguration and integration of DER, integrating power electronics-based controllers and other technologies to improve reliability and system performance, and optimizing system performance. Software combines information flowing from the automated substations with SCADA data points throughout the distribution system to analyze and recommend reconfiguration of the distribution system for optimum performance.

10.7.4 Advanced Distribution Management System

As distribution networks become more complex, the need to deliver electricity reliably, safely, and securely drive the adoption of new and advanced distribution management systems (ADMS), which represent the most effective architecture for distribution management.

ADMS is the software platform that supports the full suite of distribution management and optimization. An ADMS includes functions that automate outage restoration and optimize the performance of the distribution grid. ADMS functions include fault location, isolation, and restoration; volt/volt-ampere reactive optimization; conservation through voltage reduction; peak demand management; and support for microgrids and EVs [ADMS].

Related to DA is advanced distribution automation (ADA) in which intelligent control methods are employed to the distribution and beyond. ADA can be used for real-time adjustment to changing loads, generation, and failure conditions of the distribution system, usually without operator intervention. This necessitates control of field devices, which implies enough IT development to enable automated decision making in the field and relaying of critical information to the utility control center. The IT infrastructure includes real-time data acquisition and communication with utility databases and other automated systems. Figure 10.12 depicts examples of DMS applications (see more information in [Schneider ADMS], [Schneider DMS]). ADMS includes several DMS applications.

ADMS was evaluated by Gartner in 21 March 2012 as the product that creates an industry benchmark from a functional viewpoint [Rhodes 2013]. The functionality of ADMS is developed as a Smart Grid solution for real-time power operations and decision making integrated with business needs. From a cybersecurity point of view, the system is deployed into three zones:

- Real-time zone (also called Secure Zone).
- Demilitarized (DMZ) (also called Decision Support Zone).
- Data Entry, Test, and Staging Zone.

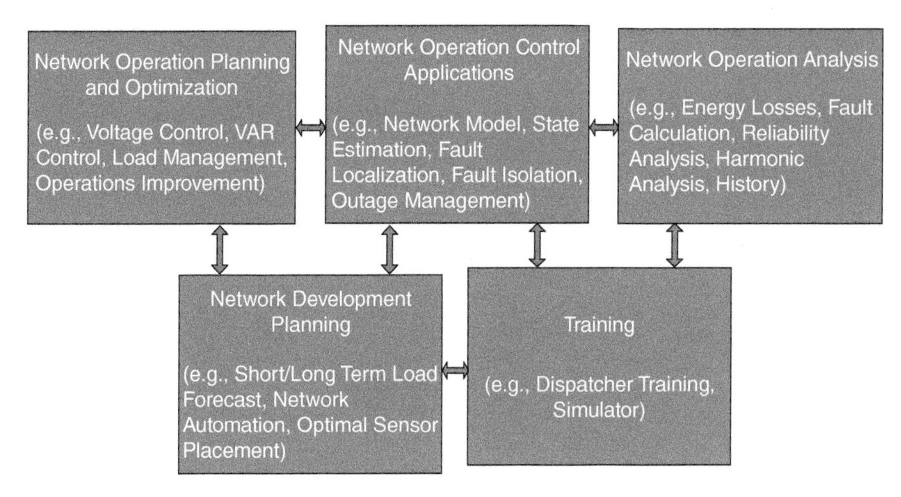

Figure 10.12 Examples of DMS analytical applications.

10.7.5 Smart Home

The concept of the smart home (or smart house) is proposed in order to use energy more efficiently, economically, and reliably in residential areas. EES technologies are expected to play an important role. Load leveling by EES can suppress the peak demand; consumers may be able to reduce electricity costs by optimizing EES operation and to use their own renewable energy sources. EES can reduce the mismatch between their power demand and their own power generation. Figure 10.13 illustrates the concept of the smart home (house).

The illustration shows home with renewable energy generation (wind and solar), storage, appliances, and plug-in hybrid electric vehicles (PHEV). One objective of the Smart Grid is demand management that would play a key role in increasing the efficiency of the load. Controlling demand-side load is known as demand response, and it is already implemented in the traditional power grid for large-scale consumers although it is not fully automated yet [Erol-Kantarci 2011]. Consumers can decrease their demand following utility instructions, and it is generally handled by the utility or an aggregator company. The subscribed consumers are notified by phone calls, for example, to turn off or to change the setpoint of their HVAC systems for a certain amount of time to reduce the load.

Examples of applications available in a smart home vary (e.g. see more about smart home in [IEC 2011]). Figure 10.14 shows an example of communications to smart home and communications to smart meter that interfaces with the home area network that connects several appliances within the house.

One of the promising demand management techniques is employing wireless sensor networks (WSNs) in demand management. A WSN is a group of small, low-cost devices that are able to sense some phenomena in their surroundings, perform limited processing on the data, and transmit the data to a sink node by communicating with their peers using the wireless medium.

The aim of demand management schemes is to schedule the appliance cycles so that the use of electricity from the grid during peak hours is reduced, which consequently reduces the need for the power from the power plants and reduces the carbon footprint of the household.

Regulations and standardization for implementing these kinds of applications are the major challenges. Around the globe, various and numerous governmental agencies, alliances, committees, and groups are working to provide standards so that Smart Grid implementations are effective, interoperable, and cost efficient. Security is another significant challenge since the smart home and power grid are becoming digitized, integrating with the Internet, and generally using open media for data transfer. Therefore, security measures for protection of smart home users and devices have become the norms of life in an information society.

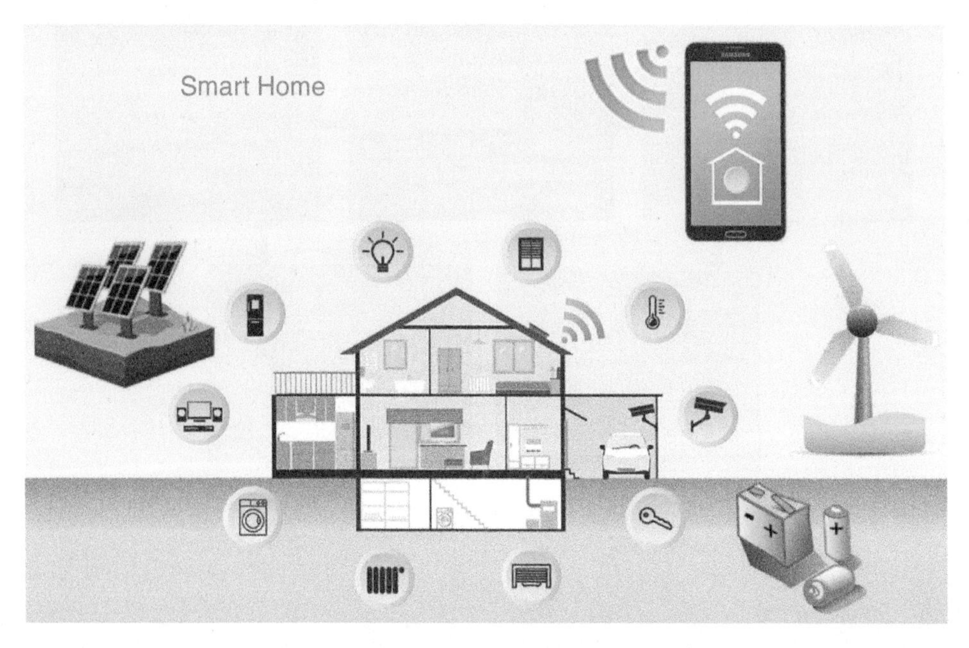

Figure 10.13 A smart home infographic. *Source:* Adapted from [SmartHome].

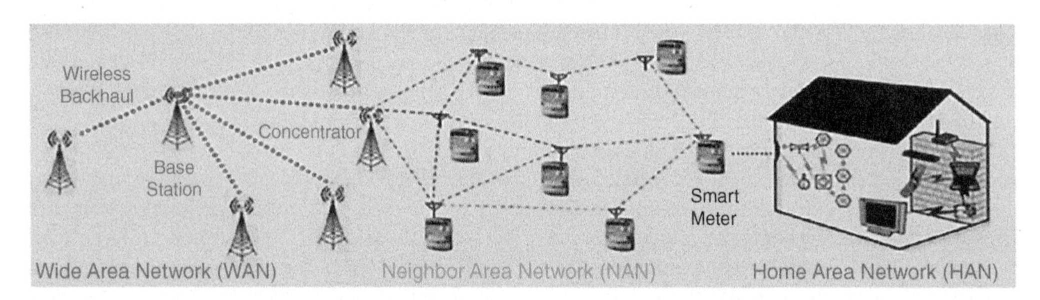

Figure 10.14 Example of communications to smart home. *Source:* [Tsampasis 2016]. Licensed under CC BY-SA 4.0.

Figure 10.15 depicts an architecture view of the smart home communication that includes entities for both the internal and external environments. As defined in [Komninos 2014], the external environment is represented by the energy services interface (ESI) entity, while internal environment is represented by the EMS environment.

The green dashed dotted lines represent the entities connected to the ESI, whereas the red dotted lines represent the entities connected to the EMS. The blue dashed line represents the communication between the smart home and its external environment.

Therefore, ESI is the interface between the smart home and the Smart Grid. It enables the remote control of devices, the support of DR programs, the monitoring of DERs such as wind turbines belonging to premises, the forwarding of consumption data to the neighborhood collection points (if it acts as a meter), plug-in electric vehicles/plug-in electric hybrid vehicles (PEV/PHEV) charging, etc. Due to cost considerations, the authors [Komninos 2014] recommend the integration of ESI and smart meter functionalities in one physical device, although logically they are separated.

The EMS enables the management of various appliances and systems within the Smart Home and supports smart home to adapt its energy profile to suit the Smart Grid capabilities. EMS controls appliances, air-conditioning systems, thermostats, light switches, pool pumps, and PEVs. The ESIv and EMS are in continuous two-way communication ensuring that the internal environment is acting in accordance with the external environment's requirements and capabilities.

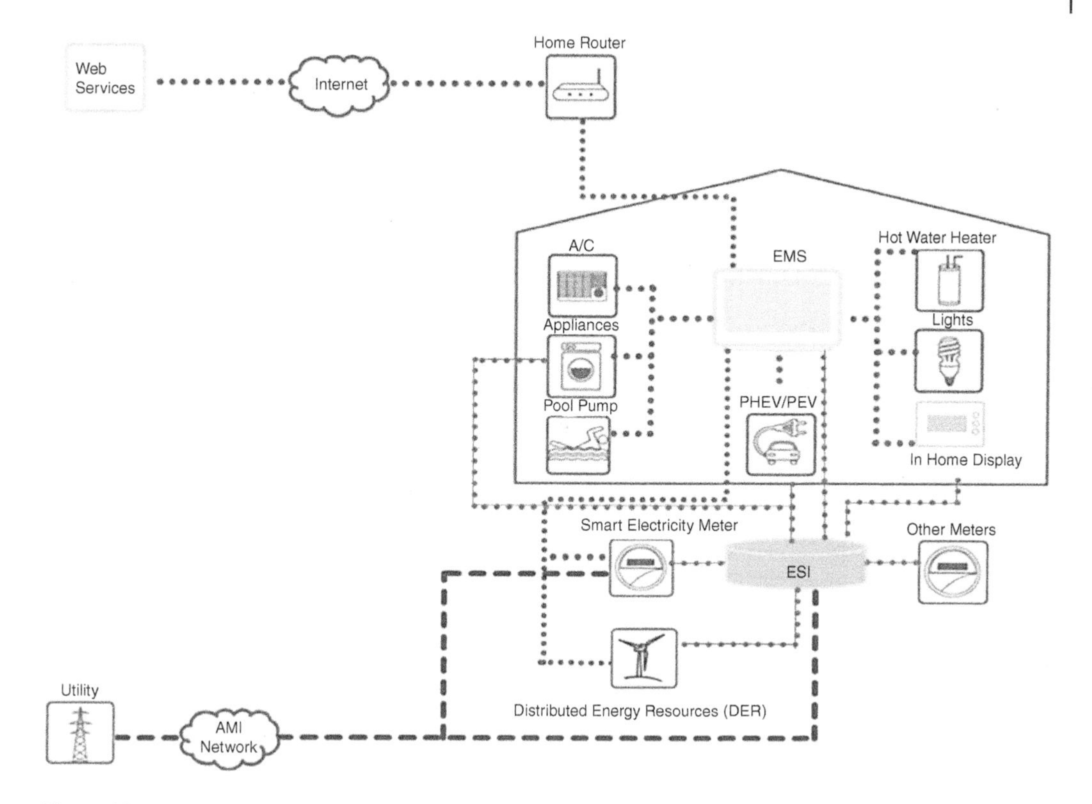

Figure 10.15 An overview of a smart home architecture. *Source:* [Komninos 2014]. © 2014, IEEE.

As illustrated in this architecture (Figure 10.15), the smart home can communicate with its external environment in two ways: either through the smart meter or through the ESI. In the former case the smart meter communicates with the NAN aggregator to report household consumption. In the latter, the ESI is responsible to enable various other interactions between the smart home and the utility such as remote load control. Through the EMS, Web Services are made available to the smart home, enabling, among other things, the remote configuration of HAN devices.

Therefore, the integration between smart home and Smart Grid enables many benefits such as DR programs, load shedding programs, effective feedback, peak shaving capabilities, and energy exchanges. As the connectivity increases, the security challenges also increase especially those challenges relative to system security. Commonly adopted security goals for Smart Grid/smart home include confidentiality, integrity, availability, authenticity, authorization, and non-repudiation, although confidentiality is not so critical as it is the customer privacy, another goal for the protection.

This work [Komninos 2014] provides a more comprehensive view of security for the smart home, taking into account its persistent interaction with the Smart Grid and focusing on the entire communication network, not only some specific subsystems that are often the focus in many published papers.

10.7.6 Smart Microgrid

A microgrid is a scaled-down version of the centralized power system that generates, distributes, and regulates the flow of electricity. Around the globe, it is observed a growth of the smart microgrid deployments. A Smart Microgrid is offering greater flexibility in meeting the power needs and is serving local and critical loads. Like the bulk power grid, smart microgrids generate, distribute, and regulate the flow of electricity to consumers but do so locally. Smart microgrids are an ideal way to integrate renewable resources on the community level and allow for customer participation in the electricity enterprise. Figure 10.16 is a representation of the components related to microgrid concept. A microgrid can operate either grid connected or islanded and, if required, can switch between the two.

Smart microgrids can be developed for households, villages, industry sites, hospitals, cities, or a university campus. A smart microgrid can either be connected to the backbone grid or to other microgrids or it can run in a so-called island mode. Dynamic islanding is one of the main solutions to overcome faults and voltage sags [Sobe 2013]. [Loreasl Blog] discusses the concept of the integrated microgrids (see also the infographic that depicts this concept). Figure 10.17 depicts a graphic of renewable resources and battery in

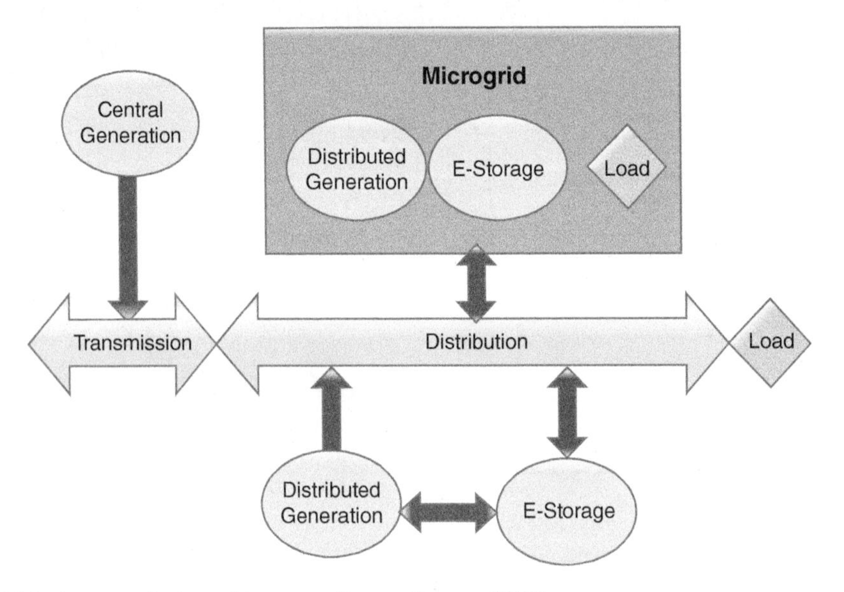

Figure 10.16 Integrated microgrid concept. *Source:* [Bossart 2012].

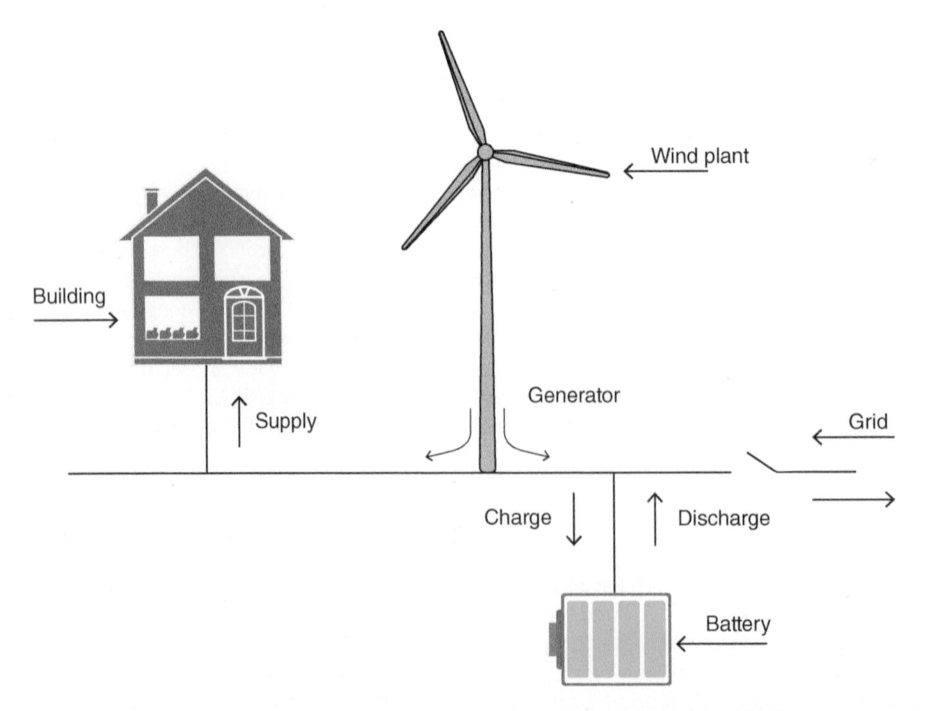

Figure 10.17 A microgrid with renewables connected to power grid. *Source:* [Dao 2016]. Licensed under CC BY-SA 4.0.

a microgrid connected to the power grid. The power generated by renewable resources is used to supply residential consumers and the grid. The storage battery can discharge energy for residential consumer or to the grid. Although the figure shows the electricity flow, there we have to consider also the information flow between different devices and entities.

For flexibility in resisting outages caused by disasters, it is very important to deploy smart microgrids as an element in constructing Smart Grids.

10.8 Standards, Guidelines, and Recommendations

The Smart Grid will ultimately require hundreds of standards. Some are more urgently needed than others. Examples include the following.

10.8.1 NIST Roadmap, Standards, and Guidelines

To prioritize its work, NIST chose to focus on seven key functionalities plus cybersecurity and network communications [NIST SP1108r3]. These functionalities are especially critical to ongoing and near-term deployments of Smart Grid technologies and services, and they include the priorities recommended by [FERC 2009a]. The nine priority areas include [NIST SP1108r3]:

- DR and consumer energy efficiency – Mechanisms and incentives for utilities, business, industrial, and residential customers to modify energy use during times of peak demand or when power reliability is at risk.
- Wide-area situational awareness – Monitoring and display of power system components and performance across interconnections and over large geographic areas in near real time.
- DER – Generation and/or electric storage systems that are interconnected with distribution systems, including devices that reside on a customer premise, behind the meter.
- Energy storage – Means of storing energy, directly or indirectly; new storage capabilities, especially for distributed storage, would benefit the entire grid, from generation to end use.
- Electric transportation – Primarily to enabling large-scale integration of PEVs.
- Network communications – A variety of public and private communication networks, both wired and wireless, that will be used for Smart Grid domains and subdomains; implementation, and maintenance of appropriate security and access controls is critical to the Smart Grid.
- AMI – Near-real-time monitoring of power usage; these advanced metering networks could also be used to implement residential DR including dynamic pricing and create a two-way network between advanced meters and utility business systems and other parties.
- Distribution grid management – Maximizing performance of various components of networked distribution systems and integrating them with transmission systems and customer operations including improved capabilities for managing distributed sources of renewable energy.
- Cybersecurity – Measures to ensure the confidentiality, integrity, and availability of the electronic information communication systems and the control systems necessary for the management, operation, and protection of the Smart Grid's energy, IT, and telecommunication infrastructures [FERC 2009a].

The three-volume document [NISTIR 7628r1], a revised version of [NISTIR 7628], presents an analytical framework that organizations can use to develop effective cybersecurity strategies tailored to their particular combinations of Smart Grid-related characteristics, risks, and vulnerabilities.

The guidelines document is a companion document to the [NIST SP1108r3], a revised version of the first document issued in 2010. The framework and roadmap report describes a high-level conceptual reference model for the Smart Grid, identifies standards that are applicable (or likely to be applicable) to the ongoing development of an interoperable Smart Grid, and specifies a set of high-priority standard-related gaps and issues. Cybersecurity is recognized as a critical, crosscutting issue that must be addressed in all standards developed for Smart Grid applications.

NIST catalog of standards for the Smart Grid includes about 56 standards (20 individual standards plus 5 series such as IEC 61850, which contain 36 additional standards) [NIST CoS].

10.8.2 NERC CIP Standards

There are many requirements documents that may be applicable to the Smart Grid. Currently, only NERC critical infrastructure protection (CIP) standards are mandatory in the United States for the bulk electric system (BES). The Version 5 CIP cybersecurity standards (CIP-002-5 through CIP-009-5, CIP-010-1, and CIP-011-1, the associated implementation plan, and the associated definitions) were approved by FERC in an order issued 22 November 2013. In that order, FERC directed further modifications to the standards [NERC 2014].

A stated goal of the NERC CIP Version 5 standards is to provide a roadmap for utilities to adopt a culture of security and due diligence, as opposed to a culture of compliance. While no system is ever completely secure, NERC CIP Version 5 makes progress not only in raising the bar for securing the safety and reliability of the BES but also in encouraging a change of approach in CIP programs – from compliance to security.

These standards are either subject to enforcement or subject to future enforcement or in development or pending regulatory filing. However, questions as to the future of CIP remain and continue to be debated. In addition, there are many unresolved technical issues including protection of low-impact cyber systems and controls that should be applied to mobile equipment such as laptops or smartphones, which could be used remotely to control BES cyber systems.

While prior versions of CIP standards were focused only on the security of critical assets of bulk power system, the newest version, Version 5 CIP cybersecurity standards, includes requirements and definitions such as BES cyber assets and cyber assets as distinct assets because of issues related to protection. However, the categorization of cyber assets among many other definitions may create confusion even with the experienced security professionals. Many entities and terms included in these standards have to be well understood before planning any activity for the security management of the Smart Grid. Therefore, it is recommended to review the definitions as provided in the most current dictionaries of terms and acronyms.

A dictionary of terms specific to operating and engineering power grid is maintained by NERC [NERC Glossary]. The cybersecurity terms are being defined by NERC and approved by FERC for the period of use. However, this issue creates ambiguity and misunderstanding when dealing with partners such as computing hardware and software manufacturers, supply chain, financial organizations, health organizations, etc. Also, many acronyms used for the Smart Grid can be found at [SGIP 2015].

Recently, FERC suggested that a dedicated organization direction should be established to be focused on handling the security standards and the cyber threats for utilities. It is believed that a new independent organization could bring together the disparate interests in the power sector to help manage cybersecurity for the nation's electric grid [Rascoe 2014], [Patel 2014].

More compliance with NERC cybersecurity standards will not assure security. It is required that utilities, large and small, to consider the application of the industry standards and best practices, including emerging security standards like NIST's Smart Grid Interoperability Standards Framework and AMI-SEC System Security Requirements, for end-to-end security of the Smart Grid, use a system-of-systems approach to cybersecurity by deploying standards published by International Organization for Standardization and International Electrotechnical Commission (ISO/IEC), National Security Agency InfoSec Assessment Methodology (NSA IAM), Information Systems Audit and Control Association (ISACA), and International Information Systems Security Certification Consortium (ISC2).

10.8.3 Security Standards Governance

Most organizations are still basing their information security on the old BS 17799 security standard that was developed several decades ago and are struggling with an increase in threats and vulnerabilities not addressed by current security standards. ISO/IEC 270001 adds a security policy life cycle approach to security management, in the hope that a more mature information security management will lead to a better information security. Already many standards of ISO/IEC 27000 series of security standards have been revised or replaced by new standards.

The recent standard [ISO/IEC 27014] refers to governance for information security as an integral part of the organization's corporate governance with strong links to IT governance, and applies principles from ISO standard [ISO/IEC 38500] to information security, and considers the relationship between information security governance and other governance and management disciplines. This standard is also adopted by ITU-T as ITU-T recommendation X.1054 with identical text.

ISO/IEC 38500 standard provides a framework for effective governance of IT to assist those at the highest level of organizations to understand and fulfill their legal, regulatory, and ethical obligations in respect of their organizations' use of IT.

ISO/IEC 27014 provides guidance on concepts and principles for the governance of information security, by which organizations can evaluate, direct, monitor, and communicate the information security-related activities within the organization. The standard identifies a governing body and executive management; both are parts of top management. The governing body is defined by [ISO/IEC 27000] standard as the person or group of people who directs and controls an organization at the highest level [ISO/IEC 27000]. Also, guidance on information security governance and planning is provided in [NIST SP800-100] standard.

COBIT (Control Objectives for Information and Related Technologies) version 5 is the only business framework for the governance and management of enterprise IT. COBIT 5 builds and expands on prior version by integrating other major frameworks, standards, and resources. An add-on for COBIT 5 related to information security was released on December 2012, and one related to assurance was released in June 2013 [COBIT 5].

Information Technology Infrastructure Library (ITIL) includes a set of practices for IT service management focusing on aligning IT services with the needs of business. Also, it includes the process for security management. It is based on the code of practice for information security management system (ISMS), known as [ISO/IEC 27002]. One goal of security management is to ensure adequate information security. The primary goal of information security, in turn, is to protect information assets against risks and thus to maintain their value to the organization [ITIL 2011].

ISO/IEC 27002:2013 provides best practice recommendations on information security management for use by those responsible for initiating, implementing, or maintaining ISMS [ISO/IEC 27002]. One issue is that information security is defined within the standard in the context of the CIA triad. The ISO/IEC TR 27019:2013 is a guide based on ISO/IEC 27002 for process control systems specific to the energy utility industry.

The standard [ISO/IEC 20000] is the first international standard for IT service management. A recent update of 2015 provides guidance on the use of ISO/IEC 20000-1:2011 for service providers delivering cloud services. Cloud services are more used by utilities and service providers.

Although these frameworks and standards are distinct, there are relationships among them. Therefore, a good understanding of the business needs and use of most recent versions is necessary.

References Part 4

[Abi-Samra 2013] Abi-Samra, N. (2013, December 13). *Distribution systems automation & optimization, Part 3.* EEWeb Pulse Electrical Engineering Community. https://www.eeweb.com/distribution-systems-automation-optimization-part-3/

[Abi-Samra 2014] Abi-Samra, N. (2014). *Utility X.0* (February 2014). [Newsletter article]. IEEE SmartGrid Resource Center. https://resourcecenter.smartgrid.ieee.org/publications/newsletters/SGNL0149.htmlhttp://smartgrid.ieee.org/february-2014/1031-utility-x-0

[Adams 2013] Adams, W.C. (2013). IEEE Standards Association smart grid strategic focus. *Wireless World Research Forum Meeting*, Vancouver, Canada, (October 2013). https://sgc2013.ieee-smartgridcomm.org/sites/ieee-smartgridcomm.org/files/WilbertAdams.pdf

[ADMS] Gartner. (n.d.). *Gartner IT Glossary*. http://wwwt.gartner.com/it-glossary/advanced-distribution-management-systems-adms

[Allen 2007] Allen, J.H., Westby, J.R. (2007). Characteristics of effective security governance (May 2007) [Newsletter]. *The EDP Journal, Audit, Control, and Security*, *35*(5), 1–17.

[Ali 2015] Ali, S., Qaisar, S.B., Saeed, H., Khan, M.F., Naeem, M., Anpalagan, A. (2015). Network Challenges for Cyber Physical Systems with Tiny Wireless Devices: A Case Study on Reliable Pipeline Condition Monitoring. *Sensors 15*(4), 7172–7205. CC BY-SA 4.0. https://doi.org/10.3390/s150407172

[Amin 2001] Amin, M. (2001). Toward self-healing energy infrastructure systems. *IEEE Computer Applications in Power*, *14*(1), 20–28. https://doi.org/10.1109/67.893351

[Amin 2008] Amin, M., Stringer, J. (2008). The electric power grid: today and tomorrow. *MRS Bulletin 33*(4), 399–407. https://doi.org/10.1557/mrs2008.80

[Arghandeh 2016] Arghandeh, R., von Meier, A., Mehrmanesh, L., Mili, L. (2016). On the definition of cyber-physical resilience in power systems. *Renewable and Sustainable Energy Reviews*, *10*(58), 1060–1069. https://doi.org/10.1016/J.RSER.2015.12.193

[Arindam 2009] Maitra, A., Sundaram, A., Gandhi, M., Bird, S., Doss, S. (2009). Intelligent Universal Transformer design and applications. *CIRED 2009 – 20th International Conference and Exhibition on Electricity Distribution - Part 1* (pp. 1–7), Prague, Czech Republic, (June 2009)

[Badal 2019] Badal, F.R., Das, P., Sarker, S.K., Das, S.K. (2019). A survey on control issues in renewable energy integration and microgrid. *Protection and Control of Modern Power Systems 4*, Article 8. https://doi.org/10.1186/s41601-019-0122-8

Building an Effective Security Program for Distributed Energy Resources and Systems: Understanding Security for Smart Grid and Distributed Energy Resources and Systems, Volume 1, First Edition. Mariana Hentea.

[Banu 2014] Banu, I.V., Istrate, M. (2014). Study on three-phase photovoltaic systems under grid faults. *2014 International Conference and Exposition on Electrical and Power Engineering (EPE)* (pp. 1132–1137), Iasi, Romania, (October 2014). https://doi.org/10.1109/ICEPE.2014.6970086

[Behr 2014] Behr, P. (2014, November 21). *Cyber attackers have penetrated U.S. infrastructure systems—NSA chief* [News]. Environment & Energy Daily. http://www.eenews.net/energywire/stories/1060009391

[Berdner 2014] Berdner, J. (2014, January 6). *California smart inverter integration needs standards development* [News]. RenewableEnergyWorld. http://www.renewableenergyworld.com/rea/news/article/2014/01/california-smart-inverter-integration-needs-standards-development

[Bologna 2000] Bologna, S., Lambiase, F., Ratto, E. (2000). Large scale electric power distribution and telecommunication systems survivability. [Paper presentation]. *Third Information Survivability Workshop* (ISW-2000), Boston, Massachusetts, USA, (October 2000).

[Borlase 2013] Borlase, S. (2013). *Smart Grids Infrastructure, Technology, and Solutions* (First Edition, edited by Stuart Borlase). CRC Press.

[Bossart 2012] Bossart, S. (2012). DOE perspective on microgrids. [Paper presentation]. *Advanced Microgrid Concepts and Technologies Workshop*. National Energy Technology Laboratory, Beltsville, Maryland, USA, (June 2012). https://www.slideserve.com/yeva/doe-perspective-on-microgrids

[Bossart 2013] Bossart, S., (2013). *Results from DOE's ARRA smart grid program*. DOE NETL. http://www.netl.doe.gov/File%20Library/research/energy%20efficiency/smart%20grid/articles/Results-from-DOE-ARRA.pdf

[Brown 2002] Brown, R.E. (2002). *Electric Power Distribution Reliability*. CRC Press.

[Bryant 2018] Bryant, S., Straker, K., Wrigley, C. (2018). The typologies of power: energy utility business models in an increasingly. *Journal of Cleaner Production 195*, 1032–1046. https://doi.org/10.1016/j.jclepro.2018.05.233

[Budka 2014] Budka, K.C., Deshpande, J.G., Thottan, M. (2014). *Communication Networks for Smart Grids Making Smart Grid Real*. Springer-Verlag. https://doi.org/10.1007/978-1-4471-6302-2

[Byres 2013] Byres, E. (2013, November 7). DNP3 vulnerabilities part 1 of 2 – NERC's electronic security perimeter is Swiss cheese [Blog]. *Tofinosecurity*. https://www.tofinosecurity.com/blog/dnp3-vulnerabilities-part-1-2-nerc%E2%80%99s-electronic-security-perimeter-swiss-cheese

[Campbell 2015] Campbell, R.J. (2015). *Cybersecurity Issues for the Bulk Power System*. CRC Press.

[Caralli 2004a] Caralli, R., Wilson, W.R. (2004a). *The Challenges of Security Management* (July 2014). Software Engineering Institute.

[Caralli 2004b] Caralli, R. (2004b). *The critical success factor method: establishing a foundation for enterprise security management* (CMU/SEI-2004-TR-010, July 2004). Software Engineering Institute.

[CEDEC] CEDEC. (n.d.). *Smart Grids for Smart Markets*. http://cedec.com/files/default/cedec_smart_grids_position_paper-2.pdf

[CEDEC 2012] CEDEC. (2012). *Annual Report 2011* (January 6, 2012). http://www.cedec.com/files/documents/267/RA%20EN%202011-final.pdf

[CENELEC 2012] EC. (2012). *CEN-CENELEC-ETSI Smart Grid Coordination Group Smart Grid Reference Architecture Version 3* (November 2012). CEN-CENELEC-ETSI Smart Grid Coordination Group. https://ec.europa.eu/energy/sites/ener/files/documents/xpert_group1_reference_architecture.pdf

[CERT Governance] CERT. (n.d.). *Governing for enterprise security implementation guide*. http://www.cert.org/historical/governance/implementation-guide.cfm?

[Cisco 2010] Cisco. (2010). *Internet protocol architecture for the Smart Grid*. White Paper. http://www.cisco.com/

[Cleanpng SGP-I] Cleanpng. (n.d.). *Smart grid integrated view*. https://www.cleanpng.com/png-electrical-grid-smart-grid-electric-power-system-e-3123111

[CNSSI 4009] CNSSI. (2010). *Committee on National Security Systems (CNSS) Information Assurance (IA) Glossary*. Committee on National Security Systems (CNSS) Instruction No. 4009. *Revised in 2015.

[COBIT 5] ISACA. (2012). *COBIT 5 for Information Security*. http://www.isaca.org/cobit/pages/info-sec.aspx

[Comer 2001] Comer, D. (2001). *Internetworking with TCP/IP – Principles, Protocols, and Architectures* (Fourth Edition). Pearson Prentice Hall.

[COMSAR] COMSAR. (n.d.). *Smart Grids*. http://comsar.com/projects-technologies/smart-grids

[Congressional] FAS. (n.d.). *Congressional Research Service Reports*. https://fas.org/sgp/crs/

[Connolly 2010] Connolly, D. (2010). *A Review of energy storage technologies for the integration of fluctuating renewable energy* (Version 4.1, December 2010). https://vbn.aau.dk/en/publications/a-review-of-energy-storage-technologies-for-the-integration-of-fl

[CORDIS] CORDIS. (n.d.). *Community research and development information service*. http://cordis.europa.eu/home_en.html at http://cordis.europa.eu/search/result_en?q=smart+grids

[DA] SmartGridToday. (n.d.). *Glossary distribution automation (DA)*. https://www.smartgridtoday.com/public/Glossary-2.cfm#D

[Dao 2016] Dao, L.A. (2016). *A microgrid with renewables connected to grid*. CC BY-SA 4.0. https://commons.wikimedia.org/wiki/File:Microgrid_with_RES_BESS_GRIDconnected.png

[Davis 2014] Davis, A. (2014, July 24). Resilience is both a technical and a business responsibility [Opinion]. *ComputerWeekly*. https://www.computerweekly.com/opinion/Security-Think-Tank-Resilience-is-both-a-technical-and-a-business-responsibility

[DeepResource 2012] DeepResource. (2012). *Smart grids in Europe* (October 2012). https://deepresource.files.wordpress.com/2012/04/smart-grid-concept.png

[Denton 2014] Denton, R., Piacentini, R. (2014, March 6). *Smart grid evolution: a new generation of intelligent electronic devices* [News]. RenewableEnergy. http://www.renewableenergyworld.com/rea/news/article/2014/03/smart-grid-evolution-a-new-generation-of-intelligent-electronic-devices

[DESERTEC 2015] DESERTEC. (2015). *Vision*. http://www.desertec.org/

[DHS Initiative] DHS. (n.d.). *Smart Grid* [News]. Homeland Security News Wire. http://www.homelandsecuritynewswire.com/topics/smart-grid

[Dignan 2013] Dignan, L. (2013, October 3). *Internet of things: $8.9 trillion market in 2020, 212 billion connected things* [News]. ZDNet. http://www.zdnet.com/article/internet-of-things-8-9-trillion-market-in-2020-212-billion-connected-things

[Dignan 2014] Dignan, L. (2014, January 16). *Proofpoint uncovers Internet of Things (IoT) cyberattack* [News]. News Wire. https://www.proofpoint.com/us/proofpoint-uncovers-internet-things-iot-cyberattack

[DOE 2009a] Energy. (2009a). *Smart Grid System Report* (July 2009). U.S. Department of Energy. http://energy.gov/sites/prod/files/2009%20Smart%20Grid%20System%20Report.pdf

[DOE 2009b] SmartGrid. (2009b). *Recovery Act Smart Grid Program*. U.S. Department of Energy. https://www.smartgrid.gov/recovery_act/

[DOE 2010] DOE. (2010). *Multi-Year Program Plan (MYPP) 2010–2014 February 2010*. September 2012 Update. U.S. Department of Energy Office of Electricity Delivery & Energy Reliability Smart Research &Development. https://www.smartgrid.gov/document/smart_grid_research_development_multi_year_program_plan_mypp_2010_2014

[DOE 2012] DOE. (2012). *Electricity Subsector Cybersecurity Capability Maturity MODEL (ES-C2M2) Version 1.0* (May 2012). U.S. Department of Energy. *Replaced by Version 1.1 February 2014. https://www.energy.gov/sites/prod/files/2014/02/f7/ES-C2M2-v1-1-Feb2014.pdf

[DOE 2014] DOE. (2014). *Smart grid investments improve grid reliability, resilience, and storm responses* (November 2014). http://energy.gov/oe/downloads/smart-grid-investments-improve-grid-reliability-resilience-and-storm-responses-november

[DOE 2016] DOE. (2016). Cyber Threat and Vulnerability Analysis of the U.S. Electric Sector Mission Support Center Analysis Report (August 2016). Idaho National Laboratory. https://www.energy.gov/sites/prod/files/2017/01/f34/Cyber%20Threat%20and%20Vulnerability%20Analysis%20of%20the%20U.S.%20Electric%20Sector.pdf

[DOE Comm] Energy. (2010). *Communications Requirements of Smart Grid Technologies* (October 5, 2010). U.S. Department of Energy. https://www.energy.gov/sites/prod/files/gcprod/documents/Smart_Grid_Communications_Requirements_Report_10-05-2010.pdf

[DOE Distributed] DOE. (n.d.). *Distributed energy*. http://energy.gov/oe/technology-development/smart-grid/distributed-energy

[DOE/OEDER 2008] Energy. (2008). *The Smart Grid: An Introduction*. U.S. Department of Energy Office of Electricity Delivery and Energy Reliability. Litos Strategic Communication. http://energy.gov/sites/prod/files/oeprod/DocumentsandMedia/DOE_SG_Book_Single_Pages%281%29.pdf

[DOE MYPP-2015] DOE. (2015). *Grid Modernization Multi-Year Program Plan November 2015*. U.S. Department of Energy. https://www.energy.gov/sites/prod/files/2016/01/f28/Grid%20Modernization%20Multi-Year%20Program%20Plan.pdf

[Dong 2010] Dong, Z., Zhang, P., Ma, J., Zhao, J., Ali, M., Meng, K., Yin, X. (2010). *Emerging Techniques in Power System Analysis*. Springer-Verlag.

[EAC 2012] EAC. (2012, October 17). *Recommendations on development of the next generation grid operating system (Energy Management System)*. Electricity Advisory Committee. https://www.energy.gov/sites/prod/files/EAC%20Paper%20-%20Recommendations%20on%20Development%20of%20the%20Next%20Gen%20Grid%20Operating%20System%20-%20Final%20-%2025%20Oct%202012.pdf

[EAC 2014] EAC. (2014, March 12). *Implementing effective enterprise security governance – March 2014*. Electricity Advisory Committee. http://energy.gov/oe/downloads/eac-recommendations-doe-action-regarding-implementing-effective-enterprise-security

[EC 2003] EC. (2003). *New ERA for electricity in Europe Distributed Generation: Key Issues, Challenges and Proposed Solutions*. European Commission. https://op.europa.eu/en/publication-detail/-/publication/c06bd1e6-6329-4eaa-a8c7-6c300479caa4

[EC 2011] EC. (2011). *M/490 Smart Grid Mandate, Standardization Mandate to European Standardisation Organisations (ESOs) to support European Smart Grid deployment* (Brussels, Belgium, March 2011). European Commission. https://ec.europa.eu/energy/sites/ener/files/documents/2011_03_01_mandate_m490_en.pdf

[EC DG] EC. (2010). *Distributed generation in Europe - the European regulatory framework and the evolution of the distribution grids towards smart grids*. European Commission. https://ec.europa.eu/jrc/en/publication/contributions-conferences/distributed-generation-europe-european-regulatory-framework-and-evolution-distribution-grids-towards

[EEI] EEI. (n.d.). *Smart Grid*. Edison Electric Institute. http://smartgrid.eei.org/Pages/EEI.aspx

[EISA 2007] EISA. (2007). *U.S. Energy Independence and Security Act of 2007, Public Law No: 110-140 Title XIII, Sec. 1301*. http://www.gpo.gov/fdsys/pkg/BILLS-110hr6enr/pdf/BILLS-110hr6enr.pdf

[Elsafdi 2019] Elsafdi, O., Khan, A.M. (2019). Developing methodological framework to enhance dynamic resilience of transportation network. [Paper presentation]. *CITE 2019 Annual Conference*, Ottawa, Canada, (June 2019). https://www.cite7.org/wpdm-package/cite-2019-compendium

[Encyclopedia] Mc Graw-Hill. (2005). *Sci-Tech Encyclopedia* (Edition 5). McGraw-Hill Encyclopedia of Science and Technology.

[EON Generation] EON. (n.d.). *Renewables and generation*. https://www.eonenergy.com/business/renewables-and-generation.

[EPRI 2009] EPRI. (2009). *Report to NIST on the Smart Grid Interoperability Standards Roadmap—Post Comment Period Version August 2009*. http://www.nist.gov/smartgrid/upload/Report_to_NIST_August10_2.pdf

[EPRI 2011] EPRI. (2011). *Estimating the Costs and Benefits of the Smart Grid a Preliminary Estimate of the Investment Requirements and the Resultant Benefits of a Fully Functioning Smart Grid Technical Report* (1022519). https://www.epri.com/research/products/000000000001022519

[EPRI 2013] EPRI. (2013). *Cyber security for DER systems 1.0*. EPRI. https://smartgrid.epri.com/doc/der%20rpt%2007-30-13.pdf

[EPRI SmartGrid] EPRI. (n.d.). *Smart Grid Resource Center*. http://smartgrid.epri.com/

[Erol-Kantarci 2011] Erol-Kantarci, M.E., Moufta, H.T. (2011). Demand management and wireless sensor networks in the smart grid. In P.G. Kini (Ed.), *Energy Management Systems* (pp. 253–274). IntechOpen. https://doi.org/10.5772/17297

[ETP 2006] EC. (2006). *Vision and strategy for Europe's electricity networks of the future* (EUR 22040). http://ec.europa.eu/research/energy/pdf/smartgrids_en.pdf

[Fang 2012] Fang, X., Misra, S., Xue, G., Yang, D. (2012). Smart grid – the new and improved power grid: a survey. *IEEE Communications Surveys and Tutorials*, *14*(4), 944–980. https://doi.org/10.1109/SURV.2011.101911.00087

[Farhangi 2014] Farhangi, H. (2014). A road map to integration. *IEEE Power & Energy Magazine*, *12*(3), 52–63. https:/doi.org/10.1109/MPE.2014.2301515

[Feng 2012] Feng, X., Stoupis, J., Mohagheghi, S., Larsson, M. (2012). Introduction to smart grid applications. In L.T. Berger, I. Krzysztof (Eds.), *Smart Grid Applications, Communications, and Security* (pp. 3–48). Wiley.

[FERC 2009a] FERC. (2009a). *Smart Grid Policy 128 FERC 61,060 Docket No. PL09-4-000 July 16, 2009*. Federal Energy Regulatory Commission. https://www.ferc.gov/sites/default/files/2020-08/E-3-Fercs-Smart-Grid-Policy.pdf

[FERC 2009b] FERC. (2009b). *FERC Standards for Business Practices and Communication Protocols for Public Utilities 128 FERC 61,263 September 17, 2009*. Federal Energy Regulatory Commission. Federal Energy Regulatory Commission. https://www.naesb.org/pdf4/ferc091709_dr_nopr.pdf

[FERC 2012] FERC. (2012). *Report on Enforcement Docket No. AD07-13-005 November 15, 2012*. Federal Energy Regulatory Commission. http://www.ferc.gov/legal/staff-reports/11-15-12-enforcement.pdf

[Fernandez 2012] Fernandez, J.J.G., Berger, L.T., Armada, A.G., Fernandez-Getino Garcia, M.I., Jimenez, V.P.G, Sorensen, T.B. (2012). Wireless communications in smart grids. In L.T. Berger, I. Krzysztof (Eds.), *Smart Grid Applications, Communications, and Security* (pp. 145–190). Wiley.

[Free Dictionary] FARLEX. (n.d.). *The Free Dictionary*. http://encyclopedia2.thefreedictionary.com/Electric+power+systems

[Galloway 2013] Galloway, B., Hancke, G.P. (2013). Introduction to Industrial Control Networks. *IEEE Communications Surveys & Tutorials*, *15*(2), 860–880. https://doi.org/10.1109/SURV.2012.071812.00124

[GAO 2011] GAO. (2011). *Critical Infrastructure protection cybersecurity guidance is available, but more can be done to promote its use* (GAO-11-117). United States Government Accountability Office. http://www.gao.gov/assets/590/587529.pdf

[GAO 2019] GAO. (2019). *Actions needed to address significant cybersecurity risks facing the electric grid* (GAO-19-332). United States Government Accountability Office. https://www.gao.gov/assets/710/701079.pdf

[Gao 2012] Gao, J., Xiao, Y., Liu, J., Liang, W., Chen, C.L.P. (2012). A survey of communication/networking in smart grids. *Future Generation Computer Systems*, *28*(2), 391–404. https://doi.org/10.1016/j.future.2011.04.014

[Gartner 2013] Firstbrook, F.P., Girard, J., MacDonald, N. (2013). *Magic Quadrant for Endpoint Protection Platforms* (G00239869). Gartner Information Technology Research. https://www.gartner.com/en/documents/2292216/magic-quadrant-for-endpoint-protection-platforms

[Gellings 2015] Gellings, C.W. (2015). *Let's build a global power grid* (28 July 2015). IEEE Spectrum. http://spectrum.ieee.org/energy/the-smarter-grid/lets-build-a-global-power-grid

[Grid Interoperability] GWAC. (n.d.). *Smart grid interoperability maturity model summary*. GridWise Architecture Council. https://www.gridwiseac.org/about/imm.aspx

[GridWise] GridWise. (n.d.). *Architecture Council*. https://www.gridwiseac.org/about/mission.aspx

[GridWise 2008] GridWise. (2008). *Interoperability context- setting framework v1.1* (March 2008). GridWise Architecture Council. https://www.gridwiseac.org/pdfs/interopframework_v1_1.pdf

[GridWise 2009] GridWise. (2009). *Reliability benefits of interoperability*. GridWise Architecture Council. http://www.gridwiseac.org/pdfs/reliability_interoperability.pdf

[GridWise 2014] GridWise. (2014). *The future of the grid: evolving to meet America's needs Final report an industry-driven vision of the 2030 grid and recommendations for a path forward* (December 2014). GridWise Alliance. http://energy.gov/sites/prod/files/2014/12/f19/Future%20of%20the%20Grid%20December%202014.pdf

[Grose 2009] Grose, T.K. (2009). *The cyber grid*. *ASSE Prism*, *19*(2), 26–31.

[Grundvig 2013] Grundvig, J. (2013, April 15). *Detecting power theft by sensors and the cloud: Awesense smart system for the grid* [News]. HUFFPOST. http://www.huffingtonpost.com/james-grundvig/detecting-power-theft-by-_b_3078082.html

[GTM 2016] GTM. (2016, June 1). *US microgrid growth beats estimates: 2020 capacity forecast now exceeds 3.7 gigawatts* [News, Research SPOTLIGHT by. O. Chen]. GTM. https://www.greentechmedia.com/articles/read/u-s-microgrid-growth-beats-analyst-estimates-revised-2020-capacity-project

[Gudzius 2011] Gudzius, S., Gecys, S., Markevicius, L.A., Miliune, R., Morkvenas, A. (2011). The model of smart grid reliability evaluation. *Elektronika Ir Elektrotechnika*, *116*(10), 25–28. CC BY-SA 4.0. https://doi.org/10.5755/j01.eee.116.10.873

[Hale 1998] Hale, J., Bose, A. (1998). Information survivability in the electric utility industry. *1998 Information Survivability Workshop (ISW'98)*, Orlando, FL, USA, (October 1998).

[Hawk 2014] Hawk, C., Kaushiva, A. (2014). Cybersecurity and the smarter grid. *The Electricity Journal*, *27*(8), 84–95. https://doi.org/10.1016/j.tej.2014.08.008

[Hines 2014] Hines, P., Veneman, J., Tivnan, B. (2014). *Smart Grid: Reliability, Security, and Resiliency*. http://www.uvm.edu/~phines/publications/2014/hines_2014_terms.pdf

[Ho 2013a] Ho, Q-D., Le-Ngoc, T. (2013a). Smart grid communications networks: wireless technologies, protocols, issues and standards. In M. Obaidat, A. Anpalagan, I. Woungang (Eds.), *Handbook of on Green Information and Communication Systems* (1st Edition, pp. 115–146). Elsevier. https://doi.org/10.1016/B978-0-12-415844-3.00005-X

[Ho 2013b] Ho, Q-D., Gao, Y., Le-Ngoc, T. (2013b). Challenges and research opportunities in wireless communication networks for smart grid. *IEEE Wireless Communications*, *20*(3), 89–95. https://doi.org/10.1109/MWC.2013.6549287

[Ho 2014] Ho, Q-D., Gao, Y., Rajalingham, G., Le-Ngoc, T. (2014). Wireless communications technologies for the SGCN. In *Wireless Communications Networks for the Smart Grid* (31–49). Springer. https://doi.org/10.1007/978-3-319-10347-1

[Hollnagel 2015] Hollnagel, E., Woods, D.D., Leveson, N. (Eds.). (2006). *Resilience Engineering: Concepts and Precepts*. CRC Press.

[Hull 2012] Hull, J., Khurana, H., Markham, T., Staggs, J. (2012). Staying in control: cybersecurity and the modern electric grid. *IEEE Power & Energy Magazine*, *10*(1), 41–48. https://doi.org/10.1109/MPE.2011.943251

[ICS-CERT 2012a] ICS-CERT. (2012a). *ICS-CERT incident response summary report 2009–2011.* https://us-cert.cisa.gov/sites/default/files/documents/ICS-CERT%20Incident%20Response%20Summary%20Report%20%282009-2011%29_S508C.pdf

[ICS-CERT 2012b] ICS-CERT. (2012b). *Malware infections in the control environment.* Monitor (ICS-MM201212) October–December 2012. https://us-cert.cisa.gov/sites/default/files/Monitors/ICS-CERT_Monitor_Oct-Dec2012.pdf

[ICS-CERT 2014a] ICS-CERT. (2014a). *Internet accessible control systems at risk.* Monitor (ICS-MM201404) January—April. https://us-cert.cisa.gov/sites/default/files/Monitors/ICS-CERT_Monitor_Jan-April2014.pdf

[ICS-CERT 2014b] ICS-CERT. (2014b). *ICS-CERT Advisory ICSA-13-291-01B – DNP3 implementation vulnerability* (Update B). https://us-cert.cisa.gov/ics/advisories/ICSA-13-291-01B

[IEA 2011] IEA. (2011). *Technology roadmap – smart grid Technology report – April 2011.* International Energy Agency. https://www.iea.org/reports/technology-roadmap-smart-grids

[IEA 2013] IEA. (2013). *Tracking clean energy progress Technology report – April 2013.* International Energy Agency. https://www.iea.org/reports/tracking-clean-energy-progress-2013

[IEC 2011] IEC. (2011). *Electrical energy storage* [White Paper]. http://www.iec.ch/whitepaper/pdf/iecWP-energystorage-LR-en.pdf

[IEC 61850] *IEC 61850:2020 SER Series Communication networks and systems for power utility automation – ALL PARTS.*

[IEC 62264] *IEC 62264-4:2015 Enterprise-control system integration – Part 4: Objects models attributes for manufacturing operations management integration.*

[IEC 62357] *IEC TR 62357-1:2016 Power systems management and associated information exchange – Part 1: Reference architecture.*

[IEC Map] IEC. (n.d.). *Smart grid standards map.* http://smartgridstandardsmap.com/

[IEEE 1547] *1547-2018 – IEEE Standard for Interconnection and Interoperability of Distributed Energy Resources with Associated Electric Power Systems Interfaces.*

[IEEE Consumer] IEEESA. (2012, January 3). *Enabling consumer connectivity through consensus building.* http://standardsinsight.com/ieee_company_detail/consensus-building

[IEEE PES-TR22] IEEE-PES. (2017). *Contribution to bulk system control and stability by distributed energy resources connected at distribution network* [Technical Report PES-TR22, 15 January 2017]. https://resourcecenter.ieee-pes.org/technical-publications/technical-reports/PESTRPDFMRH0022.html

[IEEE Smart Grid] IEEE. (n.d.). *Smart Grid.* http://smartgrid.ieee.org

[IEEE Std 493] *493-2007 – IEEE Recommended Practice for the Design of Reliable Industrial and Commercial Power Systems.*

[IEEE Std 610] *IEEE Std 610.12-1990 – IEEE Standard Glossary of Software Engineering Terminology. * superseded by ISO/IEC/IEEE 24765.*

[IEEE Std C37-2] *C37.2-2008 – IEEE Standard Electrical Power System Device Function Numbers, Acronyms, and Contact Designations.*

[India Blackout] Wikipedia. (2012). *India blackout.* https://en.wikipedia.org/wiki/2012_India_blackouts.

[Infosecurity 2014] Infosecurity. (2014). *How to govern information security.* https://spaces.internet2.edu/display/2014infosecurityguide/Information+Security+Governance

[INL 2016] INL. (2016). *Microgrids Can Enhance Diversity, Reliability, Resilience* (April 2016). https://inl.gov/article/grid-stabilization

[ISACA 2006] ISACA. (2006). *Information security governance best practices information security governance: guidance for boards of directors and executive management.* https://www.isaca.org/bookstore/it-governance-and-business-management/w3itg

[ISACA Glossary] ISACA. (n.d.). *ISACA Glossary*. https://www.isaca.org/resources/glossary

[Ishimuro 2015] Ishimuro, T. (2015). *Evolution of the grid edge: pathways to transformation* [White Paper]. www.greentechmedia.com/sponsored/resource-center/

[ISO 9000] *ISO 9000:2015 Quality Management Systems – Fundamentals and Vocabulary.*

[ISO IT] ISO. (n.d.). *Freely Available Standards*. http://standards.iso.org/ittf/PubliclyAvailableStandards/index.html

[ISO/IEC 20000] *ISO/IEC 20000-1:2018 Information Technology – Service Management – Part 1: Service Management System Requirements.*

[ISO/IEC 27000] *ISO/IEC 27000:2018 Information technology — Security techniques — Information security management systems – Overview and vocabulary (fifth edition).*

[ISO/IEC 27002] *ISO/IEC 27002:2013 Information Technology — Security Techniques — Code of Practice for Information Security Controls.*

[ISO/IEC 27014] *ISO/IEC 27014:2013 Information Technology — Security Techniques — Governance of Information Security.*

[ISO/IEC 38500] *ISO/IEC 38500:2015 Information Technology – Governance of IT for the organization.*

[ISO/IEC 42010] *ISO/IEC/IEEE 42010:2011 Systems and software engineering — Architecture description.*

[ISO/IEC TR27019] *ISO/IEC TR 27019:2013 Information technology – Security techniques – Information security management guidelines based on ISO/IEC 27002 for process control systems specific to the energy utility industry.* * revised by *ISO/IEC 27019:2017 Information technology — Security techniques — Information security controls for the energy utility industry.*

[ITGI 2003] ITGI. (2003). *Board briefing on IT governance*. IT Governance Institute.

[ITIL 2011] ITSMF. (2012). *An Introductory Overview of ITIL® 2011. IT Service Management Forum.* TSO. https://www.tsoshop.co.uk/gempdf/itSMF_An_Introductory_Overview_of_ITIL_V3.pdf

[Kim 2015] Kim, J., Filali, F., Ko, Y.-B. (2015). Trends and potentials of the smart grid infrastructure: from ICT sub-system to SDN-enabled smart grid architecture. *Applied Sciences*, *5*(4), 706–727. CC BY-SA 4.0. https://doi.org/10.3390/app5040706

[Kok 2010] Kok, J.K., Scheepers, M.J.J., Kamphuis, I.G. (2010). Intelligence in electricity networks for embedding renewables and distributed generation, intelligent infrastructures. In R.R. Negenborn, Z.Lukszo, H. Hellendoorn (Eds.), *Intelligent Systems, Control and Automation: Science and Engineering* (Vol. *42*, pp. 179–209). Springer. https://doi.org/10.1007/978-90-481-3598-1_8

[Komninos 2014] Komninos, N., Philippou, E., Pitsillides, A. (2014). Survey in smart grid and smart home security: issues, challenges and countermeasures. *IEEE Communications Surveys & Tutorials*, *16*(4), 1933–1954. https://doi.org/10.1109/COMST.2014.2320093

[Korosec 2013] Korosec, K., (2013, March 27). *The no. 1 business targeted by hackers* [News]. ZDNet. http://www.zdnet.com/article/the-no-1-business-targeted-by-hackers/

[Kundur 2004] Kundur, P., Paserba, J., Ajjarapu, V., Andersson, G., Bose, A., Canizares, C.A., Hatziargyriou, N.D., Hill, D.J., Stankovic, A.M., Taylor, C.W., Van Cutsem, T., Vittal, V. (2004). Definition and classification of power system stability IEEE/CIGRE joint task force on stability terms and definitions. *IEEE Transactions on Power Systems*, *19*(3), 1387–1401. https://doi.org/10.1109/TPWRS.2004.825981

[Lacey 2013] Lacey, S. (2013, October). *The grid edge: A New way to define the changing power sector.* Greentech Media Inc. http://www.greentechmedia.com/articles/read/the-grid-edge-a-new-way-to-define-the-changing-power-sector

[Law 2014] Law, Y.W., Pota, H.R., Jin, J., Man, Z., Palaniswami, M. (2014). Control and communication techniques for the smart grid: an energy efficiency perspective. *IFAC Proceedings*, *47*(3), 987–998. Elsevier. https://doi.org/10.3182/20140824-6-ZA-1003.01736

[Liu 2001] Liu, Y., Mili, L., De La Ree, J., and Nuqui, R.F. (2001). *State estimation and voltage security monitoring using synchronized phasor measurement* [Research Paper]. From work sponsored by

American Electric Power, ABB Power T&D company, and Tennessee Valley Authority. Virginia Polytechnic Institute and State University, 2001-07-12.

[Logan 2010] Logan, D. (2010, January 11). What is information governance? And why is it so hard? [Blog]. *Gartner*. http://blogs.gartner.com/debra_logan/2010/01/11/what-is-information-governance-and-why-is-it-so-hard/

[Loreasl Blog] LoreaslBlog. (2010, February 12). Smart Grids [Blog]. *Wordpress*. http://loreasl.wordpress.com/2010/02/12/redes-inteligentes-o-smart-grids/

[Martin 2014] Martin, R. (2014, March 15). Direct Current Power Supply Equipment for Commercial Buildings Will Reach $2.8 Billion in Annual Revenue by 2020 [Blog]. *WTE*. https://wteinternational.com/direct-current-power-supply-equipment-for-commercial-buildings-will-reach-2-8-billion-in-annual-revenue-by-2020/

[Masica 2007] Masica, K. (2007). *Recommended practices guide for securing ZigBee wireless networks in process control system environments* (April 2007). http://energy.gov/sites/prod/files/oeprod/DocumentsandMedia/Securing_ZigBee_Wireless_Networks.pdf

[Masiello 2014] Masiello, R. (2014). *Integrating distributed resources into wholesale markets and grid operations*. http://smartgrid.ieee.org/resources/ieee-in-the-news/256-gridwise-alliance-ieee-pes-enter-into-agreement

[McLaughlin 2010] McLaughlin, S., Podkuiko, D., McDaniel, P. (2010). *Energy theft in the advanced metering Infrastructure*. In E. Rome, R. Bloomfield (Eds.), *Critical Information Infrastructures Security. CRITIS 2009. Lecture Notes in Computer Science* (Vol. *6027*, pp. 176–187). Springer. https://doi.org/10.1007/978-3-642-14379-3_15

[Mekkanen 2012] Mekkanen, M., Virrankoski, R., Elmusrati, M., Antila, E. (2012). *Reliability and availability investigation for next-generation substation function based on* IEC 61850. *3rd Workshop on Wireless Communication and Applications (WoWCA2012)*, Vaasa, Finland, (April 2012).

[Meslier 2000] Meslier, F. (2000). Historical background and lessons for the future. In J.A. Casazza, G.C. Loehr (Eds.), *The Evolution of Electric Power Transmission Under Deregulation* (pp. 28–31). IEEE.

[Moslehi 2010] Moslehi, K., Kumar, R. (2010). A reliability perspective of the smart grid. *IEEE Transactions on Smart Grid*, *1*(1), 57–64. https://doi.org/10.1109/TSG.2010.2046346

[NIAC 2010] NIAC. (2010). *A framework for establishing critical infrastructure resilience goals final report and recommendations by the council* (October 2010). National Infrastructure Advisory Council. https://www.dhs.gov/xlibrary/assets/niac/niac-a-framework-for-establishing-critical-infrastructure-resilience-goals-2010-10-19.pdf

[NCOIC 2014] NCOIC. (2014). *What is Interoperability*? Network Centric Operations Industry Consortium. https://www.ncoic.org/what-is-interoperability

[NERC 2014] NERC. (2014). *Project 2008-06 Cyber Security Order 706 Version 5 CIP Standards*. North America Electric Reliability Corporation. http://www.nerc.com/pa/Stand/Pages/Project_2008-06_Cyber_Security_Version_5_CIP_Standards.aspx

[NERC Glossary] NERC. (2014). *Glossary of Terms Used in NERC Reliability Standards*. http://www.nerc.com/files/glossary_of_terms.pdf

[NETL 2007] NETL. (2007). *The NETL Modern Grid Initiative v2.0 A Systems View of the Modern Grid Conducted by the National Energy Technology Laboratory for the U.S. Department of Energy Office of Electricity Delivery and Energy Reliability, (January 2007)*. https://netl.doe.gov/sites/default/files/Smartgrid/ASystemsViewoftheModernGrid_Final_v2_0.pdf

[NETL 2008] NETL. (2008). *The Modern Grid Strategy: Characteristics of the Modern Grid*. National Energy Technology Laboratory. http://www.netl.doe.gov

[NETL 2009] NETL. (2009). *West Virginia Smart Grid implementation Plan* (September 2009). https://netl.doe.gov/sites/default/files/Smartgrid/Stakeholders-Assessment--2-15-09-edited-clean.pdf

[Neumann 2006] Neumann, S., Sioshansi, F., Vojdani, A., Yee, G. (2006). How to get more response from demand response. *The Electricity Journal, 19*(8), 25–31.

[Newmann 2007] Newmann, S., Sioshansi, F., Vojdani, A., Yee, G. (2007). The Missing Link. *Public Utilities Fortnightly* (March 2007, pp. 52—66).

[NIST 2012] NIST. (2012). *Wireless Networks for Smart Grid Applications.* https://www.nist.gov/publications/wireless-networks-smart-grid-applications

[NIST 2014] *NIST Framework for Improving Critical Infrastructure Cybersecurity Version 1.0* (February 2014). https://www.nist.gov/system/files/documents/cyberframework/cybersecurity-framework-021214.pdf

[NISTIR 7628] *NIST Interagency Report (NISTIR) 7628* Guidelines *for Smart Grid Cyber Security* (August 2010).

[NISTIR 7628r1] *NIST Interagency Report (NISTIR) 7628r1 Guidelines for Smart Grid Cybersecurity* (Revision 1, September 2014). http://dx.doi.org/10.6028/NIST.IR.7628r1

[NISTIR 7761r1] *NIST Interagency Report (NISTIR) 7761r1 NIST Smart Grid Interoperability Panel Priority Action Plan 2: Guidelines for Assessing Wireless Standards for Smart Grid Applications,* (June 2014, Updated November 2018). https://doi.org/10.6028/NIST.IR.7761r1

[NIST CoS] NIST. (n.d.). *NIST Catalog of Standards.* NIST Smart Grid Collaboration Wiki for Smart Grid Interoperability Standards. http://collaborate.nist.gov/twiki-sggrid/bin/view/SmartGrid/SGIP CoSStandardsInformationLibrary

[NIST SmartGrid] NIST. (n.d.). *What is the Smart Grid?* http://www.nist.gov/smartgrid/beginnersguide.cfm

[NIST SP1108r1] *NIST Special Publication (SP) 1108r1 NIST Framework and Roadmap for Smart Grid Interoperability Standards* (Release 1.0, January 2010). https://doi.org/10.6028/NIST.sp.1108

[NIST SP1108r3] *NIST Special Publication (SP) 1108r3 NIST Framework and Roadmap for Smart Grid Interoperability Standards* (Release 3.0, September 2014). http://dx.doi.org/10.6028/NIST.SP.1108r3

[NIST SP800-100] *NIST Special Publication (SP) 800-100 Information Security Handbook* Recommendations of the National Institute of Standards and Technology (October 2006). Computer Security Division Information Technology Laboratory.

[NREL 2014] NREL. (2014). *2014 Renewable Energy Data Book.* National Renewable Energy Laboratory (NREL). https://www.nrel.gov/docs/fy16osti/64720.pdf

[NSTC 2013] NSTC. (2013). *A Policy Framework for the 21st Century Grid: A Progress Report* (February 2013). Executive Office of the President National Science and Technology Council. https://obamawhitehouse.archives.gov/sites/default/files/microsites/ostp/2013_nstc_grid.pdf

[OECD 2004] OECD. (2004). *OECD Principles of Corporate Governance.* OECD Publishing. http://www.oecd.org/corporate/ca/corporategovernanceprinciples/31557724.pdf

[OECD 2009] OECD. (2009). *Smart Sensor Networks: Technologies and Applications for Green Growth* (December 2009). https://doi.org/10.1787/5kml6x0m5vkh-en

[OECD 2010] OECD. (2010). Greener and Smarter: ICTs, the Environment and Climate Change. In *OECD Information Technology Outlook 2010* (pp. 191–226). OECD Publishing. https://doi.org/10.1787/it_outlook-2010-7-en

[OECD 2012] OECD. (2012). *ICT Applications for the Smart Grid: Opportunities and Policy Implications.* OECD Publishing. https:doi.org/10.1787/5k9h2q8v9bln-en

[OpenEI] OpenEI. (n.d.). *Smart grid resources.* https://openei.org/wiki/Smart_Grid_Resources

[Opstal 2012] van Opstal, D. (2012). *Supply chain solutions for smart grid security: Best practices U.S. Resilience Project* 2012. https://usresilienceproject.org/wp-content/uploads/2014/09/report-Supply_Chain_Solutions_for_Smart_Grid_Security.pdf

[Osborn 2001] Osborn, J., Kawann, C. (2001). *Reliability of the US electric system: recent trends and current issues.* Electricity Markets & Policy. Berkeley Lab, California, USA.

[Ouyang 2012] Ouyang, M., Dueñas-Osorio, L., Min, X. (2012). A three-stage resilience analysis framework for urban infrastructure systems. *Structural Safety* (Vol. *36–37*, pp. 23–31). Elsevier. https://doi.org/10.1016/j.strusafe.2011.12.004

[P2030 2011] *IEEE Std 2030-2011- IEEE Guide for Smart Grid Interoperability of Energy Technology and Information Technology Operation with the Electric Power System (EPS), End-Use Applications, and Loads.*

[Patel 2014] Patel, S. (2014, March 13). *FERC directs NERC to develop physical security reliability standards* [News]. Powermagazine. https://www.powermag.com/ferc-directs-nerc-to-develop-physical-security-reliability-standards/

[Pironti 2006] Pironti, J.P. (2006). Information security governance: motivations, benefits and outcomes. *ISACA Journal Online*, *4*, 1–4. ISACA.

[Policy 2005] GPO. (2005). *Energy Policy Act of 2005, 08 August 2005 Public Law 109-58*. http://www.gpo.gov/fdsys/pkg/PLAW-109publ58/html/PLAW-109publ58.htm

[PWC 2016] PwC. (2016). *Power & Utilities Flipping the switch on disruption to opportunity Top 6 focus areas in 2016* (February 2016). https://www.pwc.se/sv/pdf-reports/flipping-the-switch-on-disruption-opportunity.pdf

[QuadTrangle] NCOIC. (2014). *NCOIC addresses the different dimensions of interoperability*. Network Centric Operations Industry Consortium. CC BY-SA 4.0. https://www.ncoic.org/ncoic-quadtrangle/

[Rascoe 2014] Rascoe, A. (2014, February 28). *U.S. utilities need industry group focused on cyber defense: report* [News]. Reuters. https://lta.reuters.com/article/idCABREA1R21620140228

[Rastler 2010] Rastler, D. (2010). *Electricity energy storage technology options: A white paper primer on applications, costs, and benefits* (EPRI, 1020676). http://large.stanford.edu/courses/2012/ph240/doshay1/docs/EPRI.pdf

[Ray 2010] Ray, P.D., Harnoor, R., Hentea, M. (2010). Smart power grid security: a unified risk management approach. In D.A. Pritchard, L.D. Sanson (Eds.), *44th Annual 2010 IEEE International Carnahan Conference on Security Technology Proceedings* (pp. 276–285), San Jose, California, USA, (October 2010). IEEE. https://doi.org/10.1109/CCST.2010.5678681

[RFC 4949] *IETF Request for Comments (RFC) 4949 Internet Security Glossary* (Version 2, August 2007).

[Rieger 2014] Rieger, C.G. (2014). Resilient control systems practical metrics basis for defining mission impact. *7th International Symposium on Resilient Control Systems* (pp. 1–10). Denver, Colorado, USA, (August 2014). https://doi.org/ 10.1109/ISRCS.2014.6900108

[Rieger 2016] Rieger, C.G., Grosshans, R.R. (2016). *Smart is not enough: Resilience and securing the power grid*. Critical National Need Idea. https://www.nist.gov/system/files/documents/2017/05/09/120_smart_is_not_enough_resilience_securing_power_grid2.pdf

[Rihan 2013] Rihan, M., Ahmad, M., Beg, M.S. (2013). Vulnerability analysis of wide area measurement system in the smart grid. *Smart Grid and Renewable Energy*, *4*(6A), 1–7. https://doi.org/10.4236/sgre.2013.46A001

[Robertson 2010] Robertson, J. (2010, March 26). *'Smart' meters have security holes* [News]. NBCNEWS. http://www.nbcnews.com/id/36055667

[Ropuszyoska-Surma] Ropuszyoska-Surma, E., Borgosz-Koczwara, M. (2019). The virtual power plant – a review of business models. *E3S Web of Conferences Issue 108* (Article 01006). Energy and Fuels 2018. CC BY 4.0. http://doi.org/10.1051/e3sconf/201910801006

[Rhodes 2013] Rhodes, R., Sumic, Z. (2013, May 24). *MarketScope for advanced distribution management systems* (G00248928). Gartner. https://www.gartner.com/en/documents/2496115/marketscope-for-outage-management-systems

[Scaini 2012] Scaini, V. (2012). *Grid support stability for reliable, renewable power understand why energy storage is fundamental to large-scale solar projects* [White Paper WP083002EN]. http://www.eaton.ca

[Schneider ADMS] Schneider. (n.d.). *Advanced Distribution Management System*. http://www.schneider-electric.com/b2b/en/solutions/for-business/s4/electric-utilities-advanced-distribution-management-system-adms/

[Schneider DMS] Schneider. (n.d.). *Smart Grid Solution for Electricity Distribution Networks Unlock the potential of the new digital grid*. http://www.schneider-electric-ms.com/media/Advanced%20Distribution%20Management%20System%20Short%20Description%20June%202012.pdf

[SGAC 2014] NIST. (2014). *NIST smart grid advisory committee report*. Committee Meeting on 3–4 June 2014. http://www.nist.gov/smartgrid/updatejuly14.cfm#3

[SGIP 2015] NIST. (2015). *IEC 61850 information model concepts and updates for distributed energy resources (DER)*. Smart Grid Interoperability Panel. https://www.nist.gov/programs-projects/smart-grid-national-coordination/smart-grid-interoperability-panel-sgiph

[Shi 2012] Shi, W., Wong, V.W.S. (2012). Introduction to smart grid communications. In L.T. Berger, I. Krzysztof (Eds.), *Smart Grid Applications, Communications, and Security* (pp. 121–144). Wiley.

[Silberman 2001] Silberman, S. (2001, July 1). *The energy web* [News]. Wired. http://archive.wired.com/wired/archive/9.07/juice.html

[SmartHome] Pixabay. (n.d.). *SmartHome*. https://pixabay.com/illustrations/smart-home-house-technology-2005993/

[Smrcka 2010] Smrcka, K. (2010, May 21). *Interconnected energy systems key to energy efficiency* [News]. Creamer Media Engineering News. https://www.engineeringnews.co.za/login.php?url=/article/intercontinnected-energy-systems-key-to-energy-efficiency-2010-05-21

[Sobe 2013] Sobe, A., Elmenreich, W. (2013). Smart Microgrids: Overview and Outlook. *ITG INFORMATIK 2012, Workshop on Smart Grids,* Braunschweig, Germany, (September 2012). http://smartmicrogrid.blogspot.com/2012/07/smart-microgrids-overview-and-outlook.html

[SPP 2010] DOE. (2010). *Energy Sector Specific Plan 2010 An Annex to the National Infrastructure Protection Plan*. http://energy.gov/sites/prod/files/oeprod/DocumentsandMedia/Energy_SSP_2010.pdf

[SPP Dictionary] DHS. (2010). *Dictionary of Key Terms*. Appendix 1 (SPP 2010). http://www.dhs.gov/xlibrary/assets/nipp-ssp-energy-2010.pdf

[Stamper 2003] Stamper, D.A., Case, T.L. (2003). *Business Data Communications* (Sixth Edition). Prentice Hall.

[Steeley 2011] Steeley, W. (May 2011). *Functional Requirements for Electric Energy Storage Applications on the Power System Grid* [Report 1022544] (Technical Update, May 2011). https://studylib.net/doc/18848099/functional-requirements-for-electric-energy-storage-appli...

[Steven 2006] Steven, J. (2006). Adopting an enterprise software security framework. *IEEE Security & Privacy*, *4*(2), 84–87. https://doi.org/10.1109/MSP.2006.33

[Storm 2013] Storm, D. (2013, July 2). Brute-force cyberattacks against critical infrastructure, energy industry, intensify [Opinion]. *Computerworld*. http://www.computerworld.com/article/2473941/cybercrime-hacking/brute-force-cyberattacks-against-critical-infrastructure--energy-industry--intens.html

[Storm 2014] Storm, D. (2014, January 15). Hackers exploit SCADA holes to take full control of critical infrastructure [Opinion]. *Computerworld*. http://www.computerworld.com/article/2475789/cybercrime-hacking/hackers-exploit-scada-holes-to-take-full-control-of-critical-infrastructure.html

[Sun 2014] Sun, M., Mohan, S., Sha, L., Gunter, C. (2014). *Addressing Safety and Security Contradictions in Cyber-Physical Systems*. https://citeseerx.ist.psu.edu/viewdoc/download?doi=10.1.1.296.3246&rep=rep1&type=pdf

[Tariq 2013] Tariq, F., Dooley, L.S. (2013). Smart grid communication and networking technologies: recent developments and future challenges. In A.B.M.S. Ali (Ed.), *Smart Grids Opportunities,*

Developments, and Trends (pp. 199–213). Green Energy and Technology. Springer. https://doi.org/10.1007/978-1-4471-5210-1_9

[Taylor 2012] Taylor, J. (2012). *Part I – What we discovered: bulk system reliability assessment and the smart grid*. EE ONLINE. https://electricenergyonline.com/energy/magazine/681/article/PART-I-WHAT-WE-DISCOVERED-Bulk-System-Reliability-Assessment-and-the-Smart-Grid.htm

[Thompson 2013] Thompson, R. (2013, October 7). *The Grid Edge: How Will Utilities, Vendors and Energy Service Providers Adapt?*. Greentechmedia. http://www.greentechmedia.com/articles/read/the-grid-edge-how-will-utilities-vendors-regulators-and-energy-service-prov

[Ton 2010] Ton, D.T. (2010). *Smart Grid Activities by the US Department of Energy* (August 31, 2010). Asia Pacific Clean Energy Summit and Expo. http://www.ct-si.org/events/APCE/sld/pdf/25.pdf

[Tong 2013] Tong, Y., Deyton, J., Sun, J., Li, F. (2013). S^3A: a secure data sharing mechanism for situational awareness in the power grid. *IEEE Transactions on Smart Grid*, *4*(4), 1751 – 1759. http://doi.org/10.1109/TSG.2013.2251016

[Tsampasis 2016] Tsampasis, E., Bargiotas, D, Elias, C., Sarakis, L. (2016). Communication challenges in smart grid. *MATEC Web Conferences 41* (Article 01004). EDP Sciences 2016. CC BY-SA 4.0. http://dx.doi.org/10.1051/matecconf/20164101004

[UN 2006] UN. (2006). Technical aspects of grid interconnection. In *Multi Dimensional Issues in International Electric Power Grid* (pp. 15–49). United Nations Publication Economic & Social Affairs. http://www.un.org/esa/sustdev/publications/energy/chapter2.pdf

[Uslar 2013] Uslar, M. (2013). Introduction and smart grid basics. In M. Uslar, M. Specht, C. Dänekas, J. Trefke, S. Rohjans, J.M. González, C. Rosinger, R. Bleiker (Eds.), *Standardization in Smart Grids Introduction to IT-Related Methodologies, Architectures and Standards* (pp. 3–12). Springer. https://doi.org/10.1007/978-3-642-34916-4_1

[Venkata 2013] Venkata, S.S., Eremia, M., Toma, L. (2013). Background of power system stability. In M. Eremia, M. Shahidehpour (Eds.), *Handbook of Electrical Power System Dynamics: Modeling, Stability, and Control* (pp. 453–475). IEEE Press. Wiley. https://doi.org/10.1002/9781118516072.ch8

[VirginiaTech 2007] VirginiaTech. (2007). *Distributed generation*. Educational Module, Consortium on Restructuring Energy, Virginia Tech. http://www.dg.history.vt.edu/ch1/introduction.html

[Vojdani 2006] Vojdani, A., Neumann, S. (2006). How to Get More Response from Demand Response. *The Electricity Journal*, *19*(8), 24–31.

[Vos 2009] Vos, A. (2009). *Demand response management systems: the next wave of smart grid software*. EE ONLINE. http://www.electricenergyonline.com/show_article.php?article=444

[Wang 2013] Wang, Z., Tong, J., Yang, K, Wang, Z. (2013). Privacy-preserving energy scheduling in microgrid systems. *IEEE Transactions on Smart Grid*, *4*(4), 1810 – 1820. https://doi.org/10.1109/ACCESS.2020.2983110

[Westby 2012] Westby, J.R. (2012). Governance of enterprise security: CyLab 2012 report how boards & senior executives are managing cyber risks (16 May 2012). http://www.rsa.com/innovation/docs/CMU-GOVERNANCE-RPT-2012-FINAL.pdf

[Wiki DG] Wikipedia. (n.d). *Distributed Generation*. https://en.wikipedia.org/wiki/Distributed_generation#cite_note-DG-virginia-tech-1

[Wiki SuperGrid] Wikipedia. (n.d.). *European SuperGrid*. CC BY-SA 2.5. https://en.wikipedia.org/wiki/European_super_grid

[Wolsink 2012] Wolsink, M. (2012). The research agenda on social acceptance of distributed generation in smart grids: renewable as common pool resources. *Renewable and Sustainable Energy Reviews*, *16*(1), 822–835. Elsevier. https://doi.org/10.1016/j.rser.2011.09.006

[Wueest 2014] Wueest, C. (2014). *Targeted Attacks Against the Energy Sector Version 1.0* (13 January 2014). http://www.symantec.com/content/en/us/enterprise/media/security_response/whitepapers/targeted_attacks_against_the_energy_sector.pdf

[Wueest Blog] Wueest, C. (2014, January 13). Attacks against the energy sector [Blog]. *Symantec.* http://www.symantec.com/connect/blogs/attacks-against-energy-sector

[Zhang 2013] Zhang, Z., Zhang, Y., Chow, M-Y. (2013). Distributed energy management under smart grid plug-and-play operations. *2013 IEEE Power & Energy Society General Meeting, Shaping the Future Energy Industry* (pp. 1–5), Vancouver, BC, Canada, (July 21–25). https://doi.org/10.1109/PESMG.2013.6672509

[Zhang 2018] Zhang, J., Hasandka, A., Wei, J., Alam, S.M.S., Elgindy, T., Florita, A.R., Hodge, B.-M. (2018). Hybrid communication architectures for distributed smart grid applications. *Energies 2018 11*(4), 871–876. https://doi.org/10.3390/en11040871

[Zonouz 2012] Zonouz, S.A., Rogers, K.M., Berthier, R., Bobba, R.B., Sanders, W.H. Overbye, T.J. (2012). SCPSE: Security-Oriented Cyber-Physical State Estimation for Power Grid Critical Infrastructures. *IEEE Transactions on Smart Grid*, *3*(4), 1790–1799. https://doi.org/10.1109/TSG.2012.2217762

Part V

Security Program Management

11

Security Management

11.1 Security Managements Overview

Security is an important management matter to be considered by any organization. Security management is a broad field of management that includes asset management, physical security, cybersecurity, and human resources safety functions. It entails the identification of an organization's information assets and the development, documentation, and implementation of policies, standards, procedures, and guidelines. Security management evolved over the years because of changes in technologies, networks, environments, computers, devices, computing systems, user's needs, information processing and structure, and information architecture, information systems, threats, legal, and organizational trends.

Security became a complex matter because information is not anymore centralized within one "glass-house." The information is found everywhere on servers, workstations, personal devices, networks, and sensors. In addition, Internet and Web technologies make security much more complex and even more critical. The real issue is that security of information is not an information technology (IT) problem but a risk facing the business. Business is ultimately responsible for information management including identification of sensitive content and how it needs to be controlled within the organization. This should be specifically included in the business strategy.

Besides technical security controls (firewalls, passwords, intrusion detection systems, disaster recovery plans, encryption, virtual private networks, etc.), security of an organization includes other issues that are typically process and people issues such as policies, training, habits, awareness, procedures, and a variety of other less technical and nontechnical issues [Heimerl 2005], [Tassabehji 2005]. All these factors make security a complex system [Volonino 2004] and a process that is based on interdisciplinary techniques [Maiwald 2004], [Mena 2004].

11.1.1 Information Security

A basic concept of security management is the information security. Information security means protecting information and information systems from unauthorized access, use, disclosure, disruption, modification, perusal, inspection, recording, or destruction. It is a general term that can be used regardless of the form the data may take either electronic or physical. Related to information security is the concept of information security management (ISM).

During the years, several definitions for information security concepts evolved and published in standards, dictionaries, and other publications such as [NIST SP800-27rA], [ISO/IEC 27000], [CNSSI 4009], [ISACA Glossary], [Pipkin 2000], [McDermott 2001], [Anderson 2003], [Venter 2003], [ISC2 SEC], [Perrin 2012].

The broad or outdated definitions that are used sometimes affect the understanding of the objectives and strategies of the information security. However, a universal definition may not be available because of the developments in the information security and evolution of standards and other factors. Therefore, it

Building an Effective Security Program for Distributed Energy Resources and Systems: Understanding Security for Smart Grid and Distributed Energy Resources and Systems, Volume 1, First Edition. Mariana Hentea.

may be necessary to identify the most recent definitions and understand the current trends in information security industry and energy sector.

11.1.2 Security Management Components

Security management includes components such as [Harris 2013]:

- Security process.
- Risk management and analysis.
- Security controls implementation and management.
- Organizational roles and responsibilities.
- Information classification.
- Information security plans and policies, procedures, standards, guidelines, baselines, information classification, and documentation.
- Personnel security and privacy.
- Security awareness and continuous education and training.

These components serve as the foundation of an organization's security program. A security program is the overall combination of technical, operational, and procedural measures and management structures implemented to provide for the confidentiality, integrity, and availability of information based on business requirements and risk analysis [ISACA Glossary].

Other components of the security program may include [Harris 2013]:

- Facilities management and information and communication technologies (ICT) continuity management programs.
- Business continuity management programs.
- Health and safety policies and procedures.
- Certification requirements, such as the ISO 27000 series or the new service auditor standard (SSAE 16/ISAE 3402).
- Related activities such as monitoring, external security services, and outsourcing arrangements.

In addition, supply chain risk requires management (SCRM) control in the organization from the perspective of information and implicitly security perspective (see definition [SCRM]). Product and supply chain data run on top of business software that connects supply chains, and weak links abound globally that can create huge disruptions [Sales 2016]. Cybersecurity and supply chain risks are drawing more attention from senior management and board members, but many companies fall short with accountability.

The term information security program is used to describe the activities to contain the risks to the information assets. Some organizations may prefer to use the information security program as defined in ISO 27001 standard that refers to as an information security management system (ISMS) [ISO/IEC 27001]. Although standards and guidelines are identified to support the implementation of minimum security measures that set a baseline for cybersecurity across energy sector, there are many security challenges that require solutions based on an effective security program [P2030 2011].

11.1.3 Management Tasks

Management must initiate the security program and must integrate it into the organization's business environment. Management's support is one of the most important pieces of a security program [Harris 2005].

Management must develop a program that effectively addresses the key needs of the organization and supports the organizations' mission objectives, the needed levels of information asset protection, regulatory and industry standards compliance, and how security must fit into the overall IT governance model. Security should play an integral part of the organization's IT governance process and the selection of IT and how it is used.

Security management has to undertake a program that provides resources for security education, information system use, training, and threat awareness to all of the organization's employees. This training should address the use, management, and maintenance of the organization's information systems. This is an important activity that is needed to support the operational viability of the organization's computing resources and to continue to meet its mission objectives.

Technology, management, organizations, and their mission objectives change within and across organizations. Since a standard set of performance metrics are not always available, another task is to build a set of systems that generates metrics that measures the security management's performance. With good data and baseline performance metrics, trending can be performed, and the activities can be tuned to yield maximum benefits. Data collection points, including but not limited to items such as call response times, duration between outages, system availability, number of incidents over a given time period, can all be used as reference points for baselining and then measuring the performance of the security program.

Incident response is another area of management's responsibility. The term incident is usually an organizational event that is associated with information system activity that violates, or appears to violate, a security policy or standard that governs that system's use as an operational IT resource. The management must establish a team to act as a central consolidation point to perform overall incident response, coordinate the investigation of incidents, facilitate incident containment and support the remediation activity, and gather forensic data for further analysis or action. Forensic data may include obtaining copies of malware or viruses and reverse engineering them and performing detailed log analysis for the determination of the specific threat and vector of attack for damage assessment and the implementation of future counter measures. Liaison activities with law enforcement and other security groups may also be expected in current IT and ICS security practices. Part of incident response's team or a separate team could have the responsibility for disaster recovery and business continuity management activities.

The operational goal is for the organization to develop the ability to reconstitute itself within a reasonable amount of time and provide operational services. The details of disaster recovery and continuity of operations must be addressed through a thorough analysis of the organization's needs, planning, and frequent testing to validate the plan. Revisions to the plan are made after each test to include lessons learned and avoid repeating mistakes. At a minimum, system configurations, capabilities, and use evolve over time, and each requires that enhancements be made to the plan. Planning is designed to address the most likely threats, or group of threats, identified by the risk assessment activity and developing sequential plans to contain, remediate, and restore system availability.

The management's role is to determine the objectives, scope, policies, priorities, and strategies to implement the security program. The primary goal of the cybersecurity strategy on Smart Grid applications should be on prevention. However, it also requires that a resilient response and recovery strategy be developed in the event of a cyber attack. The management team is expected to create and implement a security program that will reduce the possibilities that such events occur and, if so, that their impact to the organization is minimized. This program must be based on repeatable processes and not on single-based actions of an individual to solve or remediate a problem.

The primary goal of the cybersecurity strategy should be on prevention. The main objective of ISM is the protection of assets that must be done in a manner that reflects a measured and appropriate amount of the security to balance the value of the information and systems being protected against the risks to them. The most effective ways to achieve this balance are through the use of a continuous process that generates data to measure the output of the process and create feedback to the process.

Therefore, management must define the processes and steps to ensure that all components of a security program are properly addressed to ensure integrity, confidentiality, availability, and non-repudiation of the information.

11.2 Security Program

A security program describes the structure and organization of the effort related to information security and physical security. Some organizations may use the term information security program. A security program is an integrated group of activities designed and managed to meet security objectives for the organization and/or the function. A security program may be implemented at either the organization or the function level, but a higher-level implementation and enterprise viewpoint may benefit the organization by integrating activities and leveraging resource investments across the entire enterprise. The security program management comprises objectives such as the following: establish the program's strategy, establish and maintain cybersecurity architecture, perform secure software development, and management activities.

Organizing the structure of a security program is challenging, and it is determined by organizational culture, size, security personnel budget, and security capital budget [Whitman 2014]. The objective of the security program is the protection of people, assets, and organization.

11.2.1 Security Program Functions

Table 11.1 includes functions required to implement a successful security program.

These functions can be implemented following the best practices, recommendations, and guidelines provided by information security industry. Many details can be found in other published works such as [Harris 2013], [Whitman 2011], [Wadlow 2000], [Krutz 2004].

Crucial to risk management, risk assessment, and incident reporting functions is commitment from management to make these functions useful at all levels from low-level operations to top actions and

Table 11.1 Security program functions.

Function	Description
Risk assessment	Maintains the process of identifying, prioritizing, and estimating risks
Risk management	Maintains the total process of identifying, controlling, and mitigating information system-related risks. It includes risk assessment, cost–benefit analysis, and the selection, implementation, test, and security evaluation of safeguards
Legal assessment	Maintains awareness of potential laws and regulations and their impact on security of the organization
Planning	Supports identification and creation of information security plans based on a project management approach as contrasted with strategic planning for the whole organization
Policy	Establishes and reviews the security policies
Measurement	Uses specialized data collection systems to measure aspects of information security
Incident reporting and response	Handles the collection of security events and response to potential incidents
Auditing	Reviews of system's records and activities to determine the adequacy of system controls, ensure compliance with established security policy and procedures, detect breaches in security services, and recommend any changes that are indicated for countermeasures
Systems security administration	Administers the configuration of systems
Training	Trains users and personnel in various information security topics including management and operations
Vulnerability assessment in the product	Examines systematically the vulnerabilities of an information system or product to determine the adequacy of security measures; a system can have three types of vulnerabilities: (a) vulnerabilities in design or specification, (b) vulnerabilities in implementation, and (c) vulnerabilities in operation and management
System testing	Supports a process intended to reveal flaws in the security mechanisms of an information system that protect data and maintain functionality as intended

decisions. For example, incident reporting function is not enough only to report security incidents, but it must be used as a problem-solving tool, involving the entire organization. Events and data collected should be collected, aggregated, and filtered to provide support for forensic analysis, feedback control, and actions for improvement. Adequate models and systems to support these functions are needed.

11.2.2 Building a Security Program: Which Approach?

Basically, a security program is a framework that includes logical, administrative, and physical protection mechanisms, policies, procedures, business processes, and people working together to provide adequate protection level for an environment. A plan is needed to follow for building a security program. To create or maintain a secure environment, an organization needs to design a working security plan and then implement a management model to execute and maintain that plan. A framework is the outline of the more thorough blueprint, which serves as the basis for the design, selection, and implementation of all subsequent security controls, including information security policies, security education, and training programs [Whitman 2014]. Different organizations, consulting companies, and information security professionals follow different approaches for setting up an information security program. The following includes a brief overview of different models and approaches.

Many organizations follow the [ISO/IEC 27001] standard, formerly known as ISO 17799 that evolved from BS 17799, or morph it to fit their organization's specific needs. The standard ISO 27001 (formally known as ISO/IEC 27001:2005) is a specification for an ISMS. A management system as defined in [ISO/IEC 27000]:

- Uses a framework of resources to achieve an organization's objectives.
- Includes organizational structure, policies, planning activities, responsibilities, practices, procedures, processes, and resources.
- Allows an organization to support information security and goals such as:
  - Satisfy the information security requirements of customers and other stakeholders
  - Improve an organization's plans and activities
  - Meet the organization's information security objectives
  - Comply with regulations, legislation, and industry mandates
  - Manage information assets in an organized way that facilitates continual improvement and adjustment to current organizational goals

Basically, an ISMS is a framework of policies and procedures that includes all legal, physical, and technical controls involved in an organization's information risk management processes.

The [ISO/IEC 27000] provides the overview of ISMS and terms and definitions commonly used in the ISMS family of standards. It describes an ISMS consisting of policies, procedures, guidelines, and associated resources and activities, collectively managed by an organization, in the pursuit of protecting its information assets.

An ISMS is a systematic approach for establishing, implementing, operating, monitoring, reviewing, maintaining, and improving an organization's information security to achieve business objectives. It is based upon a risk assessment and the organization's risk acceptance levels designed to effectively treat and manage risks. Analyzing requirements for the protection of information assets and applying appropriate controls to ensure the protection of these information assets, as required, contributes to the successful implementation of an ISMS.

The [ISO/IEC 27001] standard provides broad advice for each of the following areas of an information security program:

- Risk assessment and treatment.
- Security policy.
- Organizational security.
- Asset classification and control.
- Personnel security.

- Physical and environmental security.
- Communications and operations management.
- Access control.
- System development and maintenance.
- Information security incident management.
- Business continuity management.
- Compliance.

The standard [ISO/IEC 27001] has 10 short clauses, plus a long annex (list of controls and their objectives), which cover:

- Scope of the standard and generic ISMS requirements.
- How the document is referenced.
- Reuse of the terms and definitions in [ISO/IEC 27000].
- Organizational context and stakeholders.
- Information security leadership and high-level support for policy.
- Planning an ISMS, risk assessment, and risk treatment.
- Supporting an ISMS.
- Making an ISMS operational.
- Performance evaluation.
- Improvement and corrective action.

11.2.3 Security Management Process

Security process is the method an organization uses to implement and achieve its security objectives. The process is designed to identify, measure, manage, and control the risks to system and data availability, integrity, and confidentiality and ensure accountability for system actions.

Security process efforts are probably the most prolific area of investment over the past ten years. As organizations woke up to the information security problem, they scrambled to define security processes. Best practices, standards, policies, procedures, manuals, guidance, consulting, implementation, and corporate governance documents abound. As a result, a massive array of resources is available to organizations in the area of security process. However, this has had both a positive and negative effect. It is good that organizations have invested in this important area, but all too often security process ends up being a checkbox on a company process form plus many boxes of documents left on shelves and unused.

Security process concepts are at the heart of the National Strategy to Secure Cyberspace.

Some of the most effective approaches to security process have come from the establishment of best practices.

The security management process consists of activities that are carried out by the security management itself or activities that are controlled by the security management. Although there are a differences among known security management processes from objectives to reporting activities, the basic process is a continuous process that describes key activities structured to fit in the management of the organization.

As with all management processes, an ISMS must remain effective and efficient in the long term, adapting to changes in the internal organization and external environment. Although no specific approach is mandated in the current ISO 27000 suite of standards, one approach called Plan–Do–Check–Act (PDCA), or Deming cycle, is defined in [ISO/IEC 27001] (version of 2005). The recent revised version of this standard, version of 2013, is not mandating a specific approach. Therefore, organizations choose different options that support processes with emphasis on managerial security or technical security. Examples include any management process (improvement) approach like PDCA or Six Sigmas (continuous efforts to achieve stable and predictable process results (e.g. by reducing process variation) [Six Sigmas]) and DMAIC (Define, Measure, Analyze, Improve and Control), a data-driven cycle used for improving, optimizing, and stabilizing business processes and designs [DMAIC].

All these approaches are based on the concept of continuous cycle that is the basis of other evolving approaches. Figure 11.1 depicts the concept of continuous process with the basic phases:

- Plan – Identifying and analyzing the problem; for example, designing the ISMS, assessing information security risks, and selecting appropriate controls.
- Do – Developing and testing a potential solution; for example, implementing and operating the controls.
- Check – Measuring how effective the test solution was and analyzing whether it could be improved in any way; for example, reviewing and evaluating the performance (efficiency and effectiveness) of the ISMS.
- Act – Implementing the improved solution fully; for example, changing and acting when necessary to bring the ISMS back to peak performance.

There can be any number of iterations of the Do and Check phases, as the solution is refined, retested, re-refined, and retested again. Different activities may require their own process such as security planning process (e.g. quarterly), security budgeting process, information security monitoring process, security enforcement process, security engineering process, systems engineering process, risk management process, etc.

11.3 Asset Management

An asset is defined as physical or logical object owned by or under the custodial duties of an organization, having either a perceived or actual value to the organization [IEC 62443-1-1]. Asset management is a systematic process of operating, maintaining, upgrading, and disposing of assets cost-effectively. Asset management is a management paradigm and a body of management practices. In the engineering environment, asset management is the practice of managing assets to achieve the greatest return and to provide the best possible service to users. Asset management applies to both tangible (physical) assets such as buildings and equipment and intangible concepts such as data, software, applications, reputation, and intellectual property.

Essential to any enterprise business is information, an asset that is like other important business assets. Information can exist in many forms. It can be printed or written on paper, stored electronically, transmitted by post or by using electronic means, shown on films, or spoken in conversation. Asset management is supported by an asset management infrastructure.

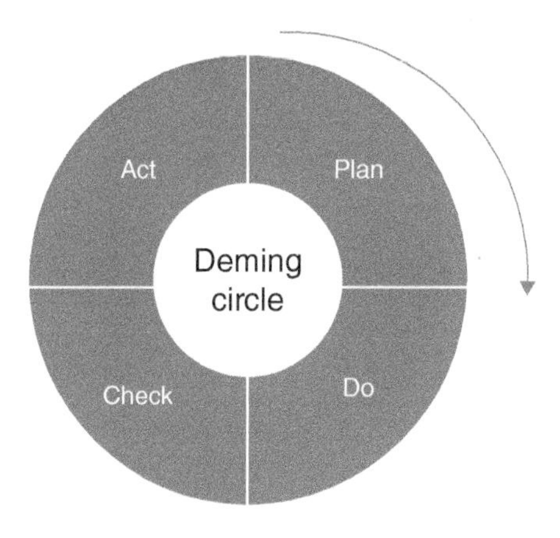

Figure 11.1 Plan–do–check–act cycle.

11.3.1 Asset Management for Power System

The electric power subsector accounts for 40% of all energy consumed in the United States. Electricity system facilities and assets are dispersed throughout the North American continent. Although most assets are privately owned, no single organization represents the interests of the entire subsector [DOE 2014]. An abstract topology of the electric grid energy delivery system showing the power system (primary equipment) alongside IT and OT systems (information management) is shown in Figure 11.2.

In the electricity subsector, the security of various types of cyber assets (e.g. cyber asset, BES cyber asset, BES critical cyber asset, BES cyber information system, etc.) including definitions is controlled by NERC, FERC, and DOE. Therefore, a security engineer needs to understand the most current terms and definitions provided in NERC vocabulary in order to design the appropriate protections and asset management applications.

11.3.2 Asset Management Perspectives

The asset management is a multidisciplinary approach with different perspectives. The asset management is the combination of management, financial, economic, engineering, and other practices applied to assets with the objective of providing the required level of service in the most cost-effective manner.

It includes the management of the whole life cycle (design, construction, commissioning, operating, maintaining, repairing, modifying, replacing, and decommissioning/disposal) of assets (physical and infrastructure assets).

Operating and sustainment of assets in a constrained budget environment require some sort of prioritization scheme for developing physical infrastructures. For economical reasons, big data and analytics needs may lead asset management applications to be implemented via cloud services: infrastructure, platform, or software. However, from a security management perspective, there are differences on visibility and control of assets security information from none to limited and more. Therefore, the use of cloud services for asset management has to be balanced with the enterprise business objectives. An overview of benefits of asset management for power system is provided in the following section.

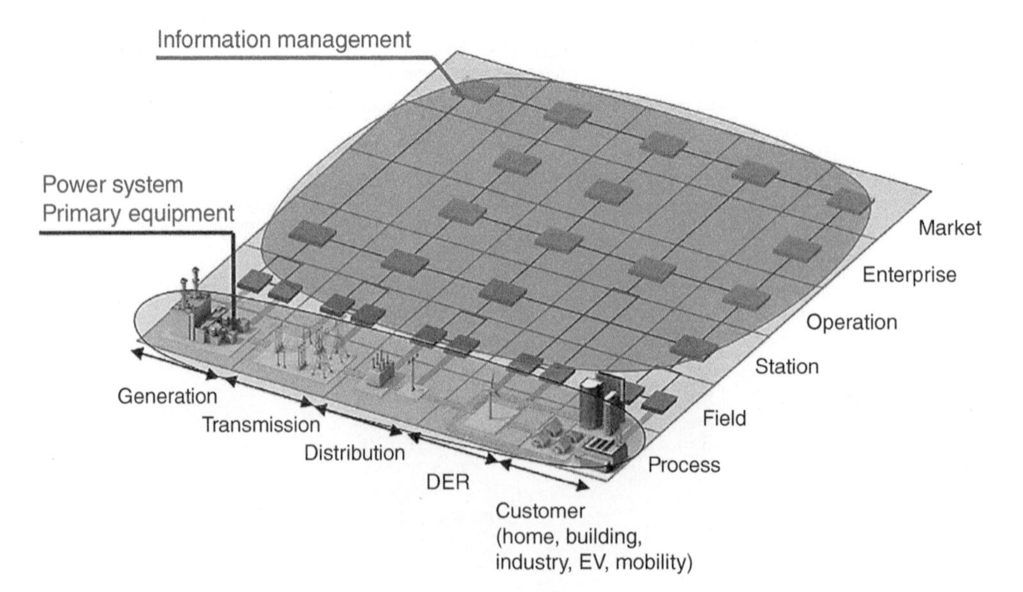

Figure 11.2 An abstract topology of the electric grid energy delivery system. *Source:* [DOE 2014]. Public Domain.

11.3.3 Benefits of Asset Management

Since electric infrastructure is aging, utilities and enterprises are looking for ways to get the most from their assets and maximize the return of investment (ROI). With Smart Grid systems, their operations are more complex than ever before. Therefore, organizations want to improve efficiency and enhance maintenance of their assets so effectively balancing cost and risk. Optimal asset management in distribution systems leads to the increase of energy efficiency, which results in the increase of benefits for distribution systems' enterprises [Dashti 2010].

Besides real-time benefits such as updating and maintaining the asset inventory, as assets are acquired and/or disposed of throughout the asset life cycle, the latest technological advances that are influencing asset management practices include successful integration of well-known technologies (telecommunications, faster and parallel computation), lab experiments, and statistical techniques and analytics intended to predict failures of assets, detect trends, patterns, and unusual events. For example, inventories of cyber assets assist in ensuring that effective asset protection takes place and are key to the risk assessment process. Given the numerous assets of the power system, we focus on more specific issues related to DER asset management.

11.3.3.1 DER Assets Classification

An inventory and a classification of assets have the potential to help an enterprise to efficiently maintain the assets and design applications [Rastler 2012]. In DER systems, information required for different activities include distributed resources and asset management, registration and certification of a distributed resource, updating and removal of a distributed resource or asset, event notification, submission of emergency signals, etc.

For example, a wide range of technologies are used for energy storage systems (ESS). Therefore, a classification based on technologies includes categories such as the following:

- Electrical such as capacitors, supercapacitors, and superconducting magnetic ESS.
- Electrochemical such as battery systems, flow batteries, and hydrogen with fuel cells.
- Mechanical such as pumped hydroelectric energy storage, compressed air energy storage, flywheel energy storage, and hydraulic accumulators.

11.3.3.2 DER Asset Data

The various data collected and recorded about an asset depends on energy storage technology and the services that asset management will provide for an enterprise. In addition, the asset data may be collected from the perspective of type of application. Understanding the different categories of applications is essential on determining the asset data. At minimum, the energy storage asset data should include:

- ID number
- Type of technology
- Description of asset
- Make or manufacturer
- Model
- Serial number
- Owner
- Location of physical or logical asset
- Installation date
- Expected service life
- Controls and protection schemes associated with each asset (if applicable)
- Operating mode
- Last failure type
- Last failure date
- Planned maintenance date
- Last maintenance start date

- Last maintenance end date
- Failure date
- Last audit date
- Energy storage parameters (power capacity, energy storage capacity, efficiency, response time, round-trip efficiency)
- Maintenance provider
- Comments.

The asset data can be used for a rapid online state-of-health assessment for batteries. The asset management information combined with data collected by an energy storage monitoring system can be used for detecting various off-normal degradation effects and estimating battery life. One particular issue is the memory effect when aging over different temperatures or degradation due nonuniform utilization. The aging behavior could directly impact applications [DOE 2011]. For the purpose of state-of-health assessment, the asset data for electrochemical battery storage system should include both actual and manufacturer' parameters for charge-to-discharge ratio, depth of discharge, and memory effect.

11.3.3.3 Asset Management Analytics

The asset management analytics can be used for the development of operations, maintenance, and capital improvement projects. The asset management analytics can make a difference in the development of operations, maintenance, and capital improvement. For example:

- Operational – Analytics can forecast a battery storage system short-term failure by using information from probes correlated with their use, health state, and history to inform the control center of pending failure avoiding a blackout. Analytics can be used to predict capacity utilization due to current distribution associated with size and location and predict optimum energy capacity in terms of electrical performance, cooling requirements, life, safety, and cost. For example, operating personnel and local fire personnel should be notified of particular safety issues and the appropriate response in case of an emergency.
- Maintenance – Analytics can help determine if a maintenance scheme has impact on asset health and can help pinpoint precisely which maintenance procedure to focus on for each asset, which results in lower maintenance costs, yielding the potential of longer asset life. In some situations, subsystems such as power electronic modules or energy storage banks (AC or DC connections) may be identified as non-serviceable by the supplier in the field, so these subsystems may be removed and replaced by maintenance team. Also, all consumable or degradable parts, such as air filters, should be classified by replacement interval.
- Capital improvement – Analytics can help determine the best replacement scheme for an asset by factoring variables like use growth, increased charging time, decreasing voltage, risk of failure, and degradation to ensure that replacements are neither too soon nor too late. Combined with improved maintenance strategy, this will help extend equipment life cycles and cut costs significantly.

For example, ESS must be able to realize multiple operations across the energy value chain and the costs of storage (e.g. battery cost, power electronics, balance of plant, etc.). These tasks can be tracked with asset management applications.

11.3.3.4 Applications

One requirement is that all components of a battery storage system must be safe, reliable, low cost, and seamlessly integrated. One benefit of asset management is support for this requirement. Another benefit could be support for proper sizing and location of equipment. Detailed and accurate data of the technical characteristics may support installation requirements as well as design and implementation of specific applications. To address each application, ESS will vary in cost, depending on size, location, and system power-to-energy ratio. Patterns of connects/disconnects frequency and charges/discharges time support and maintenance program and could help in identifying repairs, depletion of battery, upgrades, replacements, etc.

Also, it could support operational managers when defending their requests for capital investment in distribution equipment. Executive management could easily see the impact in terms of increased debt/capital spending. Also, asset management provides information that can be used to model the benefits and risks of not investing. The benefits of using the information for scheduling appropriate maintenance schedules, developing models to discover reliability trends, and avoiding failures of the battery storage systems outweigh the costs of implementing and maintaining an asset management application.

Another key asset management benefit provided is the ability for enterprises to more efficiently monitor and maintain the battery storage equipment necessary to reliably deliver power to customers.

Other benefits of leveraging the asset information and value-added services include:

- Cyber risk management, threat detection, and control.
- Outage management.
- Remote connect/disconnect.
- Load forecasting.
- Demand and response.
- Enhanced customer service.
- Power quality monitoring.
- Pricing event notification capability.
- Financial.

In addition, the asset management improves the capability of the grid to incorporate variable generation. The most notable improvement comes from asset management is the information regarding the availability of energy storage and renewable and the ability to adjust maintenance timed to a generation schedule that can change very quickly.

11.3.3.5 Asset Management Metrics

Effective asset management methods help ensure successful adherence to enterprise reliability metrics like SAIDI (System Average Interruption Duration Index), CAIDI (Customer Average Interruption Duration Index), and SAIFI (System Average Interruption Frequency Index).

CAIDI measures the average interruption duration per customer interrupted during a given period of time, typically one year.

SAIFI is the average number of times that an average customer experiences a service interruption during a year. SAIFI is an indicator of utility network performance.

SAIDI is the average total amount of time that an average customer does not have power during a year. SAIDI generally measures the operating performance of the utility in restoring customer interruptions.

MAIFI is the momentary average interruption frequency index (MAIFI).

At the distribution level, the metrics for system reliability are related to the number of customers affected, frequency, and duration: CAIDI, SAIFI, SAIDI, and MAIFI are well-understood metrics with a public utility [EPRI 2001]. Examples of reports on performance and how utilities monitor reliability and performance based on SAIDI and SAIFI indicators include [Utilities 2014], [Hawaiian Electric].

Although these reports show performance of the service, implicitly the data used for the calculation of these metrics is about specific outages due to equipment. However, data may include human errors and other failures. Grid equipment reliability metrics can be calculated using asset data collected consistently.

Typically, the traditional maintenance approach focuses on assets and then tries to consolidate the benefits. Strategic asset management uses a top-down approach, meaning it starts with the corporate responsibility goals and then finds the assets that contribute to them and proposes specific action for those assets.

11.3.3.6 Asset Management Services

With a broad range of asset management services for ESS, enterprises can optimize performance, identify opportunities for change and ways to reduce asset life cycle costs and extend asset life, prioritize their investments to make sure they earn their allowed rate of return, identify business process improvements, and enhance operations via information management and communications.

Services can be provided at each stage of the asset life cycle. The services range from business transformation and change management at the strategic level to asset condition assessment and deterioration modeling at the tactical level. Examples of asset management benefits from a strategic perspective include [Berst 2012]:

- Achieving lower total life cycle costs.
- Improving the enterprise performance.
- Optimizing use of assets.
- Improving response to emergencies by eliminating non-value-added activities.

In [Hoffman 2013], the ROI can be measured by five primary areas:

- Failure prevention and enhanced reliability (less downtime for a battery storage system and more utilization).
- Enabling operational excellence (more accurate assessment of asset conditions).
- Risk mitigation (the ability to identify critical assets with probably high failure rates).
- Solid investment choices (forecast future capital projects based on asset importance, age, and system impact).
- Eliminating guesswork (forecast short-term and long-term failures based on testing and inspections).

11.4 Physical Security and Safety

Two major components of Smart Grid security can be viewed as cybersecurity and physical security, although the demarcation between these domains is blurring. NERC defines a cybersecurity incident as a malicious act or suspicious event that compromises, or was an attempt to compromise, the electronic security perimeter or physical security perimeter or disrupts, or was an attempt to disrupt, the operation of a BES Cyber System [NERC Glossary].

Between 2011 and 2014, electric utilities reported 362 physical and cyber attacks that caused outages or other power disturbances to the US Department of Energy. Of those, 14 were cyber attacks, and the rest were physical in nature [Reilly 2015]. The threats to DER systems can be categorized in a broad set of categories such as:

- Natural environmental threats (floods, earthquakes, storms and tornadoes, fires, extreme temperature conditions, chemical contamination, etc.)
- Supply system threats (power distribution outages, communications interruptions, water interruptions, gas interruptions, etc.)
- Power controls and equipment threats (cyber–physical systems, equipment automation, electromagnetic pulses, chemical spills, etc.)
- Human threats (unintentional or intentional, e.g. unauthorized access [internal and external], explosions, vandalism, fraud, theft, etc.)
- Politically motivated threats (strikes, terrorist attacks, bombings, etc.)

Identifying and reducing potential risks from cyber attacks and physical threats such as electromagnetic pulses are mandated in [USPolicy 2013]. The protection depends on the risk acceptance level, and it is also derived from laws and regulations.

Physical security has a different set of vulnerabilities, threats, and countermeasures from that of computer and information security. Physical security provides protection for the entire organization or facility, from the outside perimeter to the inside office space, including people and all information system resources from harm, damage, and loss.

11.4.1 Physical Security Measures

Physical security mechanisms protect people, data, equipment, systems, facilities, and many other assets. Physical security involves the use of multiple layers of mechanisms and interdependent systems that include surveillance, security guards, protective barriers, locks, access control protocols, and many other techniques. In the electrical enterprise, physical and environmental controls could be a subset of measures that physically protect the facilities of an organization by implementing physical access controls, conditioning power lines, providing backup power, and establishing reuse and data use policies.

Thus, an organization to adequately protect itself from increasingly sophisticated threats is critical to leverage advancements in technologies that cater to both physical and IT environments [Lapolito 2007]. All of these converged security technologies can produce a greater amount of information from which to make security decisions. Surveillance technologies communicate in real time to deliver operations center personnel true situational awareness. However, it is crucial to be able to manage, correlate, and analyze the information to drive the right response. Also, managing all the information effectively can help in making decisions quickly.

Physical security measures aim to either prevent a direct assault on premises or reduce the potential damage and injuries that can be inflicted should an incident occur. For most organizations the recommended response will involve a sensible mix of general good housekeeping alongside appropriate investments in closed-circuit television (CCTV), intruder alarms, and lighting that deter as well as detect – measures that will also protect against other criminal acts such as theft and vandalism and address general health and safety concerns. In some locations these measures may already be in place to some degree. However, external and internal threats to organizations (and their staff) will constantly evolve, and so all procedures and technology should be kept under constant review. CCTV is a TV system in which signals are not publicly distributed but are monitored, primarily for surveillance and security purposes.

Guidelines and regulations are available for physical security. Crime prevention through environmental design (CPTED) is also a planned way for entire neighborhoods as well as specific subsystems such as transportation [Fennelly 2012]. Physical security of power grid should be also included in such plans to prevent damages of equipment. Also, the strategy should be of implementing controls that discourage attackers by convincing them that the cost of attacking is greater than the value received from the attack.

In many enterprises there are links between information security and general (physical) security that should be considered for inclusion and managed appropriately. Also, security of the personnel plays a central role in all layers of security. A security program should also comprise safety and physical security functions. Safety functions should include protection mechanisms to avoid harm to people's health and approaches to assess and manage safety risks.

All of the technological systems that are employed to enhance physical security are useless without a security force that is trained in their use and maintenance and that knows how to properly respond to breaches in security. Security personnel perform many functions such as patrols and at checkpoints to administer electronic access control, to respond to alarms, and to monitor and analyze video. Security personnel play a central role in all layers of security. Many technological systems are employed to enhance physical security. When a gunmen opened fire on a substation in California causing big damages (17 transformers were taken out), physical security of the power grid and development of specific standards was called by FERC and NERC agencies.

11.4.2 Physical Security Evolution

Focusing on physical security is an opportunity to add more sensors and software, and cybersecurity must be evaluated in conjunction with physical security [News 2013]. The emerging smart camera networks capture data in both private and public environments. The technology advances – computer vision, image sensors, embedded computing, and sensor networks – require security engineers and security personnel for more specialized skills to maintain, use the information efficiently, and deal with big data [Reisslein 2014], [Chen 2014].

The security professionals have to address more issues such as video analysis, system design (security and privacy of surveillance application and surveilled assets), protection of data collected and transmitted, integration with the existing surveillance systems, sensor networks design and monitoring, protection of smart camera networks, trade-off cost/performance, training, and management of specific activities. The traffic load generated by the emerging smart camera networks shape new decisions about the merging of physical space and cyberspace [Abas 2014], but performing multiple tasks requires strategies to find a configuration that maximizes the network's situational awareness [SanMiguel 2014].

With the integration of consumer smart cameras into camera networks, security and privacy are increasing concerns that need to be assessed and managed [Prati 2014]. Figure 11.3 shows a view of

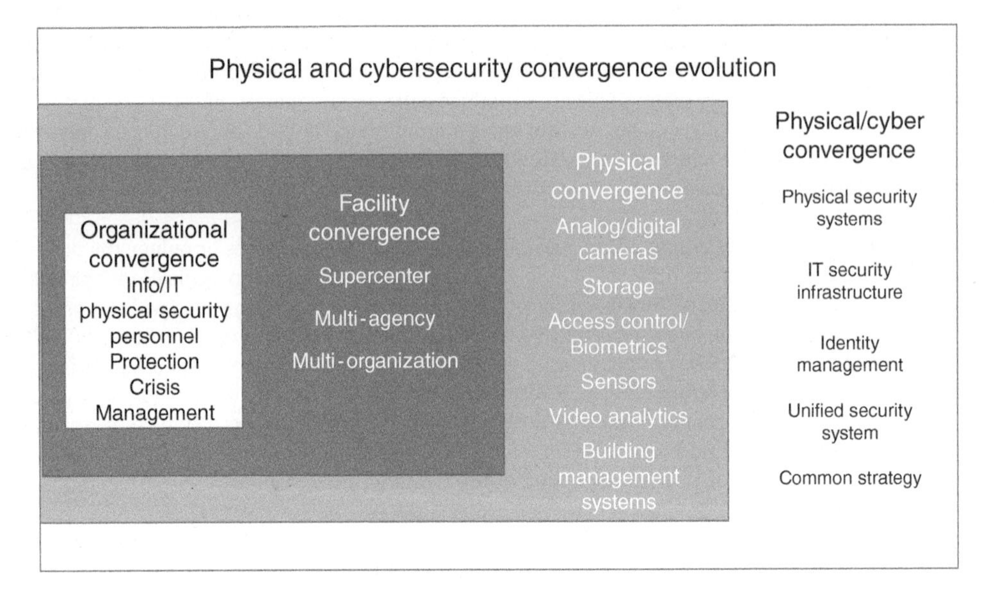

Figure 11.3 Security convergence evolution.

physical and IT security convergence evolution. Security convergence refers to the convergence of two historically distinct security functions – physical security and information security – within enterprises; both are integral parts of any coherent risk management program. Security convergence is motivated by the recognition that corporate assets are increasingly information based.

All of these levels are converging upon each other to create a new centralized view of security across an organization, bringing with it a number of technology challenges that must be acknowledged and addressed. To illustrate, consider these key technology convergence issues security organizations face as they strive to meet their primary mission of protecting people, assets, and infrastructure:

- Proprietary systems – Every product and system has a proprietary standalone management console.
- Overload – There are too many cameras to monitor and too many data sources.
- Heterogeneous systems – Systems such as surveillance, access control, information security, etc. to date have been built in silos and do not interoperate.
- Reactive – Forensics only helps solve the crime and does not accomplish the primary mission of preventing it.
- Upgrades – Upgrades for legacy equipment and networks are unrealistic.
- Data retrieval – Legacy archiving and storage are not suitable for event correlation that requires fast data retrieval.
- Lack of standards – Control plane systems do not operate on known standards.

Until now, there has been a void in the physical security market at the intersection of IT security. Despite the staggering number of cameras, alarms, and sensors feeding data to an operations center, there lacked a physical security information management (PSIM) platform that could take in all the data, correlate it with data from IT security systems, and provide security personnel the insight to make effective decisions and respond to security events.

Companies are now applying these concepts, security event management (SEM) and security incident management (SIM), to the physical security field, producing greater insight into what is going on around the facility and how to best handle a given situation.

SEM and SIM are synonymous in the IT security field and have long served as the correlation engine that enables an IT security manager to spare false alarms and false positives from true events that require attention and response. SEM enables organizations to pull data from firewalls, intrusion detection and prevention systems, antivirus software, and log files to create a clearer view of network activity and how to respond quickly and effectively.

By correlating the data feeds and alarms from various sources, a PSIM solution can provide the context that enables the right response to real threats and security events and disregard those that are false alarms or not real threats. The next chapter includes a section that describes a perspective for more rigorous and suitable solutions to Smart Grid – toward more efficient and effective systems.

In addition, safety impacts can be monitored and analyzed to establish plans that should be available for avoiding safety impacts due to cyber incidents. An incident reporting system that collects information about security and safety incidents has to be designed and managed within a security program.

However, convergence has its own challenges from human resources, cultural, and management perspectives [Lalonde 2018]. As discussed by this author, convergence aims to unified security system, which is a better way to detect, respond to, recover from, and/or prevent security incidents. Cyber and physical convergence is increasingly being recognized as organizations realize that keeping the two separate can have consequences. A unified security system is a way to eliminate silos and strengthen security, but it is essential that organizations first design a security policy with convergence in mind such as building convergence through policy and skills. Therefore, it is highlighted that convergence is widely accepted as the way of the future.

11.4.3 Human Resources and Public Safety

Human factors have been referred to as human–computer interaction (HCI), the man–machine interface, and ergonomics. In practice, much of the work on user-friendliness and usability relates to human factors. Human factors are an essential ingredient to the enabling technology and support of Smart Grid systems. Therefore, an understanding of the use and relationships between technology and people is crucial for identifying the impacts to the security of any organization.

11.5 Human and Technology Relationship

Technology is driven by the people element and the human factors. Besides technology, culture and human factors influence the functioning of the security in any organization. People influence information security through their decisions and interaction with the corporate environment, reflected in its corporate strategies and processes or in other people.

11.5.1 Use Impacts

Decision making under uncertainty and human factors' actions can influence the security risk to increase instead of reducing. Many behavior situations can occur; they are discussed elsewhere (e.g. BMIS model). A few examples on this matter include the following:

- People may be used to ignore certain procedures to access information and systems; for example, perception of "having a password" and their reaction to keep up with the requirement to change passwords frequently; if a password has to be often changed, people can be frustrated or ignore the procedure.
- Often the norm for control engineer is easy and flexible access without rigorous authorization controls.
- People may have been hired because they demonstrated certain abilities to use technology. But if information security is implemented within the organization's technological infrastructure in a manner that is contrary to the prevailing corporate culture, those same people may have difficulty in complying with the security policy, regardless of their technical skills.
- Use of encryption for email can be ignored.

The workforce of an organization is the key enablers of any security program. In the establishment of an ongoing process of effective security awareness across any organization, security practitioners should include areas such as the following (also defined in some models):

- Safety.
- Human error.
- Communication.
- Job task analyses and usability analyses, including functional requirements and resource allocation.
- Job descriptions and functions, job-related procedures, and utilization of these procedures.
- Knowledge, skills, and abilities.
- Control and display design.
- Stress.
- Visualization of data.
- Individual differences.
- Aging.
- Accessibility.

The safety and security considerations at all levels of technology are expressed by the human factors, which determine how technology is perceived, adopted, and used. The human resources and people (users of power applications) are human factors that interact with technology and the development of tools that facilitate the achievement of information security objectives.

In a major organization, technology is not limited to IT only; it covers an unusually wide range of things. Basic infrastructures (such as water, electricity, and roads) are seen as security relevant where the enterprise needs to take part or all of the responsibility for these infrastructures. At the other end of the technology range, core financial applications are also seen as security relevant, for the obvious reasons. The different layers of infrastructure, IT, and applications are managed through several technology departments that have defined links and interfaces (organization element).

Technology is a pervasive element with a wide scope and covers more than traditional IT. The safety and security considerations at all levels of technology need to be analyzed and expressed by human factors, which determines how technology is perceived, adopted, and used. The systemic view allows security management to understand technology dependencies and effects on the overall system when changes are made.

For example, the basic electrical infrastructure may be provided externally, but it should be seen as process enablers with a high level of criticality. The basic electrical infrastructure influences all processes and people. In the same time, the IT considered as middle layer in a security model – networks, hardware, and platforms – may be internal or outsourced, but it is dependent on the basic layer (electrical infrastructure). They systemically influence both high-level technology solutions and processes. In addition, a high level of IT applications, services, etc. depends on the middle layer of IT, including the security provisioning from that layer. They influence both processes and strategy as well as people (direct users). Similarly, pervasive IT dispersed and decentralized applications, devices, and processes are dependent on the high-level IT services provided by the enterprise. They systemically influence all other elements of the model and may reshape strategy and organizational design (paradigm changes in how people work). Also, the use of external security services – such as certification auditors or electronic surveillance – should be analyzed to ensure full knowledge and a detailed view of the existing security program. It should be carefully analyzed the impact of any command via IT commands to electrical equipment to avoid any events that may cause hazards and safety issues.

11.5.2 DER Systems Challenges

In the context of DER systems, safety concerns arise mainly because of power grid equipment, DER devices and installations, ICT, and security. Technology in the form of the increasing variety of options available affects everyone's daily life – from use of the camera and the car to the television and video recorder or by way of the automatic turn-on/turn-off of lights and switches.

A cybersecurity incident or attack can have impacts on safety of the personnel operating the power grid and public safety. In general, public safety involves the prevention of and protection from events that could endanger the safety of the general public. Both cyber incidents or power events could impact the normal flow of power causing blackouts or other damages that could affect the safety of people.

Therefore, the complexity of Smart Grid applications and wide scope of the technology element need a detailed approach to analyze the interconnections and ensure that dependencies of safety and security to technology including effects are identified and properly managed.

11.5.3 Security vs. Safety

Many experts contributing to current process control system security efforts have a background in safety and got involved in the security topic when they realized that a security incident could have safety-related consequences in the plant. Because the design of safe and safety systems is rather well understood today and generally accepted standards now exist, some people think that plant security should be treated like plant safety. However, this overlooks the fact that, for safety, the opponent is nature, which has a static, statistically described behavior. For security, the opponents are intelligent humans who can communicate, learn, and change their behavior.

Statistical assessments are therefore much less mature, and it is not sure whether they will ever be practical. Related to this aspect is the issue of security levels. Again, in safety, the standard [IEC 61508] describes safety integrity levels that are well-established measures for system safety (or failure probability). The standard covers safety-related systems when one or more of such systems incorporates mechanical, electrical, electronic, and programmable electronic devices. These devices can include anything from ball valves, solenoid valves, electrical relays, and switches through to complex programmable logic controllers (PLCs).

Engineering practices for the application of safety instrumented systems (SIS) in the process sector are provided in the standard [IEC 61511]. The process industry sector includes many types of manufacturing processes, such as refineries, petrochemical, chemical, pharmaceutical, pulp and paper, and power. This standard is also adopted by European countries. In the United States, the standard [ANSI/ISA 84.00.01] mirrors [IEC 61511] in content with the exception that it leaves a gap. It contains a clause that allows the owner/operator to determine and document safety aspects for the existing SIS that were designed and constructed prior to the issuance of this standard. These standards include different parts that were continuously revised since their initial publication dates. Therefore, a precise inclusion of the publication year may be obsolete.

Although the transfer of the concept of safety levels to security is argued, the practitioners may find useful to distinguish the meaning of security levels in the context. As specified in [RFC 4949], the term is usually understood to involve sensitivity to disclosure, but it also is used in many other ways and could easily be misunderstood.

11.6 Information Security Management

Related to information security are the concepts of ISMS and ISM.

ISM is the management of an organization's security program. The ISM addresses areas of security management practice including development of security solutions. ISM describes controls that an organization needs to implement to ensure that it is sensibly managing the security risks and other issues. The risks to these assets can be calculated by assessing threats, vulnerabilities, and impacts. ISM requires the security professional to possess and demonstrate managerial, technical, and political skills on a consistent and ongoing basis. Their success is measured on their ability to understand the security posture of the organization, optimize security, and facilitate the delivery of reliable IT services while helping to ensure that the security attributes of the information needed by the organization's end-user communities are met [Lewis 2010].

From one point of view, ISM evolved on application of published standards and used various security technologies promoted by the security industry. Quite often, these guidelines conflict with each other, or they target only a specific type of organizations (e.g., National Institute for Standards and Technology (NIST) standards are better suited to government organizations). However, building a security control framework focused only on compliance to standards does not allow an organization to achieve the appropriate security controls to manage risk. Besides security program management, other tasks such as security provision are described to include system development, software assurance and security engineering, system security architecture, technology research and development, system requirements planning, testing, and evaluation [Stephenson 2016].

An ISMS is a set of policies concerned with ISM- or IT-related risks. These concepts were first used in BS 7799 standard originally published by British Standards Institution (BSI) Group in 1995. The standard was later adopted and refined in ISO/IEC 27000 series of standards. Different ISMS models are available, but their continuous evolution requires management's attention.

11.6.1 Information Security Management Infrastructure

The ISM is based on its own infrastructure built on top of network management or integrated with the network management infrastructure. Therefore, the infrastructure is mostly affected by the current security technologies that lack integration and rely on human for analysis of huge data collected [Hentea 2005a].

The security management infrastructure includes system components and activities that support security policy by monitoring and controlling security services and mechanisms, distributing security information, and reporting security events [RFC 4949]. Security policy is a definite goal, course, or method of action to guide and determine present and future decisions concerning security in a system [RFC 4949]. It a set of principles that direct how a system (or an organization) provides security services to protect sensitive and critical system resources.

The ISM infrastructure is greatly affected by the network management developments (architectures), distributed real-time monitoring, data analysis and visualization, network security, ontologies, economic aspects of management, uncertainty, and probabilistic approaches, as well as understanding the behavior of managed systems [Pras 2007], [Pavlou 2011] including the most recent paradigms of autonomic computing and later, autonomic networking.

Managing the security of any organizations requires the applications of new paradigms based on process control methods [Kadrich 2007], autonomic computing [Kephart 2003], [Kephart 2005], autonomic networking [Strassner 2004], [Jennings 2007], and machine learning approach [Hentea 2005b], [Hentea 2007], [Hentea 2008]. Therefore, novel developments based on the above may trigger impetuous advances in the quality of the security products including efficiency and effectiveness of ISM.

11.6.2 Enterprise Security Model

Figure 11.4 is an example of the enterprise security governance model in the Smart Grid context. The components are aligned to emerging needs for content security implemented by an organization. It shows the security leadership on top level. Once the organization has the security strategy at a macro level, it must assign responsibility and authority for implementing and monitoring it. This responsibility for the security program is given to security leaders (e.g. information security office or officer who has the endorsement and support of the highest levels of management). Policies govern the information security objectives.

This model emphasizes that security leadership is key to the initiation and implementation of a security program. Key issues regarding an enterprise model that are discussed in [TechNews 2002a], [TechNews 2002b] include:

- Each organization requires a comprehensive information security framework of policies, which must include content security.
- Business is ultimately responsible for information management including identification of sensitive content and how it needs to be controlled within the organization. This should be specifically included in the business strategy described in a security program.

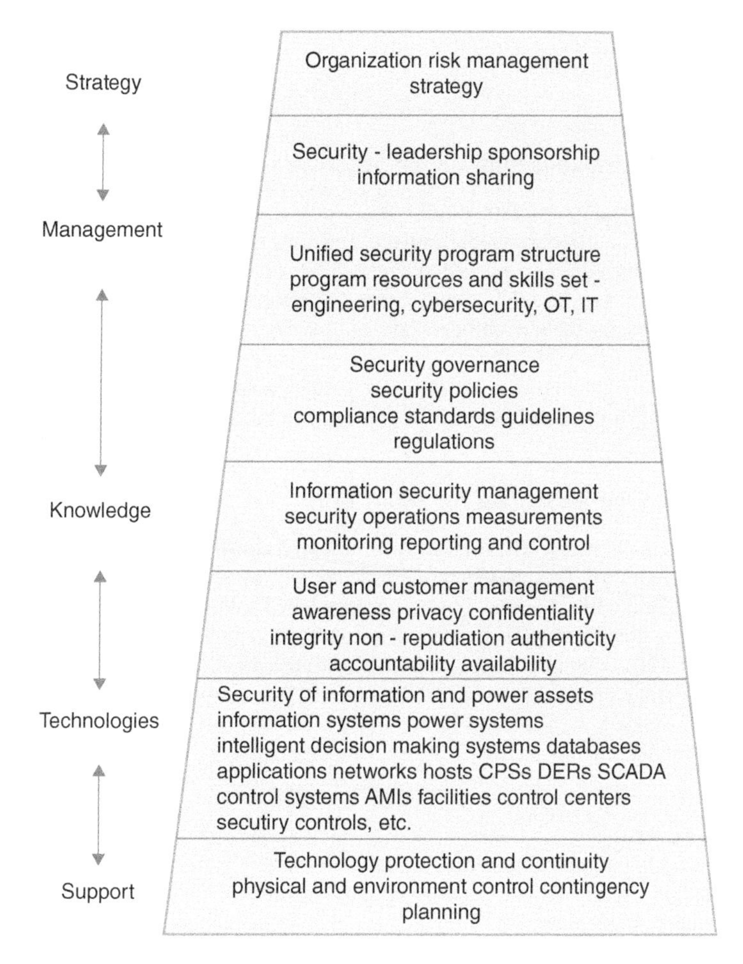

Figure 11.4 An enterprise security governance model in the Smart Grid context.

- Most organizations address e-business risk management at the applications level and then create functional and integrity controls that are not included in their overall business strategies.
- An organization must assign responsibility and authority for implementing and monitoring it. Also, it must define security policies that assist business in understanding and addressing the risks relating to information management.
- Security policies should govern the information security objectives. Key policies and procedures for end users would include those of content management.
- Organization must create business processes and operations as well as a method to monitor the security policies.
- The applications infrastructure must be designed so that content security is controlled in a secure manner. This can only be done through a security governance structure and clear security information policies.

Any energy organization needs to identify a cross-functional team that must support a culture for protecting industrial control systems. The team should include individuals at executive level manager for leadership and guidance, security and operations management at the corporate level, and full participation from control system engineers and managers. The team should be trained on the key aspects of cybersecurity for the industrial control systems and be fully aware of the present security challenges and risks that the organization needs to address in regard to its own industrial control systems infrastructure. The team should be responsible for developing policies and procedures that will increase the security capability and protection of industrial control systems.

Among other tasks, the first task on establishing an ISMS involves establishing the necessary process and management framework [ENISA Factors].

11.6.3 Cycle of the Continuous Information Security Process

Security process is the method an organization uses to implement and achieve its security objectives. The process is designed to identify, measure, manage, and control the risks to system and data availability, integrity, and confidentiality and ensure accountability for system actions. The process includes areas that serve as the framework of ISM. Security process efforts are probably the most prolific area of investment, and concepts of security process are at the heart of the National Strategy to Secure Cyberspace [Kiely 2006b].

ISM is based on a continuous process similar to information management defined as the planning, budgeting, manipulating, and controlling of information throughout its life cycle [CNSSI 4009].

11.6.4 Information Security Process for Smart Grid

The IEEE guidelines [P2030 2011], [IEEE 1547] recommend a continuous information security process based on [IEC 62351] standard. The process is composed of key activities such as risk assessment, policy, deployment, training, and audit. Each activity is accomplished following its own process so five high-level processes that are needed as part of a robust security strategy. However, some recommendations provided in [Cleveland 2012] are not entirely validated with the practices in the security industry or with the constraints of control systems. These have to be assessed before applying to them to any application. Figure 11.5 shows the continuous cycle of information security process that is dynamic and evolving. The cycle activities include the following [ISO/IEC TS 62351-1].

11.6.4.1 Risk Assessment

An assessment is used to determine the value of the information assets of an organization, the threats they are exposed to and vulnerabilities they offer, and the importance of the overall risk to the organization. The assessment is accomplished by following the risk management approach.

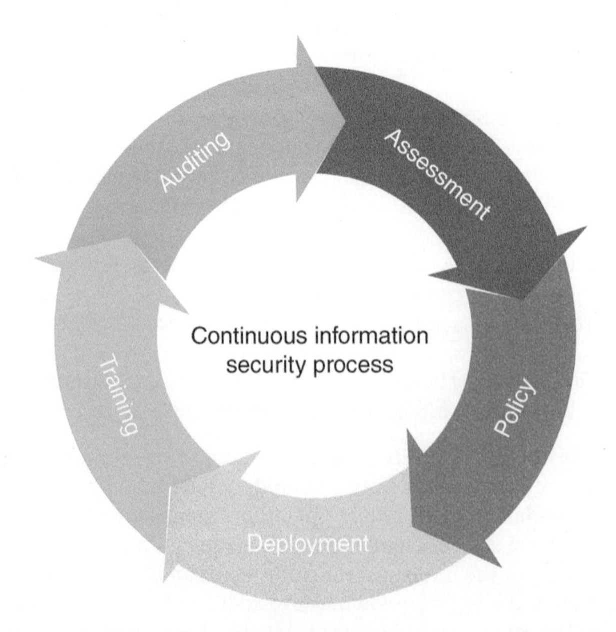

Figure 11.5 Continuous cycle of the information security process. *Source:* Adapted from [CS Odessa]. Licensed under CC BY 4.0.

11.6.4.2 Policy

A security policy is a stated objective or intent regarding the protection of the organization's assets, business, or other entity. Policies should be structured to address higher-level concerns and should not be confused with detailed procedures that are designed to implement a policy. They are the basis for the organization's security governance. A security policy can be an organizational policy, an issue specific policy, or a system policy [Harris 2013]. Examples of policies include organization security policy, information security policy, user security policy, physical security policy, access control policy, remote access policy, access control policy, virtual security policy, network security policy, etc. Writing information security policies requires proper planning and skills [Barman 2002].

11.6.4.3 Deployment

Security policies, standards, and measures to be effective should be implemented and deployed by an organization practicing due care and due diligence.

11.6.4.4 Training

Awareness and education training is the mechanism to provide necessary information to employees.

11.6.4.5 Audit

An information security audit is a systematic evidence-based evaluation of how well the organization conforms to established criteria. Audits are generally conducted by independent auditors, which implies that the auditor is not responsible for, benefited from, or in any way influenced by the audit target. Auditing ensures that controls are configured and monitored correctly with regard to policy. Functions include policy adherence audits, periodic and new assessments, and penetration testing.

The focus of management in the electric sector should be based on a sense of responsibility for security issues and a steadfast commitment to continuous improvement. The key factors that enable security processes to be effective in any organization are planning, communication, and measurement.

11.6.5 Systems Engineering and Processes

The International Council on Systems Engineering (INCOSE) supports the development of methodology called Systems Engineering Handbook (SEBoK), based on the standard [ISO/IEC/IEEE 15288]. Establishing a common framework for system developers and users, the standard covers processes and life cycle stages. Since there are a large number of life cycle process models, the choice of a specific model from any category (e.g. sequential, evolutionary, unconstrained) may include criteria such as use of organization's existing processes, people's skills, and many other aspects of system development and constraints. However, the stages of system process include specific processes that deal with the security of the system and information. Examples of such processes include information management and ISM. Also, security aspects are defined in system-level requirements that describe the system functions to satisfy stakeholder needs and nonfunctional requirements expressing the necessary levels of safety, security, reliability, etc. Related to security is also the systems security engineering process defined in [ISO/IEC 21827] standard.

11.7 Models and Frameworks for Information Security Management

Managing information security is facilitated by using established ISMS models and frameworks.

11.7.1 ISMS Models

The best-known model to follow to build and maintain a security program (called information security management system) is described in [ISO/IEC 27001], [ISO/IEC 27002] and many related standards published jointly by ISO and IEC organizations. The suite of standards, called ISO/IEC 27000 family, includes documents such as ISMS requirements, establishing a process, code of practice for ISM, and guidelines for the accreditation of organizations offering ISMS certification, sector-specific guidelines, and several other guidelines for implementation, maintenance, and improvement. The ISMS family of standards is intended to assist organizations of all types and sizes to implement and operate an ISMS. The ISO/IEC 27000 series serve as best practices for the management of security controls in a holistic manner.

The [ISO/IEC 27001] standard is focused on measuring and evaluating how well an organization's ISMS is performing and includes a new section on outsourcing, which reflects the fact that many organizations rely on third parties to provide some aspects of IT. Overall, this standard is designed to fit better alongside other management standards [ISO 9000], [ISO/IEC 20000-1], and it has more in common with them. An introduction about ISMS family of standards and relationships is available in [ISO/IEC 27000] (last revised version of 2016). This standard is considered absolutely essential to users of [ISO/IEC 27001] standard; the remaining ISO 27000 standards are optional.

While ISM presents fundamentally the same risk management challenges in all contexts, the real-time nature of process control systems and the safety and environmental criticality make some of the challenges particularly extreme for organizations in the energy industry. The standard [ISO/IEC TR 27019] is intended to help organizations in the energy industry interpret and apply this standard [ISO/IEC 27002] in order to secure their electronic process control systems. Basically, this technical report provides additional more specific guidance on ISM than the generic advice provided in the [ISO/IEC 27002] standard.

The ISMS such as described by ISO/IEC 27000 series of standards have emerged in many organizations in the past decade. In the same time, organizations have adopted knowledge management (KM), another management discipline that aims to foster a more effective management of knowledge creation for innovations. Therefore the integration of ISMS and KM into a system allows organizations to improve their security programs and information security protection [Fung 2008]. This work proposes a research initiative to integrate KM and InfoSec together into a knowledge-centric InfoSec (KCIS) System. By adopting KCIS in phases, organizations can improve their InfoSec maturity level. More details about benefits and design of a such integrated system should be investigated by each organization and further applied where it is recommended.

Different KM tools can be employed to deal with information security issues including knowledge acquisition and knowledge sharing [Mittal 2010].

Similar models are supported by other organizations. Besides ISMS based on ISO/IEC 27000 series of standards, organizations could use other ISMS models promoted by different forum, associations, and consortium. Although there is no clear information how much these models are being currently used, another known ISMS model is the Information Security Management Maturity Model (known as ISM-cubed or ISM3).

11.7.2 Information Security Management Maturity Model (ISM3) Model

The ISM3 is another form of ISMS that builds on standards such as [ISO 20000], [ISO 9001] and general information governance and security concepts. The standard [ISO/IEC 21827] supports the Capability Maturity Model (CMM) for System Security Engineering (SSE-CMM). The ISM3 model can be used as a template for an ISO 9001-compliant ISMS. While ISMS defined in the [ISO/IEC 27001] standard is controls based, ISM3 model is process based and includes process metrics. ISM3 model is a standard for security management (how to achieve the organizations mission despite of errors, attacks, and accidents with a given budget). The difference between ISM3 and SSE-CMM is that ISM3 is focused on management, while SSE-CMM is focused on engineering.

ISM3 builds on successful principles from the field of quality management and applies these ideas to the field of information security. Also, an ISM3 model builds on international standard [ISO 20000] that supports IT service management. Figure 11.6 shows the relationships between IT service management and other frameworks. Security management can be provided as a service by a third-party service provider.

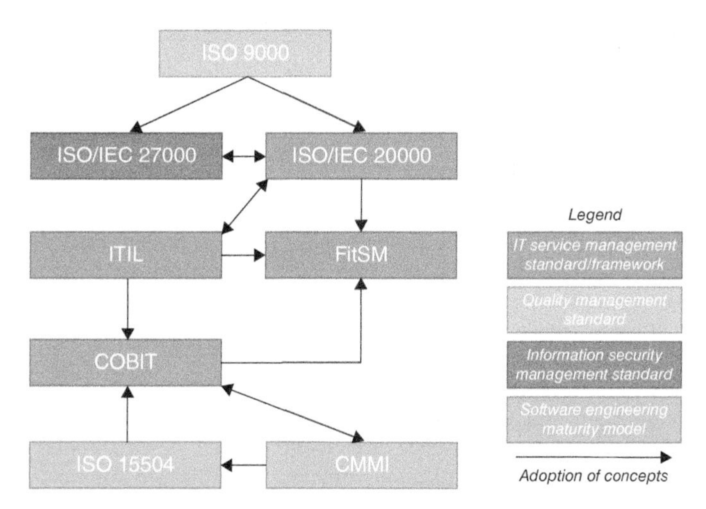

Figure 11.6 Relationships between ITSM frameworks and other standards. *Source:* [Brenner 2015]. Licensed under CC BY 4.0.

The ISM3 model is promoted by a consortium with members from many countries. It aims to provide a comprehensive approach toward effective management of information security. The ISM personnel have to continually improve their internal security management using metrics and maturity models.

The ISM3 defines security as context dependent. Traditionally, to be secure means to be resilient to any possible attack. Security in context means to be reliable, in spite of attacks, accidents, and errors. Traditionally, an incident is any loss of confidentiality, availability, or integrity. Under security in context, an incident is a failure to meet the organization's business objectives. There are three types of security objectives: the ones derived directly from business needs, the ones that are consequence of the regulatory environment, and the ones derived from the use of information systems.

The idea is that it should be a balance between business, compliance, and technical needs and limitations, like cost, functionality, privacy, liability, and risk [ISM3 2007]. The last version ISM3 v2.3 published in 2009 is based on a new approach that defines security maturity objectively as a direct result of the metrics used to manage information security processes. It looks at defining levels of security that are appropriate to the business mission and render a high return on investment. Implementations of ISM3 are compatible with the [ISO/IEC 27001] standard, which establishes control objectives for each process.

Also, the ISM3 model has a potential use in managing outsourced security processes. Many businesses find outsourcing of security management, either as a complete service or as selected services within it, to be an appropriate solution for their business model. For example, service-level agreements (SLAs) that use an ISM3 approach to operational metrics objectives and targets are specific and measurable.

11.7.3 BMIS Model

The Information Systems Audit and Control Association (ISACA) promotes a business-oriented approach to managing information security and a common language for information security and business management to talk about information protection. ISACA developed an approach based on the concepts of modeling and called Business Model for Information Security (BMIS) [BMIS]. While the original BMIS model was published some time ago by the research community [Kiely 2006a], [Kiely 2006b], the work is being taken forward by ISACA.

The model enables security professionals to examine security from a systems perspective, creating an environment where security can be managed holistically, allowing actual risks to be addressed. It addresses the security program at the strategic or business level. The security program exists not only to protect business information but also – and primarily – to support the business in reaching its objectives.

As an alternative to applying controls to apparent security symptoms in a cause-and-effect pattern, the BMIS model examines the entire enterprise system, allowing management to address the true source(s) of problems while maximizing elements of the system that can most benefit the enterprise [ISACA 2009], [BMIS]. Figure 11.7 is a simplified view of the BMIS model.

The diagram shows the three traditional elements of people, process, and technology and then adds a fourth node of organizational strategy and design to create a three-dimensional working model, best visualized as a pyramid. The connections between the nodes are shown as six dynamic interconnections (DIs), called tensions. These tensions are governance, culture, architecture, enabling emergency, and human factors. In order to advance the security issues, all of these interactions need to be further assessed, better measured, and much better understood.

Therefore, the model addresses the three traditional elements considered in IT management (people, process, and technology) and adds a critical fourth element (organization). In terms of the information security program, the flexibility and influence of elements and DI vary. Examples of DIs are culture (people, organization), governing (organization, process), architecture (organization, technology), emergence (people, process), enabling and support (process, technology), and human factors (people, technology). To obtain the maximum value from this model, it is important to understand that these DIs may be affected directly or indirectly by changes imposed on any of the other components within the model, not just the two elements at either end.

This model provides an in-depth explanation to a holistic business model that examines security issues from a systems perspective. The BMIS approach could be useful for utilities developing Smart Grid, which is a system of systems. The security problem is discussed as a real or potential compromise of safety/security and/or damage to tangible or intangible assets. These problems or breaches may be systemic (caused by the system or organizational design or culture), internal (coming from internal personnel), and/or external (coming from competitors, customers, press, terrorists, natural disasters, etc.).

Security needs to investigate not only the three-dimensional traditional concept (people, technology, process) with the added node of organizational design but also requires understanding of how people, process, technology, and organizational design all interact among themselves to create the complex mix of elements and issues.

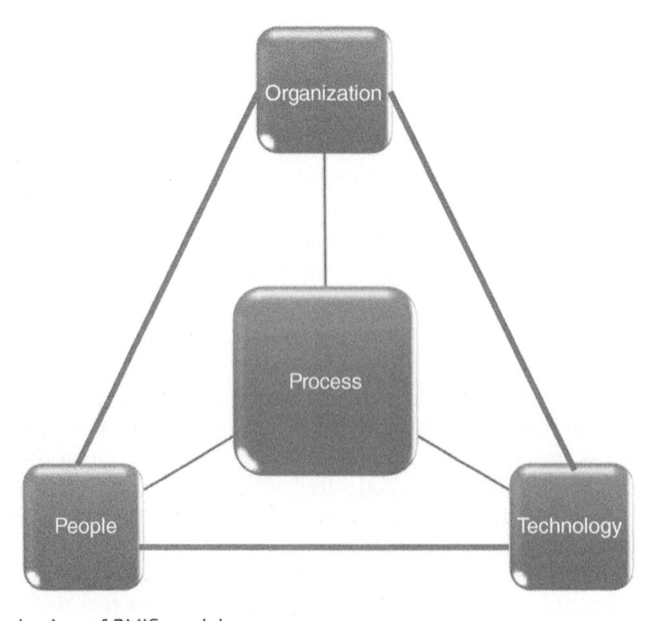

Figure 11.7 A simple view of BMIS model.

However, BMIS is primarily a model that must be supported by additional standards and frameworks. BMIS model has been designed to have the flexibility to adapt to most standards in use in information security.

11.7.4 Systems Security Engineering Capability Maturity Model (SSE-CMM)

The standard [ISO/IEC 21827] is developed by the International Systems Security Engineering Association (ISSEA). It is based on the Systems Security Engineering Capability Maturity Model (SSE-CMM) [Ferraiolo 2000]. The standard provides a roadmap for establishing and maturing security practices that include process areas (identify a comprehensive set of base security practices) and SSE-CMM model (e.g. capability levels define maturity).

Also, the standard is a process-driven framework and roadmap for assurance. It can help:

- Identify security goals.
- Assess security posture.
- Support security life cycle (identify risks, establish security requirements, implement controls, determine effectiveness).

The SSE-CMM describes the characteristics essential to the success of an organization's security engineering process and is applicable to all security engineering organizations including government, commercial, and academic. The standard does not prescribe a particular process or sequence, but captures practices generally observed in industry. The model is a standard metric for security engineering practices, activities, and project life cycles, including development, operation, maintenance, and decommissioning activities. It describes concurrent interactions with other disciplines, such as system software and hardware, human factors, test engineering, system management, operation, and maintenance. It also describes interactions with other organizations, including acquisition, system management, certification, accreditation, and evaluation.

The [ISO/IEC 21827] is a content-independent standard, which facilitates implementation of a good process for any set of security practices. Figure 11.8 depicts three life cycle processes (system, security, assurance) that share common activities.

One advantage of these processes is the presence of common activities among different processes that ensures consistency and avoidance of gaps. A system is designed, purchased, programmed, developed,

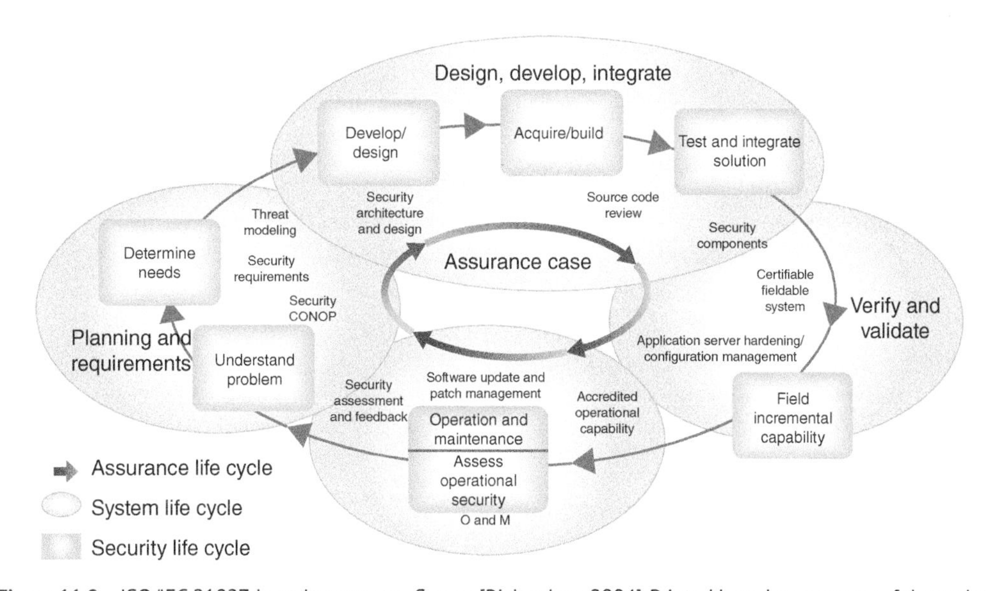

Figure 11.8 ISO/IEC 21827-based processes. *Source:* [Richardson 2004]. Printed based on courtesy of the author.

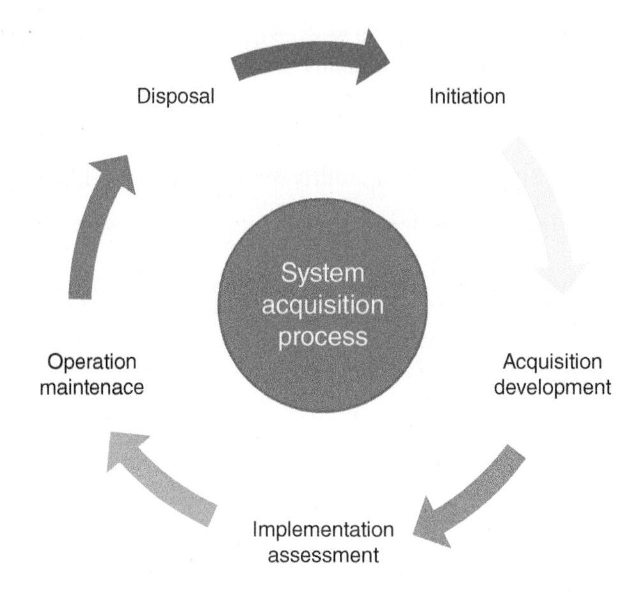

Figure 11.9 System acquisition/development and maintenance process.

or otherwise constructed. The [ISO/IEC 27002] standard provides specific information for the system acquisition, development, and maintenance. Figure 11.9 shows the cycle of this process.

In addition, a system development process may include other defined processes such as system acquisition of products (e.g. security technology controls, hardware, software, etc.) and maintenance. Before the system is actually developed or purchased, several activities should take place to ensure the end result meets the organization's needs. Examples of activities include risk assessment, security requirements analysis and design, security plan, security test, and evaluation plan.

11.7.5 Standard of Good Practice (SoGP)

One ISMS based on standard, the Standard of Good Practice (SoGP), is promoted by Information Security Forum (ISF) (https://www.securityforum.org). The standard can be used to build a comprehensive and effective ISMS. It is more best practice based as it comes from industry experiences. Best practices are used to maintain quality as an alternative to mandatory legislated standards and can be based on self-assessment or benchmarking. Best practice is a feature of accredited management standards such as [ISO 9000], [ISO 14001] standards.

The recent version is aligned with the requirements for an ISMS set described in ISO/IEC 27000 series standards and provides coverage of control topics, as well as cloud computing, consumer devices, and security governance. This standard aligns with ISO 27000 series of standards and other frameworks such as COBIT.

In addition to providing a tool to enable ISO 27001 certification, the standard provides full coverage of other frameworks (e.g. COBIT v5 topics, NIST cybersecurity framework) and offers alignment with other relevant standards and legislation for financial systems (e.g. Sarbanes–Oxley Act). The most recent standard update of 2014 covers current information security topics for critical infrastructure.

11.7.6 Examples of Other Frameworks

Different organizations support their own framework to develop a security program. Examples include COBIT 5, O-ISM3, SABSA, etc.

11.7.6.1 COBIT 5

Other methodologies that can be used to develop a security program (if one is not in place) include ISACA COBIT (Control Objectives for Information and Related Technology) [Lincke 2009].

COBIT framework facilitates alignment between general IT management and BMIS model as a convenient tool. Also, this model can be used to manage security of critical information infrastructure. This framework provides an international set of generally accepted IT control objectives for day-to-day use by business managers, IT professionals, and assurance professionals. Also, COBIT 5 builds on older version COBIT 4.1 by integrating other major frameworks, standards, and resources, including models for information security and information assurance.

In August 2016, COBIT 5 is the latest edition of ISACA's globally accepted framework, providing an end-to-end business view of the governance of enterprise IT that reflects the central role of information and technology in creating value for enterprises [COBIT 5]. The principles, practices, analytical tools, and models found in COBIT 5 embody thought leadership and guidance from business, IT, and governance experts around the world.

COBIT 5 builds and expands on older COBIT version by integrating other major frameworks, standards, and resources (e.g. ITIL and ISO/IEC 27001) including ISACA's BMIS.

11.7.6.2 Open Information Security Management Maturity Framework (O-ISM3)

The Open Information Security Management Maturity Model (O-ISM3) is based on standard that was published in 2011 by the Open Group [O-ISM3]. Building on ISM3 and other standards, this framework describes templates for managing information security processes in a wider context. It focuses on the common processes of information security, which to some extent all organizations share. The set of processes an organization may choose to use for their implementation depends on its security policy, reconciled with the resources they have available to invest in security controls, and operate their security management function. Every business has a unique context and resources.

O-ISM3 is an alternative to COBIT V5, and it aligns well with ITIL, the essential IT service delivery framework. White papers published via Web site (ISM3, http://www.ism3.com) suggest optimizing ISMS based on [ISO/IEC 27001] using O-ISM3 framework or using O-ISM3 with TOGAF or with SABSA and how to use O-ISM3 to implement the CPNI 20 Critical Security Controls for Effective Cyber Defense. Complementing the TOGAF model for enterprise architecture, O-ISM3 defines operational metrics and their allowable variances. Also, O-ISM3 Standard complements and extends ISO/IEC 27001 by adding further security management controls and applying security performance metrics [OpenGroup 2011]. These controls extend the capability of the ISO/IEC 27001 ISMS so that it will deliver specific measurements on ISMS performance against target business security objectives, so optimizing informed decision making on cost-effective ISMS investment that aligns.

11.7.6.3 Information Technology Infrastructure Library (ITIL)

In addition to specific information security standards, many enterprises have adopted and implemented more generic IT management guidance such as Information Technology Infrastructure Library (ITIL). The ITIL is a set of practices for information technology service management (ITSM) that focuses on aligning IT services with the needs of business. While these standards are much broader than just security, they often contain important components that need to be considered when working on information security. The security management process is based on the practices for ISM defined by ISO/IEC 27002 standard.

11.7.6.4 Sherwood Applied Business Security Architecture

Other organizations follow Sherwood Applied Business Security Architecture (SABSA) approach [Burkett 2012]. There are differences on these two approaches [Harris 2006a].

As discussed in this work, the main difference between former ISO/IEC 17799 standard that evolved in [ISO/IEC 27001] standard and SABSA framework is the focus and granularity. The ISO 27001 standard is security oriented on high level and provides best practices that can be used to determine what should be in a security program. SABSA is more business process oriented and detailed oriented [Harris

2006a]; therefore it is more suitable for designing the security of a system. Since this work was published, the 2005 version of [ISO/IEC 27001] standard was replaced by a revised version in 2014, which was revised again in 2016.

The SABSA framework does not work in the construct of individual areas defined by [ISO/IEC 27000] standards. It approaches all of these security issues through the specific levels of an organization. For example, when using access control, the standard [ISO/IEC 27002] lists the components that should be in an access control program (user registration, password management, node authentication, event logging, etc.). Instead of listing what should be in these individual areas, SABSA looks at access control through the following perspectives:

- Contextual – In the context of a company, how should one practice access control?
- Conceptual – What does access control program need to look like, and what is security engineer trying to accomplish?
- Logical – What pieces and parts are going to make up the access control program?
- Physical – What are the necessary standards, procedures, baselines, and process steps that the parts of the access program need to follow?
- Component – What products, tools, and personnel can be used in the access control program?
- Operational – Who is going to maintain the access control program and how?

The SABSA methodology uses different roles that work with these different perspectives such as:

- Business owner is equivalent to contextual.
- Architecture is equivalent to conceptual.
- Designer is equivalent to logical.
- Builder is equivalent to physical.
- Tradesman is equivalent to component.
- Facilities manager is equivalent to operational.

Basically, the SABSA framework is a methodology for developing risk-driven enterprise information security and information assurance architectures and for delivering security infrastructure solutions that support critical business initiatives. This methodology can be used both for both IT and operational technology (OT) environments [SABSA], although no information is available to support this statement.

Another benefit of SABSA framework is support for security governance. If a security program is only made up of physical, component, and operational mechanisms, there is no way that security governance is in place. Security governance requires all levels to be working in concert.

As the need to integrate security into business processes increases, and as governance and oversight is demanded from laws and regulations, then SABSA would be more used. Information security governance is about all of the tools, personnel, and business processes that ensure that security is carried out to meet an organization's specific needs. It requires organizational structure, roles and responsibilities, performance measurement, defined tasks, and oversight mechanisms. Information security governance is a coherent system of integrated security components (products, personnel, training, processes, policies, etc.) that exist to ensure that the organization survives and hopefully thrives.

Basically, SABSA is a framework and methodology for enterprise security architecture and service management. It was developed independently from the Zachman Framework but has a similar structure. SABSA methodology seems to be theoretical and academic in nature, which means that it can be difficult for many in the security field to understand and use [Harris 2006b]. Most people in management are not accustomed to working from an architectural level down to the component level. Also, most security professionals only work the physical, component, and operational levels when establishing a security program. These practices are causing gaps in security program management because many security issues related to other domains are not appropriately managed and addressed. However, a security program for Smart Grid systems should be managed with a higher level of quality and completeness, and the SABSA framework could be used to meet these expectations.

11.7.7 Combining Models, Frameworks, Standards, and Best Practices

A model is a schematic description of a system, theory, or phenomenon that accounts for its known or inferred properties and may be used for further study of its characteristics. Models are often used by enterprises to foster innovation and maximize the value generated through innovation or change. They can be used within an enterprise to translate strategy and mission into concepts and steps applying to processes or organizational entities.

Frameworks provide structure on how to build and manage a sound security program and ensuring its continued success via monitoring. In contrast to models, frameworks are usually normative rather than descriptive.

According to the BSI, a standard is an agreed, repeatable way of doing something. It is a published document that contains technical specifications or other precise criteria designed to be used consistently as a rule, guideline, or definition. An additional definition indicates that a standard is a basis for comparison, a reference point against which other things can be evaluated. These definitions align with how standards are seen in the information security community, providing information security professionals with a sense of direction and a way of benchmarking an enterprise's progress toward best practices [Murdoch 2006].

Best practices include a comprehensive set of approaches, solutions, rules, activities, policies, procedures, and plans; in traditional IT environment, best practices provide a balance of the need for information access and the need for adequate protection.

The standards most commonly used in the information security arena include the International Organization for Standardization (ISO) [ISO/IEC 27001] and the wider [ISO/IEC 27000] series (see [ISO 27k] for the most current versions), the NIST [NIST SP800-53r4], and the Payment Card Industry Data Security Standard (PCI DSS). The latter is an example of specific information security required for some of the processes that a financial services organization may be using. However, it often requires further context to enable and inform a cost-effective integrated implementation of PCI DSS [Murdoch 2006]. Also, the standard [NIST SP800-53Ar4] is focusing on federal information systems. NIST is also working with many public and private sector entities to establish mappings and relationships between the security standards and guidelines developed by NIST and the International Organization for Standardization and International Electrotechnical Commission (ISO/IEC).

While the options of approaches are quite reach, many definitions for information security supporting various perspectives may require to use a combination of approaches and best practices. As discussed in [Murdoch 2006], definitions of security in the literature vary according to the types of failure that are of concern. A wide range of technical and management activities are directed at managing security faults and failures. Measurement concepts are needed that bridge between specialist technical domains and integrated system management. The application of measurement principles has to address this diversity in an integrated such that a security measurement is closing the loop between decision makers and the effects of actions they undertake.

The security field is diverse and evolving rapidly to meet the dual challenges of net-centric systems and increasingly capable threat agents. Measurement concepts are needed that bridge between specialist technical domains and integrated system management. Management needs indicators of security properties that support the types of decision that have to be made at aggregated levels of systems and services. At the same time, security properties have to be evaluated as particular concerns and threats evolve.

An integrated approach to security measurement has been proposed, drawing on measurement, risk management, and systems concepts. The PSM TWG is bringing forward practical guidance materials on security measurement, informed by the concepts developed in this report.

Collaborative work is also in hand with the Measurement Working Group of ISSEA, in the context of the SSE-CMM, recent NIST measurement guidance [NIST SP800-55r1], and the ongoing development of the [ISO/IEC 27004].

The SGOP standard supports security management at the enterprise level. Enterprise is an organization with a defined mission/goal and a defined boundary, using information systems to execute that mission, and with responsibility for managing its own risks and performance. An enterprise may consist of all or some of the following business aspects: acquisition, program management, financial management (e.g. budgets), human resources, security and information systems, and information and mission management.

Organization is an entity of any size, complexity, or positioning within an organizational structure (e.g. a federal agency or, as appropriate, any of its operational elements) [FIPS 200].

Although best practices may help organizations to reduce some security efforts and activities, these may not be the best in every area of Smart Grid applications including DER systems. In fact, analysis and security requirements writing should never be bypassed.

For industrial control systems, there are good practices as the set of best industry practices, which have been shown to be effective through research and evaluation. In [CPNI 2008], one can find guidelines for applying best practices for security of industrial control systems. This work identifies good practice as being focused on key activities such as understand the business risks, implement secure architecture, and establish ongoing governance. Also good practices are provided for SCADA systems in [CPNI SCADA].

Even with the use of frameworks and standards, security professionals face challenges such as senior management's understanding of and commitment to information security initiatives, the involvement of information security in planning prior to the implementation of new technologies, integration between business and information security, alignment of information security with the enterprise's objectives, and executive and line management ownership and accountability for implementing, monitoring, and reporting on information security.

Management needs to define the security program based on common and effective methodologies. Different options may be available, and the decision could be made based on current practices in an organization or current methodologies that are commonly used and proven to be effective if there is no security program in place.

Another general issue is that combining so many tools there is room for confusion. This issue may occur when understanding or applying all these terms: model, architecture, standard, methods, methodology, and framework. Appendix A includes definitions as provided by different sources; often no universal definitions are available.

Usually, an organization may need to adapt or modify several frameworks to satisfy the needs of the business. Therefore, the chosen framework serves as the basis for the design, selection, and implementation of all subsequent security controls, including information security policies, security education and training programs, and technological controls. The continuous changes to these standards including new standards and norms have to be considered before any reference and material is used.

11.8 Standards, Guidelines, and Recommendations

Although there are no unique approaches dedicated to ISM at this time, there are many standards, best practices, and frameworks developed by various profit and nonprofit organizations. Many enterprises are using a number of recognized standards, models, and frameworks in information security. In contrast to models, frameworks are usually normative rather than descriptive. These are often seen as a subset of the wider range of general IT standards for governance, risk, compliance, or IT operations. Adherence to a standard usually means that significant effort has been put into following the structure and requirements, particularly in technology. Standards and guidelines that are relevant to ISMS developments are provided by organizations such as ISO, IEC, ISA, ISF, NIST, and ENISA.

Depending on which information security standards are used at the enterprise level, there should be an alignment of standards and frameworks periodically at regular intervals, as when a standard is updated or extended by the issuing organization. This could be the case where a series of standards feeds into one or two high-level norms, such as [ISO/IEC 27000].

12

Security Management for Smart Grid Systems

12.1 Strategic, Tactical, and Operational Security Management

One of the most important responsibilities of the security management team is security planning. This is a process of identifying an organization's immediate and long-term objectives and formulating and monitoring specific strategies to achieve them. It also entails staffing and resource allocation.

12.1.1 Unified View of Smart Grid Systems

Power grid information security and protection has aspects of both industrial control systems (ICS) and information technology (IT) systems. As discussed earlier, the security program for an organization in the electrical sector should support a unified view of Smart Grid systems that is based on an integration of activities for both traditional IT and control systems.

Although both ICS and IT systems require information security services to combat malicious attacks, the specifics of how these services are used for the power grid depend upon appropriate risk assessment and risk control methods. Distinct types of attacks targeting ICS and IT systems as well as different performance requirements of these systems determine a specific priority order of the security services implemented for each system.

Threat profiles of the power transmission and distribution management functions, where availability is paramount to all other security services, differ significantly from threat profiles of IT functions such as utility customer billing where privacy and integrity are a greater concern – hence warranting different security posturing. In addition, a structured approach should be applied to integration between security and safety in one domain model. The structured approach can act as an interface for active interactions in risk and hazard management in terms of universal coverage, finding solutions for differences, and contradictions between these two domains.

Implementation of cybersecurity guidance can occur through a variety of mechanisms, including enforcement of regulations and voluntarily in response to business incentives. Energy sector, specifically electrical sector organizations, can use several sources for designing and implementation of security programs and protection of information.

In [NSF 2011], it is mandated that cybersecurity practices must address not only the threats and vulnerabilities of traditional information systems (IS) but also issues unique to electric grid technology. These include the lengthy life expectancy of energy control systems, low-latency communications needed for real-time control, and differing requirements and regulatory frameworks among grid stakeholders. However, a long life expectancy could be an impossible goal for security technologies because of emerging needs due to increasing number of evolving vulnerabilities and threats and increasing sophistication of cyber attack methods. Therefore, Smart Grid requires a new level of cybersecurity, especially for these aspects.

Building an Effective Security Program for Distributed Energy Resources and Systems: Understanding Security for Smart Grid and Distributed Energy Resources and Systems, Volume 1, First Edition. Mariana Hentea.

Thus, a small shift in perception (from viewing data as a cost to regarding it as an asset) can dramatically change how an organization manages the data [Aiken 2007]. In addition, organizations need to implement greater legislative requirements to push software vendors into making their products more secure [Foley 2004], greater due diligence in transactions and business alliances, and coherent management strategies [Trope 2007].

Organizations are seeking to control information security, but they are applying this process in significantly different ways. First, many organizations are managing security in a somewhat inconsistent and superficial manner. Rather than taking a calculated or rational approach, there are situations emphasizing certain controls while leaving others, though no less important, poorly maintained.

Technical approaches alone cannot solve security problems for the simple reason that information security is not only a technical problem. It is also a social and organizational problem.

Organizations have to realize that many of information security problems extend far beyond technology and learn to appreciate the role that less technical controls, such as policy development, play in minimizing security breaches' impact on mission-critical operations.

Differences in perceptions of the value of policies and other management controls might explain some of the substantial disparity in control quality among organizations.

12.1.2 Organizational Security Model

Therefore, management needs to establish an organizational security model. An organizational security model is a framework made up of many entities; protection mechanisms; logical, administrative, and physical components; procedures; business processes; and configurations that all work together to provide a security level for an environment.

Each organizational security model is built to satisfy the needs and business objectives of an organization. The organizational model works in layers such that one layer provides support for the layer above it and protection for the layer below it. Figure 12.1 depicts a hierarchical view of the security model components that can be used by an organization in the electrical subsector. As shown in this figure, these components can make an organizational security model for the Smart Grid.

As shown in the hierarchical model, there are relationships such precedence among these components. For example, security requirements are derived based on assessment of vulnerabilities, risks, and threats. This model has various layers (one layer provides support for the layer above it and protection for

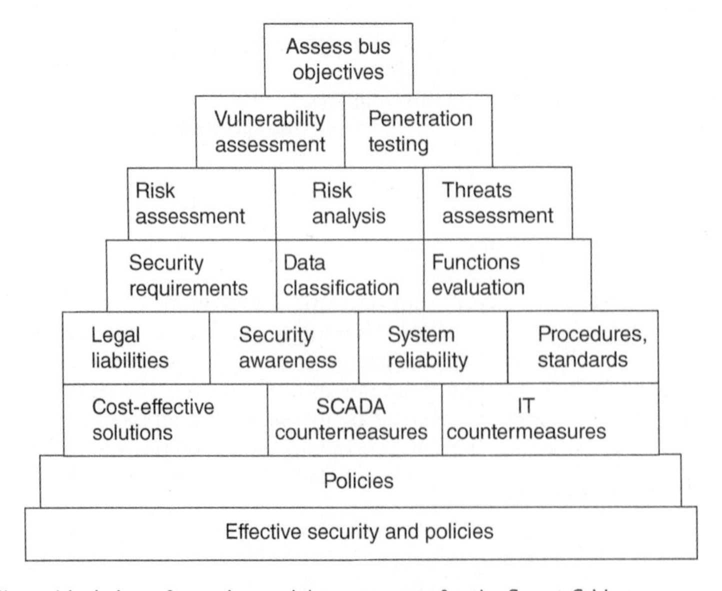

Figure 12.1 Hierarchical view of security model components for the Smart Grid.

the layer below it), but it also has different types of goals to accomplish in different time frames. The goals are operational (daily), tactical (short term), and strategical (long term). Thus, security planning should be broken into three different areas: operational, tactical, and strategical. This approach to planning is called the planning horizon. The many aspects of managing information security can also be classified as being strategically, tactically, or operationally oriented. Therefore, security is a business issue.

12.2 Security as Business Issue

A security program organized in layers should address security of assets from a strategic, tactical, and operational view as shown in Figure 12.2.

Whenever possible, management should combine multiple layers of security to ensure adequate coverage. At the same time, the plan needs to be unified [Whitman 2014]. That is, the separate elements of

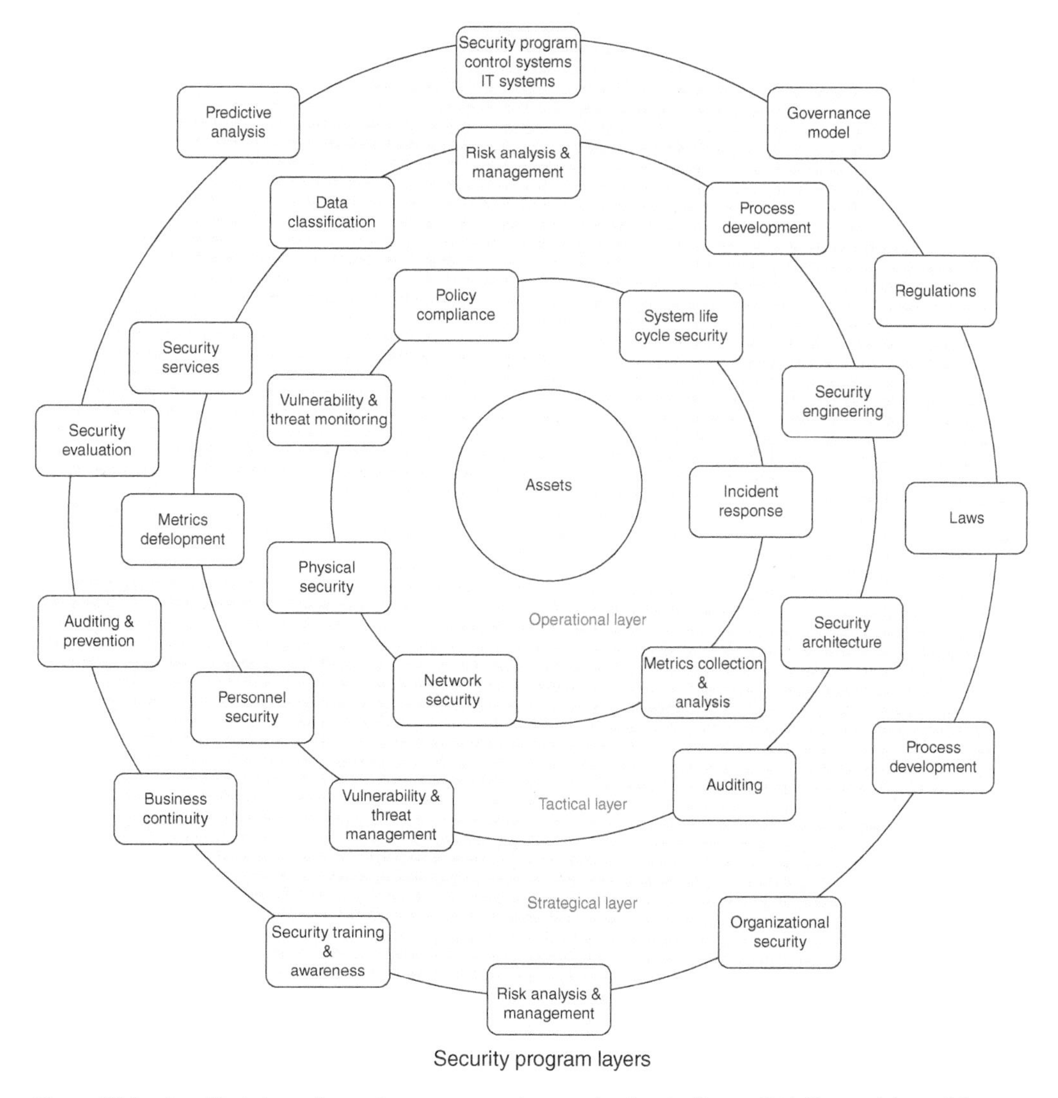

Figure 12.2 A unified view of security program and strategies for the Smart Grid. *Source:* Adapted from [Harris 2013].

the information security plan must be managed as a single effort or strategy based on risk management approach. Risk is the potential of an undesirable or unfavorable outcome resulting from a given action, activity, and/or inaction.

A successful information security management (ISM) plan must contain policies and procedures that cover multiple dimensions: hardware, software, and people as the primary components. A unified security management plan for Smart Grid systems and DER systems should include specific policies and procedures for IT and control systems.

12.2.1 Strategic Management

Strategic ISM addresses the role of information resources and information security infrastructure over the long term. It is focused on ensuring the organization has the infrastructure that it needs to achieve its long-range business goals and objectives. Data management, risk management, and contingency planning are other strategically oriented activities. Strategic management is accountable to stakeholders for the use of resources through governance arrangements. The customers of strategic management are therefore external (and possibly internal) stakeholders.

The practice of strategic management includes specific goals to fulfill responsibilities in respect of security that provides leadership and coordination of:

- Information security.
- Physical security.
- Workplace security.
- Interaction with organizational units.
- Improvement of information security management system (ISMS).
- Appointment of managers and internal and external auditors.
- Relationships with other organizations, such partners, vendors, and contractors.
- Providing resources for information security.
- Defining security objectives consistent with business goals and objectives, protecting stakeholders interests.

Essential to any business are new strategies [Johnson 2007] that should be applied to DER systems include:

- Protecting intellectual property.
- Use of security metrics that are shared across organization to help in better decision making.
- Investing in security from reactive add-ons to proactive initiatives that are aligned with the organization's strategic goals.
- Building a secure culture based on education and ongoing discussions about the security requirements for the Smart Grid and DER systems.

12.2.2 Tactical Management

Tactical ISM includes the translation of strategic security plans into more detailed actions. Tactical ISM involves the development of implementation plans and schedules for implementing new security controls. Other important tactical management functions include selecting vendors, training users and administrators, and developing follow-up evaluation and maintenance plans.

12.2.3 Operational Management

Operational ISM concerns the activities associated with managing day-to-day security operations of an organization. Best business practice models including log analysis do not always provide the greatest level of performance in the protection of the information.

12.3 Systemic Security Management

An approach to security management called systemic security management (SSM) is proposed in [Kiely 2006a]. This is a management approach to security that serves the extended enterprise but also partners, suppliers, customers, and communities. It is argued that often security issues have been studied too simplistically, as a three-dimension (process, technology, and people) or four-dimension (with added organization) concept, in a static or somewhat static collection of three independent dimensional issues. This work recommends investigating and understanding of how people, process, technology, and organizational design all interact among themselves to create that complex mix of elements and issues that the question of security really is.

The SSM approach is built around a set of core principles whose intent is to ensure an optimal balance of protection while maintaining the ability to share information and develop innovations among strategic partners. It is an approach designed to make it possible to do business in a highly integrated way, yet ensuring that digital assets, intellectual property, and proprietary technologies are protected [Kiely 2006b]. The primary focus of SSM approach is on prevention of potentially catastrophic security incidents through advanced technology, proactive planning, open communication, and workforce commitment. Organizations need to establish fully integrated policies where process is part of their culture, as well as the tensions of architecture, enablement, emergence, and human factors.

12.3.1 Comparison and Discussion of Models

The success of a security program is dependent on the implementation of the ISMS model and framework. Therefore, the selection of the appropriate security methodology and model is an important business objective for any organization. Given the multitude of models and frameworks for an ISMS, an organization needs to analyze and compare the benefits and drawbacks of current solutions including the promises of new trends.

While [ISO/IEC 27001] is controls based, ISM3 is process based and includes process metrics. Metrics are used to promote improvements that increase the value added by a process. ISM3 is a standard for security management on how to achieve organization's mission with a given budget despite of the occurrence of undesirable events such as errors, attacks, and accidents. Aspects of security engineering process and relations to other processes are defined by [ISO/IEC 21827] standard. The difference between ISM3 and ISO/IEC 21827 is that ISM3 is focused on management, while ISO/IEC 21827 is focused on engineering. ISM3 model is the basis of O-ISM3 framework, the most recent updated version.

Other frameworks such as ITIL and COBIT cover security issues but are mainly geared toward creating a governance framework for information and IT more generally. Both frameworks cover security issues.

There are some benefits and drawbacks that have to be understood when choosing a model. For example, when deciding about the ISM3 model and ISMS based on [ISO/IEC 27001] standard, there are differences (model maturity levels, organizational model, etc.). The following may be useful [ISM3 COMPARE]:

- If the optimum level of investment requires less far fewer controls than those required by [ISO/IEC 27001] standard, the organization cannot show the security assurance because it is not accreditable when using [ISO/IEC 27001].
- If an organization has limited resources, then it cannot show efforts in security by using [ISO/IEC 27001] standard because it does not include maturity levels.
- The ISM3 maturity levels allow to show progress toward better security management.
- The ISM3 maturity levels enable prioritizing investment, as processes are required in order of importance.
- With ISM3 the distribution of responsibilities is granular and specific (e.g. customer, process owner, supervisor, owner of assets), division of duties and reporting among strategic, tactical and operational levels, etc.
- The standard [ISO/IEC 27001] allows to detail the ownership and responsibilities of assets.

In the past few years, ISM has further evolved to include trust issues and other socio-cognitive phenomena. Therefore, ISM is increasingly relying on qualitative and quantitative models that primarily address the human factor.

12.3.2 Efficient and Effective Management Solutions

To deliver protection against the latest generation of cyber threats, the rules of preemptive protection have to meet criteria for effectiveness, performance, and protection. Effectiveness is a property of a target of evaluation (TOE) representing how well it provides security in the context of its actual or proposed operational use [RFC 4949].

ISM is the framework for ensuring the effectiveness of information security controls over information resources to ensure no repudiation, authenticity, confidentiality, integrity, and availability of the information [Hentea 2005b]. As ISMS is the tool used in ISM, critical success factors and capabilities for ISMS to be effective are summarized as follows [ENISA Factors]:

- Continuous and visible support and commitment of the organization's top management.
- Centrally managed based on a common strategy and policy across the entire organization.
- ISMS should be an integral part of the overall management of the organization related to and reflecting the organization's approach to risk management, the control objectives and controls, and the degree of assurance required.
- Security objectives and activities be based on business objectives and requirements and led by business management.
- Undertake only necessary tasks and avoiding over control and waste of valuable resources.
- Fully comply with the organization philosophy and mindset by providing a system that instead of preventing people from doing what they are employed to do, it will enable them to do it in control and demonstrate their fulfilled accountabilities.
- Be based on continuous training and awareness of staff and avoid the use of disciplinary measures and police or military practices.
- Be a never-ending process.

Effectiveness of security management system is also determined by the intelligence of the system, defined as the ability to detect unknown attacks with accuracy, along with enough time to strategically take action against intruders [Wang 2004], [Wang 2005a].

While information security practices are improving across the corporate world, not too many organizations implement a model that successfully protects while doing so in an efficient manner. Various international technical standards and frameworks are available as guidelines for managing information security, but there are no guidelines how to do efficiently in terms of costs and time. For example, a lack of clarity about how to operate efficiently means that many companies are wasting critical resources in lower-priority areas, resulting in insufficient resources to do the critical work of protecting the information and intellectual capital. Often, the efficiency of delivery of business applications and the choice of information technologies are in opposition to effective and efficient information security [Tipton 2006]. Study indicates that companies are emphasizing the effectiveness of their systems and processes, but few use consistent metrics to determine whether they are doing the job efficiently. Seventy-nine percent of respondents say their companies do not have a clear idea about how to define and, therefore, measure the efficiency of their information security processes [ATKERANEY].

Managing information security is not just restricted to maintaining confidentiality, integrity and availability (CIA) [Dhillon 2001a]. In addition to CIA, the responsibility, integrity, trust, and ethicality principles hold the key for successfully managing information security in the next millennium [Dhillon 2001b]. Several challenges need to be addressed toward an effective and efficient security management. Examples of these challenges include [Dhillon 2001a], [Dhillon 2000]:

- Often, solutions designed in a reactive manner still dominate.
- Controls tend to ignore other existing controls and their contexts; often controls have dysfunctional effects because isolated solutions are proposed for specific problems.

- Formal models for managing information security fall short of maintaining their completeness and validity in the commercial environment.
- Formal security models for maintaining the CIA of information cannot be applied to commercial organizations on a grand scale; models based on the military domain are bound to be inadequate for the commercial organizations.
- A security model enables enforcement of a security policy for particular situations.
- Organization structures are not the same for all enterprises; this means that the stated security policy for one organization is bound to be different from that of the other organizations.
- Organizations relying exclusively on risk analysis as a means to ensure information security tend to ignore all the other organizationally grounded IS security vulnerabilities and problems [Dhillon 2006].

While few cybersecurity maturity models exist that specifically address ICS, the Electricity Subsector Cybersecurity Capability Maturity Model (ES-C2M2) model specifically targets the energy critical infrastructure sector and ICS. C2M2 was developed by US Department of Energy (DOE). Other sectors could benefit from the development of similar maturity models tailored to their specifics [Luiijf 2015]. Maturity models are commonly used to review and benchmark an organization's IT capabilities and processes. A cybersecurity maturity model serves two purposes:

- Obtain an objective metric for their actual cybersecurity posture (level).
- Define milestones in an organization's roadmap toward improved cybersecurity capabilities.

Another model proposed by the World Economic Forum [WEF 2014] defines five stages of cybersecurity maturity as follows:

- Stage 1 – Unaware; organization is not managing cyber risk.
- Stage 2 – Fragmented; limited insight in risk management practices; siloed approach to cyber risk.
- Stage 3 – Top-down; CEO has initiated a top-down risk response program, but the organization does not view cyber risk management as a competitive advantage.
- Stage 4 – Pervasive; organization's management has understood the vulnerabilities, threats, controls, and interdependencies with third party and has developed policies and frameworks.
- Stage 5 – Networked; organization is managing cyber risk, highly connected to its peers and partners, sharing information, and mitigates cyber risk as part of their day-to-day operations.

These stages graphically represent the hyperconnection readiness curve, or the extent to which organizations are ready to address cybersecurity challenges in a hyperconnected world. The term hyperconnectivity not only pertains to technical interfaces but also mean the need for cooperation and information exchange with peers, suppliers, and operators of other dependent infrastructures.

Therefore, the hyperconnection readiness curve can be used for the ICS domain to assess whether an organization makes conscious decisions about the desired level of connectedness, and what the maturity level of the organization is, in terms of identifying and managing the related risk [Luiijf 2015].

12.3.3 Means for Improvement

Organizations must enhance the effectiveness and efficiency of their ISM. Solutions should be based on embedding professional information security measures into all business activities. Information security cannot be created through separate information security modules. The only tenable approach is its natural implementation into all business processes. Thus, integrating information security into the improved activities of the organization leads to a better balance between humans and technology [Anttila 2005].

Efficiency is the next step in the evolution of ISM. Efficient information security requires broad, cross-business collaboration that engages the whole company. It requires teamwork, decisive leadership, effective communication, and a culture of continuous improvement.

New norms and principles based on organizational theory, management science, and information science are recommended in managing information security [Dhillon 2000]. This work recommends solutions to the problem of managing information security in the new millennium need to shift emphasis from technology to business and social process. Also, organizations need to develop a focus on the pragmatic aspects in managing IS security. The way forward is to create newer models for particular aspects of the business for which information security needs to be designed. This would mean that micro-strategies be created for unit or functional levels.

Also, efficient ISM requires an intelligent system that supports security event management approach with enhanced real-time capabilities, adaptation, and generalization to predict possible attacks and to support human's actions [Hentea 2007]. There is also a growing need to extract and highlight the unusual traffic and unusual traffic patterns, real-time analysis and visualization, to reduce detection and reaction time.

Management is fundamentally about deciding and delivering behavior. Researchers want to model and manage the behaviors of hardware, software, and even users with a system. Behavior implies the ability to predict changes in a system, either changes made autonomously or in response to input (events or programming). However, behavior can be understood empirically or theoretically. In [Pras 2007], the authors argue that more research is required to investigate the relationship between behavior, economics, and uncertainty. The existing challenges of ISM combined with the lack of scientific understanding of organizations' behaviors call for better computational systems.

Several paradigms are needed to meet the requirements of the ISMS. This is broad and requires an intelligent approach. Collaborative efforts with academia, government, and commercial organizations could facilitate the implementation of the most promising paradigms for the development of ISMS. In addition, security industry has to give adequate attention on solving many problems related to software, hardware, and standardization [Hentea 2008]. New approaches based on intelligent techniques require implementing functions such as follows [Hentea 2007]:

- Management of heterogeneous devices and security technologies
  It is imperative to harness information models and ontologies to abstract away vendor-specific functionality to facilitate a standard way of aggregating and viewing the data.
- Adaptability
  One of the promises of autonomic operation is the capability to adapt the functionality of the system in response to changes in policies, requirements, business rules, and/or environmental conditions.
- Learning and reasoning capabilities to support intelligent decisions
  Statistics can be gathered and analyzed to determine if a given device is experiencing a cyber attack. This information must be inferred using a security knowledge base and other data and retained for future reference. There is a need to incorporate sophisticated, state-of-the-art learning and reasoning algorithms into ISMS.
- Control model
  Using closed-loop process control methods, we can more accurately set an acceptable limit of risk, build trust, and thus protect the organization more effectively.
- Decision making
  Using monitoring data, we can automate decision making and taking actions to control the behavior of the device, applications, or system, thus preventing cyber attacks.
- Intelligent assistant
  Security personnel can use the system to identify key areas where human intervention is needed or the human requires advice on making decisions. Human intervention will be required for the refinement of policies and also to resolve policy conflicts never before encountered by the system.
- Building knowledge
  The vision of intelligent systems for ISM is that of a self-managing security infrastructure that itself can access, or generate, the knowledge it requires to enable it to optimally react to changing of policies or operational contexts.

Intelligent systems emerged as new software systems to support complex applications. The architecture for an intelligent system for information security management (ISISM) is described [Hentea 2007]. Intelligent systems may include intelligent agents that exhibit a high level of autonomy and function successfully in situations with a high level of uncertainty. For example, the innovative solutions include new approaches for spam filtering using multiple filters and dynamic statistical analysis and evaluation of the email messages [Kim 2007]. There is progress on tracking the root of botnets, and traffic analysis tools can be used to identify other types of attacks, such as spam or click fraud [Gaudin 2007]. In the situation of automated threats, security researchers cannot combat malicious software using manual methods of disassembly or reverse engineering. Hatton states that in bounding software errors free, security vendors will have to do more than just use code signatures to recognize and stop malware [Hatton 2007]. Therefore, analysis tools must analyze malware automatically, effectively, and correctly without user intervention.

Also, key to the progress of ISM is addressing prevention of software vulnerabilities and developing a reliable infrastructure that can mitigate current security problems without end-user intervention [Shannon 2007].

12.4 Security Model for Electrical Sector

A security model and the method for applying the model (e.g. appraisal method) are intended to be used as a:

- Tool for engineering organizations to evaluate their security engineering practices and define improvements.
- Method by which security engineering evaluation organizations such as certifiers and evaluators can establish confidence in the organizational capability as one input to system or product security assurance.
- Standard mechanism for customers to evaluate a provider's security engineering capability.

A Capability Maturity Model (CMM) is used to evaluate and document process maturity for a given area. The term maturity relates to the degree of formality and structure, ranging from ad hoc to optimized processes. Developed in the mid-1980s, the model has since been adopted for subjects as diverse as information security, software engineering, systems engineering, project management, risk management, system acquisition, IT services, and personnel management. It is sometimes combined with other methodologies such as ISO 9001, Six Sigma, Extreme Programming (XP), and DMAIC [Greene 2014].

12.4.1 Electricity Subsector Cybersecurity Capability Maturity Model (ES-C2M2)

The ES-C2M2 can help electricity subsector organizations of all types evaluate and make improvements to their cybersecurity programs. The ES-C2M2 is part of the DOE Cybersecurity Capability Maturity Model (C2M2) program and was developed to address the unique characteristics of the electricity subsector. The ES-C2M2 model uses concepts of CMM.

A framework for electricity subsector is provided in [DOE 2014]. This document, along with several others, supports organizations in the effective use of the ES-C2M2 model. It introduces the model and provides the ES-C2M2's main structure and content.

The ES-C2M2 comprises a maturity model, an evaluation tool, and DOE-facilitated self-evaluations that are described in [DOE 2014].

The model was developed as a result of the administration's efforts to improve electricity subsector cybersecurity capabilities and to understand the cybersecurity posture of the energy sector. The ES-C2M2 includes the core C2M2 as well as additional reference material and implementation guidance specifically tailored for the electricity subsector.

A maturity model consists of a set of characteristics, attributes, indicators, or patterns that represent capability and progression in a particular discipline. The model thus provides a benchmark against which an organization can evaluate the current level of capability of its practices, processes, and methods and set goals and priorities for improvement.

To measure progression, maturity models typically have levels along a scale. The ES-C2M2 model uses a scale of maturity indicator levels (MILs) 0–3, which are described in the document. A set of attributes defines each level. If an organization demonstrates these attributes, it has achieved both that level and the capabilities that the level represents. Based on the measurable transition states between the levels, it is possible to:

- Define an organization's current state.
- Determine its future, more mature state.
- Identify the capabilities it must attain to reach that future state.

Each of the model's 10 domains contains a structured set of cybersecurity practices. Each set of practices represents the activities an organization can perform to establish and mature capability in the domain. The domains are risk management, situational awareness, information sharing and communications, event and incident response, continuity of operations, supply chain and external dependencies management, supply chain and external dependencies management, workforce management, and cybersecurity program management. For example, the risk management domain is a group of practices that an organization can perform to establish and mature cybersecurity risk management capability. Cybersecurity program management is the function to establish and maintain an enterprise cybersecurity program that provides governance, strategic planning, and sponsorship for the organization's cybersecurity activities in a manner that aligns cybersecurity objectives with the organization's strategic objectives and the risk to critical infrastructure.

Basically, the model is a common set of industry cybersecurity practices, grouped into ten domains and arranged according to maturity level. The ES-C2M2 evaluation tool allows organizations to evaluate their cybersecurity practices against ES-C2M2 cybersecurity practices. Based on this comparison, a score is assigned for each domain. Scores can then be compared with a desired score, as determined by the organization's risk tolerance for each domain.

The model builds on a number of existing cybersecurity resources and initiatives within the electrical sector. The model provides descriptive, not prescriptive, guidance to help organizations develop and improve their cybersecurity capabilities. As a result, the model practices tend to be abstract so that they can be interpreted for utilities of various structures, functions, and sizes. Many ES-C2M2 practices refer to assets that include both traditional and emerging enterprise IT assets and any ICS in use, including process control systems, supervisory control and data acquisition (SCADA) systems, and other operations technology (OT) systems.

The ES-C2M2 model is described in [DOE 2014] document. This model is a sector-specific version that includes the core C2M2 as well as additional reference material and implementation guidance specifically tailored for the electricity sector. The C2M2 model has already been adopted by many energy sector organizations. The model enables organizations to voluntarily share knowledge and effective practices using common terminology.

A framework on how to use C2M2 model is described in [DOE 2015b] guidance. Basically, this framework is based on ES-C2M2 model, a DOE and industry developed C2M2 model described in [DOE 2014].

The framework guide [DOE 2014] recommends to perform analysis and activities for the specification of security needs such that a common understanding of desired security posture is reached between all parties including consumers. Therefore, it is required that all members of the project team (e.g. management, power engineers, security professionals) have an understanding of security issues so they can perform their functions.

To gain understanding of a utility's security needs, security engineer has to:

- Identify applicable laws, policies, and constraints.
- Identify system security context.
- Capture security view of system operation.
- Capture security high-level goals.
- Define security related requirements (e.g. activities, processes, etc.).
- Obtain agreement from management and all stakeholders.
- Identify security inputs to include all system issues (e.g. security constraints, security alternatives, operational issues, monitor security posture, analysis of security events, configuration requirements, review security posture, manage security incident response, manage security program, protect security monitoring artifacts, etc.).

Figure 12.3 summarizes the recommended continuous cycle approach for using the model.

An organization performs an evaluation against the model, uses that evaluation to identify gaps in capability, prioritizes those gaps and develops plans to address them, and finally improves the model. Each of the model's 10 domains contains a structured set of cybersecurity practices. Each set of practices represents the activities an organization can perform to establish and mature capability in the domain.

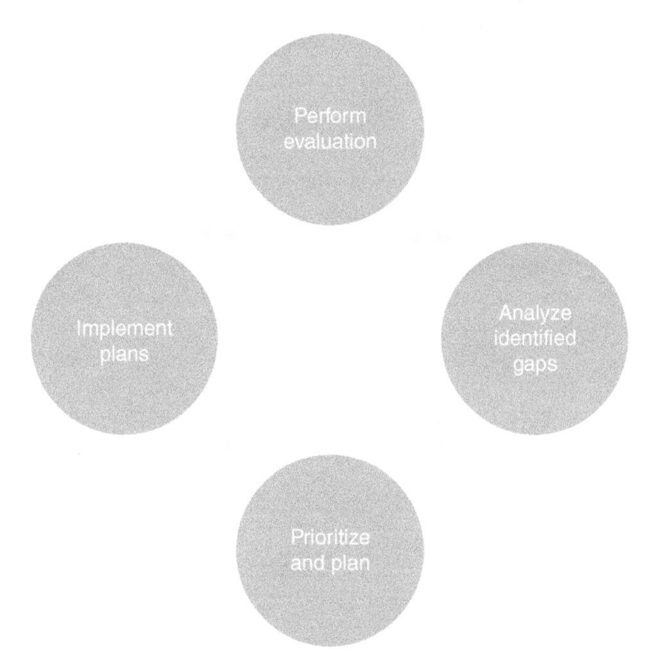

Figure 12.3 Recommended approach for using the RM model. *Source:* [DOE 2014]. Public Domain.

For example, the Risk Management domain is a group of practices that an organization can perform to establish and mature cybersecurity risk management capability.

The practices within each domain are organized into objectives, which represent achievements that support the domain. For example, the Risk Management domain comprises three objectives:

- Establish cybersecurity risk management strategy.
- Manage cybersecurity risk.
- Management practices.

The guidance on cybersecurity program supports the ongoing development and measurement of cybersecurity capabilities within the electricity subsector, and the model can be used to:

- Strengthen cybersecurity capabilities in the electricity subsector.
- Enable utilities to effectively and consistently evaluate and benchmark cybersecurity capabilities.
- Share knowledge, best practices, and relevant references within the subsector as a means to improve cybersecurity capabilities.
- Enable utilities to prioritize actions and investments to improve cybersecurity.

The ES-C2M2 model is designed for use with a self-evaluation methodology and toolkit (available by request) for an organization to measure and improve its cybersecurity program. The model provides descriptive rather than prescriptive industry focused guidance. Therefore, the content is presented at a high level of abstraction so that it can be interpreted by subsector organizations of various types, structures, and sizes.

This guidance is intended to address only the implementation and management of cybersecurity practices associated with IT and OT assets and the environments in which they operate. This guidance is not part of any regulatory framework and is not intended for regulatory use. The guidance is not intended to replace or subsume other cybersecurity-related activities, programs, processes, or approaches that electricity subsector organizations have implemented or intend to implement, including any cybersecurity activities associated with legislation, regulations, policies, programmatic initiatives, or mission and business requirements. The guidance is intended to complement a comprehensive enterprise cybersecurity program. Additionally, the model can inform the development of a new cybersecurity program.

12.4.2 Which Guidance and Recommendations Apply in Electrical Sector?

While no simple answer can be found to this question, the multitude of security issues that organizations face have to be addressed by following analysis and design principles. Therefore, several paradigms are needed to meet the requirements of ISM for DER systems. As with all management processes, an ISMS for DER systems must remain effective and efficient in the long term, adapting to changes in the utility center and external environment. In addition, comprehensive solutions based on guidance and recommendations could benefit the security programs in the electrical sector.

ISM should begin with the creation and validation of a security framework, followed by the development of an information security blueprint [Whitman 2014]. Frameworks provide structure on how to build and manage a sound security program that ensures continued success via monitoring and control. The framework serves as the basis for the design, selection, and implementation of all subsequent security controls, including information security policies, security education and training programs, and technological controls. The framework is the result of the design and validation of a working security plan that is then implemented and maintained using a management model such as ISMS described in ISO/IEC 27000 family of standards. Although frameworks are helpful, they are also at high level such that a security engineer needs to develop security blueprints.

The need for cybersecurity program and cybersecurity program management in electrical sector is documented in [DOE 2014]. Also, in response to administrative strategic call of 12 February 2013 [EO 13636], the Office of Electricity Delivery and Energy Reliability (OE) initiated a comprehensive approach to cybersecurity for the grid [DOE 2015a] and implementation of the National Institute of Standards and Technology (NIST) framework for improving the cybersecurity of critical infrastructure [NIST 2014]. Critical infrastructure consists of systems and assets, whether physical or virtual, so vital to the United States that the incapacity or destruction of such systems and assets would have a debilitating impact on security, national economic security, national public health or safety, or any combination of those matters.

12.4.3 Implementing ISMS

As defined in [ISO/IEC 27000] standard, an ISMS is a systematic approach for establishing, implementing, operating, monitoring, reviewing, maintaining, and improving an organization's information security to achieve business objectives. It is based upon a risk assessment and the organization's risk acceptance levels designed to effectively treat and manage risks. Analyzing requirements for the protection of information assets and applying appropriate controls to ensure the protection of these information assets, as required, contribute to the successful implementation of an ISMS. The following fundamental principles also contribute to the successful implementation of an ISMS:

- Awareness of the need for information security.
- Assignment of responsibility for information security.
- Incorporating management commitment and the interests of stakeholders.
- Enhancing societal values.
- Risk assessments determining appropriate controls to reach acceptable levels of risk.
- Security incorporated as an essential element of information networks and systems.
- Active prevention and detection of information security incidents.
- Ensuring a comprehensive approach to ISM.
- Continuous reassessment of information security and making of modifications as appropriate.

12.4.4 NIST Framework

DOE and the private sector stakeholders recognize that many organizations operate in multiple critical infrastructure sectors and as a result need alignment between the guidance developed by overlapping sector-specific agencies and associated cybersecurity approaches. To advance risk management strategies and to improve decision making, organizations of electrical sector are advised to identify risks and gaps in their current security posture. Guidance on the activities and the implementation of NIST framework in the electrical sector are described in [DOE 2015b]. This guide is based on a self-evaluation scoring method described in [DOE 2014] and NIST framework.

NIST framework is recommended to the critical infrastructure community that includes public and private owners and operators and other entities with a role in securing the nation's critical infrastructure. Members of each critical infrastructure sector perform functions that are supported by IT and ICS. This document is not a prescriptive guide, although the reliance on technology, communication, and the interconnectivity of IT and ICS has changed and expanded the potential vulnerabilities and increased potential risk to operations of critical infrastructure.

The framework enables organizations to apply the principles and best practices of risk management to improving the security and resilience of critical infrastructure. It provides an approach to cybersecurity by assembling standards, guidelines, and practices that are working effectively in industry today. Because it references globally recognized standards for cybersecurity, the NIST framework can also be used by organizations located outside the United States and can serve as a model for international cooperation on strengthening critical infrastructure cybersecurity.

By using the NIST framework, organizations can determine activities that are important to critical service delivery and can prioritize investments to maximize the impact of costs. Although not a one-size-fits-all approach, the framework is aimed at reducing and better managing cybersecurity risks for organizations that are part of the critical infrastructure. However, organizations could have unique risks, different threats, different vulnerabilities, and different risk tolerances, so the implementation of the practices in the framework may vary.

The framework is based on a risk-based approach to reducing cybersecurity risk. It is composed of three parts [NIST 2014]:

- Framework core – A set of cybersecurity activities and references that are common across critical infrastructure sectors and are organized around particular outcomes; it includes four types of elements identified as functions, categories, subcategories, and informative references.
- Framework profile – A representation of the outcomes that a particular system or organization has selected from the framework categories and subcategories.
- Framework implementation tiers – How an organization views cybersecurity risk and the processes in place to manage that risk.

Basic functions include identify, protect, detect, respond, and recover. Examples of categories include specific activities such as asset management and asset control. Examples of subcategories (specific outcomes of technical and/or management activities) include external IS that are catalogued and IDS notifications that are investigated.

However, the NIST framework is just a descriptive document of activities to be performed for a set of functions: identify, detect, protect, respond, and recover. The security compliance is left with each organization's desire and need to support cybersecurity of the assets. DOE implementation guide shows how to use a mapping of one tool – the C2M2 – to the framework. Other tools and processes are in active use, or in development, which may provide similar cybersecurity risk management capabilities.

While the framework provides broad coverage of the cybersecurity and risk management domains, an organization may have deployed standards, tools, methods, and guidelines that achieve outcomes not defined by or referenced in the framework leading to the opportunity to share the information with other organization within the sector or across sectors. The current profile should identify these practices as well.

Also, NIST document suggests organizations to use the framework as a guideline to setup some functions of new cybersecurity program or used for improving an existing program. However, it has a limited spectrum of functions that are required by an ISMS, and it does not have explicit functions for prevention, assurance, and measuring of the effectiveness of the program. NIST recommends the framework to be used as a means of expressing cybersecurity requirements to business partners and customers. However, the framework does not provide any approaches or methods on how to implement the security requirements. Although the framework provides a general set of considerations and processes for considering privacy and civil liberties implications in the context of a cybersecurity program, there could be more specific aspects that should be enforced. The framework suggests that technical privacy standards, guidelines, and additional best practices may need to be developed to support improved technical implementations of privacy programs. It is possible that cybersecurity activities may result in [NIST 2014]:

- Over-collection or over-retention of personal information.
- Disclosure or use of personal information unrelated to cybersecurity activities.
- Cybersecurity mitigation activities can result in denial of service or other similar potentially adverse impacts.
- Activities such as some types of incident detection or monitoring may impact freedom of expression or association.

Although the framework can help on identifying gaps in an organization's cybersecurity practices, finding the improvements is left with the organization.

12.4.5 Blueprints

Blueprints are important tools to identify, develop, and design security requirements for specific business needs [Harris 2013]. To design a security blueprint, most organizations follow established models and practices, after the framework is established. The model can be offered by a service organization. The model could be proprietary or based on open standards. Some of these models are available for high fees; others are relatively inexpensive, such as ISO standards, and some are free. Free models are available from NIST and other sources.

These blueprints must be customized to fulfill the organization's security requirements, which are based on its regulatory obligations, business drivers, and legal obligations. Blueprints are used to identify, develop, and design security requirements for a particular business solution. A blueprint lays out the security solutions, processes, and components the organization uses to match its security business needs. Blueprints must be applied to different business units within an organization. Not all aspects of a particular blueprint will apply, but all should be considered.

One example of a blueprint is identity management that should outline roles, registration management, authoritative source, identity repositories, single sign-on solutions, etc. For example, the identity management practiced in each of the different departments of an energy storage aggregator or service provider should follow the crafted blueprint. The blueprint allows for standardization, easier metric acquisition and collection, and governance. Blueprints can be tight together using standards.

Figure 12.4 shows blueprint solutions for a privacy program. It includes privacy plan and other related plans such as identity management plan, application integrity plan, infrastructure plan, business continuity management plan, and management plan. Infrastructure blueprint includes individual blueprints that support tailored requirements meeting the organization's specific requirements. The infrastructure blueprints are influenced by legal, regulatory, business, and IT drivers.

Organizations such as utilities need a systematic approach for ISMS that addresses security consistently at every level. They need systems that support optimal allocation of limited security resources on the basis of predicted risk rather than perceived vulnerabilities. Security cannot be viewed in isolation from the larger organizational context and only based on technology. Aspects of strategic, tactical, and operational activities should be employed and balanced to meet the power grid reliability and security goals of each organization. DER systems could be integrated with a utility information management system or managed by a third-party provider. ISM for DER systems is broad and requires an intelligent approach as well.

12.4.6 Control Systems

While ISMS requirements focus on business systems, the need to support the specific requirements of control systems is provided by ISO/IEC technical report [ISO/IEC TR 27019]. This report provides guiding principles based on ISO/IEC 27002 standard for ISM applied to process control systems as used in the energy utility industry.

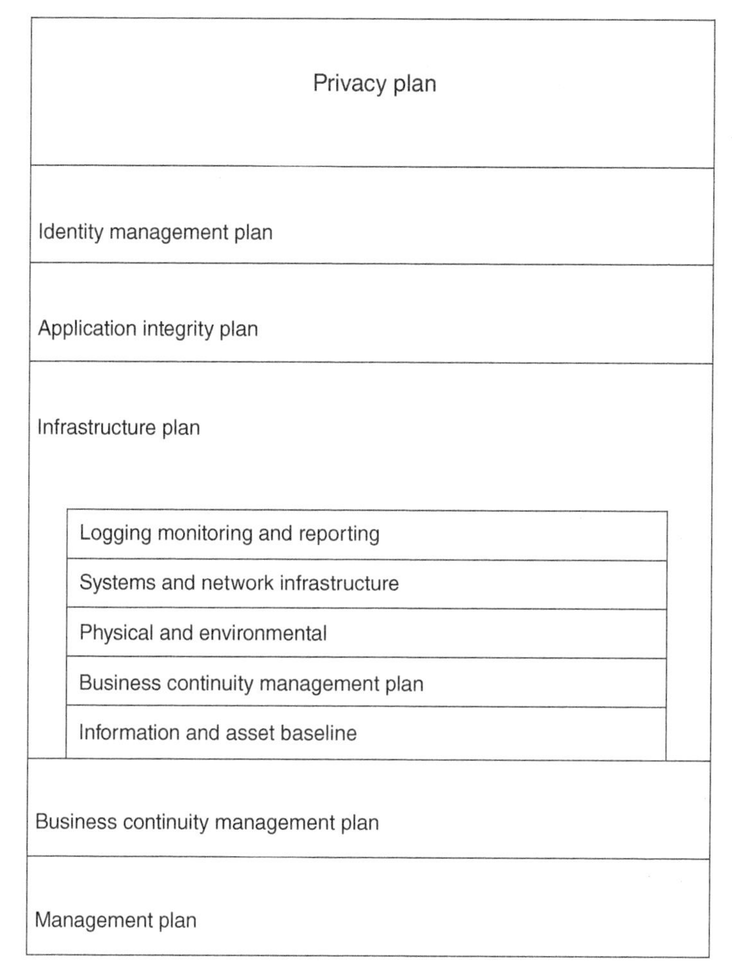

Figure 12.4 Example of blueprint solutions for privacy program.

The aim is to extend the ISO/IEC 27000 set of standards to the domain of process control systems and automation technology, thus allowing the energy utility industry to implement a standardized ISMS in accordance with ISO/IEC 27001 that extends from the business to the process control level. It covers process control systems used by the energy utility industry for controlling and monitoring the generation, transmission, storage, and distribution of electric power, gas, and heat in combination with the control of supporting processes.

12.5 Achieving Security Governance

Organizations today are facing constant and often profound change – from the marketplace, competitors, advancing technologies, and growing client expectations. In addition, global changes such as corporate governance reform, security concerns arising from terrorism, and increased malicious Internet activity have required organizations to be resilient in times of competition and uncertainty.

While the approach to information security may vary between organizations due to a difference in resources and business objectives, there is an underlying set of requirements that all organizations must follow in order to ensure the security of their information assets [Wang 2005b].

12.5.1 Security Strategy Principles

The seven basic principles of information security that must underpin the organization's strategy for protecting and securing its information assets that are recommended include the following [Wang 2005b]:

1) Information security is integral to enterprise strategy.
2) Information security impacts on the entire organization.
3) Enterprise risk management defines information security requirements.
4) Information security accountabilities should be defined and acknowledged.
5) Information security must consider internal and external stakeholders.
6) Information security requires understanding and commitment.
7) Information security requires continual improvement.

These principles allow organizations to better meet their obligations in achieving corporate governance requirements for information security, including legal and regulatory compliance. In addition, there is a need for a governance model and framework. These principles have been developed in line with global and national information security best practice and have been thoroughly reviewed and endorsed by the Australian IT Security Expert Advisory Group (ITSEAG*) [TISN 2007].

These principles and their recommendations form a baseline that allow critical infrastructure organizations to structure their information security governance program to ensure information security risks are consistently and appropriately addressed. However, the application of these principles may vary from organization to organization and industry to industry.

12.5.2 Governance Definitions and Developments

Governance is the process of managing, directing, controlling, and influencing organizational decisions, actions, and behaviors. Corporate governance has become the global benchmark, accepted in OECD and non-OECD countries. The Organization for Economic Cooperation and Development (OECD) states that governance should include the structure through which the objectives of the enterprise are set and the means of attaining those objectives and monitoring performance are determined [OECD 2015a]. First released in May 1999 and subsequently revised in 2004, the OECD Corporate Governance Committee conducted a further review of the OECD Principles of Corporate Governance. The review process started in 2014 and concluded in 2015 with a new revised document.

A subsequent OECD document [OECD 2015b] provides a coherent framework of eight interrelated, interdependent, and complementary high-level principles on digital security risk management and encourages organizations and governments to adopt these principles in their approach to digital security risk management. Also, principles are addressing more specifically the highest level of leadership to work toward an appropriate digital security risk management governance framework. A key aspect that the framework can address is the modalities to ensure that the business and information and communication technology (ICT) leadership within the organization work hand in hand to manage digital security risk.

A subset of corporate governance is information security governance. It is the process of establishing and maintaining a framework, supporting management structures and processes to provide assurance that information security strategies are aligned with and support business objectives.

It is argued that information security governance characterizes the fourth wave of information security development [von Solms 2006], which started in 2005. The fifth wave, which is called the cybersecurity wave, started in 2006 and is discussed in [von Solms 2010]. The waves represent new developments that started in a certain period and placed new emphasis on aspects related to information security and should therefore be seen as existing in parallel with each other.

The first three waves are characterized in [von Solms 2000]:

- Technical (up to about early 1980s), mainly characterized by a very technical approach to information security.
- Management (from about early 1980s to mid-1990s), characterized by a growing management realization of and involvement with the importance of information security, supplementing the technical wave.

- Institutional (best practices, from the last few years of the 1990s up to 2005), characterized by aspects like best practices and codes of practice for ISM, international information security certification, cultivating information security as a corporate culture, and dynamic and continuous information security measurement.

The fifth wave is characterized by concerns about the security of many Internet-based systems. While both fourth wave and fifth wave currently exist in parallel, the author argues that the fifth wave is the one that will really challenge information security specialist to start acting as information security professional, acting more like an engineer than a technician to install software or apply patches.

Security engineering should be established as a valid profession in the minds of the public and policy makers [Schneier 2012]. The author argues that amateurs produce amateur security, which costs more in dollars, time, liberty, and dignity while giving us less – or even no – security.

The purpose of security engineering is to solve engineering problems involving security. Examples of goals include:

- Plan and develop a security program.
- Determine customer security needs.
- Develop solutions and guidance on security issues.
- Coordinate activities with other engineering groups.
- Monitor and control security posture.
- Improve security program.

Figure 12.5 shows the diagram of the overall security engineering area.As emphasized in [Bayuk 2011], the trend should be to escape from best practices checklists and return to core systems engineering methods, processes, and tools.

Security engineering is practiced for:

- Preconcept definition.
- Concept exploration and definition.
- Demonstration and validation.
- Engineering, development, and manufacturing.
- Production and deployment.
- Operations and support.
- Disposal.

Security as a job focus started in corporate security. Most books, classes, and training emphasize the corporate side, and too few security professionals emphasize product security [Watson 2013]. This

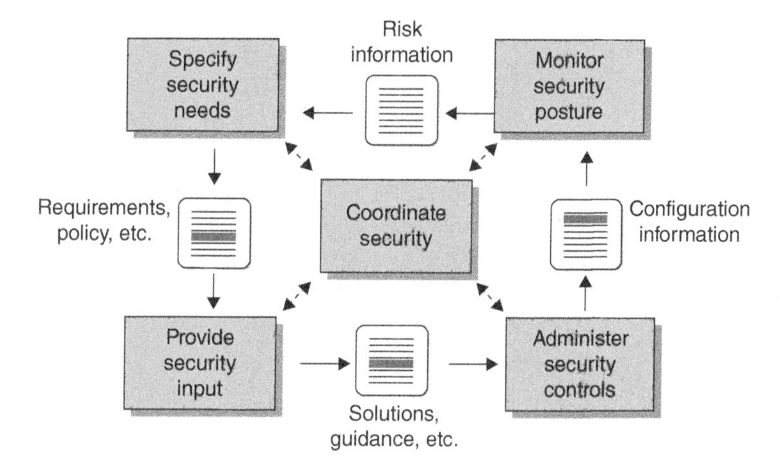

Figure 12.5 Security engineering model. *Source:* [Ferraiolo 2000]. Public Domain.

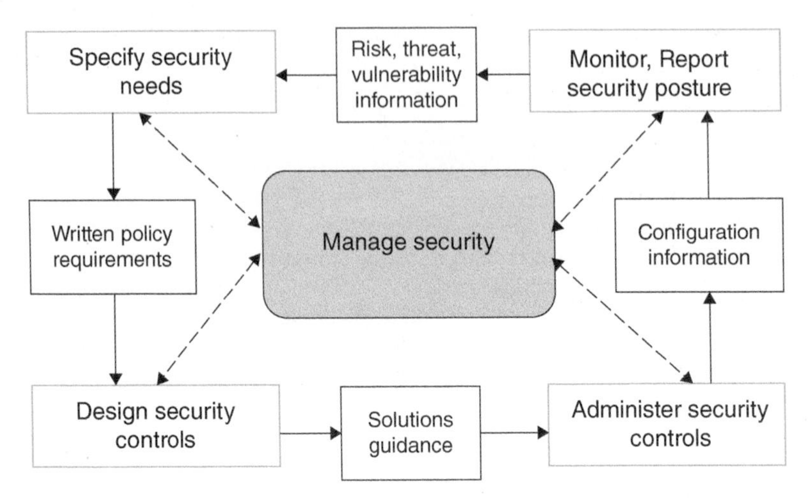

Figure 12.6 Security engineering activities. *Source:* Adapted from [Ferraiolo 2000].

requires that the product security and security engineering should rise to the front. The author argues that it should not be assumed that all security work is the same, because it is likely that one will overshadow the other, leaving either the product or the corporate information insecure. There could be situations where the product and the corporate realms may overlap, so responsibility should be designated accordingly. Figure 12.6 depicts key security engineering activities.

A framework for achieving information security governance is described in [TISN 2007].

A successful governance program enables the organization to effectively manage security during periods of change. A successful governance structure must define key security principles, accountabilities, and actions that an organization must follow to ensure their objectives are achieved [TISN 2007].

12.5.3 Information Security Governance

The business software providers argue that the lack of progress in security is due in part to the absence of a governance framework [BSA 2003]. A survey of existing governance and frameworks was initiated by Business Software Alliance as part of the project to define a governance framework. The report highlights issues such as the following [BSA 2003]:

- A management framework that instructs personnel at different levels about how to implement solutions is crucial, if progress is to be accelerated.
- No single document provides the necessary governance framework for information security.
- Existing guidance is either too detailed or not actionable in a comprehensive manner from the top to bottom of an organization.
- Various initiatives, both in the private sector and in government, have addressed various issues of security program management; these initiatives describe proposed management structures, give security checklists, and offer best practices and, in the case of government, legislation, but they do not offer a governance framework.
- Documents covering information security governance topics are classified into three categories:
  - Information security as a fundamental governance issue
  - Organizing for information security – essential program components
  - Governance documents under development

While the categories of documents that were examined in [BSA 2003] may cover these topics, the examined documents are not an exhaustive list of known published material on the topic. For example, guidance on the implementation of information security governance is provided in other sources such as [Brotby 2009]. Specific aspects of security governance related to SCADA and ICSs are discussed in [Haegley 2016]. Although there are many definitions for governance, the author clarifies that governance refers to interacting processes and decision-making processes among the actors who are collectively aiming to ensure and maintain the security of an ICS. Such processes ensure that benefits of ICS are delivered in a controlled manner and are aligned with long-term goals of the enterprise. However, this definition is kind of poor adaptation of the one definition – IT governance, provided in [Howe 2010].

Therefore, Haegley's definition of governance is not complete. It is limited to the discussion of the processes and documentation that includes policies, standards, guidelines, and procedures.

Therefore, other issues that are also relevant to information security governance are described further in this section. In this document [Allen 2007], the guidelines provide guidance on the implementation of an effective program to govern IT and information security. The document provides information that can help to make well-informed decisions about many important components of governing for enterprise security, such as adjusting organizational structure, designating roles and responsibilities, allocating resources (including security investments), managing risks, measuring results, and gauging the adequacy of security audits and reviews.

A successful governance framework is essential for the development of a "culture of security" within the enterprise [TISN 2007]. A governance structure must define key security principles, accountabilities, and actions that an organization must follow to ensure their objectives are achieved. Examples of recommendations describe that organizations are [Greene 2014]:

- Manage security risk.
- Implement and maintain security policies.
- Establish security roles and responsibilities.
- Educate staff members.

A governance framework is important because it provides a roadmap for the implementation, evaluation, and improvement of information security practices.

12.5.4 Implementation Challenges

It is observed a growing gap between the speed of technology adoption and that of security control implementation (see Figure 12.7). The governance framework described in [TISN 2007] provides a number of strategies for achieving strong security governance given this gap.

One of the most important features of a governance framework is that it defines the roles of different members of an organization. By specifying who does what, it allows organizations to assign specific tasks and responsibilities [BSA 2003].

Within IT governance, information security governance becomes a very focused activity, with specific value drivers: integrity of information, continuity of services, and protection of information assets.

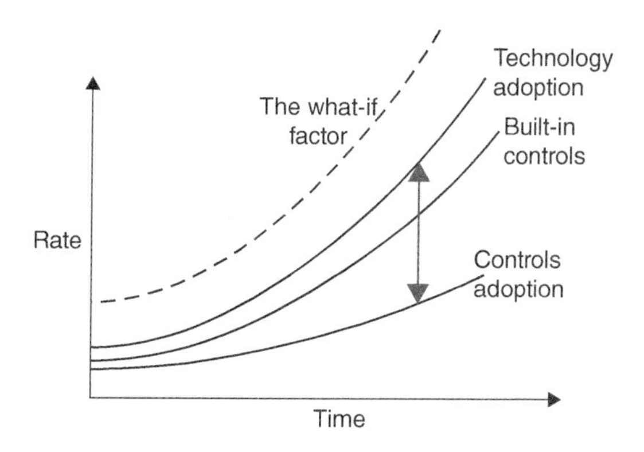

Figure 12.7 IT adoption vs. controls adoption. *Source:* [TISN 2007]. Licensed under CC BY 3.0.

Information security governance, when properly implemented, should provide basic outcomes such as [TISN 2007]:

- Strategic alignment.
- Security requirements driven by enterprise requirements.
- Security solutions fit for enterprise processes.
- Investment in information security aligned with the enterprise strategy and agreed-upon risk profile.
- Value delivery.
- Complete solutions, covering organization and process as well as technology.
- A continuous improvement culture.
- Performance measurement and defined set of metrics.

Also, the governing principle behind an ISMS is that an organization should design, implement, and maintain a coherent set of policies, processes, and systems to manage risks to its assets, thus ensuring acceptable levels of information security risk. As with all management processes, an ISMS must remain effective and efficient in the long term, adapting to changes in the internal organization and external environment.

However, for information security to be properly addressed, greater involvement of boards of directors, executive management, and business process owners is required. For information security to be properly implemented, skilled resources such as IS auditors, security professionals, and technology providers need to be utilized. All interested parties should be involved in the process. Sharing of information with those responsible for governance is critical to success.

12.5.5 Responsibilities and Roles

Governing for enterprise security means viewing adequate security as a nonnegotiable requirement of being in business. To achieve a sustainable capability, organizations must make enterprise security the responsibility of leaders at a governance level, not of other organizational roles that lack the authority, accountability, and resources to act and enforce compliance [Greene 2014]. Guidance and training for board members to reduce the risk of security abuses are also provided in this document [IT 2001].

To achieve a sustainable capability, organizations must make enterprise security the responsibility of leaders at a governance level, not of other organizational roles that lack the authority, accountability, and resources to act and enforce compliance [Greene 2014]. It is also needed to distinguish the skill and task requirements needed for different security topics. The skill and task requirements for designing in security may be vastly different from what is needed by a typical corporate security team. Therefore, a separate security team should be dedicated to product security [Watson 2013] because product security has a different focus than corporate security. Forrester recommends even more [Kark 2010], splitting into four security groups: security oversight, IT risk, security engineering, and security operations.

Organizational roles and responsibilities are defined in standards, guidelines, and several published materials (e.g. [Greene 2014], [von Solms 2010]) including regulations. Examples of regulations requiring organizations to assign security specialists for handling specific security activities include Gramm–Leach–Bliley Act (GLBA), Health Insurance Portability and Accountability Act/Health Information Technology for Economic and Clinical Health Act (HIPAA/HITECH), Payment Card Industry Data Security Standard (PCI DDS), and Standards for the Protection of Personal Information of the Residents of the Commonwealth (CMR). The regulation GLBA, known as the Financial Modernization Act of 1999, is a federal law enacted in the United States to control the ways that financial institutions deal with the private information of individuals; HIPAA Privacy Rule establishes national standards to protect individuals' medical records and other personal health information and applies to health plans, healthcare clearinghouses, and those healthcare providers that conduct certain healthcare transactions electronically. Also, the Sarbanes–Oxley Act of 2002 (SOX), which mandates strict reforms to improve financial disclosures from corporations and prevent accounting fraud, outlines requirements for IT departments regarding electronic records.

The implementation of privacy management programs, called for by in the revised Privacy Guidelines by OECD [OECD 2013], could benefit from being integrated into existing risk management frameworks and governance structures. Besides the privacy program concept, these guidelines introduce other concepts: a multifaceted national privacy strategy coordinated at the highest levels of the government and data security breach notification. Basic privacy principles were first published in [OECD 1980].

Risk management, governance, and information policy are the basis of an information security program. Policies related to these domains include the following policies: information security policy, information security policy authorization and oversight, CISO, information security steering committee, information security risk management oversight, information security risk assessment, and information security risk management. Security should be given the same level of respect as other fundamental drivers and influencing elements of the business [Greene 2014].

The objectives to safeguard critical energy infrastructure, satisfy regulatory requirements, and protect customer data and corporate reputations can only be accomplished with the full support and leadership of senior executives and a new approach to enterprise security governance in the energy sector. Recommendations for implementation of security governance in the energy sector are outlined in [EAC 2014]. The EAC recommends that DOE promote the establishment of enterprise security governance as a corporate norm in the energy sector in the near term.

Also, effective security requires the active involvement, cooperation, and collaboration of stakeholders, decision makers, and the user community. The factors that influence information security decision making and policy development include:

- Guiding principles.
- Regulatory requirements.
- Risks related to achieving their business objectives.

In global governance, the fundamental consideration is the control model.

12.5.6 Governance Model

A governance model is needed for managing a global IT organization that supports clear decision making, oversight, and visibility into what's happening across time zones and continents. The question is where the authority should reside, should IT authority reside centrally, locally, or in combination? [Pastore 2008]. The author suggests that criteria for the selection of a model are efficiency and effectiveness.

Although a centralized management model supports efficiency and a distributed model supports effectiveness, these models have their own drawbacks, so a better choice is a hybrid model that includes both models. Strategy and core processes must be the same globally. However, the strategy should always be executed in a way that makes sense for the location – a trademark of the hybrid model. The foundation of a distributed governance model is the principle that stewardship is an organizational responsibility.

At the base of ISM is the evaluation of information security posture and the improvement process. It is recommended to evaluate the enterprise information security maturity. This can be done by using a variation of the CMM [Greene 2014]. The CMM is a methodology used to develop and refine an organization's software development process. The model describes a five-level evolutionary path of increasingly organized and systematically more mature processes. The maturity level or capability level of an organization provides a way to characterize its capability and performance. Greene's model includes an extra level.

Figure 12.8 illustrates the functions of the security program and assessment results. After completing this table, a graph can be plotted to convey the state of the information security program on a per-domain basis [Greene 2014]. The results of evaluation are defined using a scale-based model shown in Table 12.1.

This author argues that this kind of assessment is an effective mechanism for reporting to those responsible for oversight, such as the board of directors or executive management. Process improvement objectives are a natural outcome of a CMM assessment.

A CMM assessment is an evaluation of process maturity for a given area. In contrast to an audit, the application of a CMM is generally an internal process. Audits and maturity models are good indicators of policy acceptance and integration.

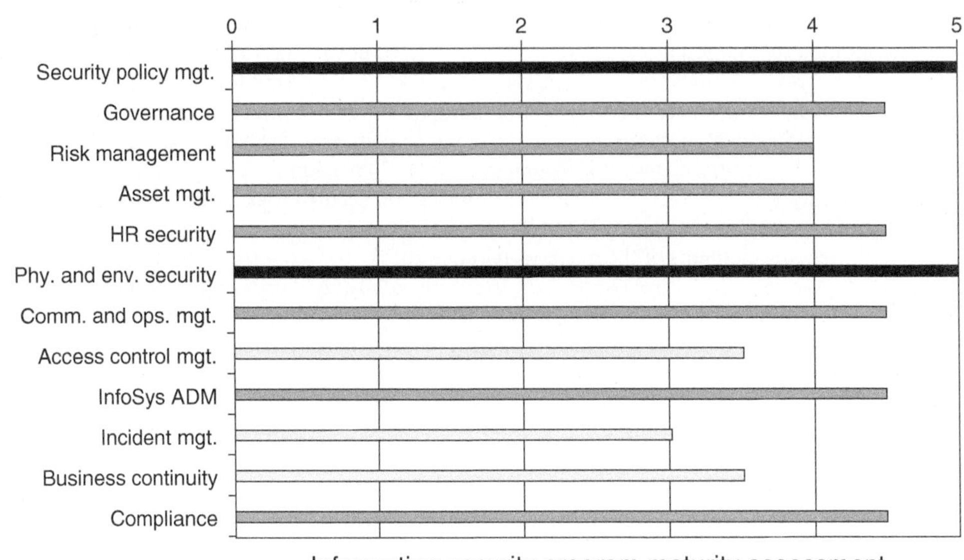

Figure 12.8 Example of information security program assessment. *Source:* [Greene 2014]. © 2014, Pearson.

Table 12.1 Capability maturity model (CMM) scale.

Level	State	Description
0	Nonexistent	The organization is unaware of the need for policies or processes
1	Ad hoc	There are no documented policies and processes; there is sporadic activity
2	Repeatable	Policies and processes are not fully documented; however, the activities occur on a regular basis
3	Defined process	Policies and processes are documented and standardized; there is an active commitment to implementation
4	Managed	Policies and processes are well defined, implemented, measured, and tested
5	Optimized	Policies and processes are well understood and have been fully integrated into the organizational culture

Source: [Greene 2014]. © 2014, Pearson.

The recommendations included in [EAC 2014] emphasize the need for the enterprise security program effectiveness for both physical and cyber assets. It should seek to drive a single holistic view of the security and regulatory posture.

The characteristics of effective security governance in energy industry are defined as follows [EAC 2014]:

- Clearly defined responsibilities from the board of directors to senior leadership to employees.
- Presence of an active security governance board composed of senior stakeholders from across the company.
- An executive owner of enterprise security: with purview over IT, OT, and physical security policy, designated CSO or similar.
- Striving for 100% alignment with of security with business/mission.
- Using measurement of key indicators to increase awareness and drive improvement (with maturity tools like the C2M2).

In addition, effective security governance must address issues of safety and assurance. To address the issues of safety, the scope of information security governance must be considerably broader than either IT security or IS. It becomes obvious that many of the functions that deal with aspects of safety must be integrated into the governance framework. Also, a strategy to increasingly integrate assurance functions over time allows for improving security, lowering costs, reducing losses, and helping to ensure the preservation of the organization and its ability to operate.

Alternatively, for most organizations, failure to implement effective information security governance will result in the continued chaotic, increasingly expensive, and marginally effective firefighting mode of operation typical of most security departments today [Brotby 2009]. One way of increasing good governance and trust is ensuring information assurance (IA).

12.6 Ensuring Information Assurance

IA includes measures that protect and defend information and IS by ensuring their availability integrity, authentication, confidentiality, and non-repudiation. These measures include providing for restoration of IS by incorporating protection, detection, and reaction capabilities [CNSSI 4009], [RFC 4949]. Assurance is the degree of confidence that security needs are satisfied.

IA is a continuous crisis in the digital world. The attackers are winning, and efforts to create and maintain a secure environment are proving not very effective. IA is challenged by the application of ISM, which is the framework for ensuring the effectiveness of information security controls over information resources. A summary of security management topics to ensure IA includes the following:

- Core security principles for the attributes such as confidentiality, integrity, availability, non-repudiation, etc.
- Functional requirements that define the security behavior of the IT or control system.
- Assurance requirements that establish confidence that the security function will be performed as intended.
- Information security blueprints that provide proven models for establishing cost-efficient, security-effective, and sustainable security business practices and solutions.
- Elevating the governance of the typical myriad assurance functions to the highest levels of the organization.
- Developing an assurance governance framework that integrates assurance functions with cybersecurity functions and other critical functions under a common strategy tightly aligned with and supporting business objectives.
- Developing and implementing a cybersecurity/IA measurements framework.

With the increasing dependence of the power grid applications on ICT, the need to protect information is increasingly important for electrical industry. The demand for products, systems, and services with which to manage and maintain information is therefore increasing, and the realization of superficial security controls is not sufficient. Therefore, it is necessary to apply a rigorous approach to the assessing and improvement of the security of products and processes that take place in the context of ICT.

As with any other process, security cannot be managed if it cannot be measured. The need for metrics is important for assessing the current security status, to develop operational best practices, and also for guiding future security research [Patriciu 2009].

While a number of cybersecurity/IA strategies, methods, and tools exist for protecting IT assets, there are no universally recognized, reliable, and scalable methods to measure the security of those assets. The security program's success in protecting and defending the organization is dependent on achievement of accurate cybersecurity and IA measurements. Also, it is needed to collect real-time measurements that would make possible to understand, improve, and predict the state of cybersecurity/IA.

For federal agencies, a number of existing laws, rules, and regulations cite IT performance measurement in general, and IT security performance measurement in particular, as a requirement. In the United States, several initiatives for security measurement programs were supported by the government and private organizations. The benefits of using measures, types of measures, and the relationship of the types of measures to the maturity of security program being measured are discussed in [NIST SP800-55r1].

Efficient measurement means automating metric production, consolidation, analysis, and presentation [Ayoub 2006]. Examples and components of cybersecurity/IA measurement tools typically fall into the following four categories:

- Integration (frameworks/platforms).
- Collection/storage.
- Analysis/assessment.
- Reporting.

While some organizations may choose to design, implement, and deploy their own security measures, the adoption of standards and guidelines for security measurement greatly improves the quality of an organization's measurement program and allows organizations to better share and improve their security postures. Commonly, security assurance is the degree of confidence that security needs are satisfied.

Several standards and guidelines documents have emerged over the years to address the challenge that many organizations face in developing cybersecurity/IA measurement programs. Generally, these standards and guidelines fall into the following categories [Bartol 2009]:

- Processes for developing information security measures to assess effectiveness of enterprise or system-level security controls and implementing the measures.
- Maturity model frameworks that provide a framework for assigning a maturity level to a grouping of security processes, based on specific criteria.
- Product evaluation frameworks that assess the level of assurance the products provide against a specific criteria and assigning a product evaluation level based on this criteria.

Security programs that do not have established processes and procedures, and where data needs to be collected manually, are likely to have greater difficulty collecting effectiveness/efficiency and impact measures. These programs are advised to focus on implementation measures.

As measurement programs mature, old measures that are no longer useful can be phased out and new measures can be introduced to continue monitoring and improving the status of cybersecurity/IA.

Programs with a greater number of institutionalized processes and some level of automated data collection tools are likely to be more successful in leveraging effectiveness/efficiency measures. These programs are also better equipped to move toward the business impact measures, which are more sophisticated than other types of measures.

Several standards, guidelines, and best practices documents that provide processes, frameworks, and metamodels for cybersecurity/IA measurement are described in [Bartol 2009]. They can be used by organizations to structure their programs and processes in a robust and repeatable way to facilitate long-term viability and success. The most common standards are [NIST SP800-55r1], [ISO/IEC 27004].

12.6.1 NIST SP800-55

The standard [NIST SP800-55r1] describes the benefits of using measures, types of measures, and the relationship of the types of measures to the maturity of security program being measured. It focuses on three key measurement categories:

- Implementation measures.
- Effectiveness/efficiency measures.
- Impact measures.

Also, the standard describes two primary processes:

- Measures implementation process (depicted in Figure 12.9).
- Measures development process (depicted in Figure 12.10), which serves as the first phase of measures implementation process.

Effectiveness/efficiency measures address two aspects of security control implementation results: the robustness of the result itself, referred to as **effectiveness**, and the timeliness of the result, referred to as **efficiency**. Effectiveness/efficiency measures help:

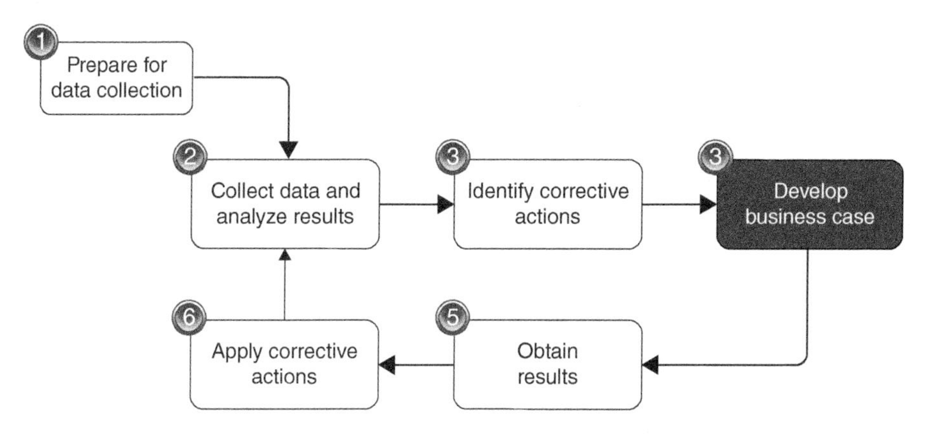

Figure 12.9 Information security measurement program implementation process. *Source:* [NIST SP800-55r1]. Public Domain.

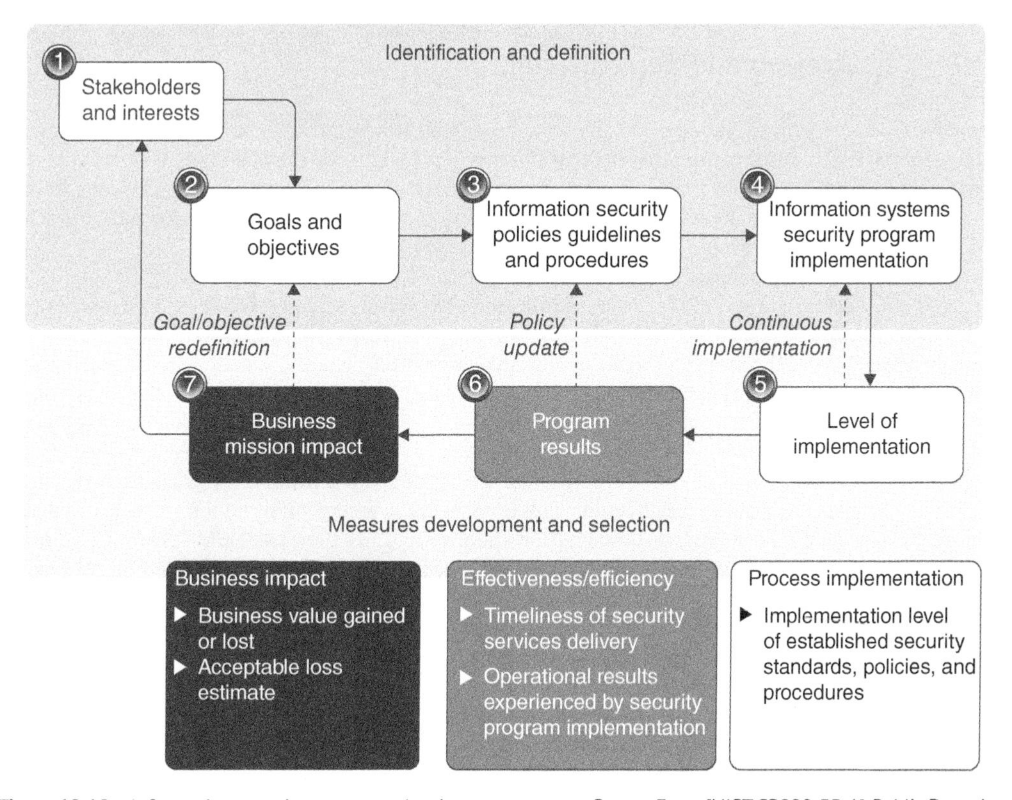

Figure 12.10 Information security measures development process. *Source:* From [NIST SP800-55r1]. Public Domain.

- Provide key information for information security decision makers about the results of previous policy and acquisition decisions.
- Offer insight for improving performance of information security programs.
- Determine the effectiveness of security controls.

12.6.2 ISO/IEC 27004

The standard [ISO/IEC 27004] does not contain requirements; rather, it contains recommendations. As an international standard, [ISO/IEC 27004] is being developed by a group of experts from across the globe. Many national standards bodies have provided contributions and inputs to amalgamate into a comprehensive solution for measuring the effectiveness of ISMS controls and of ISMSs. Among these inputs, ISO/IEC 27004 incorporates materials from [NIST SP800-55r1] standard and other nations' standards and guidelines.

The measures development and implementation processes used by [ISO/IEC 27004] are very similar to processes detailed in [NIST SP800-55r1], but the document uses ISMS terminology, rather than NIST terminology. The standard [ISO/IEC 27004] is also harmonized with [ISO/IEC 15939]; it uses the overall measurement process and the measurement model originally published in [ISO/IEC 15939].

The measurement process used in [ISO/IEC 27004] consists of these steps:

- Developing measures.
- Operating measurement program.
- Analyzing and reporting results.
- Evaluating and improving the measurement program itself.

The evaluation and improvement of the measurement program ensures that the program continues to be effective, and it is refreshed regularly or when the needs or operating environment changes.

12.7 Certification and Accreditation

Certification is a technical evaluation of the security capabilities and their compliance for the purpose of the accreditation. Certification ensures that the product is right for the customer. The certification provides documentation about the good and bad performance of the product. Certification is the technical testing of a system or product. Established verification procedures are followed to ensure the effectiveness of the security controls.

Accreditation is the formal acceptance of the adequacy of the overall security and functionality by management. Accreditation is the formal authorization given by management to allow a system to operate in a specific environment. By doing this, management confirms the understanding of the level of protection provided by the security product in its current environment and the security risks associated with installing and maintaining this product. The accreditation decision is based upon the results of the certification process. Security products have to continue to be certified and accredited, due to continuous changes of software, systems, and environments.

An ISMS may be certified compliant with [ISO/IEC 27001] by a number of accredited registrars worldwide. The ISO/IEC 27001 certification is like other ISO management system certifications and usually involves an auditing process. Some nations publish and use their own ISMS standards, e.g. the Department of Defense (DoD) Information Technology Security Certification and Accreditation Process (DITSCAP) of the United States, the Department of Defense Information Assurance Certification and Accreditation Process (DIACAP) of the United States, the German IT baseline protection, ISMS of Japan, ISMS of Korea, and Information Security Check Service (ISCS) of Korea.

Since software, systems, and environments continually change and evolve, the certification and accreditation should also continue to take place [Harris 2013]. Given the reality that an organization may be using different types of ISMSs, evaluation process can become cumbersome and expensive. Therefore, an approach such as an information security management evaluation system (ISMES) described in [Jo 2011] could help to reduce costs of information security evaluation and improve information security of the organization. The most known standards for certification and accreditation are Common Criteria (CC) [ISO/IEC 15408], [ISO/IEC 27001] standards.

12.7.1 Common Criteria

Some best-known ISMSs for computer security certification are the CC, International Standard [ISO/IEC 15408], and its predecessors such as Information Technology Security Evaluation Criteria (ITSEC) and Trusted Computer System Evaluation Criteria (TCSEC). This is an earlier standard supporting computer security certification. It consists of three parts (general model, security functional components, and security assurance components).

CC is a framework in which computer system users can specify their security functional and assurance requirements (SFRs and SARs, respectively) through the use of protection profiles (PPs), vendors can then implement and/or make claims about the security attributes of their products, and testing laboratories can evaluate the products to determine if they actually meet the claims. For purposes of security design, acquisition, and management of the IT products, it is recommended to use the specifications provided by all three parts (general model, security functional components, and security assurance components) of this standard.

CC provides assurance that the process of specification, implementation, and evaluation of a computer security product has been conducted in a rigorous and standard and repeatable manner at a level that is commensurate with the target environment for use.

For evaluating security attributes of computer systems and products, CC standard (ISO 15408) is being used in the United States and more globally. An evaluation is carried out on a product, and an Evaluation Assurance Level (EAL) is assigned. The range of assurance levels spans from EAL1, where functionality testing takes place, to EAL7, where system is formally verified and tested. The CC uses PPs in its evaluation process.

The PP contains the set of requirements, environmental assumptions, functions, and the assurance rating (EAL) level that the intended product will require. When the functionality of a system's protection mechanisms is being evaluated, the services that are provided to the subjects are examined and measured.

Assurance is the degree of confidence in the protection mechanisms and their effectiveness and capability to perform consistently. Once a security product achieves a specific rating, it only applies to that particular version and only to certain configurations of that product. Thus if an organization buys a firewall with a high assurance rating, the organization has no guarantee the next version of that software will have that rating. The next version will need to be evaluated. When the product is implemented in the real environment, factors other than its rating need to be addressed and assessed to ensure it is properly protecting resources and the environment.

CC can be perceived by a customer as metrics to evaluate trust; it can be perceived by a vendor as security capabilities to build in a product.

12.7.2 ISO/IEC 27001

An ISMS may be certified compliant with [ISO/IEC 27001] standard by a number of accredited registrars (also called accredited certification bodies or other names in some countries) worldwide. The ISO/IEC 27001 certification, like other ISO management system certifications, usually involves a three-stage audit process:

1) Informal review, for example, checking the existence and completeness of key documentation such as the organization's information security policy and risk treatment plan; it allows auditors to know the organization.
2) Detailed and formal compliance audit, testing the ISMS against the requirements specified in [ISO/IEC 27001]. The auditors will seek evidence to confirm that the management system has been properly designed and implemented and it is in fact in operation. Certification audits are usually conducted by ISO/IEC 27001 lead auditors. Passing this stage results in the ISMS being certified compliant with ISO/IEC 27001.
3) Follow-up reviews to confirm that the organization remains in compliance with the standard. Certification maintenance requires periodic reassessment to confirm that the ISMS continues to operate as specified and intended. These should happen at least annually but (by agreement with management) are often conducted more frequently, particularly while the ISMS is still maturing.

The [ISO/IEC 27001] standard is an internationally recognized standard for ISMS published worldwide in 2005 and revised in September 2013, which is known as [ISO/IEC 27001]. The revised [ISO/IEC 27001] new standard emphasizes more on measuring and evaluating how well an organization's ISMS is performing. Also, it includes information security related controls along with other requirements. Basically, the [ISO/IEC 27001] standard includes:

- A comprehensive set of controls comprising best practices in information security.
- A structure for continuous improvement for management and information security controls.

This standard is applicable in many types of industry and few areas where certified organizations in [ISO/IEC 27001] are demanded. These include:

- Finance and insurance.
- Manufacturing companies having research and development and critical information.
- Software development.
- Data processing.
- Banks and hospitals.
- Telecommunications.
- IT service sectors.
- Retail sectors.
- Manufacturing sector.
- Various service industries.
- Transportation sector.
- Government.

The certification follows a process [ISO Certification] as shown in Figure 12.11. The diagram explains the logical flow of the process itself including the ISMS development framework. The certification process starts after an organization makes the decision to embark upon the exercise. However, other standards should be used with the ISMS process. Examples of standards include:

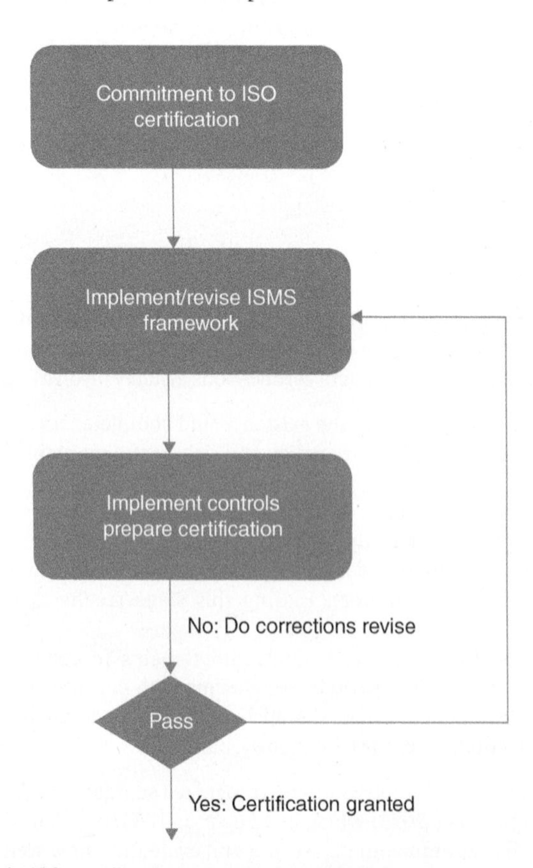

Figure 12.11 Key steps in ISO certification process with ISMS framework.

- [ISO/IEC 27006] is the accreditation standard that guides certification bodies on the formal processes they must follow when auditing their clients' ISMSs against [ISO/IEC 27001] standard in order to certify or register them compliant. The accreditation processes laid out in the standard give assurance that ISO/IEC 27001 certificates issued by accredited organizations are valid.
- [ISO/IEC 27007] provides guidance for auditing an ISMS against [ISO/IEC 27001] standard, in addition to the guidance contained in [ISO 19011] standard.
- [ISO/IEC 27008] provides guidelines for ISM auditing with respect to security controls. This differs from [ISO/IEC 27007] in that the latter is focused upon the management system (ISMS) itself, rather than specific controls. The objective of the standard itself is to provide requirements for establishing, implementing, maintaining, and continuously improving an ISMS.
- [ISO/IEC 27021] is a draft that plans to cover the knowledge, skills, and competencies requirements for ISM professionals that are certifying ISMS.
- [ISO 19011] standard provides guidance on auditing management systems, including the principles of auditing, managing an audit program, and conducting management system audits, as well as guidance on the evaluation of competence of individuals involved in the audit process, including the person managing the audit program, auditors, and audit teams.
- [ISO/IEC 27003] standard focuses on the critical aspects needed for successful design and implementation of an ISMS in accordance with [ISO/IEC 27001] standard. It describes the process of ISMS specification and design from inception to the production of implementation plans. It describes the process of obtaining management approval to implement an ISMS, defines a project to implement an ISMS (referred to in ISO/IEC 27003:2010 as the ISMS project), and provides guidance on how to plan the ISMS project, resulting in a final ISMS project implementation plan.
- [ISO/IEC 27011] provides guidelines supporting the implementation of information security controls in telecommunications organizations to meet baseline information security management requirements of confidentiality, integrity, availability, and any other relevant security property.
- [ISO/IEC 27015] provides guidelines in addition to the guidance given in the ISO/IEC 27000 family of standards for initiating, implementing, maintaining, and improving information security within organizations providing financial services. This technical report is a special supplement to [ISO/IEC 27001], [ISO/IEC 27002] standards for use by organizations providing financial services to support them in implementing ISMS and designing controls.
- [ISO/IEC 27017] provides guidelines for information security controls applicable to the provision and use of cloud services by providing additional implementation guidance and additional controls with implementation guidance that specifically relate to cloud services. This guidance serves both cloud service providers and cloud service customers.
- [ISO/IEC 27018] establishes commonly accepted control objectives, controls, and guidelines for implementing measures to protect personally identifiable information (PII) in accordance with the privacy principles in [ISO 29100] for the public cloud computing environment.
- [ISO/IEC TR 27019] provides guidelines for information security controls to be implemented in process control systems used by the energy utility industry for controlling and monitoring the generation, transmission, storage, and distribution of electric power, gas, and heat in combination with the control of supporting processes.

As certification is granted, there should be a continuous monitoring of changes in the security posture. More details about each step can be found in [ISO Certification].

12.7.3 ISMS Accreditation

Although not a legal requirement, accreditation is often stipulated and supported by local governments. A certification body applying for the accreditation of ISMS based on [ISO/IEC 27001] standard must conform to [ISO/IEC 17021] and other additional international requirements as detailed in specific requirements for accreditation for ISMS scheme. Several documents have to be submitted with the application for the accreditation to United Accreditation Foundation (UAF), a not-profit organization [UAF].

12.8 Standards, Guidelines, and Recommendations

Standards and guidelines that are relevant to ISMS developments are provided by organizations such as ISO, IEC, ISA, ISF, NIST, and ENISA.

12.8.1 ISO/IEC Standards

The ISO standards are developed by groups of experts, within technical committees (TCs). TCs are made up of representatives of industry, NGOs, governments, and other stakeholders, who are put forward by ISO's members. Each TC deals with a different subject, for example, there are TCs focusing on different topics such as Joint Task Committee 1 focusing on IT (ISO/IEC JTC 1 IT) for business and consumer applications. In accordance with ISO/IEC JTC 1 and the ISO and IEC councils, a list of freely available international standards is provided at [ISO freely]. Included in the list is [ISO/IEC 27000] standard.

The standard [ISO/IEC 27000] gathers in one place all the essential terminology used in the ISO/IEC 27001 family. The standard [ISO/IEC 27000] has been revised and replaced with newer editions.. It has been updated and extended to align with the revised version of ISO/IEC 27001 and other standards of the family that are currently under review. Relationships between the ISMS family of standards are illustrated in Figure 12.12.

12.8.2 ISA Standards

ISA99 remains the name of the Industrial Automation and Control System Security Committee of the International Society of Automation (ISA). Since 2002, the committee has been developing a multipart

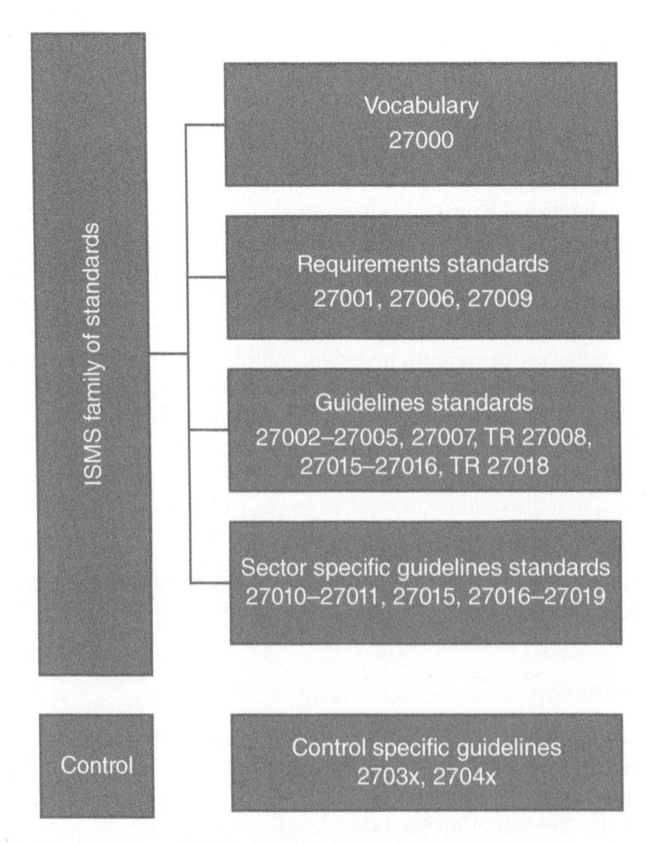

Figure 12.12 ISO/IEC 2700 series – ISMS family of standards relationships.

series of standards and technical reports on the subject. These work products are then submitted to the ISA approval and publishing under the American National Standards Institute (ANSI). They are also submitted to IEC for review and approval as standards and specifications in the IEC 62443 series [ISA99].

12.8.2.1 ISA/IEC 62443 Standards

Formerly ISA-99, the ISA/IEC 62443 is a series of standards, technical reports, and related information that define procedures for implementing electronically secure industrial automation and control systems (IACS). This guidance applies to end users (e.g. asset owner), system integrators, security practitioners, and control systems manufacturers responsible for manufacturing, designing, implementing, or managing IACS.

These documents were originally referred to as ANSI/ISA-99 or ISA99 standards, as they were created by the ISA and publicly released as ANSI documents. In 2010, they were renumbered to be the ANSI/ISA-62443 series. This change was intended to align the ISA and ANSI document numbering with the corresponding International Electrotechnical Commission (IEC) standards.

Although the ISA work products are using the convention ISA-62443-x-y and previous ISA99 nomenclature is maintained for continuity purposes only, corresponding IEC documents are referenced as IEC 62443-x-y. The approved IEC and ISA versions are generally identical for all functional purposes. All ISA-62443 standards and technical reports are organized into four general categories called General, Policies and Procedures, System, and Component.

The Technical Committee of IEC, TC 65, prepares and coordinates the international standards for industrial process measurement and control concerning continuous and batch processes of the systems operating with electrical, pneumatic, hydraulic, mechanical, or other systems of measurement and/or control.

The status of the various work products in the ISA/IEC 62443 series of IACS standards and technical reports are available on the [ISA99] Web site.

The standards are organized in groups and many are under development. Specific standards of the IEC 62443 series include [ISA SCADA]:

- IEC/TS 62443-1-1 provides general model concepts and definitions.
- IEC 62443-1-2 is a master glossary of terms used by ISA committee.
- IEC 62443-1-3 identifies a set of compliance metrics for IACS security.
- IEC 62443-2-2 addresses how to operate an IACS security program.
- IEC 62443-3-1 addresses how to setup an IACS security program.
- IEC 62443-3-2 addresses how to define security assurance levels using the zones and conduits concept.
- IEC 62443-4-1 addresses the requirements for the development of secure IACS products and solutions.
- IEC 62443-4-2 series address detailed technical requirements for IACS components.
- IEC TR 62443-3-1 is a technical report on the subject of suitable technologies for IACS security.

12.8.2.2 ISA Security Compliance Institute (ISCI)

Related to the work of ISA 99 is the work of the ISA Security Compliance Institute (ISCI), a not-for-profit automation controls industry consortium. The ISCI has developed compliance test specifications for ISA99 and other control system security standards.

They have also created an ANSI-accredited certification program called ISASecure for the certification of industrial automation devices such as programmable logic controllers (PLC), distributed control systems (DCS), and safety instrumented systems (SIS). These types of devices provided automated control of industrial processes such as those found in the oil and gas, chemical, electric utility, manufacturing, food and beverage, and water/wastewater processing industries. There is growing concern from both governments and private industry regarding the risk that these systems could be intentionally compromised by hackers, disgruntled employees, organized criminals, terrorist organizations, or even state-sponsored groups.

ISASecure independently certifies industrial automation and control (IAC) products and systems to ensure that they are robust against network attacks and free from known vulnerabilities. The ISASecure program is based upon the IAC security life cycle as defined in [IEC 62443-1-1] standard. At this time, the scope of the ISASecure certifications includes assessment of off-the-shelf IAC products and IAC product development security life cycle practices [ISASecure].

12.8.3 National Institute of Standards and Technology (NIST)

The NIST provides several standards and guidelines for security management:

- [NIST SP800-12] provides a broad overview of computer security and control areas. It also emphasizes the importance of the security controls and ways to implement them. Initially this document was aimed at the federal government although most practices in this document can be applied to the private sector as well. Specifically, it was written for those people in the federal government responsible for handling sensitive systems.
- [NIST SP800-53Ar4] provides advice on how to manage IT security; it emphasizes the importance of self-assessments as well as risk assessments; it provides recommendations regarding security and privacy controls for federal information systems and organizations.
- [NIST SP800-37r1] provides a risk management approach to federal information systems.
- [NIST SP800-53r4] provides recommendations regarding security and privacy controls for federal information systems and organizations.

12.8.4 Internet Engineering Task Force (IETF)

The Internet Engineering Task Force (IETF) provides recommendations for developing security policies and procedures for IS connected on the Internet:

- [RFC 2196] provides a general and broad overview of information security including network security, incident response, or security policies. The document is very practical and focusing on day-to-day operations.

12.8.5 ISF Standards

The Information Security Forum (ISF) is an independent, not-for-profit association of organizations from around the world. It is dedicated to investigating, clarifying, and resolving key issues in information security and developing best practices methodologies, processes, and solutions that meet the business needs of its members.

Among other programs, the ISF offers its member organizations a comprehensive benchmarking program based on the [SOGP] standard.

12.8.6 European Union Agency for Network and Information Security Guidelines

The European Union Agency for Network and Information Security (ENISA) is the center of expertise for cybersecurity in Europe. ENISA is contributing to a high level of network and information security (NIS) within the European Union, by developing and promoting a culture of NIS in society to assist in the proper functioning of the internal market. Examples of ENISA published guidelines and recommendations for the security of ICS include:

- [ENISA 2011] provides an overview of existing methods, procedures, and guidelines in the area of control system (cyber) security. It takes into account activities of international organizations and important national activities in Europe and the United States (as far as the consortium was aware of these activities).
- [ENISA 2015a] is a study that reveals the current maturity level of ICS-SCADA cybersecurity in Europe and identifies good practices used by European member states to improve this area.
- [ENISA 2015b] is a report that proposes a series of recommendations for the development of cybersecurity certifications for ICS/SCADA professionals.

12.8.7 Information Assurance for Small Medium Enterprise (IASME)

The Information Assurance for Small Medium Enterprise (IASME) is a UK-based consortium and an accreditation body for assessing and certifying against the government's Cyber Essentials Scheme. IASME provides criteria and certification for small to medium business cybersecurity readiness. It also allows small to medium business to provide potential and existing customers and clients with an accredited measurement of the cybersecurity posture of the enterprise and its protection of personal/business data. The audited IASME certification is also seen as showing compliance to ISO27001 by an increasing number of companies. The consortium developed the IASME Governance standard.

12.8.7.1 IASME Governance Standard

The IASME Governance standard, based on international best practice, is risk-based and includes aspects such as physical security, staff awareness, and data backup. The IASME standard was recently recognized as the best cybersecurity standard for small companies by the UK government [IASME]. The IASME standard follows the same implementation pattern used by the international standards community including the ISMS, which provides a management framework.

The standard demonstrates baseline compliance with the international standard [ISO/IEC 27001]. Although the international standard is comprehensive, it is extremely challenging for a small organization to achieve and maintain an ISMS. The IASME standard is following the same methodology as the [ISO/IEC 27001] standard but specifically for small organizations.

References Part 5

[Abas 2014] Abas, K., Porto, C., Obraczka, K. (2014). Wireless smart camera networks for the surveillance of public spaces. *Computer, 47*(5), 37–44. https://doi.org/10.1109/MC.2014.140

[Allen 2007] Allen, J.H., Westby, J.R. (2007). *Governing for Enterprise Security (GES) Implementation Guide* [Technical Note, CMU/SEI-2007-TN-020]. Software Engineering Institute. https://doi.org/10.1184/R1/6574010.v1 http://www.cert.org/historical/governance/implementation-guide.cfm

[Anderson 2003] Anderson, J. M. (2003). Why we need a new definition of information security. *Computers & Security, 22*(4), 308–313. Elsevier. https://doi.org/10.1016/S0167-4048(03)00407-3

[Aiken 2007] Aiken, P., Allen, M.D., Parker, B., Mattia, A. (2007). Measuring data management practice maturity: a community's self-assessment. *Computer, 40*(4), 42–50. https://doi.org/10.1109/MC.2007.139

[ANSI/ISA 84.00.01] *ANSI/ISA 84.00.01-2004 Part 1 Functional Safety: Safety Instrumented Systems for the Process Industry Sector - Part 1: Framework, Definitions, System, Hardware and Software Requirements.*

[Anttila 2004]Anttila, J., Kajava, J., Varonen, R. (2004). Balanced integration of information security into business management. *Proceedings of 30th Euromicro Conference* (pp. 558-564), Rennes, France, (September 2004). https://doi.org/10.1109/EURMIC.2004.1333422

[ATKERANEY] ATKERANEY. (n.d.). *The golden rules of operational excellence in information security management.* IT Strategic Management. http://www.ATKEARNEY.com

[Ayoub 2006] Ayoub, R. (2006). Analysis of business driven metrics: measuring for security value [White paper]. *DM Review* (March 2006). Frost & Sullivan. http://www.intellitactics.com

[Barman 2002] Barman, S. (2002). *Writing Information Security Policies.* New Riders.

[Bartol 2009] Bartol, N., Bates, B., Mercedes Goertzel, K., Winograd, T. (2009). *Measuring cyber security and information assurance state-of-the-art report* (SOAR) (May 8, 2009). Information Assurance Technology Analysis Center (IATAC). https://www.csiac.org/wp-content/uploads/2016/02/cybersecurity.pdf

[Bayuk 2011] Bayuk, J.L. (2011). Systems security engineering. *IEEE Security & Privacy, 9*(2), 72–74. https://doi.org/10.1109/MSP.2011.41

[Berst 2012] Berst, J. (2012). *Strategic Asset Management a primer for electric utilities.* [News]. SmartGrid News. http://assets.fiercemarkets.net/public/smartgridnews/SAM_ebook_V4_1.pdf

[BMIS] ISACA. (n.d.). *The business model for information security (BMIS).* https://www.isaca.org/bookstore/it-governance-and-business-management/wbmis1

[Brenner 2015] Brenner, M. (2015). *Relationships between ITSM frameworks* (January 2015). CC BY 4.0. Wikimedia. https://commons.wikimedia.org/wiki/File:Itsm-context.png

[Brotby 2009] Brotby, K. (2009). *Information Security Governance.* Wiley.

Building an Effective Security Program for Distributed Energy Resources and Systems: Understanding Security for Smart Grid and Distributed Energy Resources and Systems, Volume 1, First Edition. Mariana Hentea.

[BSA 2003] BSA. (2003). *Information security governance: Toward a framework for action*. Business Software Alliance. http://www.globaltechsummit.net/press/ISGPaper_2003.pdf

[Burkett 2012] Burkett, J.S. (2012). Business security architecture: weaving information security into your organization's enterprise architecture through SABSA®. *Information Security Journal: A Global Perspective, 21*(1), 47–54. https://doi.org/10.1080/19393555.2011.629341

[CS Odessa] CS Odessa. (2014). *Continuous cycle of the information security process*. CC BY 4.0. Wikimedia. https://commons.wikimedia.org/wiki/File:Circular-Arrows-Diagram-BPM-Life-Cycle.png

[Chen 2014] Chen, C-H., Favre, J., Kurillo, G., Andriacchi, T.P., Bajcsy, R., Chellappa, R. (2014). Camera networks for healthcare, teleimmersion, and surveillance. *Computer, 47*(5), 26–36. http://10.1109/MC.2014.112

[Cleveland 2012] Cleveland, F. (2012). *IEC TC57: IEC 62351 Security Standards the Power System Information Infrastructure*. http://iectc57.ucaiug.org/wg15public/Public%20Documents/White%20Paper%20on%20Security%20Standards%20in%20IEC%20TC57.pdf

[CNSSI 4009]CNSSI. (2010). *Committee on National Security Systems (CNSS) Information Assurance (IA) Glossary*. Committee on National Security Systems (CNSS) Instruction No. 4009. *Revised in 2015.

[COBIT 5] ISACA. (n.d.). *COBIT 5 Framework*. https://www.isaca.org/resources/cobit

[CPNI 2008] CPNI. (2008). *Good Practice Guide Process Control and SCADA Security*. Centre for Protection of National Infrastructure. http://osgug.ucaiug.org/conformity/security/Shared%20Documents/Reference/UK%20-%20CPNI%20-%20GPG%20Process%20Control%20and%20SCADA%20Security.pdf

[CPNI SCADA] CPNI. (n.d.). *Protecting my asset SCADA*. https://www.cpni.gov.uk/protecting-my-asset

[Dashti 2010] Dashti, R., Afsharnia, S., Bayat, B., Barband, A. (2010). Energy efficiency based asset management infrastructures in electrical distribution system. *2010 IEEE International Conference on Power and Energy (PECon)* (pp. 880-885), Kuala Lumpur, Malaysia, (January 2010). https://doi.org/10.1109/PECON.2010.5697703

[Dhillon 2000]Dhillon, G., Backhouse, J. (2000). Information system security management in the new millennium, technical Opinion. *Communications of the ACM, 43*(7), 125–128.

[Dhillon 2001a] Dhillon, G. (2001a). Challenges on managing information security in the new millennium. In G. Dhillon (Ed.), *Information Security Management: Global Challenges in the New Millennium* (pp. 1–8). IGI Global. https://doi.org/10.4018/978-1-878289-78-0.ch001

[Dhilon 2001b] Dhillon, G. (2001b). Principles for managing information security. In G. Dhillon (Ed.), *Information Security Management: Global Challenges in the New Millennium* (pp. 173–177). IGI Global. https://doi.org/10.4018/978-1-878289-78-0.ch012

[Dhillon 2006] Dhillon, G., Torkzadeh, G. (2006). Value-focused assessment of information system security in organizations. *Information Systems Journal, 16*(3), 293–314. https://doi.org/10.1111/j.1365-2575.2006.00219.x

[DMAIC] Wikipedia. (n.d.). *Define, Measure, Analyze, Improve and Control (DMAIC)*. https://en.wikipedia.org/wiki/DMAIC

[DOE 2011] DOE. (2011). *Energy storage research and development FY 2011 progress report*. https://www.energy.gov/eere/vehicles/energy-storage-resesearch-and-development-fy-2011-progress-report

[DOE 2014] DOE. (2014). *Electricity subsector cybersecurity capability maturity model (ES-C2M2) Version 1.1* (February 2014). http://energy.gov/sites/prod/files/2014/02/f7/ES-C2M2-v1-1-Feb2014.pdf

[DOE 2015a] DOE. (2015a). *Cybersecurity*. https://www.energy.gov/national-security-safety/cybersecurity

[DOE 2015b] DOE. (2015b). *Energy sector cybersecurity framework implementation guidance January 2015*. http://energy.gov/sites/prod/files/2015/01/f19/Energy%20Sector%20Cybersecurity%20Framework%20Implementation%20Guidance_FINAL_01-05-15.pdf

[EAC 2014] DOE. (2014). *Implementing effective enterprise security governance Outline for energy sector executives and boards*. Electricity Advisory Committee. http://energy.gov/sites/prod/files/Mar2014EAC_Recs-CyberGovernance.pdf

[ENISA 2011] ENISA. (2011). *Annex III. ICS security related standards, guidelines and policy documents* (December 2011). https://www.enisa.europa.eu/activities/Resilience-and-CIIP/critical-infrastructure-and-services/scada-industrial-control-systems/annex-iii

[ENISA 2015a] ENISA. (2015a). *Analysis of ICS-SCADA cyber security maturity levels in critical sectors* (December 2015). https://www.enisa.europa.eu/publications/maturity-levels

[ENISA 2015b] ENISA. (2015b). *Certification of cyber security skills of ICS/SCADA professionals* (February 2015). https://www.enisa.europa.eu/publications/certification-of-cyber-security-skills-of-ics-scada-professionals

[ENISA Factors] ENISA. (n.d.). *Critical success factors Critical success factors for ISMS*. https://www.enisa.europa.eu/topics/threat-risk-management/risk-management/current-risk/risk-management-inventory/rm-isms/critical-success-factors

[EPRI 2001] EPRI. (2001). *Grid Equipment Reliability Study* (1001971). http://www.epri.com/abstracts/Pages/ProductAbstract.aspx?ProductId=000000000001001971&Mode=download

[EO 13636] WhiteHouse. (2013, February 12). *Executive Order (EO) 13636 Improving Critical Infrastructure Cybersecurity*. https://www.whitehouse.gov/the-press-office/2013/02/12/executive-order-improving-critical-infrastructure-cybersecurity

[Fennelly 2012] Fennelly, L.J. (2012). *Effective Physical Security* (Fourth Edition). Butterworth-Heinemann.

[Ferraiolo 2000] Ferraiolo, K. (2000). The systems security engineering capability maturity model (SSE-CMM) [Paper presentation]. *23rd National Information Systems Security Conference (NISSC)*, Baltimore, Maryland, USA, (October 2000). http://csrc.nist.gov/nissc/2000/proceedings/papers/916.pdf

[FIPS 200] NIST. (2006). *FIPS PUB 200 Minimum security requirements for federal information and information systems* (March 2006). Federal Information Processing Standard (FIPS). http://csrc.nist.gov/publications/fips/fips200/FIPS-200-final-march.pdf

[Foley 2004] Foley, J., Hulme, G.V., Marlin, S. (2004, July 4). Applying pressure. *InformationWeek*, 20–22.

[Fung 2008] Fung, W.S.L., Fung, R.Y.K. (2008). Knowledge-centric information security. *International Conference on Security Technology (SECTECH '08) Proceedings* (pp. 27-34), Sanya, Hainan Island, China, (December 2008). https://doi.org./10.1109/SecTech.2008.9

[Gaudin 2007] Gaudin, S. (2007, August 7). Storm turns into a hurricane; is a botnet attack brewing? *Information Week*, 38.

[Greene 2014]. Greene, S.S. (2014). *Security Program and Policies: Principles and Practices* (Second Edition). Pearson IT Certification.

[Haegley 2016] Haegley, D. (2016). Governance and assessment strategies for industrial control systems. In E.J.M Colbert, A. Kott (Eds.), *Cyber-security of SCADA and Other Industrial Control Systems* (pp. 279–304). Springer. https://doi.org/10.1007/978-3-319-32125-7_14

[Harris 2005] Harris, S. (2005). *All in One CISSP Exam Guide* (Third Edition). McGraw-Hill/Osborne.

[Harris 2006a] Harris, S. (2006a). *Developing an information security program using SABSA., ISO 17799.* https://searchsecurity.techtarget.com/tip/Developing-an-information-security-program-using-SABSA-ISO-17799

[Harris 2006b] Harris, S. (2006b). *What is information security governance?* http://security.networksasia.net/content/information-security-governance-guide

[Harris 2013] Harris, S. (2013). *All in One CISSP Exam Guide* (Sixth Edition). McGraw-Hill Education.

[Hatton 2007] Hatton, L. (2007). The chimera of software quality. *IEEE Computer*, *40*(8), 102–104.

[Hawaiian Electric] Hawaiianelectric. (2016). *Key performance metrics Service reliability.* https://www.hawaiianelectric.com/about-us/key-performance-metrics/service-reliability

[Heimerl 2005] Heimerl, J.L., Voight, H. (2005). Measurement: the foundation of security program design and management. *Computer Security Journal*, *XXI*(2), 1–20.

[Hentea 2005a] Hentea, M. (2005a). Use of reconnaissance patterns for intelligent monitoring model. *Proceedings of Information Resources Management Association International Conference* (pp. 160–163), San Diego, California, USA, (May 2005).

[Hentea 2005b] Hentea, M. (2005b). Information security management. In M. Pagani (Ed.), *Encyclopedia of Multimedia Technology and Networking* (*2*-volume set, pp. 390–395). IGI Global. https://doi.org/10.4018/978-1-59140-561-0.ch056

[Hentea 2007] Hentea, M. (2007). Intelligent system for information security management: architecture and design issues. *Journal of Issues in Informing Science and Information Technology*, *4*, 29–43. https://doi.org/10.28945/930

[Hentea 2008] Hentea, M. (2008). Information security management. In M. Pagani (Ed.) *Encyclopedia of Multimedia Technology and Networking* (Second Edition, *3*-volume set, pp. 675–681). IGI Global. https://doi.org/10.4018/978-1-60566-014-1.ch091

[Hoffman 2013] Hoffman, P. (2013). *Investment Today for Tomorrow's Grid.* IEEE Power & Energy Society. https://resourcecenter.ieee-pes.org/conference-videos-and-slides/isgt/PESVID1233.html

[Howe 2010] Howe, D. (2010). *The Free On-line Dictionary of Computing Information technology governance.* https://foldoc.org/information+technology+governance

[IASME] IASME. (n.d.). *IASME governance.* https://iasme.co.uk/iasme-governance/

[IEC 61508] *IEC 61508:2010 CMV Commented version Functional safety of electrical/electronic/programmable electronic safety-related systems – Parts 1 to 7 together with a Commented version.*

[IEC 61511] *IEC 61511-1:2016 Functional safety – Safety instrumented systems for the process industry sector – Part 1: Framework, definitions, system, hardware and application programming requirements.*

[IEC 62443-1-1] *IEC TS 62443-1-1:2009 Industrial communication networks – Network and system security – Part 1-1: Terminology, concepts and models.*

[IEC 62351] *IEC 62351:2020 SER Series Power systems management and associated information exchange – Data and communications security -- ALL PARTS.*

[IEEE 1547] *1547-2018 – IEEE Standard for Interconnection and Interoperability of Distributed Energy Resources with Associated Electric Power Systems Interfaces.*

[ISA99] ISA. (n.d.). *ISA99 industrial automation and control systems security.* https://www.isa.org/intech-plus/2019/may/new-isa-iec-62443-standard-specifies-security-capa

[ISA SCADA] ISA. (n.d.). *Overview of ISA 62443 series of standards.* https://www.isa.org/search?query=Overview%20of%20ISA%2062443%20Series%20of%20Standards

[ISACA 2009] ISACA. (2009). *An introduction to the business model for information security.* http://media.techtarget.com/Syndication/SECURITY/BusiModelforInfoSec.pdf

[ISACA Glossary] ISACA. (2008). *Glossary of terms.* https://www.isaca.org/resources/glossary

[ISASecure] ISASecure. (n.d.). *Certification.* http://www.isasecure.org/en-US/Certification.

[ISC2 SEC] ISC2. (n.d.). *(ISC)^2 International Academic Program (IAP).* https://www.isc2.org/Development

[ISM3 2007] ISM3. (2007). *ISM^3 Handbook* v2.00 – *Information security management maturity model.* ISM3 Consortium. CC BY-ND 3.0. https://www.academia.edu/8289662/v2_00_INFORMATION_SECURITY_MANAGEMENT_MATURITY_MODEL_CONTACT_INFORMATION

[ISM3 COMPARE] Slideshare. (n.d.). *ISM3 compared to ISO27001.* https://www.slideshare.net/vaceituno/oism3-vs-iso27001

[ISO 9000] *ISO 9000:2015 Quality Management Systems – Fundamentals and Vocabulary.*

[ISO 9001] *ISO 9001:2015 Quality Management Systems – Requirements.*

[ISO 14001] *ISO 14001:2015 Environmental management systems — Requirements with guidance for use.*

[ISO 19011] *ISO 19011:2018 Guidelines for auditing management systems.*

[ISO 20000] *ISO 20000 Information technology — Service management. * suite of parts*

[ISO 29100] *ISO/IEC 29100:2011 Information technology — Security techniques — Privacy framework.*

[ISO 27k] ISO. (n.d.). *About the ISO27k Standards.* International Standards Organization. https://www.iso27001security.com/html/iso27000.html

[ISO Certification] ISO. (n.d.). *The ISO27001 certification process.* http://www.27000.org/ismsprocess.htm

[ISO Freely] ISO. (n.d.). *Freely Available Standards.* http://standards.iso.org/ittf/PubliclyAvailableStandards/

[ISO/IEC 15408] *ISO/IEC 15408 Information Technology – Security Techniques – Evaluation Criteria for IT Security, Parts 1--3.*

[ISO/IEC 15939] *ISO/IEC/IEEE 15939:2017 Systems and software engineering — Measurement process.*

[ISO/IEC 17021] *ISO/IEC 17021-1:2015 Conformity assessment — Requirements for bodies providing audit and certification of management systems — Part 1: Requirements.*

[ISO/IEC 20000-1] *ISO/IEC 20000-1:2018 Information technology — Service management — Part 1: Service management system requirements.*

[ISO/IEC 21827] *ISO/IEC 21827:2008 Information technology — Security techniques — Systems Security Engineering — Capability Maturity Model® (SSE-CMM®).*

[ISO/IEC 27000] *ISO/IEC 27000:2018 Information Technology — Security Techniques — Information Security Management Systems — Overview and Vocabulary* (Edition 5).

[ISO/IEC 27001] *ISO/IEC 27001:2013 Information technology — Security techniques — Information security management systems — Requirements.*

[ISO/IEC 27002] *ISO/IEC 27002:2013 Information Technology — Security Techniques — Code of Practice for Information Security Controls.*

[ISO/IEC 27003] *ISO/IEC 27003:2017 Information technology — Security techniques — Information security management systems — Guidance.*

[ISO/IEC 27004] *ISO/IEC 27004:2016 Information technology — Security techniques — Information security management — Monitoring, measurement, analysis and evaluation.*

[ISO/IEC 27006] *ISO/IEC 27006:2015 Information technology — Security techniques — Requirements for bodies providing audit and certification of information security management systems.*

[ISO/IEC 27007] *ISO/IEC 27007:2020 Information security, cybersecurity and privacy protection — Guidelines for information security management systems auditing.*

[ISO/IEC 27008] *ISO/IEC TS 27008:2019 Information technology — Security techniques — Guidelines for the assessment of information security controls.*

[ISO/IEC 27011] *ISO/IEC 27011:2016 Information technology — Security techniques — Code of practice for Information security controls based on ISO/IEC 27002 for telecommunications organizations.*

[ISO/IEC 27015] *ISO/IEC TR 27015:2012 Information technology — Security techniques — Information security management guidelines for financial services.*

[ISO/IEC 27017] *ISO/IEC 27017:2015 Information technology — Security techniques — Code of practice for information security controls based on ISO/IEC 27002 for cloud services.*

[ISO/IEC 27018] *ISO/IEC 27018:2019 Information technology — Security techniques — Code of practice for protection of personally identifiable information (PII) in public clouds acting as PII processors.*

[ISO/IEC TR 27019] *ISO/IEC TR 27019:2013 Information technology - Security techniques - Information security management guidelines based on ISO/IEC 27002 for process control systems specific to the energy utility industry.* * Revised by [ISO/IEC 27019]

[ISO/IEC 27021] *ISO/IEC 27021:2017 Information technology — Security techniques — Competence requirements for information security management systems professionals.*

[ISO/IEC/IEEE 15288] *ISO/IEC/IEEE 15288:2015 Systems and software engineering — System life cycle processes.*

[ISO/IEC TS 62351-1] *IEC TS 62351-1:2007 Power systems management and associated information exchange - Data and communications security - Part 1: Communication network and system security - Introduction to security issues.*

[IT 2001] ISACA. (2001). *Information Security Governance Guidance for Boards of Directors and Executive Management.* Information Systems Audit and Control Foundation (ISACF). https://citadel-information.com/wp-content/uploads/2010/12/isaca-information-security-governance-guidance-for-boards-of-directors-and-executive-management-2001.pdf

[Jennings 2007] Jennings, B., van der Meer, S., Balasubramaniam, S., Botvich, D., Donnelly, W., Strassner, J. (2007). Towards autonomic management of communications networks. *IEEE Communications Magazine, 45*(10), 112–121. https://doi.org/10.1109/MCOM.2007.4342833

[Jo 2011] Jo, H., Kim, S., Won, D. (2011). Advanced information security management evaluation system. *KSII Transactions on Internet and Information Systems, 5*(6), 1192–1213. https://doi.org/10.3837/tiis.2011.06.006

[Johnson 2007] Johnson, M.E., Goetz, E. (2007). Embedding information security into the organization. *IEEE Security & Privacy, 5*(3), 16–24. https://doi.org/10.1109/MSP.2007.59

[Kadrich 2007] Kadrich, M.S. (2007). *Endpoint Security.* Addison-Wesley.

[Kark 2010] Kark, K., Dines, R.A. (2010). *Security organization 2.0: Building a robust security organization.* Forrester Research. htttp://www.forrester.com

[Kephart 2003] Kephart, J.O., Chess, D.M. (2003). The vision of automatic computing. *Computer, 36*(1), 41–50. https://doi.org/10.1109/MC.2003.1160055

[Kephart 2005] Kephart, O. (2005). Research challenges of autonomic computing. *Proceedings of 27th International Conference Software Engineering (ICSE '05)* (pp. 15-22), St. Louis, Missouri, USA (May 2005). ACM Press. https://doi.org/10.1145/1062455.1062464

[Kiely 2006a] Kiely, L., Benzel, T. (2006a). Systemic security management: A new conceptual framework for understanding the issues, inviting dialogue and debate, and identifying future research needs. https://www.academia.edu/9918325/Systemic_Security_Management

[Kiely 2006b] Kiely, L., Benzel, T. (2006b). Systemic security management. *IEEE Security & Privacy, 4*(6), 74–77. https://doi.org/10.1109/MSP.2006.167

[Kim 2007] Kim, J., Chung, K., Choi, K. (2007). Spam filtering with dynamically updated URL statistics. *IEEE Security & Privacy, 5*(4), 33–39. https://doi.org/10.1109/MSP.2007.95

[Krutz 2004] Krutz, R.L., Vines, R.D. (2004). *The CISSP Prep Guide* (Second Edition). Wiley.

[Lalonde 2018] Lalonde, M. (2018). Combining strengths: cyber and physical security convergence. *The Conference Board of Canada 2018*, Ottawa, Canada, (February 2018). https://doi.org/10.13140/RG.2.2.31955.02083

[Lapolito 2007] Lapolito, T. (2007, February 6). *The rise of physical security information management* [News]. TechNewsWorld. http://www.technewsworld.com/story/55581.html

[Lewis 2010] Lewis, E. (2010). Information security management. In H. Bidgoli (Ed.), *The Handbook of Technology Management* (Second Edition, *3*-volume set, pp. 940–956). Wiley.

[Lincke 2009] Lincke, S.J. (2009). *Information Security Program Development.* http://www.cs.uwp.edu/Classes/Cs490/notes/SecurityPgmDev.ppt

[Luiijf 2015] Luiijf, H.A.M., Paske, B.J.t. (2015). *Cyber Security of Industrial Control Systems.* TNO Publishing. https://doi.org/10.13140/RG.2.1.3797.4566

[Maiwald 2004] Maiwald, E. (2004). *Fundamentals of Network Security.* McGraw-Hill/Technology Education.

[McDermott 2001] Blackley, B., McDermott, E., Geer, D. (2001). Information security is information risk management. *Proceedings of the 2001 Workshop on New Security Paradigms NSPW '01* (pp. 97 –104). ACM. https://doi.org/10.1145/508171.508187

[Mena 2004] Mena, J. (2004). Homeland security connecting the DOTS. *Software Development, 12*(5), 34–41.

[Mittal 2010] Mittal, Y.K., Roy, S., Saxena, M. (2010). Role of knowledge management in enhancing information security. *International Journal of Computer Science Issues, 7*(6), 320–324. www.IJCSI.org.

[Murdoch 2006] Murdoch, J. (2006). *Security Measurement Version 3.0* (13 January 2006) [Whitepaper]. PSM Safety and Security Technical Working Group. http://www.psmsc.com/Downloads/TechnologyPapers/SecurityWhitePaper_v3.0.pdf

[NERC Glossary] NERC. (n.d.) *Glossary of Terms Used in NERC Reliability Standards.* http://www.nerc.com/files/Glossary_of_Terms.pdf

[News 2013] Tweed, K. (2013). Bulletproofing the grid A gun attack on a Silicon Valley substation has utilities looking to boost physical security. *IEEE Spectrum, 49*(5), 13.

[NIST 2014] *NIST Framework for Improving Critical Infrastructure Cybersecurity Version 1.0* (February 2014). https://www.nist.gov/system/files/documents/cyberframework/cybersecurity-framework-021214.pdf

[NIST SP800-12] *NIST Special Publication (SP) 800-12r1 An Introduction to Computer Security* (Revision 1, June 2017). https://doi.org/10.6028/NIST.SP.800-12r1

[NIST SP800-27rA] *NIST Special Publication (SP) 800-27 Rev A Engineering Principles for Information Technology Security (A Baseline for Achieving Security) Revision A* (June 2004). * superseded by NIST SP800-160 (November 2016).

[NIST SP800-37r1] *NIST Special Publication (SP) 800-37 Guide for Applying the Risk Management Framework to Federal Information Systems: A Security Life Cycle Approach Revision 1* (June 2014). *Superseded by NIST SP800-37r2 (December 2018). https://doi.org/10.6028/NIST.SP.800-37r2

[NIST SP800-53Ar4] *NIST Special Publications (SP) 800-53Ar4 Assessing Security and Privacy Controls in Federal Information Systems and Organizations: Building Effective Assessment Plans* (Revision 4, December 2014). Computer Security Division Information Technology Laboratory. http://dx.doi.org/10.6028/NIST.SP.800-53Ar4

[NIST SP800-53r4] *NIST Special Publication (SP) 800-53 Security and Privacy Controls for Federal Information Systems and Organizations Revision 4*, (April 2013). Computer Security Division Information Technology Laboratory. http://dx.doi.org/10.6028/NIST.SP.800-53r4 * Superseded by [NIST SP800-53r5] in September 2020; Withdrawal Date September 23, 2021.

[NIST SP800-55r1] *NIST Special Publication (SP) 800-55 Performance Measurement Guide for Information Security Revision 1* (July 2008). Computer Security Division Information Technology Laboratory.

[NSF 2011] NSF. (2011, June). *A Policy framework for the 21st century grid: enabling our secure energy future*. U.S. National Science Foundation. https://www.ourenergypolicy.org/wp-content/uploads/2011/12/2011_06_WhiteHouse_PolicyFramework21stCenturyGrid.pdf

[OECD 1980] OECD. (1980). *Guidelines on the Protection of Privacy and Transborder Flows of Personal Data*. Organisation for Economic Co-operation and Development. https://www.oecd.org/internet/ieconomy/oecdguidelinesontheprotectionofprivacyandtransborderflowsofpersonaldata.htm

[OECD 2013] OECD. (2013). *The OECD Privacy Framework*. Organisation for Economic Co-operation and Development. https://www.oecd.org/sti/ieconomy/oecd_privacy_framework.pdf

[OECD 2015a] OECD. (2015a). *G20/Principles of Corporate Governance*. OECD Publishing. http://dx.doi.org/10.1787/9789264236882-en

[OECD 2015b] OECD. (2015b). *Digital Security Risk Management for Economic and Social Prosperity: OECD Recommendation and Companion Document*. OECD Publishing. http://dx.doi.org/10.1787/9789264245471-en

[O-ISM3] OpenGroup. (2011). *Open information security management maturity model (O-ISM3)*. https://www2.opengroup.org/ogsys/catalog/C102

[OpenGroup 2011] OpenGroup. (2011). *Optimizing ISO/IEC 27001:2013 using O-ISM3*. https://publications.opengroup.org/g125

[P2030 2011] *IEEE Std 2030-2011- IEEE Guide for Smart Grid Interoperability of Energy Technology and Information Technology Operation with the Electric Power System (EPS), End-Use Applications, and Loads.*

[Pastore 2008] Pastore, R. (2008). *Models for global IT governance* (March 3, 2008). IT Strategy Leadership and Management. http://www.cio.com/article/2437034/it-organization/models-for-global-it-governance.html

[Patriciu 2009] Patriciu, V.V., Priescu, I., Nicolaescu, S. (2009). Security metrics for enterprise information systems. *Journal of Applied Quantitative Methods (JAQM)*, *1*(2), 151–158.

[Pavlou 2011] Pavlou, G., Pras, A. (2011). Topics in network and service management. *IEEE Communications Magazine*, *49*(7), 78–79.

[Perrin 2012] Perrin, C. (2012, June 30). The CIA Triad [Blog]. *TechRepublic*. http://www.techrepublic.com/blog/security/the-cia-triad/488.

[Pipkin 2000] Pipkin, D. (2000). *Information Security: Protecting the Global Enterprise* (First Edition). Pearson.

[Pras 2007] Pras, A., Schonwalder, J., Burgess, M., Festor, O., Perez, G.M., Stadler, R., Burkhard, S. (2007). Key research challenges in network management. *IEEE Communications Magazine*, *45*(10), 104–110. https://doi.org/10.1109/MCOM.2007.4342832

[Prati 2014] Prati, A., Qureshi, F.Z. (2014). Integrating consumer smart cameras into camera networks: opportunities and obstacles. *Computer*, *47*(5), 45–51. https://doi.org/10.1109/MC.2014.125

[Rastler 2012] Rastler, D. (2012). Electricity Energy Storage Technology Options: System Cost Benchmarking. *IPHE Workshop Hydrogen - A competitive Energy Storage Medium for large scale integration of renewable electricity*, Seville, Spain, (November 2012). https://moam.info/2012-powerpoint-template-version-20-international-partnership-_59df90631723dd66c747eb0d.html

[Reilly 2015] Reilly, S. (2015, March 24). *Bracing for a big power grid attack: 'One is too many'* [News]. USA Today. https://eu.usatoday.com/story/news/2015/03/24/power-grid-physical-and-cyber-attacks-concern-security-experts/24892471/

[Reisslein 2014] Reisslein, M., Rinner, B., Roy-Chowdhury, A. (2014). Smart camera networks [Guest editors' introduction]. *Computer*, *47*(5), 23–25. https://doi.org/ 10.1109/MC.2014.134

[RFC 2196] *IETF Request for Comments (RFC) 2196 Site Security Handbook (September 1997).*

[RFC 4949] *IETF Request for Comments (RFC) 4949 Internet Security Glossary* (Version 2, August 2007).

[Richardson 2004] Richardson, J.F., Bartol, N., Moss, M. (2004). *ISO/IEC 21827 Systems security engineering capability maturity model (SSE-CMM).* https://www.academia.edu/15324011/3-ISOIEC-21827-CMMIAnd_Assurance_Aug2-Moss-Richardson

[SABSA] Wikipedia. (n.d.). *Sherwood Applied Business Security Architecture (SABSA).* http://en.wikipedia.org/wiki/Sherwood_Applied_Business_Security_Architecture

[Sales 2016] Sales, F. (2016, March 18). *Managing cybersecurity and Supply chain risks: The board's role* [News]. TechTarget. https://searchcompliance.techtarget.com/feature/Managing-cybersecurity-and-supply-chain-risks-The-boards-role

[SanMiguel 2014] SanMiguel, J.C., Shoop, K., Cavallero, A., Micheloni, C., Foresti, G.L. (2014). Self-reconfigurable smart camera networks. *Computer, 47*(5), 67–73. https://doi.org/10.1109/MC.2014.133

[Schneier 2012] Schneier, B. (2012, September 15). The Importance of Security Engineering [Blog]. *Schneir.* https://www.schneier.com/crypto-gram-1209.html

[SCRM] Techtarget. (n.d.). *Supply chain risk management (SCRM).* IT Encyclopedia. http://whatis.techtarget.com/definition/supply-chain-risk-management-SCRM

[Shannon 2007] Shannon, C. (2007). Current network security threats: DoS, viruses, worms, botnets [Paper presentation]. *TERENA Networking Conference*, San Diego, California, USA, (May 2007). Cooperative Association for Internet Data Analysis. http://www.caida.org/publications/presentations/2007/terena_security/terena_security.pdf

[Six Sigmas] Wikipedia. (n.d.). *Six Sigmas.* https://en.wikipedia.org/wiki/Six_Sigma

[SOGP] Securityforum. (2020). *Standard of Good Practice (SOGP).* Information Security Forum. https://www.securityforum.org/tool/standard-of-good-practice-for-information-security-2020/

[Stephenson 2016] Stephenson, P. (2016). *Information Security Management Handbook* (7th Edition). Auerbach Publications.

[Strassner 2004] Strassner, J. (2004). Autonomic networking – theory and practice. *2004 IEEE/IFIP Network Operations and Management Symposium (IEEE Cat. No.04CH37507)* (Vol. 1, pp. 927), Seoul, Korea, (April 2004). https://doi.org./10.1109/NOMS.2004.1317811

[Tassabehji 2005] Tassabehji, R. (2005). Information security threats. In M. Pagani (Ed.), *Encyclopedia of Multimedia Technology and Networking* (2-volume set, pp. 404–410). IGI Global. https://doi.org/10.4018/978-1-59140-561-0.ch058

[TechNews 2002a] Technews. (2002a, March). *Security transformation – The A-Z of content security: Part 2* [News]. Computer Business. http://www.cbr.co.za/article.aspx?pklarticleid=1689

[TechNews 2002b] TechNews. (2002b, August). *Protecting Your Company's Valuable Assets: Data* [News]. Computer Business. August 2002. http://www.cbr.co.za/article.aspx?pklarticleid=1987

[Tipton 2006] Tipton, H.F., Krause, M. (2006). *Information Security Management Handbook* (Fifth Edition). Auerbach Publications.

[TISN 2007] TISN. (2007). *Secure your information: information security principles for enterprise architecture trusted information Sharing Network for Critical Infrastructure Protection Report* (June 2007). www.tisn.gov.au

[Trope 2007] Trope, R.L., Power, E.M., Polley, V.I., Morley, B.C. (2007). A coherent strategy for data security through data governance. *IEEE Security & Privacy, 5*(3), 32–39. https://doi.org/10.1109/MSP.2007.51

[UAF] UAF. (n.d.). *Accreditation for information security management system (ISMS) scheme based on ISO/IEC 27001.* https://uafaccreditation.org/accreditation/information-security-management-systems-isms-scheme-based-on-iso-iec-27001/

[USPolicy 2013] Whitehouse. (2013). *A Policy Framework for the 21st Century Grid: A Progress Report. Executive Office of the President National Science and Technology Council, February 2013.* https://obamawhitehouse.archives.gov/sites/default/files/microsites/ostp/2013_nstc_grid.pd

[Utilities 2014] Oregon. (2014, June). *Oregon investor-owned utilities Seven-Year electric service reliability statistics summary 2007-2013.* Oregon Public Utility Commission. https://www.oregon.gov/puc/forms/Forms%20and%20Reports/Seven-Year_Electric_Service_Reliability_Statistics_Summary_(2011-2017).pdf

[Venter 2003] Venter, H.S., Eloff, J.H.P. (2003). A taxonomy for information security technologies. *Computers & Security, 22*(4), 299–307. https://doi.org/10.1016/S0167-4048(03)00406-1

[Volonino 2004] Volonino, L., Robinson, S.R. (2004). *Principles and Practice of Information Security.* Pearson Prentice Hall.

[von Solms 2000] von Solms, S.H. (2000). Information security – the third wave?. *Computers & Security, 19*(7), 615–620. https://doi.org/10.1016/S0167-4048(00)07021-8

[von Solms 2006] von Solms, S.H. (2006). Information security – the fourth wave. *Computers & Security, 25*(3), 165–168. https://doi.org/10.1016/j.cose.2006.03.004

[von Solms 2010] von Solms, S.H. (2010). The 5 waves of information security – from Kristian Beckman to the present. In K. Rannenberg, V. Varadharajan, C. Weber (Eds.), *IFIP International Federation for Information Processing 2010, SEC 2010, IFIP Advances in Information and Communication Technology* (Vol *330*, pp. 1–8). Springer. https://doi.org/10.1007/978-3-642-15257-3_1

[Wadlow 2000] Wadlow, T.A. (2000). *The Process of Network Security.* Addison-Wesley.

[Wang 2004] Wang, W. (2004). *The intelligent, proactive information assurance and security technology* [White Paper]. Cybershield Networks. http://infosecwriters.com/text_resources/pdf/Intelligent_IPDM.pdf

[Wang 2005a] Wang, W. (2005a). The intelligent proactive information assurance and security technology. *IEEE Intelligent Systems, 19*(5), 92–96.

[Wang 2005b] Wang, G. (2005b). Strategies and influence for information security. *Information System Control Journal, 1.* Information Systems Audit and Control Association.

[Watson 2013] Watson, P. (2013). *Corporate vs. product security* [White paper]. SANS. https://www.sans.org/reading-room/whitepapers/leadership/corporate-vs-product-security-34237

[WEF 2014] Weforum. (2014). *Risk and responsibility in a hyperconnected world (WEF Principles), Geneva, Switzerland.* http://www.weforum.org/reports/risk-and-responsibilityhyperconnected-world-pathways-global-cyber-resilience

[Whitman 2011] Whitman, M.E., Mattord, H.J. (2011). *Principles of Information Security.* Cengage Learning.

[Whitman 2014] Whitman, M.E., Mattord, H.J. (2014). *Management of Information Security.* Cengage Learning.

A

Cybersecurity Concepts

Cybersecurity

Appendix A includes definitions for cybersecurity terms and concepts that are recommended including information about various glossaries that include cybersecurity definitions. The wide spread of the cybersecurity glossaries with not unique definitions is creating often confusion and room for the misinterpretation at the expense of a sound guide to provide for adequate security controls.

Table A.1 shows a comparison of definitions for a selected set of security and privacy terms as they are found in known glossaries: IETF RFC 4949; CNSSI 4009; NIST SP 800-53r4; and American Dictionary. Also, included are comments about these definitions.

IETF glossary includes these types of definitions:

"I": Recommended definitions of Internet origin

"N": Recommended definitions of Non-Internet origin

"O": Other terms and definitions to be noted

"D": Deprecated terms and definitions.

Table A.2 includes definitions for a selected set of security and privacy terms that are recommended.

Table A.3 includes examples of glossaries that define terms for IT, SCADA, and cybersecurity.

Building an Effective Security Program for Distributed Energy Resources and Systems: Understanding Security for Smart Grid and Distributed Energy Resources and Systems, Volume 1, First Edition. Mariana Hentea.

Table A.1 Comparison of security terms defined in different glossaries or standards.

Term/glossary	CNSSI 4009 Glossary	IETF RFC 4949 Glossary	NIST SP 800-53r4	American Dictionary
Information	"Any communication or representation of knowledge such as facts, data, or opinions in any medium or form, including textual, numerical, graphic, cartographic, narrative, or audiovisual" Comments: Same as NIST, references NIST SP800-53	"Facts and ideas, which can be represented (encoded) as various forms of data" Comments: Incomplete; refers only to digital data	Same as CNSSI 4009, references CNSSI 4009 Comments: A vicious circle; each glossary points to other glossary	"A collection of facts or data"
Information security	"The protection of information and information systems from unauthorized access, use, disclosure, disruption, modification, or destruction in order to provide confidentiality, integrity, and availability" Comments: Incomplete; no reference to protection against repudiation	"Measures that implement and assure security services in information systems, including in computer systems and in communication systems (see: COMPUSEC) and in communication systems (see: COMSEC)." Comments: It refers to security services	Same as CNSSI 4009	No definition
Cyberspace	"A global domain within the information environment consisting of the interdependent network of information systems infrastructures including the Internet, telecommunications networks, computer systems, and embedded processors and controllers" Comments: Incomplete; no inclusion of facilities, people, and their interactions	Term not included in the glossary	Same as CNSSI 4009, references CNSSI 4009	No definition

Cyber attack	"An attack, via cyberspace, targeting an enterprise's use of cyberspace for the purpose of disrupting, disabling, destroying, or maliciously controlling a computing environment/infrastructure; or destroying the integrity of the data or stealing controlled information" Comments: Incomplete; refers to cyberspace and enterprise only	1) (I) An intentional act by which an entity attempts to evade security services and violate the security policy of a system. That is, an actual assault on system security that derives from an intelligent threat. (See: penetration, violation, vulnerability.) 2) (I) A method or technique used in an assault (e.g. masquerade). (See: blind attack, distributed attack.) Comments: It uses the term "attack" and defines the meaning of the attack; the definition relates to some other basic security terms (threat, vulnerability, countermeasures) and information security	Same as CNSSI 4009, references CNSSI 4009	No definition
Cybersecurity	"The ability to protect or defend the use of cyberspace from cyber attacks" Comments: Incomplete; it does not imply protection from human errors or natural disasters	Not included in the glossary	Same as CNSSI 4009, references CNSSI 4009	Incomplete; no reference to protection of data and information "Measures taken to protect a computer or computer system (as on the Internet) against unauthorized access or attack"
Privacy	No definition	"The right of an entity (normally a person), acting in its own behalf, to determine the degree to which it will interact with its environment, including the degree to which the entity is willing to share its personal information with others" Comments: Relevant; refers to PII	No definition	"1. The condition of being secluded from others; 2. Secrecy" Comments: Not relevant; no reference to personal information

(Continued)

Table A.1 (Continued)

Term/glossary	CNSSI 4009 Glossary	IETF RFC 4949 Glossary	NIST SP 800-53r4	American Dictionary
Security	"A condition that results from the establishment and maintenance of protective measures that enable an enterprise to perform its mission or critical functions despite risks posed by threats to its use of information systems. Protective measures may involve a combination of deterrence, avoidance, prevention, detection, recovery, and correction that should form part of the enterprise's risk management approach" Comments: Incomplete; refers to enterprise only	"1a. (I) A system condition that results from the establishment and maintenance of measures to protect the system. 1b. (I) A system condition in which system resources are free from unauthorized access and from unauthorized or accidental change, destruction, or loss. (Compare: safety.) 2. (I) Measures taken to protect a system." Comments: Definitions 1a and 2 are incomplete and refer to system only; definition 1b appears to be more appropriate	References CNSSI 4009	"Measures adopted to guard against attack, theft, or disclosure" Comments: Not relevant

Each entry in RFC 4949 is preceded by a character – I, N, O, or D – enclosed in parentheses, to indicate the type of definition:
– "I" for a RECOMMENDED term or definition of Internet origin.
– "N" if RECOMMENDED but not of Internet origin.
– "O" for a term or definition that is NOT recommended for use in IDOCs but is something that authors of Internet documents should know about.
– "D" for a term or definition that is deprecated and SHOULD NOT be used in Internet documents.

Table A.2 Recommended definitions of most common security terms.

Term	Definition	Reference
Acceptable risk	A risk that is understood and tolerated by a system's user, operator, owner, or accreditor, usually because the cost or difficulty of implementing an effective countermeasure for the associated vulnerability exceeds the expectation of loss. (See: adequate security, risk, "second law" under "Courtney's laws")	RFC 4949
Access	The ability and means to communicate with or otherwise interact with a system to use system resources either to handle information or to gain knowledge of the information the system contains	RFC 4949
Accountability	The property of a system or system resource that ensures that the actions of a system entity may be traced uniquely to that entity, which can then be held responsible for its actions	RFC 4949
Accountability service	A security service that records information needed to establish accountability for system events and for the actions of system entities that cause them	RFC 4949
Attack	1. (I) An intentional act by which an entity attempts to evade security services and violate the security policy of a system. That is, an actual assault on system security that derives from an intelligent threat. (See: penetration, violation, vulnerability) 2. (I) A method or technique used in an assault (e.g. masquerade)	RFC 4949
Authentication	The process of verifying a claim that a system entity or system resource has a certain attribute value. (See: attribute, authenticate, authentication exchange, authentication information, credential, data origin authentication, peer entity authentication, "relationship between data integrity service and authentication services" under "data integrity service," simple authentication, strong authentication, verification, X.509)	RFC 4949
Authentication information	Information used to verify an identity claimed by or for an entity. (See: authentication, credential, user. Compare: identification information)	RFC 4949
Authentication service	A security service that verifies an identity claimed by or for an entity. (See: authentication)	RFC 4949
Authenticity	The property of being genuine and able to be verified and be trusted. (See: authenticate, authentication, validate vs. verify)	
Automated security monitoring	Use of automated procedures to ensure security controls are not circumvented or the use of these tools to track actions taken by subjects suspected of misusing the information system	CNSSI 4009

(Continued)

Table A.2 (Continued)

Term	Definition	Reference
Availability	The property of a system or a system resource being accessible, or usable or operational upon demand, by an authorized system entity, according to performance specifications for the system; i.e. a system is available if it provides services according to the system design whenever users request them. (See: critical, denial of service. Compare: precedence, reliability, survivability)	RFC 4949
Availability service	A security service that protects a system to ensure its availability	RFC 4949
Confidentiality	See: data confidentiality	RFC 4949
Countermeasure	An action, device, procedure, or technique that meets or opposes (i.e. counters) a threat, a vulnerability, or an attack by eliminating or preventing it, by minimizing the harm it can cause, or by discovering and reporting it so that corrective action can be taken	RFC 4949
Critical	/system resource/A condition of a system resource such that denial of access to, or lack of availability of, that resource would jeopardize a system user's ability to perform a primary function or would result in other serious consequences, such as human injury or loss of life. (See: availability, precedence. Compare: sensitive)	RFC 4949
Critical infrastructure	The systems and assets, whether physical or virtual, so vital to society that the incapacity or destruction of such may have a debilitating impact on the security, economy, public health or safety, environment, or any combination of these matters	NICCS
Cyber ecosystem	The interconnected information infrastructure of interactions among persons, processes, data, and information and communications technologies, along with the environment and conditions that influence those interactions	NICCS
Cyber infrastructure	An electronic information and communications systems and services and the information contained therein Extended definition: An electronic information and communications systems and services and the information contained therein. Extended definition: The information and communications systems and services composed of all hardware and software that process, store, and communicate information, or any combination of all of these elements: • Processing includes the creation, access, modification, and destruction of information • Storage includes paper, magnetic, electronic, and all other media types • Communications include sharing and distribution of information	NICCS

Table A.2 (Continued)

Term	Definition	Reference
Cybersecurity plan	Formal document that provides an overview of the cybersecurity requirements for an IT and ICS and describes the cybersecurity controls in place or planned for meeting those requirements	DOE RMP
Cybersecurity policy	A set of criteria for the provision of security services	DOE RMP
Cybersecurity program strategy	A plan of action designed to achieve the performance targets that the organization sets to accomplish its mission, vision, values, and purpose	CERT RMM
Cybersecurity program management	Provides governance, strategic planning, and sponsorship for the organization's cybersecurity activities in a manner that aligns cybersecurity objectives with the organization's strategic objectives and the risk to critical infrastructure	ES-C2M2
Cyberspace	The interdependent network of information technology infrastructures, and includes the Internet, telecommunications networks, computer systems, and embedded processors and controllers in critical industries. Common usage of the term also refers to the virtual environment of information and interactions between people	CPR CPR definition is based on National Security Presidential Directive 54/Homeland Security Presidential Directive 23 (NSPD-54/HSPD23)
Cyber–physical systems	Integrations of computation, networking, and physical processes	http://cyberphysicalsystems.org cyber–physical systems
Cybersecurity	The ability to protect or defend the use of cyberspace from cyber incident	CNSSI 4009
Data availability	The property that data is available when it is needed	RFC 4949
Data confidentiality	The property that data is not disclosed to system entities unless they have been authorized to know the data. (See: Bell–LaPadula model, classification, data confidentiality service, secret. Compare: privacy)	RFC 4949
Data confidentiality service	A security service that protects data against unauthorized disclosure (See: access control, data confidentiality, datagram confidentiality service, flow control, inference control)	RFC 4949
Data integrity	The property that data has not been changed, destroyed, or lost in an unauthorized or accidental manner. (See: data integrity service. Compare: correctness integrity, source integrity)	RFC 4949
Data integrity service	A security service that protects against unauthorized changes to data, including both intentional change or destruction and accidental change or loss, by ensuring that changes to data are detectable. (See: data integrity, checksum, datagram integrity service)	RFC 4949
Data origin authentication	The corroboration that the source of data received is as claimed [I7498-2] (See: authentication)	RFC 4949

(Continued)

Table A.2 (Continued)

Term	Definition	Reference
Data origin authentication service	A security service that verifies the identity of a system entity that is claimed to be the original source of received data. (See: authentication, authentication service)	RFC 4949
Data security	The protection of data from disclosure, alteration, destruction, or loss that either is accidental or is intentional but unauthorized	RFC 4949
Domain	An environment or context that (i) includes a set of system resources and a set of system entities that have the right to access the resources and (ii) usually is defined by a security policy, security model, or security architecture. (See: CA domain, domain of interpretation, security perimeter. Compare: COI, enclave)	RFC 4949
Effectiveness	/ITSEC/A property of a TOE representing how well it provides security in the context of its actual or proposed operational use	RFC 4949
Event	An observable occurrence in an information system or network	NICCS
Fail safe	A mode of termination of system functions that prevents damage to specified system resources and system entities (i.e. specified data, property, and life) when a failure occurs or is detected in the system (but the failure still might cause a security compromise). (See: failure control)	RFC 4949
Fail secure	A mode of termination of system functions that prevents loss of secure state when a failure occurs or is detected in the system (but the failure still might cause damage to some system resource or system entity). (See: failure control. Compare: fail safe)	RFC 4949
Governance	An organizational process of providing strategic direction for the organization while ensuring that it meets its obligations, appropriately manages risk, and efficiently uses financial and human resources. Governance also typically includes the concepts of sponsorship (setting the managerial tone), compliance (ensuring that the organization is meeting its compliance obligations), and alignment (ensuring that processes such as those for cybersecurity program management align with strategic objectives)	Adapted from CERT RMM
Identity	The collective aspect of a set of attribute values (i.e. a set of characteristics) by which a system user or other system entity is recognizable or known. (See: authenticate, registration. Compare: identifier)	RFC 4949
Incident	See: security incident	RFC 4949

Table A.2 (Continued)

Term	Definition	Reference
Information assurance	/US Government/Measures that protect and defend information and information systems by ensuring their availability, integrity, authentication, confidentiality, and non-repudiation. These measures include providing for restoration of information systems by incorporating protection, detection, and reaction capabilities	RFC 4949
Information security (INFOSEC)	Measures that implement and assure security services in information systems, including in computer systems (see: COMPUSEC) and in communication systems (see: COMSEC)	RFC 4949
Information security governance	The set of responsibilities and practices exercised by the board and executive management with the goal of providing strategic direction, ensuring that objectives are achieved, ascertaining that risk is managed appropriately and verifying that the enterprise's resources are used responsibly	ISACA glossary
Information security program	The overall combination of technical, operational, and procedural measures and management structures implemented to provide for the confidentiality, integrity, and availability of information based on business requirements and risk analysis	OECD 2002
Information system	An organized assembly of computing and communication resources and procedures – i.e. equipment and services, together with their supporting infrastructure, facilities, and personnel – that create, collect, record, process, store, transport, retrieve, display, disseminate, control, or dispose of information to accomplish a specified set of functions. (See: system entity, system resource. Compare: computer platform.)	RFC 4949
Information system	A discrete set of information resources organized for the collection, processing, maintenance, use, sharing, dissemination, or disposition of information. Note: Information systems also include specialized systems such as industrial/process control systems, telephone switching and private branch exchange (PBX) systems, and environmental control systems	CNSSI 4009
Integrity	See: data integrity, datagram integrity service, correctness integrity, source integrity, stream integrity service, system integrity	RFC 4949
Integrity	The property whereby information, an information system, or a component of a system has not been modified or destroyed in an unauthorized manner	NICCS
Integrity service	A security service that protects against unauthorized changes to data, including both intentional change or destruction and accidental change or loss by ensuring that changes to data are detectable	RFC 4949

(Continued)

Table A.2 (Continued)

Term	Definition	Reference
Intellectual property (IP)	A number of distinct types of legal monopolies over creations of the mind, both artistic and commercial, and the corresponding fields of law. Under intellectual property law, owners are granted certain exclusive rights to a variety of intangible assets, such as musical, literary, and artistic works; discoveries and inventions; and words, phrases, symbols, and designs. Common types of intellectual property include copyrights, trademarks, patents, industrial design rights, and trade secrets in some jurisdictions	
Intrusion	1. (I) A security event, or a combination of multiple security events, that constitutes a security incident in which an intruder gains, or attempts to gain, access to a system or system resource without having authorization to do so. (See: IDS) 2. (I) A type of threat action whereby an unauthorized entity gains access to sensitive data by circumventing a system's security protections. (See: unauthorized disclosure)	RFC 4949
Intrusion detection	Sensing and analyzing system events for the purpose of noticing (i.e. becoming aware of) attempts to access system resources in an unauthorized manner. (See: anomaly detection, IDS, misuse detection. Compare: extrusion detection.) [IDSAN, IDSSC, IDSSE, IDSSY]	RFC 4949
Intrusion detection system	1. (N) A process or subsystem, implemented in software or hardware, that automates the tasks of (i) monitoring events that occur in a computer network and (ii) analyzing them for signs of security problems. [SP31] (See: intrusion detection) 2. (N) A security alarm system to detect unauthorized entry [DC6/9]	RFC 4949
Intrusion prevention system	System that can detect an intrusive activity and can also attempt to stop the activity, ideally before it reaches its targets	CNSSI 4009
Legacy system	A system that is in operation but will not be improved or expanded while a new system is being developed to supersede it	RFC 4949
Level of concern	/US DoD/A rating assigned to an information system that indicates the extent to which protective measures, techniques, and procedures must be applied. (See: critical, sensitive, level of robustness)	RFC 4949
Level of robustness	/US DoD/A characterization of a) the strength of a security function, mechanism, service, or solution and b) the assurance (or confidence) that it is implemented and functioning. [Cons, IATF] (See: level of concern)	RFC 4949
Management controls	Actions taken to manage the development, maintenance, and use of the system, including system-specific policies, procedures and rules of behavior, individual roles and responsibilities, individual accountability, and personnel security decisions	CNSSI 4009

Table A.2 (Continued)

Term	Definition	Reference
Measure (noun)	Variable to which a value is assigned as the result of measurement Note: "Measures" is used to refer collectively to base measures, derived measures, and indicators	ISO/IEC 15939:2007
Measurement	Set of operations having the object of determining a value of a measure	ISO/IEC 15939:2007
Mission critical	A condition of a system service or other system resource such that denial of access to, or lack of availability of, the resource would jeopardize a system user's ability to perform a primary mission function or would result in other serious consequences. (See: critical. Compare: mission essential)	RFC 4949
Non-repudiation service	A security service that provide protection against false denial of involvement in an association (especially a communication association that transfers data)	RFC 4949
Non-repudiation with proof of origin	A security service that provides the recipient of data with evidence that proves the origin of the data and thus protects the recipient against an attempt by the originator to falsely deny sending the data. (See: non-repudiation service)	RFC 4949
Non-repudiation with proof of receipt	A security service that provides the originator of data with evidence that proves the data was received as addressed and thus protects the originator against an attempt by the recipient to falsely deny receiving the data. (See: non-repudiation service)	RFC 4949
Operational controls	The security controls (i.e. safeguards or countermeasures) for an information system that are primarily implemented and executed by people (as opposed to systems)	CNSSI 4009 4009
Operational controls	The security controls (i.e. safeguards or countermeasures) for an information system that are primarily implemented and executed by people (as opposed to systems)	
Personal information	Information that reveals details, either explicitly or implicitly, about a specific individual's household dwelling or other type of premises. This is expanded beyond the normal "individual" component because there are serious privacy impacts for all individuals living in one dwelling or premise. This can include items such as energy use patterns or other types of activities. The pattern can become unique to a household or premises just as a fingerprint or DNA is unique to an individual	NISTIR 7628 Vol. 3, Glossary
Physical security	Tangible means of preventing unauthorized physical access to a system. Examples: fences, walls, and other barriers; locks, safes, and vaults; dogs and armed guards; sensors and alarm bells. [FP031, R1455] (See: security architecture)	RFC 4949

(Continued)

Table A.2 (Continued)

Term	Definition	Reference
Quality measure	A measurement function of two or more values of quality measure elements	ISO/IEC 25021
Quality measure element	Measure defined in terms of an attribute and the measurement method for quantifying it, including optionally the transformation by a mathematical function	ISO/IEC 25021
Quality model	A set of characteristics, and of relationships between them, that provides a framework for specifying quality requirements and evaluating quality	ISO/IEC 25000:2005
Quality of service	The measurable end-to-end performance properties of a network service, which can be guaranteed in advance by a service-level agreement between a user and a service provider, so as to satisfy specific customer application requirements. Note: These properties may include throughput (bandwidth), transit delay (latency), error rates, priority, security, packet loss, packet jitter, etc.	CNSSI 4009
Quality plan	Document that specifies the procedures and resources that will be needed to carry out a project, perform a process, realize a product, or manage a contract	ISO 9000:2005
Quality planning	Setting quality objectives and then specifying the operational processes and resources that will be needed to achieve those objectives. Quality planning is one part of quality management	ISO 9000:2005
Quality policy	Management's commitment to quality, it describes an organization's general quality orientation and clarify its basic intentions (based on the ISO 9000 quality management principles)	ISO 9000:2005
Quality property	Measurable component of quality	ISO/IEC 25000:2005
Reliability	The ability of a system to perform a required function under stated conditions for a specified period of time. (Compare: availability, survivability)	RFC 4949
Repudiation	1. (I) Denial by a system entity that was involved in an association (especially a communication association that transfers data) of having participated in the relationship. (See: accountability, non-repudiation service) 2. (I) A type of threat action whereby an entity deceives another by falsely denying responsibility for an act. (See: deception)	RFC 4949
Residual risk	The portion of an original risk or set of risks that remains after countermeasures have been applied. (Compare: acceptable risk, risk analysis)	RFC 4949
Risk	1. (I) An expectation of loss expressed as the probability that a particular threat will exploit a particular vulnerability with a particular harmful result. (See: residual risk) 2. (O) /SET/The possibility of loss because of one or more threats to information (not to be confused with financial or business risk) [SET2]	RFC 4949

Table A.2 (Continued)

Term	Definition	Reference
Safeguards	Protective measures prescribed to meet the security requirements (i.e. confidentiality, integrity, and availability) specified for an information system. Safeguards may include security features, management constraints, personnel security, and security of physical structures, areas, and devices. Synonymous with security controls and countermeasures	CNSSI 4009
Safety	The property of a system being free from risk of causing harm (especially physical harm) to its system entities. (Compare: security)	RFC 4949
Secure state	1a. (I) A system condition in which the system is in conformance with the applicable security policy. (Compare: clean system, transaction) 1b. (I) /formal model/A system condition in which no subject can access any object in an unauthorized manner. (See: secondary definition under "Bell–LaPadula model")	RFC 4949
Security	1a. (I) system condition that results from the establishment and maintenance of measures to protect the system 1b. (I) system condition in which system resources are free from unauthorized access and from unauthorized or accidental change, destruction, or loss. (Compare: safety) 2. Measures taken to protect a system	RFC 4949
Security architecture	A plan and set of principles that describe: a) the security services that a system is required to provide to meet the needs of its users, b) the system components required to implement the services, and c) the performance levels required in the components to deal with the threat environment (e.g. [R2179]). (See: defense in depth, IATF, OSIRM security architecture, security controls, tutorial under "security policy")	RFC 4949
Security assurance	1. (I) An attribute of an information system that provides grounds for having confidence that the system operates such that the system's security policy is enforced. (Compare: trust) 2. (I) A procedure that ensures a system is developed and operated as intended by the system's security policy	RFC 4949
Security by obscurity	Attempting to maintain or increase security of a system by keeping secret the design or construction of a security mechanism	RFC 4949
Security compromise	A security violation in which a system resource is exposed, or is potentially exposed, to unauthorized access. (Compare: data compromise, exposure, violation)	RFC 4949

(*Continued*)

Table A.2 (Continued)

Term	Definition	Reference
Security controls	The management, operational, and technical controls (safeguards or countermeasures) prescribed for an information system that, taken together, satisfy the specified security requirements and adequately protect the confidentiality, integrity, and availability of the system and its information. [FP199] (See: security architecture)	RFC 4949
Security domain	See: domain	RFC 4949
Security engineering	An interdisciplinary approach and means to enable the realization of secure systems. It focuses on defining customer needs, security protection requirements, and required functionality early in the systems development life cycle, documenting requirements, and then proceeding with design, synthesis, and system validation while considering the complete problem	CNSSI 4009
Security environment	The set of external entities, procedures, and conditions that affect secure development, operation, and maintenance of a system (See: "first law" under "Courtney's laws")	RFC 4949
Security event	An occurrence in a system that is relevant to the security of the system. (See: security incident)	RFC 4949
Security incident	A security event that involves a security violation. (See: CERT, security event, security intrusion, security violation)	RFC 4949
Security intrusion	A security event, or a combination of multiple security events, that constitutes a security incident in which an intruder gains, or attempts to gain, access to a system or system resource without having authorization to do so	RFC 4949
Security management infrastructure	System components and activities that support security policy by monitoring and controlling security services and mechanisms, distributing security information, and reporting security events	RFC 4949
Security mechanism	A method or process (or a device incorporating it) that can be used in a system to implement a security service that is provided by or within the system. (See: tutorial under "security policy." Compare: security doctrine)	RFC 4949
Security model	A schematic description of a set of entities and relationships by which a specified set of security services are provided by or within a system. Example: Bell–LaPadula model, OSIRM. (See: tutorial under "security policy")	RFC 4949

Table A.2 (Continued)

Term	Definition	Reference
Security service	1. (I) A processing or communication service that is provided by a system to give a specific kind of protection to system resources (See: access control service, audit service, availability service, data confidentiality service, data integrity service, data origin authentication service, non-repudiation service, peer entity authentication service, system integrity service) Tutorial: Security services implement security policies and are implemented by security mechanisms 2. (O) A service, provided by a layer of communicating open systems, [that] ensures adequate security of the systems or the data transfers [I7498-2]	RFC 4949
Sensitive information	1. (I) Information for which (i) disclosure, (ii) alteration, or (iii) destruction or loss could adversely affect the interests or business of its owner or user. (See: data confidentiality, data integrity, sensitive. Compare: classified, critical) 2. (O) /US Government/Information for which (i) loss, (ii) misuse, (iii) unauthorized access, or (iv) unauthorized modification could adversely affect the national interest or the conduct of federal programs, or the privacy to which individuals are entitled under the Privacy Act of 1974, but that has not been specifically authorized under criteria established by an executive order or an Act of Congress to be kept classified in the interest of national defense or foreign policy	RFC 4949
Sensitivity	A condition of a system resource such that the loss of some specified property of that resource, such as confidentiality or integrity, would adversely affect the interests or business of its owner or user. (See: sensitive information. Compare: critical)	RFC 4949
Signal intelligence	The science and practice of extracting information from signals. (See: signal security)	RFC 4949
System	A combination of interacting elements organized to achieve one or more stated purposes	ISO/IEC 15288:2008
System integrity	An attribute or quality that a system has when it can perform its intended function in an unimpaired manner, free from deliberate or inadvertent unauthorized manipulation. [C4009, NCS04] (See: recovery, system integrity service)	RFC 4949
System integrity service	A security service that protects system resources in a verifiable manner against unauthorized or accidental change, loss, or destruction. (See: system integrity)	RFC 4949
Technical security	Security mechanisms and procedures that are implemented in and executed by computer hardware, firmware, or software to provide automated protection for a system. (See: security architecture. Compare: administrative security)	RFC 4949

(Continued)

Table A.2 (Continued)

Term	Definition	Reference
Threat	1a. (I) A potential for violation of security, which exists when there is an entity, circumstance, capability, action, or event that could cause harm. (See: dangling threat, INFOCON level, threat action, threat agent, threat consequence. Compare: attack, vulnerability) 1b. (N) Any circumstance or event with the potential to adversely affect a system through unauthorized access, destruction, disclosure, or modification of data, or denial of service. [C4009] (See: sensitive information)	RFC 4949
Threat action	A realization of a threat, i.e. an occurrence in which system security is assaulted as the result of either an accidental event or an intentional act. (See: attack, threat, threat consequence)	RFC 4949
Threat agent	A system entity that performs a threat action, or an event that results in a threat action	RFC 4949
Threat consequence	A security violation that results from a threat action	RFC 4949
Trust	1. (I) /information system/ A feeling of certainty (sometimes based on inconclusive evidence) either (i) that the system will not fail or (ii) that the system meets its specifications (i.e. the system does what it claims to do and does not perform unwanted functions). (See: trust level, trusted system, trustworthy system Compare: assurance.) 2. (I) /PKI/A relationship between a certificate user and a CA in which the user acts according to the assumption that the CA creates only valid digital certificates	RFC 4949
Trusted system	1. (I) /information system/A system that operates as expected, according to design and policy, doing what is required – despite environmental disruption, human user and operator errors, and attacks by hostile parties – and not doing other things [NRC98]. (See: trust level, trusted process. Compare: trustworthy) 2. (N) /multilevel secure/A [trusted system is a] system that employs sufficient hardware and software assurance measures to allow its use for simultaneous processing of a range of sensitive or classified information. [NCS04] (See: multilevel security mode)	RFC 4949
Trustworthy system	A system that not only is trusted but also warrants that trust because the system's behavior can be validated in some convincing way, such as through formal analysis or code review. (See: trust. Compare: trusted)	RFC 4949
Vulnerability	A flaw or weakness in a system's design, implementation, or operation and management that could be exploited to violate the system's security policy. (See: harden)	RFC 4949
Vulnerability assessment	Systematic examination of an information system or product to determine the adequacy of security measures, identify security deficiencies, provide data from which to predict the effectiveness of proposed security measures, and confirm the adequacy of such measures after implementation	CNSSI 4009

Table A.3 Examples of cybersecurity glossaries provided by different organizations.

Organization	Name	Address
ANSI/ISA	ANSI/ISA-62443-1-1 (99.01.01)-2007, Manufacturing and Control Systems Security, Part 1: Concepts, Models and Terminology	
ASQ	Quality	http://asq.org/learn-about-quality/six-sigma/overview/overview.html
Bank of America	Privacy & Security Glossary	https://www.bankofamerica.com/privacy/privacy-policy-glossary.go
CDSE	Handbook	Glossary of Security Terms, Definitions, and Acronyms, http://www.cdse.edu/documents/cdse/Glossary_Handbook.pdf
CNSSI 4009	National Information Assurance (IA) Glossary (CNSSI 4009 Instruction No. 4009	https://www.CNSSI 4009.gov/CNSSI 4009/issuances/Instructions.cfm
Computer Security	Glossary	http://www.computer-security-glossary.org
DHS	The Risk Management Process for Federal Facilities: An Interagency Security Committee Standard 2nd Edition, November 2016, Glossary of Terms	https://www.dhs.gov/sites/default/files/publications/isc-risk-management-process-2016-508.pdf
EN ISO	EN ISO 9000:2015, Quality management systems – Fundamentals and vocabulary	
ENISA	ENISA, 2012, Smart Grid Security, Annex I General concepts and dependencies with ICT	http://www.enisa.europe.eu
ENISA	ENISA, 2013, Proposal for a list of security measures for Smart Grids, Annex I – Glossary	http://www.enisa.europe.eu
ENISA	Risk Management Glossary	https://www.enisa.europa.eu/activities/risk-management/current-risk/risk-management-inventory/glossary
Gartner	IT Glossary Smart Grid	http://www.gartner.com/it-glossary/smart-grid
Global Data Management Association	Dictionary	The DAMA Dictionary of Data Management (2nd Edition)
IEC	IEC TS 62351-2:2008, Power systems management and associated information exchange – Data and communications security – Part 2: Glossary of terms	https://webstore.iec.ch/publication/6905

(*Continued*)

Table A.3 (Continued)

Organization	Name	Address
IETF	RFC 4919, Internet Security Glossary, V2	https://tools.ietf.org/html/rfc4949
Industrial Internet Consortium (IIC)	Industrial Internet Vocabulary	https://workspace.iiconsortium.org/kws/public/download/1267/IICVocabulary_v1.0_approved_20150625.pdf
Internet Society	Glossary	http://www.internetsociety.org/publications/ietf-journal/glossary
ISA	Dictionary	http://isa99.isa.org/ISA99%20Wiki/Master-Glossary.aspx
ISACA	Glossary	http://www.isaca.org/knowledge-center/documents/glossary/cybersecurity_fundamentals_glossary.pdf
ISO	ISO 9000, Glossary – Guidance on selected words used in the ISO 9000 family of standards	http://www.iso.org/iso/03_terminology_used_in_iso_9000_family.pdf
ISO	ISO Guide 73:2009, Risk Management – Vocabulary	https://www.iso.org/standard/44651.html
ISO	ISO/IEC 2700 third edition, 2014, Information technology – Security techniques – Information security management systems – Overview and vocabulary	http://www.oso.org
ISO	ISO 2700 Definitions and Terms	http://www.praxiom.com/iso-27000-definitions.htm
ISO/IEC	27000:2014, Information technology – Security techniques – Information security management systems – Overview and vocabulary	https://www.iso.org/standard/63411.html
ISO/IEC	ISO/IEC 15408, Common Criteria for Information Technology Evaluation, Part 1: Introduction and general model, 2012,V3.1,R4	https://www.commoncriteriaportal.org/files/ccfiles/CCPART1V3.1R4.pdf
ISO/IEC	ISO/IEC 2382:2015, Information technology – Vocabulary	https://www.iso.org/standard/63598.html
ISO/IEC	ISO/IEC TR 15443-1:2012, Information technology – Security techniques – Security assurance framework – Part 1: Introduction and concepts	https://www.iso.org/standard/59138.html
National Initiative for Cybersecurity Careers and Studies (NICCS)	A Glossary of Common Cybersecurity Terminology	https://niccs.us-cert.gov/glossary
NERC	Glossary of Terms Used in Reliability Standards	http://www.nerc.com/pa/Stand/Glossary%20of%20Terms/Glossary_of_Terms.pdf

NetSmartz	Internet Safety Terms	http://www.netsmartz.org/safety/definitions
NIST	NISTIR 7298r2, Glossary of Key Information Security Terms	http://nvlpubs.nist.gov/nistpubs/ir/2013/NIST.IR.7298r2.pdf
NIST	NIST SP800-53r4, Security and Privacy Controls for Federal Information Systems and Organizations, Appendix B Glossary	https://nvd.nist.gov/800-53/Rev4
NIST	NISTIR 7628r1, Guidelines to Smart Grid Cybersecurity, Appendix J – Glossary and Acronyms	https://csrc.nist.gov/publications/detail/nistir/7628/rev-1/final
NIST	NIST SP800-160, Systems Security Engineering: Considerations for a Multidisciplinary Approach in the Engineering of Trustworthy Secure Systems, Vol. 1, Glossary Common Terms and Definitions, Appendix B	https://csrc.nist.gov/publications/detail/sp/800-160/vol-1/final
NIST	NIST SP800-160, Systems Security Engineering: Cyber Resiliency Considerations for the Engineering of Trustworthy Secure Systems, Vol. 2, Draft, Glossary Common Terms and Definitions, Appendix B	https://csrc.nist.gov/publications/detail/sp/800-160/vol-2/draft
OECD	Glossary of Statistical Terms	http://stats.oecd.org/glossary/detail.asp?ID=3108
PC Magazine Encyclopedia	Definitions on common technical and computer related terms	https://www.pcmag.com/encyclopedia
Radware	Glossary	http://www.radware.com/Glossary
SANS	Glossary of Terms	http://www.sans.org/security-resources/glossary-of-terms
Techopedia	IT Dictionary	http://www.techopedia.com/it-dictionary
TechTarget	Dictionary	http://whatis.techtarget.com
US Energy Information Administration (EIA)	EIA Glossary	http://www.eia.gov/tools/glossary
US DOE	Cybersecurity Procurement Language	http://energy.gov/sites/prod/files/2014/04/f15/CybersecProcurementLanguage-EnergyDeliverySystems_040714_fin.pdf
US DOE	Electricity Subsector Cybersecurity Capability Maturity Model, Annex B	http://energy.gov/sites/prod/files/2014/02/f7/ES-C2M2-v1-1-Feb2014.pdf
US DOE	Electricity subsector cybersecurity risk management process, May 2012, Appendix B	https://www.energy.gov/sites/prod/files/Cybersecurity%20Risk%20Management%20Process%20Guideline%20-%20Final%20-%20May%202012.pdf
US DOE	Risk Management Guide DOE G 413.3-7A Attachment 15 1-12-2011	https://www.directives.doe.gov/directives-documents/400-series/0413.3-EGuide-07a/@@images/file
US-CERT	Glossary	http://niccs.us-cert.gov/glossary

References

[American Dictionary] Houghton Mifflin Company. (1994). *The American Heritage dictionary* (third edition). Dell Publishing.

[CNSSI 4009] CNSSI. (2010). *Committee on National Security Systems (CNSS)* Information Assurance (IA) Glossary. Committee on National Security Systems (CNSS) Instruction No. 4009.

[IETF RFC 4949] *IETF Request for Comments (RFC) 4949 Internet Security Glossary* (Version 2, August 2007).

[NIST SP800-53r4] *NIST Special Publication (SP) 800-53 Security and Privacy Controls for Federal Information Systems and Organizations Revision 4*, (April 2013). Computer Security Division Information Technology Laboratory. http://dx.doi.org/10.6028/NIST.SP.800-53r4 * Superseded by [NIST SP800-53r5] in September 2020; Withdrawal Date September 23, 2021.

B

Power Grid Concepts

Power

Appendix B includes definitions of basic terms and concepts used in power industry, an overview of US electric power industry, and examples of dictionaries with definitions of terms for power grid (see Table B.2).

B.1 Basic Terms

B.1.1 Ancillary Services

All services necessary for the operation of a transmission or distribution system. It comprises compensation for energy losses, frequency control (automated, local fast control, and coordinated slow control), voltage and flow control (reactive power, active power, and regulation devices), and restoration of supply (black start, temporary island operation). These services are required to provide system reliability and power quality. They are provided by generators and system operators.

B.1.2 Bulk Electric System

Bulk power system (BPS) has been revised and changed to bulk electric system (BES) that is the electrical generation resources, transmission lines, interconnections with neighboring systems, and associated equipment, generally operated at voltages of 100 kV or higher; radial transmission facilities serving only load with one transmission source are generally not included in this definition.

B.1.3 Bulk Power System (BPS)

A bulk power system (BPS) is a large interconnected electrical system made up of generation and transmission facilities and their control systems. A BPS does not include facilities used in the local distribution of electric energy. If a BPS is disrupted, the effects are felt in more than one location. In the United States, BPS are overseen by the North American Electric Reliability Corporation (NERC) (BPS has changed to BES).

Building an Effective Security Program for Distributed Energy Resources and Systems: Understanding Security for Smart Grid and Distributed Energy Resources and Systems, Volume 1, First Edition. Mariana Hentea.

B.1.4 Centralized Historian

A historian centralizing data from multiple supervisor consoles (SCADA). Data retention time is often long, but with a coarser granularity than in the local historian.

Management can use this historian for data analysis, statistics, etc. concerning the entire production unit.

B.1.5 Circuit Breakers

Devices used to open or close electric circuits. If a transmission or distribution line is in trouble, a circuit breaker can disconnect it from the rest of the system.

B.1.6 Cooperative

A cooperative (also known as co-op, co-operative, or coop) is an autonomous association of people united voluntarily to meet their common economic, social, and cultural needs and aspirations through a jointly owned and democratically controlled business.

Cooperatives include nonprofit community organizations and businesses that are owned and managed by the people who use their services (a consumer cooperative), by the people who work there (a worker cooperative), by the people who live there (a housing cooperative); hybrids such as worker cooperatives that are also consumer cooperatives or credit unions; multi-stakeholder cooperatives such as those that bring together civil society and local actors to deliver community needs; and second- and third-tier cooperatives whose members are other cooperatives.

Cooperatives frequently have social goals that they aim to accomplish by investing a proportion of trading profits back into their communities.

B.1.7 Control System

System of components that act together to maintain actual system performance close to a desired set of performance specifications. The type of control needed by applications determines the architecture of a control system: open loop and closed loop (or feedback control). These are the basic strategies for the implementation of automation schemes.

Open-loop control systems are those in which the output has no effect on the input; it supports prior known response to the control inputs (see Figure B.1).

Closed-loop control system includes a way to measure its output to sense changes and take feedback correction to input (see Figure B.2a,b). The response or the output is continuously compared with the desired result, and the control output to the process is modified to reduce the deviation and aim for the reference.

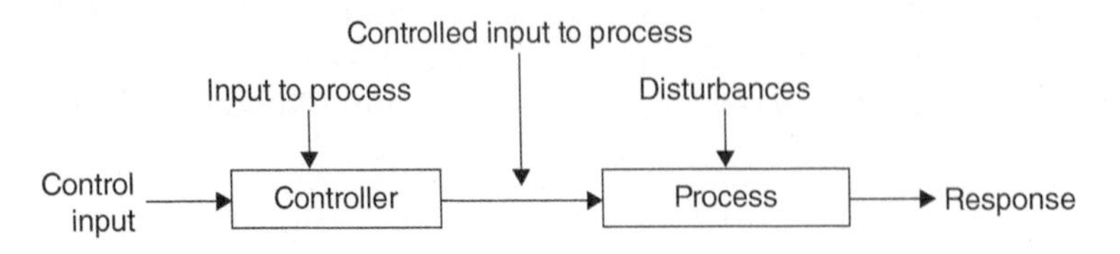

Figure B.1 Open-loop control. *Source:* [Sharma 2011]. © 2011, Elsevier.

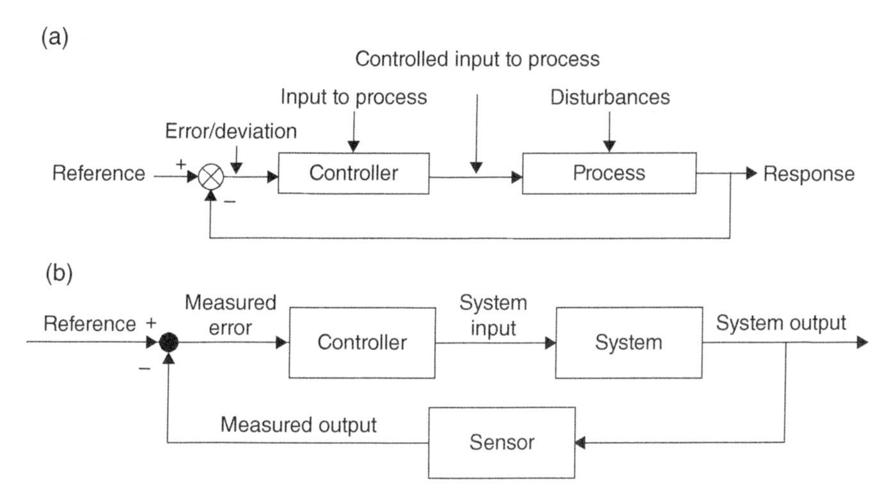

Figure B.2 Examples of closed-loop control diagram. (a) Closed-loop control diagram. *Source:* [Sharma 2011]. © 2011, Elsevier, and (b) closed-loop control with measured output. *Source:* [Orzetto]. Licensed under CC BY-SA 4.0.

One advantage of feedback systems is that the effect of distortion, noise, and unwanted disturbances can be effectively reduced. Of primary interest is the transient response performance that characterizes the stability of the system.

A stable system is defined as a system with a bounded system response. If the system is subjected to a bounded input or disturbance and the response is bounded in magnitude, the system is called to be stable.

When power engineers talk about security, they are thinking of system stability, not information security. In North America, the term "secure state" is used for a stable state that is not a potentially unstable state.

B.1.8 Conservation of Energy

Energy can neither be created (produced) nor destroyed by itself. It can only be transformed. The total inflow of energy into a system must equal the total outflow of energy from the system, plus the change in the energy contained within the system. Energy is subject to a strict global conservation law; that is, whenever one measures (or calculates) the total energy of a system of particles whose interactions do not depend explicitly on time, it is found that the total energy of the system always remains constant. The total energy of a system can be calculated by adding up all forms of energy in the system.

B.1.9 Critical Characteristic

Any feature throughout the life cycle of a Critical Safety Item, such as dimension, tolerance, finish, material or assembly, manufacturing or inspection process, operation, field maintenance, or depot overhaul requirement that if nonconforming, missing, or degraded may cause the failure or malfunction of a Critical Safety Item.

Often, the term Critical Characteristic is synonymous with Critical Safety Characteristic and Flight Safety Characteristic.

B.1.10 Critical Safety Item

An item (part, assembly, installation, or production system) that, if missing or not conforming to the design data, quality requirements, or overhaul and maintenance documentation, would result in an unsafe condition per the established risk acceptance criteria.

Critical Safety Items (CSIs) include items determined to be "life limited," "fracture critical," "fatigue sensitive," etc. The determining factor in CSIs is the consequence of failure, not the probability that the failure or consequence would occur.

B.1.11 Distributed Information System

Distributed information system is an information system interconnecting multiple sites. A system falls into this category when it is not possible to set up a closed perimeter with access control around the entire system. This applies in particular to the cables and optical fiber for the network supporting the system.

B.1.12 Engineering Station

Engineering station computer devices with software packages for configuring, designing, programming, or administrating industrial devices such as PLCs and SCADAs. This device is connected to the industrial network and made available to different teams (e.g. maintenance, engineering, support).

B.1.13 Historian Database

Historian database containing logs of alarms and process values collected by the supervision software (SCADA). These historians are often local or centralized.

B.1.14 Energy

A property of objects that can be transferred to other objects or converted into different forms.

B.1.15 Electrical Energy

Energy newly derived from electric potential energy or kinetic energy. When loosely used to describe energy absorbed or delivered by an electrical circuit (for example, one provided by an electric power utility), electrical energy is about energy that has been converted from electric potential energy. This energy is supplied by the combination of electric current and electric potential that is delivered by the circuit. Thus, all electrical energy is potential energy before it is delivered to the end use. Once converted from potential energy, electrical energy can always be called another type of energy (heat, light, motion, etc.).

Electricity is a secondary energy source that means that we get it from the conversion of other sources of energy, like coal, natural gas, oil, nuclear power, and other natural sources, which are called primary sources. The energy sources we use to make electricity can be renewable or nonrenewable, but electricity itself is neither renewable or nonrenewable.

B.1.16 Energy Forms

All of the many forms of energy are convertible to other kinds of energy. In Newtonian physics, there is a universal law of conservation of energy that says that energy can be neither created nor be destroyed; however, it can change from one form to another. Common energy forms include kinetic, elastic, chemical, radiant, and thermal energy.

B.1.17 Electricity Generation

The process of generating electrical energy from other forms of energy. The fundamental principle of electricity generation was discovered during the 1820s and early 1830s by the British scientist Michael Faraday. His basic method is still used today: electricity is generated by the movement of a loop of wire or disc of copper between the poles of a magnet.

For electric utilities, it is the first step in the delivery of electricity to consumers. Other processes such as electricity transmission, distribution, and electrical power storage and recovery using pumped storage methods are normally carried out by the electric power industry.

Electricity is most often generated at a power station by electromechanical generators, primarily driven by heat engines fueled by chemical combustion or nuclear fission but also by other means such as the kinetic energy of flowing water and wind. There are many other technologies that can be and are used to generate electricity such as solar photovoltaics and geothermal power.

B.1.18 Electric Generator

A device for converting mechanical energy into electrical energy. The process is based on the relationship between magnetism and electricity. When a wire or any other electrically conductive material moves across a magnetic field, an electric current occurs in the wire.

An electric utility power station uses either a turbine, engine, water wheel, or other similar machine to drive an electric generator or a device that converts mechanical or chemical energy to generate electricity. Steam turbines, internal combustion engines, gas combustion turbines, water turbines, and wind turbines are the most common methods to generate electricity.

B.1.19 Electric Circuit Components

From the standpoint of electric power, components in an electric circuit can be divided into two categories:

- **Passive devices** or **loads**
  When electric charges move through a potential difference from a higher to a lower voltage, that is, when conventional current (positive charge) moves from the positive (+) terminal to the negative (−) terminal, work is done by the charges on the device. The potential energy of the charges due to the voltage between the terminals is converted to kinetic energy in the device. These devices are called passive components or loads; they consume electric power from the circuit, converting it to other forms of energy such as mechanical work, heat, light, etc. Examples are electrical appliances, such as light bulbs, electric motors, and electric heaters. In alternating current (AC) circuits, the direction of the voltage periodically reverses, but the current always flows from the higher potential to the lower potential side.
- **Active devices** or **power sources**
  If the charges are moved by an exterior force through the device in the direction from the lower electric potential to the higher (so positive charge moves from the negative to the positive terminal), work will be done on the charges, and energy is being converted to electric potential energy from some other type of energy, such as mechanical energy or chemical energy. Devices in which this occurs are called active devices or power sources, such as electric generators and batteries.

Some devices can be either a source or a load, depending on the voltage and current through them. For example, a rechargeable battery acts as a source when it provides power to a circuit but as a load when it is connected to a battery charger and is being recharged.

B.1.20 Electric Grid

The term refers to the electricity infrastructure that lies between the generation sources and the consumer (e.g. transmission and distribution or electricity delivery).

B.1.21 Electric Power

Electric power is the rate at which electrical energy is transferred by an electric circuit. The SI unit of power is the watt, 1 J per second. Electric power is usually sold by the kilowatt hour, which is the product of power in kilowatts multiplied by running time in hours. Electric utilities measure power using an electricity meter, which keeps a running total of the electric energy delivered to a customer.

Electric power is usually produced by electric generators but can also be supplied by sources such as electric batteries. It is usually supplied to businesses and homes by the electric power industry through an electric power grid. Electric power is transmitted on overhead lines and also on underground high-voltage cables.

B.1.22 Electric Power Network

Network consisting of installations, substations, lines, or cables for the transmission and distribution of electric energy. The boundaries of the different parts of an electric power network are defined by appropriate criteria, such as geographical situation, ownership, voltage, etc.

B.1.23 Electric Power Versus Electric Energy

Electric power is the rate at which electricity does work – measured at a point in time, that is, with no time dimension. The unit of measure for electric power is a watt. The maximum amount of electric power that a piece of electrical equipment can accommodate is the capacity or capability of that equipment.

Electric energy is the amount of work that can be done by electricity. The unit of measure for electric energy is a watt-hour. Electric energy is measured over a period of time and has a time dimension as well as an energy dimension. The amount of electric energy produced or used during a specified period of time by a piece of electrical equipment is referred to as generation or consumption.

B.1.24 Electric Power System

Electric system refers to the entire system of generation, transmission, distribution, storage, and end use. An electric power system is a group of generation, transmission, distribution, communication, and other facilities that are physically connected and operated as a single unit under one control. The flow of electricity within the system is maintained and controlled by dispatch centers that can buy and sell electricity based on system requirements.

B.1.25 Electricity Supply System

System consisting of all installations and plant provided for the purpose of generating, transmitting, and distributing electric energy.

B.1.26 Electric Utility

An electric utility is a company in the electric power industry (often a public utility) that engages in electricity generation and distribution of electricity for sale generally in a regulated market. The electrical utility industry is a major provider of energy in most countries.

The utilities can be engaged in all or only some aspects of the industry. Electricity markets are also considered electric utilities – these entities buy and sell electricity, acting as brokers, but usually do not own or operate generation, transmission, or distribution facilities. Utilities are regulated by local and national authorities. Electric utilities are facing increasing demands including aging infrastructure, reliability, and regulation.

B.1.27 Human–Machine Interface

Human–machine interface allowing an intervener to interact with and control the operation of an ICS.

B.1.28 Industrial Control Network (ICS)

A system of interconnected equipment used to monitor and control physical equipment in industrial environments. These networks differ quite significantly from traditional enterprise networks due to the specific requirements of their operation. Despite the functional differences between industrial and enterprise networks, a growing integration between the two has been observed. The technology in use in industrial networks is also beginning to display a greater reliance on Ethernet and Web standards, especially at higher levels of the network architecture. This has resulted in a situation where engineers involved in the design and maintenance of control networks must be familiar with both traditional enterprise concerns, such as network security, and traditional industrial concerns, such as determinism and response time.

As described in [Amin 2008], the continental-scale grid is configured as a multiscale, multilevel hybrid system consisting of vertically integrated hierarchical networks including the generation layer and three basic levels: transmission level, sub transmission level, and distribution level. The power system adaptation to disturbances can be characterized as having multiple states, or modes, during which specific operational and control actions and reactions are taking place. In addition to these spatial, energy, and operational levels, power systems are also multiscale in the time domain, from nanoseconds to decades, as shown in Table B.1.

B.1.29 Industrial Control System

Set of human and material resources designed to control or operate a group of sensors and actuators.

B.1.30 Instrumentation and Control (I&C) System

Instrumentation and control (I&C) system is the collection of devices that monitor, control, and protect the power system. The I&C devices built using microprocessors are commonly referred to as intelligent electronic devices (IEDs). Microprocessors are single-chip computers that allow the devices into which they are built to process data, accept commands, and communicate information like a computer. Automatic processes can be run in the IEDs. Examples of devices are:

- Remote terminal unit (RTU) used to transfer collected data to other devices and receive data and control commands from other devices; a user programmable RTU is referred to as smart RTU.
- Meter used to record measurements of power system current, voltage, and power values.
- Digital fault recorder used to record power system disturbance information such as harmonics, frequency, and voltage.

- Programmable logic controller (PLC) used to can be programmed to perform logical control and working upon the command given by their master.
- Protective relay used to sense power system disturbances and automatically perform control actions on the I&C system and the power system to protect personnel and equipment.
- Controlling (output) devices include load tap changer and recloser controller: load tap changer used for automatically changing the tap position on transformers, either locally or remotely; recloser controller remotely monitors and controls the operation of automated reclosers and switches.
- Communication devices include a communications processor and a substation controller incorporating the functions of many other I&C devices into one IED, performs data acquisition and control of the other substation IEDs, and concentrates the data it acquires for transmission to one or many masters inside and outside the substation.

B.1.31 Intelligent Electronic Device (IED)

IED is a microprocessor-based controller of power system equipment, such as circuit breakers, transformers, and capacitor banks. In addition to controlling a device, an IED may have connections as a client, or as a server, or both, with computer-based systems including other IEDs. An IED is, therefore, any device incorporating one or more processors, with the capability to receive data from an external sender or to send data to an external receiver.

B.1.32 Key Characteristic

A feature of a material, process, or part (includes assemblies) whose variation within the specified tolerance has a significant influence on product fit, performance, service life, or manufacturability.

B.1.33 Local Historian

Historian located near the industrial devices for which records data. Data retention time is often limited, but with a very fine granularity. This historian allows operators to perform detailed analyses when production incidents occur.

B.1.34 Management Information System

Information system including services and applications designed for management purposes (e.g. office applications, human resources, customer service).

B.1.35 Measuring Electricity

Electricity is measured in units of power called watts. It was named to honor James Watt, the inventor of the steam engine. One watt is a very small amount of power. It would require nearly 750 W to equal one horsepower. A kilowatt represents 1000 W. A kilowatt-hour (kWh) is equal to the energy of 1000 W working for one hour. The amount of electricity a power plant generates or a customer uses over a period of time is measured in kilowatt-hours (kWh). Kilowatt-hours are determined by multiplying the number of kW's required by the number of hours of use.

B.1.36 Moving Electricity

To solve the problem of sending electricity over long distances, George Westinghouse developed a device called a transformer. The transformer allowed electricity to be efficiently transmitted over long distances (see Figure B.3). This made it possible to supply electricity to homes and businesses located far from the electric generating plant.

The electricity produced by a generator travels along cables to a transformer, which changes electricity from low voltage to high voltage. Electricity can be moved long distances more efficiently using high voltage. Transmission lines are used to carry the electricity to a substation. Substations have transformers that change the high-voltage electricity into lower-voltage electricity. From the substation, distribution lines carry the electricity to homes, offices, and factories, which require low-voltage electricity.

Figure B.4 shows the control center with the typical communication networks among different elements of the power grid.

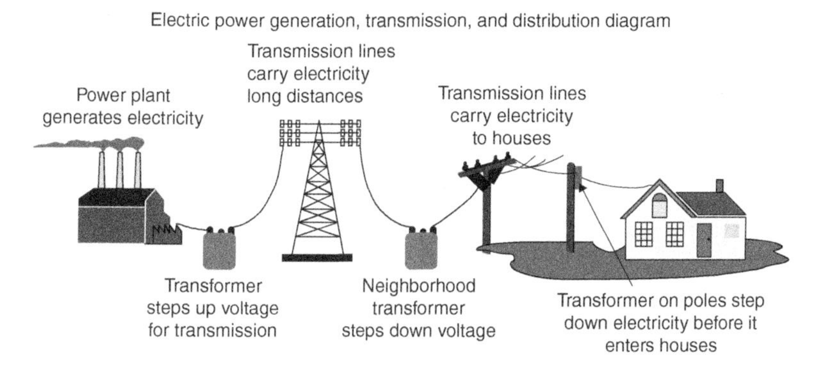

Figure B.3 Transporting electricity from generator to customer. *Source:* [Education]. Public Domain.

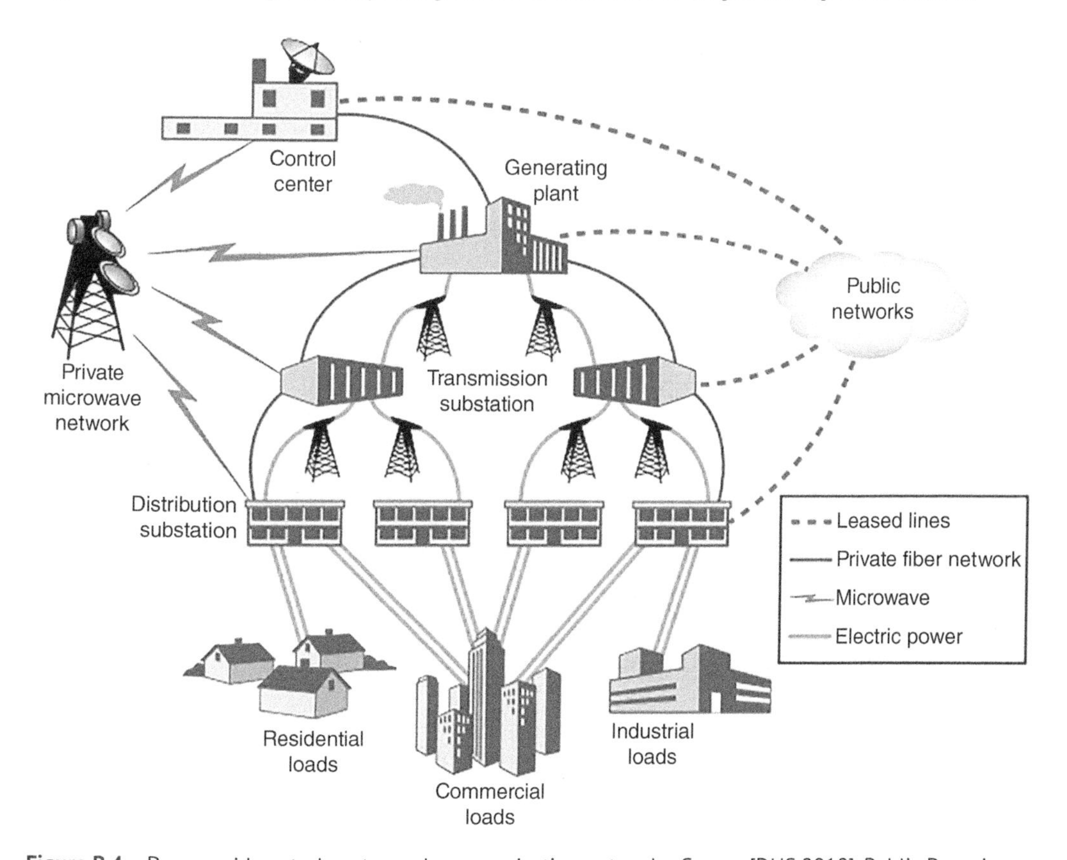

Figure B.4 Power grid control center and communication networks. *Source:* [DHS 2010]. Public Domain.

B.1.37 Passive Sign Convention

Since electric power can flow either into or out of a component, a convention is needed for which direction represents positive power flow. Electric power flowing out of a circuit *into* a component is arbitrarily defined to have a positive sign, while power flowing into a circuit from a component is defined to have a negative sign. Thus, passive components have positive power consumption, while power sources have negative power consumption. This is called the passive sign convention.

B.1.38 Power Plant Efficiency

Most power plants are about 35% efficient. This means that for every 100 units of energy that go into a plant, only 35 units are converted to usable electrical energy.

B.1.39 Power System

The collection of devices that make up the physical systems that generate, transmit, and distribute power.

B.1.40 Power System Automation

The act of automatically controlling the power system via instrumentation and control devices.

B.1.41 Power Delivery System

Power delivery system includes the busbar located at the generating plant (where the power delivery system begins) and extends to the energy-consuming device or appliance at the end user.

This means that the power delivery system encompasses generation step-up transformers; the generation switchyard; transmission substations, lines, and equipment; distribution substations, lines, and equipment; intelligent electronic devices; communications; distributed energy resources (DERs) located at end users; power quality mitigation devices and uninterruptible power supplies; sensors; energy storage devices; and other equipment.

Inadequacies in the power delivery system are manifested in the form of poor reliability, excessive occurrences of degraded power quality, vulnerability to mischief or terrorist attack, the inability to integrate renewables, and the inability to provide enhanced services to consumers.

B.1.42 Substation Automation

Substation automation includes use of data from Intelligent electronic devices (IEDs), control and automation capabilities within the substation, and control commands from remote users to control power system devices.

Power system automation includes processes associated with generation and delivery of power. Monitoring and control of power delivery systems in the substation and on the pole reduce the occurrence of outages and shorten the duration of outages that do occur.

The IEDs, communication protocols, and communication methods work together as a system to perform power system automation. Since full substation automation relies on substation integration, these terms – substation automation and power system automation – are often used interchangeably.

B.1.43 Reliability

Reliability, in a bulk power electric system, is the degree to which the performance of the elements of that system results in power being delivered to consumers within accepted standards and in the amount desired. The degree of reliability may be measured by the frequency, duration, and magnitude of adverse effects on consumer service.

B.1.44 Reliable Operation

Reliable operation means operating the elements of the bulk power system (BPS) within equipment and electric system thermal, voltage, and stability limits so that instability, uncontrolled separation, or cascading failures of such system will not occur as a result of a sudden disturbance, including a cybersecurity incident or unanticipated failure of system elements.

B.2 US Electric Power Industry Overview

Electricity is an integral part of life in the United States. It is indispensable to factories, commercial establishments, homes, and even most recreational facilities. Lack of electricity causes not only inconvenience but also economic loss due to reduced industrial production. Various aspects of the electric power industry are provided in this overview.

The vision of electric power grid will move from an electromechanically controlled system to an electronically controlled network in the next two decades as described in [Amin 2008] is one goal of Smart Grid too. While challenges still need to be addressed, the unprecedented "developments in both information technology and materials science and engineering promise significant improvements in the security, reliability, efficiency, and cost effectiveness of electric power delivery systems" [Amin 2008].

B.2.1 Traditional Electric Utilities

In the United States, electric utilities include investor-owned, publicly owned, cooperatives, and federal utilities.

Power marketers are also considered electric utilities – these entities buy and sell electricity, but usually do not own or operate generation, transmission, or distribution facilities. Utilities are regulated by local, state, and federal authorities.

Generally, interstate activities (those that cross state lines) are subject to federal regulation, while intrastate activities are subject to state regulation.

Wholesale rates (sales and purchases between electric utilities), licensing of hydroelectric facilities, questions of nuclear safety and high-level nuclear waste disposal, and environmental regulation are federal concerns. Approval for most plant and transmission line construction and retail rate levels are state regulatory functions.

State public service commissions have jurisdiction primarily over the large, vertically integrated, investor-owned electric utilities that own more than 75% of the nation's generating and transmission capacity and serve about 75% of ultimate consumers. There are 239 investor-owned electric utilities, 2009 publicly owned electric utilities, 912 consumer-owned rural electric cooperatives, and 10 federal electric utilities. Approximately 20 states regulate cooperatives, and 7 states regulate municipal electric utilities; many state legislatures, however, defer this control to local municipal officials or cooperative members.

B.2.2 Meters

Electric utilities use electric meters installed at customers premises to measure electric energy delivered to their customers for billing purposes. They are typically calibrated in billing units; the most common is the kilowatt-hour [kWh]. The electricity consumptions are usually read once each billing period.

An electricity meter, electric meter, or energy meter is a device that measures the amount of electric energy consumed by a residence, business, or an electrically powered device.

B.2.3 Consumer Sectors

Utility service territories are geographically distinct from one another. Each territory is usually composed of many different types of consumers. Electricity consumers are divided into classes of service or sectors based on the type of service they receive such as residential, commercial, industrial, and others.

Sectorial classification of consumers is determined by each utility and is based on various criteria such as:

- Demand levels.
- Rate schedules.
- North American Industry Classification System (NAIC) codes.
- Distribution voltage.
- Accounting methods.
- End-use applications.
- Other social and economic characteristics.

Electric utilities use consumer classifications for planning (for example, load growth and peak demand) and for determining their sales and revenue requirements (costs of service) in order to derive their rates. Utilities typically employ a set of rate schedules for a single sector. The alternative rate schedules reflect consumers' varying consumption levels and patterns and the associated impact on the utility's costs of providing electrical service. Reclassification of consumers, usually between the commercial and industrial sectors, may occur from year to year due to changes in demand level, economic factors, or other factors.

B.2.4 Energy Sources

Various sources of energy can be converted into electric energy or electricity. The major or dominant sources include fossil fuels, uranium, and water.

Fossil fuels supply about 70% of the energy sources for the generation requirements of the nation. Coal, petroleum, and gas are currently the dominant fossil fuels used by the industry. Other sources of energy can also be converted into electricity, including geothermal, solar thermal, photovoltaic, wind, and biomass.

B.2.5 Electric Power Transactions and the Interconnected Networks

B.2.5.1 Power Transactions

The flow of electricity with the power system is maintained and controlled by dispatch centers. It is the responsibility of the dispatch center to match the supply of electricity with the demand for it. In order to carry out its responsibilities, the dispatch center is authorized to buy and sell electricity based on system requirements. Authority for those transactions has been preapproved under interconnection agreements signed by all the electric utilities physically interconnected or with coordination agreements among utilities that are not connected.

B.2.5.2 The Interconnected Networks

The US bulk power system has evolved into three major networks (power grids), which also include smaller groupings or power pools.

The major networks consist of extra-high-voltage connections between individual utilities designed to permit the transfer of electrical energy from one part of the network to another. These transfers are restricted, on occasion, because of a lack of contractual arrangements or because of inadequate transmission capability. The three networks are:

- The Eastern Interconnected System
- The Western Interconnected System
- The Texas Interconnected System

The Texas Interconnected System is not interconnected with the other two networks (except by certain direct current lines). The other two networks have limited interconnections to each other.

Both the Western and the Texas Interconnects are linked with different parts of Mexico.

The Eastern and Western Interconnects are completely integrated with most of Canada or have links to the Quebec Province power grid.

Virtually all US utilities are interconnected with at least one other utility by these three major grids. The exceptions are in Alaska and Hawaii.

The interconnected utilities within each power grid coordinate operations and buy and sell power among themselves. The bulk power system makes it possible for utilities to engage in wholesale (for resale) electric power trade. Wholesale trade has historically played an important role, allowing utilities to reduce power costs, increase power supply options, and improve reliability.

Historically, almost all wholesale trade was within the National Electric Reliability Council (NERC) regions, but utilities are expanding wholesale trade beyond those traditional boundaries. US international trade is mostly imports of electricity. Normally, most imports are from Canada and the remainder are from Mexico.

B.2.5.3 Reliability Planning

Overall reliability planning and coordination of the interconnected power systems are the responsibility of NERC, which was voluntarily formed in 1968 by the electric utility industry as a result of the 1965 power failure in the Northeast.

NERC's 9 regional councils cover the 48 contiguous states, part of Alaska, and portions of Canada and Mexico. The councils are responsible for overall coordination of bulk power policies that affect the reliability and adequacy of service in their areas. They also regularly exchange operating and planning information among their member utilities.

Steady progress toward competitive wholesale markets for electric power recently has been accelerated by FERC Order 888, which opens access to transmission lines and encourages greater wholesale trade.

B.2.5.4 The Changing Electric Power Industry

The electric power industry is evolving from a highly regulated, monopolistic industry with traditionally structured electric utilities to a less regulated, competitive industry. EPACT removed some constraints on ownership of electric generation facilities and encouraged increased competition in the wholesale electric power business.

The EPACT amended the Federal Power Act (FPA) such that any electric utility can apply to the FERC for an order requiring another electric utility to provide transmission services (wheeling). Prior to EPACT, the FERC could not mandate that an electric utility provide wheeling services for wholesale electric trade. This change in the law permits owners of electric generating equipment to sell wholesale power (sales for resale) to noncontiguous utilities.

In April 1996, the FERC issued two final rules, 888 and 889, implementing EPACT's provisions for open access to transmission lines.

Rule 888 addresses equal access to the transmission grid for all wholesale buyers and sellers, transmission pricing, and the recovery of stranded costs. Stranded costs are investments, mostly in generation, made by utilities under the regulated environment that are presently recovered in cost-based rate structures and may not be recoverable in a competitive environment with market-based rates.

Rule 889 requires jurisdictional utilities that own or operate transmission facilities to establish electronic systems to post information about their available transmission capacities.

B.2.5.5 Time Hierarchy of Power Systems

Table B.1 Multiscale time hierarchy of power systems.

Action/operation	Time frame
Wave effects (fast dynamics, lightning caused over voltages)	Microseconds to milliseconds
Switching over voltages	Milliseconds
Fault protection	100 ms or a few cycles
Electromagnetic effects in machine windings	Milliseconds to seconds
Stability	60 cycles or 1 s
Stability augmentation	Seconds
Electromechanical effects of oscillations in motors and generators	Milliseconds to minutes
Tie line load frequency control	1–10 s; ongoing
Economic load dispatch	10 s to 1 h; ongoing
Thermodynamic changes from boiler control action (slow dynamics)	Seconds to hours
System structure monitoring (what is energized and what is not)	Steady state; ongoing
System state measurement and estimation	Steady state; ongoing
System security monitoring	Steady state; ongoing
Load management, load forecasting, generation, scheduling	1 h to 1 day or longer; ongoing
Maintenance scheduling	Months to 1 year; ongoing
Expansion planning	Years; ongoing
Power plant site selection, design, construction, environmental impact, etc.	2–10 years or longer

Source: [Amin 2008]. © 2008, Cambridge University Press.

B.3 Examples of Power Grid Glossaries

Table B.2 Power grid glossaries.

Organization	Glossary	Address
EIA	US Energy Information Administration Glossary	http://www.eia.gov/tools/glossary
Electropedia	The World's Online Electrotechnical Vocabulary	www.electropedia.org/
FERC	Federal Energy Regulatory Commission (FERC) Glossary	http://www.ferc.gov/market-oversight/guide/glossary.asp
IEC	Quality Attributes for Electrical Systems and Components	IEC electropedia definitions, http://dom2.iec.ch/iev/iev.nsf/index?openform&part=191
NERC	North American Energy Reliability Corporation Glossary of Terms	http://www.nerc.com/pa/Stand/Glossary%20of%20Terms/Glossary_of_Terms.pdf

References

[Amin 2008] Amin, M., Stringer, J. (2008). The electric power grid: today and tomorrow. *MRS Bulletin 33*(4), 399–407. Cambridge University Press. https://doi.org/10.1557/mrs2008.80

[DHS 2010] DHS. (2010). *Energy Sector-Specific Plan An Annex to the National Infrastructure Protection Plan*. https://www.dhs.gov/xlibrary/assets/nipp-ssp-energy-2010.pdf

[Education] DOE. (2013). *Education, Transporting Electricity to Consumer*. https://twitter.com/ENERGY/status/382226112836681728/photo/1

[Orzetto] Orzetto. (2008). *Closed loop control with measured output*. CC BY-SA 4.0. https://commons. wikimedia.org/wiki/File:Feedback_loop_with_descriptions.svg

[Sharma 2011] Sharma, KLS. (2011). Automation strategies. In *Overview of Industrial Process Automation* (53–62). Elsevier. https://doi.org/10.1016/B978-0-12-415779-8.00006-1

C

Critical Infrastructures Concepts

Critical Infrastructures and Energy Infrastructure

Appendix C provides highlights of US critical infrastructures plans, regulations, frameworks, and programs with a focus on energy sector and cybersecurity. Related terms and concepts can be found in [DHS Glossary].

C.1 Critical Infrastructures: Plans, Regulations, Frameworks, Programs

Cybersecurity-specific authorities and various federal strategies, directives, policies, and regulations provide the basis for actions and activities associated with implementing the cyber-specific aspects of the national plans. However, the regulations and plans change frequently. Figure C.1 is a summary of evolving threats to critical infrastructures.

In the United States, several directives were published and updated, and programs were established and innovated. For example:

- PDD-63, Presidential Decision Directive 63 of May 1998 [PDD-63], established the framework to protect the critical infrastructure, and the Presidential document of 2003 established the national strategy to secure cyberspace and stated that securing SCADA systems is a national priority.
- PPD-21, Critical Infrastructure Security and Resilience [PPD-21], advances a national policy to strengthen and maintain secure, functioning, and resilient critical infrastructure; it identifies the energy sector as uniquely critical because it provides an "enabling function" across all critical infrastructure sectors. More than 80% of the US energy infrastructure is owned by the private sector, supplying fuels to the transportation industry, electricity to households and businesses, and other sources of energy that are integral to growth and production across the nation.

C.1.1 Critical Infrastructure Protection Framework

The national security program (CERT, DHS, US national labs) provides coordination and analysis of control system vulnerabilities and improvement of security. Issue that is still under discussion is how vulnerabilities information should be disclosed, distributed, and shared. There are also international aspects because the same technology may be used differently in different countries with different regulations and business models.

Building an Effective Security Program for Distributed Energy Resources and Systems: Understanding Security for Smart Grid and Distributed Energy Resources and Systems, Volume 1, First Edition. Mariana Hentea.

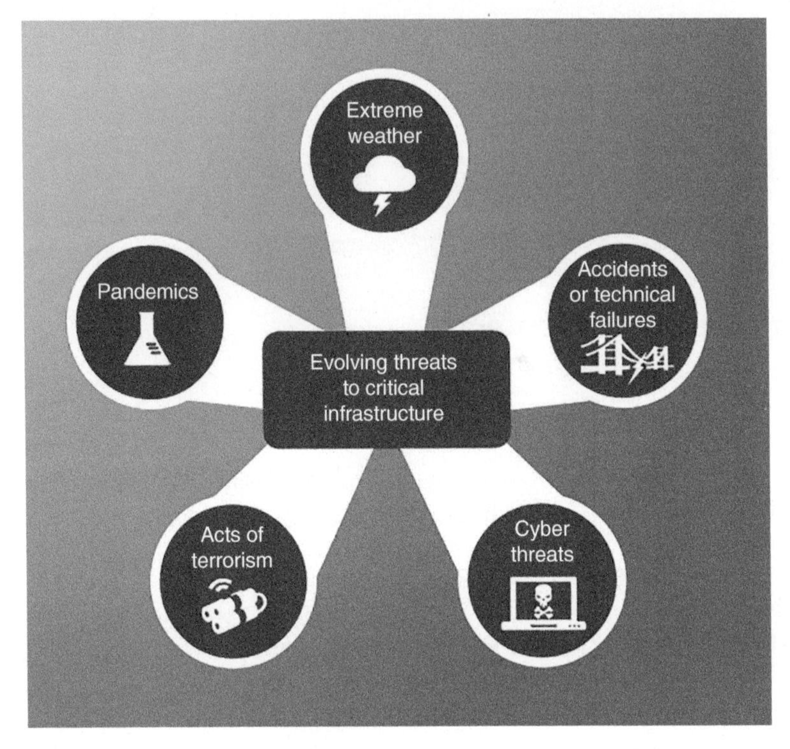

Figure C.1 Evolving threats to critical infrastructure [NIPP 2013]. *Source:* [NIPP 2013]. Public Domain.

C.1.2 Critical Infrastructure Protection (CIP) Program

PDD-63 set up a national critical infrastructure protection (CIP) program was established. CIP is a concept that relates to the preparedness and response to serious incidents that involve the critical infrastructure of a region or nation. The US CIP and several bills issued since 1998 are programs to assure the security of vulnerable and interconnected infrastructures of the United States.

Similar CIP programs are established in other countries. The European Union (EU) established in 2006 European Program for Critical Infrastructure Protection (EPCIP) that designates European critical infrastructure that, in case of fault, incident, or attack, could impact both the country where it is hosted and at least one other European member state. Member states are obliged to adopt the 2006 directive into their national statutes.

C.1.3 Critical Infrastructure Security and Resilience

PPD-8 (presidential directives on national preparedness in March 2011 [PPD-8]) and PPD-21 (Critical Infrastructure Security and Resilience) promote an all-hazards approach; stress the need to anticipate cascading impacts, especially due to dependencies and interdependencies; highlight the shared responsibility of CIP and resilience to all levels of government, the private sector, and individual citizens; and define, as a strategic imperative, the implementation of an integrated analysis function to inform planning and operations decisions regarding critical infrastructure.

The Executive Order 13636 – Improving Critical Infrastructure Cybersecurity – issued in February 2013 [EXEC 13636], reinforces the need to "enhance the security and resilience of the Nation's critical infrastructure and to maintain a cyber-environment that encourages efficiency, innovation and economic prosperity while promoting safety, security, business confidentiality, privacy and civil liberties."

C.1.4 Control Systems Security Program

The Control Systems Security Program (CSSP) coordinates efforts among federal, state, local, tribal, and territorial governments, as well as control system owners, operators, and vendors to improve control system security within and across all critical sectors.

The program coordinates activities to reduce the likelihood of the success and severity of a cyber attack against critical infrastructure control systems through risk mitigation activities. These activities include assessing and managing control system vulnerabilities, assisting the US-CERT Control Systems Security Center with control system incident management, and providing control system situational awareness through outreach and training initiatives.

The National Cyber Security Division (NCSD) of DHS implements the program. NCSD also performs vulnerability assessments of operational control systems and vendor equipment to improve their security posture.

C.1.5 National Infrastructure Protection Plan

The National Infrastructure Protection Plan (NIPP), first released in 2006, then subsequently revised – last revised in 2013 – integrates the concepts of resilience and protection and broadens the focus of NIPP-related programs and activities to an all-hazards environment.

The critical infrastructure risk management framework supports a decision-making process that critical infrastructure partners collaboratively undertake to inform the selection of risk management actions.

The NIPP framework (2009) calls for critical infrastructure and key resource (CIKR) partners to assess risk from any scenario as a function of consequence, vulnerability, and threat [DHS Glossary]:

$$R = f\left(C, V, T\right), \text{where}$$

C.1.5.1 Risk

The potential for an unwanted outcome resulting from an incident, event, or occurrence as determined by its likelihood and the associated consequences; risk is influenced by the nature and magnitude of a threat, the vulnerabilities to that threat, and the consequences that could result.

C.1.5.2 Consequence

The effect of an event, incident, or occurrence reflects the level, duration, and nature of the loss resulting from the incident. For the purposes of the NIPP, consequences are divided into four main categories: public health and safety (i.e. loss of life and illness), economic (direct and indirect), psychological, and governance/mission impacts.

C.1.5.3 Vulnerability

Physical feature or operational attribute that renders an entity open to exploitation or susceptible to a given hazard. In calculating the risk of an intentional hazard, a common measure of vulnerability is the likelihood that an attack is successful, given that it is attempted.

C.1.5.4 Threat

Natural or man-made occurrence, individual, entity, or action that has or indicates the potential to harm life, information, operations, the environment, and/or property. For the purpose of calculating risk, the threat of an intentional hazard is generally estimated as the likelihood of an attack being attempted by an adversary; for other hazards, threat is generally estimated as the likelihood that a hazard will manifest itself. In the case of terrorist attacks, the threat likelihood is estimated based on the intent and capability of the adversary.

C.1.5.5 All Hazards

A threat or an incident, natural or man-made, that warrants action to protect life, property, the environment, and public health or safety and to minimize disruptions of government, social, or economic activities. It includes natural disasters, cyber incidents, industrial accidents, pandemics, acts of terrorism, sabotage, and destructive criminal activity targeting critical infrastructure.

C.1.5.6 Asset

Person, structure, facility, information, material, or process that has value.

CIKR-related risk assessments consider all three components of risk to be conducted on assets, systems, or networks, depending on the characteristics of the infrastructure being examined. Once the three components of risk have been assessed for one or more given assets, systems, or networks, they must be integrated into a defensible model to produce a risk estimate.

CIKR provide the essential services that underpin American society. Their exploitation or destruction by terrorists could cause catastrophic health effects or mass casualties comparable with those from the use of a weapon of mass destruction or could profoundly affect national prestige and morale. CIKR is an interdependent network of vital physical and information facilities, networks, and assets, including the telecommunications, energy, financial services, water, and transportation sectors, that private business and the government rely upon (including for the defense and national security of the United States).

The NIPP risk management framework (see Figure C.2) is tailored toward and applied on an asset, system, network, or functional basis depending on the fundamental characteristics of the individual CIKR sectors. Each sector must pursue the approach that produces the most effective use of resources for the sector and contributes to cross-sector comparative risk analyses conducted by DHS.

This document [NIPP 2013] provides a unifying framework for critical infrastructure based on a risk management framework. This framework was developed through a collaborative process involving stakeholders from all 16 critical infrastructure sectors, all 50 states, and from all levels of government and industry. It provides a call to action to leverage partnerships, innovate for risk management, and focus on outcomes. A call to action refers to collective actions through joint planning efforts that also references the cybersecurity framework under development by the National Institute of Standards and Technology (NIST), stating that a new round of updated sector council plans should describe current and planned cybersecurity efforts, including, but not limited to, use of the NIST framework [NIST 2014] (also, see the updated document [NIST 2018]).

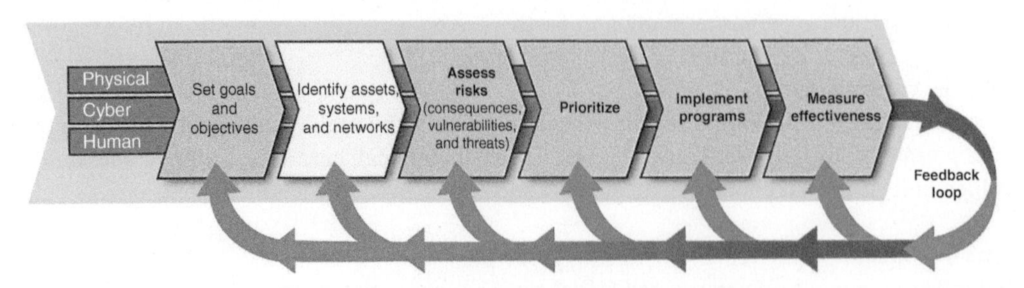

Figure C.2 NIPP 2013 risk management framework. *Source:* [NIPP 2013]. Public Domain.

A Sector-Specific Plan (SSP) details the application of the NIPP concepts to the unique characteristics and conditions of the critical sector and a Sector-Specific Agency (SSA) leads the process for CIP within each infrastructure [PPD-21].

The Department of Energy (DOE) is the SSA for the energy sector.

The new NIPP calls on national level councils to jointly issue multiyear priorities based on multiple information sources, including results of state and regional Threat and Hazard Identification and Risk Assessment (THIRA) process. This process should be used as a method to integrate human, physical, and cyber elements of critical infrastructure risk management.

In 2017 the Cybersecurity and Infrastructure Security Agency's Infrastructure Security Division and the National Institute for Hometown Security announced the 2017 NIPP Security and Resilience Challenge plan. This provides an opportunity for the critical infrastructure community to help develop technology, tools, processes, and methods that address immediate needs and strengthen the long-term security and resilience of critical infrastructure.

C.2 Energy Sector

The energy infrastructure is divided into three interrelated segments: electricity, oil, and natural gas. The US electricity segment contains more than 6413 power plants (this includes 3273 traditional electric utilities and 1738 nonutility power producers) with approximately 1075 GW of installed generation. Approximately 48% of electricity is produced by combusting coal (primarily transported by rail), 20% in nuclear power plants, and 22% by combusting natural gas. The remaining generation is provided by hydroelectric plants (6%), oil (1%), and renewable sources (solar, wind, and geothermal) (3%). The heavy reliance on pipelines to distribute products across the nation highlights the interdependencies between the energy and transportation systems sector (see [Overview CISA]).

The reliance of virtually all industries on electric power and fuels means that all sectors have some dependence on the energy sector. The energy sector is recognizing its vulnerabilities and is leading a significant voluntary effort to increase its planning and preparedness. Cooperation through industry groups has resulted in substantial information sharing of best practices across the sector. Many sector owners and operators have extensive experience abroad with infrastructure protection and have more recently focused their attention on cybersecurity.

The Energy SSP details how the NIPP risk management framework is implemented within the context of the unique characteristics and risk landscape of the energy sector. Each SSA develops an SSP through a coordinated effort involving its public and private sector partners. The DOE is designated as the SSA for the energy sector.

The purpose of the Energy SSP is to help guide and integrate the sector's continuous effort to improve the security and resilience of its critical infrastructure and to describe how the energy sector contributes toward the national critical infrastructure security and resilience goals.

The 2015 Energy SSP updates and augments the prior versions of the SSP in accordance with the NIPP 2013.

C.2.1 2015 Energy Sector Goals

- Assess and analyze threats to, vulnerabilities of, and consequences to critical infrastructure to inform risk management activities.
- Secure critical infrastructure against human, physical, and cyber threats through sustainable efforts to reduce risk while accounting for the costs and benefits of security investments.

- Enhance critical infrastructure resilience by minimizing the adverse consequences of incidents through advance planning and mitigation efforts, as well as effective responses to save lives and ensure the rapid recovery of essential services.
- Share actionable and relevant information across the critical infrastructure community to build awareness and enable risk-informed decision making.
- Promote learning and adaptation during and after exercises and incidents.

C.2.2 Electricity Sector Priorities

- Tools and technology – Deploying tools and technologies to enhance situational awareness and security of critical infrastructure.
- Deploying proprietary government technologies on utility systems that enable machine-to-machine information sharing and improved situational awareness of threats to the grid.
- Implementing the NIST Cybersecurity Framework.
- Information flow is to making sure actionable intelligence and threat indicators are communicated between the government and industry in a time-sensitive manner.
- Improving the bidirectional flow of threat information.
- Coordinating with interdependent sectors.
- Incident response includes planning and exercising coordinated responses to an attack.
- Developing playbooks and capabilities to coordinate industry government response and recovery efforts.
- Ongoing assessments of equipment sharing programs.

C.2.3 Electricity Subsector Risks and Threats

The document [NIPP 2013] defines a risk as the potential for an unwanted outcome resulting from an incident, event, or occurrence, as determined by its likelihood and the associated consequences. Many organizations conduct a wide variety of risk assessments of the electricity subsector. For example, the North American Electric Reliability Corporation (NERC) assesses risks in terms of the potential impact to the reliability of the bulk power system (e.g. did an event result in the loss or interruption of service to customers?), while private companies and utilities examine risks and threats as they relate to the operational and financial security of each company (e.g. could a threat negatively impact the organization's financial health?).

Based on a review by some of the largest US electric utilities (in terms of revenue) as well as the analysis by NERC, a wide variety of issues were considered threats in the electricity subsector. Despite the differences in what constitutes risk, the electricity subsector identified several issues as the key risks and threats to its infrastructure and/or continuity of business in 2012 and 2013 to include:

- Cyber and physical security threats.
- Natural disasters and extreme weather conditions.
- Workforce capability ("aging workforce") and human errors.
- Equipment failure and aging infrastructure.
- Evolving environmental, economic, and reliability regulatory requirements.
- Changes in the technical and operational environment, including changes in fuel supply.

In addition, state, local, tribal, and territorial (SLTT) governments are crucial stakeholders in providing a secure and reliable energy infrastructure for the nation. They are responsible for emergency planning and response, developing energy security and reliability policies and practices, and facilitating energy sector protection activities. In times of crisis, citizens turn to these organizations, which play a significant role in preparing for and responding to energy supply events and mitigating the impacts of emergencies that do arise.

State and local governments are required under federal homeland security funding guidance to implement the NIPP, as well as the NRF and National Incident Management System.

C.2.4 Addressing Cybersecurity

The DOE also works with industry to develop new cybersecurity solutions for energy delivery systems through an integrated planning and research and development (R&D) effort. DOE's Cybersecurity for Energy Delivery Systems is one such program, which emphasizes collaboration among the government, industry, universities, national laboratories, and end users to advance R&D in cybersecurity [DOE R&D]. The aim of the program is to reduce the risk of energy disruptions due to cyber incidents as well as survive an intentional cyber assault with no loss of critical function. This program is helping to increase the security of energy delivery systems around the country.

In addition to government programs, various industry partners, including trade associations, have been carrying out numerous cybersecurity-related activities. They include the establishment of three Information Sharing and Analysis Centers (ISACs) and efforts under various cyber working groups through trade associations, the development of enterprise-wide cybersecurity guidance, and the development of the NERC CIP Reliability Standards.

However, cross-sector interdependency during the last half of the twentieth century and technical innovations and developments in digital information and telecommunications dramatically increased interdependencies among the nation's critical infrastructures. The energy infrastructure provides essential functions to all critical infrastructure sectors, and without energy, none of them can operate properly.

Thus, the energy sector serves one of the four lifeline functions, which means that its reliable operation is so critical that a disruption or loss of energy function will directly affect the security and resilience of other critical infrastructure sectors. In turn, the energy sector depends on many other critical infrastructure sectors, such as transportation, information technology (IT), communications, water, financial services, and government facilities. A disruption in a single facility of capability can generate disturbances within other infrastructure or sectors and over long distances. A series of related interconnections can extend or amplify the effects of a disruption. Figure C.3 is a simplified illustration of the interdependencies among 16 critical infrastructure sectors, including the four critical lifeline sectors – communications, energy, transportation systems, and water – that provide lifeline functions to all critical infrastructure sectors.

Over time, cyber/IT dependencies have increased dramatically. For example, electricity and natural gas suppliers rely heavily on data collection systems to ensure accurate billing. Energy control systems and the information and communication technologies on which they rely play a key role in the North American energy infrastructure. These cyber/IT components are essential in monitoring and controlling the production and distribution of energy. They have helped to create the highly reliable and flexible energy infrastructure in the United States; however, the reliance of energy infrastructure on cyber infrastructure can also present vulnerability. Another aspect is the international interdependencies.

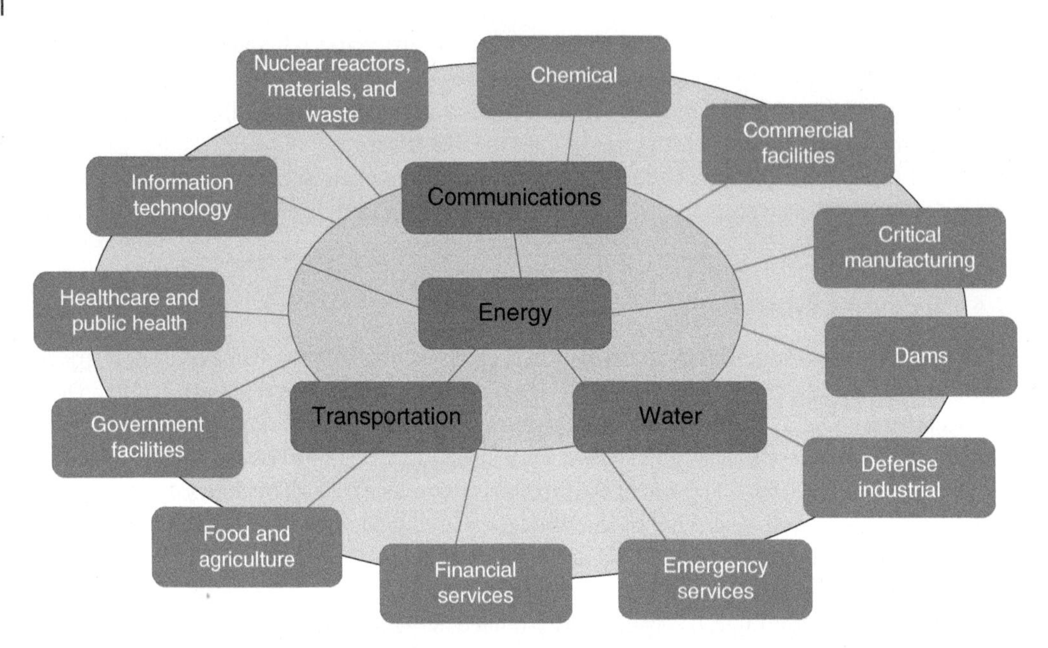

Figure C.3 Critical infrastructure interdependencies. *Source:* [Overview CISA]. Public Domain.

C.2.5 International Interdependency and Coordination

Energy infrastructure interdependencies cross international borders in many ways. The United States depends on cross-border flows of energy resources to meet its total energy requirements. In addition, the energy system is supported by and critically dependent upon global flows of information, knowledge, and investment capital.

The United States relies on the import of critical technologies and equipment as well as many key raw materials that are essential to the manufacturing of certain electrical infrastructure. Further, many oil companies have facilities in various global locations that operate under different foreign sovereignties.

North America has an integrated system of oil and natural gas pipelines and electric transmission lines. Pipeline interconnections between the United States and Canada and between the United States and Mexico move considerable volumes of natural gas (and also oil in case of Canada) between the countries. This also requires coordination to ensure that protective measures across borders provide adequate risk reduction across the full length of these systems.

In 2011, the administration released the 2011 International Strategy for Cyberspace, which highlighted the need to develop a US government position for an international cybersecurity policy framework and to strengthen international partnerships to create initiatives that would address cybersecurity activities (see most current activities at [DOE Cyber]).

Energy sector stakeholders from public and private sector collaboratively developed the 2011 roadmap [DOE 2011] to achieve Energy Delivery Systems Cybersecurity. The Roadmap articulates the sector's vision that "by 2020, resilient energy delivery systems are designed, installed, operated, and maintained to survive a cyber-incident while sustaining critical functions." It also presents a set of milestones that guide energy sector cybersecurity activities in five strategic areas: building a culture of security, assessing and monitoring cybersecurity risks, developing and implementing new protective measures to reduce risks, managing incidents, and sustaining security improvements.

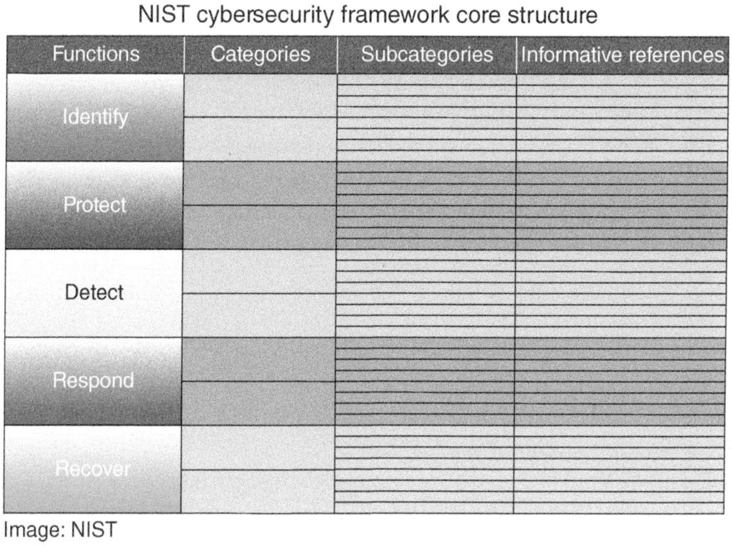

NIST cybersecurity framework core structure

Functions	Categories	Subcategories	Informative references
Identify			
Protect			
Detect			
Respond			
Recover			

Image: NIST

Figure C.4 Framework core structure. Source: [NIST 2014]. Public Domain.

C.3 NIST Cybersecurity Framework

Protection of critical infrastructure constantly evolves to anticipate and reflect changes in the risk environment. This NIST document [NIST 2014] describes a framework that focuses on using business drivers to guide cybersecurity activities and considers cybersecurity risks as part of the organization's risk management processes. Figure C.4 shows the functions and related concepts of NIST framework.

This framework focuses on using business drivers to guide cybersecurity activities and considering cybersecurity risks as part of the organization's risk management processes. It outlines a maturity model of four tiers against which organizations can benchmark the efficiency of their cybersecurity program. However, the final document lacks provision for privacy protection.

Recognizing the drawbacks of the framework, NIST released a companion document – Roadmap for Improving Critical Infrastructure Cybersecurity on 12 February 2014, which discusses NIST's next steps for improving the framework and identifies key areas of development, alignment, and collaboration (see more information in [NIST Roadmap]).

C.4 Privacy

The Fair Information Practice Principles (FIPPs) are a widely accepted framework that is at the core of the Privacy Act of 1974 in the United States [Privacy Act] and is mirrored in the laws of many US states, as well as many foreign nations and international organizations. FIPPs are principles for balancing the need for privacy and other interests. The FIPPs were first discussed in 1973, and consequently different variants were released by various agencies [IT Privacy]. Thus, several practices are known, but no legal means for enforcement of privacy principles.

Major reports setting forth the core FIPPs are provided by the Organisation for Economic Cooperation and Development (OECD) [OECD 1980], advocate groups (e.g. Information Infrastructure Task Force, Information Policy Committee, Privacy Working Group, Privacy and the

National Information Infrastructure: Principles for Providing and Using Personal Information 1995, The Asia-Pacific Economic Cooperation (APEC) Privacy Framework), and governments of many countries (United States, Canada, EU, Australia, etc.).

The most recent document, the European General Data Protection Regulation (EU) 2016/679 (GDPR), is a regulation in EU law on data protection and privacy for all individuals within the EU and the European Economic Area (EEA). It also addresses the export of personal data outside the EU and EEA areas. The regulation aims to reshape the way data is handled across every sector from healthcare to banking and beyond. This is the most important change in data privacy regulation in 20 years.

References

[DHS Glossary] DHS. (2013). *Glossary of Terms*. https://www.dhs.gov/sites/default/files/publications/National-Infrastructure-Protection-Plan-2013-508.pdf.

[DOE 2011] DOE. (2011). *Roadmap to Achieve Energy Delivery Systems Cybersecurity*. Energy Sector Control Systems Working Group, September 2011.

[DOE Cyber] DOE. (n.d.). *Cybersecurity for Critical Energy Infrastructure*. https://www.energy.gov/ceser/activities/cybersecurity-critical-energy-infrastructure.

[DOE R&D] DOE. (2019). *Cybersecurity for Energy Delivery Systems Program*. https://www.energy.gov/sites/prod/files/2020/08/f77/Cybersecurity%20for%20Energy%20Delivery%20Systems.pdf

[EXEC 13636] WhiteHouse. (2013, February 19). *Executive Order 13636 – Improving Critical Infrastructure Cybersecurity*. Presidential Documents. Federal Register *78*(33).

[IT Privacy] Wiki. (n.d.). *Fair Information Privacy Principles*. http://itlaw.wikia.com/wiki/Fair_Information_Practice_Principles.

[NIPP 2013] DHS. (2013). *Partnering for Critical Infrastructure Security and Resilience*. https://www.dhs.gov/sites/default/files/publications/National-Infrastructure-Protection-Plan-2013-508.pdf

[NIST 2014] NIST. (2014). *Framework for Improving Critical Infrastructure Cybersecurity* (Version 1.0, February 12, 2014). National Institute of Standards and Technology. https://www.nist.gov/system/files/documents/cyberframework/cybersecurity-framework-021214.pdf

[NIST 2018] NIST. (2018). *Framework for Improving Critical Infrastructure Cybersecurity* (Version 1.1, April 16, 2018). https://doi.org/10.6028/NIST.CSWP.04162018

[NIST Roadmap] NIST (2014). *Roadmap for Improving Critical Infrastructure Cybersecurity* (February 12, 2014). http://www.nist.gov/cyberframework/upload/roadmap-021214.pdf

[OECD 1980] OECD. (1980). *OECD Guidelines on the Protection of Privacy and Transborder Flows of Personal Data*. http://www.oecd.org/document/18/0,3343,en_2649_34255_1815186_1_1_1_1,00.html.

[Overview CISA] CISA. (2015). *Energy Sector Overview. 2015 Energy Sector-Specific Plan*. https://www.cisa.gov/sites/default/files/publications/nipp-ssp-energy-2015-508.pdf

[PDD-63] WhiteHouse. (1998, May). *Presidential Decision Directive 63 Critical Infrastructure Protection*. https://www.fas.org/irp/offdocs/pdd/pdd-63.pdf

[PPD-21] WhiteHouse. (2013). *Presidential Policy Directive/PPD-21 – Critical Infrastructure Security and Resilience February 12, 2013*. http://www.whitehouse.gov/the-press-office/2013/02/12/presidential-policy-directive-critical-infrastructure-security-and-resil

[PPD-8] WhiteHouse. (2011). *Presidential Policy Directive/PPD-8 – National Preparedness 30 March 2011*. http://www.dhs.gov/xlibrary/assets/presidential-policy-directive-8-national-preparedness.pdf

[Privacy Act] Wiki. (n.d.). *Privacy Act Law*. http://itlaw.wikia.com/wiki/Privacy_Act_of_1974

[Strategy 2003] WhiteHouse. (2003). *The National Strategy to Secure Cyberspace February 2003*. https://www.us-cert.gov/sites/default/files/publications/cyberspace_strategy.pdf.

D

Smart Grid Concepts

Smart Grid: Policy, Concepts, and Technologies

Appendix D includes information related to US Smart Grid policy, examples of Smart Grid definitions and US vision of future grid, examples of technologies and concepts, examples of glossaries defining Smart Grid terms, and NIST conceptual model.

D.1 US Smart Grid Policy

It is the policy of the United States [42 U.S. CODE § 17381 – STATEMENT OF POLICY ON MODERNIZATION OF ELECTRICITY GRID (Pub. L. 110–140, title XIII, § 1301, Dec. 19, 2007, 121 Stat. 1783)] to support the modernization of the nation's electricity transmission and distribution system to maintain a reliable and secure electricity infrastructure that can meet future demand growth and to achieve each of the following, which together characterize a Smart Grid:

1) Increased use of digital information and control technology to improve reliability, security, and efficiency of the electric grid.
2) Dynamic optimization of grid operations and resources, with full cybersecurity.
3) Deployment and integration of distributed resources and generation, including renewable resources.
4) Development and incorporation of demand response (DR), demand-side resources, and energy-efficiency resources.
5) Deployment of "smart" technologies (real-time, automated, interactive technologies that optimize the physical operation of appliances and consumer devices) for metering, communications concerning grid operations and status, and distribution automation.
6) Integration of "smart" appliances and consumer devices.
7) Deployment and integration of advanced electricity storage and peak-shaving technologies, including plug-in electric and hybrid electric vehicles, and thermal storage air conditioning.
8) Provision to consumers of timely information and control options.
9) Development of standards for communication and interoperability of appliances and equipment connected to the electric grid, including the infrastructure serving the grid.
10) Identification and lowering of unreasonable or unnecessary barriers to adoption of Smart Grid technologies, practices, and services.

Building an Effective Security Program for Distributed Energy Resources and Systems: Understanding Security for Smart Grid and Distributed Energy Resources and Systems, Volume 1, First Edition. Mariana Hentea.

D.2 Smart Grid Definitions and Vision

There are too many definitions, perspectives, and goals for Smart Grid. Examples of definitions follow.

Smart Grid is the electric delivery network, from electrical generation to end-use customer, integrated with the latest advances in digital and information technology to improve electric system reliability, security, and efficiency [DHS 2010].

Smart Grid is the integration of power, communications, and information technologies for an improved electric power infrastructure serving loads while providing for an ongoing evolution of end-use applications [IEEE P2030].

A Smart Grid is an electricity network that can intelligently integrate the actions of all users connected to it – generators, consumers, and those that do both – in order to efficiently deliver sustainable, economic, and secure electricity supplies [EC 2006].

The fundamental challenge in the Smart Grid is to ensure balance of generation and demand/consumption when integrating all those new technologies that are aimed at addressing in a sustainable manner energy independence and modernization of the aging power grid:

- Utility-scale renewable energy sources (RES) feeding into the transmission system.
- Distributed energy resources (DER) feeding into the distribution system.
- Plug-in (hybrid) electric vehicles (PHEV).
- Demand-side management (DSM).
- Consumer participation.
- Storage to compensate for the time-varying nature of some renewables.

D.2.1 DOE Future Smart Grid Vision

Examples of features related to DERs and Smart Grid include the following [DOE 2014]:

- Energy storage will leverage energy storage to optimize operation of the Smart Grid and will allow for mitigating the variability in non-dispatchable generation sources such as solar photovoltaic and wind generation, as well as provide some additional ancillary services such as spinning reserve requirements.
- Multiple-customer and single-customer microgrid operations will complement the future grid providing services to utilities and vice versa. Aggregating distributed generation (DG) into microgrids, it is possible to optimize generation and reduce load on peak demand days or for other operational efficiencies, including reducing the impacts of outages.
- A mix of regulated and competitive services and a mixture of regulated and competitive services will emerge.
- A retail market exchange that supports the true value of renewable energy, traditional energy, capacity, distribution, transmission, aggregation, and other services provided by market participants will develop in some jurisdictions.
- Microgrids used for managing multiple DERs in some areas will operate either in parallel or in island mode as needed. Microgrids can be private, owned by utility or community.
- Advanced analytics that leverage exponential growth in data will play a critical role.
- Safeguards will mitigate and protect against cyber, physical, and other threats using Smart Grid technologies, trained utility personnel, improved business processes, and upgraded information technology and communications infrastructure.

Smart Grid offers improved capabilities and new features such as [Wiki SG]:

- Reliability – Use of technologies, such as state estimation and prediction, can improve fault detection and allow self-healing of the network without the intervention of technicians and finally ensure more reliable supply of electricity and reduce vulnerability to natural disasters or attack. Source: Wikipedia, Smart Grid.
- Flexibility in network topology – Better handle of two-direction energy flows, allowing for DG such as from photovoltaic panels on building roofs, but also the use of fuel cells, charging to/from the batteries of electric cars, wind turbines, pumped hydroelectric power, and other sources.
- Efficiency – Overall improvement of the efficiency of energy infrastructure is anticipated from the deployment of Smart Grid technology, in particular including DSM, less redundancy in transmission and distribution lines, and greater utilization of generators, leading to lower power prices.
- Load adjustment/load balancing – Using mathematical prediction algorithms, it is possible to predict the load and how many standby generators need to be used, to reach a certain failure rate.
- Peak curtailment/leveling and time of use pricing – To reduce demand during the high-cost peak usage periods, communications and metering technologies inform smart devices in the home and business when energy demand is high and track how much electricity is used and when it is used.
- Sustainability – The improved flexibility of the Smart Grid permits greater penetration of highly variable RES such as solar power and wind power, even without the addition of energy storage.
- Market-enabling – Only the critical loads will need to pay the peak energy prices, and consumers will be able to be more strategic in when they use energy.
- DR support – Allows generators and loads to interact in an automated fashion in real time, coordinating demand to flatten spikes.
- Platform for advanced services – As with other industries, use of robust two-way communications, advanced sensors, and distributed computing technology will improve the efficiency, reliability, and safety of power delivery and use. It also opens up the potential for entirely new services or improvements on existing ones, such as fire monitoring and alarms that can shut off power, make phone calls to emergency services, etc.
- Provision megabits, control power with kilobits, sell the rest – The amount of data required to perform monitoring and switching one's appliances off automatically is very small compared with that already reaching even remote homes to support voice, security, Internet, and TV services. Many Smart Grid bandwidth upgrades are paid for by over-provisioning to also support consumer services and subsidizing the communications with energy-related services or subsidizing the energy-related services, such as higher rates during peak hours, with communications.

D.3 Examples of Smart Grid Technologies

D.3.1 Active Distribution Network

Active distribution networks are capable of handling bidirectional power and information flows based on the latest automation, information, and communication technologies, as well as on corresponding metering services. The term is used to express the shift from the conventional and more passive operation approach (without any monitoring and control in lower distribution levels for the integration of DER toward more advanced Smart Grid concepts).

D.3.2 Advanced Metering Infrastructure (AMI)

Systems that measure, collect, and analyze energy usage and communicate with metering devices such as electricity meters, gas meters, heat meters, and water meters, either on request or on a schedule are called advanced metering infrastructure (AMI). These systems include hardware, software, communications, consumer energy displays and controllers, customer-associated systems, meter data management software, and supplier business systems.

In settings when energy savings during certain periods are desired, meters may measure demand, the maximum use of power in some interval. "Time of day" metering allows electric rates to be changed during a day to record usage during peak high-cost periods and off-peak, lower-cost, periods. Also, meters may have relays for DR load shedding during peak load periods.

Power companies often install remote reporting meters specifically to enable remote detection of tampering and specifically to discover energy theft. The change to smart power meters is useful to stop energy theft.

AMI differs from traditional automatic meter reading (AMR) in that it enables two-way communications with the meter. Systems only capable of meter readings do not qualify as AMI systems.

D.3.3 Aggregation and Aggregator

Aggregation is the process of organizing small groups of industrial, commercial, or residential customers into a larger, more effective bargaining unit with a view to strengthening their participation in electricity trading. Aggregation can link demand (DR) and/or generation and, if applicable, storage units.

An aggregator is a legal organization that consolidates or aggregates a number of individual customers and/or small generators into a coherent group of business players.

Aggregators aim at optimizing energy supply and consumption both technically and economically. This optimization process improves the coherent operation of DER to support system balancing.

An aggregator is therefore a facility manager able to design and offer energy services downstream to energy customers (at the micro level – a large number of contracts) and upstream to several key players (at the macro level – system operators, electricity traders, etc.).

Such aggregators can be considered as power plants that use distributed resources together with other generation solutions to supply their aggregated customers. Simultaneously, aggregators can also be viewed as entities that bring a group of consumers together to buy electricity.

Any marketer, supplier/retailer, broker, public agency, and municipality, which combine the loads of multiple end-use customers and facilitate the sale and purchase of electric energy and other services on behalf of these customers, can be considered as an aggregator. A public aggregator, for instance, is an organization set up by a municipality to purchase electricity in bulk for its citizens in order to increase their purchasing power. Participation is voluntary: consumers can opt out, if they wish, and go back to the standard service offer within a given period of time.

D.3.4 Automatic Meter Reading

AMR is the technology of automatically collecting consumption, diagnostic, and status data from water meter or energy metering devices (gas, electric) and transferring that data to a central database for billing, troubleshooting, and analyzing. AMR technologies include handheld, mobile, and network technologies based on telephony platforms (wired and wireless), radio frequency (RF), or powerline transmission.

D.3.5 Backhaul Communications

The backhaul (otherwise referred to as back-net or backbone or transport network), in cellular networks, is the network that connects the nodes (e.g. cellular radio stations) to the core network and consists mostly of dedicated fiber, copper, microwave, and occasionally satellite links.

Utilities refer to backhaul communications as the infrastructure used to connect the AMI head-end system to the AMI data collectors or access points. Backhaul communications typically utilize fiber-optic cables, high-speed wireless connections, or other networks that can handle large amounts of data. Backhaul communications can utilize utility-owned infrastructure or third-party communications providers.

D.3.6 Balance

Balance is the compliance of the total amount of electricity produced by the generating plants with the electrical consumption in the system, including the power required for auxiliaries' services, and the losses in the transmission and distribution networks, considering the scheduled power exchanges.

D.3.7 Balancing market

A balancing market is that part of the overall electricity market that provides for meeting requirements for balancing electrical power in the electrical power system operation. A balancing market consists generally of two important parts:

- Balancing mechanisms and defining features of the balancing market, e.g. the way to bid, constraints/requirements on the balancing market participants, methods of payment to the bidders, constraints on the TSOs, who/how makes the merit order, etc.

Imbalance arrangements and pricing, where the cost-reflective and transparent prices for the users (e.g. balance responsible parties) emerge according to the predefined, transparent, and agreed rules and regulatory framework; these rules also include the way TSOs determine the imbalance prices for the balance responsible parties. Different ways of calculating imbalances exist, e.g. only one imbalance for generation and demand or separated imbalance accounts.

D.3.8 Building Area Networks (BANs)

A building area network (BAN) is a local area network (LAN) that covers an entire building. It may be a collection of smaller LANs. For example, if each floor is considered a single LAN, then the combination of each per floor LAN is considered a BAN. A BAN is also known as a basement area network (BAN).

D.3.9 Building Automation

The automatic centralized control of a building's lighting, heating, ventilation, and air conditioning (HVAC), and other systems through a building automation system (BAS). The objectives of building automation are improved occupant comfort, efficient operation of building systems, and reduction in energy consumption and operating costs.

Building automation is a distributed control system of devices designed to monitor and control the mechanical, security, fire and flood safety, lighting (especially emergency lighting), HVAC, and humidity control and ventilation systems in a building.

A BAS should reduce building energy and maintenance costs compared with a noncontrolled building. Most commercial, institutional, and industrial buildings built after 2000 include a BAS. Many older buildings have been retrofitted with a new BAS, typically financed through energy and insurance savings and other savings associated with preemptive maintenance and fault detection.

A building controlled by a BAS is often referred to as an intelligent building, smart building, or (if a residence) a smart home. Commercial and industrial buildings have historically relied on robust proven protocols (like BACnet), while proprietary and poorly integrated purpose-specific protocols (like X-10 or those from Johnson Controls, Honeywell, Siemens, or other major manufacturers of smart thermostats, etc.) were used in homes.

Therefore, commercial, industrial, military, and other institutional users now use systems that differ from home systems mostly in scale.

D.3.10 Content of Information Exchange Between DER and Aggregator

Depending on the type of aggregation, different information needs to be exchanged between the commercial aggregator and the flexible DER unit, either in a unilateral or bilateral direction. It is not possible to aggregate DER without an appropriate exchange of data.

The unidirectional flow of information comprises the monitoring of DER units by the aggregator, e.g. with smart metering, or the sending of information from the aggregator to the DER unit, e.g. in the form of price information.

Bidirectional communication allows data exchange in both directions and allows the aggregator to send out signals (info or set points) to the DER unit based on the information received by the unit (such as metered information, information about availability).

D.3.11 Distributed Energy Resources (DER)

DER comprises DG, the storage of electrical and thermal energy and/or flexible loads. DER units are operated either independently of the electrical grid or connected to the low or medium voltage distribution level of the main network. They are located close to the point of consumption, irrespective of the technology, but are smaller than 10 MWe of electrical power.

D.3.12 Distributed Generation (DG)

DG is defined as an electrical generation unit connected to the electrical distribution system at or close to the point of consumption. It includes all types of generation technologies based on fossil fuels and RES.

D.3.13 Distribution System Operator (DSO) and Transmission System Operator (TSO)

A system operator is a natural or legal person responsible for operating, ensuring the maintenance of and, if necessary, development of the transmission and distribution system in a given area and, where applicable, its interconnections with other systems. The mission is also to ensure the long-term capability of the system to meet reasonable demands for electricity transmission and distribution. However, both DSO and TSO are usually regulated monopolies, which are unbundled from the energy market functions.

D.3.14 Flexible DER

Flexible DER is defined as a DER with the two following key characteristics:

- Controllability through an existing communication and control infrastructure: the DER is able to react to remote signals, such as schedules and set points sent through communication channels. This type of DER can be remotely controlled.
- Adaptability means a DER is able to adapt (up/down) its generation/load profile according to incentives, instructions, or price signals; also, the output is variable.

Both characteristics are needed in order to allow a remote operator to benefit from DER flexibility and thus make full use of the aggregation business.

D.3.15 Home Area Networks (HANs)

A home network or home area network (HAN) is a type of LAN with the purpose to facilitate communication among digital devices present inside or within the close vicinity of a home. Devices capable of participating in this network, for example, smart devices such as network printers and handheld mobile computers, often gain enhanced emergent capabilities through their ability to interact. These additional capabilities can be used to increase the quality of life inside the home in a variety of ways, such as automation of repetitious tasks, increased personal productivity, enhanced home security, and easier access to entertainment.

Basically, HAN is a communication network of appliances and devices within a home. Resources such as electricity, gas, water, heat, solar panels, etc. can be equipped with smart meters. These meters also connect smart appliances (e.g. smart dishwashers, dryers, ovens, washers, etc.) with communications and remote control functions, and finally these meters connect to a metering gateway.

D.3.16 Information and Communication Infrastructure

Recent developments in information and communication technologies (ICT) allowing for secure, cost-efficient, and reliable data exchange with an appropriate performance between key players constitute an important enabling technology for the aggregation business.

The ICT infrastructure requires the definition of the communication functions, technologies, architectures, media, and data requirements. Performance, interoperability, security, and reliability also need to be defined. Some applications (e.g. control systems, SCADA) require real-time response.

For data transmission over a geographical distance, different communication media can be used depending on performance requirements:

- Powerline carrier (PLC) transmission.
- Fiber-optic cables (fiber-optic converters).
- Telephone network (dedicated lines or dial-up lines).
- Mobile (and satellite) transmission (GSM, UMTS, GPRS, messaging, etc.).
- Internet.

D.3.17 LonWorks

Local operating network is a platform to address the needs of control applications. The platform is built on a protocol created by Echelon Corporation for networking devices over media such as twisted pair, powerlines, fiber optics, and radio frequencies. It is used for the automation of various functions within buildings such as lighting and HVAC.

D.3.18 Market

There is a multiplicity of markets, such as spot market, balancing market, or ancillary services market, where stakeholders involved in the electricity supply sector can actively trade different energy products. This document focuses on the spot and balancing markets as they currently provide most the profitable business for aggregation, even though short-term operational reserves may bring more to the system.

D.3.19 Metropolitan Area Network (MAN)

A metropolitan area network (MAN) is the network that spans an entire city or campus. MANs are formed by connecting multiple LANs. Thus, MANs are larger than LANs but smaller than wide area networks (WAN). MANs are extremely efficient and provide fast communication via high-speed carriers, such as fiber-optic cables. MAN using the former standard [IEEE 802.6]. Distributed Queue Dual Bus (DQDB)] extends up to 30–40 km or 20–25 miles.

D.3.20 Mesh Network

A mesh network is a network topology in which each node relays data for the network. All nodes cooperate in the distribution of data in the network. Every node in a mesh network is called a mesh node. A fully mesh network is where each node is connected to every other node in the network. Mesh networks dynamically self-organize and self-configure, which can reduce installation overhead. In the event of a hardware failure, many routes are available to continue the network communication process. Mesh networks can relay messages using either a flooding technique or a routing technique. The ability to self-configure enables dynamic distribution of workloads, particularly in the event a few nodes should fail. This in turn contributes to fault tolerance and reduced maintenance costs.

Fully connected wired networks have the advantages of security and reliability: problems in a cable affect only the two nodes attached to it. However, in such networks, the number of cables, and therefore the cost, goes up rapidly as the number of nodes increases.

D.3.21 Microgrid

A localized grouping of electricity generation, energy storage, and loads that normally operates connected to a traditional centralized grid (*macrogrid*). This single point of common coupling with the macrogrid can be disconnected. The microgrid can then function autonomously. Generation and loads in a microgrid are usually interconnected at low voltage. From the point of view of the grid operator, a connected microgrid can be controlled as if it were one entity.

Microgrid generation resources can include fuel cells, wind, solar, or other energy sources. The multiple dispersed generation sources and ability to isolate the microgrid from a larger network would provide highly reliable electric power. Produced heat from generation sources such as microturbines could be used for local process heating or space heating, allowing flexible trade-off between the needs for heat and electric power.

D.3.22 Neighborhood Area Networks

Neighborhood area network (NAN) or field area network (FAN) is a network of multiple HANs or BANs to deliver the metering data to data concentrators and to deliver control date to HANs and

BANs. The network allows devices in a small area to communicate with each other. For example, all the smart meters in a neighborhood may communicate with each other and with a router to form an interconnected mesh of smart devices. Many metering gateways of home areas connect each other to form a possible wireless mesh network. A WAN connects smart metering gateways with utility and the distribution control system.

NAN is one of the most important sections in Smart Grid communications. It connects residential customers as part of a two-way communication infrastructure responsible for transmitting power grid sensing and measuring status as well as the control messages.

D.3.23 Personal Area Network

Personal area network (PAN) refers to the interconnection of information technology devices or gadgets within the environment of an individual user (typically within 10 m). These interconnected devices might include laptop computers, PDAs, cellphones, printers, PCs, or other wearable computer devices – also known as a wireless personal network (WPAN).

D.3.24 Power Exchange

A power exchange, also known as spot market, is a commercial entity that establishes a competitive spot market for electrical power through day- and/or hour-ahead auctioning of generation and demand bids, thus facilitating the development of transparent spot prices for energy.

It determines the market clearing prices and defines those generator units that have bid at or below the clearing price and are scheduled for generation. Depending on the closing time of the power exchange, it is called a day-ahead market, an intraday market, or real-time market if it closes just before delivery.

D.3.25 Renewable Energy (Re) Sources (RES)

RES means renewable energy sources (wind, solar, geothermal heat, wave, tidal, hydro [falling water], ocean energy, biomass, landfill gas, sewage treatment plant gas, and biogases). They are generally not subject to depletion. They are virtually inexhaustible in duration but limited in the amount of energy that is available per unit of time.

D.3.26 Resilience/Resiliency

Resilience/resiliency is the ability to resist, absorb, recover from, or successfully adapt to adversity or a change in conditions. In the context of energy security, resilience is measured in terms of robustness, resourcefulness, and rapid recovery [DHS 2010].

There are more definitions of resilience from different disciplines such as:

- The ability to reduce the magnitude and duration of disturbances; it depends upon the system's ability to predict, absorb, and adapt to disturbances and recover rapidly (discipline: infrastructure systems).
- The ability of an organization to identify risks and to handle perturbations that affect its competencies, strategies, and coordination (discipline: organizational systems).
- The ability of community to withstand stresses and disturbances caused by social, political, and economic changes (discipline: social systems).
- The ability to provide and maintain an acceptable level of service in the face of faults and challenges to normal operation. Threats and challenges for services can range from simple misconfiguration over large-scale natural disasters to targeted attacks (discipline: computer networking).

- The ability of the system (system resilience) to withstand a major disruption within acceptable degradation parameters and to recover within an acceptable time (discipline: systems engineering).
- The ability of failover that distributes redundant implementations of IT resources across physical locations (called also resilient computing); IT resources can be preconfigured so that if one becomes deficient, processing is automatically handed over to another redundant implementation (discipline: cloud computing).
- The ability to keep operating during a detected attack or incident, to keep operating under the assumption that an undetected compromise has occurred, to operate with reduced capability or capacity, and to provide graceful degradation and recovery during and after an incident (called also cyber resilience) (discipline: information security).

D.3.27 Smart Device

A smart device is an electronic device, generally connected to other devices or networks via different wireless protocols such as Bluetooth, Wi-Fi, 3G, etc., that can operate to some extent interactively and autonomously. Examples of smart devices are smart phones, tablets, smart watches, and smart key chains. The term can also refer to a ubiquitous computing device: a device that exhibits some properties of ubiquitous computing including artificial intelligence.

Smart devices can be designed to support a variety of form factors, a range of properties pertaining to ubiquitous computing and to be used in three main system environments: physical world, human-centered environments, and distributed computing environments.

D.3.28 Smart Grid Communications

Smart Grid applications require communication networks supporting reliable information transfer between the various entities in the electric grid, but there are many issues related to network performance, suitability, interoperability, and security that need to be resolved. One need is network traffic control to provide quality of service (QoS) to Smart Grid applications and to manage power flows in the Smart Grid between traditional and renewable generation sources and between utility-owned and customer-owned assets.

For example, wide area measurement system traffic is structurally different from Internet traffic. The delay and loss requirements for Smart Grid applications vary widely; some are very tolerant of long delays or lost information (metering), while others demand near-instantaneous data delivery with virtually no loss (wide area measurement systems). Also, the amount of data exchanged can grow very large as in the case of wide area measurement systems. As these systems scale up to a large number of phasor measurement units (PMU), the centralized super-phasor data collector (PDC) architecture becomes untenable.

D.3.29 Smart Meter

A smart meter is an electronic device that records consumption of electric energy and communicates the information to the electricity supplier for monitoring and billing. Smart meters typically record energy hourly or more frequently and report at least daily.

D.3.30 Smart Metering

Smart metering, advanced metering, smart metering infrastructure, and advanced metering infrastructure are often used as synonyms in different parts of the world. They all emphasize the metering infrastructure aspect in contrast to acronyms AMR (automatic meter reading) and AMM (four different commonly used meanings such as advanced metering management) that are limited to the internal issues of metering.

However, there could be differences that need to be accounted for when designing security measures. For example, advanced metering provides more control. Participating in residential energy management and other energy efficiency programs is completely optional. Customers who participate can use the information they receive to manage their energy usage day by day.

Smart metering, based on its corresponding metering services, has the following features: automatic processing, transfer, management and use of metering data, automatic management of meters, two-way data communication with meters, provision of meaningful and timely consumption information to the relevant actors and their systems including the energy consumer, and support services that improve the energy efficiency of the energy consumption and energy system (generation, transmission, distribution, and especially end use). Smart metering supports the aggregation of flexible distributed energy.

D.3.31 Virtual Power Plant

A virtual power plant (VPP) is a cluster of DG installations (such as wind turbines, solar panels, small hydro, etc.) that are collectively run by a central control entity. The concerted operational mode delivers extra benefits such as the ability to deliver peak load electricity or load aware power generation at short notice. Such a VPP can replace a conventional power plant while providing higher efficiency and more flexibility. Note that more flexibility allows the system to react better to fluctuations. However, a VPP is also a complex system requiring a complicated optimization, control, and secure communication methodology.

D.3.32 Wide Area Network

A WAN is the network that exists over an extended, large-scale geographical area. A WAN connects different smaller networks, including LAN and metro area networks (MAN). This ensures that computers and users in one location can communicate with computers and users in other locations. WAN implementation can be done either with the help of the public transmission system or a private network. Typically, TCP/IP is the protocol used for a WAN in combination with devices such as routers, switches, firewalls, and modems.

D.3.33 Smart Grid Glossaries

A report [CENELEC 2011] includes standardization recommendations regarding terminology, object identification, and classification such as:

- Harmonization of glossaries and establishing of a process for harmonizing Smart Grid vocabulary over different domains.
- Alignment of data model glossaries as much as possible with electronic data models and graphical representations of the equipment (Table D.1).

Table D.1 Examples of Smart Grid glossaries.

Organization	Glossary	Address
Clean Energy Solutions	Clean Energy Policy Glossary	https://cleanenergysolutions.org/policy-briefs/glossary
IEEE	Definitions of the IEEE Smart Grid Domains	https://smartgrid.ieee.org/images/files/domains/IEEE_Smart_Grid_Domains_and_SubDomains_Definitions.pdf
ISGAN	International Smart Grid Action Network (ISGAN) Smart Grid Glossary	https://openei.org/wiki/ISGAN_Smart_Grid_Glossary
IEEE	IEEE Standards Dictionary: Glossary of Terms and Definitions	http://shop.ieee.org/
ITU	ITU Smart Grid Terminology Deliverable, Smart-O-30Rev.6, December 2011	http://www.itu.int
PC Magazine Encyclopedia	Smart sensor	http://www.pcmag.com/encyclopedia/term/59600/smart-sensor
SGIP	SGIP Terminology and Definitions	http://sgip.org/Terms-Definitions
SmartGrid	SmartGrid Lexicon	https://www.smartgrid.gov/lexicon/6/letter_d
SmartGridToday	Glossary	https://www.smartgridtoday.com/public/Glossary-2.cfm
TechTarget	Smart Grid Glossary	www.whatissmartgrid.org/smart-grid-101/smart-grid-glossary
US DOE	US Department of Energy Office of Energy Efficiency and Renewable Energy	http://energy.gov/eere/energybasics/articles/glossary-energy-related-terms
US DOE	The Smart Grid: An Introduction, Glossary: coming to terms with the Smart Grid	https://www.energy.gov/sites/prod/files/oeprod/DocumentsandMedia/DOE_SG_Book_Single_Pages(1).pdf
What is Smart Grid	Smart Grid Glossary	http://www.whatissmartgrid.org/smart-grid-101/smart-grid-glossary

Also, this report [CENELEC 2012] includes information about security standards.

D.4 A Smart Grid Diagram: Past, Present, and Future

See Figure D.1.

D.5 Smart Grid Conceptual Model

A conceptual model is a model made of the composition of concepts, which are used to describe or simulate a subject the model represents. Some models may consist of physical objects. Conceptual models are often abstractions of things in the real world, whether physical or social. Semantics

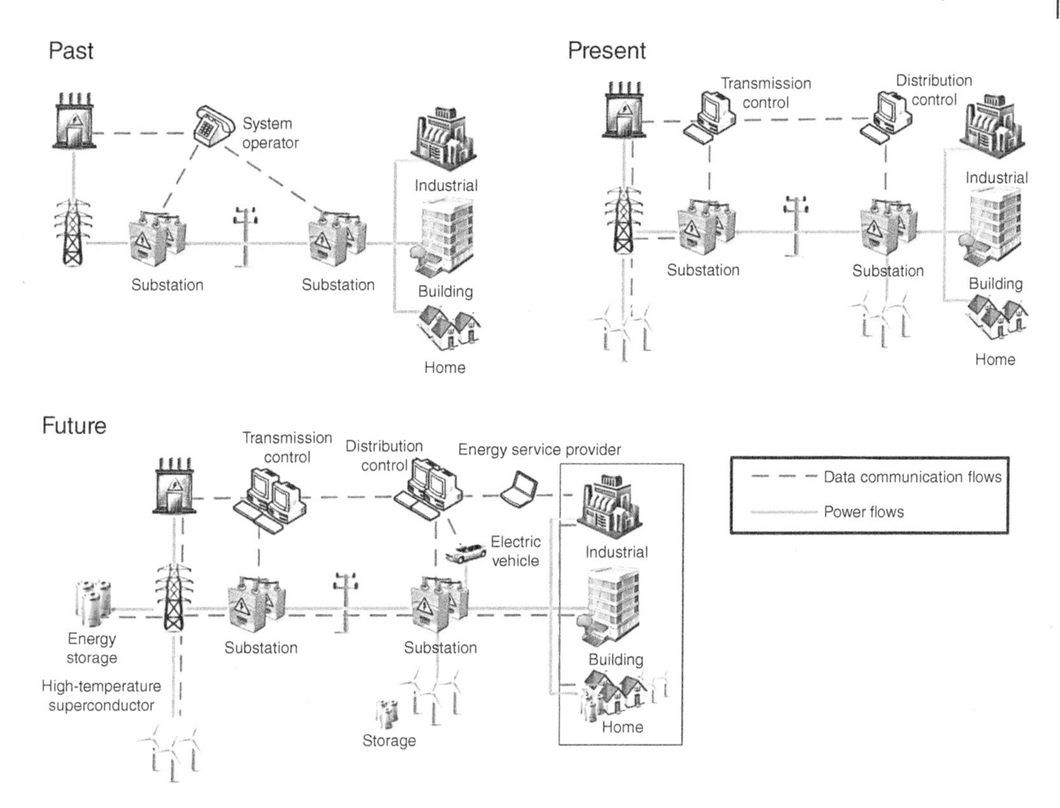

Figure D.1 Past, present, and future of the Smart Grid. *Source:* [ENISA 2012]. Public Domain.

studies are relevant to various stages of concept formation, and use of semantics is basically about concepts, the meaning that thinking beings give to various elements of their experience. An NIST document [NIST SP1108r1] defines the conceptual model (see Figure D.2) of the Smart Grid to include seven domains: generation, transmission, distribution, operations, market, customer, and service provider.

Each of the seven Smart Grid conceptual domains is a high-level grouping of physical organizations, buildings, individuals, systems, devices, or other actors that have similar objectives and that rely on – or participate in – similar types of services. Communications among roles and services in the same domain may have similar characteristics and requirements. Domains contain subdomains. Moreover, domains have many overlapping functionalities, as in the case of the transmission and distribution domains. Because transmission and distribution often share networks, they are represented as overlapping domains.

The Smart Grid conceptual model is presented as successive diagrams of increasing levels of detail, as shown in the following diagrams.

D.5.1 Bulk Generation

The bulk generation domain of the Smart Grid generates electricity from renewable and nonrenewable energy sources in bulk quantities (Figure D.3). These sources can also be classified as renewable, variable sources, such as solar and wind; renewable, non-variable, such as hydro, biomass, geothermal, and pump storage; or nonrenewable, non-variable, such as nuclear, coal, and gas. Energy that is stored for later distribution may also be included in this domain (Figure D.4).

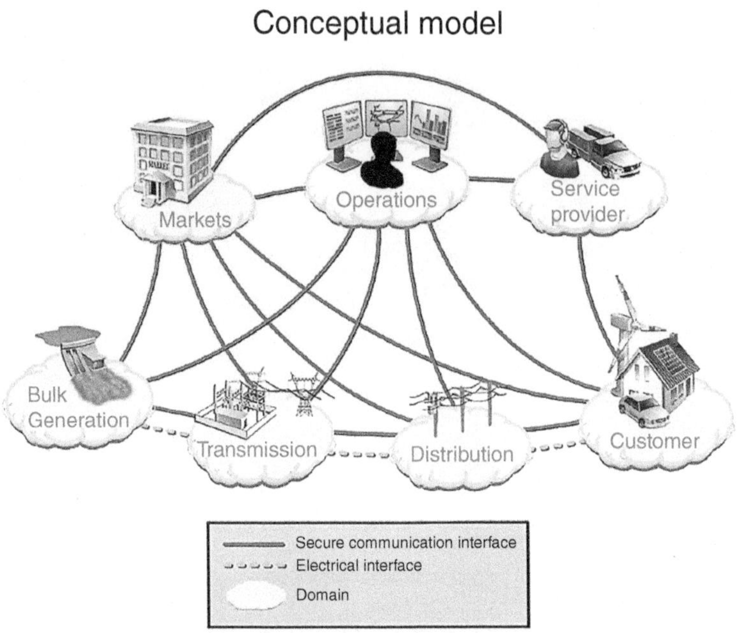

Figure D.2 Overview of Smart Grid conceptual model. *Source:* [NIST SP1108r3]. Public Domain.

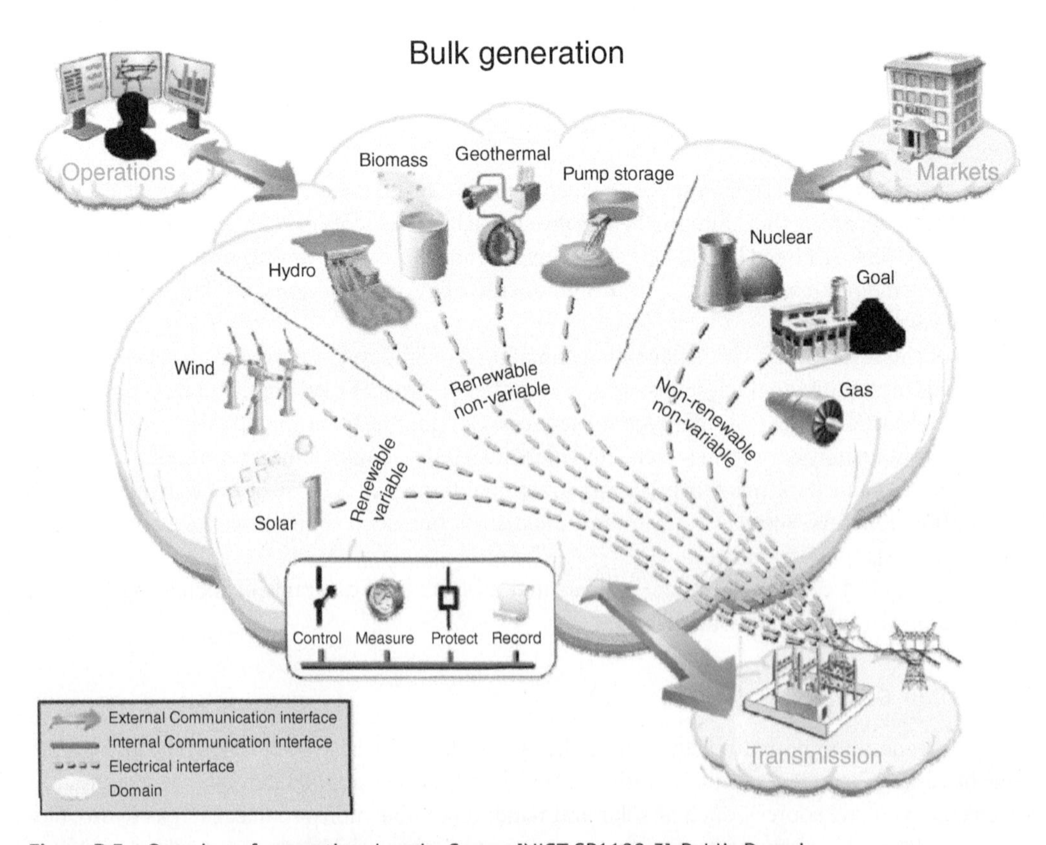

Figure D.3 Overview of generation domain. *Source:* [NIST SP1108r3]. Public Domain.

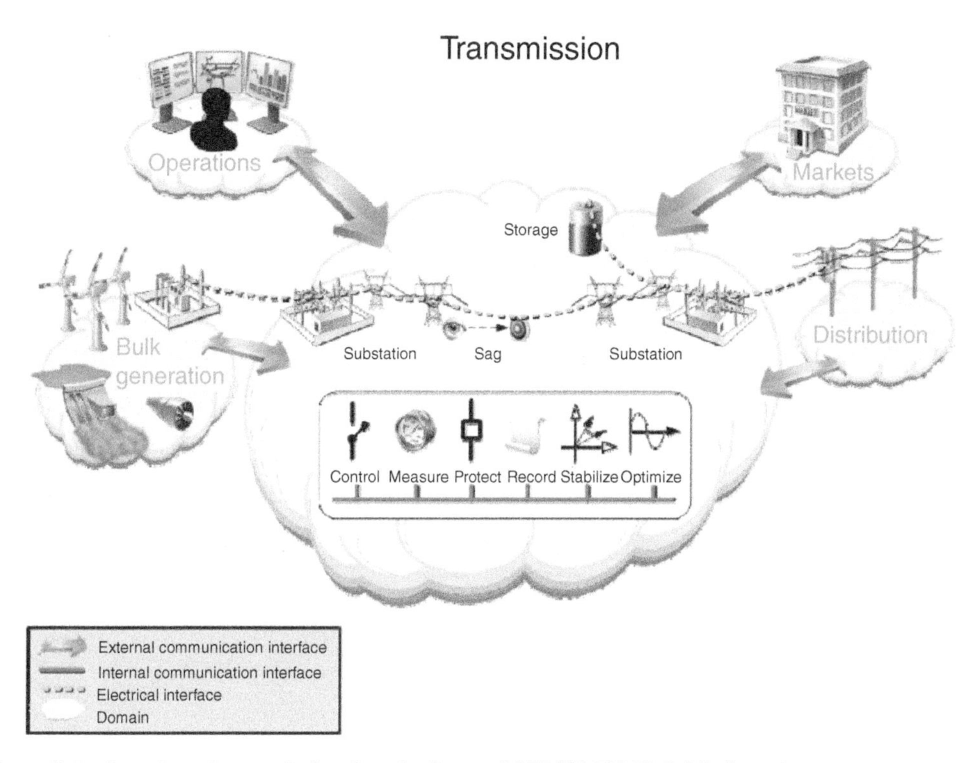

Figure D.4 Overview of transmission domain. *Source:* [NIST SP1108r3]. Public Domain.

Transmission is the bulk transfer of electrical power from generation sources to distribution through multiple substations. A transmission network is typically operated by a regional transmission operator or independent system operator (RTO/ISO) whose primary responsibility is to maintain stability on the electric grid by balancing generation (supply) with load (demand) across the transmission network.

The transmission domain may contain DER such as electrical storage or peaking generation units. Energy and supporting ancillary services (capacity that can be dispatched when needed to stabilize the grid) are procured through the markets domain and scheduled and operated from the operations domain and finally delivered through the transmission domain to the distribution system and finally to the customer domain (Figure D.5).

D.5.2 Distribution

The distribution domain distributes the electricity to and from the end customers in the Smart Grid. The distribution network connects the smart meters and all intelligent field devices, managing and controlling them through a two-way wireless or wireline communication network. It may also connect to energy storage facilities and alternative DER at the distribution level.

D.5.3 Customer

The customer domain of the Smart Grid is where the end users of electricity (home, commercial/building, and industrial) are connected to the electric distribution network through the smart meters (Figure D.6). The smart meters control and manage the flow of electricity to and from the

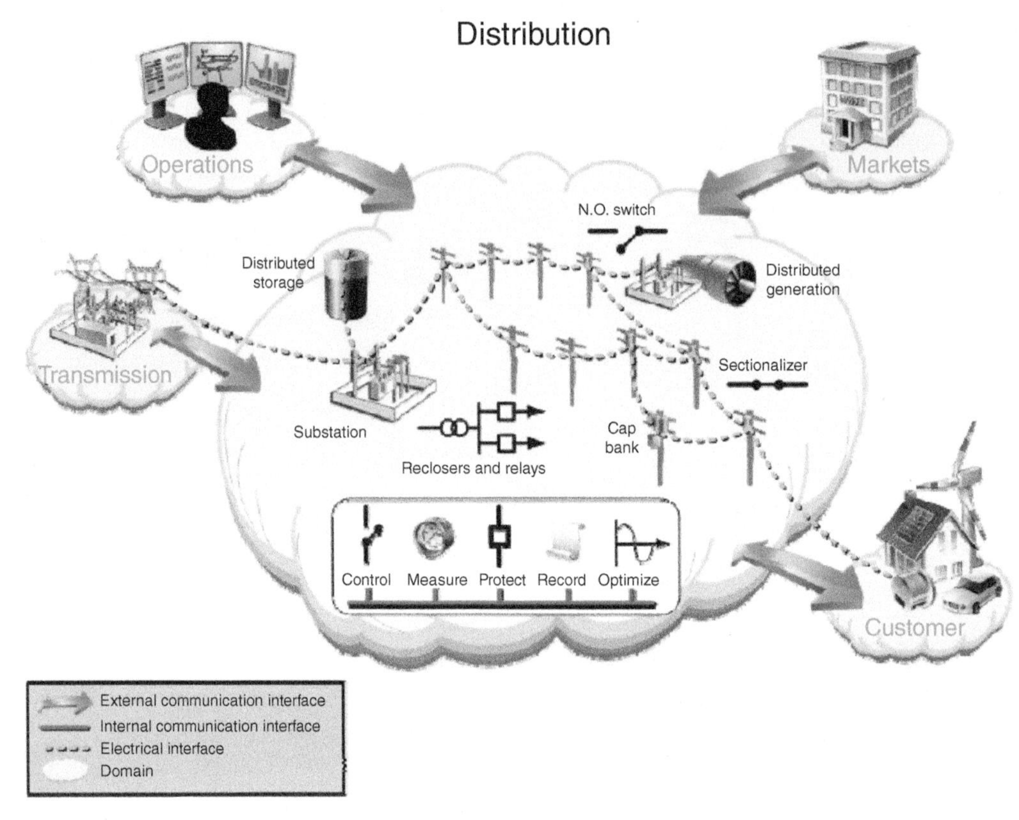

Figure D.5 Overview of distribution domain. *Source:* [NIST SP1108r3]. Public Domain.

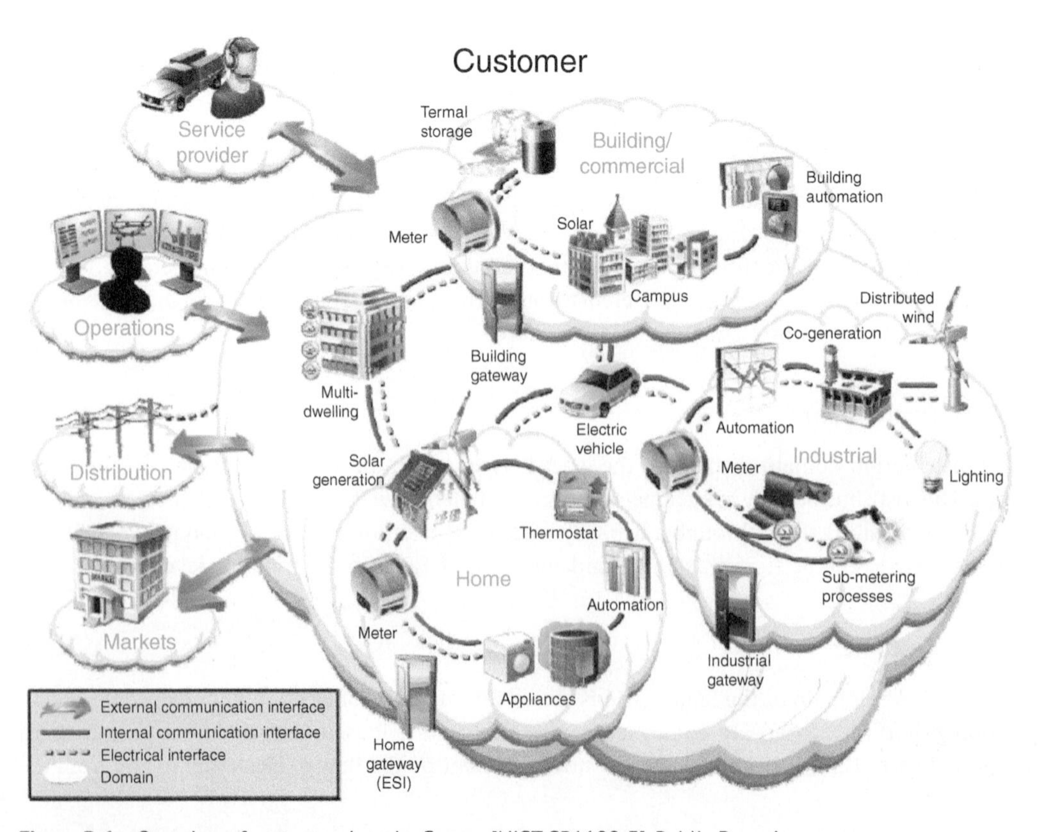

Figure D.6 Overview of customer domain. *Source:* [NIST SP1108r3]. Public Domain.

customers and provide energy information about energy usage and patterns. Each customer has a discrete domain composed of electricity premise and two-way communication networks. A customer domain may also generate, store, and manage the use of energy, as well as the connectivity with plug-in vehicles.

D.5.4 Operations

The operations domain manages and controls the electricity flow of all other domains in the Smart Grid (Figure D.7). It uses a two-way communication network to connect to substations, customer premises networks, and other intelligent field devices. It provides monitoring, reporting, controlling, and supervision status and important process information and decisions. Business intelligence processes gather data from the customer and network and provide intelligence to support the decision making.

D.5.5 Markets

The markets domain operates and coordinates all the participants in electricity markets within the Smart Grid (Figure D.8). It provides the market management, wholesaling, retailing, and trading of energy services. The markets domain interfaces with all other domains and makes sure they are coordinated in a competitive market environment. It also handles energy information clearinghouse operations and information exchange with third-party service providers. For example, roaming billing information for inter-utility plug-in-vehicles falls under this domain.

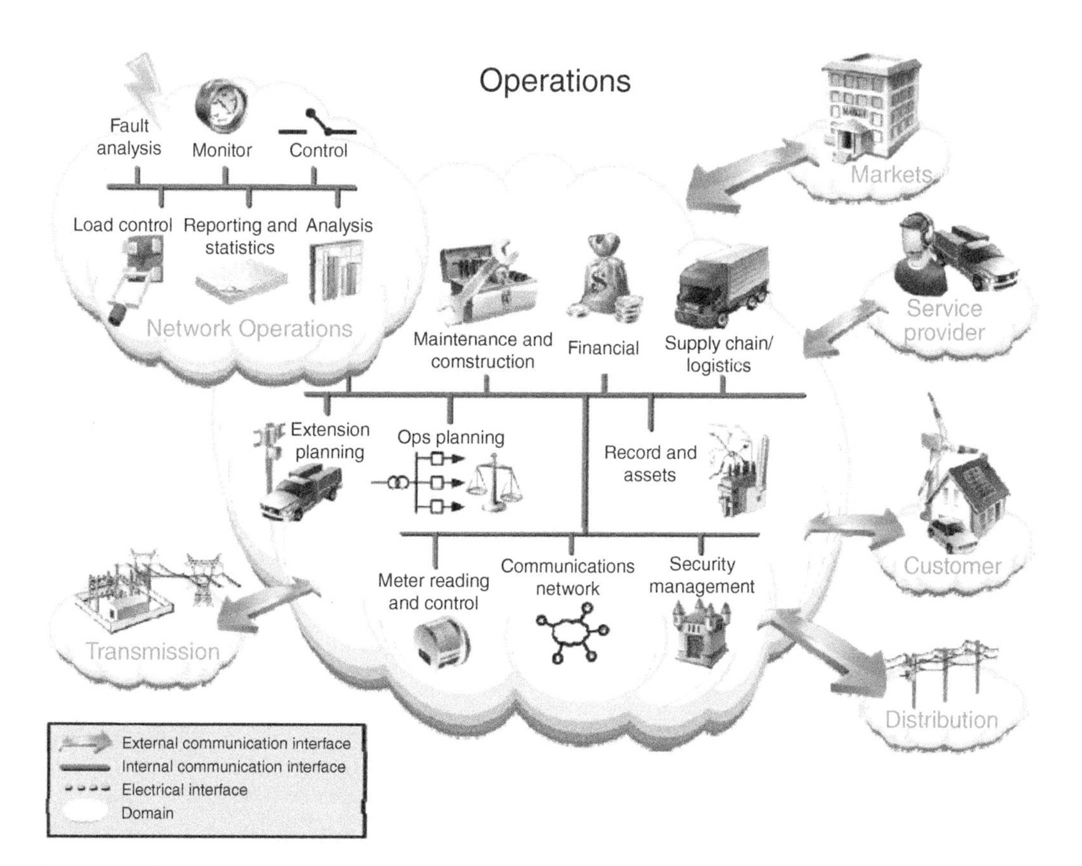

Figure D.7 Overview of operations domain. *Source:* [NIST SP1108r3]. Public Domain.

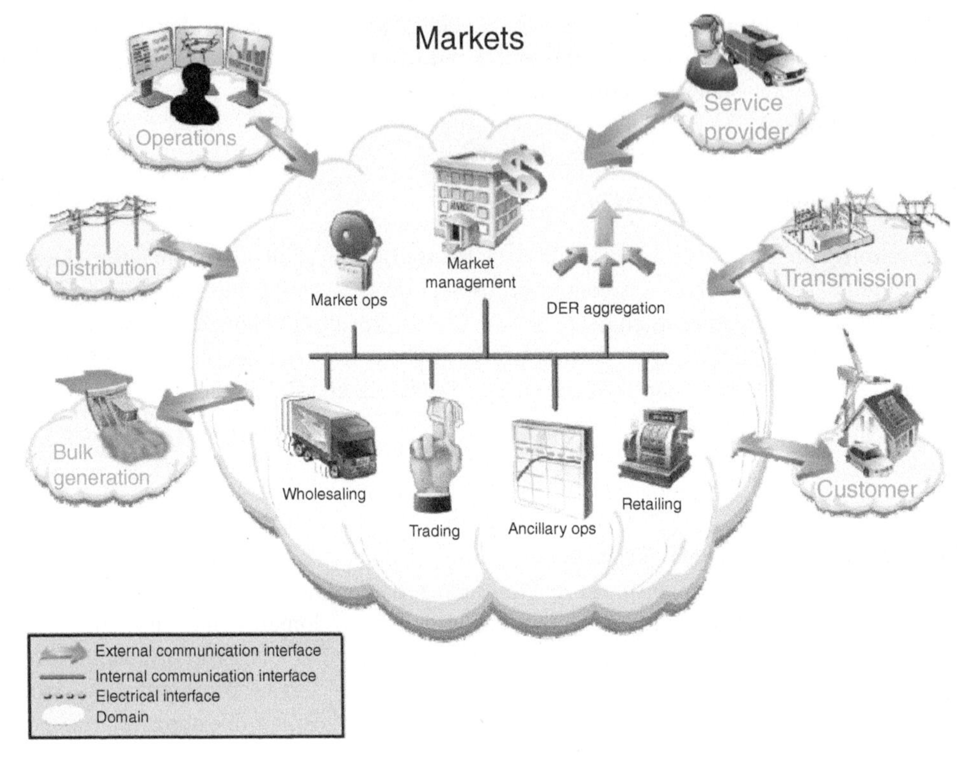

Figure D.8 Overview of markets domain. *Source:* [NIST SP1108r3]. Public Domain.

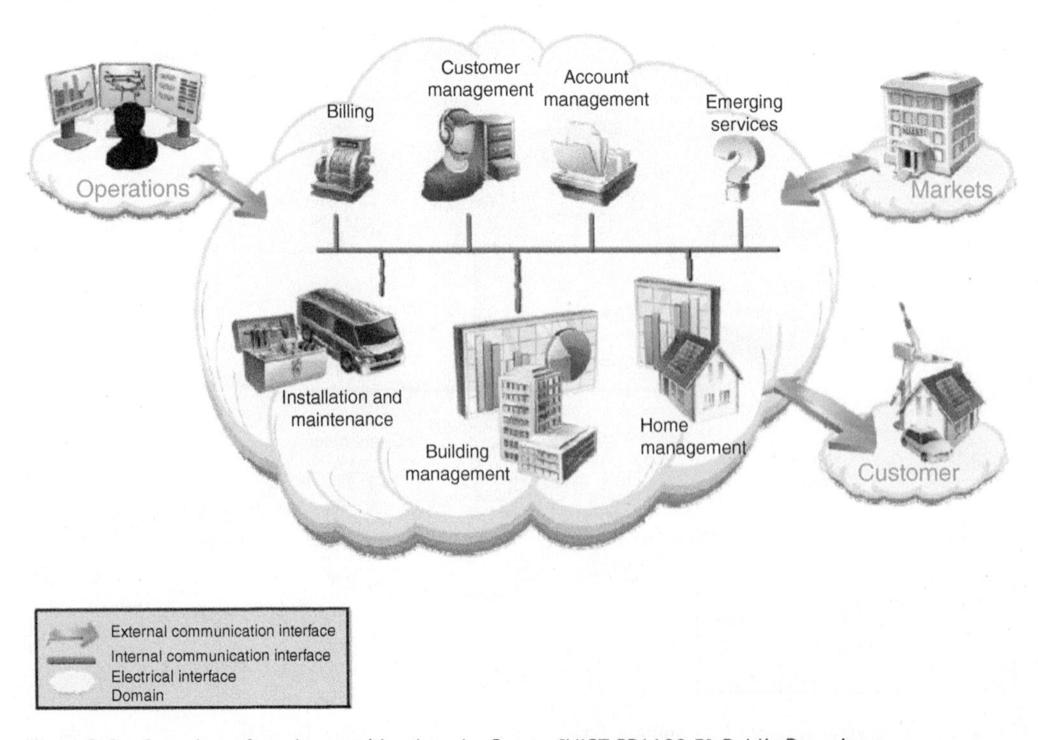

Figure D.9 Overview of service provider domain. *Source:* [NIST SP1108r3]. Public Domain.

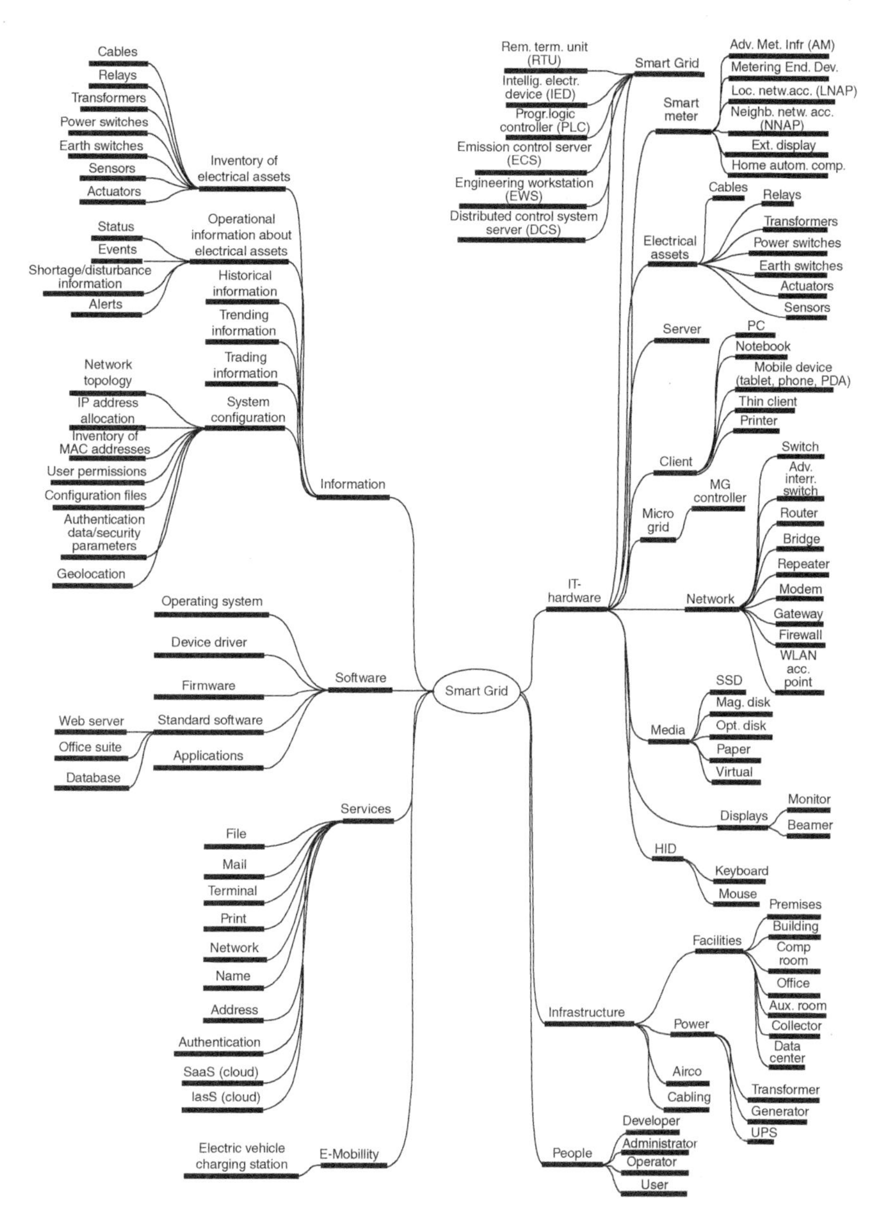

Figure D.10 Overview of Smart Grid assets. *Source:* [ENISA 2014]. Public Domain.

D.5.6 Service Provider

The service provider domain of the Smart Grid handles all third-party operations among the domains (Figure D.9). These might include web portals that provide energy efficiency management services to end customers and data exchange between the customer and the utilities regarding energy management and regarding the electricity supplied to homes and buildings. It may also manage other processes for the utilities, such as DR programs, outage management, and field services.

A diagram of assets of different nature (information, electric equipment, computer hardware, people, etc.) requiring security protection is depicted in Figure D.10. Besides these assets, composite assets related to Smart Grid Architecture Model (SGAM) model are described in [ENISA 2014].

References

[CENELEC 2011] CEN-CENELEC-ETSI. (2011). *Final Report of the CEN/CENELEC/ETSI Joint Working Group on Standards for Smart Grids May 2011.*

[CENELEC 2012] CEN-CENELEC-ETSI. (2012). *SG-CG/M490/H_Smart Grid Information Security December 2012.*

[DHS 2010] DHS. (2010). *Appendix 1 Dictionary of Key Terms, SPP 2010, Energy Sector Specific Plan 2010 An Annex to the National Infrastructure Protection Plan.* http://www.dhs.gov/xlibrary/assets/nipp-ssp-energy-2010.pdf

[DOE 2014] DOE. (2014). *Future Smart Grid.* http://energy.gov/sites/prod/files/2014/12/f19/Future%20of%20the%20Grid%20December%202014.pdf

[EC 2006] EC. (2006). *SmartGrids European Technology Platform Vision and Strategy for Europe's Electricity Networks of the Future.* http://www.smartgrids.eu/documents/vision.pdf

[ENISA 2012] ENISA. (2012). *Smart Grid Security Annex I General concepts and dependencies with ICT.*

[ENISA 2014] ENISA. (2014). *Proposal for a List of Security Measures for Smart Grids December 2013.* Security measures for Smart Grids - Dissemination workshop. https://www.enisa.europa.eu/news/enisa-news/security-measures-for-smart-grids-dissemination-workshop

[IEEE 802.6] *IEEE 802.6 1990 Local and Metropolitan Area Networks: Distributed Queue Dual Bus (DQDB) Subnetwork of a Metropolitan Area Network (MAN)* * Withdrawn in 2003.

[IEEE P2030] *IEEE Std 2030 ™-2011 IEEE Guide for Smart Grid Interoperability of Energy Technology and Information Technology Operation with the Electric Power System (EPS) End-Use Applications, and Loads.* http://www.ieee.org

[NIST SP1108r3] NIST. (2014). *NIST Special Publication 1108 NIST Framework and Roadmap for Smart Grid Interoperability Standards Release 3.0 September 2014.*

[Wiki SG] Wiki. (n.d.). *Smart Grid.* http://en.wikipedia.org/wiki/Smart_grid

J

Acronyms

Appendix J includes the most common acronyms for organizations, standards, technologies, protocols, and business terms used in the development of Smart Grid applications.

1G, 2G, 3G, 4G, 5G	First, Second, Third, Fourth, Fifth Generation Wireless
3DES	Triple DES
3GPP	3rd Generation Partnership Project
6LoWPAN	IPv6 over Low-Power Wireless Personal Area Network
AAA	authentication, authorization, and auditing
ABR	available bit rate
AC	access control
AC	alternating current
ACID	atomicity, consistency, isolation, durability
ACK	Acknowledgment
ACL	access control list
AD	Active Directory
ADA	Advanced Distribution Automation
ADR	automated demand response
ADSL	asymmetric digital subscriber line
AES	Advanced Encryption Standard
AGA	American Gas Association
AH	Authentication Header
AHJ	Authority Having Jurisdiction
AI	artificial intelligence
ALDS	Application Level Detection System
ALE	Annualized Loss Expectancy
AMC	adaptive modulation and coding
AMI	advanced metering infrastructure
AMR	automatic meter reading
AN	access network
ANN	artificial neural network

Building an Effective Security Program for Distributed Energy Resources and Systems: Understanding Security for Smart Grid and Distributed Energy Resources and Systems, Volume 1, First Edition. Mariana Hentea.

ANSI	American National Standards Institute
AP	access point
API	application programming interface
APP	Application
ARO	annualized rate of occurrence
ARP	Address Resolution Protocol
AS	autonomous system
ASCII	American Standard Code for Information Interchange
ASHRAE	American Society of Heating, Refrigerating, and Air-Conditioning Engineers
AT	awareness and training
ATM	Asynchronous Transfer Mode
AU	Audit and Accountability
AV	autonomous vehicle
AWL	application whitelisting
AWWA	American Water Works Association
B2B	business to business
B2C	business to consumer
BAN	building area network
BC	business continuity
BCM	business capability model
BCM	business continuity management
BCP	business continuity plan
BGP	Border Gateway Protocol
BEMS	Building Energy Management System
BIOS	basic input/output system
BLE	Bluetooth Low Energy
BMIS	Business Model for Information Security
BOOTP	Bootstrap Protocol
BPL	Broadband over Power Line
BT	Bluetooth
C2M2	Cybersecurity Capability Maturity Model
C2S	client to site
CA	certificate authority
CA	Security Assessment and Authorization
CAES	compressed air energy storage
CAG	Consensus Audit Guidelines
CAPEX	capital expenditures
CASE	computer-aided software engineering tools
CBC	cipher block chaining
CBK	Common Body of Knowledge
CBR	constant bit rate

CC	Critical Control
CC	Common Criteria
CCHP	combined cooling, heat, and power
CCTV	closed-circuit television
CD	compact disc
CDF	cumulative distribution function
CDMA	code division multiple access
CEN	Comité Européen de Normalisation
CEN	European Committee for Standardization
CENELEC	Comité Européen de Normalisation Electrotechnique
CENELEC	European Committee for Electrotechnical Standardization
CEO	chief executive officer
CFB	cipher feedback mode
CHAP	Challenge Handshake Authentication Protocol
CHP	combined heat and power
CIA	confidentiality, integrity, and availability
CIO	chief information officer
CISO	chief information security officer
CDI	constrained data item
CIKR	critical infrastructures and key resources
CIM	Common Information Model
CIP	critical infrastructure protection
CIR	committed information rate
CIS	Customer Information Systems
CISSP	Certified Information Systems Security Professional
CM	configuration management
CMAC	Cipher-Based Message Authentication Code
CMM	Capability Maturity Model
CMMI	Capability Maturity Model Integration
CNSS	Committee on National Security Systems
CNSSI	Committee on National Security Systems Instruction
COBIT	Control Objectives for Information and Related Technology
COM	Component Object Model
CORBA	Common Object Request Broker Architecture
COTS	commercial off the shelf
CP	contingency panning
CPO	chief privacy officer
CPP	critical peak pricing
CPTED	Crime Prevention Through Environmental Design
CPU	central processing unit
CRUD	create, read, update, delete

CSI	Computer Security Institute
CSL	coordinated sampled listening
CSMA/CA	Carrier Sense Multiple Access with Collision Avoidance
CSP	concentrated solar power
CSRF	Cross-Site Request Forgery
CT-IAP	Communications Technology Interoperability Architectural Perspective
CTS	Clear to Send
CVR	conservation voltage reduction
CWE	Common Weakness Enumeration
DA	distribution automation
DAC	discretionary access control
DAC	Dual Attached Concentrator
DAG	directed acyclic graph
DAP	data aggregation point
DAR	Design Architecture Review
DASD	direct access storage device
DC	direct current
DCE	data circuit terminating equipment
DCL	data control language
DCOM	Distributed Component Object Model
DCS	distributed control system
DDOS	distributed denial of service
DEA	Data Encryption Algorithm
DER	distributed energy resource
DES	Data Encryption Standard
DFS	Distributed File Service
DG	distributed generation
DHCP	Dynamic Host Configuration Protocol
DHS	US Department of Homeland Security
DIFS	Distributed Coordination Function Interframe Spacing
DIO	DODAG Information Object
DIS	DODAG Information Solicitation
DLC	direct load control
DLR	dynamic line rating
DMS	distribution management system
DMZ	demilitarized zone
DNP3	Distributed Network Protocol Version 3
DNS	Domain Name Service
DOCIS	Data Over Cable Interface Specifications
DODAG	Destination-Oriented Directed Acyclic Graph
DOE	US Department of Energy

DOS	denial of service
DR	demand response
DRM	digital rights management
DRMS	Demand Response Management System
DS	distributed storage
DSL	digital subscriber line
DSM	demand-side management
DSN	Data Source Name
DSO	distribution system operator
DSS	Digital Signature Standard
DSSS	direct sequence spread spectrum
DSU	Data Service Unit
DTLS	Datagram Transport Layer Security
DTP	Dynamic Trunking Protocol
DVD	digital video disc
EAL	Evaluation Assurance Level
EAM	enterprise asset management
EAP	Extensible Authentication Protocol
ebIX	(European forum for) energy Business Information eXchange
ECB	Electronic Code Book mode
ECC	Elliptic Curve Cryptography algorithm
ECU	electronic control unit
EDGE	Enhanced Data rates for GSM Evolution
EDR	Enhanced Data Rate
EER	equal error rate
EF	exposure factor
EGP	Exterior Gateway Protocol
EGx	EU Smart Grid Task Force Expert Group x(1–3)
EIA	Electronic Industries Association
EIFS	Extended Interframe Spacing
EIGRP	Enhanced Interior Gateway Routing Protocol
EMI	electromagnetic interference
EMS	energy management system
ENTSO-E	European Network of Transmission System Operators for Electricity
EPM	Energy Portfolio Management
EPS	electric power system
ERP	enterprise resource planning
ES-C2M2	Electricity Subsector Cybersecurity Capability Maturity Model
ESCO	energy service company
eTOM	extended Telecom Operations Map
ETSI	European Telecommunications Standards Institute

ETX	expected transmission count
EU DPD	European Union Data Protection Directive
EV	electrical vehicle
EVO	electrical vehicle operator
FACTS	flexible alternating current transmission systems
FAN	field area network
FDMA	frequency division multiple access
FERC	Federal Energy Regulatory Commission
FHSS	frequency hopping spread spectrum
FIPS	Federal Information Processing Standards
FISMA	Federal Information Security Management Act
FLIR	Fault Location, Isolation, and Restoration
FLISR	Fault Location, Isolation, and Service Recovery
FLISR	Fault Location, Isolation, and System Restoration
FMEA	failure modes and effects analysis
FTP	File Transfer Protocol
G2V	grid to vehicle
GDPR	General Data Protection Regulation
GEO	geographic routing
GF	greedy forwarding
GIF	Graphics Interchange Format
GIS	geographical information system
GLBA	Gramm–Leach–Bliley Act
GPS	global positioning system
GPSR	Greedy Perimeter Stateless Routing
GRE	Generic Routing Encapsulation
GSM	Global System for Mobile
GSM	Global System for Mobile Communications
GW	gigawatt
GWAC	GridWise Architecture Council
HAN	home area network
HART	Highway Addressable Remote Transducer
HDSL	High-Level Data Link Control
HDSL	high-bit-rate digital subscriber line
HEMS	home energy management system
HFRT	High-frequency ride-through
HIDS	host-based intrusion detection system
HIP	Host Identity Protocol
HIPAA	Health Insurance Portability and Accountability Act
HMI	human–machine interface
HSDPA	High-Speed Downlink Packet Access

HSPA	High-Speed Packet Access
HTTP	Hypertext Transfer Protocol
HTTPS	HTTP Secure
HVAC	heating, ventilation, and air conditioning
HVDC	high-voltage direct current
HVRT	high-voltage ride-through
IA	identification and authenticity
IaaS	Infrastructure as a Service
IAB	Internet Architecture Board
IAM	identity and access management
IAN	industrial area network
IAP	Interoperability Architectural Perspective
IC	internal combustion
ICMP	Internet Control Message Protocol
ICMPv6	Internet Control Message Protocol version 6
ICS	industrial control system
ICS-CERT	Industrial Control Systems Cyber Emergency Response Team
ICT	information and communication technology
ICV	Integrity Check Value
ID	identification
ID	identity
IdM	identity management
IDEA	International Data Encryption Standard
IDS	intrusion detection systems
IDS/IPS	intrusion detection system/intrusion prevention system
IEA	International Energy Agency
IEC	International Electrotechnical Commission
IED	intelligent electronic device
IEEE	Institute of Electrical and Electronics Engineers
IETF	Internet Engineering Task Force
IGMP	Internet Group Management Protocol
IGP	Interior Gateway Protocol
IGRP	Interior Gateway Routing Protocol
IKE	Internet Key Exchange
IOCE	International Organization on Computer Evidence
IoT	Internet of Things
IP	Internet Protocol
IPS	Intrusion Prevention System
IPsec	Internet Protocol Security
IPv4	Internet Protocol version 4
IPv6	Internet Protocol version 6

IR	incident response
IRENA	International Renewable Energy Agency
ISACA	Information Systems Audit and Control Association
ISDN	Integrated Services Digital Network
IS	information system
ISA	International Society of Automation
ISAE	International Standard on Assurance Engagements
ISM	industrial, scientific, and medical
ISMS	information security management system
ISO	independent system operator
ISO	International Organization for Standardization
ISP	Internet Service Provider
ISSA	Information Systems Security Association
ISSAP	Information Systems Security Architecture Professional
IT	information technology
IT-IAP	Information Technology Interoperability Architectural Perspective
ITIL	Information Technology Infrastructure Library
ITSEC	Information Technology Security Evaluation Criteria
ITU	International Telecommunication Union
ITU-T	International Telecommunication Union for the Telecommunication Standardization Sector
IV	initialization vector
IVVC	Integrated Volt/VAR Control
JWG	Joint Working Report for Standards for the Smart Grids
KDC	key distribution center
Key	cryptographic key
kW	kilowatt
kWh	kilowatt-hour
kWp	kilowatt peak
L2TP	Layer 2 Tunneling Protocol
LAN	local area network
LCP	Link Control Protocol
LDAP	Lightweight Directory Access Protocol
LEAP	Lightweight Extensible Authentication Protocol
LFRT	low-frequency ride-through
LIDAR	Light Detection and Ranging
Li-ion	lithium ion
LLC	Logical Link Control
LR-WPAN	Low-Rate Wireless Personal Area Network
LTE	Long Term Evolution
LTE-A	LTE Advanced

LVRT	low-voltage ride-through
M2M	machine-to-machine
MA	system development and maintenance
MAC	Medium Access Control
MAC	Message Authentication Code
MAN	metropolitan area network
MD5	Message Digest algorithm series 5
MFA	multi-factor authentication
MFR	Most Forwarding Progress Within Radius
MIME	Multipurpose Internet Mail Extensions
MIMO	multiple-input multiple-output
MITM	man in the middle
MP	Media Protection
MP2P	multipoint-to-point
MPDU	MAC Protocol Data Unit
MPLS	Multiprotocol Label Switching
MPLS-TP	MPLS Transport Profile
MQTT	Message Queue Telemetry Transport
MTBF	mean time between failures
MTC	machine-type communications
MTTR	mean time to repair
MTX	mean transmission time
MTU	maximum transmission unit
NAN	neighborhood area network
NAT	Network Address Translation
NAVV	Network Architecture Validation and Verification
NCCIC	National Cybersecurity & Communications Integration Center
NCP	Network Control Protocol
NEC	National Electrical Code
NERC CIP	North American Electric Reliability Corporation Critical Infrastructure Protection
NERC	North American Electric Reliability Corporation
NESCOR	National Electric Sector Cybersecurity Organization Resource
NFC	near-field communication
NFP	Nearest with Forwarding Progress
NIC	network interface card
NIST	National Institute of Standards and Technology
NRC	Nuclear Regulatory Commission
NVD	National Vulnerability Database
NWP	Numerical Weather Prediction
O&M	operations and maintenance
OCTAVE	Operationally Critical Threat, Asset, and Vulnerability Evaluation

OECD	Organization for Economic Cooperation and Development
OEM	original equipment manufacturer
OF	objective function
OFDM	orthogonal frequency division multiplexing
OFDMA	orthogonal frequency division multiple access
OLE	Object Linking and Embedding
OMS	outage management system
OPC	OLE for Process Control
OpenADR	Open Automated Demand Response
OPEX	operational expenditure
O-QPSK	offset quadrature phase-shift keying
OS	operating system
OSA	Open Security Architecture
OSHA	Occupational Safety and Health Administration
OSI	Open Systems Interconnection
OSPF	Open Shortest Path First
OT	operational technology
OTN	Optical Transport Network
OTP	one-time password
OWASP	Open Web Application Security Project
OWL	Web Ontology Language
P2MP	point-to-multipoint
P2P	point-to-point
PAP	Priority Action Plan
PAP	Password Authentication Protocol
PBX	private branch exchange
PCI DSS	Payment Card Industry Data Security Standard
PCS	process control system
PDC	phasor data concentrator
PDCA	plan–do–check–act cycle
PDF	Portable Document File
PDR	packet delivery ratio
PE	Physical and Environmental Protection
PEV	plug-in electric vehicle
PGP	Pretty Good Privacy
PHEV	plug-in hybrid electric vehicle
PHY	physical layer
PIFS	Point Coordination Function Interframe Spacing
PIN	personal identification number
PIPEDA	Personal Information Protection and Electronic Documents Act
PIV	Personal Identity Verification

PKI	public key infrastructure
PL	planning
PLC	powerline carrier
PLC	powerline communications
PLC	programmable logic controller
PM	program management
PMU	phasor measurement unit
PON	passive optical network
PPP	Point-to-Point Protocol
PPS	Proactive Parent Switching
PPTP	Point-to-Point Tunneling Protocol
PS	personnel security
PSH	pumped storage hydropower
PS-IAP	Power Systems Interoperability Architectural Perspective
PSTN	public switched telephone network
PUC	public utility commission
PV	photovoltaic
QoS	quality of service
QoSS	quality of security service
RA	risk assessment
RA	Request Authority
RAD	Rapid Application Development
RADIUS	Remote Authentication Dial-In User Service
RAID	Redundant Array of Independent Disks
RAM	random access memory
RARP	Reverse Address Resolution Protocol
RBAC	role-based access control
RE	renewable energy
RF	radio frequency
RFID	radio-frequency identification
RG	regulatory guide
RMP	risk management process
RNC	Radio Network Controller
RoE	rules of engagement
ROLL	Routing over Low-Power and Lossy Networks
RPC	remote procedure call
RPL	Routing Protocol for Low-Power and Lossy Networks
RTP	real-time pricing
RTS	Request to Send
RSVP	Resource Reservation Protocol
S/MIME	Secure MIME

SA	System and Services Acquisition
SaaS	Software as a Service
SABSA	Sherwood Applied Business Security Architecture
SAE	System Architecture Evolution
SAFECode	Software Assurance Forum for Excellence in Code
SAIDI	System Average Interruption Duration Index
SAML	Security Assertion Markup Language
SAN	storage area network
SANS	SysAdmin, Audit, Networking, and Security Institute
SAS	Substation Automation System
SASL	Simple Authentication and Security Layer
SC	System and Communications Protection
SCADA	supervisory control and data acquisition
SCAP	Security Content Automation Protocol
SC-FDMA	single-carrier frequency division multiple access
SD	Secure Digital memory card
SDH	synchronous optical networking
SDN	software-defined networking
SDO	standards developing organization
SEP	Smart Energy Profile
SG	Smart Grid
SGAM	European Smart Grid Architecture (Reference) Model
SGCG	Smart Grid Coordination Group
SGCN	Smart Grid Communications Network
SGIP	Smart Grid Interoperability Panel
SGIRM	Smart Grid Interoperability Reference Model
SET	Secure Electronic Transaction
SI	System and Information Integrity
SIEM	Security Information and Event Management
SIFS	Short Interframe Spacing
SIG	Special Interest Group
SIP	Session Initiation Protocol
SIS	safety instrumented systems
SKIP	Simple Key Management Protocol
SLA	service-level agreement
SM	smart meter
SMTP	Simple Mail Transfer Protocol
SNMPv3	Simple Network Management Protocol
SOA	Service-Oriented Architecture
SOC	service organization control
SOCKS	circuit-level proxy firewall

SOAP	Simple Object Access Protocol
SODAR	Sonic Detection and Ranging
SOX	Sarbanes–Oxley Act
SP	Special Publication
SQL	Structured Query Language
SQL/SQLi	Structured Query Language/Structured Query Language Injection
SSDLC	Secure Software Development Life Cycle
SSH	Secure Shell
SSL	Secure Sockets Layer
ST	slot time
STATCOM	static synchronous compensator
SUN	smart utility network
SVC	static VAR compensator
T&D	transmission and distribution
TCB	trusted computing base
TCP	Transmission Control Protocol
TCP/IP	Transmission Control Protocol/Internet Protocol
TDM	time division multiplexing
TDMA	time division multiple access
TKIP	Temporal Key Integrity Protocol
TLS	Transport Layer Security
TMF	TeleManagement Forum
TOGAF	The Open Group Architecture Framework
TOU	time-of-use
TPM	Trusted Platform Module
TSA	Transportation Security Administration
TSO	Transmission System Operator
TTL	time to live
US	United States
UBR	unspecified bit rate
UDDI	Universal Description Discovery and Integration
UDP	User Datagram Protocol
UHF	ultra high frequency
UL	Underwriters Laboratories
UMTS	Universal Mobile Telecommunications System
UPS	uninterruptible power supply
URL	Uniform Resource Locator
USB	Universal Serial Bus
US-CERT	US Computer Emergency Readiness Team
USIM	Universal Subscriber Identity Module
V2G	vehicle to grid

VAR	volt-ampere reactive
VHF	Very high frequency
VLAN	virtual LAN
VM	virtual machine
VoIP	Voice over Internet Protocol
VPN	virtual private network
VPP	virtual power plant
WAMS	wide area management systems
WAN	wide area network
WASA	wide area situational awareness
WEB	World Wide Web
WEP	Wired Equivalent Privacy
Wi-Fi	Wireless Fidelity
WiMAX	Worldwide Interoperability for Microwave Access
WMN	wireless mesh network
WPA	Wi-Fi Protected Access
WPA2	Wi-Fi Protected Access version 2
WPAN	wireless personal area network
WSAN	wireless sensor and actuator network
WSDL	Web Services Description Language
WSN	wireless sensor network
XACML	Extensible Access Control Markup Language
xDSL	digital subscriber line
XML	Extended Markup Language
XOR	exclusive OR
XSS	Cross-Site Scripting

Index

d

e

k

l

m

r

s

t